NATURAL HISTORY
UNIVERSAL LIBRARY

U0215794

西方博物学大系

主编：江晓原

THE ILLUSTRATED
NATURAL HISTORY

自然图志

[英] 约翰·乔治·伍德 著

华东师范大学出版社

图书在版编目（CIP）数据

自然图志 = The Illustrated Natural History：英文 /（英）约翰·乔治·伍德著. — 上海：华东师范大学出版社，2018
（寰宇文献）
ISBN 978-7-5675-7718-3

Ⅰ.①自… Ⅱ.①约… Ⅲ.①动物-图集 Ⅳ.①Q95-64

中国版本图书馆CIP数据核字(2018)第095418号

自然图志
The Illustrated Natural History
（英）约翰·乔治·伍德著

特约策划　黄曙辉　徐　辰
责任编辑　庞　坚
特约编辑　许　倩
装帧设计　刘怡霖

出版发行　华东师范大学出版社
社　　址　上海市中山北路3663号　邮编 200062
网　　址　www.ecnupress.com.cn
电　　话　021-60821666　行政传真　021-62572105
客服电话　021-62865537
门市（邮购）电话　021-62869887
地　　址　上海市中山北路3663号华东师范大学校内先锋路口
网　　店　http://hdsdcbs.tmall.com/

印 刷 者　虎彩印艺股份有限公司
开　　本　16开
印　　张　152
版　　次　2018年6月第1版
印　　次　2018年6月第1次
书　　号　ISBN 978-7-5675-7718-3
定　　价　2600.00元（精装全三册）

出 版 人　王　焰

（如发现本版图书有印订质量问题，请寄回本社客服中心调换或电话021-62865537联系）

《西方博物学大系》总序

江晓原

《西方博物学大系》收录博物学著作超过一百种，时间跨度为 15 世纪至 1919 年，作者分布于 16 个国家，写作语种有英语、法语、拉丁语、德语、弗莱芒语等，涉及对象包括植物、昆虫、软体动物、两栖动物、爬行动物、哺乳动物、鸟类和人类等，西方博物学史上的经典著作大备于此编。

中西方"博物"传统及观念之异同

今天中文里的"博物学"一词，学者们认为对应的英语词汇是 Natural History，考其本义，在中国传统文化中并无现成对应词汇。在中国传统文化中原有"博物"一词，与"自然史"当然并不精确相同，甚至还有着相当大的区别，但是在"搜集自然界的物品"这种最原始的意义上，两者确实也大有相通之处，故以"博物学"对译 Natural History 一词，大体仍属可取，而且已被广泛接受。

已故科学史前辈刘祖慰教授尝言：古代中国人处理知识，如开中药铺，有数十上百小抽屉，将百药分门别类放入其中，即心安矣。刘教授言此，其辞若有憾焉——认为中国人不致力于寻求世界"所以然之理"，故不如西方之分析传统优越。然而古代中国人这种处理知识的风格，正与西方的博物学相通。

与此相对，西方的分析传统致力于探求各种现象和物体之间的相互关系，试图以此解释宇宙运行的原因。自古希腊开始，西方哲人即孜孜不倦建构各种几何模型，欲用以说明宇宙如何运行，其中最典型的代表，即为托勒密（Ptolemy）的宇宙体系。

比较两者，差别即在于：古代中国人主要关心外部世界"如何"运行，而以希腊为源头的西方知识传统（西方并非没有别的知识传统，只是未能光大而已）更关心世界"为何"如此运行。在线性发展无限进步的科学主义观念体系中，我们习惯于认为"为何"是在解决了"如何"之后的更高境界，故西方的分析传统比中国的传统更高明。

然而考之古代实际情形，如此简单的优劣结论未必能够成立。例如以天文学言之，古代东西方世界天文学的终极问题是共同的：给定任意地点和时刻，计算出太阳、月亮和五大行星（七政）的位置。古代中国人虽不致力于建立几何模型去解释七政"为何"如此运行，但他们用抽象的周期叠加（古代巴比伦也使用类似方法），同样能在足够高的精度上计算并预报任意给定地点和时刻的七政位置。而通过持续观察天象变化以统计、收集各种天象周期，同样可视之为富有博物学色彩的活动。

还有一点需要注意：虽然我们已经接受了用"博物学"来对译 Natural History，但中国的博物传统，确实和西方的博物学有一个重大差别——即中国的博物传统是可以容纳怪力乱神的，而西方的博物学基本上没有怪力乱神的位置。

古代中国人的博物传统不限于"多识于鸟兽草木之名"。体现此种传统的典型著作，首推晋代张华《博物志》一书。书名"博物"，其义尽显。此书从内容到分类，无不充分体现它作为中国博物传统的代表资格。

《博物志》中内容，大致可分为五类：一、山川地理知识；二、奇禽异兽描述；三、古代神话材料；四、历史人物传说；五、神仙方伎故事。这五大类，完全符合中国文化中的博物传统，深合中国古代博物传统之旨。第一类，其中涉及宇宙学说，甚至还有"地动"思想，故为科学史家所重视。第二类，其中甚至出现了中国古代长期流传的"守宫砂"传说的早期文献：相传守宫砂点在处女胳膊上，永不褪色，只有性交之后才会自动消失。第三类，古代神话传说，其中甚至包括可猜想为现代"连体人"的记载。第四类，各种著名历史人物，比如三位著名刺客的传说，此三名刺客及所刺对象，历史上皆实有其人。第五类，包括各种古代方术传说，比如中国古代房中养生学说，房中术史上的传说人物之一"青牛道士封君达"等等。前两类与西方的博物学较为接近，但每一类都会带怪力乱神色彩。

"所有的科学不是物理学就是集邮"

在许多人心目中，画画花草图案，做做昆虫标本，拍拍植物照片，这类博物学活动，和精密的数理科学，比如天文学、物理学等等，那是无法同日而语的。博物学显得那么的初级、简单，甚至幼稚。这种观念，实际上是将"数理程度"作为唯一的标尺，用来衡量一切知识。但凡能够使用数学工具来描述的，或能够进行物理实验的，那就是"硬"科学。使用的数学工具越高深越复杂，似乎就越"硬"；物理实验设备越庞大，花费的金钱越多，似乎就越"高端"、越"先进"……

这样的观念，当然带着浓厚的"物理学沙文主义"色彩，在很多情况下是不正确的。而实际上，即使我们暂且同意上述"物理学沙文主义"的观念，博物学的"科学地位"也仍然可以保住。作为一个学天体物理专业出身，因而经常徜徉在"物理学沙文主义"幻影之下的人，我很乐意指出这样一个事实：现代天文学家们的研究工作中，仍然有绘制星图，编制星表，以及为此进行的巡天观测等等活动，这些活动和博物学家"寻花问柳"，绘制植物或昆虫图谱，本质上是完全一致的。

这里我们不妨重温物理学家卢瑟福（Ernest Rutherford）的金句："所有的科学不是物理学就是集邮（All science is either physics or stamp collecting）。"卢瑟福的这个金句堪称"物理学沙文主义"的极致，连天文学也没被他放在眼里。不过，按照中国传统的"博物"理念，集邮毫无疑问应该是博物学的一部分——尽管古代并没有邮票。卢瑟福的金句也可以从另一个角度来解读：既然在卢瑟福眼里天文学和博物学都只是"集邮"，那岂不就可以将博物学和天文学相提并论了？

如果我们摆脱了科学主义的语境，则西方模式的优越性将进一步被消解。例如，按照霍金（Stephen Hawking）在《大设计》（*The Grand Design*）中的意见，他所认同的是一种"依赖模型的实在论（model-dependent realism）"，即"不存在与图像或理论无关的实在性概念（There is no picture- or theory-independent concept of reality）"。在这样的认识中，我们以前所坚信的外部世界的客观性，已经不复存在。既然几何模型只不过是对外部世界图像的人为建构，则古代中国人干脆放弃这种建构直奔应用（毕竟在实际应用中我们只需要知道七政"如何"运行），又有何不可？

传说中的"神农尝百草"故事，也可以在类似意义下得到新的解读："尝百草"当然是富有博物学色彩的活动，神农通过这一活动，得知哪些草能够治病，哪些不能，然而在这个传说中，神农显然没有致力于解释"为何"某些草能够治病而另一些则不能，更不会去建立"模型"以说明之。

"帝国科学"的原罪

今日学者有倡言"博物学复兴"者，用意可有多种，诸如缓解压力、亲近自然、保护环境、绿色生活、可持续发展、科学主义解毒剂等等，皆属美善。编印《西方博物学大系》也是意欲为"博物学复兴"添一助力。

然而，对于这些博物学著作，有一点似乎从未见学者指出过，而鄙意以为，当我们披阅把玩欣赏这些著作时，意识到这一点是必须的。

这百余种著作的时间跨度为 15 世纪至 1919 年，注意这个时间跨度，正是西方列强"帝国科学"大行其道的时代。遥想当年，帝国的科学家们乘上帝国的军舰——达尔文在皇家海军"小猎犬号"上就是这样的场景之一，前往那些已经成为帝国的殖民地或还未成为殖民地的"未开化"的遥远地方，通常都是踌躇满志、充满优越感的。

作为一个典型的例子，英国学者法拉在（Patricia Fara）《性、植物学与帝国：林奈与班克斯》（*Sex, Botany and Empire, The Story of Carl Linnaeus and Joseph Banks*）一书中讲述了英国植物学家班克斯（Joseph Banks）的故事。1768 年 8 月 15 日，班克斯告别未婚妻，登上了澳大利亚军舰"奋进号"。此次"奋进号"的远航是受英国海军部和皇家学会资助，目的是前往南太平洋的塔希提岛（Tahiti，法属海外自治领，另一个常见的译名是"大溪地"）观测一次比较罕见的金星凌日。舰长库克（James Cook）是西方殖民史上最著名的舰长之一，多次远航探险，开拓海外殖民地。他还被认为是澳大利亚和夏威夷群岛的"发现"者，如今以他命名的群岛、海峡、山峰等不胜枚举。

当"奋进号"停靠塔希提岛时，班克斯一下就被当地美丽的土著女性迷昏了，他在她们的温柔乡里纵情狂欢，连库克舰长都看不下去了，"道德愤怒情绪偷偷溜进了他的日志当中，他发现自己根本不可能不去批评所见到的滥交行为"，而班克斯纵欲到了"连嫖妓都毫无激情"的地步——这是别人讽刺班克斯的说法，因为对于那时常年航行于茫茫大海上的男性来说，上岸嫖妓通常是一项能够唤起"激情"的活动。

而在"帝国科学"的宏大叙事中，科学家的私德是无关紧要的，人们关注的是科学家做出的科学发现。所以，尽管一面是班克斯在塔希提岛纵欲滥交，一面是他留在故乡的未婚妻正泪眼婆娑地"为远去的心上人绣织背心"，这样典型的"渣男"行径要是放在今天，非被互联网上的口水淹死不可，但是"班克斯很快从他们的分离之苦中走了出来，在外近三年，他活得倒十分滋润"。

法拉不无讽刺地指出了"帝国科学"的实质："班克斯接管了当地的女性和植物，而库克则保护了大英帝国在太平洋上的殖民地。"甚至对班克斯的植物学本身也调侃了一番："即使是植物学方面的科学术语也充满了性指涉。……这个体系主要依靠花朵之中雌雄生殖器官的数量来进行分类。"据说"要保护年轻妇女不受植物学教育的浸染，他们严令禁止各种各样的植物采集探险活动。"这简直就是将植物学看成一种"涉黄"的淫秽色情活动了。

在意识形态强烈影响着我们学术话语的时代，上面的故事通常是这样被描述的：库克舰长的"奋进号"军舰对殖民地和尚未成为殖民地的那些地方的所谓"访问"，其实是殖民者耀武扬威的侵略，搭载着达尔文的"小猎犬号"军舰也是同样行径；班克斯和当地女性的纵欲狂欢，当然是殖民者对土著妇女令人发指的蹂躏；即使是他采集当地植物标本的"科学考察"，也可以视为殖民者"窃取当地经济情报"的罪恶行为。

后来改革开放，上面那种意识形态话语被抛弃了，但似乎又走向了另一个极端，完全忘记或有意回避殖民者和帝国主义这个层面，只歌颂这些军舰上的科学家的伟大发现和成就，例如达尔文随着"小猎犬号"的航行，早已成为一曲祥和优美的科学颂歌。

其实达尔文也未能免俗，他在远航中也乐意与土著女性打打交道，当然他没有像班克斯那样滥情纵欲。在达尔文为"小猎犬号"远航写的《环球游记》中，我们读到："回程途中我们遇到一群黑人姑娘在聚会，……我们笑着看了很久，还给了她们一些钱，这着实令她们欣喜一番，拿着钱尖声大笑起来，很远还能听到那愉悦的笑声。"

有趣的是，在班克斯在塔希提岛纵欲六十多年后，达尔文随着"小猎犬号"也来到了塔希提岛，岛上的土著女性同样引起了达尔文的注意，在《环球游记》中他写道："我对这里妇女的外貌感到有些失望，然而她们却很爱美，把一朵白花或者红花戴在脑后的髣鬓上……"接着他以居高临下的笔调描述了当地女性的几种发饰。

用今天的眼光来看，这些在别的民族土地上采集植物动物标本、测量地质水文数据等等的"科学考察"行为，有没有合法性问题？有没有侵犯主权的问题？这些行为得到当地人的同意了吗？当地人知道这些行为的性质和意义吗？他们有知情权吗？……这些问题，在今天的国际交往中，确实都是存在的。

也许有人会为这些帝国科学家辩解说：那时当地土著尚在未开化或半开化状态中，他们哪有"国家主权"的意识啊？他们也没有制止帝国科学家的考察活动啊？但是，这样的辩解是无法成立的。

姑不论当地土著当时究竟有没有试图制止帝国科学家的"科学考察"行为，现在早已不得而知，只要殖民者没有记录下来，我们通常就无法知道。况且殖民者有军舰有枪炮，土著就是想制止也无能为力。正如法拉所描述的："在几个塔希提人被杀之后，一套行之有效的易货贸易体制建立了起来。"

即使土著因为无知而没有制止帝国科学家的"科学考察"行为，这事也很像一个成年人闯进别人的家，难道因为那家只有不懂事的小孩子，闯入者就可以随便打探那家的隐私、拿走那家的东西、甚至将那家的房屋土地据为己有吗？事实上，很多情况下殖民者就是这样干的。所以，所谓的"帝国科学"，其实是有着原罪的。

如果沿用上述比喻，现在的局面是，家家户户都不会只有不懂事的孩子了，所以任何外来者要想进行"科学探索"，他也得和这家主人达成共识，得到这家主人的允许才能够进行。即使这种共识的达成依赖于利益的交换，至少也不能单方面强加于人。

博物学在今日中国

博物学在今日中国之复兴，北京大学刘华杰教授提倡之功殊不可没。自刘教授大力提倡之后，各界人士纷纷跟进，仿佛昔日蔡锷在云南起兵反袁之"滇黔首义，薄海同钦，一檄遥传，景从恐后"光景，这当然是和博物学本身特点密切相关的。

无论在西方还是在中国，无论在过去还是在当下，为何博物学在它繁荣时尚的阶段，就会应者云集？深究起来，恐怕和博物学本身的特点有关。博物学没有复杂的理论结构，它的专业训练也相对容易，至少没有天文学、物理学那样的数理"门槛"，所以和一些数理学科相比，博物学可以有更多的自学成才者。这次编印的《西方博物学大系》，卷帙浩繁，蔚为大观，同样说明了这一点。

最后，还有一点明显的差别必须在此处强调指出：用刘华杰教授喜欢的术语来说，《西方博物学大系》所收入的百余种著作，绝大部分属于"一阶"性质的工作，即直接对博物学作出了贡献的著作。事实上，这也是它们被收入《西方博物学大系》的主要理由之一。而在中国国内目前已经相当热的博物学时尚潮流中，绝大部分已经出版的书籍，不是属于"二阶"性质（比如介绍西方的博物学成就），就是文学性的吟风咏月野草闲花。

要寻找中国当代学者在博物学方面的"一阶"著作，如果有之，以笔者之孤陋寡闻，唯有刘华杰教授的《檀岛花事——夏威夷植物日记》三卷，可以当之。这是刘教授在夏威夷群岛实地考察当地植物的成果，不仅属于直接对博物学作出贡献之作，而且至少在形式上将昔日"帝国科学"的逻辑反其道而用之，岂不快哉！

2018 年 6 月 5 日
于上海交通大学
科学史与科学文化研究院

《自然图志》

约翰·乔治·伍德

（1827-1889）

英国著名博物学家、作家约翰·乔治·伍德（John George Wood），1827年生于伦敦，是外科医生约翰·弗里曼·伍德的儿子。由于体质羸弱，他年少时不得不在家中受教育，并于1830年随父母搬到牛津居住。在牛津，小伍德常去乡间野地游玩，对他日后的博物学研究造成决定性影响。

十一岁时，伍德开始跟随叔父生活，在埃施波恩文法学校过了六年寄宿生活。十七岁时他考入牛津大学默顿学院，学业有成。1851年，弱冠二十四岁的伍德出版了自己的第一本著作《自然图志》，本书厚达2400页，配有大量精美铜版画，内容涉及的动物既有乡间常见的品类，也有当时对英国人而言远在天边的珍奇物种，引起轰动，至今仍是博物学史的名作。

他本人也于三年后成为牧师，但对博物学的热爱使他割舍了教区服侍，很快辞去神职，开始专门从事博物学研究与写作。

伍德一生笔耕不辍，至1889年在考文垂去世为止，出版过十多部博物学专著，以及相关的文学作品，包括著名的"小生灵"三部曲、《自然的教诲：仿生学》、《去远行》、《大众博物学》和《蛮野之民：人类的历史》等。伍德的著作文字秀美，显著有别于那些艰涩的科研论文，以细腻的观察与富于情感的笔触，结合对故土与自然生灵的热爱，成就了一部又一部博物学史上的奇迹丰碑。

本次影印底本为1853年出版的《自然图志》豪华版。

THE

ILLUSTRATED

NATURAL HISTORY.

BY THE REV.

J. G. WOOD, M.A. F.L.S.

AUTHOR OF "ANECDOTES OF ANIMAL LIFE," "COMMON OBJECTS OF THE COUNTRY,"
"MY FEATHERED FRIENDS," ETC. ETC.

WITH NEW DESIGNS

BY WOLF, ZWECKER, WEIR, COLEMAN, HARVEY, ETC. ETC.

ENGRAVED BY THE BROTHERS DALZIEL.

VOL. I.

MAMMALIA.

LONDON:

ROUTLEDGE, WARNE, AND ROUTLEDGE, FARRINGDON STREET.

NEW YORK: 56, WALKER STREET.

LONDON:
PRINTED BY RICHARD CLAY,
BREAD STREET HILL.

PREFACE.

In the present Volume I have endeavoured to carry out, on a more extended scale, the principle which has been partially indicated in several of my smaller works; namely, to present to the reader the outlines of zoologic knowledge in a form that shall be readily comprehended, while it is as intrinsically valuable as if it were couched in the most repellent vocabulary of conventional technicalities. In acting thus, an author must voluntarily abnegate the veneration which attaches itself to those who are the accredited possessors of abstruse learning, and must content himself with the satisfaction of having achieved the task which has been placed in his hands. In accordance with this principle, the technical language of scientific zoology has been carefully avoided, and English names have been employed wherever practicable in the place of Greek or Latin appellatives.

The body of the work has been studiously preserved in a simple and readable form, and the more strictly scientific portions have been removed to the "Compendium of Generic Distinctions" at the end of the volume. In this Compendium the reader will find a brief notice of the various characteristics which are employed by our best systematic naturalists, such as Owen, Gray, Van der Hoeven, and others, for the purpose of separating the different genera from each other; and by its aid he will be enabled to place every animal in that position which it is at present supposed to occupy. Even in that Compendium simplicity of diction has been maintained. For example: the word "five-toed" has been substituted for "pentedactylous;" "pointed" for "acuminate;" "ringed" for "annulate;" together with innumerable similar instances which need no separate mention.

Owing to the inordinate use of pseudo-classical phraseology, the fascinating study of animal life has been too long considered as a profession or a science restricted to a favoured few, and interdicted to the many until they have undergone a long apprenticeship to its preliminary formulæ. So deeply rooted is this idea, that the popular notion of a scientific man is of one who possesses a fund of words, and not of one who has gathered a mass of ideas. There is really not the least reason why any one of ordinary capabilities and moderate memory should not be acquainted with the general outlines of zoology, and possess some knowledge of the representative animals, which serve as types of each group, tribe, or family; for when relieved of the cumbersome diction with which it is embarrassed, the study of animal life can be brought within the comprehension of all who care to examine the myriad varieties of form and colour with which the Almighty clothes His living poems.

The true object of Zoology is not, as some appear to fancy, to arrange, to number, and to ticket animals in a formal inventory, but to make the study an inquiry into the Life-nature, and not only an investigation of the lifeless organism. I must not, however, be understood to disparage the outward form, thing of clay though it be. For what wondrous clay it is, and how marvellous the continuous miracle by which the dust of earth is transmuted into the glowing colours and graceful forms which we most imperfectly endeavour to preserve after the soul has departed therefrom. It is a great thing to be acquainted with the material framework of any creature, but it is a far greater to know something of the principle which gave animation to that structure. The former, indeed, is the consequence of the latter. The lion, for example, is not predacious because it possesses fangs, talons, strength, and activity; on the contrary, it possesses these qualities because its inmost nature is predacious, and it needs these appliances to enable it to carry out the innate principle of its being; so that the truest description of the lion is that which treats of the animating spirit, and not only of the outward form. In accordance with this principle, it has been my endeavour to make the work rather anecdotal and vital than merely anatomical and scientific. The object of a true zoologist is to search into the essential nature of every being, to investigate, according to his individual capacity, the reason why it should have been placed on earth, and to give his personal service to his Divine Master in developing that nature in the best manner and to the fullest extent.

What do we know of Man from the dissecting room? Of Man, the warrior, the statesman, the poet, or the saint? In the lifeless corpse there are no records of the burning thoughts, the hopes, loves, and fears that once animated that now passive form, and which constituted the very essence of the being. Every nerve, fibre, and particle in the dead bodies of the king and the beggar, the poet and the boor, the saint and the sensualist, may be separately traced, and anatomically they shall all be alike, for neither of the individuals is there, and on the dissecting table lies only the cast-off attire that the spirit no longer needs. What can an artist learn even of the outward form of Man, if he lives only in the dissecting room, and studies the human frame merely through the medium of scalpel and scissors? He may, indeed, obtain an accurate muscular outline, but it will be an outline of a cold and rigid corpse, suggestive only of the charnel-house, and devoid of the soft and rounded form, the delicate tinting, and breathing grace which invest the living human frame. A feeling eye will always discover whether an artist has painted even his details of attire from a lay figure instead of depicting the raiment as it rests upon and droops from the breathing form of a living model ; for such robes are not raiment, but a shroud. So it is with the animal kingdom. The zoologist will never comprehend the nature of any creature by the most careful investigation of its interior structure or the closest inspection of its stuffed skin, for the material structure tells little of the vital nature, and the stuffed skin is but the lay figure stiffly fitted with its own cast coat.

The true study of Zoology is of more importance than is generally conceived, for although "the proper study of mankind is Man," it is impossible for us to comprehend the loftiness and grandeur of humanity, or even its individual and physical nature, without possessing some knowledge of the earlier forms of God's animated organizations. We must follow the order of creation, and as far as our perceptions will permit, begin where the Creator began. We shall then find that no animal leads an isolated existence, for the minutest atom of animated life which God has enfranchised with an individual existence, forms, though independent in itself, an integral and necessary portion of His ever-changing yet eternal organic universe. Hence every being which draws the breath of life forms a part of one universal family, bound together by the ties of a common creaturehood. And as being ourselves members of that living and breathing family, we learn to view with clearer eyes and more reverent hearts those beings which, although less godlike than ourselves in their physical or moral natures, demand for that very reason our kindliest sympathies and most indulgent care. For we, being made in the image of God, are to them the visible representations of that Divine Being who gave the Sabbath alike for man and beast, and who takes even the sparrows under His personal protection.

INTRODUCTION.

IN order to understand any science rightly, it needs that the student should proceed to its contemplation in an orderly manner, arranging in his mind the various portions of which it is composed, and endeavouring, as far as possible, to follow that classification which best accords with nature. The result of any infringement of this rule is always a confusion of ideas, which is sure to lead to misapprehension. So, in the study of living beings, it is necessary to adhere to some determinate order, or the mind becomes bewildered among the countless myriads of living creatures that fill earth, air, and water.

That some determinate order exists is evident to any thinking mind, but the discovery of the principle on which this order is founded is a problem that as yet has received but a partial solution. We already know some of the links of that wondrous chain that connects Man with the microscopic animalcule, but the one plan on which the Animal Kingdom is formed, has yet to be made known.

It is impossible to contemplate the vast mass of animal life without the conviction that the most supreme harmony has been observed in their creation, and the most perfect order exists in their connexion one with the other. Whatever may be the key to this enigma,—and it is of a certainty a very simple one, possibly eluding us from its very simplicity—from the days of Aristotle to the present time zoologists have been diligently seeking for the true system of animated nature; and until that auspicious discovery be achieved, we must be content with making as near an approximation as possible.

As a general arranges his army into its greater divisions, and each division into regiments and companies, so does the naturalist separate the host of living beings into greater and smaller groups. The present state of zoological science gives five as the number of divisions of which the animal kingdom is composed, the highest of which is that in which Man himself is, by some, placed. These are called Vertebrates, Molluscs, Articulates, Radiates, and Protozoa. Of each of these divisions a slight description will be given, and each will be considered more at length in its own place.

1st. The VERTEBRATES include Man and all the Mammalia, the Birds, the Reptiles, and the Fish.

The term Vertebrate is applied to them because they are furnished with a succession of bones called "vertebræ," running along the body and forming a support and protection to the nervous cord that connects the body with the brain by means of numerous branches The Vertebrates, with one or two known exceptions, have red blood and a muscular heart.

2d. The MOLLUSCA, or soft-bodied animals, include the Cuttle-fish, the Snails, Slugs, Mussels, &c. Some of them possess shells, while others are entirely destitute of such defence. Their nervous system is arranged on a different plan from that of the Vertebrates. They have no definite brain, and no real spinal cord, but their nerves issue from certain masses of nervous substance technically called ganglia.

3d. The ARTICULATES, or jointed animals, form an enormously large division, comprising the Crustaceans, such as the Crabs and Lobsters, the Insects, Spiders, Worms, and very many creatures so different from each other, that it is scarcely possible to find any

1.   B

common characteristics. It is among these lower animals that the want of a true classification is most severely felt, and the present arrangement can only be considered as provisional.

4th. The next division, that of the RADIATED animals, is so named on account of the radiated or star-like form of the body, so well exhibited in the Star-fishes and the Sea-anemones. Their nervous system is very obscure, and in many instances so slight as to baffle even the microscope. Many of the Radiates possess the faculty of giving out a phosphorescent light, and it is to these animals that the well-known luminosity of the sea is chiefly owing.

5th. The PROTOZOA, or primitive animals, are, as far as we know, devoid of internal organs or external limbs, and in many of them the signs of life are so feeble, that they can scarcely be distinguished from vegetable germs. The Sponges and Infusorial Animal-cules are familiar examples of this division.

VERTEBRATES.

The term Vertebrate is derived from the Latin word *vertere*, signifying to turn ; and the various bones that are gathered round and defend the spinal cord are named vertebræ, because they are capable of being moved upon each other in order to permit the animal to flex its body. Were the spinal cord to be defended by one long bone, the result would be that the entire trunk of the animal would be stiff, graceless, and exceedingly liable to injury from any sudden shock. In order, therefore, to give the body latitude of motion, and at the same time to afford effectual protection to the delicate nerve-cord, on which the welfare of the entire structure depends, the bony spine is composed of a series of distinct pieces, varying in form and number according to the species of animal, each being affixed to its neighbour in such a manner as to permit the movement of one upon the other. The methods by which these vertebræ are connected with each other vary according to the amount of flexibility required by the animal of which they form a part. For example, the heavy elephant would find himself prostrate on the ground if his spine were composed of vertebræ as flexible as those of the snakes ; while the snake, if its spine were stiff as that of the elephant, would be unable to move from the spot where it happened to lie. But in all animals there is some power of movement in the spinal column, although in many creatures it is very trifling.

Anatomy shows us that, in point of fact, the essential skeleton is composed of vertebræ, and that even the head is formed by the development of these wonderful bones. The limbs can but be considered as appendages, and in many Vertebrated animals, such as the common snake of our fields, the lamprey, and others, there are no true limbs at all.

The perfect VERTEBRA consists of three principal portions. Firstly, there is a solid, bony mass, called the centre, which is the basis of the whole vertebra. From this centre springs an arch of bone, through which runs the spinal cord, and directly opposite to this arch a second arch springs, forming the guardian of the chief blood-vessel of the body. Each arch is called by a name significative of its use ; those through which the spinal cord runs being termed the neural, or nerve arch, and that for the passage of the blood-vessel is named the hæmal, or blood arch. There are other portions of the vertebræ which are developed into the bones, called "processes," some of which we can feel by placing a hand on any part of the spine.

It will be seen that, strictly speaking, the vertebræ are not of so much importance in the animal as the spinal cord, of which the vertebræ are but guardians, and that the division should rather have been defined by the character of the nerve than by that of the bone which is built around it.

Indeed, wherever the chief nervous column lies, it seems to gather the bony particles, and to arrange them round itself as its clothing or armour. This may be seen in a very young chicken, if the egg in which it is formed is opened during the first few days of incubation.

The position of the spinal cord is always along the back in every Vertebrate animal The insects, the lobster, and other invertebrate animals exhibit the principal nerve-cords running along the abdomen; the position, therefore, of the chief nervous cord settles the division to which the animal belongs. This rule is of great importance in classification, because in every group of animals there are some in whom the distinguishing character-istics are so slight that they hardly afford a real criterion by which to judge. In the lower divisions the number of these enigmatical animals is very considerable, and even in the highest of all, namely, the Vertebrates, there are one or two individuals whose position is but dubious. The best known of these creatures is the Amphioxus, a small, transparent fish, not uncommon on sandy coasts. In this curious animal the vertebral column is composed of, or rather represented by, a jelly-like cord, on which the divisions of the vertebræ are indicated by very slight markings. The spinal cord lies on the upper surface of this gelatinous substance, and there is no distinct brain, the nervous cord simply terminating in a rounded extremity. The blood is unlike that of the generality of Vertebrate animals, being transparent like water, instead of bearing the red hue that is so characteristic of their blood. Neither is there any separate heart, the circulation seeming to be effected by the contraction of the arteries.

On account of these very great divergencies from the usual vertebrate characteristics, its claim to be numbered among the Vertebrates appears to be a very hopeless one. But the spinal cord is found to run along the *back* of the creature, and this one fact settles its position in the Animal Kingdom.

It must be remembered that the Amphioxus is to be considered an exceptional being, and that when the anatomy of Vertebrate animals is described, the words " with the exception of the Amphioxus" must be supplied by the reader. The character of the nerves, bones, blood, and other structures, will be shown, in the course of the work, in connexion with the various animals of which they form a part.

MAMMALIA.

The Vertebrated animals fall naturally into four great classes, which are so clearly marked that, with the exception of a few singularly constructed creatures, such as the Lepidosiren, or Mud-fish of the Gambia, any vertebrate animal can be without difficulty referred to its proper class. These four classes are termed MAMMALS, BIRDS, REPTILES, and FISHES,—their precedence in order being determined by the greater or less develop-ment of their structure.

Mammals, or Mammalia, as they are called more scientifically, comprise Man, the Monkey tribes, the Bats, the Dogs and Cats, all the hoofed animals, the Whales and their allies, and other animals, amounting in number to some two thousand species, the last on the list being the Sloth. The name by which they are distinguished is derived from the Latin word *mamma*, a breast, and is given to them because all the species belonging to this class are furnished with a set of organs, called the MAMMARY GLANDS, secreting the liquid known as milk, by which the young are nourished.

The number of the mammæ varies much, as does their position. Many animals that produce only one, or at the most two, young at the same birth, have but two mammæ, such as the monkey, the elephant, and others; while some,—such as the cat, the dog, and the swine,—are furnished with a sufficient number of these organs to afford sustenance to their numerous progeny. Sometimes the mammæ are placed on the breast, as in the monkey tribe; sometimes by the hind legs, as in the cow and the horse; and sometimes, as in the swine, along the abdomen.

The glands that supply the mammæ with milk lie under the skin, and by the microscope are easily resolvable into their component parts. Great numbers of tiny cells, or cellules, as they are named, are grouped together in little masses, something like bunches of minute grapes, and by means of very small tubes pour their secretions into vessels of a larger size. As the various tube-branches join each other they become larger, until they unite in five or six principal vessels, which are so constructed as to be capable of enlargement according to the amount of liquid which they are called upon to hold. In some animals, such as the cow, these reservoirs are extremely large, being capable of containing at least a quart of milk. The reservoirs are much smaller towards the mamma itself, and serve as tubes for the conveyance of the milk into the mouth of the young. Of the milk itself we shall speak in another part of the work.

The BLOOD of the Vertebrate animals is of a light red colour when freshly drawn from the arteries. This wondrous fluid, in which is hidden the life principle that animates the being, is of a most complex structure, as may be imagined when it is remembered that all the parts of the body are formed from the blood; and therefore to give a full description of that fluid would occupy more space than can be afforded to one subject. It is, however, so important a substance that it demands some notice.

When it is freshly drawn, the blood appears to be of an uniform consistence, but if poured into a vessel and suffered to remain undisturbed it soon begins to change its aspect. A comparatively solid and curd-like mass, of a deep red colour, rises to the surface, and there forms a kind of cake, while the liquid on which it floats is limpid and almost colourless. The solid mass is called the clot, and the liquid is known by the name of serum. The whole time consumed in this curious process is about twenty minutes. While thus coagulating the blood gives out a peculiar odour, which, although far from powerful, can be perceived at some distance, and to many persons is inexpressibly revolting.

The upper part of the clot is covered with a thick film of an elastic and tenacious nature, which can be washed free from the red colouring substance, and then appears of a yellowish white tint. It can be drawn out and spread between the fingers, as if it were an organic membrane; and, as its particles arrange themselves into fibres, the substance is called fibrin. When a portion of fibrin is drawn out until it is much lengthened, the fibres are seen crossing each other in all directions, sometimes forming themselves into regular lines.

The red mass, which remains after the fibrin and serum have been removed, is almost wholly composed of myriads of small rounded bodies, called corpuscules, which can be readily seen by spreading a drop of blood very thinly on glass, and examining it with a

BLOOD CORPUSCULES OF MAN.

microscope. The general appearance of the blood corpuscules of man is seen in the accompanying illustration. Some of the disc-like corpuscules are seen scattered about, while others have run together and adhered by their flat sides, until they look somewhat like rouleaux of coin. There is sufficient distinction between the blood corpuscules of the various Mammalia to indicate to a practised eye the kind of animal from which they were taken; while the blood of the four great divisions of the Vertebrates is so strongly marked, that a casual glance will detect the ownership of the object under the microscope. The specimen represented above is magnified about two hundred diameters. The blood corpuscules of the Mammalia are circular, while those of the other three divisions are more or less elliptical.

That the blood contains within itself the various substances of which the body is composed, is evident to the intellect, although as yet no investigator has discovered the mode of its operation.

How the blood corpuscules are generated from the vegetable and animal substances taken into the stomach, we know not; but we do know that each globule possesses life, passing through its regular stages of birth, development, age, and death. When yet in their first stages of existence, the blood corpuscules are colourless, not taking the well-known ruddy tint until they have attained their full development. The living current that passes through our bodies is truly a fathomless ocean of wonders! Even the material formation of this fluid is beyond our present sight, which cannot penetrate through the veil which conceals its mysteries. Much less can we explain the connexion of the blood with the mind, or know how it is that one thought will send the blood coursing through the frame with furious speed, crimsoning the face with hot blushes; or another cause the vital fluid to recoil to the heart, leaving the countenance pallid, the eyes vacant, and the limbs cold and powerless, as if the very life had departed from the body.

Not without reason do the earlier Scriptures speak so reverently of the blood, accepting the outpoured life of beasts as an atonement for the sin, and witness of the penitence of man, and forbid its use for any less sacred office. Nor was it without a still mightier meaning that the later Scriptures endue the blood with a sacramental sense, giving even to its vegetable symbol, the blood of the grape, a dignity greater than that of the former sacrifices.

A few words must also be given to the mode by which the blood is kept continually running its appointed course through the animal frame. This process, commonly called CIRCULATION, takes place in the following manner, Man being an example :—

In the centre of the breast lies the heart, an organ composed of four chambers, the two upper being termed auricles, and the two lower being distinguished by the title of ventricles. These are only conventional terms, and do not express the office of the parts. The auricles are comparatively slight in structure, but the ventricles are extremely powerful, and contract with great force, by means of a curiously spiral arrangement of the muscular fibres. These latter chambers are used for the purpose of propelling the blood through the body, while the auricles serve to receive the blood from these vessels, and to throw it into the ventricles when they are ready for it.

By the systematic expansion and contraction of the heart-chambers, the blood is sent on its mission to all parts of the body, through vessels named arteries, gradually diminishing in diameter as they send forth their branches, until they terminate in branchlets scarcely so large as hairs, and which are therefore called "capillaries," from the Latin word *capillus*, a hair. The formation of the capillary system is well shown

CAPILLARY.

CAPILLARIES IN SKIN OF
HUMAN FINGER.

CAPILLARIES OF HUMAN
TONGUE.

by the accompanying sketches. The first figure exhibits a portion of capillaries which are found in the fatty tissues, while the second and third are examples of the corresponding vessels in the finger and the tongue.

In the capillaries the blood corpuscles would end their course, were they not met and welcomed by a second set of capillaries, which take up the wearied and weakened globules, carrying them off to the right-hand chambers of the heart, which impel them through vessels known by the name of "veins," to be refreshed by the air which is supplied to them in the beautiful structure known as the lungs. Meeting there with fresh vitality—if it may so be called—the blood corpuscles throw off some of their effete portions, and so, brightened and strengthened, are again sent from the heart to run their round of existence.

It is indeed a marvellous system, this constant circular movement, that seems to be inherent in the universe at large, as well as in the minute forms that inhabit a single orb. The planets roll through their appointed courses in the macrocosmal universe, as the blood globules through the veins of the microcosm, man : each has its individual life, while it is inseparably connected with its fellow-orbs, performing a special and yet a collective work in the vast body to which it belongs ; darkening and brightening in its alternate night and day until it has completed its career.

In order to prevent other organs from pressing on the heart, and so preventing it from playing freely, a membranous envelope, called from its office the "pericardium," surrounds the heart and guards it.

The various operations which are simultaneously conducted in our animal frame are so closely connected with each other that it is impossible to describe one of them without trenching upon the others. Thus, the system of the circulatory movement, by which the blood passes through the body, is intimately connected with the system of RESPIRATION, by which the blood is restored to the vigour needful for its many duties.

In order to renew the worn-out blood, there must be some mode of carrying off its effete particles, and of supplying the waste with fresh nourishment. For this purpose the air must be brought into connexion with the blood without permitting its escape from the vessels in which it is confined. The mode by which this object is attained, in the Mammalia, is briefly as follows :—

A large tube, appropriately and popularly called the "windpipe," leads from the back of the mouth and nostrils into the interior of the breast. Just as it enters the chest it divides into two large branches, each of which subdivides into innumerable smaller branchlets, thus forming two large masses, or lobes. In these lobes, or lungs, as they are called, the air-bearing tubes become exceedingly small, until at last they are but capillaries which convey air instead of blood, each tube terminating in a minute cell. The diameter of these cells is very small, the average being about the hundred and fiftieth of an inch. Among these air-bearing capillaries the blood-bearing capillaries are so intermingled that the air and blood are separated from each other

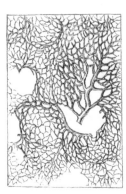

Fig. 1.
AIR-TUBES OF THE LUNGS.

Fig. 2
CAPILLARIES OF THE LUNGS.

only by membranes so delicate that the comparatively coarse substance of the blood cannot pass through, although the more ethereal gases can do so. So, by the presence of the air, the blood is renewed in vigour, and returns to its bright florid red, which had been lost in its course through the body, while the useless parts are rejected, and gathered into the air-tubes, from whence they are expelled by the breath.

The accompanying illustrations will give a good idea of the capillary structure. Fig. 1 represents the air-tubes of the lungs, and fig. 2 exhibits the capillaries through which the blood is conveyed.

The heart is placed between the two lobes of the lungs, and is in a manner embraced by them. The lungs themselves are enclosed in a delicate membrane called the "pleura." These two great vital organs are situated in the breast, and separated from the digestive and other systems by a partition, which is scientifically known by the name of "diaphragm," and in popular language by the term "midriff." This structure does not exist in the Birds ; and its presence, together with that of the freely-suspended lungs, is an unfailing characteristic of the Mammalian animal.

Thus the entire structure bears the closest resemblance to a tree, growing with its root upwards and its leaves downward,—the trachea being the trunk, the branchial tubes the limbs, the smaller tubes are the branches, and the air-cells the leaves. A similar idea runs through the nerve system and that of the blood ; all three being interwoven with each other in a manner most marvellous and beautiful.

The ORGANS OF NUTRITION occupy the greater part of the space between the diaphragm and the lower limbs, and are composed of the following parts. The mouth receives and, in most cases, grinds the food until it is sufficiently soft to be passed onwards into the general receptacle, called the stomach. Here begins the process of digestion, which is chiefly carried on by means of a liquid called the gastric juice, which is secreted by glands within the stomach, and dissolves the food until it is of an uniform soft consistency. In this state the food is called "chyme," and passes from the stomach into a tube called the "duodenum." Here the chyme begins to separate into two portions; one, an indigestible and useless mass, and the other, a creamy kind of liquid, called "chyle." The former of these substances is propelled through the long and variously-formed tube, called the intestinal canal, and rejected at its outlet; while the chyle is taken up by numerous vessels that accompany the intestines, and is finally thrown into one of the large veins close by the heart, and there mixes with the blood.

There is another curious system called the "lymphatic," on account of the limpid appearance of the liquid which is conveyed through the lymphatic vessels. These are analogous to the lacteals, but instead of belonging to the intestines, they are spread over the whole frame, being thickly arranged just under the skin. They are curiously shaped, being studded with small knotty masses, and fitted with valves which keep the contained liquid in its proper course. Both the lacteal and lymphatic vessels pour their contents into one large trunk, called from its position the thoracic duct. This vessel is about twenty inches in length, and when distended, is in its widest part as large as a common lead pencil.

All these wonderful forms and organs would, however, be but senseless masses of matter, differing from each other by the arrangement of their component parts, but otherwise dead and useless. It needs that the being which is enshrined in this bodily form (whether it be man or beast) should be able to move the frame at will, and to receive sensations from the outer world.

More than this. As all vertebrated animals are forced at short intervals to yield their wearied bodies to repose, and to sink their exhausted minds in the temporary oblivion of sleep, there must of necessity be a provision for carrying on the vital functions without the active co-operation of the mind. Were it otherwise, the first slumber of every being would become its death-sleep, and all the higher classes of animals would be extirpated in a few days. The mind would be always on the stretch to keep the heart to its constant and necessary work; to watch the play of the lungs in regenerating the blood; to aid the stomach in digesting the food, and the intestinal canal in sifting its contents; together with many other duties of a character quite as important.

Supposing such a state of things to be possible, and to be put in practice for one single hour, how terrible would be the result to humanity! We should at once degenerate into a mass of separate, selfish individuals, each thinking only of himself, and forced to give the whole of his intellectual powers to the one object of keeping the animal frame in motion. Society would vanish, arts cease from the face of the earth, and the whole occupation of man would be confined to living an isolated and almost vegetable life.

This being the case with man, the results to the lower portions of the animal kingdom would be still more terrible. For their intellect is infinitely below that of the dullest of the human race, and they would not even possess the knowledge that any active exertion would be necessary to preserve their lives. And for all living beings the wandering of the mind but for a few seconds would cause instantaneous death.

All these difficulties are removed, and the animal kingdom preserved and vivified, by means of certain vital organs, known by the name of nerves.

It is clear enough that mind does not act directly upon the muscles and the various organs of the material body, but requires a third and intermediate substance, by which it is enabled to convey its mandates and to receive information. The necessarily multitudinous channels through which this substance is conveyed are called "nerves," and are of a consistency more delicate than that of any other portions of the animal frame. There is a rather striking and close analogy between the mode in which the three systems of mind, nerve, and muscle act together, and the working of a steam-engine. In the engine we may take the fire as the analogue of the mind ; the water, of the nervous substance—the water-tubes representing the nerves ; and the iron and brass machine as the representative of the bone and muscle. Thus we may make as large a fire as we like, heap on coals, and urge a fierce draught of air through the furnace, until the grate is filled with a mass of glowing white-hot matter. But the fire cannot act on the wheels without the intermediate substance, the water. This medium being supplied, the fire acts on the water, and the water on the metallic bars and wheels, so that the three become one harmonious whole.

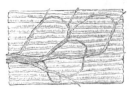

TERMINATION OF NERVE-LOOPS
IN MUSCULAR FIBRES.

The great nerve mass, called by the name of "brain," is the general source of all the nerve-cords that supply the body with vital energy, and seems to be the nerve-heart, so to speak. From the brain, a cord of nervous matter, called the "spinal cord," runs along the back, under the guardian-ship of the vertebræ, continually giving off branches of various sizes, according to the work which they have to fulfil. These branches ramify into smaller twigs, subdividing until they become so small that they almost even baffle the micro-scope. A familiar proof may be given of the wonderfully minute subdivision of the nerves, by trying to probe the skin with the point of a fine needle, and to discover any spot so small that the needle-point does not meet with a nerve.

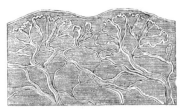

NERVES OF FINGER TIPS.

The cause of the peculiarly delicate sensibility of the finger tips is shown by the accompanying engraving, which exhibits the mode in which the nerve-loops are distributed. The object is greatly magnified, the two ridges being the enlarged representations of the minute raised lines which appear on the tips of the fingers and thumbs.

That the nerves all find their way to the brain and issue from thence, is plainly shown by the well-known fact that if the spinal cord be injured all sensation ceases in the parts of the body that lie below the injury. And it is possible to deprive any limb of sensation by dividing the chief nerve that supplies that member with nerve-fibres.

There seem to be two sets of nerves for the two purposes of conveying motive power to the body and of bringing to the nervous centres the sensations of pain or pleasure felt by any part of the body. These are appropriately known as nerves of motion and nerves of sensation.

Connected with these nerves is a second system of a very curious nature, known by the name of the "sympathetic nerve." The greater portion of the sympathetic nerve in the human frame "communicates with the other nerves immediately at their exit from the cranium and vertebral canal. It is called the ganglionic nerve, from being constituted of a number of ganglia, and from the constant disposition which it evinces in its distribution to communicate and form small knots or ganglia."* It is wonderfully interwoven with the vital organs, and from this disposition it is sometimes termed the "organic nerve." Its functions are closely connected with the phenomena of organic life, and it seems to be especially sensitive to emotional disturbances. There are several aggregations of the ganglia in various portions of the body ; the largest, which is known by the name of the "solar plexus," is placed in the pit of the stomach or "epigastrium." Its importance may be easily inferred from the extreme agony that is caused by the slightest blow near the region of that group of ganglia. A concussion that would hardly be felt upon any other portion of the body, will, if it takes place on the epigastrium, at once cause the injured person to fall as if shot,

* Wilson.

bring on collapse, deprive him of breath for some time, and leave him gasping and speechless on the ground ; while a tolerably severe blow in that region causes instantaneous death.

Anxiety seems to fix its gnawing teeth chiefly in the solar plexus, causing indigestion and many other similar maladies, and deranging the system so thoroughly that even after the exciting cause is removed the effects are painfully evident for many a sad year.

By means of this complicated system of nerves the entire body, with its vital organs, is permeated in every part by the animating power that gives vitality and energy to the frame so long as the spirit abides therein.

This is the portion of the nervous system that never slumbers nor sleeps, knowing no rest, and never ceasing from its labours until the time comes when the spirit finally withdraws from the material temple in which it has been enshrined. It is the very citadel of the nerve forces, and is the last stronghold that yields to the conquering powers of death and decay.

Thus it will be seen that each animal is a complex of many animals, interwoven with each other, and mutually aiding each other. In the human body there is, for example, the nerve-man, which has just been described ; there is a blood-man, which, if separated from the other part of the body, is found to present a human form, perfect in proportions, and composed of large trunk-vessels, dividing into smaller branches, until they terminate in their capillaries. A rough preparation of the blood-being may be made by filling the vessels with wax, and dissolving away the remaining substances, thus leaving a waxen model of the arteries and veins with their larger capillaries.

Again, there is the fibrous and muscular man, composed of forms more massive and solid than those which we have already examined.

Lastly, there is the bone-man, which is the least developed of the human images, and which, when stripped of the softer coverings, stands dense, dry, and lifeless ;—the grim scaffolding of the human edifice. Although the bones are not in themselves very pleasing objects, yet their mode of arrangement, their adaptation to the wants of the animal whose frame they support, and the beautiful mechanism of their construction, as revealed by the microscope, give a spirit and a life, even to the study of dry bones.

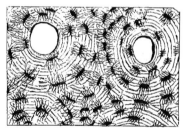

SECTION OF HUMAN BONE

The accompanying illustration represents the appearance of a transverse section of human bone, as seen under a tolerably powerful microscope.

The larger hollows are caused by the minute blood-vessels which penetrate the bone throughout its substance, and serve to deposit new particles, and to remove those whose work is over. They are, in fact, a kind of lungs of the bones, through which the osseous system is regenerated in a manner analogous to the respiration which regenerates the blood. In order to supply a sufficient volume of blood to these various vessels, several trunk vessels enter the bones at different parts of their form, and ramify out into innumerable branchlets, which again separate into the hair-like vessels that pass through the above-mentioned canals. These are termed, from their discoverer, C. Havers, the Haversian canals, and their shape and comparative size are most important in determining the class of beings which furnished the portion of bone under examination.

In the human bone these canals run so uniformly, that their cut diameters always afford a roundish outline. But in the bird-bone, the Haversian canals frequently turn off abruptly from their course, and running for a short distance at right angles, again dip and resume their former direction.

The reptiles possess very few Haversian canals, which, when they exist, are extremely large, and devoid of that beautiful regularity which is so conspicuous in the mammalia, and to a degree in the birds.

The fish-bone is often totally destitute of these canals, while, in other cases, the bone is thickly pierced with them, and exhibits also a number of minute tubes, white and delicate, as if made of ivory.

Returning to the human bone, the Haversian canals are seen to be surrounded with a number of concentric bony rings, varying much in number and shape, on which are placed sundry little black objects that somewhat resemble ants or similar insects. These latter objects are known by the name of bone-cells ; and the little dark lines that radiate from them are the indications of very minute tubes, the number and comparative dimensions of which are extremely various in different animals.

Thus, it will be seen, how easily the observer can, in a minute fragment of bone, though hardly larger than a midge's wing, read the class of animal of whose framework it once formed a part, as decisively as if the former owner were present to claim his property ; for each particle of every animal is imbued with the nature of the whole being. The life-character is enshrined in and written upon every sanguine disc that rolls through the veins ; is manifested in every fibre and nervelet that gives energy and force to the breathing and active body ; and is stereotyped upon each bony atom that forms part of its skeleton framework.

Whoever reads these hieroglyphs rightly is truly a poet and a prophet ; for to him the "valley of dry bones" becomes a vision of death passed away, and a prevision of a resurrection and a life to come. As he gazes upon the vast multitude of dead, sapless memorials of beings long since perished, "there is a shaking, and the bones come together" once again ; their fleshly clothing is restored to them ; the vital fluid courses through their bodies ; the spirit of life is breathed into them ; "and they live, and stand upon their feet." Ages upon ages roll back their tides, and once more the vast reptile epoch reigns on earth. The huge saurians shake the ground with their heavy tread, wallow in the slimy ooze, or glide sinuous through the waters ; while winged reptiles flap their course through the miasmatic vapours that hang dank and heavy over the marshy world. As with them, so shall it be with us,—an inevitable progression towards higher stages of existence, the effete and undeveloped beings passing away to make room for new, and loftier, and more perfect creations. What is the volume that has thus recorded the chronicles of an age so long past, and prophecies of as far distant a future? Simply a little fragment of mouldering bone, tossed aside contemptuously by the careless labourer as miners' "rubbish."

Not only is the past history of each being written in every particle of which its material frame is constructed, but the past records of the universe to which it belongs, and a prediction of its future. God can make no one thing that is not universal in its teachings, if we would only be so taught ; if not, the fault is with the pupils, not with the Teacher. He writes his ever-living words in all the works of his hand ; He spreads this ample book before us, always ready to teach, if we will only learn. We walk in the midst of miracles with closed eyes and stopped ears, dazzled and bewildered with the Light, fearful and distrustful of the Word !

It is not enough to accumulate facts as misers gather coins, and then to put them away on our bookshelves, guarded by the bars and bolts of technical phraseology. As coins, the facts must be circulated, and given to the public for their use. It is no matter of wonder that the generality of readers recoil from works on the natural sciences, and look upon them as mere collections of tedious names, irksome to read, unmanageable of utterance, and impossible to remember. Our scientific libraries are filled with facts, dead, hard, dry, and material as the fossil bones that fill the sealed and caverned libraries of the past. But true science will breathe life into that dead mass, and fill the study of zoology with poetry and spirit.

QUADRUMANA;

OR, THE MONKEY TRIBE.

THE QUADRUMANOUS, or Four-handed animals, are familiarly known by the titles of Apes, Baboons, and Monkeys. There is another family of Quadrumana, called Lemurs, which bear but little external resemblance to their more man-like relations, are comparatively little known, and have even been popularly termed "rats," "cats," or "dogs," by travellers who have come in contact with them.

With the exception of a few small species, such as the marmosets and the lemurs, the Quadrumana are not very pleasing animals in aspect or habits; while the larger apes and baboons are positively disgusting. The air of grotesque humanity that characterises them is horribly suggestive of human idiocy; and we approach an imprisoned baboon with much the same feeling of repugnance that would be excited by a debased and brutal maniac. This aversion seems to be caused not so much by the resemblance that the ape bears to man, but by the horror lest man should degenerate until he resembled the ape. It is true that the naturalist learns to see wonder or beauty in all things of nature, and therefore looks with lively interest on such animals as the shark, the toad, the viper, the vulture, the hyæna, or the ape. But still, these creatures are less pleasing in his sight than many others which may be not so highly developed; and in truth there are few who, if the choice lay between the two fates, would not prefer to suffer from the fangs and claws of the lion, than from the teeth and hands of the ape.

Although these animals are capable of assuming a partially erect position, yet their habitual attitude is on all fours, like the generality of the mammalia. Even the most accomplished ape is but a bad walker when he discards the use of his two upper limbs, and trusts for support and progression to the hinder legs only. There are many dogs which can walk, after the biped manner, with a firmer step and a more assured demeanour than the apes, although they do not so closely resemble the human figure.

We are all familiar with the small monkeys that are led about the streets in company with a barrel-organ, or seated, in equestrian fashion, upon a bear or dog. These poor little creatures have been trained to stand upon their hind feet, and to shuffle along at a slow and awkward pace. But if they are startled, and so forget for a moment their acquired art, or if they wish to hurry their pace, they drop down on all fours, and scamper off with an air of easy comfort that contrasts forcibly with their former constrained and vaccillating hobble. The difficulty seems to increase almost proportionately with the size of the animal, and the largest apes, such as the orang-outan, are forced to balance themselves with outstretched arms.

However carefully a monkey may be eductaed, yet it never can assume an attitude truly erect, like that of man. The construction of its whole frame is such, that its knees are always bent more or less, so that a firm and steady step is rendered impossible. When in the enjoyment of liberty among their native haunts, none of the monkey tribes seem to use their hind legs exclusively for walking, although they often raise themselves in a manner similar to that of the bears, and other animals, when they wish to take a more extended view of the surrounding localities.

On account of the structure of the limbs, the term "hand" is given to their extremities; but hardly with perfect fitness. It must be borne in mind that the thumb is not invariably found on the fore extremities of these animals. In several genera of the monkeys, the fore-paws are destitute of effective thumbs, and the hand-like grasp is limited to the hinder feet. The so-called hands of the monkey tribes will not bear comparison with those of man. Although the thumb possesses great freedom of motion, and can be opposed to the fingers in a manner resembling the hand of man, yet there is no intellectual power in the monkey hand; none of that characteristic contour which speaks of the glorious human soul so strongly, that an artist can sketch a single hand, and in that one member exhibit the individuality of its owner! The monkey's "hand" is a paw—a thieving, crafty, slinking paw, and not a true hand. So is his foot but a paw, and not a true foot, formed for grasping and not for walking. Man seems to be the only earthly being that possesses true feet and hands. Some animals patter along upon their paws, some trot and gallop upon hoofs, others propel themselves with paddles; but Man alone can walk. Man is never so much Man as when erect, whether standing or walking. It is no mere figure of speech to say that man walks with God.

In order to bring this point more clearly before the eyes of the reader, the skeleton of a man is contrasted with that of the gorilla, the most highly organized of all the apes. The heavy, ill-balanced form of the ape; its head sunk upon its shoulders; its long, uncouth arms, with those enormous paws at their extremities; its short, bowed, and tottering legs, unable to support the huge body without the help of the arms; the massive

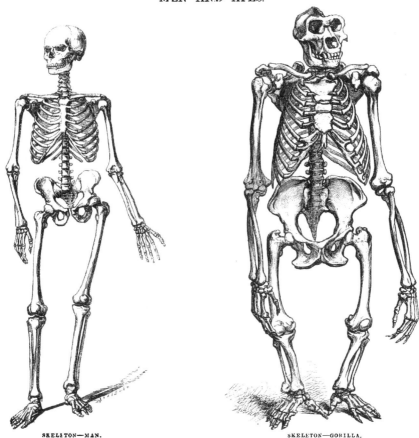

SKELETON—MAN. SKELETON—GORILLA.

jaw-bones and protruding face, put the creature at an unappreciable distance from humanity, even though it is represented in an attitude as similar to that of the human being as the organization of the bones will permit. Any one who could fancy himself to be descended, however remotely, from such a being, is welcome to his ancestry.

Contrast with the skeleton of the gorilla, that of man. Light in structure, and perfectly balanced on the small and delicate feet; the slender arms, with their characteristic hands; the smooth and rounded skull; the small jaw-bones and regular teeth, all show themselves as the framework of a being whose strength is to lie in his intellect, and not in the mere brute power of bone and muscle. There seems to be a strange eloquence in *form*, which speaks at once to the heart in language that can only be felt, and is beyond the power of analysis to resolve. Thus, the contrasted shapes of these two frames speak more forcibly of the immeasurable distance between the two beings of which they form a part, than could be expressed in many pages of careful description. Strength for strength, the ape is many times the man's superior, and could rend him to pieces in single combat. But that slender human frame can be so intellectually strengthened, that a single man could destroy a troop of apes, if he so desired, and without offering them the possibility of resistance.

One great cause of the awkward bipedal walk of the monkey tribes, is the position of the orifice in the skull, through which the spinal cord enters the brain. In the human skull this orifice is so placed that the head is nearly equally balanced, and a considerable portion of the skull projects behind it; but in the lower animals, this orifice—called the "occipital foramen"—is set so far back, that the whole weight of the brain and skull is thrown forwards, and so overbalances the body.

Another cause is seen in the structure of the hind limbs. These members are intended for progression among the branches of trees, and are so formed that, when the animal uses them for terrestrial locomotion, it is forced to tread, not upon their soles, but upon their sides. The muscular calves, which brace the foot and limb, are wanting in the Quadrumanous animals; and even when they are standing as uprightly as possible, the knees are always partially bent. The monkeys, then, are just quadrupeds, although their paws are more perfectly developed than those of the generality of animals.

We will now proceed to our example of the Quadrumanous animals.

SIMÍADÆ, OR APES.

The Apes are at once distinguished from the other Quadrumana by the absence of those cheek-pouches which are so usefully employed as temporary larders by those monkeys which possess them; by the total want of tails, and of those callosities on the hinder quarters which are so conspicuously characteristic of the baboons. Besides these external differences there are several distinctions to be found in the interior anatomy both of the bones and the vital organs.

The first in order, as well as the largest of the Apes, is the enormous ape from Western Africa, the Gorilla, the skeleton of which has already been given. This animal is comparatively new to modern zoologists, and very little is at present known of its habits. The first modern writer who brought the Gorilla before the notice of the public, seems to be Mr. Bowdich, the well-known African traveller; for it is evidently of the Gorilla that he speaks under the name of Ingheena. The natives of the Gaboon and its vicinity use the name Gina, when mentioning the Gorilla. The many tales, too, that are told of the habits, the gigantic strength, and the general appearance of the Ingheena, are precisely those which are attributed to the Gorilla.

Of the Ingheena, Mrs. Lee (formerly Mrs. Bowdich) speaks as follows:—"It is in equatorial Africa that the most powerful of all the Quadrumana live, far exceeding the orang-outan, and even the pongo of Borneo.

"Mr. Bowdich and myself were the first to revive and confirm a long-forgotten and vague report of the existence of such a creature, and many thought that, as we ourselves had not seen it, we had been deceived by the natives. They assured us that these huge creatures walk constantly on their hind feet, and never yet were taken alive; that they watch the actions of men, and imitate them as nearly as possible. Like the ivory hunters, they pick up the fallen tusks of elephants, but not knowing where to deposit them, they carry their burdens about until they themselves drop, and even die from fatigue; that they built huts nearly in the shape of those of men, but live on the outside; and that when one of their children dies, the mother carries it in her arms until it falls to pieces; that one blow of their paw will kill a man, and that nothing can exceed their ferocity."

Its existence was evidently known to some adventurous voyagers more than two thousand years ago, and a record has been preserved of these travels.

Somewhere about the year 350 B.C., the Carthaginians, then a most powerful and flourishing nation, organized a naval expedition for the purpose of examining the coasts and of founding colonies. The command of the fleet, which consisted of sixty large vessels containing nearly thirty thousand men and women, together with provisions and other necessaries, was entrusted to Hanno, who wrote memoirs of the voyage in a small work that is well known by the title of the "Periplus," or the Circumnavigation of Hanno. In the course of this voyage he founded seven colonies, and after advancing as far as the modern Sierra Leone, was forced to return for want of provisions.

The whole treatise is one of great interest, especially in the present day, when travels of discovery in Africa have been prosecuted with so much energy. The passage, however, which bears on the present question is briefly as follows. After narrating the meeting with these creatures on an island off the west coast of Africa, he proceeds to say :—" There were many more females than males, all equally covered with hair on all parts of the body. The interpreters called them GORILLAS. On pursuing them we could not succeed in taking a single male; they all escaped with astonishing swiftness, and threw stones at us; but we took three females, who defended themselves with so much violence that we were obliged to kill them, but we brought their skins stuffed with straw to Carthage." It is evident that Hanno (or Annon, as his name is sometimes given) considered these

GORILLA.—*Troglodytes Gorilla.*

Gorillas to be the veritable savage human inhabitants of the island; perhaps rather more savage and powerful than ordinary, and rather less given to clothing; which latter deficiency, however, was supplied by the natural covering of hair.

Imperfect as is his description, yet it is of much interest, as it proves the existence of extraordinarily huge apes hitherto unknown even to the Carthaginians, the stuffed skins being visible tests.

For two thousand years nothing was heard of the Gorilla except certain floating rumours of satyr-haunted woods, and of wild men who used to make their appearance at distant intervals and then to disappear; "of which kind," it is said, "there are still in Ethiopia." But by degrees the truthfulness of the narrative was made clear; detached bones were discovered and sent to Europe, and at last the complete animal made its appearance. Indeed, we are much indebted to this straightforward and simple-minded sailor, for his unadorned narrative, which forms such a favourable contrast to the travellers' tales of later voyagers, who on some small substratum of truth raised such enormous fictions as the monopods, the pigmies and cranes, the acephali, and other prodigies. For a vivid description, and graphic though rude figures of these and many other monsters, the reader is referred to the " Nuremberg Chronicle."

Perhaps it may be of this animal that the following history is narrated :—

" A certain ape after a shipwreck, swimming to land, was seen by a countryman, and thinking him to be a man in the water, gave him his hand to save him, yet in the mean-time asked him what countryman he was, who answered he was an *Athenian.* 'Well,' said the man, 'dost thou know *Piræus?*' (which is a port in *Athens*).

" 'Very well,' said the ape, 'and his wife, friends, and children;' whereat the man being moved, did what he could to drown him."

At present we have but a very slight acquaintance with the mode of life adopted by the Gorilla in a wild state, or even with its food. For a knowledge of the habits of animals is only to be gained by a long residence in their vicinity, and by careful watching. With some creatures this is an easy task, but there are some which are so wary, so active, and so fierce, that a close inspection is almost an impossibility. Among the worst of such objects is the Gorilla. In the first place, it is only to be found in the thickest jungles of the Gaboon, far from man and his habitations. Then, it is wary, as are all the apes, and is said to be so ferocious, that if it sees a man, it immediately attacks him, so that there would be little time for gaining any knowledge of the creature's domestic habits, and scarcely any likelihood of surviving to tell the result of the investigation.

To judge by the structure of the skeleton, and of the entire form, the strength of an adult male must be prodigious. The teeth are heavy and powerful, and the great canines or tusks are considerably more than an inch in their projection from the jaw. The jaw-bone, too, is enormously developed, and the strength of the muscles that move it, is indicated by the deep bony ridges that run over the top of the skull, and in different parts of the head. As is usual among such animals, the tusks of the male Gorilla are nearly double the size of those of the female ape.

Although the body is comparatively small, as are the hinder legs, yet the breadth of shoulder and length of arm are singularly great; while an ordinary human hand placed on that of the ape, dwindles down to insignificance before the huge muscular paw. The thumb of the hinder paws is enormously large, as is well shown in the engraving.

There is a treacherous and cruel aspect about this hind foot, with its enormous thumb; and if all tales be true, the foot belies not its character. The natives of the Gaboon country hold the Gorilla in great dread, fearing it even more than the lion itself, on account of its furtively murderous disposition.

Concealed among the thick branches of the forest trees, the Gorilla, itself unseen, watches the approach of the unsuspecting negro. Should he pass under the tree, woe betide him; for the Gorilla lets down its terrible hind foot, grasps its victim round the throat, lifts him from the earth, and finally drops him on the ground, dead.

Sheer malignity must prompt the animal to such a deed, for it cares not to eat the dead man's flesh; but finds a fiendish gratification in the mere act of killing. It is a kind of sporting; though the game is of a better quality than that which is usually chased over the fields, shot in the air, or hooked out of the water; not to be eaten, but for the sport.

Such a deed as the capture of an adult Gorilla has never been attempted, and much less achieved, by the human inhabitants of the same land. There are many reasons for this circumstance.

In the first place, the negroes seeing that the Gorilla is possessed of strength, ferocity, and pitiless cruelty, conceive that the animal must be inspirited by the soul of one of their kings; for in the lower stages of man's progress he does honour to physical force alone, and values his ruler in proportion to his power, brutality, and heartlessness. It is the best boast of a savage chief, no matter of what nation and of what country, that he has "no heart." The savage crouches in terror before the imaged incarnation of the evil principle, and adores, for he can only worship the object of his fears. His belief is truly that of the demons—"he believes and trembles." Reverence for the inborn royalty of the Gorilla does not save the animal from the fate of being eaten whenever it falls a victim to the weapons of its negro assailants. Perhaps the very feeling of reverence may incite to the act, in a manner analogous to the filial piety of the Scythians, which was best shown by killing their aged parents and dutifully eating them.

But putting aside the terrors of diabolism, which are engrained in the native African mind, the task of capturing a living and full-grown Gorilla is well calculated to appal the heart of any man. The strength, the activity, and the cunning and sanguinary malevolence of the animal are so great, that the uncivilized Africans may well be excused for their dread of its powers.

Yet it does not follow that although the Africans have failed, Europeans should not succeed. The native Africans have not dared to attempt the capture of the elephant, although Europeans have succeeded in that endeavour, and have subdued the terrible foe, converting it into a docile servant, and even making it an attached and intelligent friend.

Once or twice, the young Gorillas have been captured, in spite of the furious resistance which is made by their male friends ; but from some reason they have always died in a very short time.

Cunning as is the Gorilla, and ingenious in some things to a striking degree, its intelligence is but limited, and the animal exhibits such unexpected instances of fatuity, that it well shows the distinction between cunning and wisdom, and proves itself to be but an animal, and nothing more.

If it finds the remnant of a fire which has been relinquished by the persons who kindled it, the Gorilla is greatly charmed with the novel sensation produced by artificial warmth, and sits by the bright wonder with much satisfaction. As the fire fails, and the glowing brands sink into white ashes, the animal draws closer to the expiring embers, and does not leave them until all heat has left the spot. But it never thinks of keeping up the fire by placing fresh fuel upon it, and does not even learn to imitate that action, which it may often have seen performed by the hunters who kindled the fire, and kept it well supplied with fuel during the night. It is most providential that the beast is devoid of this faculty, for, with the usual perseverance of the monkey race in such cases, it would probably continue to heap fuel until the forest itself was ablaze.

It is said also, that when the Gorilla makes an incursion into a sugar plantation, it has sufficient sense to bite off a number of the canes, and to twist them into a bundle for better conveyance. But it frequently includes several of the growing canes in its faggot, and then feels woefully discomfited because it cannot carry away the parcel which had cost so much trouble in making.

The natives of Africa have an idea that these, and other large apes, are really men ; but that they pretend to be stupid and dumb, in order to escape impressment as slaves. Work, indeed, seems to be the *summum malum* in the African mind, and a true African never works if he can help it. As to the necessary household labours, and the task of agriculture, he will not raise a finger, but makes his wives work, he having previously purchased them for that purpose. In truth, in a land where the artificial wants are so few—unless the corruptions of pseudo-civilization have made their entrance—and where unassisted nature is so bountiful, there is small need of work. The daily life of a "black fellow" has been very graphically described in a few words. He gets a large melon ; cuts it in two and scoops out the inside ; one half he puts on his head, he sits in the other half, and eats the middle.

It is rather singular that this legendary connexion of apes and indolence should prevail on the continents of Africa and Asia.

The outline of the Gorilla's face is most brutal in character, and entirely destroys the slight resemblance to the human countenance, which the full form exhibits. As in the Chimpansee, an ape which is placed in the same genus with the Gorilla, the colour of the hair is nearly black; but in some lights, and during the life of the animal, it assumes a lighter tinge of greyish brown, on account of the admixture of variously coloured hairs. On the top of the head, and the side of the cheeks, it assumes a grizzly hue. The length of the hair is not very great, considering the size of the animal, and is not more than two or three inches in length. On the arms it is arranged in a rather curious manner, the hair from the shoulder to the elbow points downwards, while that from the elbow to the fingers points upwards, so that the two sets of hairs meet at the elbow, and make a pendent tuft. A similar structure is found in other large apes, but the object of so curious a disposition is not yet known. One reason for this arrangement of the hair, may be that if their long hairs were to hang along the arm and wrist, they would get into the hand, and interfere with the grasp, while by their reverted growth such an embarrassment is removed. The colour of the eye is dark brown, glowing with a baleful emerald light, when the fierce passions are roused.

It will be seen, on referring to the two engravings, which represent the skeleton of this animal, and the living creature itself, that the paws of the four extremities are not precisely alike in their development. On the two fore-paws, the fingers are enormous, the thumbs being comparatively trifling in dimensions; while the corresponding members of the hinder paws are just reversed in their size. The figure of the Gorilla, on p. 15, marks these peculiarities with great fidelity, and in the action of the creature shows the reason for the extraordinary and gigantic thumbs of the hinder limbs.

As to the size of a full-grown Gorilla, accounts vary much. The specimen which is best known in England is five feet six inches high, when placed erect. From shoulder to shoulder it measures nearly three feet, while the body is only two feet four inches, measured from the hip-joint. It is possible, however, that there may be much larger individuals. Independent, however, of the impression made on the minds of the spectators by the sight of an infuriated animal, it is a fact that the feeling of anger does dilate the form, whether of man or beast. And as one effect of anger is to cause the hair to bristle up (as indeed is seen familiarly in dogs, cats, and other animals), the ape while under the influence of that fiery rage to which these animals are so subject, would in reality present a larger outline than if it were calmly engaged in its usual pursuits. Six, or even seven feet of height, have been attributed to these creatures. But it must be remembered that a wild, fierce animal always looks very much larger when living and in motion, than when lying dead and still on the ground, or even "set up" in a museum, with glass eyes, and straw-distended skin. Elephants of sixteen feet high, have shrunk to eleven and ten feet under the application of the measuring rod, and it is proverbial among anglers, that the fish which they do *not* catch, are finer and heavier than those which they can subject to scales and foot-measure. So it is likely enough, that a wild and savage Gorilla, with his fury-flashing eyes, his fierce gestures, and enormous arms, would impress the mind of his opponent with an idea of a very much larger animal. It is not only upon Gadshill that two men in buckram multiply unto eleven.

But granting that the Gorilla does not attain to any much greater height than five feet, even then it is an animal much to be dreaded as an enemy, and capable of doing vast mischief, if so inclined. But it is a most merciful provision, and one that seems to be universal among creatures of such a stamp, that in proportion as their bodily powers increase, their mental powers degenerate. The larger apes are, in their period of childhood, so to speak, teachable and tolerably docile; while when they attain to years of maturity, the animal attributes assume strength, gradually gain dominion over the mental, until at last the reasoning capacities seem to degenerate into a mere contracted cunning.

It seems that this degeneration is intended to prevent the animal from passing beyond the bounds to which it is confined, and by the very laws of its being to prevent it from using its vast strength for bad purposes. The ape evidently does not know his strength, nor how terrible an enemy he could be, if he only knew how to use the singular power and activity which he possesses. These huge apes seem to live apart from each other, and not to band together in large herds as do the baboons and other quadrumanous animals. If they were to unite, and to understand the principle of combination, they could speedily depopulate any country that was inhabited by men who were not possessed of fire-arms, and were unable to construct defences.

But, fortunately for those human beings who are within reach of these terrible animals, the adult ape is one of the most dull and stupid creatures imaginable; sulky, ferocious, and given solely to its own animal appetites.

Here is a sketch of one of the lowest and least developed of human beings, probably the very lowest of the human race. This little man, who belongs to the same country as the Gorilla, hardly attains even to the same stature, and in muscular proportions is a very pigmy. Yet that in mere animal form the Bushman is infinitely higher than the ape, is evident from the contrast displayed by the two figures; while, if the comparison be extended to the mental endowments, the impassable barrier that exists between the two beings, exhibits itself in the most unmistakeable manner.

Modern zoologists have done rightly in refusing to admit mankind into the same order with beings so infinitely below them, as are even the very highest of the apes. The unprogressive animal is restricted to a narrow circle of thought and reason, and is totally devoid of that great privilege of human nature which we call by the name of aspiration. Man ever proceeds onwards and upwards, anticipating something beyond that which he possesses, while the brute creation remain in the same course of life in which they were originally placed. The records of geological experience, show that Simiadæ of gigantic stature existed on earth ages before the creation of human beings. Relics of these creatures have been found in various parts of the globe, and even in the tertiary formations of our own island.

BUSHMAN.

Apes were, therefore, at least contemporary with mankind; but while men have progressed, the apes have stood still, and always will stand still as long as they remain upon earth. The ape which saw the light in the year B.C. 4,000, was not a whit behind its descendant of the year A.D. 1859 in intellect or civilization; and if the order were to be continued for twenty thousand years longer, the last ape would be not a step nearer civilization than the primeval pair. Within its own little circle of life, many of its bodily senses are far more acute than those of man, and its bodily powers greater; but there ends the advantage. The animals are only partial and individual in their existence, restricted to a small sphere of life, and often confined within a very limited portion of the earth. These very limits place the animals at an immeasurable distance from man, who spreads himself over the entire earth, enduring with equal ease the fierce rays of the tropical sun, or the icy blasts of the arctic gales, and accommodating himself, through the agencies which his intellect projects, to these totally dissimilar modes of life.

CLOSELY connected with the preceding animal is the large black ape, which is now well known by the name of CHIMPANSEE.

This creature is found in the same parts of Western Africa as the gorilla, being very common near the Gaboon. It ranges over a considerable space of country, inhabiting a belt of land some ten or more degrees north and south of the torrid zone.

THE CHIMPANSEE.—*Troglodytes Niger.*

For some little time it was supposed that the gorilla was simply an adult Chimpansee, but zoologists now agree in separating it from that animal, and giving it a specific name of its own.

The title *niger*, or black, sufficiently indicates the colour of the hair which envelops the body and limbs of the Chimpansee. The tint of the hair is almost precisely the same as that of the gorilla, being nearly entirely black; the exception being a few whiter hairs scattered thinly over the muzzle. Age seems to give the hair of the animal a greyish tint in many places. As in the gorilla, the hair of the fore-arm is turned towards the elbow, where it meets the hair from the upper arm, and forms a pointed tuft. On the chest and abdomen it is rather thinner than on the remainder of the body, and permits the skin to be seen between the hairs, but on the arms and other parts it is sufficiently thick and long to hide the skin altogether. There is a small beard on the chin and face, which has a Chinese kind of aspect about it.

With very few exceptions, the nostrils of the Quadrumana are placed almost flat upon the face, and are devoid of that projecting character which gives such expression to the human countenance. Even in that very large-nosed animal, the Proboscis Monkey, the nostrils are only oval orifices for the conveyance of air, and seem as devoid of character as those of a wax doll.

Just as man is the only being that possesses two hands and feet, so is he the only inhabitant of earth who can lay claim to a nose. All the Mammalia have nostrils, and some species are endowed with wonderful powers of scent, such as the dogs, the deer, and others. Some of them carry a proboscis more or less elongated, such as the elephants and the tapirs. Then there are some, such as those of the porcine group, which possess snouts; but not one of them has a nose.

So in the Chimpansee and its relatives, the muzzle projects exceedingly, and the nostrils lie almost flatly upon the projecting mass. Herein lies one of the chief characteristics of the simian countenance, which is not so conspicuous when the face is viewed directly from the front, as when it is turned with the profile towards the observer. In front, the flattened and divergent nostrils, together with the projecting muzzle, are not forced on the notice, and might escape a hasty observation; but if the animal turns its head, then the simian character shows itself in all its repulsive brutality.

Even in the young Chimpansee, this preponderance of the face and jaws over the brain-skull is very considerable, and, as we have already seen, continues to increase as the animal draws nearer to maturity. The accompanying sketch exhibits the general characteristics of the Chimpansee skull, and shows how radically it differs from that of the human being. The distinction is even more clearly shown if the lower jaw be removed, and the skull examined from below; for then, the disproportion between the animal and reflective parts shows itself most forcibly.

SKULL OF CHIMPANSEE.

In its native country, the Chimpansee lives in a partly social state, and at night the united cries of the community fill the air with their reiterated yells. If we may credit the reports given by the natives of Western Africa, the Chimpansees weave huts for themselves, and take up their residence in these dwellings. Now it is a well-known fact that the orang-outan, which comes next in our list, can rapidly frame a kind of platform of interwoven branches, and so it is not beyond the bounds of credibility that the Chimpansee may perform a work of similar character. Only, the chief difference between the customs of the two animals seems to be, that the one lives *upon* the structure or roof, if it may so be called, and the other beneath it. Some travellers say, that although the huts are actually inhabited, yet that only the females and young are permitted to take possession of the interior, and that the male takes up his position on the roof.

The latter supposition derives more force from those habits of the Chimpansees with which we are acquainted, and which have induced naturalists to give to the entire genus, the name of *troglodytes*. This term is compounded from two Greek words, signifying a "diver into caverns," and was applied to this ape, because it seems to prefer rocky and broken ground to the forest branches, which form the refuge of nearly all quadrumanous animals.

This compound word is not of modern invention; for in the works of Aristotle, Pliny, and other writers on the subject of natural history, much mention is made of a race of men who lived in rocky caverns, and who earned, by their burrowing habits, the title above mentioned. The language and costume of these people were as barbarous as their habitations, for the former characteristic was said to resemble the hissing of serpents, rather than to bear any likeness to articulate speech, and in the latter accomplishment they were totally deficient in the hotter months. It is possible that the Bushman tribes may have given rise to these descriptions, which, indeed, would not be very erroneous if they had been used in depicting the "Digger" Indians of the New World.

Be this as it may, it is a remarkable fact that the Chimpansees are groundlings, and are not accustomed to habitual residence among the branches of trees. Although these apes do not avail themselves of the protection which would be afforded by a loftier habitation, yet they are individually so strong, and collectively so formidable, that they dwell in security, unharmed even by the lion, leopard, or other members of the cat tribes, which are so dreaded by the monkey tribes generally. Even the elephant yields to these active and ferocious animals, and leaves them undisturbed. Yet a Chimpansee would not dare to meet a panther in single combat, and depends for safety upon the assistance that would be afforded by its companions. This is shown by a curious and rather absurd incident that occurred on board a ship, where a young and docile Chimpansee suddenly came in sight of a caged panther, which had taken voyage in the same vessel.

The unexpected sight of the panther entirely overcame his feelings, and with a fearful yell he dashed along the deck, knocking over sundry of the crew in his passage. He then dived into the folds of a sail which was lying on deck, covered himself up with the sail-cloth, and was in such an agony of terror, that he could not be induced to come out of his retreat for a long time.

His fright was not groundless, for the panther was as much excited as the ape, only with eager desire, and not with fear. It paced its cage for hours afterwards, and continued to watch restlessly, much as a cat may be seen to watch the crevice through which a mouse has made good its escape.

There are also strange reports, which are still credited, that the Chimpansees carry off negresses, and detain them in the woods for years, sometimes until they are released by death from their terrible captivity.

The food of these creatures appears to be almost entirely of a vegetable nature, and they are very unprofitable neighbours to anyone who has the misfortune to raise crops of rice, or to plant bananas, plantains, or papaus, within an easy journey of a Chimpansee settlement. As is the case with many of the monkey tribes, the animal will eat food of a mixed character, when it is living in a domesticated state.

Many specimens have been brought to Europe, and some to England; but this insular climate seems to have a more deleterious effect on the constitution of this ape than even on that of the other Quadrumana. In this country, our worst, most insidious, and most irresistible malady fastens upon the apes with relentless hand. The lungs of these creatures are accustomed to the burning suns which heat and rarefy the air of the tropical climates, and are peculiarly sensitive to cold and damp. Few members of this family live to any length of years, after they have once crossed the Channel. They are, after a while, seized with a short hacking cough, the sure sign that consumption has begun that work which it is so sure to accomplish.

It may be, that the atmosphere of so small an island as England, is loaded with marine and saline exhalations which prove too irritative to the lungs of the ape. Be this as it may, the free use of food which supplies a large amount of carbon, is the best preventive of this lethal ailment. Cod liver oil will be found very efficacious; and I know of one successful cure, where the animal was treated to a glass of wine daily. It seems to supply, internally, the heat principle, which is poured in fiery vehemence from the vertical sun of the tropics, and which our temperate zones can only afford in moderate proportions.

A monkey, when afflicted with this disease, is a truly pitiful sight. The poor animal sits in such a woeful attitude, coughing at intervals, and putting its hands to its chest in a way terribly human. And it looks so mournfully and reproachfully out of its dark brown eyes, just as if it were rebuking the spectator for his part in bringing it from its native land, where it was happy among its friends, to die a solitary death of cold and consumption, behind the bars of its prison.

The climate of France seems to be better suited to these animals than that of England.

In the Jardin des Plantes, in Paris, there was a remarkably fine specimen of the Chimpansee. Black, sleek, and glossy, he was *facile princeps* in the establishment, and none dared to dispute his authority.

He was active enough, and displayed very great strength, and some agility, as he swung himself from side to side of the cage, by means of the ropes that are suspended from the roof; but he preserved a dignified air as became the sole ruler.

There was a kind of aristocratic calmness about the animal, and he would, at intervals, pause in his airy promenade, and, seating himself on a convenient spot, deliberately scan the large assembly that generally surrounded the monkey-house. His survey completed, he would eat a nut or a piece of biscuit, and recommence his leisurely gambols. His health seemed to be perfectly good, as was shown by the alertness of his movements, and the full, open look of his eyes.

A sad contrast to this animal was presented by a wretched little Chimpansee which I saw in England. It was still possessed of sufficient strength to move about its cage, but executed all its movements in a slow, listless manner, that would have told its own tale, had not the frequent hacking cough spoken so plainly of the malady that was consuming its vitals. The countenance of the poor creature was very sad, and it did not appear to take the least interest in anything that occurred.

I have seen many monkeys with this sad aspect, and was always haunted by their piteous looks for days afterwards.

The ravages which this disease can make in the delicate formation of a monkey's lung, before the creature finally succumbs, must be seen to be appreciated. The whole organ is so eaten up, and its colour and substance so changed, that the spectator marvels that the creature's life could have been sustained for an hour under such circumstances.

As long, however, as they resist the untoward influence of our climate, the specimens which we have known, have always been extremely gentle and docile. Taught by the instinctive dread of cold, they soon appreciate the value of clothing, and learn to wrap themselves in mats, rugs, or blankets, with perfect gravity and decorum. Dress exercises its fascinations even over the ape, for one of these animals has been known to take such delight in a new and handsome costume, that he repudiated the previous dress, and in order to guard against the possibility of reverting to the cast-off garment, tore it to shreds. Whether the natives of Western Africa speak rightly in asserting that the Chimpansee is capable of using weapons, is at present rather a doubtful point. The negroes say that the "Baboos," as they call the animals (the name evidently being a corruption from our own word Baboon), make use of clubs, staves, and other rude weapons, and that they can use them with great address. Certain it is, that the adult Chimpansee has been known to snap with a single effort branches so thick, that the united strength of two men could hardly bend them. But whether the animal would possess sufficient intellectual power to make use of a weapon thus obtained, is not so certain.

It is said that they have a sufficient amount of knowledge to be aware that the strength of a man lies in his weapons, and not in his muscles only; and that if a hunter should draw on himself the vengeance of the troop, by wounding or killing one of their number, he can escape certain death by flinging down his gun. The enraged apes gather round the object that dealt the fatal stroke, and tear it to pieces with every mark of fury. While they are occupied with wreaking their vengeance on the senseless object, the owner of the fatal weapon escapes unnoticed.

The strength of arm with which this animal is endowed, has already been shown. But although the hinder limbs are not possessed of that gigantic muscular strength which is given to the arms, yet they are powerful to a degree that would be remarkable in any animal less athletic than the Chimpansee. One of these creatures has been seen to lower itself backwards from the bar on which it was sitting, and to draw itself up again, merely by the grasp of the hinder feet.

The age to which the Chimpansee attains in its wild state, is as yet unknown. But to judge by the length of time that elapses before the animal reaches maturity, its life cannot be very much less than that of the human inhabitants of the same land. Nine or ten years are spent by the Chimpansee before it has reached the perfection of its development; and it is well known that the inhabitants of the tropical regions attain to maturity at a very early age indeed.

A peculiarly fine specimen of the Chimpansee, which was tamed and domesticated in its native country, lived to the age of twenty-one years. This animal was possessed of gigantic strength, and on one occasionwas intercepted in the act of carrying a soldier into the tree to which he was chained. This ape might, however, have been a specimen of the gorilla.

One great and almost radical objection to the weapon-using powers of the Chimpansee, may be found in the difficulty which these animals experience in standing erect. In order to use a weapon effectively, the hands and arms must be at liberty, and the feet planted firmly on the ground. A defect in either of these conditions, is fatal to the right handling of the weapon. Now, as the Chimpansee has much difficulty in preserving even a semi-erect position, and is forced to aid itself by placing the backs of its hands on the ground, it will be at once seen that a club would not give very much assistance to the creature. It might certainly launch stones with force and effect; but a weapon that requires the full and independent use of both sets of limbs, would be of small benefit.

Besides, the creature is already so terribly armed by nature with formidable fangs, and limbs of Herculean strength, that it needs no artificial means of offence, and would probably be rather embarrassed by them than otherwise.

Still, it is not improbable that these inquisitive animals have seen their human neighbours armed with sticks, and in that irresistible spirit of imitation to which monkey nature seems to be a victim, have armed themselves in similar manner, though with certain detrimental results. Should they really have recourse to these artificial and useless weapons, when brought into collision with human foes, it may be a providential means of depriving them of those terrible natural weapons, which would be truly formidable, and so causing them to be the more easily overcome by man. Judging from the familiar instances of their imitative nature, we may safely allow that the Chimpansees do carry sticks, although we may infer that such weapons would be worse than useless to their bearers.

In common with the orang-outan, and several other members of the same family, the Chimpansee is possessed of extremely mobile lips. In the lips, indeed, the whole expression of the face seems to be concentrated; and by the lips, the animal expresses the various emotions of fear, astonishment, hatred, rage, or pleasure, that agitate the ape's brain. Those lips can be protruded until they assume an almost snout-like aspect; they can be moulded into the strangest forms; they can be withdrawn, and almost obliterated from the countenance, when the creature extends its mouth into the grin of anger, exhibiting its sharp teeth, and uttering its furious cries. There are in the face of the ape none of those delicate lines that render the human countenance an index of the mind within; and, therefore, the animal makes the most of the limited means which it possesses. Articulate voice it has none, although it can be taught to comprehend the commands of its instructor; but it is a proficient in natural language of action, and by gesture can make itself understood without difficulty.

Though the language of the ape be not articulate, according to our ideas, yet in their wild state the Chimpansees can talk well enough for their own purposes. One proof of this, is the acknowledged fact that they can confer with one another sufficiently to act in unison, at the same time and place, and with a given object.

Strong and daring as they are, they do not appear to seek a contest with human beings, but do their best to keep quietly out of the way. Like most animals that herd together, even in limited numbers, the Chimpansees have ever a watchful sentinel posted on the look-out, whose duty it is to guard against the insidious approach of foes, and to give warning if he sees, hears, or smells, anything of a suspicious character.

Should the sentinel ape perceive a sign of danger, he sets up a loud cry, which has been likened to the anguished scream of a man in sore distress. The other apes know well enough the meaning of that cry, and signify their comprehension by answering cries. If the danger continues to threaten, then the ape-conversation becomes loud, shrill, and hoarse, and the air is filled with the various notes of the simian language, perfectly understood by themselves, although to human ears it consists of nothing but discordant yells and barks.

On reference to the engraving on p. 20, it will be seen that the arms of this animal, of the gorilla, and the orang-outan, are of considerably greater length than might be inferred from the height of the animal. It will also be seen that each creature is represented with the knuckles of one hand resting on the tree-trunk on which they are supported. This peculiar action has been thus noted, because, when these creatures aid their steps by placing the hands on the ground, they have the curious habit of resting the knuckles on the ground, instead of the palms of the hands, as might have been supposed. From this peculiarity, the three apes have received the appropriate title of " knuckle-walkers."

The head of the Chimpansee is remarkable for the large development of the ears, which stand prominently from the sides of the head, and give a curiously peculiar expression to the contour of the head and face.

We should probably have seen many more specimens of this ape imported into this country, had not the superstitious fears of the natives kept them aloof from meddling with these animals. Probably on account of the weird resemblance to the human form, which is one characteristic of their race, or on account of their cunning, the inhabitants of the Gaboon and the neighbourhood labour under the dread of being bewitched by the Chimpansees, and so very prudently let them alone. Certainly, they would be " no canny " to deal with, and the discretion exercised is not to be blamed.

THE ORANG-OUTAN.

THE title of Satyrus, or Satyr, is very rightly applied to the huge ape which is known by the name of ORANG-OUTAN.

For, saving that the long-eared Satyrs of the classic authors were more intellectual in countenance, and usually wore hoofs instead of hands at the extremities of the lower limbs, there is no small resemblance between the veritable and the imaginary wild man of the woods.

An ancient proverb tells us that there is no smoke without fire ; and we generally find that even the wildest travellers' tales have some foundation in fact. The ruddy colour of the hair of these Satyrs is especially noticed, and the reader will remark that the Orang-outan is at once distinguishable from the two preceding animals, by the reddish, chestnut colour of its hair. The goat-legs with which Satyrs were generally furnished, do not seem to be indispensable, for I have now before me two curious old wood-engravings of Satyrs, neither of which creatures possess the hircine leg.

One of them is represented with a flute in his hand, and legs and feet of a human form, while the other is a composite animal altogether. On the top of his head is a huge fleshy comb, like that of a cock ; two ibex horns curl over his shoulders, and his ears are those of an ass, dressed and pointed. Three large skin pouches hang from the throat to the middle of the breast, concealed at their origin by an enormous beard that curves upwards until its tip is on a level with the nose. The body and limbs are those of a man, fringed and studded with tufts of long hair, and the tail is that of a wolf. The hands are replaced by four-clawed paws, and the feet are modelled from those of the chameleon.

The account which is affixed to the portrait, avers the colour of the nondescript to be a "yellowish carnation," and states that it was seen in a forest belonging to the Bishop of Saltsburg, in the year 1530. The date of the print is 1658.

In connexion with this subject we may mention the curiously similar legends of Brazil, as told by Dr. Lund. With the exception of colour, and of several added peculiarities, the native accounts of the Caypore, as they call the creature, differ but very slightly from the tales told of the Ingheena of Africa.

The animal is said to be equal in stature to the human form, to be in the habit of walking in the erect posture, to be quiet and harmless when young, but when aged to become fierce and dangerous, and to attack mankind with the formidable tusks that grow from its jaws. So much for the points of similarity, which are sufficiently striking. The additional properties are as follow :—

The Caypore, or "Dweller of the Wood," is covered with long curling hair of a brown colour, so thick as to be invulnerable except in a single white spot on the abdomen. Its feet are each furnished with two heels, one in the usual position, and the other in the place where the toes are generally placed. On account of this peculiarity, its footmarks, although they cannot be mistaken for those of any other animal, cannot be tracked, as no one can tell in which direction the footprints proceed. It is the lord of the wild hogs, and if any of its subjects be killed, its angry voice warns the slayer to make his escape. The upper portion of its body is that of an ape, and from the waist downwards, that of a pig. It may be seen in the middle of the wild swine, riding upon the largest of the herd.

THE ORANG-OUTAN.—*Simia Sátyrus.*

It is most remarkable that there should be similar legends in Western Africa, in Borneo, and in Brazil; and the elucidation of the mystery would in all probability bring to light some curious physical facts.

The Orang-outan is a native of Asia, and only to be found upon a small portion of that part of the globe. Borneo and Sumatra are the lands most favoured by the Orang-outan, which inhabits the woody districts of those islands, and there rules supreme, unless attacked by man.

There seem to be at least two species of this animal, that are found in Borneo, and some zoologists consider the Sumatran ape to be a third species.

The natives distinguish the two Bornean species by the name of Mias-kassar, and Mias-pappan, the latter of which animals is the *Simia satyrus*, so well represented in the engraving.

The Pappan is a truly terrible animal when roused to anger, and would be even more formidable than is the case, were it endowed with a less slothful disposition. Its length of arm is very great; for when the animal stands erect, and permits the arms to hang by its sides, its hands can nearly touch the ground. The muscular power of these arms is proportionate to their length, and it is chiefly by means of the upper limbs that the ape makes progress among the boughs of the trees on which it loves to live.

So powerful, indeed, are the arms, that a female Orang has been known to snap a strong spear like a reed, and this after she had been weakened by many wounds and loss of blood. In attack the Orang-outan is not sparing of teeth as well as hands; and uses to the utmost the weapons with which it has been endowed. The teeth of an adult Orang are truly formidable weapons, and it is said that even the leopard cares not to prove their power. So strong are even the front teeth, that they are capable of gnawing through and tearing away the dense fibrous covering in which the cocoa-nut is enveloped, and possibly can cut through the hard shell itself. Besides these teeth, the Orang is furnished with enormous canines, or tusks, the object of which is probably to act as offensive weapons; for the Orang is a vegetable-feeding animal, and the canine teeth can hardly be given merely for the purpose of cutting vegetable food.

Although the hind limbs are not so largely developed as the arms, yet they possess great power, and are perfectly adapted to the purpose which they serve. For terrestrial locomotion they are anything but fitted, as the animal is unable to plant the sole, or rather the palm, flat upon the ground, and rests upon the outside edges of the feet.

The walk of the Orang-outan is little better than an awkward hobble, and the creature shuffles along uneasily by help of its arms. The hands are placed on the ground, and are used as crutches in aid of the feet, which are often raised entirely from the ground, and the body swung through the arms. Sometimes it bends considerably backwards, and throwing its long arms over its head, preserves its equilibrium by their means.

This attitude is caused by the peculiar structure of the hind limbs, which, besides their comparative shortness, are only loosely jointed to the hip-bones. The Orang-outan is destitute of the short, but very strong ligament, that binds the thigh-bone to the hip-joint, and which is called the *ligamentum teres*. This ligament is very powerful in man, and plays an important part in giving him that steady tread, which alone is sufficient to distinguish the human species from the apes.

But the Orang-outan is intended for an arboreal life, and requires limbs that can adapt themselves to the boughs. Therefore the legs are so twisted inwards, that the feet can grasp the branches freely, and hold the body in its position, while the long arms are stretched out to take a fresh hold.

Among the trees the Orang-outan is in its element, and traverses the boughs with an ease and freedom that contrasts strongly with its awkward movements when on the ground. It has a curious habit of making for itself a temporary resting-place, by weaving together the branches so as to make a rude platform or scaffold on which it reposes. The powerful limbs of the animal enable it to execute this task in a very short time. Rajah Brooke of Sarāwak narrates an interesting tale of a female Orang-outan, which when severely wounded ceased her attempts to escape, and weaving together a branch-platform, seated herself upon it, and quietly awaited her end. The poor animal received several more shots before she expired, and as she fell dead upon her extemporary edifice, the hunters were put to some trouble before they could dislodge the dead body. The whole process of weaving the branches and seating herself did not occupy more than a minute.

When the hunters desire to capture an adult Orang-outan, they hem him in by felling the trees around that on which he is seated, and so deprive him of the means of escape.

Having thus cut off his retreat, they apply the axe to the tree of refuge, and endeavour to secure the ape before he has recovered from the shock of the fall.

The adult male animal is singularly hideous in aspect, owing much of its repulsiveness to the great projection of the jaws and the callosities that appear on the cheeks. As is the case with all the larger apes, it becomes sullen and ferocious as it approaches its adult state, although in the earlier years of its life it is docile, quiet, and even affectionate. Several young specimens have been brought to Europe, and were quite interesting animals, having many curious tricks, and exhibiting marks of strong affection to anyone who treated them kindly. One of these animals learned to take its meals in a civilised maner, using a spoon, or a cup and saucer, with perfect propriety.

When brought to colder climates than that of its native land, the animal covets warmth, and is fond of wrapping itself in any woollen clothes, or blankets that it can obtain. On board ship it has been known to rob the sailors or passengers of their bedding, and to resist with much energy any attempt to recover the stolen property.

Though sufficiently docile and good-tempered when it has its own way, the young Orang is rather subject to sudden gusts of passion when crossed in its wishes, and in such cases puts forth its powers with much effect. But the angry passion soon passes away, and the creature seems to be ashamed of its conduct.

One of these animals which I watched for some little time, had a curiously wistful and piteous expression of countenance, and although very young, its face was wrinkled like that of an old man of eighty. The creature sat and looked out of its deeply set eyes, as if the cares of the nation rested on its shoulders. It was not very lively, but moved about among the branches with great ease. The form was not at all symmetrical, for the long arms, and feet, and hands seemed strangely out of proportion with its round, weakly-looking body, so that it involuntarily reminded the spectator of those long-legged, round-bodied spiders that are so common about old walls.

The lips were very mobile, and the animal moved them when agitated by any emotions; sometimes shooting them forward like the poutings of a petulant child, and sometimes drawing them together in strange wrinkles. The neck was but slightly indicated, and the whole animal presented an uncouth, goblin-like aspect.

One of these animals that was brought to England by Dr. Abel, exhibited many curious habits.

It had been taught to walk in an erect position, without supporting itself by extraneous help, but the erect posture was so ill adapted to its structure, that it could only preserve its balance by raising the arms over its head, and throwing them behind it, as has already been mentioned. The mode in which the head is united to the neck renders the equilibrium uncertain.

This animal was tolerably omnivorous in appetite, for although its usual food consisted of fruits and bread, it was exceedingly fond of raw eggs, and would eat almost any kind of meat, whether dressed or raw. It would drink water, or milk, or beer, preferring the two latter liquids to any other. But it was also fond of wine, and was partial to mixtures of a still more potent character. Coffee and tea were favourite beverages with the animal, so that it displayed a decidedly civilized taste.

As might be expected, while it was on board ship the sailors petted their companion after their wont, and it was quite familiar with them, showing no fear, and even occasionally indulging in a sham fight. But it was struck with unaccountable fright at some very harmless creatures that became inmates of the same vessel. They were only common turtles, perfectly incapable of doing damage, and destined for soup. But the mere sight of them terrified the Orang-outan to such an extent that it ran away to the mast-head, and, protruding its lips, uttered a series of strange sounds. A land tortoise affected the animal in a similar manner, as also did the sight of a number of men bathing and floating in the water. Perhaps there was some connexion in the mind of the ape between the turtle and the cayman, which supposition is strengthened by the alarm caused by the bathers. I have known a common snail cause a great turmoil in a cage of monkeys, and there may possibly be some instinctive antipathy between monkeys and crawling animals.

This singular emotion is worthy of notice, because it proves the fallacy of judging any animal to be the natural enemy of another, merely because the latter is terrified at its approach. Granting that the apes might occasionally have been prompted by their mischievous nature to meddle with the turtles, and to have been half-blinded by a sand-shower thrown from the turtle's flippers, or have suffered a painful wound from the snap of a turtle's sharp jaws, yet the little land-tortoise could not do damage. As we have just mentioned, even the presence of a poor garden-snail is a terror to many members of the monkey race.

It is therefore evident that the antipathy does not exist only in some individuals which may have suffered by the reptiles, but that it is the common propensity of these strange animals. We can easily understand that an ape should display an agony of terror at the sight of a leopard, or a snake, for the one has teeth and claws, being also very fond of ape-flesh, and the other has fangs. But that the same animal should be just as frightened when it sees a turtle, a tortoise, or a man bathing, is indeed remarkable.

Our best insight into the habits of animals is generally gained by watching the actions of a single individual, and these biographies are usually found to be most interesting. An admirable description has been given by Dr. Abel of the young Orang-outan, which has been already mentioned.

At first the ape was put into a cage, but he broke the bars and got out. Then he was chained, but he detached the chain from the staple, and finding that the heavy links incommoded him, he coiled the chain round his shoulder, and to prevent it from slipping, held the end in his mouth. As he always succeeded in escaping from his bonds, his keepers made a virtue of necessity, and permitted him to enjoy the full range of the vessel. Among the ropes he was quite at home, and, trusting to his superior activity, was accustomed to take liberties with the sailors, and then escape among the ropes. One very curious trait in his character must be given in the words of the narrator.

" Although so gentle when not exceedingly irritated, the Orang-outan could be excited to violent rage, which he expressed by opening his mouth, showing his teeth, and seizing and biting those who were near him.

" Sometimes, indeed, he seemed almost driven to desperation ; and on two or three occasions committed an act which in a rational being would have been called the threatening of suicide. If repeatedly refused an orange when he attempted to take it, he would shriek violently and swing furiously about the ropes, then return and endeavour to obtain it. If again refused, he would roll for some time like an angry child upon the deck, uttering the most piercing screams ; and then, suddenly starting up, rush furiously over the side of the ship and disappear.

" On first witnessing this act, we thought that he had thrown himself into the sea ; but on a search being made, found him concealed under the chains."

He learned artificial tastes of civilization, and preferred tea and coffee to water. Tastes less natural and more to be regretted soon followed, for he took to drinking wine, and was so fond of spirituous liquids, that he was detected in stealing the captain's brandy-bottle. This interesting animal survived the English climate for about eighteen months, and then succumbed to the usual foe of the monkey race. The fatal issue of the disease was probably promoted by the shedding of his teeth.

In its native woods, the Orang-outan seems to be an unsocial animal, delighting not in those noisy conversaziones which rejoice the hearts of the gregarious monkeys and deafen the ears of their neighbours. It does not even unite in little bands of eight or ten as do many species, but leads a comparatively eremitical existence among the trees, sitting in dreamy indolence on the platform which it weaves, and averse to moving unless impelled by hunger, anger, or some motive equally powerful. When it does move, it passes with much rapidity from tree to tree, or from one branch to another by means of its long limbs, and launches itself through a considerable distance, if the space between the branches be too great for its reach of arm.

It has already been mentioned that the adult Orang is a sullen and ferocious animal, and if the reader will refer to the accompanying sketch of the skull, he may form an opinion of the nature that belonged to the animal that owned such a skull. It is almost totally animal in character; there is hardly any space for the brain; the head is surmounted with heavy ridges of bone, showing the great strength of the muscles that are attached to them; the lower part of the face and the jaws projects greatly, and, in fine, the skull is almost wholly made up of face, jaws, and bony ridges. The teeth, too, are very formidable.

SKULL OF ORANG-OUTAN.

The hair of the Orang-outan is of a reddish chestnut hue, deepening here and there into brown. The texture of the hair is coarse, and its length varies according to the part of the body on which it is placed. Over the face, back, breast, shoulders, and arms, it falls in thick profusion, becoming especially long at the elbow-joint, where the hairs of the upper and fore-arm meet. The face is partly covered with a beard, which seems to increase in size as the animal grows older. The hair of the face takes a lighter tinge of red than that of the body, and merges the red or auburn tint in the brown, on the inside of the limbs.

At a little distance, the face appears to be black; but if examined closely is found to present a bluish tint.

The Mias-kassar is similar to the Mias-pappan in general appearance, and colour of hair; but is evidently a different species from the Pappan, and not the young of that animal. Of this ape, Sir J. Brooke says, that it is "a small, slight animal; by no means formidable in its appearance; with hands and feet proportioned to the body. They do not approach the gigantic extremities of the Pappan either in size or power; and, in short, a moderately strong man could readily overpower one; when he would not stand a shadow of a chance with the Pappan."

The height of a full-grown Pappan does not seem to be quite so great as has been supposed. Credible informants, however, tell us that they usually grow to the height of five feet, or even more, which, taking into consideration the extreme length of the arms, and the general muscular development, gives us a very large ape indeed. Sir J. Brooke was deceived into the belief that one of these animals which he killed was nearly six feet in stature; but was surprised to find when the animal was dead that the height was very much overrated.

Many of the quadrumanous animals, among which are the large apes, the siamang, many of the tailed monkeys, and the baboons, are furnished with a singular appendage to the throat, which has been carefully investigated by M. Vrolik. This appendage consists of a pouch, varying in form and size, which is connected with the lungs by an opening into the windpipe, and can be dilated with air at the pleasure of the animal.

The result of his researches is, that the air-pouch is not connected with the voice; but that it is intended to reduce the specific gravity of the animal, and to assist it in climbing or leaping. The pouch is not a mere hollow sac; but is furnished with many subordinate receptacles, something like a badly made glove, with three or four additional fingers or thumbs. These prolongations lie between the muscles of the throat. They are larger in the male than in the other sex, and increase together with the growth of the animal. In the Orang-outan, these pouches are very largely developed; much more so than in the chimpansee. The siamang possesses them of a large size, while the gibbons are without them.

The generic name *Simia*, which is applied to these apes, and which serves to distinguish the entire family, is derived from the Greek word *Simos*, signifying "flat-nosed."

THE SIAMANG.

THE accounts of this ape vary extremely. Some authors pronounce the Siamang to be a dull and stupid animal, caring not to distinguish between friends and foes; never moving until forced to do so, and hardly even taking the trouble to put food into its mouth. Others give to the Siamang the character of being a lively and affectionate creature, soon tamed, and attaching itself strongly to those with whom it has made acquaintance, and who behave kindly to it. As the latter character has been borne by the Siamang when in the possession of those who treated it well, and studied its habits, it is but justice to the creature to give it the credit of good behaviour.

The SIAMANG is a Sumatran animal, and, as far as is known, is found in no other spot on the globe. The colour of the hair is black, and it is so thickly planted, that, although it is but short, it conceals the skin, except in one or two spots, such as the upper part of the breast, where the skin can be seen through the woolly covering. It is a large animal, measuring some three feet in height, when it has attained to its full growth. The arms are long, and the hands narrow, with slender fingers covered with the woolly black hair as far as the roots of the nails. The term *Syndactyla*, or "joined-fingers," is applied to this ape because the first and second fingers of the hinder limbs are united as far as

THE SIAMANG.—*Siamanga Syndáctyla.*

the middle of the second joint. This union of the members is by means of a membrane that runs between the fingers, and does not extend to the bones, which when stripped of their fleshy coverings are found to be as distinct as those of any other animal.

There is a curious structure of the throat which is worth notice. This consists of a double pouch under the chin and throat, formed by the loose folds of skin. When the animal is excited either by anger or pleasure, it inflates these pouches to such a degree, that their exterior surface becomes quite glossy. The pouches are without hair.

At sunrise and sunset, the Siamangs assemble in great numbers, under the command of a chief who is thought by the natives to be weapon-proof, and, being assembled, utter most hideous yells, each striving to outdo the other in their cries. It is supposed by some writers that the peculiar resonance of the animal's cry, is in a great measure to be attributed to the throat-pouches above mentioned. M. Vrolik, however, seems to be of a different opinion, as has been already noticed in the account of the Orang-outan. Except at the beginning and end of the day, the Siamangs are comparatively quiet.

There is not a very great development of the combative nature in this animal, which is timid, unless urged by those feelings which inspire even the weakest and mildest creatures with reckless courage. The poor animal has no notion how to inflict or avoid a blow; but in defence of its young, when threatened with danger, or in revenge for their loss, if slain, the mother Siamang dauntlessly flings herself upon the enemy, caring nothing for her own life in comparison with that of her offspring.

When permitted to range unmolested in the woods, the care of the mother Siamang for her young affords a pleasing, and sometimes an amusing spectacle. But the father must not be passed over without the tribute of honour due to his paternal virtues. Those who have watched the Siamangs as they wandered unrestrainedly, say that the parents divide the care of the family between them; the father taking care of the male offspring, and the mother of the females. They are properly solicitous about the cleanliness of their young charge, and duly wash them, rub and dry them, in spite of the screams and struggles of the little ones.

It seems to be a general rule, that when an animal is peculiarly adapted for one mode of life, displaying singular powers therein, it is quite at a loss when placed in an uncongenial condition. The bats, for example, are awkward and helpless animals when placed on a level surface ; so are many of the swift-winged birds, such as the albatross, the frigate-bird, and others, while the diving-birds are just as clumsy on land as they are agile in the water. So it is with the Siamang, for its great length of limb, that gives it such powers of locomotion among trees, forms a serious impediment to its progress on level ground. Among the trees the Siamang is unapproachable ; and although not quite so active as the gibbons, is yet sufficiently so to be perfectly secure from pursuit. But let the creature once descend to earth, and it is so embarrassed by its long limbs that it can be overtaken and captured with ease. Indeed, those specimens that have been taken unhurt, have almost invariably been made prisoners while struggling to regain the shelter of the trees.

One of these animals was for some time an inmate of a ship, where it became quite companionable, and gained the affections of passengers and crew. So far from exhibiting the sullen and sluggish demeanour which has been attributed to this ape, the Siamang displayed great activity and quickness, skipping about the ropes, and given to harmless tricks. It took a fancy to a little Papuan girl who was on board, and would sit with its arms round her neck, eating biscuit with her. It was of an inquisitive nature, running up the rigging, and watching from its elevated position a passing vessel, and remaining there until the ship was out of sight. In temper it was rather uncertain, and apt to fly into a passion if opposed in any wish.

When thus excited, it would fling itself down, just like a naughty, spoiled child, roll about the deck with great contortion of limbs and face, strike at everything which came in its way, and scream incessantly, with a sound like "Ra! ra! ra!"

It had a strange predilection for ink, and in order to procure this remarkable dainty, would drain the ink-bottle whenever there was an opportunity of so doing, or suck the pens in default of the liquid itself. Being itself destitute of a tail, and feeling no fear of reprisals in that direction, the Siamang used to make very free with the tails of some monkeys that lived on board of the same vessel. Catching an unfortunate monkey by its caudal appendage, away went Ungka, as the ape was named, dragging the monkey after him along the deck, until the wretched animal writhed itself free from its tormentor. At another time, Ungka would carry the monkey by the tail up the rigging, in spite of its squeaks and struggles, and then quietly let it drop.

It was sensitive to ridicule ; and when its feelings were hurt, it used to inflate its throat until it resembled a huge wen, and looked seriously at the offenders, uttering hollow barks at intervals. This sound seemed to be used for the purpose of expressing irritation. Anger was expressed by the shrieking "Ra! ra!" and pleasure by a kind of mixture between a squeak and a chirp.

For the account of this animal we are indebted to Mr. Bennett, who has related many other traits indicative of its character. Sir S. Raffles possessed several specimens of this ape, and describes them as being social in their manners, and of an intelligent nature. Although they were powerful animals, they were gentle, and showed themselves to be pleased with the society of those persons to whom they were attached.

THE GIBBONS.

ALTHOUGH in their physical characters the GIBBONS bear much resemblance to the apes which have already been described, yet there are some peculiarities in form and anatomy which show them to be a link of transition between the great apes, and the lesser monkeys and baboons.

They possess, although in a small degree, those singular callosities on the hinder quarters which are so conspicuous in the baboon family, and assume such strange

tints. The gorilla, chimpansee, and the orangs, are entirely destitute of these peculiarities, but the Gibbons are found to possess them, although the callosities are very
small, and hidden by the fur from a casual view.

As in the great apes, the arms of the Gibbons are of enormous length, and endowed
with exceeding power of muscle, though the strength which resides in these largely
developed limbs is of a different character.

If the gigantic and powerful gorilla be compared to Hercules, then the light and
active Gibbons may find their type in Mercury, the swift aerial messenger of the Olym-
pian deities. The ponderous weight of the larger apes binds them to earth; and even the
orangs, which are more active than the chimpansee, are no very great adepts at leaping
through great intervals of space. But the Gibbons seem to pass nearly as much time
in the air as on the branches, shooting from one resting-place to another, with such
rapid movements, that the eye can hardly follow their course—the very swallows of the
monkey race.

THE LAR GIBBON.—*Hylóbates Lar.*

From their wonderful agility in flinging themselves from branch to branch, or from
tree to tree, naturalists have given to these animals the generic name of *hylobates*,
signifying, "tree traverser." And carrying out the mythological comparison which has
just been mentioned, the name Lar has been attributed to this species.

According to the legends of antiquity, it appears that a very beautiful and very
loquacious Naiad, named Lara, indiscreetly acquainted Juno with one of the many
causes for jealousy for which her husband gave occasion. Jupiter, being greatly incensed
at her conduct, deprived her of the offending tongue, and sent her off to Hades under
the charge of Mercury. That faithless messenger, however, found that pity melted
the heart to love, and instead of obeying the order of his master, became enamoured
of the beautiful criminal, saved her from the punishment to which she had been
destined, and married her instead of delivering her to Pluto.

From this union sprang the Lares, twin demigods, who took on themselves the
guardianship of domestic hearths, and the peace of families. The Romans symbolised
these protecting deities under the form of monkeys clothed with the skins of dogs, and
placed their images around the hearths which they protected, and behind the doors which
they guarded from evil.

1.   D

These children of the eloquent and swift deity, Mercury, and the Naiad offspring of the waters, were supposed to combine the space-traversing attributes of both parents, and so the name of "Lar" is sufficiently appropriate for this most agile of animals.

The derivation of the name Gibbon is rather doubtful, although it is of great antiquity. The opinion which seems to be most in accordance with probability is, that the term is a corruption of *Kophin*, a Chaldaic word, signifying an ape. Delachamp thinks that it may be derived from *Keipos*, which in Strabo's version of the well-known word *Kephos*, signifies an ape or monkey. The difficulty in the latter case appears to be that the *Keïpon* resides in Ethiopia, while the Gibbons are Asiatic animals.

The present species is sometimes called the "White-handed Gibbon," because the hands and feet are of a much paler tint than the rest of the body and limbs.

There are slight differences in the colour of the fur in different individuals, but the prevailing tint is a darkish brown, changing to a creamy hue about the hands, and the face is quite black. Some specimens have the fur nearly black, while others assume a whitish tint along the throat and abdomen, and several specimens have the fur of the hind quarters rather paler than that of the remainder of the body.

In all the Gibbons, the hair is thicker and finer than in any of the preceding animals. It is short, being only an inch or so in length, and has more of a woolly appearance than is seen in most of the monkey tribe.

Many animals exhibit great differences of form and colour in the various periods of life, and in the two sexes. It has often happened that the greatest confusion has been caused by these changes of form, so that the young, and the two sexes of an animal have been described as several distinct species. We are the more liable to error when we cannot watch the entire development of the creature, and therefore such animals as the monkey tribes are very embarrassing to the systematic naturalist.

The Lar Gibbon seems to be one of these animals, and is probably identical with the Little Gibbon; this latter animal appears to be only a smaller specimen than usual, and its disparity of colour to be of little importance. The proportions are precisely the same as those of the Lar Gibbon, and although the general tints are so unlike those of the Gibbons as to earn from Cuvier the name of "Variegated Orang," yet we have already seen that the tint of the fur is extremely capricious, and can form no true criterion, unless accompanied by other distinctions.

The Lar, or White-handed Gibbon is an inhabitant of Malacca and Siam.

On looking at a living specimen of this animal, or indeed at any of the same genus, the hands are seen to differ much from those of the large apes, and especially in the shape and direction of the thumb. As we have already seen, the thumb of the chimpansee is very large, and is so formed that it can be opposed to the fingers in order to grasp any object between them. But the thumb of these tree-traversing apes is comparatively small, is hardly opposable to the fingers, and is placed in the same direction as the fingers themselves. Moreover, the bones of the hand are so formed, that the thumb appears to take its origin from the wrist, and not to be set on after the usual manner. Sometimes it is found that the first and second fingers of the hinder paws or hands are fixed together.

The reason of this arrangement is evident to anyone who has practised gymnastic exercises. In order to grasp a pole in the firmest manner, and with the least expenditure of strength, the fingers must be set close to each other, the thumb placed against the forefinger, and the hand hooked over the pole. In this position the muscles of the fore-arm are not subjected to the exhausting grasp of the thumb, and the power of the limbs is applied in precisely the right direction.

So it is with these apes, the most accomplished gymnasts in the world. If a monkey be watched while dancing about the bars and poles of his cage (not on hanging ropes, for then the thumb is wanted), it will be seen that the animal seldom or never grasps a horizontal bar, except occasionally with the hinder paws. The hands are always just hooked over the bars, and by their aid the animal flings itself from one place to another, using the grasp of the hinder feet to check itself when it wishes to sit still for a time.

This mode of employing the two sets of limbs is well shown in the Gibbons, and in order to fit them in the best manner for their arboreal existence, the thumb of the fore-hands is found to be almost destitute of the muscular prominence which is popularly known as the "ball" of the thumb, is therefore incapable of grasping, and can only follow the direction of the fingers; while the corresponding member on the hinder hands is very large and powerful in proportion.

All the Gibbons are gifted with voices as powerful as their limbs, and the creatures seem to lose few opportunities of exercising lungs or limbs. The cry which these animals utter is a singular one, loud, and piercing, and has been represented by the syllables "wou-wou," which duplex combination of intonations is often used as a general name common to the whole family. Some writers express the sound by the words "oa-oa," and others as "woo-woo," among which the reader is left to choose.

The several species of Gibbon do not seem to inhabit the same localities, although they all, without an exception, live among trees. Some reside among the mountainous ranges and their forests of fir-trees, while others prefer the lower regions of the wooded plains and valleys. All, however, agree in their exceeding activity and noisy voices, thus proving themselves in every way to be worthy types of their mythological lineage.

All animals which are destined to move with great rapidity, bear a sure sign of their destiny in the configuration of their bodies. Active exertions cause the heart to beat so fiercely, and the blood to circulate with such rapidity, that a provision must be made to give the blood a sufficiency of air to refresh it after its hard labour. This can only be done by the gift of very large lungs with plenty of room for their free action. Accordingly, the frames of all swift animals are found to be made on a similar model, although necessarily modified according to the description of animal.

Thus, among the well-known living creatures with which all are familiar, we may cite the greyhound and the racehorse. Or if we turn to the birds, the falcons and swallows are good examples of this formation of body. The chest and fore-part of the body are wide and capacious, in order to accommodate the large lungs which are necessary for the creature. The limbs which aid the progress of the animal, whether it be bird or beast, are very largely developed, while the subordinate parts of the body and limbs are reduced to the smallest size compatible with the well-being of their possessor. A greyhound in proper health, and ready for the course, has not an ounce of superabundant weight about it; neither has the racehorse when at the post. So with the falcons and swallows, until we come to the humming-birds, which exhibit this modification of limb and body in singular perfection.

The Gibbons are formed on a model of a similar nature, their enormously long arms and broad shoulders contrasted with the smaller hinder limbs and thin flank showing that they are capable of rapid movement, while the deep and capacious chest gives indication that they can endure a long continuance of labour without being exhausted by it.

Of the habits of the Gibbons in a wild state, very little is known, as they are shy in their nature, and by means of their wonderful agility escape among the trees in a manner that baffles pursuit or observation. As to the species which is represented in the accompanying engraving, it seems to be the

AGILE GIBBON.—*Hylóbates Ágilis.*

most active of this agile family, and well deserves the name that has been given to it. Rather more has been noticed of this wonderful creature, and a further insight into its habits has been gained by means of a female specimen, which was captured and brought safely to London, where it lived for some time.

In their native woods, these animals are most interesting to the observer, if he is only fortunate enough to get near them without being seen by the vigilant creatures. A good telescope affords an excellent mode of watching the customs of animals that are too timid to permit a human being to come near their haunts.

When startled, the Agile Gibbon flits at once to the top of the tree, and then, seizing the branch that seems best adapted to its purpose, it swings itself once or twice to gain an impetus, and launches itself through the air like a stone from a sling, gaining its force very much on the same principle. Seizing another branch, towards which it had aimed itself, and which it reaches with unerring certainty, the creature repeats the process, and flings itself with ease through distances of thirty or forty feet, flying along as if by magic. Those who have seen it urging its flight over the trees, have compared its actions and appearance to those of a bird. Indeed, these creatures seem to pass a life that is more aerial than that of many birds, putting out of question the heavy earth-walking birds which have not the power to raise themselves from the ground, even if they had the will.

The colour of this species is extremely variable, and as may be seen by reference to the figure, the offspring is not necessarily of the same colour as the parent. This difference in tint is not solely caused by age, for it frequently happens that a cream-coloured mother has a dark infant, and *vice versâ*. Of the specimens in the British Museum, hardly any two are alike in the tint of their soft woolly fur. Some are nearly black, some are brown, and some are of a light cream colour. It is worthy of remark that one of the black specimens was brought from the Himalayas; the brown and the cream-coloured examples being from Malacca.

The natives of Sumatra, where the Agile Gibbon is found in the greatest plenty, call it the Ungka-puti, or sometimes Ungka-etam. Sometimes the Siamang goes by the same name of Ungka, being called the Black Ungka Ape.

The singularly active manners of this animal were exhibited by the ape above-mentioned as being a visitor to our shores. A large apartment was prepared for it, and branches set up at some distance from each other, so as to give it as much room as possible for its wonderful evolutions. Eighteen feet appears to have been the farthest distance between the branches, and this space was cleared with consummate ease, as would probably be the case with an animal which was accustomed to launch itself through a space nearly double the eighteen feet. The animal, however, was hindered by many drawbacks. Putting aside the disadvantages of a strange climate and the want of the usual food, she had been subjected to the inconvenience of a long sea voyage, had suffered from confinement and the deprivation of its natural atmosphere. Even with all these drawbacks, the Gibbon exhibited such singular feats of agility, that the spectators were lost in astonishment.

She was accustomed to fling herself, without the least warning or apparent pre-paration, from the branch on which she might be sitting, towards another branch, which she invariably succeeded in catching with her outstretched hand. From branch to branch the Gibbon would continue her flight, for so it might be aptly termed, without cessation, until checked. The most curious part of the performance was, that she did not seem to require any further impulse after her first swing, but was content just to touch the branches as she passed from one to the other. So easy was this exercise, and of such quick eye and hand was the animal possessed, that the spectators were accustomed to amuse themselves by throwing fruits or other objects in the air, which she would adroitly catch as she passed along, without thinking it needful to stop for that purpose.

Swift as was its flight, the equilibric powers of the animal were so perfect, that even in its most rapid course it could arrest itself in a moment, catching a branch with the hands, and then suddenly drawing up the hinder feet to the same level. The firm grasp of the hinder feet then came into play, and the creature sat on the branch as quietly as if it had never stirred,

Some idea of the proportion of limbs and body of this ape may be gained by contrasting them with those of the human form. An ordinary man, when standing erect, permitting the arms to hang freely by his sides, finds that the tips of his fingers reach to the middle of the thigh. But when the Gibbon assumes the erect attitude, its finger-tips reach as far as the ankle-joint. Again, if a well-proportioned man stands perfectly erect, and stretches his arms out in a horizontal direction, the distance between the extended finger-tips is as nearly as possible equal to the height of the body, measured from the top of the head to the ground. But if the Agile Gibbon extends its limbs in a similar manner, the measurement between the fingers is just double that of the entire height of the animal.

On account of this great preponderance of the arms over the legs, the Agile Gibbon is not a very good walker on its hinder feet, but waddles along in an awkward fashion. While thus employed, the animal sways its long arms as balancers after the fashion of a rope-dancer, and now and then helps itself along the level surface with the hands on the ground. The Gibbon, though so marvellously light and active among trees, is totally out of its element when it is deprived of the branches, and forced to traverse the flat ground. All its elegance and exquisite address are lost, and the creature becomes as clumsy as it was formerly graceful. A swan while awkwardly hobbling over dry land, with a gait like that of a lame Silenus, affords no greater contrast to the same bird when proudly sailing on the water with arched neck and gliding movement, than does the Gibbon when stranded on unfamiliar earth to the same animal disporting itself among the congenial branches.

This species does not appear to love society as much as do many of the apes and monkeys, but lives in pairs, contented with the society of its own family.

The voice of this ape is of a very peculiar character, and its powers are put forth with the greatest intensity while the animal is performing its wonderful feats of agility. The time of day seems to have some influence upon the creature and its cry, for in its native state the Gibbon is most noisy in the early mornings,—the loud, strange cry being probably a call-note to its companions. Even in the open air, this call-note is exceedingly loud, and can be heard at great distances, so that when the animal is confined in a room, and exerts its voice, the ears of the bystanders suffer somewhat from its deafening resonance.

In themselves, the notes of this curious cry are rather musical than otherwise, but they are uttered with such vigour, that they become painful to the ears.

To judge by the cry of the female Gibbon, it is quite a musical performance, capable of being set to musical notes, and coming to an abrupt conclusion, by a couple of barks in octaves. The animal achieves the chromatic scale admirably, effecting the descent (no easy task even to the practised human vocalist) with a precision and rapidity that renders the vocal gymnastics as remarkable as those of the limbs. The note on which the creature began was E; and starting from this note, she began a series of chromatics, first ascending to the upper octave, and then descending in the same way, but always sounding the lower E almost simultaneously with the upper note, whatever that note might be. These musical efforts seemed to excite the creature greatly, for her whole frame appeared strung to a pitch of great intensity, her body dilated and quivered with excitement while she uttered her rapid cry, and at its conclusion she shook with all her strength the object to which she was clinging.

This individual was pleasing in manners, gentle and caressing to those whom she favoured. With delicate discrimination, she at once admitted ladies into her confidence, and would come to them voluntarily, shake hands, and permit herself to be stroked. But when gentlemen tried to gain her affection, she deliberated on the matter, and did not allow of a nearer acquaintance without further investigation. But when her scruples were once overcome, she was perfectly affectionate and confiding.

The SILVERY GIBBON derives its name from the silver-grey colour which generally pervades the fur. In some parts of the body, however, there is a browner tinge, and the face and palms of the hands are quite black. The sides of the face are covered with white, furry hair, which is so plentiful, that although the ears are tolerably large, they are nearly hidden among the luxuriant hairy fringe that encircles the head. The eyes of this and of the other Gibbons are deeply sunk in the head. The size of the Silvery Gibbon is little different from that of Gibbons generally, the adult animal measuring about three feet or so in height. Active, as are all its relatives, it lives among the branches and tall

SILVERY GIBBON.—*Hylóbates Leuciscus.*

canes of the Malaccas, and displays in these congenial habitations the same sportive agility that is so peculiar to the Gibbons.

A very different group of animals now comes before us, separated even by the outer form from the apes.

The chief distinction which strikes the eye, is the presence of a tail, which is of some length, and in several species, among which we may mention the SIMPAI itself, is extremely long and slender in proportion to the body. The arms of these animals are not of that inordinate length which is seen in the limbs of the apes, but are delicate and well proportioned. The hinder paws, or hands, are extremely slender, their thumbs being short, and, as will be seen by reference to the engraving, are twice the length of the fore-paws.

Some of these monkeys are furnished with small cheek-pouches, while others appear to be destitute of these natural pockets. The callosities of the hinder quarters are well shown.

In this group of the Quadrumana, the characteristics of the apes disappear, and the animals betray more clearly their quadrupedal nature. Very seldom do they assume the erect attitude, preferring to run on all fours like a dog, that being their legitimate mode of progression. Even

SIMPAI.—*Presbýtes Melalophos.*

when they do stand on their hind feet, the long tail at once deprives them of that grotesque semblance of the human form, which is so painfully exhibited in the tail-less apes. Besides these external distinctions, there are many remarkable peculiarities in the anatomy of the internal organs, which also serve to settle the position of the animal in the order of nature. Among these internal organs, the stomach displays the most remarkable construction, being very large, and divided into compartments that bear some resemblance to those in the stomach of ruminating animals.

These monkeys are distributed through several parts of the world, the Simpai making its residence in Sumatra.

This is a beautiful little animal, and is pleasing both for elegance of shape, and the contrasting tints with which its fur is decorated. The prevailing colour of the body is a light chestnut, with a perceptible golden tinge, showing itself when the light falls obliquely on the fur. The inside of the limbs and the abdomen are not so bright as the rest of the body, but take a most sober tint of grey. At the top of the head the hair is straight, and is set on nearly perpendicularly, so as to form a narrow crest. The colour of the crest, together with that of a narrow band running over the eyes and temples, is black. From this conspicuous peculiarity, the Simpai is also called the Black-crested Monkey. The name Presbytes signifies an old man, and is given to these monkeys on account of the wizened, old-fashioned aspect of their countenances. The term "melalophos" is literally "black-crested," and therefore a very appropriate name for this species.

The length of this animal, measured from the nose to the root of the tail, is about twenty inches, and that of the tail itself is not very far from three feet. Its fur is very soft and glossy.

Several allied species are rather celebrated among furriers for the beauty of their natural garments, and suffer much from the hunters. A well-known example, the Negro Monkey, sometimes called the Moor, or the Budeng (*Presbytes Maura*), furnishes the long black monkey-fur that is put to so many uses. Jet black as is the long silky fur of an adult Budeng, it is of a very different colour when the creature is young. The fur of the very young Negro Monkey is of a yellowish red colour, and the black tint appears first on the hands, whence it spreads up the arms, across the shoulders, and by degrees creeps over the whole body.

It is a native of Java, and is a gregarious animal, being found in troops of fifty or more in number, and extremely noisy on the approach of a human being. In temper it is said to be morose and sulky, so that, in spite of its beautiful coat, it is seldom domesticated. In such a case a bad temper must be a positive blessing to a monkey.

Not only for the skins are these monkeys valuable. Their teeth are in some favour for the composition of ornaments, being pierced and curiously strung together.

There is another substance which is furnished by some individuals among this group of monkeys, but is not always found in them. This is the bezoar, a substance which was long in high esteem for the cure of disease, and even now is used for that purpose by the physicians of the East. The word bezoar is originally "bâd-zahr," or poison-expeller, and was applied to this substance as it was supposed to possess extraordinary virtue in destroying the effects of poison, whether administered internally, or applied to the bite of serpents, or the wounds caused by poisoned weapons. The bezoars are concretions, chiefly of phosphate of lime, which are found in the stomachs of many ruminating animals, the most valuable being those of the Persian wild goat. So highly valued were the last, that they were sold for ten times their weight in gold.

Those of the Asiatic monkeys are considered the most valuable of all the bezoars, as, although small in size, they are powerful in quality. It is a somewhat remarkable circumstance that these monkeys, with their approximation to the ruminant stomach, should produce the same description of substance that was formerly thought to be the special property of the ruminating animals.

A well-known example of this group of monkeys is the HOONUMAN, or ENTELLUS. This is a considerably larger animal than the Simpai, as the adult Hoonuman measures three or four feet from the nose to the root of the tail, and the tail itself rather exceeds the body in length. The colour of this monkey when young is a greyish brown, excepting a dark brown line along the back and over the loins. As the animal increases in years, the fur darkens in colour, chiefly by means of black hairs that are inserted at intervals. The face, hands, and feet are black.

It is a native of India, and fortunately for itself, the mythological religion is so closely connected with it that it lives in perfect security. Monkeys are never short-sighted in spying out an advantage, and the Entellus monkeys are no exception to the rule. Feeling themselves masters of the situation, and knowing full well that they will not be punished for any delinquency, they take up their position in a village with as much complacency as if they had built it themselves. They parade the streets, they mix on equal terms with the inhabitants, they clamber over the houses, they frequent the shops, especially those of the 'pastrycooks and fruit-sellers, keeping their proprietors constantly on the watch.

ENTELLUS.
Presbytes Entellus.

Reverencing the monkey too much to afford active resistance to his depredations, the shopkeepers have recourse to passive means, and by covering the roofs of their shops with thorn-bushes, deprive the thieving deity of his chief point of vantage. Let it not be matter of wonder that a thief can be a god, for even the civilised Romans acknowledged Mercury to be the god of thieves, and they only borrowed their mythology from a much more ancient source. Certainly the Hoonuman gives practical proof of his claims to be the representative of such a deity ; for he possesses four hands with which to steal, and neglects no opportunity of using them all.

Conscious of the impropriety of its behaviour, the monkey does not steal anything while the proprietor is looking at it, but employs various subtle stratagems in order to draw off the owner's attention while it filches his goods. Many ludicrous anecdotes of such crafty tricks are known to everyone who has visited India, and employed his eyes.

The banyan-tree is the favoured habitation of these monkeys ; and among its many branches they play strange antics, undisturbed by any foes excepting snakes. These reptiles are greatly dreaded by the monkeys, and with good reason. However, it is said that the monkeys kill many more snakes in proportion to their own loss, and do so with a curiously refined cruelty. A snake may be coiled among the branches of the banyan, fast asleep, when it is spied by a Hoonuman. After satisfying himself that the reptile really is sleeping, the monkey steals upon it noiselessly, grasps it by the neck, tears it from the branch, and hurries to the ground. He then runs to a flat stone, and begins to grind down the reptile's head upon it, grinning and chattering with delight at the writhings and useless struggles of the tortured snake, and occasionally inspecting his work to see how it is progressing. When he has rubbed away the poor animal's jaws, so as to deprive it of its poison-fangs, he holds great rejoicings over his helpless foe, and tossing it to the young monkeys, looks complacently at its destruction.

Besides the reverence in which this animal is held through its deification, it has other claims to respect through the doctrine of metempsychosis, or the transmigration of souls through the various forms of animal life. From the semblance of human form which is borne by the monkeys, their frames were supposed to be the shrines of human souls that had nearly reached perfection, and thereby made their habitations royal. Therefore, to insult the Hoonuman is considered to be a crime equivalent to that of insulting one of the royal family, while the murder of a monkey is high treason, and punished by instant death. Many times have enthusiastic naturalists, or thoughtless "griffs," en-dangered their lives by wounding or killing one of these sacred beings. The report of such a sacrilegious offence is enough to raise the whole population in arms against the offender ; and those very men who study cruelty as a science, and will inflict the keenest tortures on their fellow-beings without one feeling of compunction,—who will leave an infirm companion to perish from hunger and thirst, or the more merciful claws of the wild beasts, will be outraged in their feelings because a monkey has been wounded.

The hunters in India find these animals to be useful auxiliaries in some cases, though tiresome in the main. They collect on boughs when a tiger or similar animal of prey passes under them, and often serve to point out to the hunter the whereabouts of the quarry. A tree thus covered with monkeys is a curious sight; for the boughs are studded with them as thickly as fruit, and the pendent tails give an absurd appearance to the group.

Although each part of every animal must be formed with some definite object, there are many which seem to be devoid of use, and among them is the monkey's tail.

Some of the monkeys—the spider-monkeys of America, for example—find in their tail a most useful member, by means of which they can suspend themselves from boughs, aid their limbs in tree-climbing, or, on an emergency, pick an object out of a crevice which the hand could not enter. But the use of the tails belonging to these old-world monkeys does seem to be very obscure.

Some writers have opined that the tails are intended to balance the body in the various attitudes assumed by its owner. But when we reply that the Gibbons, although very much more agile, and, from their very form, requiring more balancing than the monkeys, yet are totally devoid of tails, this supposition falls to the ground. It cannot be for the purpose of flapping away flies that these animals are furnished with such long and slender tails, for their shape renders them useless for that occupation; and, besides, the hands of the monkey are much better fly-flappers than its tail could possibly be.

The question arises, "What does the monkey do with his tail?"

Answer. He nibbles it sometimes, when he is at a loss for occupation.

It is a curious fact that—at all events in captivity—the long-tailed monkeys *will* eat their tails, and nothing seems to deter them from this strange act. The tips of those members have been covered with plaisters, and have been tied up in bandages, but without effect. The ends of the tails have been treated with aloes, cayenne pepper, and other disagreeable substances, just as the finger-tips of a nail-biting child are dressed. But, though the creature splutters and makes strange grimaces at the horrid flavours that greet his palate, he cannot refrain from the accustomed luxury, and perseveres in his nibbling. One great charm of this habit seems to be the excitement felt by the monkey in trying how far he can nibble without smarting for it. Whatever may be the cause, the effect is that the tail is gradually eaten up, in spite of all endeavours to prevent such a consummation. Considered in a social light, the tails are calculated to promote the merriment of the company, for they are admirable handles for practical jokes, and afford mutual amusement, not unmingled with indignation.

The PROBOSCIS MONKEY, or KAHAU, as it is sometimes called, on account of its cry bearing some resemblance to that word, is an inhabitant of Borneo, and probably of several neighbouring countries. It is, as may be seen by the engraving, an animal of very unattractive features, principally on account of its enormously lengthened nose. This feature does not present itself in perfection until the Kahau has reached its maturity. When the animal is very young, there are but few indications of the singular length to which this feature will attain; for, although it is rather more prominent than in most of the monkeys, it is rather of that description of nose denominated "*retroussé*."

In size, the Kahau is about equal to the hoonuman, and seems to be an active animal, leaping from branch to branch, through distances of fifteen feet or more. The natives assert, that while leaping they take their noses in

KAHAU.—*Presbytes Larvatus.*

their hands, in order to guard that feature from being damaged by contact with branches. Whether this refinement of caution be true or not, it is certain that they do hold their outstretched hands in a manner unlike that of the generality of monkeys, and probably for the purpose just mentioned.

These monkeys are fond of society, assembling together in large troops, and howling with exceeding fervour. They observe hours, regulating themselves by the sun, at whose rising and setting they congregate together, and perform their arborial gymnastics.

For the preternatural ugliness of the countenance, the Kahau is partially compensated by the beautiful colouring of its fur, which is thick, but not woolly, nor very long. The principal colour in the body is a bright chestnut red; the sides of the face, part of the shoulders, and the under parts of the body being of a golden yellow. A rich brown tint is spread over the head and between the shoulders; the arms and legs taking a whiter tinge than the shoulders.

The nostrils of this creature do not at all resemble those of man, although the animal's nose seems to be a burlesqued edition of the corresponding feature of the human countenance. They are placed quite at the extremity of the nose, and are separated from each other by a very thin cartilage. They are therefore, as has been observed in a former page, quite devoid of that expressive character which is so strongly exhibited in the contour of the human nostril.

We will pass on to more pleasing animals; but before taking leave of this group of monkeys we must observe that they are hardly deserving of the title "Slow Monkeys," which has been applied to them. They sit quietly on the branches, with their tails hanging down, and their bodies gathered together; but they only need some exciting cause to make them throw off their seeming apathy. They then spring from branch to branch, flinging themselves towards their mark with wonderful precision, and are all life and energy.

URSINE COLOBUS.—*Colobus Ursinus.*   BLACK COLOBUS.—*Colobus Sátanas.*

THE COLOBUS.

THE scientific name which is given to this genus of monkeys, explains—as is the proper office of names—one of the leading peculiarities of the animals. The title "Colobus" is a Greek word, signifying "stunted," or "maimed," and is given to these animals because the thumbs of the two fore-limbs give but little external indication of their presence, so that the hand consists merely of four fingers. They are exclusively African animals. They are rather handsome creatures, and their hair is sufficiently long and silky to be valuable as a fur.

The Ursine, or Bear-like Colobus, is so named because the general colour of its long black fur, and the form of the monkey itself, with the exception of the tail, has something of the bearish aspect. The cheeks and chin of this animal are covered with white hair; there is a white patch on the hind legs; and, with the exception of a few inches at its root, which retain the black hue of the body, the tail is of a beautiful white, terminated with a long and full white tuft.

Another species, called the Full-maned Colobus, is rather a remarkable animal, not so much on account of its habits, of which little is known, but on account of the huge mass of long hairs which cover the head and shoulders, falling nearly as low as the middle of the breast. The colour of this mane, or "full-bottomed peruke," as it has also been called, is yellow, with black hairs intermixed. Like the Ursine Colobus, the Full-mane possesses a tail of a white colour, decorated with a snowy-white tuft.

The Black Colobus is devoid of those exquisitely white portions of the fur that are so strongly marked in the Ursine and the Full-maned Colobus. The head, body, limbs, and even the tail, are jet black, unrelieved by any admixture of a lighter tint. This uniform black hue of the long glossy fur, has earned for the animal the demoniacal title which will be found appended to the figure. Beside the sable garments that are conventionally attributed to the powers of darkness, the animal in question is probably in part indebted for its name to the black crest, that projects over the forehead and eyes with so pert and impish an air.

GUEREZA.—*Cólobus Gueréza.*

Our last example of this genus is the beautifully adorned GUEREZA. This monkey presents a singular example of contrast in colours. The back, shoulders, the crown of the head, the limbs, and part of the tail, are black. But along the sides, the black hairs have hardly run a fifth of their course, when they suddenly become of a pure white. This change is not effected by a gradual melting of the black into white, but the line of demarcation is clearly defined. There is also a fringe of white hairs that encircles the cheeks, and becoming suddenly very narrow, runs across the forehead, just above the eyes, and is boldly contrasted with the black face and black scalp. The tail ends in a whitish tuft, but not so large as that of the Ursine Colobus, nor so purely white.

Very little is known of the habits of this animal, but it is said to be a gentle creature, feeding on insects as well as on the usual vegetable food for monkeys.

It is a native of Abyssinia, and its name "Guereza" is its Abyssinian title.

The beauty of its fur causes it to be much sought after by the natives of the country, who make its skin into coverings for the curiously shaped shields which they bear. The white fringe is the part that is chiefly valued, and its appearance on a shield points out at once a person of distinction in its bearer.

GRIVET.   GREEN MONKEY.   VERVET.
Cercopithécus Engythithia.   *Cercopithécus Sabæus.*   *Cercopithécus Pygerythrus.*

We now arrive at a group of small monkeys, with exceedingly long names. The term "Cercopithécus" is composed from two Greek words, signifying "tailed ape."

The monkeys belonging to this genus are very abundant in their native forests, and the unfortunate peripatetic monkeys that parade the streets in tormenting company with barrel organs, or seated on the backs of dejected and pensive bears, are mostly members of this group. The first glance at one of these monkeys will detect a peculiar sheen of the fur, that bewilders the eye and conceals the precise colour. If, however, the hairs are examined separately, each hair will be found to be varied in colour several times, black and yellow being the principal colours. First the hair will be black for a part of its length, then yellow, then black again, and so on to the tip. As the black has something of a bluish tinge in it, the mixture of the yellow and blue gives an undefined greenish hue, which in the central figure of the engraving is so decided, as to cause the name of Green Monkey to be given to the animal.

The Cercopitheci are remarkable for the singularly large development of the cheek pouches, which seem to possess an illimitable power of extension, and to accumulate a strange medley of articles. Supply one of these monkeys with nuts or biscuit, and he will contrive to put the greater part of the food into his cheek pouches, only eating a small portion at the time.

I never knew but one instance when the pouches were quite full, and even then the monkey was a small one, and the nuts were large. The little creature was liberally gifted with nuts, with the special purpose of ascertaining the capabilities of the pouches, and after dilating its cheeks to a wonderful extent with large "cob" nuts, it was at last compelled to empty them into its hands.

These pouches have been aptly compared to the stomach of a ruminant animal, and are employed in much the same manner. By means of the possession of these natural cupboards, the monkey is enabled to make little incursions, to eat as much food as hunger demands, and to carry away sufficient nourishment for one or two meals more, without being embarrassed in its retreat by its burden.

It is worth notice that the word "monkey" is derived from the name of one of this group, the Mona. The diminutive of Mona is Monikin, the transition from which word to our "monkey" is sufficiently evident.

The GRIVET, or TOTA, as it is called by some writers, is of a sombre green colour; the green being produced, as has been already mentioned, by the black and yellow hair. The limbs and tail are of a greyer tint than the rest of the body, the yellow portion of the hair being changed to a dull white. The inside of the limbs and the abdomen are slightly tinged with white. In the male animal the canine teeth are rather protuberant, showing themselves beyond the lips. The naked skin of the face, ears, and palms, is black, dashed with that deep violet hue that is found in so many of the monkeys. At each side of the head, the white hairs stand out boldly, whisker fashion, and give a very lively character to the head. It is an African animal, and common in Abyssinia.

The centre of the group is occupied by the GREEN MONKEY, sometimes called the Callithrix, or Beautiful-haired Monkey, on account of the exquisitely delicate marking of each separate hair. The inside of the limbs is nearly white, as is the under surface of the body, and the outer side of the limbs takes a greyish tinge. The hairy fringe that grows over the side of the face is of a delicate golden yellow.

This monkey is a native of Senegal and the neighbouring parts, and is frequently brought to this country.

The VERVET is the last of the figures. This is rather a variable animal in point of colour, some specimens being decidedly pale, while others assume a blackish hue. In general, the colour of the animal is as follows. The prevailing tint of the fur is much the same as that of the Grivet, to which animal the Vervet bears a strong resemblance. The head, the throat, and breast, are of a light dun, the paws being very dark. In the male Vervet the canines are rather long, and show their points beyond the lips.

These little animals are extremely abundant in their native land, and in Senegal especially are seen among the branches in immense troops. They seem to feel their own dignity as masters of the wood, and are aggrieved by the intrusion of human beings into their special domains. They are so agile and swift in their movements, and withal so quick of sight, that they almost invariably descry an intruder before themselves are visible. There may be hundreds of little heads peering through the branches of the very tree under which the traveller is seated, and double the number of sharp little eyes glittering among the foliage; but their owners are so lithe and cautious, that their presence remains undiscovered until they choose to announce themselves in their own fashion.

Monkeys have their code of etiquette as well as men; and, as they do not possess cards, the correct mode in which a monkey announces its presence to a human visitor is by dropping a piece of stick upon him. Perhaps he may consider the stick to be only a twig fallen in the course of nature, and so take no notice of it. Down comes another stick, and if that does not cause him to look up, several more are let fall upon him until his attention is drawn to the assembly in the branches.

This point having been gained, the next object is to let the intruder know that his company is undesirable, and that the sooner he takes his departure the more agreeable it will be for all parties.

That the long-tailed party are averse to so big an animal without an inch of tail, is clearly shown by the angry chattering that is set up, and the double rows of white and sharp teeth that are freely exhibited; and that the position of the objectionable individual will become anything but agreeable, is practically proved by the riot among the branches, which are shaken with noisy violence, the constant cries and chattering, and the shower of sticks and various missiles that pour upon him from above. Whether the object of their dislike be armed or not, seems to make but little difference to these tetchy animals. Should he retreat from so unpleasant a proximity, well and good—they have achieved their point, and satisfied their pride of place. Should he retaliate, and hurl deadly leaden missiles among his persecutors in exchange for the harmless but disagreeable assaults committed on himself, they sullenly receive his fire, unterrified by the fall of their slaughtered companions, and, even when wounded, continue the unequal conflict. They evidently feel themselves in the right, and refuse to abandon their position. One traveller who had been thus treated by the monkeys, killed twenty-three of the poor animals in less than an hour—not much to his credit.

Killing a monkey is always a pitiful business, for it is so much like an act of murder committed on a human being. Many are the travellers who, urged either by anger, curiosity, scientific researches, or innate destructiveness, have destroyed these animals, and have been so stricken by remorse at the effect of their cruelty, that they have vowed never to kill another monkey as long as they lived. There are several most touching narratives of such scenes, but they are so trying to the feelings, that I can neither bring myself to write them, nor to inflict such tragical tales on my readers. It were much to be wished that men could read the effects of their cruelty in the eyes of other animals except the monkeys, and would bind themselves never to inflict one unnecessary pang upon any living creature. Surely no wounded monkey could look at its tormentor with more pitiful eyes than those of the over-laden and over-driven ass, or even the neglected and ill-treated dog. These latter animals, too, are always with us, and need not only the cessation of actual cruelty, but even the gift of human sympathies, before they can take their proper place in creation, and become the true servants and companions of man. It rests with man, who gave names to all living beings, to complete the work which God began in making them, and by stooping from his own superior nature, to be a protecting and loving providence to the beings that are placed under him. By so doing, man draws out, fosters, and develops the better nature which is inherent in every animal, and which would remain concealed, like a seed in ice-bound soil, unless it were brought into vigorous life by the genial influence of a higher being. I cannot believe that any animal is utterly untameable, and so totally brutish as to be insensible to the touch of kindness. There are many animals which are proof against the old-fashioned way of education, and which are only rendered more fierce and obstinate by the tortures and blows which were formerly so freely bestowed on animals in course of training. But these very animals have proved to be sensitive to gentle and kind treatment, and, though fierce and savage towards one who only approached in order to torment, became docile and subdued when in the hands of a tender and sympathetic owner.

The same rule holds good with human beings; and the great and beautiful truth becomes daily more apparent, that severity of punishment has an injurious rather than a beneficial effect, and that the only true rule is that of love.

The Grivets and Vervets are frequent visitors to our land; and being extremely inquisitive in character, as well as active in body, play strange pranks in their land of exile. One of these creatures which resided in London some few years ago, caused considerable annoyance to his neighbours, one of whom very kindly favoured me with the following account of some of his misdemeanours.

"A few years ago, we lived next door to a lady who had a pet monkey, which was one of the most imitative and mischievous little beings that ever existed. His imitative nature caused the servants so much trouble, that he had not a friend among those of his own house.

"One day he observed the ladies'-maid washing her mistresses' lace; and his offers of assistance having been somewhat roughly repulsed by her, chattering and scolding he went forth in search of adventures. Unfortunately, my windows were invitingly open, and he entered, with the idea of washing fresh in his head.

"His spirit of curiosity induced him to open two small drawers, from which he abstracted their whole contents, consisting of lace, ribbons, and handkerchiefs. He placed these things in a foot-pan, together with all the water and soap that happened to be in the room, and he must then have washed away with great vigour; for when I returned to my room, after an absence of an hour or so, to my astonishment, I found him busily engaged in his laundry operations, spreading the torn and disfigured remnants to dry. He was well aware that he was doing wrong, for without my speaking to him, he made off the moment he saw me, going very quickly and hiding himself in the case of the kitchen clock in his own home.

"By this act, the servants knew he had been doing mischief, as this was his place of refuge when he was in trouble or disgrace.

"One day he watched the cook while she was preparing some partridges for dinner, and I suppose that in his own mind he considered that all birds ought to be so treated, for he managed to get into the yard where his mistress kept a few pet bantam fowls, and after robbing them of their eggs, he secured one of the poor hens, with which he proceeded to the kitchen, and then commenced plucking it. The noise that the poor bird made brought some of the servants to the rescue, but they found it in such a pitiful and bleeding state, that in mercy it was at once killed.

"After this outrageous act, Mr. Monkey was chained up, which humiliated him so much that he steadily refused his food, and soon died."

In their native woods these animals are very amusing if they can be watched without exciting their anger or fears. They chase one another about the branches, screaming, chattering with delight when they have succeeded in playing off a practical joke on a comrade, and anon shrieking with anger when suffering from a joke played on themselves. Not only do they chase the members of their own race, but wage a constant war against the tail-feathers of the brilliant and noisy parrots that inhabit the same country.

The motives that incite the monkeys to pluck out these feathery trophies are twofold, each of them dear to the very soul of the mischievous creature. The first and most obvious motive is that of sheer mischief, but the second is of rather a more complex character. When an immature feather is recently drawn from a bird, its quill portion is generally soft, and filled with the material by which the feather is supplied with nourishment. The monkeys take great delight in sucking these soft feathers; and in order to procure a supply of this curious dainty, chase the poor parrots, even to the tops of the trees. At first sight, it would appear that the legs and arms of the monkey would have little chance of winning a prize defended by the beak and wings of the parrots, which sit exultantly screaming on twigs that bear their weight easily enough, but are too slender even for the monkeys to venture upon. But the restless vigilance and quick hand of the monkey often win the day; and while the parrot is shrieking defiance to an enemy in front, it is suddenly startled from its fancied security by the loss of its tail, which has been snatched away by a stealthy foe from behind. The deafening din which is occasioned by the joint voices of parrots and monkeys, may be easier imagined than described.

That the monkeys should take an interest in so singular a game, and should play it with such spirit, is no matter of wonder, inasmuch as they have nothing to lose in case of failure, and a pleasant little reward in case of success. But the parrots seem to be actuated by very strange motives when they consent to hazard so valuable a stake upon their own alertness; and even if they win the game, can gain nothing but the retention of their own tails. A stroke or two of their wings would carry them beyond the reach of the most agile monkey that ever tenanted a tree; but they prefer to measure their own agility and vigilance against that of their four-handed antagonists, and often pay the penalty of so witless a pastime.

Were the parrots capable of connected reasoning, they might sometimes find cause for alleviating the pangs of defeat, by vindictive satisfaction in seeing their foes succumb to a still worse fate than that which had been inflicted on themselves. If the monkey likes to suck the bleeding trophies snatched painfully from the bird's person, there are many animals which feel a great partiality for the monkey, not as a pleasant companion, but as an agreeable article of diet. Some of these foes, such as the leopards and snakes, have been already mentioned; but there is one enemy who is more to be dreaded than serpent or pard, and this foe is man.

Monkey flesh forms a favourite article of food with the human inhabitants of the same country, and is said to be tolerably good eating, though extremely dry and sapless. Part of this fault seems, however, to lie with the very primitive style of cooking which is prevalent in those regions, and which is achieved by running a sharp stake through the animal's body, and letting it roast before the fire.

Europeans find a difficulty in accustoming themselves to the sight of broiled monkey; for it presents an appearance so unpleasantly suggestive of a toasted child, that horrid ideas of cannibalism arise in the mind, and even a stomach sharpened by hunger revolts from the unsightly banquet.

The well-known Mona monkey belongs to the same genus as the foregoing animals. All the long-tailed African monkeys are termed Monas by the Moors. On account of its green, maroon, grey, and white fur, it is sometimes called the Variegated Monkey. Little is known of its habits in a state of nature, and accounts of its captive character vary as much as is usually found in similar cases.   On the authority of one writer, who speaks from personal experience, we are told that the adult Mona is savage and irritable ; while another, who also writes from personal observation, tells us that the Mona is gentle, and devoid of petulance or malice, its excellent disposition remaining unaltered by age.

One of these animals, which passed several years in Europe, was remarkable for its amiable temper ; and although by no means free from the little mischievous and pilfering habits that are so inextricably interwoven in the monkey nature, was so quiet and gentle as to be left at perfect liberty.   He was an adept at unlocking boxes and examining their contents, could unravel the intricacies of a knot, and was possessed of a hand dexterous and nimble at picking pockets.   The last-named occupation seemed to afford peculiar gratification, which was increased by the fact that his visitors were accustomed to carry nuts, cakes, and other delicacies in their pockets, on purpose for the monkey to find them there.

Many specimens of this animal have been brought to Europe, and their disposition seems to vary according to the temperament of their owner.   Monkeys are very sensitive animals, and take much of their tone of character from that of the person with whom they are most familiar.

They seem to be affected almost instantaneously by predilection or antipathy, and on their first interview with a stranger, will evince either a satisfaction at, or objection to, his presence, which they will maintain for ever afterwards.   I have often watched this propensity, and seen the same animal come voluntarily and offer itself to be caressed by one person, while the very approach of another would set it chattering with anger.   It may be that the animal is actuated simply by caprice ; but the more rational mode of accounting for such an action, is to suppose that the fine instincts which are implanted in its nature, enable it to discover its true friends at a glance without the trouble of testing them.

THE WHITE-NOSE MONKEY.   THE PATAS.   THE DIANA MONKEY.
  Cercopithécus Petaurista.   Cercopithécus ruber.   Cercopithécus Diana.

The three monkeys which form the subject of the accompanying engraving are all members of the same genus, although they are marked by decided differences of colour and general aspect.

The little animal which occupies the left hand of the group is the White-nose Monkey of Western Africa.   It is a curious little creature, with an air of quaint conceit, for which it is indebted to the fringe of white hairs that surrounds its face, and the conspicuous white spot on the nose, which has earned for it the title of White-nose.   As is so often the case in these animals, the under side of the body and inside of the limbs is of a much lighter tint than the upper portions.   This distinction is peculiarly well marked in the long tail, which is nearly black above, and beneath takes a greyish hue.

It is a very graceful little creature, playful, but petulant and coquettish, disliking to be touched, but fond of notice and nuts, and often balanced in curious perplexity between its coy shyness and the charms of an offered dainty. When in perfect health, it is seldom still, but flits with light grace from one spot to another, performing the most difficult muscular efforts with exquisite ease, and profoundly sensible of the admiration which its pretty antics never fail to excite in the spectators.

It is by no means a large animal, its head and body only measuring fifteen or sixteen inches, the tail being little short of two feet in length.

The central figure of the group is the PATAS, sometimes called the Red Monkey, on account of the ruddy colour of the hair. The general tint of the fur is a bright chestnut, or fawn colour, with a deep shading of red. This hue is shown very decidedly on the sides and on the outer portions of the hind legs, the legs themselves being of a darkish cream colour. The breast and the fore-limbs are covered with hair, which much resembles that of the Green Monkey.

It is an inhabitant of Western Africa, being found very commonly in Senegal. In size it is much superior to the last-mentioned animal, reaching more than three feet in length.

When left to an undisturbed life, these creatures are playful and inquisitive, but mischievous and spiteful withal. They display great courage when engaged in a fray, and if their size and strength were proportionate to their bravery and endurance, would be truly formidable antagonists. Even the fall of their comrades only seems to redouble their rage, and to stimulate them to increased exertions.

Too crafty to venture upon close combat, these monkeys retain their posts of vantage on the tree-tops, and hurling from thence every kind of offensive missile that can be procured, render their attack a matter of exceeding inconvenience, even to armed men. During the skirmish, the monkeys distort their features into strange grimaces, and rend the air with their cries of rage. They have been known to follow boats up the course of a river, keeping pace upon the overhanging trees, and becoming so troublesome from the constant shower of sticks, fruits, and other missiles, that the occupants of the boats were forced to fire at their assailants, and to kill many of the number before they could be freed from the annoyance.

This, as well as the foregoing long-tailed monkeys, belongs to that large group of quadrumanous animals called the GUENONS, nearly all of which possess similar character-istics of disposition. They are amusing and playful creatures, very active, and move with much grace of deportment. In captivity they are remarkable for their mercurial tem-perament, their ingenuity in devising and executing small malevolent pranks, and their insatiable appetite for nuts, and other similar dainties. They are curiously sensitive to ridicule, being thrown into furious excitement by any mocking gestures or sounds. Nothing seems to irritate a monkey more than a grin and a chatter, in imitation of its own habits. It will fly at the offender with furious looks and screams of rage, and, unless restrained by chains or bars, would be likely to inflict some damage by its sharp teeth. It will remember the person of its tormentor with singular tenacity of memory, and will be thrown into a state of angry agitation even by the sound of the hated voice.

Although rather tetchy and hot-tempered, and too apt to resent any supposed slight or injury, the Guenons are very capable of education, and in the hands of a kind and gentle teacher can be trained to perform many curious feats. Severity defeats its own aim, and only makes the creature fall back upon the innate obstinacy which is inherent in most animals, and of which the monkey has a large share. But a kind instructor, and one who will never lose his own temper, may take in hand even a savage monkey and reduce it to gentle obedience. As a general rule, the male monkeys are less open to higher influences than the females, and are therefore more difficult subjects for the trainer.

Nearly all the long-tailed monkeys that come to England belong to the Guenons, and the many anecdotes that are related of them may be safely attributed to this group of animals.

The monkey which is known by the name of the DIANA is remarkable not only for its quaint aspect, but for the richly variegated tints with which its fur is adorned. The most conspicuous feature in the Diana Monkey, is the long and sharply pointed beard which

1.   E

decorates its chin and face. The colour of the beard is a pure white, and the animal is extremely solicitous about the perfect spotlessness of its hue, taking every precaution to preserve the cherished ornament from stain. So careful is the monkey, that when it drinks it holds back its beard with one hand, lest it should dip into the liquid and be soiled.

It may seem rather singular that an animal which bears so masculine an adornment should be named after the bright virgin huntress of mythology, radiant in her perpetual youth. But though as Diana the beard might be scarcely appropriate, yet as Hecate it would not be so very inconsistent. The reason, however, for giving to this monkey the title of the Diana, may be found not on the chin but on the forehead : where a semi-lunar line of white hair gleams out conspicuously against the black brows, and bears a close resemblance to the silvery crescent borne by the Diana of the ancients.

The colouring of the fur is extremely diversified, and in several parts assumes a force and richness of tint that we should rather expect in the plumage of a bird than in the fur of a monkey. The back is mostly of a deep chestnut colour, and is relieved by a bright orange hue that covers the lower part of the abdomen and the inside of the thighs. The orange colour is very much the same as that of the well-known penguin feathers which are so extensively used for slippers, pouches, and other similar purposes.

A band of pure white separates the chestnut from the orange, and serves to set them off to great advantage. The remainder of the body is of a rather dark grey, and the hands are nearly black. The colour of the eye is a clear grey.

In captivity it is rather a pleasing animal ; almost fastidiously clean in habits, therein being in advantageous contrast to many of the monkey tribe. It is easily tamed, and walks deliberately forward to receive any gift at the hands of its visitors. When walking, its diverse colours produce a curious effect, especially when it is viewed from behind.

Although it is by no means a rare species, and is found in plenty in Guinea, Congo, and other places, it is not so often imported as might be expected. The total length of tail and body is about four feet and a half, of which the tail occupies rather more than the moiety.

THE SOOTY MANGABEY.—*Cercocebus fuliginosus.*

There are several species of monkeys belonging to the genus Cercocebus (*i.e. Tailed Monkey*), of which the animal that is so well depicted in the accompanying illustration is a good type. The Mangabeys, as these monkeys are called, are all inhabitants of Western Africa, and are tolerably frequent visitors to our island. They are amusing in their habits, and gentle in manner; easily domesticated, and open to instruction. Their temper does not seem to be so irritable as that of many monkeys; and even when they are roused to anger, their ire is comparatively evanescent.

On account of the white hue which marks the eyelids, the Mangabeys are sometimes termed the "White-eyelid Monkeys." The Sooty Mangabey is well named; for its general colour is nearly black, something like a half-tint chimney-sweeper. The black hue is only found in the adult animal, the colour of the young Mangabey being a fawn tint. Sometimes it goes by the name of the Negro Monkey; and under these several titles suffers somewhat from the confusion that is almost inseparable from such uncertain nomenclature. It is rather a small animal, measuring some eighteen inches or so from the nose to the root of the tail, which occupies about the same space.

Among the peculiar habits which distinguish the Mangabeys, we may especially notice the action of their lips, and the mode in which they carry the tail. They have a strange way of writhing their faces into a kind of quaint grin, in which they raise the lips, and exhibit the teeth almost as if they were laughing. When walking, they have a fashion of turning their tails over their backs, and carrying them reversed, in a line almost parallel with the direction of the spine.

Few monkeys can assume more *outré* attitudes than the Mangabeys, which seem to be, among monkeys, almost the analogues of the acrobats among mankind; and twist themselves into such strange contortions, that they seem to be able to dispense with the bones and joints with which other animals are furnished. They seem to be quite aware of their own accomplishments, and soon learn that their display will bring in a supply of nuts, cakes, and fruit to their exchequer. So they keep a vigilant eye on the visitors, and when they conceive that they have drawn attention to themselves, they execute a series of agile gambols, in the hope of meeting the reward which sweetens labour.

Their attention is soon excited by any object that is more than ordinarily glittering; jewellery of all kinds being as magnets, to which their eyes and fingers are instinctively drawn. My own fingers have more than once been endangered by the exceeding zeal manifested by the animal in its attempts to secure a ring to which it had taken a sudden liking. The monkey held out its paw as if it wanted to shake hands, seized my fingers with both its hands, and did its best to remove the object of its curiosity; fortunately, the ring fitted rather tightly, or it would probably have been lost or swallowed. As it was, a few scratches on my hands, and an outburst of disappointed anger on the part of the monkey, were the only results of the sudden attack.

MACAQUES.

THE various species of monkeys which are ranged under the common title of Macaques, are mostly well-known animals; being plentiful in their native lands, and frequently domesticated, both in their own and in foreign countries. They are all inhabitants of Asia, although the word Macaco is the name which is given to all kinds of quadrumanous animals on the coast of Guinea, and is almost synonymous with our own word monkey.

One of the best typical examples of this genus is found in the BONNET MACAQUE, or MUNGA, as it is often called. A native of Bengal and Ceylon, it is a frequent visitor to our shores; being tolerably hardy in constitution, bearing the long voyage well, and suffering less from our insular climate than many of the monkey tribe.

For the title of Bonnet it is indebted to the peculiar arrangement of the hairs on the crown of the head, which radiate in such a manner that they seem to form a kind of cap or bonnet. The general colour of the animal is a rather bright olive-grey, fading into white beneath. The skin of the face is of a leathery flesh colour.

The distinctions between the Macaques and the Cercopitheci, are not very striking; but by comparison of the two genera, sufficiently decided variations are visible. These are rather comparative than absolute. In the Macaques, the muzzle is slightly more solid than in the Guenons, the body and head are larger, and in most species the tail is shorter. The callosities are well marked, and in some instances are rendered more conspicuous by a surrounding fold of skin devoid of hair. The limbs, too, are more muscular than those of the Guenons. These peculiarities may be seen on reference to the illustration.

Whether the fault lies with its proprietor, or whether the temper of this Macaque be really uncertain, is difficult to say ; but its general disposition when in captivity is rather of a snappish and crabbed character. Those who have had much to do with the Munga, say that it is very capricious, and that its good humour cannot be depended upon, as is the case with many domesticated monkeys.

In its native land, the Munga enjoys exemption from most of the external ills to which monkey nature is liable ; for, in common with several other species, it is piously protected by the natives, on account of its importance in their myriad-deitied religion. Not content with permitting these monkeys to devastate his plantations at will, the devout Hindoo prepares a home for them in his temple, where they rule supreme, and tolerate not the intrusion of any monkeys of another caste. When old, they are of a very high caste indeed, according to the Hindoo ideas on the subject. The more fierce and savage the monkey, the higher is its caste ; and among serpents, the cobra is significantly the Brahmin.

The RHESUS, or BHUNDER MONKEY, is rather a handsome animal in point of colour; the usual olive-green and yellow being relieved by warmer tints of a very bright chestnut, almost amounting to orange. The back is of a brownish hue, while the lower part of the spine and the outside of the thighs is of the warm tint already mentioned. The arms and shoulders are lighter, and change to dun below. The eye is of a light brown colour.

As will be seen in the engraving, the Rhesus is of a short and sturdy make, and looks more like an ordinary quadruped than any of the preceding monkeys. The tail, too, is very short, and the callosities are very conspicuous ; more on account of their ruddy colour, than their size.

For cool impudence and audacity, this monkey stands unrivalled among its congeners ; surpassing even the previous animal in both these characteristics.

So excellent and spirited a description has been given by Captain Johnson, of these monkeys in their wild state, that I cannot do better than present his account in his own words.

" At Bindrabun (which name, I imagine, was originally Baunder-bund, literally signi-fying a jungle of monkeys), a town only a few miles distant from the holy city of Muttra, more than a hundred gardens are well cultivated with all kinds of fruit, solely for the support of these animals, which are kept up and maintained by religious endowments from rich natives.

" When I was passing through a street in Bindrabun, an old monkey came down to the lower branches of a tree we were going under, and pulled off my Harcarrah's turban, as he was running in front of the palanquin, decamped with it over some houses where it was impossible to follow him, and was not again seen.

" I once resided a month in that town, occupying a large house on the banks of the river, belonging to a rich native ; it had no doors, and the monkeys frequently came into the room where we were sitting, carrying off bread and other things from the breakfast-table. If we were sleeping or sitting in a corner of the room, they would ransack every other part.

" I often feigned sleep, to observe their manœuvres, and the caution with which they proceeded to examine everything. I was much amused to see their sagacity and alertness. They would often spring twelve or fifteen feet from the house to another, with one, some-times two young ones under their bellies, carrying with them also, a loaf of bread, some sugar, or other article ; and to have seen the care they always took of their young would have been a good lesson to many mothers.

" I was one of a party at Teekarry, in the Bahar district ; our tents were pitched in a large mango garden, and our horses were picqueted in the same garden at a little distance off. When we were at dinner, a Syce came to us, complaining that some of the horses had broken loose, in consequence of being frightened by monkeys on the trees ; that, with

BONNET MACAQUE.—*Macacus Sinicus.*   RHESUS.—*Macacus Rhesus.*

their chattering and breaking off the dry branches in leaping about, the rest would also get loose, if they were not driven away.

"As soon as dinner was over, I went out with my gun to drive them off, and I fired with small shot at one of them, which instantly ran down to the lowest branch of the tree, as if he were going to fly at me, stopped suddenly, and coolly put its paw to the part wounded, covered with blood, and held it out for me to see : I was so much hurt at the time, that it has left an impression never to be effaced, and I have never since fired a gun at any of the tribe.

"Almost immediately on my return to the party, before I had fully described what had passed, a Syce came to inform us that the monkey was dead ; we ordered the Syce to bring it to us, but by the time he returned, the other monkeys had carried the dead one off, and none of them could anywhere be seen.

"I have been informed by a gentleman of great respectability, on whose veracity I can rely (as he is not the least given to relating wonderful stories), that in the district of Cooch-Bahar, a very large tract of land is actually considered by the inhabitants to belong to a tribe of monkeys inhabiting the hills near it; and when the natives cut their different kinds of grain, they always leave about a tenth part piled in heaps for the monkeys. And as soon as their portion is marked out, they come down from the hills in a large body, and carry all that is allotted for them to the hills, storing it under and between rocks, in such a manner as to prevent vermin from destroying it.

"On this grain they chiefly live ; and the natives assert, that if they were not to have their due proportion, in another year they would not allow a single grain to become ripe, but would destroy it when green. In this account, perhaps, superstition has its full influence."

The natives are nearly as careful of the Rhesus, as of the Hoonuman itself ; and take sanguinary revenge on any one who wounds or kills one of these animals. On one occasion, two officers, together with their servant, lost their lives in a popular tumult caused by the death of a monkey, at which they had thoughtlessly fired. But although the monkeys may not be hurt, and are allowed to plunder the crops at their own sweet will, the Hindoo cultivators are by no means pleased to see their fields so often devastated, and would willingly preserve them from the depredators in spite of their divine, though thievish character.

To drive away the monkeys is almost an impossible act on the part of the native proprietor; for the monkeys consider themselves as quite on an equality with any dark-skinned human being, and decline to move an inch. So the only resource is to beg a European to undertake the task; and the monkeys, knowing that a white man is not so scrupulous as a black one, take the hint, and move off.

One ready-witted Englishman succeeded in keeping the monkeys away from his plantation for more than two years, and that without using any violence, or offending the prejudices of the natives.

He had planted a patch of sugar-canes, and had seen his growing crops eaten by elephants, swine, deer, monkeys, and other animals, without being able to guard the ground from the robbers. The heavier animals he excluded by means of a deep trench surrounding the cane-patch, and a strong palisading of bamboos just within the ditch. But the monkeys cared nothing for moat or wall, and carried off whole canes in their hands, eating them complacently as they proceeded to the shelter of the trees.

For a long time this state of things continued, and the planter was doomed to see the ripening canes devoured in his very presence, and the chewed fragments spit in his face by the robbers. This last insult proved too great a strain for his patience to endure, and after some thought, he hit upon a stratagem which answered even beyond his expectation.

He chased a flock of the monkeys into a tree, which he then felled; and by the help of his assistants, captured a number of the young, which he conveyed home.

He then mixed some treacle with as much tartar-emetic as could be spared from the store, and after painting all the young monkeys with this treacherous mixture set them free. Their anxious parents had been watching for their offspring, and carried them away out of danger. The liberated captives were then surrounded by the whole troop, who commenced licking the treacle from their fur. Before very long, the expected effects made their appearance, and the poor monkeys presented a most pitiful appearance.

The result of the affair was, that the monkeys were so terrified at the internal anguish which their depredations had caused them to suffer, that they fled the place, and not a monkey was seen in that locality until long afterwards.

In captivity they are most mischievous, and are always on the watch for an opportunity of exhibiting a little malice.

They tear pieces out of the dress of anybody who may happen to approach near their cage; they snatch at any ornament that strikes their quick eyes; they grin and chatter with exultation when they succeed in their mischief, and scream with rage when they are foiled. They prefer to exercise these abilities on human sufferers; but in default of man, whom they consider their legitimate game, they are not above playing practical jokes upon each other, and, better still, upon the inhabitants of neighbouring cages.

Some are of so jealous a disposition that the sight of another monkey eating a nut will throw them into a state of angry irritation, which is not always pacified even by the gift of a similar or even a better article.

The skin of this monkey is very loose about the throat and abdomen, and generally hangs in folds.

The animal which is shown in the following engraving is one of the best known of the monkey tribe; as it is tolerably hardy, it endures the changeable and chilly European climates better than most of its race.

As its name implies, it is a native of Barbary, where it is found in great numbers, but has also been naturalized upon the rock of Gibraltar. The Gibraltar MAGOTS are frequently mentioned in books of travel, and display great ingenuity in avoiding pursuit and discovering food. They keep to the most inaccessible portions of the rock, and scamper away hurriedly on the slightest alarm. But with the aid of a moderately good telescope, their movements may be watched, and are very amusing.

When in their native wilds, the Magots live in large flocks, each band seeming to be under the orders of some chosen leader. They are very intelligent, and possessed of a large share of the cunning that belongs to the monkeys, and which, when aided by their strength of muscle, agility of limb, and quickness of sight, keeps them in tolerable security from foes, and enables them to make raids upon cultivated lands without suffering the penalty due to their crimes.

The enemies which these creatures hold in greatest dread are the climbing felidæ; and on the approach of one of these animals, the colony is instantly in a turmoil. The leaders yell their cry of alarm and give the signal for retreat, the mothers snatch up their little ones, the powerful males range themselves in battle array, and the whole body seeks a place of refuge.

Open attacks are little feared by the Magots, as their combined forces are sufficiently powerful to repel almost any enemy. But at night, when they are quietly sleeping, the crafty foe comes stealing along, and climbing up the trees or rocks on which the Magots are sitting asleep, strikes down its unsuspecting prey.

When young, the Magot is tolerably gentle; and as it is sufficiently intelligent to learn many tricks, it is frequently brought to Europe, and its accomplishments exhibited before the public. But this state of comparative domesticity is only for

MAGOT, OR BARBARY APE.--*Macacus Innuus.*

a time, and as the bodily frame becomes more developed, so does the Magot lose its gentle nature, and put on a sullen and fierce deportment. Captivity seems to exert a terribly depressing influence over the animal when it becomes fitted by nature for its wild independence; and as the stimulus to the mind is removed by the restrictions under which the animal is placed, the mind loses its spring, and the creature is deserted by the apt intelligence that characterizes its wild state, and for which it has then no need.

This monkey is not very widely spread, for with the exception of the Rock of Gibraltar, it seems to be confined to Northern Africa. Some authors state that it is found in India, China, and even the entire African continent, but it seems clear that there has been some confusion of species. Indeed, the Magot has caused some little labour in placing it in its right position.

It is not a very large animal, as the full-grown males only measure about a yard in length, and the females are rather smaller. The general size of the Magot is about that of an ordinary bull-terrier dog.

The colour of the fur is tolerably uniform, differing chiefly in depth of shade, and is of a clear greyish colour. The head is strong and heavy, the eyes deeply set under the overhanging brows, the neck is short and powerful, the teeth are fully developed and sharp, the finger-nails are sufficiently strong to inflict a severe wound; so that the entire aspect of an adult male Magot is that of a fierce and dangerous animal.

Its walk on level ground is rather awkward, this animal making use of feet and hands for that purpose; but it climbs with ease and agility up trees or rocks, and in a domesticated state is fond of running up and down ropes, and swinging itself about its cage.

In captivity it will eat almost any kind of food, but in its wild state it prefers fruit, leaves and other vegetable fare, varying its diet by sundry insects which it captures. When enraged it utters a fierce harsh yell, which, when enhanced by the force of numbers, the fury-flashing eyes and warlike gestures, often suffices to intimidate a foe from venturing upon an attack. But when it is not under the influence of angry feelings, its voice is comparatively mild and gentle, being a soft and almost caressing chatter.

There is a strange grimace in which this animal habitually indulges on almost every emotion, whether it be caused by pleasure, anger, or disappointment. The cheeks are sucked in, the lips are contracted over the gums, and the teeth are freely exhibited.

Although it is popularly termed the Barbary Ape, the Magot is not a true ape, being organized after a very different fashion from the veritable Simians. Belonging to the same genus as the Munga and Rhesus, it is almost entirely destitute of the tail which is so conspicuous an adornment of these monkeys. In the Magot the tail is reduced to a mere projection, sufficient to mark the spot where that member would have been placed, but not prominent enough to be ranked among real tails. Owing to this formation, the Magot, although one of the Macaques, was placed among the apes by earlier naturalists.

When at liberty in its native lands, the Magot has a great predilection for hunting scorpions, insects, and similar creatures, and devouring them on the spot. It displays peculiar aptitude for discovering and pouncing upon its prey.

Scorpions and beetles are found in profusion under stones, logs, or in similar sheltering places, and are there secure from any ordinary foe. But the quick senses of the Magot detect them in their concealment, and the ready hands sweep away the shelter and make the insect prisoner before it recovers the sudden surprise of its violated roof. On the rock of Gibraltar these monkeys are constantly engaged in turning over the loose stones, and by their perpetual industry have, in course of years, quite altered the surface of the earth, affording, it may be, grounds for sore perplexity in the minds of future geologists.

To any ordinary animal the scorpion would be rather a dangerous prey, and would probably avenge its death most fully by a stroke of its torture-giving and swiftly-lashing tail. The Magot, however, has hands which can overmatch even the scorpion's tail, and no sooner is one of these baneful creatures brought to light, than the monkey pounces upon it, twitches off the poison-joints of the tail, and then, grasping the disarmed scorpion, eats it as composedly as if it were a carrot.

In default of such large insects as have been mentioned, the Magot turns its attention to smaller deer, and, entering into a mutual engagement with a friend of its own race, they reciprocally exterminate the parasitic insects with which monkeys generally swarm.

Small though the quarry may be, the Magot displays much excitement in the chase, and after running down its prey successfully, holds the captured insect to its eyes, contemplates it with a grimace of satisfaction, and then daintily eats it. When in captivity it continues the same pursuits, and may often be seen nestling close to a friendly cat or dog, busily engaged in a minute investigation of its fur, and ever and anon giving vent to a little complacent chuckle which proclaims a successful chase. Sometimes the Magot contracts a strong friendship for its master, and being desirous to render every service in its power, jumps on his shoulder, and examines his head with much care, though, we may hope, with little ultimate satisfaction.

It often happens that the domesticated Magot takes a fancy for some other animals that may chance to come in its way, especially if they are young and comparatively helpless. It then acts as a voluntary nurse, and performs sundry kind offices for its charge, carrying them about with it, and, like nurses in general, becomes horribly jealous if its authority be in the least infringed.

Its attitudes are rather singular. When walking or running, it goes chiefly on all-fours, but when it wishes to rest, it sits in a manner very similar to the corresponding attitude in man; when sleeping it generally lies extended at length, reclining on one side, or gathered up in a seated position, with its head drooping between its hind legs.

In the absence of a tail, and in general form, the BLACK MACAQUE bears some resemblance to the Magot, but in colour and arrangement of hair it is entirely distinct from that animal.

The tint of the fur is as deep a black as that of the Budeng, or Black Colobus, which was mentioned on p. 42. Both these monkeys are possessed of crests which give a peculiar character to the whole aspect. That of the Black Colobus, however, is reverted forward, and curves to a point over the forehead, while that of the animal before us rises from the head and bends backward over the neck in a manner not unlike that of the cockatoo.

Like the Magot, the Black Macaque has been called an ape by some writers, and a baboon by others, on account of the apology for a tail with which its hinder quarters are

BLACK MACAQUE.—*Macacus Niger.*

terminated, but not decorated. It is an inhabitant of the Phillippines and the neighbouring countries.

THERE are few races of animals which have not been impressed by their human superiors into their service. Although the bodily powers of man are often more limited than those of the inferior animals, yet the lofty human intellect can more than compensate for corporeal deficiencies by making use of these faculties which are possessed by the subservient creation.

Thus the Indian hunters take advantage of the active and stealthy chetah to capture the prey which is too vigilant of sight and too active of foot to be approached by man.

In the bird-kingdom, the falcons take the place of the chetah, and chase through the realms of air those creatures whose wings would carry them beyond the grasp of man or the range of any weapon which he could devise.

Again, the otter and the cormorant are both employed for the capture of fish in their native element, although the one is a quadruped and the other a bird.

The ponderous strength of the elephant, and the drought-enduring powers of the camel, are equally utilized by man; and indeed, throughout the whole creation, whether of animate or inanimate bodies, there is perhaps no one object that cannot, either directly or indirectly, be converted to some human use.

Some there are, which are more directly profitable than others, among which may be enumerated the long list of domesticated animals which are familiar to us from childhood. Many of these animals, such as the horse and the dog, are universally employed in all parts of the world, while others, such as the camel, are of no service except in the peculiar climate and among the peculiar circumstances for which they were created.

Among these latter animals is the monkey which is depicted in the engraving on the next page. This is the PIG-TAILED MACAQUE, sometimes called the BRUH.

An inhabitant of Sumatra and neighbouring parts, the Bruh is possessed of the activity which distinguishes the monkey tribes, and withal is endowed with a larger share of intelligence than usual, even with the quadrumanous animals. The inhabitants of Sumatra are in the habit of capturing the Pig-tailed Macaque when young, and training it to climb the lofty cocoa-nut palms for the purpose of gathering the fruit. So clever are the monkeys, and so ingenious are the teachers, that the young scholars are instructed to select the matured nuts only, leaving the others to ripen on the tree. On this account, the Bruh has been called by a name which signifies the "fruit-gatherer."

In captivity it is generally an amusing animal, displaying to the full those traits of curiosity, impertinence, petty malice, and quaint humour, for which the monkeys are celebrated, enhanced by a spice of something that is not very far removed from wit.

I have often remarked the exceeding ingenuity of this animal in planning an attack on some unsuspecting person, its patience in biding its time, and its prompt rapidity of execution.

On one occasion, a young lady happened to pass near a cage where a pair of these animals were confined, and their attention was immediately drawn to some beautiful white feathers which she bore on her hat. Now, the monkeys were far too wise to betray the least emotion, and not even by a look did they show that they had even observed the objects on which their very hearts were fixed. But any one who knew the ways of monkeys could divine, by the sudden sparkle of the eye, that there was mischief brewing.

For some time, all went on as usual. The two monkeys held out their paws for nuts, cracked them, ate the sound kernels, and flung the bad nuts at the donors, just as if they had nothing on their minds, and had no soul above nuts. Interested by the amusing pranks which the creatures were playing, the owner of the feathers incautiously approached within reach of the cage.

Almost too quickly for the eye to follow, one of the Bruhs shot down the bars, and with a single adroit movement, whipped out one of the white feathers and leaped to the back of the cage.

Seating himself on the ground, he gravely inspected his prize, turning it over in every direction, smelling it critically, and biting off little strips of the feather, in order to ascertain the flavour. Having satisfied himself on these points, he stuck the feather behind one of his ears, so that it drooped over his head in ludicrous imitation of the manner in which it had been fastened into the hat. Thus accoutred, he paraded about the floor of the cage with stately pride.

His companion now thought himself entitled to some share in the booty, and, creeping up stealthily from behind, made a sudden spring at the feather. It was quite useless, for the original thief was on the alert, and, putting the feather in his mouth, climbed up a suspended rope with wonderful agility ; and in order to guard against an attack from below, he coiled up the rope with his hinder feet as fast as he ascended, thus cutting off all communication. When he reached the ceiling, he hitched his fingers and toes through the staple to which the rope was attached, and thus remained for awhile in perfect security.

However, even a monkey's limbs will not maintain their hold for ever, and the Bruh was forced to descend. His companion was waiting for him on the floor, and, when he reached the ground, gave chase, the two monkeys leaping about the cage, climbing the bars, and swinging from the ropes in the most agile manner.

At last they seemed to be tired of the game, and, sitting on one of the bars, amicably

PIG-TAILED MACAQUE.—*Macacus nemestrinus.*

set to work at the feather, picking out each vane separately, nibbling it, and spurting the fragments on the floor.

Just at this juncture the keeper made his appearance at the door, and the very gleam of his cap was a signal for the delinquents to dive into the furthermost corner of their cage, out of reach of stick or whip. The feather was ultimately restored to its rightful owner, but as its shaft had been bitten nearly through, had lost many of its snowy vanes, and hung limp and flaccid, as if it had been mangled, there was but slight probability of its ever renewing its position upon hat or bonnet.

As to the depredators, they were incorrigible. Hardly had the excitement caused by the feather-robbery begun to subside, when a fresh storm of laughter and exclamations arose.

On my returning to the cage, the same monkey was seen perched on his bar examining leisurely a new prize in the shape of a bracelet, which he had snatched from the hand of a lady who was offering some biscuit. It was one of those bracelets that are composed of large beads, threaded on elastic cord, and the whole attention of the thief was absorbed in the amusement caused by drawing the bracelet to its full length, and letting it snap. The clatter of the beads seemed to amuse the monkey mightily, and he was so entirely charmed with this novel recreation, that he did not even see the approaching keeper. At the sound of his voice, however, down went beads, away went monkey, and the bracelet was soon in possession of its owner.

It was a very fortunate circumstance for the monkey that he was deprived of his prize. He would most certainly have pulled the bracelet until the string broke, and the beads fell on the floor; and in that case, he would inevitably have swallowed every bead that had not been seized and eaten by his companion.

The floor of the cage was strewed with fragmentary trophies of the powers of these most mischievous creatures. There were scraps of ribbon, evidently torn from feminine wrists ; there were odd fingers and thumbs of gloves, of every material and make ; there were patches of various laces and light textures, which had once formed part of summer dresses ; even to little pieces of slight walking-sticks, which had been seized and broken by the monkey in excusable avenging of insults offered by their bearers ;—there were representative fragments of man, woman, and child, lying tossed about in admirable confusion.

I never knew so excellent a show of trophies, excepting in one instance, where several monkeys were confined in the same cage, and even in that case, I fancy that the superiority was simply occasioned by the less frequency with which the cage was swept. It is quite a common sight to see the skeleton of a parasol or two lying helplessly on the floor, or hung derisively from some bar or hook that is out of reach of any hand but that of the monkey.

Tassels of all kinds fall easy victims to the monkey's quick paw, and, after being well gnawed, are thrown contemptuously on the ground. The hard knob that is usually found in the upper part of a tassel irritates the monkey exceedingly. He thinks that he has found a nut concealed in the silken threads, and expends much time and labour in trying to crack it. The fine fibres of the silk annoy him wonderfully, and the air of angry vexation with which he spits out the obnoxious threads is highly amusing.

The fur of the Pig-tailed Macaque is tolerably uniform in its hue. The colour of the greater part of the fur is a light fawn ; a dark brown tint is washed over the top of the head and along the back, spreading partly over the sides, and colouring the upper surface of the tail. The under parts of the body and tail, together with the cheeks are of a lighter tint.

WANDEROO.—*Silénus veter.*

The last of the Macaques which we shall notice in this work is the monkey which is well known under the name of WANDEROO, or OUANDEROO, as it is sometimes written.

Although the Wanderoo is by our best authorities considered to be a member of the Macaques, and is therefore placed among them in this work ; some naturalists are more inclined to give it a place at the head of the Baboons, and assert that it forms the link between them and the Macaques.

To this decision they are led by the general physiognomy of this monkey, and by the fact that the extremity of the tail is furnished with a brush. Still, the muzzle is not of that brutal character which is so repulsively exhibited in the baboons, and the nostrils are situated in their ordinary position, instead of being pierced at the extremity of the muzzle.

The Indian name of this animal is "Nilbandar," or more properly "Neel-bhunder," the word being a composite one, and signifying a black Bhunder.

This very singular animal is a native of the East Indies, and is found commonly enough in Ceylon. The heavy mass of hair that surmounts the head and envelops the entire face, gives it a rather dignified aspect, reminding the observer of the huge peruke under whose learned shade the great legal chiefs consider judgment. The hair on the top of the head is black, but the great beard that rolls down the face and beneath the chin is of a grey tint, as if blanched by the burden of many years. In some instances this beard is almost entirely white, and then the Wanderoo looks very venerable indeed.

It is not a very mischievous animal in its wild state, and withdraws itself from the habitations of men. When in captivity it is of a tetchy and capricious disposition, sometimes becoming mild in its demeanour, and presently, without the least apparent motive, bursting into a fit of passion, and indulging in all kinds of malicious tricks. But, as is the case with so many of the monkey tribe, as the creature becomes older, it loses the gentle part of its nature, and develops the brutality alone. Thus, a Wanderoo may be quiet, docile, and even affectionate at a year old, and appear quite a model of monkey nature; at two years of age the same animal will be full of lively caprice, at times playful, and at times cross and savage; while at full age, the creature will be surly, inert, savage, and revengeful.

From the form of the tail, which is of a moderate length, and decorated with a hairy tuft at its extremity, the Wanderoo is also known by the name of the Lion-tailed Baboon.

The greater part of the fur of this animal is of a fine black, but the colour assumes a lighter hue on the breast and abdomen. The callosities on the hinder quarters are of a light pink.

It is not a very large animal, being rather less than three feet from the nose to the tip of the tail.

The name Silenus is appropriate enough, for the white beard and whiskers bear some resemblance to those facial ornaments attributed to the aged companion of the youthful Bacchus. And the specific title of "Veter," signifying "old," is well earned by the veteran aspect of the animal. The eye is a bright brown, and looks knowingly out of the hairy mass, from which it peers inquisitively at the bystanders.

Probably on account of the sapient mien, for which it is indebted to the mass of circumfluous locks, the Wanderoo is considered by the inhabitants to be a personage of great distinction among its own people. All other monkeys of the same land are said to pay the most profound reverence to their bearded chief, and, in his presence, to humble themselves as subjects before an emperor.

When feeding, the Wanderoo has a discreet custom of filling its cheek pouches before it begins to eat, thus laying up a provision against future emergencies before it has begun to satisfy the actual present wants of hunger. This habit presents a curious analogy with the peculiar stomach of the ruminating animals, when in the act of eating; a portion of the food passes into a series of pockets or pouches, where it is retained until the creature is possessed of time and leisure for re-mastication.

In its earlier youth, the Wanderoo is susceptible of education, and can be trained to perform many ingenious tricks, preferring those of a grave and sedate cast to the mercurial and erratic accomplishments displayed by the generality of learned monkeys.

GROUP OF CHACMAS.

BABOONS, OR DOG-HEADED MONKEYS.

A WELL-MARKED group of animals now comes before us, popularly known by the name of BABOONS. In more learned language they are entitled "Cynocéphali," or Dog-headed animals, on account of the formation of the head and jaws, which much resemble those of the dog tribe.

One distinguishing characteristic of these creatures is that the nostrils are situated at the extremity of the muzzle, instead of lying nearly flat upon its base, and just under the eyes, as in the apes, and other quadrumanous animals. The muzzle, too, is peculiar in its form, being, as it were, cut off abruptly, leaving a round and flattened extremity, which is well shown in the engraving of the Gelada, on p. 64. This extreme projection is not so conspicuous in the young baboon as when it attains a more mature age, and, indeed, is sometimes so little developed, that the young baboons have been taken for adult Macaques.

Of all the Quadrumana, the baboons are the most morose in temper, the fiercest in character, and the most repellent in manners.

So odiously disgusting are the habits in which many of these animals continually indulge, that, as a general rule, their presence is offensive in the extreme, and excepting for purposes of scientific investigation, it is better to shun the cage that holds any specimen of these creatures.

There are now and then exceptional cases, but they are few and far between, and it is hardly possible to watch an adult baboon for many minutes without incurring a risk of some shock to the nerves. Even their exceeding cunning, and the crafty wiles which are hatched in their fertile brains, cannot atone for their habitual offences against decorum.

It is rather curious that in the preceding genera, such as the Cercopitheci, and the Cercocebi, the chief characteristic from which the genus derives its rather lengthy title is founded upon the tail; while in the baboons, the systematic naturalists leaped at one bound to the opposite extremity of the body, and took up their stand upon the head.

For the introduction to science of the GELADA, one of the most singular of these animals, we are indebted to Dr. Ruppell, who has gained so well-earned a name in the annals of natural science.

Together with all the Cynocephali, the Gelada is a native of Africa, Abyssinia being the country from which our specimens have been derived. Dr. Ruppell, in his work on the "Fauna of Abyssinia," places this animal among the Macaques. The adult animal exhibits in perfection the curious mass of hair that is seen to cover the neck and shoulders of the monkeys of this group, and sits magnificently placid under the shade of its capillary mantle.

The young Gelada is almost totally devoid of this heavy mane, if it can be so called, and only by slight indications gives promise of the future development.

The general colour of this animal is a brown tint of varying intensity. The body and mane are of a dark brown, fading into a much lighter hue on the top of the head and sides of the face. The limbs partake of the character of the body, with the exception of the fore-legs, and paws, and the hinder feet, on which the fur is nearly black.

The baboons are more quadrupedal in their gait than any of the animals hitherto described, their formation being well adapted to such a style of progression. Even in walking some three or four steps, they seldom move otherwise than on all-fours, and when at liberty in their native haunts, are almost invariably seen either to walk like a dog, or to sit in the usual monkey fashion, discarding all attempts to imitate the human attitude. Sometimes they will stand in a tolerably erect posture for a few moments if they are desirous of looking at a distant object, or of playing some of their fantastic pranks; but even in that case, they usually aid themselves by resting a paw on any convenient support.

Their paces are generally of two kinds, a walk when they are at leisure and un-interrupted in their proceedings, and a gallop when they are alarmed, or otherwise hurried. The walk is remarkable for its jaunty impertinence, and must be seen before it can be properly appreciated. There is an easy, undulating swagger of the whole person, and a pretentious carriage of the tail, that, aided by the quick cunning blink of the little deep-set eyes, imparts an indescribable air of effrontery to the animal. This characteristic action is admirably hit off by the artist in the lesser figures depicted in the engraving on p. 62. Their pace, when hurried, is a gallop, somewhat resembling that of a dog.

All the baboons are excellent climbers of trees, as well as accomplished cragsmen, and are seldom found very far from trees or rocks. As they band together in great

GELADA.—*Gélada Ruppellii.*

numbers, they are nearly invincible in their own domains, whether of forest or cliff, bidding defiance to almost every enemy but man.

Although more ready to shun an enemy than to attack, and always preserving the better part of valour, they are terrible foes when they are brought to bay, and turn upon their enemies with the furious energy of despair. Active to a degree, and furnished with powerful limbs, they would be no despicable antagonists were their means of attack limited to hands and feet alone; but when their long sharp teeth and massive jaws are thrown into the scale, it will be seen that hardly the leopard itself is a more formidable animal.

The teeth are formed in a manner which peculiarly fits them for the mode of attack that is employed by all the baboons. The great canine teeth are long and pointed at their tips, while their inner edge is sharp as that of a knife, and can cut with more effect than many a steel weapon.

Knowing well the power of the terrible armature with which he is gifted, the enraged baboon leaps upon his foe, and drawing it towards him with his hands and feet, fixes his teeth in its throat until the sharp fangs meet together. He then violently pushes the miserable aggressor from him, so that the keen-edged teeth cut their way through the flesh, and inflict a wound that is often immediately fatal.

In this manner they repel the attacks of dogs; and woe be to the inexperienced hound who is foolish enough to venture its person within grasp of the baboon's feet or hands.

Many a time have these reckless animals paid for their audacity by their life. The whole affair is the work of only a few seconds. The baboon is scampering away in hot haste, and the hound following at full speed. Suddenly the fugitive casts a quick glance behind him, and seeing that he has only one antagonist close upon him, wheels round, springs on the dog before it can check itself, and in an instant flings the dying hound on the earth, the blood pouring in torrents from its mangled throat.

Of the Dog-headed baboons, the species which is most celebrated for such feats of prowess is the well-known animal called the CHACMA, or URSINE BABOON, the latter title being given to it on account of the slightly bear-like aspect of the head and neck. The word Chacma is a corrupted, or rather a contracted form of the Hottentot name T'chakamma. The Zulu name for this baboon is Imfena, a much more euphonious word, without that odious click, so impossible of achievement by ordinary vocal organs. In the same dialect, one which is in almost every case remarkable for the rich softness of its intonation, the word "Inkau," is the synonyme for a monkey.

This animal, when it has attained its full age, equals in size a large mastiff, or an ordinary sized wolf; while, in bodily strength and prowess, it is a match for any two dogs that can be brought to attack it.

Curiously enough, although it is so ruthless an antagonist, being the certain slayer of any hound that may come to close quarters, there is no animal which is so eagerly hunted by the South African hounds. Experience seems in this case to have lost its proverbially instructive powers; and the cruel death of many comrades by the trenchant fangs of the Chacma, has no effect in deterring the ardent hound from attacking the first baboon that comes in its way.

The owners of the hounds are more careful in this matter than are the dogs themselves, and evince more caution in setting their dogs on the track of a baboon than on the "spoor" of a leopard, or even of the regal lion himself.

The Chacma is a most accomplished robber, executing his burglaries openly whenever he knows that he will meet with no formidable opposition, and having recourse to silent craft when there are dogs to watch for trespassers, and men with guns to shoot them.

With such consummate art do these animals plan, and with such admirable skill do they carry out their raids, that even the watchful band of dogs is comparatively useless; and the cunning robbers actually slip past the vigilant sentries without the stirring of a grass blade, or the rustling of a dried twig, to give notice to the open ears of the wakeful but beguiled sentries.

In such a case, the mode to which they resort is clever in the extreme.

They know full well, that if a number of their body were to enter the forbidden domain, they could hardly elude the observation or escape the hearing of dogs and men; so they commit the delicate task of entering the enemy's domains to one or two old experienced baboons. These take the lead, and gliding softly past the sentry dogs, find admission by some crevice, or by the simpler mode of climbing over the fence.

Meanwhile, the rest of the band array themselves in a long line, leading from the scene of operations to some spot where they will be out of danger from pursuit.

All being ready, the venturous leaders begin to pluck the fruit, or to bite off the stalks, as the case may be, and quietly hand the booty to the comrade who is nearest to them. He passes the fruit to a third, who again hands it to a fourth; and thus the spoil is silently conveyed to a distance, in a manner similar to that which is employed in handing water-buckets to a fire-engine. When a sufficient amount of plunder has been secured, the invading party quietly make their retreat, and revel in security on their ill-gotten goods.

Although on service for the general weal, each individual baboon is not unmindful of his personal interest; and while he hands the booty to his next neighbour, deftly slips a portion into his pouches, much on the same principle that an accomplished epicure, while busily carving for the assembled guests, never loses sight of his own particular predilection, and when he has exhausted the contents of the dish, quietly assumes the portion which he had laid aside.

When young, the Chacma is docile enough, and by its curious tricks affords much amusement to its master and those around it. Not only for amusement, however, is this

1.   F

animal detained in captivity, but its delicate natural instincts are sometimes enlisted in the service of its master. It displays great ability in discovering the various roots and tubers on which it feeds, and which can also be used as food for man; and in digging like Caliban, with his long nails, pignuts.

A more important service is often rendered by this animal than even the procuration of food; and that is, the hunting for, and almost unfailing discovery of water.

In the desert life, water loses its character of a luxury, and becomes a dread necessity; its partial deficiency giving birth to fearful sufferings, while its total deprivation, even for a day or two, causes inevitable death. The fiery sun of the tropical regions, and the arid, scorching atmosphere, absorb every particle of moisture from the body, and cause a constant desire to supply the unwonted waste with fresh material, exactly where such a supply is least attainable.

Among these climates, the want of a proper supply of water is soon felt, the longing for the cool element becomes a raging madness; the scorched and hardened lips refuse their office, and the tongue rattles uselessly in the mouth, as if both tongue and palate were cut out of dried wood.

The value of any means by which such sufferings can be alleviated is incalculable; and the animal of which we are speaking, is possessed of this priceless faculty.

When the water begins to run short, and the known fountains have failed, as is too often the sad hap of these desert wells, fortunate is the man who owns a tame Chacma, or "Babian," as it is called. The animal is first deprived of water for a whole day, until it is furious with thirst, which is increased by giving it salt provisions, or putting salt into its mouth. This apparent cruelty is, however, an act of true mercy, as on the Chacma may depend the existence of itself and the whole party.

A long rope is now tied to the baboon's collar, and it is suffered to run about wherever it chooses, the rope being merely used as a means to prevent the animal from getting out of sight. The baboon now assumes the leadership of the band, and becomes the most important personage of the party.

First it runs forward a little, then stops; gets on its hind feet, and sniffs up the air, especially taking notice of the wind and its direction. It will then, perhaps, change the direction of its course; and after running for some distance take another observation Presently it will spy out a blade of grass, or similar object, pluck it up, turn it on all sides, smell it, and then go forward again. And thus the animal proceeds until it leads the party to water; guided by some mysterious instinct which appears to be totally independent of reasoning, and which loses its powers in proportion as reason gains dominion.

The curious employment of the animal for the discovery of water, is mentioned by Captain Drayson, R. A., in his interesting work, "Sporting Scenes among the Kaffirs of . South Africa." In the course of the same work he gives many life-like illustrations of baboon habits, whether wild or tame.

Of the daily life of the baboons, the following affords a graphic and amusing description.

"During the shooting trip with the Boers, I awoke before daybreak, and as I felt very cold and not inclined to sleep, I got up, and taking my gun, walked to a little ravine, out of which a clear, murmuring stream flashed in the moonlight, and ran close past our outspan. A little distance up this kloof, the fog was dense and thick; the blue and pink streaks of the morning light were beginning to illuminate the peaks of the Draakensberg, but all immediately around us still acknowledged the supremacy of the pale moonlight. I wanted to see the sun rise in this lonely region, and watch the changing effects which its arrival would produce on the mountains and plains around.

"Suddenly I heard a hoarse cough, and on turning, saw indistinctly in the fog a queer little old man standing near, and looking at me. I instinctively cocked my gun, as the idea of bushmen and poisoned arrows flashed across my mind. The old man instantly dropped on his hands; giving another hoarse cough, that evidently told a tale of consumptive lungs; he snatched up something beside him, which seemed to leap on his shoulders, and then he scampered off up the ravine on all-fours. Before half this performance was

THE CHACMA.—*Cynocéphalus porcarius.*

completed, I had discovered my mistake; the little old man turned into an ursine baboon with an infant ditto, who had come down the kloof to drink. The 'old man's' cough was answered by a dozen others, at present hidden in the fogs; soon, however,·

> " 'Up rose the sun, the mists were curl'd
> Back from the solitary world
> Which lay around,' "

and I obtained a view of the range of mountains gilded by the morning sun.

"A large party of the old gentleman's family were sitting up the ravine, and were evidently holding a debate as to the cause of my intrusion. I watched them through my glass, and was much amused at their grotesque and almost human movements. Some of the old ladies had their olive branches in their laps, and appeared to be 'doing their hair,' while a patriarchal old fellow paced backwards and forwards with a fussy sort of look; he was evidently on sentry, and seemed to think himself of no small importance.

"This estimate of his dignity did not appear to be universally acknowledged; as two or three young baboons sat close behind him watching his proceedings; sometimes with the most grotesque movements and expressions they would stand directly in his path, and hobble away only at the last moment. One daring youngster followed close on the heels of the patriarch during the whole length of his beat, and gave a sharp tug at his tail as he was about to turn. The old fellow seemed to treat it with the greatest indifference, scarcely turning round at the insult. Master Impudence was about repeating the performance, when the pater, showing that he was not such a fool as he looked, suddenly sprang round, and catching the young one before he could escape, gave him two or three such cuffs, that I could hear the screams that resulted therefrom. The venerable gentleman then chucked the delinquent over his shoulder, and continued his promenade with the greatest coolness : this old baboon was evidently acquainted with the practical details of Solomon's proverb.

"A crowd gathered round the naughty child, who childlike, seeing commiseration, shrieked all the louder. I even fancied I could see the angry glances of the mamma, as she took her dear little pet in her arms and removed it from a repetition of such brutal treatment."

One of these animals, personally known to Captain Drayson, was a great practical jester, and was fond of terrifying the Kaffir women by rushing at them open mouthed, catching them by their ankles, and mowing at them with extravagant grimaces, as if he meant to eat them up bodily. Sometimes a dog would be set at him while thus employed, and change the aspect of affairs in a moment. The pursuer then became the pursued, and quitting his prey, made for the nearest tree, up which he scuttled, and settled himself among the branches just so high as to be out of reach of the dog's jaws, and just so low as to give hopes of success by a higher than ordinary leap. There he would sit as if there were no such being in the world as a dog, and giving himself up to the contemplation of the surrounding scenery, or the aspect of the sky, would leisurely pursue his train of thought until the dog was tired and went away.

His keenness of sight was remarkable, his eyes possessing powers of distant vision that rivalled the telescope.

In order to prove the powers of the creature's sight, his master made several experiments, by going to so great a distance that the baboon perched on its pole was barely perceptible to the naked eye, and from thence producing sundry distortions of countenance, and strange attitudes of body. By looking through a telescope, he was able to see that the animal was not only capable of discerning and imitating his gestures, but even the very changes of countenance ; so that a grimace on the part of the gallant owner was immediately reproduced, or rather, represented by a grin on the part of the baboon.

There is a well-known story of a monkey who literally " plucked a crow " which had been in the habit of stealing his food, and curiously enough, the scene was re-enacted by this very animal, with the exception of one or two slight differences.

He was chained to the pole because he was rather too mischievous to be left entirely at liberty. He had been already detected in eating a box of wafers, studying practically the interior construction of a watch, and drinking a bottle of ink—in this last exploit displaying similar tastes with the siamang described on p. 32 of this volume. His age was only two years at the time when the account of his performances was written.

Captain Drayson has very kindly furnished me with the following original anecdotes of this tame Chacma :—

"A young baboon which had been reared by his owner from infancy resided for some months near my tent, and often served to while away an idle hour.

"Sometimes a stout earthen pot, which had just been emptied of its contents of good English jam, was submitted to the mercy of 'Jacob,' as this animal was named. The neck of the pot would not admit even a hand to be inserted, and it was most amusing to watch the manœuvres which were practised to procure some of the remnants of the sweets. If a stick were near, the jam was scooped out ; but if not, the pot was elevated high above Jacob's head, and then flung to the ground with great force.

"The earthen pot was stout and strong ; but upon one occasion, by good luck, the pot struck a stone, and was fractured. Great was the delight of Jacob, but not unmixed with suspicion ; for he appeared to think that the bystanders had been merely waiting to take advantage of his skill in projectiles, and that they would now purloin his fragments. Cramming his pouches full of bits of the jam-pot, he then seized the largest remaining piece and retreated to the top of his pole to enjoy the licking,

"He was always fully occupied for some hours after these feats ; for the jam adhered to his body, and he had to contort himself to lick off all the particles.

"There is almost as much expression in the tail of a baboon, as there is in his face. The alteration of the curve in which it is usually carried, or the lowering of this appendage, having a special meaning, according to the character of the individual.

"The baboon is perfectly aware of the dangerous character of the snake, and when he approaches a clump of bushes for the purpose of feasting upon the young shoots or ripe berries, he invariably peeps suspiciously amongst the underwood in search of his dreaded foe.

"In consequence of Jacob's detestation of the serpent race, a cruel trick was frequently played upon him, but which was one that gave great amusement. This was to frighten him with a dead snake.

"Serpents of every description were here very common; and sometimes when one had been killed, it was laid across a stick and taken towards Jacob. The instant his persecutor came in sight, the snake was sure to be seen; Jacob would then wrap himself up in his blanket and turn over an old box, under which he would hide. This retreat soon failed him, as there was a small knot-hole in the box, through which the tail of the snake was insinuated.

"Finding that this artifice had failed, he would upset the box, and spring away; a little dodging would then take place, and Jacob would be hemmed in so that the snake was brought close to him. Then, indeed, things required a desperate remedy, and with great presence of mind, he would seize the *tail*—invariably the tail—of the snake, and would fling the reptile to a distance. He would then at once rush towards his persecutor, and sit down beside him, as though to intimate that he wished to be friends.

"There was only one method from which there was no escape; this was to tie the snake loosely around the upper part of Jacob's chain, and then hold it so that a little shaking caused the reptile to slide towards him.

"After several jumps and grimaces, he would appear to be convinced that escape was useless, and would then resign himself complacently to his fate.

"Lying down on his side as though perfectly prepared for the worst, he would remain as though dead. But as soon as the snake was taken away, the mercurial temperament of the creature instantly showed itself; for he would then jump on the shoulders of any person who might happen to be near, and would play off some practical joke as a retaliation.

"Although evidently alarmed whenever snakes were brought near him, he still appeared perfectly to understand that nothing more than a joke was intended.

"His treatment of small dogs was very quaint.

"If by chance a young pup came near him, he would seize hold of it and cuddle it in his arms in a most affectionate and maternal way; not being very particular, however, whether he held the animal by the ear, the tail, or a leg.

"If the pup, as sometimes happened, objected to this treatment, and endeavoured to escape or to misbehave, Jacob would catch hold of its hind leg or tail, and would swing it round at arm's length, and at last fling it from him.

"The morning of life is decidedly the period of light-heartedness with the baboon; when the weight of years has been accumulated upon the shoulders of a veteran he becomes staid and philosophic, and sometimes rather quarrelsome, objecting strongly to the presuming manners of his juniors, and taking every opportunity to punish them should they be caught taking liberties with him."

The Chacma is supposed to be rather a long-lived animal, and with some reason. For although it is not easy to follow the course of a Chacma's existence from birth to death, and there are not as yet any official registers among the quadrumanous tribe, there are certain registers which are written by Nature's hand, and not subject to erasion, forgery, or alteration. One of these official registers, is the proportion that exists between the time which is passed by an animal before it attains its adult state, and the entire term of its life. It is found that the Chacma arrives at its full development at the age of eight or nine years; and, therefore, its lease of life may be calculated at about forty years.

The chief, and most legitimate food of this baboon, is the plant which is called from this circumstance, Babiana. It affords a curious example of vegetable life existing under trying circumstances, as it only gets rain for three months in the year; and during the remainder of the twelvemonth is buried in a soil so parched, that hardly any plant except itself can exist. The portion that is eaten is the thick, round, subterraneous stem, which is neatly peeled by the more fastidious baboons, and eaten entire by the less refined and more hungry animals.

The number of species belonging to the Dog-headed Baboons is very limited. All of them seem to be possessed of very similar habits and modes of action. The species which is represented in the accompanying engraving presents characteristics that are typical of the entire race, and is therefore called the Baboon, *par excellence.* There is some difficulty about the precise distinctions between several of the species,—a circumstance which, although to be regretted, is almost inevitable from the great external changes which are occasioned by age and sex, and the impossibility of keeping a close watch on these animals in their wild state.

The most interesting portion of natural history is that which relates the habits and manners of the creatures observed; and in the majority of instances the narrations are given by persons who, although fully alive to the little traits of temper, humour, or ingenuity, are unacquainted with the more recondite details of systematic zoology.

Consequently, an act performed by a baboon is considered by them in virtue of the deed itself, rather than in relation to the particular species of the animal who achieved it; and the intellectual power displayed by the animal is thought to be of more real value than the number of projections upon its molar teeth. This uncertainty is very great among the baboons, and as long as an act of theft or cunning is performed by a baboon, the narrator seems to care little whether the species be the Chacma, the Baboon, the Papion, or any other member of the same genus.

There are many most curious and interesting anecdotes on record which admirably illustrate the baboon nature, and yet which are not to be attributed with absolute certainty to any one species.

For example, there is a well-authenticated tale of a tame baboon which used to perform all kinds of clever tricks, some for the pecuniary benefit of its master, and others for its own individual pleasure.

The animal must have been of great service to its owner, for it cost him nothing in food, being accustomed to steal its own daily supply. On one occasion this capability was put to the test; a date-seller being the unfortunate subject upon whom the talents of the baboon were tried. The performance began by a simulated fit on the part of the animal, which fell down apparently in great pain, and grovelled on the earth in a paroxysm of contortions, its eyes steadily fixed on those of the date-seller.

Apparently motiveless as this conduct might be, it was the result of much care, for every writhing twist of the body brought the creature nearer to the basket which contained the coveted dainties. When it had arrived within reach, it fixed the date-seller's attention by strange grimaces, and, with its *hind feet,* commenced emptying the basket.

The most absurd part of the story is, that its "wicked conscience smited it" for the theft, and that it perfectly understood the unjustifiable character of the deed which it had just accomplished; for, as it was retreating, after having secured its plunder, a mischievous boy gave the animal a sly tug of the tail. The baboon, fancying that the insult had come from the date-seller, in reprisal for the abstraction of his goods, turned round, flew at the man, and, if it had not been captured by its master, would probably have done him some material injury.

A very quaint story is told of the same animal, which, if true, exhibits the strangest combination of cunning, simplicity, and ready wit, that ever entered the brain of living creature. At all events, if it be not true, it ought to be so.

It appears that the baboon was so tame, and had proved so apt a pupil, that its master had taught it to watch the pot in which he prepared his dinner, and was accustomed to leave it in charge of the culinary department while he was engaged in other business. One day, he had prepared a fowl for his dinner, and, after putting it into the pot, and the pot on the fire, went away for a time, leaving the baboon in charge, as usual.

For a time all went well, and the animal kept a quiet watch over the fire. After a while, it was seized with a desire to see what might be in the pot, and so, taking off the lid, peeped in. The odour that issued from the boiled fowl was gratifying to the animal's nostrils, and induced it, after a brief mental struggle, to pick just a little bit from the fowl, and to put the bird back again. This was done accordingly, but the experiment was so very successful that it was speedily repeated. Again and again was a morsel pinched from the fowl, until the natural consummation followed—the fowl was picked quite clean, and nothing left but the bones.

Now came remorse and sudden fear, causing the wretched animal to chatter with terror at the thought of the scarifying which was sure to follow so grievous an offence.

What was the poor thing to do? Time was passing, and the master must soon return for his dinner. At last a brilliant thought flashed through the animal's brain, and it immediately acted upon the idea.

Now, in order to understand the depth of the craft which was employed, it must be remembered that the baboons are furnished, in common with very many monkeys, with two callosities on the hinder quarters, which serve them for seats, and which are, in these animals, of a light red colour.

Rolling itself over and over in the dust, it covered its body with an uniformly sombre coating, and then, gathering itself well together, and putting its head and knees on the ground, it presented an appearance marvellously resembling a rough block of stone with two pieces of raw meat laid on its top. In those climates the birds of prey absolutely swarm, and, being encouraged by their well-earned impunity, crowd round every place where cooking is going on, and where they may have a chance of securing a portion, either by lawful gift, or lawless rapine. Several of these birds, among which were some kites, being attracted by the scent of the boiling meat, came to the spot, and seeing, as they thought, some nice raw meat temptingly laid out for them, swept upon their fancied prize.

THE BABOON.—*Cynocephalus Babouin.*

In a moment the baboon had sprung to its feet, and, with a rapid clutch, seized one of the kites. The lid was again taken off the pot, and the shrieking and struggling prisoner thrust into the boiling water in spite of its beak and claws. The lid was then replaced, and the baboon resumed its post of sentry with the placid ease that belongs to a conscience void of offence.

The baboons, when in their native fastnesses, are under a very complete system of discipline, and enforce its code upon each other most strictly. Considering the daring inroads which these creatures constantly make upon their neighbours' property, and the daily dangers to which all gregarious animals are necessarily subject, the most wary vigilance and the most implicit obedience are necessary for the safety of the whole community.

The acknowledged chiefs of the association are easily recognised by the heavy mass of hair that falls over their shoulders, and which, when thick and grey with age, is a natural uniform that cannot be wrongly assumed or mistaken.

These leaders have a mode of communicating their orders to their subordinates, and they again to those placed under them, in a curiously-varied language of intonations. Short and sharp barks, prolonged howls, sudden screams, quick jabberings, and even gestures of limbs and person, are all used with singular rapidity, and repeated from one to the other. There was a system of military telegraphing, by means of attitudes and sounds, which was invented some time ago, and which really might have been copied from the baboons, so much do their natural tactics resemble the artificial inventions of mankind.

It must be remembered that, clever as are these animals, their ingenuity is quite equalled, and even surpassed, by many of the animal kingdom which are placed much lower in its system. Therefore, although these examples of their sagacity are thus placed on record, it is not to be imagined that the quadrumanous animals are put forward as the most rational of the lower creations.

In recording the known instances of the mental powers displayed by the monkey tribe, we only give to the creature its due meed of praise, and act honestly by treating of every being with equal justice. It is so sad that many writers should set about such a task, having a purpose to serve, and that, in order to give to their own theory the greatest weight, they lay the greatest stress upon those records which tell in their favour, while they suppress those facts which might tend to overthrow or modify their own peculiar views.

To resume the account of the baboons :—

Like all animals which assemble in flocks, they never rest or move without the protection of certain sentries, which are chosen out of their number, and which keep the most careful watch over the troop to which they belong. The duty is anything but an agreeable one, and its labours are equally divided among the community, each competent member taking that task upon himself in his own turn.

When they make an attack upon a field or a plantation, they always guard against surprise by posting sentries on elevated spots, and, knowing that due notice will be given if any suspicious object be seen or heard, they devote all their energies to the congenial business of theft, while the sentries remain at their posts, never daring to withdraw their attention from the important charge which is committed to them. However, the sentinels do not entirely lose the benefit of all the good things, but take their proper share of the spoil after the thievish band has returned to a place of safety ; so that their greatest trial is an exercise of patience of rather a prolonged character.

In their rocky fastnesses, their chief foe is the leopard, and so terrified are they at the very sound of their enemy's voice, that even a very poor imitation of a growl is sufficient to set them flying off as fast as their legs can carry them, while a breath of air that bears upon its wings the least taint of that rank odour which exhales so powerfully from the large Felidæ, scatters dire consternation among the assemblage. There is a story of a life saved by means of the ingenuity of a native servant, who, seeing his master beset by a party of angry baboons, quietly stepped behind a rock, and imitated the growl of a leopard with that startling fidelity that is so general an accomplishment among savage tribes.

The leopard seldom attacks an adult baboon, not caring to risk its claws and fangs against the hands and teeth of so powerful an opponent. Much less does it openly venture to assault a band of baboons in hopes of securing one of their number. Its mode of procedure is by slily creeping round their rocky domains, and whipping off one of the young baboons before an alarm is given.

Bold as are these animals, they will not dare to follow a leopard into its den ; so that, if their dreaded foe succeeds in once getting clear of their outposts, it may carry off its prey with impunity. The constant dread which the leopard seems to excite in a baboon's mind appears to be occasioned more by the stealthy craft and persevering aggression of the animal, rather than by its physical powers alone.

One of these animals, the Thoth Baboon, bore a conspicuous part in the sculptured mythology of the Egyptians, and may be seen in almost every stony document that is impressed with the hieroglyphical wisdom of that wondrous nation. Only the male seems to have been considered worthy of forming one of the symbols of that representative language, as is shown by the fact that, whenever the Thoth Baboon is engraved, the large mass of hair over the shoulders proves it to be of the male sex, and adult. The attitude is generally a sitting position.

Among the Egyptians, the god Thoth held the same place among the minor deities, as Hermes of the Greeks, and Mercury of the Romans,—being probably the prototype of them both.

Another well-known species of the Dog-headed Baboons is the PAPION, an animal of rather a more refined aspect than the Chacma, or, more properly speaking, not quite so brutal.

The face, although unattractive enough, is yet not so repulsive as that of the Chacma, and the colours are rather more bright than those of that animal.

Great reverence was paid to these creatures, and specially to certain selected individuals which were furnished with a safe home in or near their temples, liberally fed while living, and honourably embalmed when dead. Many mummied forms of these baboons have been found in the temple caves of Egypt, swathed, and spiced, and adorned, just as if they had been human beings.

Some authors say that the Thoth Baboon was an object of worship among the Egyptians, but hardly with sufficient reason. Various animal forms were used as visible living emblems of the attributes of deity,

THE PAPION.—*Cynocephalus Sphinx.*

and the qualities of the human intellect, but were no more objects of idolatrous worship than the lion of England, or the eagle of America.

The fur of the Papion is of a chestnut colour; in some parts fading into a sober fawn, and in others warmed with a wash of ruddy bay. The paws are darker than the rest of the body. When young, it is of a lighter hue, and deepens in colour until it reaches its full age. In the prime of existence its colours are the lightest, but as years begin to lay their burden on the animal, the hairs begin to be flecked with a slight grizzle, and, in process of time, the snows of age descend liberally, and whiten the whole fur with hoary hairs.

The sense of smell is very largely developed in the baboons, their wide and roomy snouts giving plenty of space for the olfactory nerve to spread its branches. Aided by this formation, they are enabled to distinguish between poisonous and wholesome food—much to the advantage of their human neighbours, who profit by their intelligence, knowing that they may safely eat any vegetable which a baboon will admit into its list of viands. What is good for baboon is good for man, say they.

As to the animal food in which these animals indulge, it might possibly be made use of under the pressure of imminent starvation, but hardly under any circumstances less distressing. It must require a very hungry man to eat a scorpion or a centipede, although ants and some other insects are said to possess quite a delicate and almond-like flavour.

As has already been mentioned, they are singular adepts at discovering the presence of water, even though the priceless element should lie concealed under sand or stony ground. In such a case of subterraneous springs, the baboons set regularly to work, and, using their hands in lieu of spades and mattocks, dig with wonderful celerity. While thus working, they divide the task among themselves, and relieve each other at regular intervals.

When the baboons move in parties, they employ an almost military mode of arranging their numbers. In the advanced guard are the young males, who keep forward, well in front of the main body, and run from side to side, for the purpose of reconnoitering the ground over which they will have to pass. The females and their young occupy the centre, while the rear is brought up by the old and experienced males.

Thus, the more active and vigilant animals lead the way, the weakest are kept under protection, and the powerful elders have the whole of their charge constantly in view. In order to ensure the utmost precision in the line of march, several trusty animals are selected as "whippers in," whose business it is to keep order, to drive stragglers back to their proper position, to moderate the exuberant playfulness of the advanced guard, to keep a watchful eye upon the weaker members of the community, and to maintain a correspondence with the venerable chiefs in the rear.

The number of individuals composing a troop is sometimes above one hundred, ten or twelve being adult males, twenty or so, adult females, and the rest of the band composed of the young of both sexes.

The specimens of baboons that have been captured and domesticated, are generally taken by a crafty stratagem. Jars of well-sweetened beer are placed near their haunts, and drugged with some of those somniferous herbs which are so well known to the Orientals.

The baboons, seeing the jars left apparently unwatched, come cautiously from their homes, and assemble round the novel articles with much grin and chatter. They first dip in a cautious finger, and taste suspiciously. Misgiving gives place to confidence, and they partake freely of the sweet treachery. The soporific liquid soon manifests its power, and the baboons fall easy victims to their captors.

The two animals with which this history of baboons is closed, are removed from the preceding species, on account of various points in their conformation, and are placed in a separate genus, under the name of Papio.

Few animals present a more grotesque mixture of fantastic embellishment and repulsive ferocity than the baboon which is known under the name of MANDRILL.

The colours of the rainbow are emblazoned on the creature's form, but always in the very spots where one would least expect to see them. A bright azure glows, not in its "eyes of heavenly blue," but on each side of its nose, where the snout is widely expanded, and swollen into two enormous masses. The surfaces of these curious and very unprepossessing projections are deeply grooved, and the ridges are bedizened with the cerulean tint above mentioned. Lines of brilliant scarlet and deep purple alternate with the blue, and the extremity of the muzzle blazes with a fiery red like Bardolph's nose.

That all things should be equally balanced, the opposite end of the body is also radiant with chromatic effect, being plenteously charged with a ruddy violet, that is permitted to give its full effect, by the pert, upright carriage of the tail.

The general colour of the fur is of an olive brown tint, fading into grey on the under side of the limbs, and the chin is decorated with a small yellow pointed beard. The muzzle is remarkable for a kind of rim or border, which is not unlike the corresponding part in a hog, and is well shown in the engraving. The ears are small, devoid of fur, and of a black colour with a tinge of blue.

As in the Diana, the colours of this animal are more of a character that we look for in the plumage of birds, than in one of the mammals. These bright tints do not, however, belong to the hair, but only are developed in the skin, fading away after death, and turning into a dingy black. The same circumstance is found to take place in many other animals, the skin colours being very fugitive.

So dependent are these tints upon the life of the animal, that unless it be in perfect health and strength, the bright colours dim their beauty, and form, by their brilliancy or faintness, a tolerable test of the state of the creature's health.

The curious cheek expansions are due, not to the muscles of the face, but to the very bones themselves, which are heavy, protuberant, and ridged in the bone skull as in the living head. This addition to the usual form of the skull, adds greatly to the brutish appearance of the animal, and gives it a less intelligent aspect than that which is seen in most of the monkey tribe.

THE MANDRILL.—*Papio Maimon.*

Only the male Mandrill possesses these strange adornments in their full beauty of size and colour, the females being only gifted with the blue tint upon the muzzle, and even that is of a much less brilliant hue than in the male. The cheek-bones are but little elevated above the face, and are without the deep furrows that give so strange an appearance to the male sex.

Even in the male animal, these ornaments do not fully develop themselves until the creature has attained maturity. Not until the task of dentition is fully accomplished does the Mandrill shine out in all the glory of his huge azure nose, his crimson mouth, and carmine termination.

Of all the baboons, the Mandrill appears to be the most hopelessly savage, though examples are not wanting of individuals which have been subjected to kind treatment, and have proved tractable and gentle—that is, for baboons.

The adult Mandrill is liable to terrible gusts of passion, during which it seems to be bereft of reason and possessed with an insane fury. That which in other monkeys is a hasty petulance, easily excited and soon passing away, becomes in this animal a paroxysm of wild and blind rage, to which the anger of an ordinary monkey is but a zephyr to a tornado.

When thus infuriated—and but small cause is needed for its excitation—the animal seems to be beside itself with fury, heedless of everything but the object of its anger.

A demon light glares from the eyes, and it seems verily possessed with a demon's strength and malignity. With such violence do its stormy passions rage, that the vital powers themselves have been known to yield before the tempest that agitates the mind, and the animal has fallen lifeless in the midst of its wild yells and struggles.

"Sudden and quick in passion" as is the Mandrill, it bears no short-lived anger, after the custom of most quick-tempered beings, but cherishes a rancorous and deeply-rooted vengeance against any one who may be unfortunate enough to irritate its froward temper. It will often call in the aid of its natural cunning, and will pretend to have forgotten the offence, in order to decoy the offender within reach of its grasp.

The power of this animal is very great, and more than might be inferred from its size alone, though its dimensions are far from trifling.

Although in a foreign land, this, in common with most of the monkey tribes, seldom reaches the stature to which it would have attained had it passed its existence among the congenial influences of its own country; even in England it has been known to reach so considerable a size, that it was looked on as a dangerous animal, and one which required strong bars and careful surveillance.

In this country, the Mandrill is seldom seen to equal a tolerably large terrier in size, but in its native land a full-grown male measures more than five feet when standing upright, a stature which equals, if not excels, that of the chacma. As with monkeys in general, the muscular power is very great in proportion to the size of the limbs, and therefore the attack of a Mandrill is a serious matter. Even an armed man would as soon encounter a leopard or a bear as a Mandrill, while a weaponless man would be quite at the animal's mercy—and mercy it has none.

Perhaps it may be on account of the repulsive look of this animal that it is held in such detestation by the natives, as much as on account of its ferocity and strength. Be this as it may, the Mandrill is thoroughly feared and hated by the inhabitants of Guinea.

Unless they travel in large numbers and well armed, the natives shrink from passing through the woods in which these animals make their residence.

For the Mandrills live in society, and their bands are so powerful in point of numbers, and so crafty in point of management, that they are about as formidable neighbours as could be imagined. It is said that wherever they take up their abode they assume supreme sway, attacking and driving from their haunts even the lordly elephant himself.

These animals are also affirmed to keep a watch over the villages, and, when their male population is dispersed to field labour, that they issue in large companies from the woods, enter the defenceless villages, and plunder the houses of everything eatable, in spite of the terrified women. Some of the female population are said to fall victims to the Mandrills, which carry them away to the woods, as has been related of the Chimpansee.

This latter assertion may be untrue, but it is strengthened by much collateral evidence. The large male baboons, when in captivity, always make a great distinction between their visitors of either sex, preferring the ladies to the gentlemen. Sometimes they are so jealous in their disposition that they throw themselves into a transport of rage if any attentions be paid to a lady within their sight.

This curious propensity was once made the means of re-capturing a large baboon— a chacma—that had escaped from its cage in the Jardin des Plantes, in Paris.

It had already baffled many attempts to entice it to its home, and when force was tried, repelled the assailants, severely wounding several of the keepers. At last a ready-witted keeper hit upon a plan which proved eminently successful.

There was a little window at the back of the cage, and when the keeper saw the baboon in front of the open door, he brought a young lady to the window, and pretended to kiss her. The sight of this proceeding was too much for the jealous feeling of the baboon, which flew into the cage for the purpose of exterminating the offending keeper. Another keeper was stationed in ambush near the cage, and the moment the infuriated animal entered the den, he shut and fastened the door.

The male Mandrills are always more ferocious and less tameable than the females, who are also comparatively free from the revolting habits that are so unfortunately found in the adult males.

There are several instances on record of Mandrills which have led a peaceful life in captivity, and learned many accomplishments—some, perhaps, rather of a dubious nature.

One of the most celebrated of these individuals, surnamed "Happy Jerry," on account of his contented disposition, was a well-known inhabitant of the menagerie at Exeter 'Change during his lifetime; and, even after his death, is still before the public who visit the British Museum.

He was accustomed to drink porter, which he liked, and to smoke a pipe, which he tolerated. He had the honour of being a royal guest, by special invitation, and seems to have passed a life as happy as could well fall to the lot of an expatriated animal.

There are several allusions to this baboon by ancient writers, although they seem to have been very undecided about the real character of the animal.

Topsel gives a really good illustration of the Mandrill, placing it among the hyænas, because preceding writers had done so. However, his own penetrative mind refused to accept this opinion, and after saying that it *might* be the Artocyon, a beast which was supposed to be the offspring of bear and dog, diffidently puts forward his own idea on the subject, which is the correct one, as is usual when men will venture to think boldly for themselves, and shake off the trammels of conventional prejudice.

"His fore-feet," says Topsel, "are divided like a man's fingers. It continually holdeth up his tail, for at every motion it turneth that as other beasts do their head. It hath a short tail, and but for that I should judge it to be a kind of ape." Many of the traits recorded by the same author are precisely applicable to the Mandrill, although, as he thought, that it ought to be a hyæna, he has intermixed with his account a few truly hyenine anecdotes.

His name for it is, "The Second Kinde of Hyæna, called Papio, or Dabuh."

In its native land, the usual food of the Mandrill is of a vegetable nature, although, in common with the rest of the baboons, it displays a great liking for ants, centipedes, and similar creatures.

Sometimes it happens that it takes a carnivorous turn, and then will capture and devour small birds, quadrupeds, and reptiles. In captivity it is tolerably carnivorous, its tastes being sufficiently universal to accommodate itself to strong drink, as well as to civilized fare. Meat of all kinds seems acceptable to the animal, as does beer and wine. Tobacco, as we have seen, it can endure, but hardly appreciate.

It drinks by shooting forward its mobile lips into the vessel, and drawing the liquid into its mouth by suction.

When it eats, it generally commences its repast by filling its pouches with food in readiness for another meal, and unless very severely pressed by hunger, never neglects this precaution.

The tail of this animal is a remarkable feature, if it may so be termed, in the general aspect of the baboon. It is short, set high on the back, and curved upwards in a manner that is most singular, not to say ludicrous, in the living animals, and conspicuously noticeable in the skeleton. The skull of an adult Mandrill is most brutal in character. The brain has but little place in the cranium, and the greater part of the surface is either composed of, or covered with, heavy ridges of solid bone that are formed for the support of the large muscles which move the jaws.

The eyes are placed extremely high in the face, leaving hardly any forehead above them, and they are deeply set beneath a pair of morosely overhanging brows. The hair on the head is rather peculiar in its arrangement, forming a kind of pointed crest on the crown, and thus giving an almost triangular outline to the head.

It is a very common animal in its own country, but on account of its great strength, cunning, and ferocity, is not so often captured as might be expected. Even when a specimen is made prisoner, it is generally a very young one, which soon loses in captivity the individuality of its being, and learns to accommodate itself to the altered circumstances among which it is placed.

The name "Maimon," which is applied to the Mandrill, is most appropriate. It is a Greek word, signifying a hobgoblin, and is therefore peculiarly applicable to so uncanny a looking animal.

The DRILL, co-native with the Mandrill, of the coast of Guinea, somewhat resembles the female or young male mandrill, and is not of quite so savage and grotesque an aspect as that animal.

Its cheek-bones are not nearly so protuberant as those of the mandrill's, nor is its skin so brilliantly coloured. The upper parts of the body are greener than those of the mandrill, the yellow rings in the hair being more frequent. Its face and ears are of a light polished black, and the palms of the hands and feet are devoid of hair, and of a coppery tinge.

Formerly the Drill was thought to be only a young mandrill, and was so named. But the fact that even after their second dentition, the male Drills do not put on the furrowed

THE DRILL.—*Pápio Leucophœus.*

cheek-bones, or the bright colouring that distinguishes the mandrill, is sufficient to prove that it is a distinct species.

Little is known of its habits when in a state of nature, as it has probably been confounded with the mandrill, and its deeds narrated as if they belonged to the last-named animal.

It is a frequent visitor to England, and lives in tolerably good health. As far as is known, it is much like the mandrill and other baboons in temper, being quiet and docile when young, but subsiding into morose apathy as it becomes older.

The little stumpy tail is very like that of the mandrill, and is covered with short and stiff hair. Its length is not more than two inches even in a full-grown male. The Drill is always a smaller animal than the mandrill, and the female much smaller than the male, from whom she differs also in the comparative shortness of her head, and the generally paler tint of her fur.

AMERICAN MONKEYS.

WE have now taken a rapid survey of the varied forms which the Quadrumana of the old world assume; forms so diversified that there hardly seems to be scope for further modifications. Yet the prolific power of nature is so inexhaustible, that the depth of our researches only brings to view objects of such infinite variety of shape that the mind is lost in wonder and admiration.

Thus it is with the Cebidæ, or American Monkeys. While preserving the chief characteristics of the monkey nature, thus proving their close relationship with the Old World monkeys, they exhibit the strangest modification of details. The four hand-like paws, and other quadrumanous peculiarities, point out their position in the animal kingdom, while sundry differences of form show that the animals are intended to pass their life under conditions which would not suit the monkeys of the Old World.

It is curious to observe how the same idea of animal life is repeated in various lands and various climates, even though seas, now impassable to creatures unaided by the light of true reason, separate the countries in which they dwell. So we have the Simiadæ of Asia and Africa represented by the Cebidæ of America. The lion, tiger, and other Felidæ of the Eastern continents, find Western representatives in the jaguar and puma. The dogs are spread over nearly the whole world, taking very diversified forms, colour, and dimensions, but still being unmistakeably dogs. The same circumstance may be remarked of nearly all the families of mammalian animals ; of the chief bird forms ; of the reptiles ; the fishes, and so on, through the entire animal kingdom.

It seems, also, as if a similar system ran through the various classes of the animal kingdom, the nature or instinct being the original creation, and the outward shape only the manifestation thereof.

Thus, taking the Destructive Idea, as an example. Among the Mammalia it takes form as a lion, a tiger, or a leopard. In the birds, it becomes an eagle or a falcon. Descending to the reptiles, we find the destructive idea more constantly developed in the crocodiles and alligators, and serpents ; while among fishes, the lowest of the vertebrated animals, the shark, pike, and indeed almost every species of fish, exhibits this same idea enshrined in outward shape.

The records of the past, written upon rocks and stones, prove that in the earlier ages of this world the destructive element was powerfully manifested and widely diffused, and that nearly every creature to whom Almighty God imparted the breath of animated life, and that moved on the earth in those strange dark times, was of a rapacious character, living almost exclusively on slaughtered animals, and waging ceaseless wars against every being less powerful than itself.

As the earth, under the moulding hand of its divine Maker, advanced towards a more perfect state of being, the old fierce creations died out, and were replaced by milder and gentler races. Thus, by slow degrees, it was made a fit residence for man, the epitome of all previous beings, combining in himself a capacity of inflicting torture more appalling than the aggregated cruelty of all the rapacious animals that belong to the material world, and a faculty of self-sacrificing love that belongs wholly to the better world to which he alone is privileged to look forward.

Even in man himself, there exists an analogy from which we may infer that the same grand system reigns. At one extreme of the human scale we see the ruthless savage, pouring out blood like water, exultant at another's suffering, and feasting with diabolical enjoyment on the banquet torn from the still breathing body of his fellow man. At the other extreme we have the man, more like what God intended that being to be when He made him in His own image, shunning to pain another even by an unkind thought, the aim of whose life is to love and to labour for all mankind.

"Be fruitful, and multiply, and replenish the earth," was the benediction pronounced upon the true humanity, and just as good is in itself its own blessed reward, and through love continually gives birth to love, so evil, being destructive, bears within its very being the doom of eternal death, and by unwilling self-annihilation prepares the way for better and higher natures. Therefore, in the earlier and less perfect races, there was greater destructiveness, because there was more evil to destroy.

Herein we may find a key to that problem that must present itself to all reflective minds, namely, the reason why rapacious animals should exist at all.

The answer to this enigma is, that all creation represents somewhat of the Creator's being, and thus the destructive animals are the visible embodiments of God's evil-destroying power. As the evil is destroyed, so will the destroyers perish, "the evil beasts shall cease out of the land," and vanish from the face of the earth as completely as the rapacious

CHAMECK.—*Ateles Chameck.*

saurians of æons long passed away, leaving but their dry and fossil remains as records of an evil time that has been, but is no longer. So with mankind. The wild beasts melt away before the savage man, and the human wild beasts die out before the resistless march of higher races; and thus the earth is gradually purified and regenerated. Imperfect though it be, the world is better than it has ever been, and it rests with each individual who is placed upon it, to aid by his own efforts the advancement of the orb on which he lives, and the progress of that vast humanity of which he forms a part.

In each embodiment of the prevailing idea there is a strong individuality, which causes great modifications in the external form, according to the time, place, and climate, in which the animal is intended to pass its existence.

The lion, tiger, wolf, bear, and weasel, are all rapacious animals, being inspired with the same prevailing principle; but each carries out that principle in its own way, and thus performs its allotted task without interfering with the work which is assigned to any other being. As with the mammals, so with the birds, the eagle, vulture, owl, and shrike, being examples of different kinds of rapacity. The same remark may be made upon the instances which have just been quoted from the reptiles and the fishes.

So, all the parts of the world are filled with endless variety, and whether by night or day, in the fierce rays of the tropical sun, or under the sunless winter of the polar regions, earth, air, and water, are peopled with infinite multitudes of living forms, each performing its allotted task in working out its individual portion of the universal principle.

It appears to be only consistent with reason, to suppose that this system is not solely confined to the animal kingdom, but reigns through the entire creation, and that even in vegetable and mineral objects we may discover the same beautiful order to prevail.

The curiously shaped monkey which is represented in the above engraving, is an excellent example of the Cebidæ, or Sapajous, as they are often called.

The name "Ateles," which is given to the entire genus to which this animal belongs, signifies "imperfect," and has been applied to the creatures because the fore-paws are devoid of useful thumbs. Sometimes that member is almost entirely absent, and in other instances it only just shows itself.

In the CHAMECK, the thumb is slightly projecting, but even in this case it has only a single joint, and is not furnished with a nail after the usual custom of thumbs and fingers. Even when the thumb reaches its greatest size, it cannot be used as the human thumb, as it is not capable of being opposed to the fingers.

1.   G

The Chameck is a native of various parts of Brazil, where it is found rather profusely. From all accounts, it seems to be a very gentle creature, and susceptible of a high amount of cultivation. It does not appear to be so capricious of temper as the monkeys of the Old World, and although playful when in the humour for sport, is not so spitefully tricky as its transatlantic relatives. It soon learns to distinguish those persons who treat it with kindness, and will often enter into playful mock combats, pretending to inflict severe injuries, but never doing any real damage.

It is not a very large animal, the length of its body being about twenty inches, and the tail just over two feet in length. The fur is tolerably long, and falls densely over the body and limbs.

On referring to the engraving, it will be seen that the hair is longer than usual by the region of the hips, and rather thickly overhangs the hinder quarters. This arrangement seems to stand the creature in place of the callosities which have so often been alluded to, and which are not possessed by the Cebidæ. These monkeys are also destitute of cheek-pouches, but, as if to compensate them for the want of these appendages, they are furnished with an additional supply of teeth, having thirty-six instead of thirty-two, which is the ordinary complement.

The nostrils are very different from those of the monkeys which have already been described, as they open at the sides instead of underneath, and are separated from each other by a wide piece of cartilage. The ear is less unlike that of man than is the case with the greater part of the monkey tribe, the greatest distinction between the two being that the ear of the monkey is destitute of that soft lower lobe, which is so characteristic of the human ear, and through which ladies barbarously hook their auricular trinkets.

If the reader will refer to the illustration of the Chameck, he will see that the tail is the most conspicuous member of the animal. For the greater part of its length it is thickly covered with long drooping fur, but the last seven or eight inches are nearly denuded of hair on the upper surface, and entirely so on the lower. Towards the base it is extremely thick, and is furnished with muscles of great strength and marvellous flexibility, destined to aid the member in the performance of those curiously active movements for which these monkeys are so renowned.

The tail of these animals is to them equivalent, and more than equivalent, to a fifth hand. The naked extremity is endowed with so sensitive a surface that it can be applied to most of the uses to which the hand can be put, while the powerful muscles that move it are so strong and lithe that they can exert a singular amount of strength, even so as to suspend the entire weight of the animal.

In ascending trees or traversing the branches, the monkeys continually aid their progress by twining the end of the tail round the neighbouring boughs. Sometimes they even suspend themselves wholly by their tails, and after giving their bodies a few oscillating movements, boldly swing themselves from one branch to another, clearing considerable spaces in the effort. On account of these capabilities, the tail is known by the name of "prehensile."

The colour of the Chameck is nearly black, and of an uniform tint over the head, body, and limbs. Its hair is rather long and thick, in some parts taking a slight curl. The head is very small in proportion to the rest of the body. During the life of the animal the face is of a deep brown colour, as are the ears, cheeks, and chin, on which some long black hairs are scattered at distant intervals. Its lips are possessed of some mobility, but not equal to those of the chimpansee or orang-outan.

The COAITA, or QUATA, as the word is frequently written, resembles the chameck in many characteristics.

It is one of the best known of this group of animals, which are called by the name of Spider Monkeys, on account of their long sprawling limbs, and their peculiar action while walking.

It is very remarkable, that although these creatures appear to be much less calculated for bipedal locomotion than the large apes, they should really be better walkers than most of the monkey tribe. When placed on a level surface and desirous to walk

COAITA.—*Ateles Paniscus.*

in an erect position, they always attempt to aid their tottering steps by means of their prehensile tails, which they twine about in every direction in the hope of grasping some object by which to help themselves along. But when they find that all chances of external support are vain, they bravely throw themselves on their own resources, and, using their tail as a balance, move along with tolerable ease.

The mode in which they apply the tail to this unexpected use is by raising it up behind until it is on a level with the head, and then curling the tip of it downwards, so as to form the figure of a letter " S."

The spider monkeys can apply the tail to uses far more remarkable than any of those which have been mentioned. With such singularly delicate sense of touch is it furnished, that it almost seems to be possessed of the power of sight, and moves about among the branches with as much decision as if there were an eye in its tip. Should the monkey discover some prize, such as a nest of eggs, or any little dainty, which lies in a crevice too small for the hand to enter, it is in nowise disconcerted, but inserts the end of its tail into the cranny, and hooks out the desired object.

It is impossible to contemplate this wonderful provision of nature without a feeling of admiration at the manner in which the most unlikely portions of an animal are developed for the purpose of performing sundry uses. There seems to be a curious parallel between the elephant's trunk, and the spider monkey's tail, being developments of the two opposite extremes of the body, the former belonging to the Old World and the latter to the New.

There is a wonderful resemblance in the use to which these members are put, excepting of course those discrepancies that must arise from the different natures of the organs, and the habits of the animals to which they belong. Even in external form the proboscis and the tail are marvellously similar ; so much so, indeed, that an outline of one would almost serve as a sketch of the other. Each is gifted with discriminating faculty of touch, and therefore able to pick up any small object; while at the same time its muscular powers are so great, that it can endure severe and prolonged exertion.

The proboscis of the elephant can seize a tree-branch and tear it from its parent trunk. The spider monkey has no such gigantic strength, but it can sling itself from a bough by its tail, and remain suspended for almost any length of time. There is a beautiful formation of the tail of this creature, by means of which the grasp of that member retains its hold even after the death of the owner. If a spider monkey is mortally wounded, and not killed outright, it curls its tail round a branch, and thus suspended yields up its life. The tail does not lose its grasp when the life has departed; and the dead monkey hangs with its head downwards for days, until decomposition sets in and the rigid muscles are relaxed.

We may here trace another curious analogy between this automatic contraction of the tail, and the well-known structure by which a bird is enabled to hold itself on its perch during sleep. If the spider monkey's tail be drawn out till it is straightened, the tip immediately curls round, and remains so until the member is suffered to return to its usual curve. Perhaps one reason for this provision may be, that it is for the purpose of retaining the animal in its arboreal residence, and guarding it against a fall.

Still, it is a curious fact, and cannot be wholly accounted for on those grounds; for the monkeys of the Old World, although not gifted with prehensile tails, are quite as arboreal as their brethren of the New, and consequently as liable to Eutychian casualties. It may be remarked, *en passant*, that there are Preacher Monkeys in America, and consequently that an especial provision against such misfortunes may be more requisite in Brazil than in Africa.

In their native country, the spider monkeys may be seen in great profusion, swinging from the tree-branches in groups, like bunches of enormous fruits.

They are very lazy animals, and will sit, swing, or recline for hours in the strangest attitudes without moving a limb; just as if they were striving to emulate the Hindoo Fakirs in their motionless penances. Such a propensity is the more curious, because the slight forms of the animals, their long and slender limbs, and above all, their wonderful tail, would lead us to anticipate the same singular swiftness and activity that are found in the gibbons. In the American monkeys, however, we do not find the capacious chest and thin flanks which mark out the character of the gibbons.

Yet, when aroused by hunger or other sufficient motive, the spider monkeys can move fast enough; and in such a manner, that nothing without wings can follow them. In their native land, the forests are so dense and so vast, that if it were not for the rivers which occasionally cut their path through the dark foliage, the monkeys could travel for hundreds of miles without once coming to the ground.

Not that the monkeys care very much for a river, provided that the distance between the banks is not very great; and as they detest going into the water, they most ingeniously contrive to get over without wetting a hair. The manner in which they are said to achieve this feat of engineering is as follows.

When a marching troop, often amounting to a hundred or more, arrives at the bank of a river, the principal body halts, while the oldest and most experienced of their band run forward, and carefully reconnoitre the locality. After mature deliberation they fix on some spot where the trees of the opposite banks incline riverwards, and approximate nearest to each other.

Running to the overhanging boughs, the most powerful monkeys twist their tails firmly round the branch, and permit themselves to hang with their heads downwards. Another monkey then slides down the body of the first, twines his tail tightly round his predecessor, and awaits his successor. In this way a long chain of monkeys is gradually formed, until the last, who is always one of the strongest of the troop, is able to plant his paws on the ground. He then begins to push the ground with his hands, so as to give the dependent chain a slight oscillating movement, which is increased until he is able to seize a branch on the opposite side of the river.

Having so done, he draws himself gradually up the branches, until he finds one that is sufficiently strong for the purpose in view, and takes a firm hold of it. The signal is then given that all is ready, and the rest of the band ascend the tree, and cross the river by means of this natural suspension bridge.

So far, so good! The monkeys run over the bridge easily enough ; but how is the bridge itself to get over. Their plight is very like that of the man who invented a system of iron doors to be closed from the interior, and who, after closing them in the most admirable and effectual manner, was obliged to open them again in order to get out.

Still, whatever may be the case with human beings, when monkeys are clever enough to make such a bridge, they are at no loss to achieve the passage of the bridge itself.

Two or three of the stoutest keep themselves in reserve for this emergency, and, attaching themselves to the last links of the living chain, relieve their comrade from his arduous task of clutching the boughs, and at the same time slightly lengthen the chain. They then clamber up the tree as high as the chain will stretch, or the boughs bear the strain, and take a firm hold of a tough branch. A second signal is now given, and the monkey on the opposite bank relaxing his hold, the entire line of monkeys swings across the river, perhaps, slightly ducking the lowermost in the passage. Once arrived, the lower monkeys drop to the ground, while the others catch at branches, and break their connexion with the much-enduring individual at the top. When the last monkey has secured itself, the leaders descend the tree, and the whole troop proceed on their march.

Those who have witnessed this curious scene, say that it is a most amusing affair, and that there is a considerable comic element in it, on account of the exuberant spirits of the younger and less staid individuals, who delight in playing off little practical jokes on the component parts of the bridge in their passage; knowing that there is no opportunity for immediate retaliation, and trusting to escape ultimately in the confusion that follows the renewal of the march.

The Coaita is by no means a large animal, measuring very little more than a foot from the nose to the root of the tail, while the tail itself is two feet in length. Its colour is very dark and glossy ; so dark, indeed, as to be almost black. The hair varies much in length and density. On the back and the outside of the limbs it hangs in long drooping locks, forming a thick covering through which the skin cannot be seen. But on the abdomen the hair is quite scanty, and is so thinly scattered that the skin is plainly visible. The skin of the face is of a dark copper colour.

The Coaita seems to be as much averse to the intrusion of strangers into its domains as the African monkeys, whose proceedings have been already narrated. Banding together in large troops, these monkeys will assault a stranger with great vigour. Their first proceeding upon the approach of any intruder, whether man or beast, is to descend to the lower branches of their trees, and to satisfy themselves by a close inspection, whether the object be a friend or a trespasser. Having decided on the latter point of view, they re-ascend to their stronghold, and commence an assault by pelting with sticks, and keep up their attacks, until they fairly worry the intruder out of their dominions.

Another example of this wonderful group of monkeys is found in the MARIMONDA ; an inhabitant, like the two last-named animals, of Central America, and found in greatest numbers in Spanish Guiana, where, according to Humboldt, it fills the place of the Coaita.

The general shape, the formation of its limbs, and the long prehensile tail, point it out at once as another of the spider monkeys. It is certainly a very appropriate name for these animals. Their heads are so small, their bodies so short, their limbs so slender, and their tail so limb-like, that the mind unconsciously draws a parallel between these monkeys and the long-legged spiders that scuttle so awkwardly over the ground, and are so indifferent respecting their complement of legs.

The resemblance holds good even when the monkey is at rest, or even when it only appears before the eye in an illustration. But when the creature begins to walk on level ground, and especially if it be hurried, its clumsy movements are so very spider-like, that the similitude is ten times more striking. Be it remarked, that both creatures are supposed to be placed in uncongenial circumstances. The spider is deft and active enough among the many threads of its air-suspended nets, as is the monkey among the slight twigs of the air-bathed branches.   But when both animals are subjected to circumstances which are directly opposed to their natural mode of existence, they become alike awkward, and alike afford subjects of mirth.

The mode by which a spider monkey walks on level ground is rather singular, and difficult to describe, being different from that which is employed by the large apes. They do not set the sole of either paw, or hand, flat upon the ground, but, turning the hinder feet inwards, they walk upon their outer sides.   The reverse process takes place with the fore-paws, which are twisted outwards, so that the weight of the animal is thrown upon their inner edges.

It will easily be seen how very awkward an animal must be which is forced to employ so complicated a means for the purpose of locomotion.   Although it has been already stated that the spider monkey has been known to walk in a manner much more steady than that of any other monkey, yet it must be remembered that this bipedal progression was only employed for a few paces, and with a haven of rest in view in the shape of a window-sill, on which the creature could rest its hands.

MARIMONDA.—*Ateles Bélzebuth.*

In captivity, the Marimonda is a gentle and affectionate animal, attaching itself strongly to those persons to whom it takes a fancy, and playing many fantastic gambols to attract their attention.   Its angry feelings, although perhaps easily roused, do not partake of the petulant malignity which so often characterizes the monkey race, and are quite free from the rancorous vengeance which is found in the baboons.   Very seldom does it attempt to bite, and even when such an event does take place, it is rather the effect of sudden terror than of deliberate malice.

On account of its amiable nature it is often brought into a domesticated state, and, if we may give credence to many a traveller, is trained to become not only an amusing companion, but an useful servant.

The colour of this animal varies much according to the age of the individual.

When adult, the leading colour is of an uniform dull black, devoid of the glossy lustre which throws back the sunbeams from the coaita's furry mantle.   On the back, the top of the head, and along the spine, the hair is of a dense, dead black, which seems to have earned for the animal the very inapposite name with which its nomenclators have thought fit to dedecorate the mild and amiable Marimonda.

MIRIKI.—*Brachyteles hypoxanthus.*

The throat, breast, inside of the limbs, and the under side of the tail are much lighter in tint, while in some individuals a large, bright chestnut patch covers the latter half of the sides.

It seems to be of rather a listless character, delighting to bask in the sun's rays, and lying in the strangest attitudes for hours without moving. One of the postures which is most in vogue is achieved by throwing the head back with the eyes turned up, and then flinging the arms over the head. The position in which this animal is depicted in the illustration is a very favourite one with most of the spider monkeys.

There are several other species belonging to this group of animals, among which may be mentioned the Cayou, or Black Spider Monkey, the Chuva, the Brown Coaita, and others. The habits, however, of all these creatures are very similar, and therefore only one more example will be described. This is the MIRIKI, or MONO, as some authors call it.

The hair of this species is very thick, short, and furry, of a tolerably uniform brown tint over the head, body, and limbs, the paws being much darker than the rest of the animal. There is a slight moustache formed by a continuation of the long black hairs which are scantily planted on the chin and face. On account of the thick coating of fur with which the skin of this animal is covered, water has but little effect upon it. Knowing this wet-repellent property, the hunters of Brazil are accustomed to make the skin of the Miriki into cases wherewith to cover the locks of their guns in rainy days.

This species is easily distinguishable from its companions by the presence of a better developed thumb on the fore-paws than falls to the lot of spider monkeys generally.

I conclude the account of the spider monkeys with a few anecdotes of one of these animals, that have been kindly narrated to me by its owner, a captain in the royal navy.

The monkey—a lady—to whom the name of Sally was given, was captured in British Guiana, and brought to the governor of Demerara, from whom it passed to its present gallant possessor. Sally seems to be a wondrous favourite, and to take in her owner's heart the place of a favourite child. There are many photographic portraits of this sable pet, three of which are at present before me, one representing Sally as lying contentedly in her master's lap, her little wrinkled face looking over his arm, and her tail twisted round his knees, while one hind-foot is grasping this appendage. A second portrait exhibits her standing on a pedestal, by the side of the captain's coxswain,—to whose care she was chiefly committed—her left arm flung lovingly round his neck, and her tail coiled several times round his right hand, on which she is partly sitting. In the third, she is shown standing by the side of the same man, with her foot upon his hand, and the tip of her tail round his neck, by way of a change.

In almost every case there is a slight blur in the monkey's form, owing to the difficulty in persuading so volatile an animal as a monkey to remain still for two seconds together. However, the proportions of the animal are well preserved, and its characteristic attitudes shown clearly enough.

She is a most gentle creature, only having been known to bite on two occasions, one of which was simply in self-defence. She had got loose in the dock yard at Antigua, and had been chased by the men for some time. At last she was hemmed into a corner, and would have been taken easily, had not the dockyard labourers rather feared her teeth. Her master, however, in order to prove that she was not dangerous, caught her, and was rewarded by a rather severe bite on his thumb. Had it not been, however, that poor Sally was terrified out of her senses by the pursuit of the labourers, she would not have behaved so badly.

So gentle was she in general, that whenever she received a slight correction for some fault, she would never attempt to retaliate, but only sidle away and accept the rebuke. Malice does not seem to be in her nature, for she soon forgets such injuries, and does not lose her kind feelings towards her corrector. Her master tells me that if any one gets bitten by her, it is entirely the fault of the sufferer, and not of the monkey.

On board ship she is not trammeled by chain or rope, but is permitted to range the vessel at her own sweet will. She revels among the rigging, and when she becomes playful, dances about a rope in such a strange manner, and flings her limbs and tail about so fantastically, that the spectators are at a loss to distinguish the arms and legs from the tail. When thus engaged, the name of spider monkey is peculiarly apposite, for she looks just like a great overgrown tarantula in convulsions. During these fits of sportiveness, she stops every now and then to shake her head playfully at her friends, and, screwing up her nose into a point, utters little, short, soft grunts at intervals. She generally becomes vivacious towards sunset.

There is a curious custom in which she is in the habit of indulging. She likes to climb up the rigging until she reaches a horizontal rope, or small spar, and then, hooking just the tip of her tail over it, will hang at full length, slowly swinging backward and forward, while she rubs each arm alternately from the wrist to the elbow, as if she were trying to stroke the hair the wrong way. She always must needs have her tail round something, and, if possible, would not venture a step without securing herself to some object by the means of that long and lithe member.

Unlike many of her relatives, who are inveterate thieves, and with the tips of their tails quietly steal objects from which their attention is apparently turned, Sally is remarkably honest, never having stolen anything but an occasional fruit or cake. She is accustomed to take her dinner at her master's table, and behaves herself with perfect decorum, not even beginning to eat until she has obtained permission, and keeping to her own plate like a civilized being. Her food is mostly composed of vegetables, fruit, and sopped bread, although she occasionally is treated to a chicken bone, and appreciates it highly.

In the matter of food she is rather fastidious, and if a piece of too stale bread be given to her, smells it suspiciously, throws it on the floor, and contemptuously ignores its existence. With true monkey instinct, she is capable of distinguishing wholesome from harmful food, and, after she had left the tropical fruits far behind, she accepted at once an apple which was offered to her, and ate it without hesitation.

At Belize, Sally was permitted to range the town at large for some days. One morning, as her master was passing along the streets, he heard high above his head a little croaking sound, which struck him as being very like the voice of his monkey ; and on looking up, there was Sally herself, perched on a balcony, croaking in pleased recognition of her friend below.

Once, and once only, poor Sally got into a sad scrape. Her master was going into his cabin, and found Sally sitting all bundled together on the door-mat. He spoke to her, and the creature just lifted up her head, looked him in the face, and sank down again in her former listless posture.

" Come here, Sally ;" said the captain.

But Sally would not move.

The order was repeated once or twice, and without the accustomed obedience.

Surprised at so unusual a circumstance, her master lifted her by the arms, and then made the shocking discovery that poor Sally was quite tipsy. She was long past the jovial stage of intoxication, and had only just sense enough left to recognise her master. Very ill was Sally that night, and very penitent next day.

The reason for such a catastrophe was as follows :—

The officers of the ship had got together a little dinner-party, and being very fond of the monkey, had given her such a feed of almonds and raisins, fruits of various kinds, biscuits and olives, as she had not enjoyed for many a day. Now of olives in particular, Sally is very fond, and having eaten largely of these dainties, the salt juice naturally produced an intense thirst. So, when the brandy and water began to make its appearance, Sally pushed her lips into a tumbler, and to the amusement of the officers, drank nearly the whole of its cool but potent contents.

Her master remonstrated with the officers for permitting the animal to drink this strong liquid ; but there was no necessity for expostulating with the victim. So entirely disgusted was the poor monkey, that she never afterwards could endure the taste or even the smell of brandy. She was so thoroughly out of conceit with the liquid that had wrought her such woe, that even when cherry-brandy was offered to her, the cherries thereof being her special luxury, she would shoot out her tongue, and with just its tip taste the liquid that covered the dainty fruits beneath, but would not venture further.

She seemed to bear the cold weather tolerably well, and was supplied with plenty of warm clothing which stood her in good stead even off the icy coasts of Newfoundland, where, however, she expressed her dislike of the temperature by constant shivering. In order to guard herself against the excessive cold, she hit upon an ingenious device. There were on board two Newfoundland dogs. They were quite young, and the two used to occupy a domicile which was furnished with plenty of straw. Into this refuge Sally would creep, and putting an arm round each of the puppies and wrapping her tail about them, was happy and warm.

She was fond of almost all kinds of animals, especially if they were small, but these two puppies were her particular pets. Her affection for them was so great, that she was quite jealous of them, and if any of the men or boys passed nearer the spot than she considered proper, she would come flying out of the little house, and shake her arms at the intruders with a menacing gesture as if she meant to bite them.

A kennel had been built for her special accommodation, but she never would go into it. She is a very nervous animal, and apparently has a great dislike to any kind of covering over her head. So she was accustomed to repudiate her kennel, and to coil herself up in the hammock nettings, where she would sleep soundly. She is rather somnolent in character, giving up her eventide gambols soon after dark, and falling into a sound slumber from which she does not awake until quite late in the morning.

She has now been in the possession of her present owner some three years, and probably is not more than four or five years of age, to judge by her teeth ; though from her old-fashioned, wrinkled face, she might be a century old. Her colour is black, but it is remarkable, that once when she was ill, her jetty coat became interspersed with hairs of a red tint, imparting an unpleasant rusty hue to her furry mantle.

She is expected to reach England in the course of the summer, and it may chance that the public will one day have the opportunity of studying the biography of Sally the spider monkey.

THE ANIMAL which is engraved on the next page, is an example of the celebrated group of HOWLING MONKEYS, or ALOUATTES as they are termed by some naturalists, whose strange customs have been so often noticed by travellers, and whose reverberating cries rend their ears. Little chance is there that the Howling Monkeys should ever fade from the memory of any one who has once suffered an unwilling martyrdom from their mournful yells.

Several species of Howling Monkeys are known to science, of which the ARAGUATO as it is called in its own land, or the URSINE HOWLER as it is popularly named in this country, is, perhaps, the commonest and most conspicuous. It is larger than any of the New World monkeys which have hitherto been noticed; its length being very nearly three feet when it is fully grown, and the tail reaching to even a greater length.

The colour of the fur is a rich reddish-brown, or rather bay, enlivened by a golden lustre when a brighter ray of light than usual plays over its surface. The beard which so thickly decorates the chin, throat, and neck, is of a deeper colour than that of the body.

Few animals have deserved the name which they bear so well as the Howling Monkeys. Their horrid yells are so loud, that they can be heard plainly although the animals which produce them are more than a mile distant; and the sounds that issue from their curiously formed throats are strangely simulative of the most discordant outcries of various other animals—the jaguar being one of the most favourite subjects for imitation. Throughout the entire night their dismal ululations resound, persecuting the ears of the involuntarily wakeful traveller with their oppressive pertinacity, and driving far from his wearied senses the slumber which he courts, but courts in vain. As if to give greater energy to the performance, and to worry their neighbours as much as possible, the Araguatos have a fashion of holding conversations, in which each member does his best to overpower the rest.

A similar custom is in vogue with many of the African and Asiatic monkeys, but with this difference. The above-mentioned animals certainly lift up their voices together, but then, each individual appears to be talking on his own account, so that the sound, although it is sufficiently loud to affect a listener's ears most unpleasantly, is disjointed and undecided.

But the Howlers give forth their cries with a consentaneous accord, that appears to be the result of discipline rather than of instinct alone.

Indeed, the natives assert that in each company, one monkey takes the lead, and acting as toast-master, or as conductor of an orchestra, gives a signal which is followed by the rest of the band. The result of the combined voices of these stentorian animals may be imagined. And when the effect of this melancholy and not at all musical inter-mittent bellow is heightened by the silence of night and the darkness that hangs over the midnight hours in the dense forests, it may easily be supposed, that but little sleep would visit the eyes of one who had not served an apprenticeship to the unearthly sounds that fill the night air of these regions.

In order that an animal of so limited a size should be enabled to produce sounds of such intensity and volume, a peculiar structure of the vocal organs is necessary.

The instrument by means of which the Howlers make night dismal with their funestral wailings, is found to be the "hyoid bone," a portion of the form which is very slightly developed in man, but very largely in these monkeys. In man, the bone in question gives support to the tongue and is attached to numerous muscles of the neck. In the Howling Monkeys it takes a wider range of duty, and, by a curious modification of structure, forms a bony drum which communicates with the windpipe and gives to the voice that powerful resonance, which has made the Alouattes famous.

It is said by those who have been able to watch the habits of these creatures, that the

ARAGUATO, OR URSINE HOWLER.—*Mycetes ursinus.*

howlings of the Alouattes are but nocturnal serenades addressed by the amorous monkeys to their arboreal lovers. It is proverbial that good taste, both in beauty and art, are dependent entirely upon race and date, and so the deafening yells of a band of howling Araguatos may be as pleasing in the ears of their listening mates as Romeo's loving words to Juliet in her balcony ; or as, to bring the matter nearer our home and sympathies, the tender plaints of our favourite Tom-cat upon the housetop to his inamorata in the neighbouring garden.

The howling monkeys are said to be less gentle than the spider monkeys, and to partake more of the baboon nature than any of their American brethren. From the fact of their large size, their formation of head and face, together with one or two other peculiarities, some naturalists have considered the Alouattes to be the Western representatives of the baboons that inhabit the Eastern continent.

There is rather an ingenious mode of capturing these monkeys, which is worthy of notice.

A certain plant, the "Lecythis," produces a kind of nut, which, when emptied of its contents, becomes a hollow vessel with a small mouth. Into one of these hollowed nuts a quantity of sugar is placed, the nut left in some locality where the monkey is likely to find it, and the monkey-catchers retreat to some spot whence they can watch unseen the effect of their trap.

So tempting an object cannot lie on the ground for any length of time without being investigated by the inquisitive monkeys. One of them soon finds out the sweet treasure of the nut, and squeezes his hand through the narrow opening for the purpose of emptying the contents. Grasping a handful of sugar, he tries to pull it out, but cannot do so because the orifice is not large enough to permit the passage of the closed hand with its prize. Certainly, he could extricate his hand by leaving the sugar and drawing out his hand empty, but his acquisitive nature will not suffer him to do so. At this juncture, the ambushed hunters issue forth and give chase to the monkey. At all times, these monkeys are clumsy enough on a level surface, but when encumbered with the heavy burden, which is often as big as the monkey's own head, and deprived of one of its hands, it falls an easy victim to the pursuers.

All these monkeys are eaten by the inhabitants of these lands, being cooked upon an extempore scaffolding of hard wood. Their flesh is very dry indeed, so much so, that a monkey's arm has been preserved for many years only by being roasted over a fire.

They are not so playful in their habits as most of the monkey tribe, even when young, preserving a solid gravity of demeanour. They are very numerous among the trees of their favourite resorts, as many as forty individuals having been seen upon one tree.

The CAPUCIN MONKEYS, two examples of which are here given, are active little animals, lively and playful. In habits, all the species seem to be very similar, so that the description of one will serve equally for any other. In consequence of their youth and sportive manners they are frequently kept in a domesticated state, both by the native Indians and by European settlers. Like several other small monkeys, the Capucin often strikes up a friendship for other animals that may happen to live in or near its home, the cat being one of the most favoured of their allies. Sometimes it carries its familiarity so far as to turn the cat into a steed for the nonce, and, seated upon her back, to perambu- late the premises. More unpromising subjects for equestrian exercise have been pressed

THE CAPUCIN.—*Cebus Apella.*   HORNED SAPAJOU.—*Cebus fatuellus.*

into the service by the Capucin. Humboldt mentions one of these creatures which was accustomed to catch a pig every morning, and, mounting upon its back, to retain its seat during the day. Even while the pig was feeding in the savannahs, its rider remained firm, and bestrode its victim with as much pertinacity as Sinbad's old man of the sea.

There is some difficulty in settling the species of the Capucins, for their fur is rather variable in tint, in some cases differing so greatly as to look like another species. The general tint of the CAPUCIN is a golden olive, a whiter fur bordering the face in some individuals, though not in all.

The HORNED CAPUCIN is much more conspicuous than the last-mentioned animal, as the erect fringe of hair that stands so boldly from the forehead points it out at once. When viewed in front, the hair assumes the appearance of two tufts or horns, from which peculiarity the creature derives its name. These horns are not fully developed until the monkey has attained maturity.

In colour, too, it is rather different from the Capucin, having a constant tinge of red in it. The fur is mostly of a deep brown, but in some individuals resembles that peculiar purple black which is obtained by diluting common black ink with water, while in others the ruddy hue prevails so strongly as to impart a chestnut tint to the hair. The fringed crest is tipped with grey.

The last example of the Capucins which will be noticed in these pages, is the WEEPER MONKEY, or SAI.

SAI.—*Cebus Capucinus.*

As is the case with the two previously-mentioned animals, it is an inhabitant of the Brazils, and as lively as any of its congeners. The tails of the Capucins are covered with hair, but are still possessed of prehensile powers. All these monkeys seem to be possessed of much intelligence, and their little quaint ways make them great favourites with those who watch their motions.

Their food is chiefly of a vegetable nature, but they are fond of various insects, sometimes rising to higher prey, as was once rather unexpectedly proved. A linnet was placed, by way of experiment, in a cage containing two Capucin monkeys, who pounced upon their winged visitor, caught it, and the stronger of the two devoured it with such avidity that it would not even wait to pluck off the feathers. Eggs are also thought to form part of the Capucin's food.

There is always much difficulty with regard to the names of various animals, as almost every systematic naturalist prefers a name of his own invention to one which has already been in use. It often happens, therefore, that the same creature has been burdened with ten or fifteen titles, given to it by as many writers. The chacma, for example, has been named "Cynocephalus porcarius" by one author, "Simia porcaria" by another, "Simia sphingiola" by a third, "Papio comatus" by a fourth, and "Cynocephalus ursinus" by a fifth. In order to avoid the great waste of valuable space that would be caused by giving a list of these various names, I only make use of the title by which each animal is designated in the catalogue of the British Museum, and under which name it may be found in that magnificent collection.

A very pretty genus of monkeys comes next in order, deriving, from the beauty of their fur, the term Callithrix, or "beautiful hair." Sometimes these animals are called Squirrel Monkeys, partly on account of their shape and size, and partly from the squirrel-like activity that characterizes these light and graceful little creatures. The TEE-TEE, or

Titi—as the name is sometimes given—is a native of Brazil, and is found in great numbers. Another name for the animal is the SAIMIRI.

The colours of the Tee-tee are very diversified. A greyish olive is spread over the body and limbs, the latter being washed with a rich golden hue. The ears are quite white, and the under surface of the body is whitish grey. The tip of the tail is black.

There are several species of Tee-tee, four of which are in the British Museum. Our engraving of the last of these monkeys, namely, the COLLARED TEE-TEE, is given opposite.

They are most engaging little creatures, attaching themselves strongly to their possessors, and behaving with a gentle intelligence that lifts them far above the greater part of the monkey race. Their temper is most amiable, and anger seems to be almost

TEE-TEE.—*Callithrix Sciureus.*

unknown to them. In the expression of their countenance, there is something of an infantine innocence, which impresses itself the more strongly when the little creatures are alarmed. Sudden tears fill the clear hazel eyes, and, by the little, imploring, shrinking gestures, they establish an irresistible claim on all kindly sympathies.

The Tee-tees have a curious habit of watching the lips of those who speak to them, just as if they could understand the words that are spoken, and when they become quite familiar, are fond of sitting on their friend's shoulder, and laying their tiny fingers on his lips. They seem to have an intuitive idea of the empire of language, and to try, in their own little way, to discover its mysteries.

A pleasant musky odour exhales from these animals. Their beautiful, furry tails have no prehensile power, but can be wrapped about any object, or even coiled round their own bodies in order to keep them warm.

The strange looking animal which is represented in the engraving on the opposite page, is no less remarkable in its character than in its looks. It is savage in its temper, and liable to gusts of furious passion, during which it is apt to be a very unpleasant neighbour, for it has long sharp teeth, and does not hesitate to use them.

On examining this animal, the attention is at once drawn to the curious manner in which both extremities of the body are decorated.

The beard is of a dull black colour, and is formed chiefly by hairs which start from the sides of the jaw and chin, and project forward in the curious fashion which gives the animal so strange an expression.

COLLARED TEE-TEE.—*Callithrix torquatus.*

Of this ornament the Cuxio is mightily careful, protecting its facial ornament with a veneration equal to that beard-worship for which the mediæval Spanish noble was world-famous. It is even more fastidious in this respect than the Diana monkey, whose beard-protecting customs have been alluded to on page 50. The Diana will hold its

CUXIO, OR BEARDED SAKI.—*Brachyurus Satanas.*

beard aside when it drinks; but the more cautious Cuxio forbears to put its face near the water. Instead of drinking a deep draught by suction, as is the custom with most monkeys, it scoops up the liquid in the palm of its hand, and so avoids the danger of wetting its beard.

This curious habit, however, is but rarely witnessed, as the animal dislikes to exhibit its fastidiousness before spectators, and only when it thinks itself unwatched will it use its natural goblet. When in the presence of witnesses it drinks as do other monkeys, wetting its beard without compunction.

The general colour of this monkey is a grizzled brown, sometimes speckled with rust-coloured hairs, and the limbs, tail, and head are black. If, however, the hair of the body be blown aside, a greyish hair takes the place of the dark brown ; for the hairs are much lighter towards their insertion, and in many cases are nearly white. The hair of the head is remarkable for the mode of its arrangement, which gives it an air as if it had been parted artificially. The long black hairs start from a line down the centre of the head, and fall over the temples so densely that they quite conceal the ears under their thick locks. The large quantity of hair that decorates the head and face increases the really great comparative size of the rounded head. The nostrils are rather large, and are separated from each other by a dividing cartilage which is larger than is usual even in the American monkeys.

The teeth are so sharp and the jaws so strong, that Humboldt has seen the animal, when enraged, drive its weapons deeply into a thick plank. When it suffers from a fit of passion, it grinds these sharp teeth, leaps about in fury, and rubs the extremity of its long beard. Even when slightly irritated, it grins with savage rage, threatening the offender with menacing grimaces, and wrinkling the skin of its jaws and face.

It is not known to live in companies, as is the wont of most American monkeys, but passes a comparatively solitary life, limiting its acquaintance to its partner and its family. The cry of this animal is rather powerful, and can be heard at a considerable distance. The colour of the female Cuxio is not so dark as that of her mate, being almost wholly of a rusty brown. It is chiefly nocturnal in its habits.

BLACK YARKE.—*Pithécia Leucocéphala.*

There are several monkeys known by the name of Sakis, among which are reckoned the Cuxio, which has just been described, and two other species, which are easily distinguished from each other by the colour of their heads. The first of these animals is the BLACK YARKE, or WHITE-HEADED SAKI, and the other the CACAJAO, or BLACK-HEADED SAKI.

The former of these Sakis is a rather elegant creature in form, and of colour more

CACAJAO.—*Pithecia Melanocephala.*

varied than those of the Cuxio. As will be seen from the accompanying engraving, the head is surrounded with a thick and closely-set fringe of white hair, which is rather short in the male, but long and drooping in the female. The top of the head is of a deep black, and the remainder of the body and tail is covered with very long and rather coarse hair of a blackish-brown. Under the chin and throat the hairs are almost entirely absent, and the skin is of an orange hue.

Beside the difference of length in the facial hairs of the female Yarke, there are several distinctions between the sexes, which are so decided as to have caused many naturalists to consider the male and female to belong to different species. The hair of the female Yarke is decorated near the tip with several rings of a rusty brown colour, while the hair of the male is entirely devoid of these marks.

The natural food of these animals is said to consist chiefly of wild bees and their honeycombs. Perhaps the long furry hair with which the Sakis are covered, may be useful for the purpose of defending them from the stings of the angry insects. On account of the full and bushy tail with which the members of this group are furnished, they are popularly classed together under the title of Fox-tailed Monkeys.

The two animals which have just been noticed are marked by such decided peculiarities of form and colour that they can easily be distinguished from any other monkeys. The Cuxio is known by its black beard and parted hair, the Black Yarke by its dark body and white head-fringe, while the Cacajao is conspicuous by reason of its black head and short tail.

When this animal was first discovered, it was thought that the tail had been docked either by some accident, or by the teeth of the monkey itself, as is the custom with so many of the long-tailed monkeys of the Old World. But the natives of the country where it lives assert that its brevity of tail is a distinctive character of the species.

1. H

Indeed, among the many names which have been given to the Cacajao, one of them, "Mono Rabon," or short-tailed Mono, refers to this peculiarity. On account of the very short tail, and the general aspect of the animal, the Cacajao is supposed by some naturalists to be the American representative of the Magot.

The head of the creature is not only remarkable for its black hue, but for its shape, which instead of being rounded, as is the case with most monkeys, is slightly flattened at the temples. The general colour of the fur is a bright yellowish-brown, the only exceptions being the head and the fore-paws, which are black. The ears are devoid of hair, are very large in proportion to the size of the animal, and have something of the human character about them. The length of the head and body is said to reach nearly two feet in full-grown animals, and the tail is from three to five inches long, according to the size of the individual.

Very little is known of the habits of the Cacajao in a wild state, but in captivity it bears the character of being a very inactive and very docile animal. Fruits seem to be its favourite diet, and when eating them it has a habit of bending over its food in a very peculiar attitude. It is not so adroit in handling objects as are the generality of monkeys, and seems to feel some difficulty in the management of its long and slender fingers, so that its manner of eating is rather awkward than otherwise.

Among the names by which this monkey is known, we may mention, "Mono-feo," or Hideous Monkey, Chucuto, Chucuzo, and Caruiri. The term "Melanocephala" signifies Black-headed, while the word "Leucocephala," which is applied to the Yarke, signifies White-headed.

It seems to be a timid, as well as a quiet animal, as a Cacajao which had been domesticated displayed some alarm at the sight of several small monkeys of its own country, and trembled violently when a lizard or a serpent was brought before its eyes.

The localities where it is most generally found are the forests which border the Rio Negro and the Cassiquiare, but it does not seem to be very plentiful even in its own land.

The term "Nyctipithecus," or Night-monkey, which is used as the generic title of the DOUROUCOULI, refers to its habits, which are more strictly nocturnal than those of the animals heretofore mentioned. The eyes of this little creature are so sensitive to light, that it cannot endure the glare of day, and only awakes to activity and energy when the shades of night throw their welcome veil over the face of nature.

In its wild state, it seeks the shelter of some hollow tree or other darkened place of refuge, and there abides during the hours of daylight, buried in a slumber so deep, that it can with difficulty be aroused, even though the rough hand of its captor drag it from its concealment. During sleep, it gathers all its four feet closely together, and drops its head between its fore-paws. It seems to be one of the owls of the monkey race.

DOUROUCOULI.—*Nyctipithecus Trivergâtus.*

The food of this Douroucouli is mostly of an animal nature; and consists chiefly of insects and small birds, which it hunts and captures in the night season. After dark, the Douroucouli awakes from the torpid lethargy in which it has spent the day, and shaking off its drowsiness, becomes filled with life and spirit. The large dull eyes, that shrank from the dazzling rays of the sun, light up with eager animation at eventide; the listless

MARMOSET.—*Jacchus Vulgáris.*

limbs are instinct with fiery activity, every sense is aroused to keen perception, and the creature sets off on its nightly quest. Such is then its agile address, that it can capture even the quick-sighted and ready-winged flies as they flit by, striking rapid blows at them with its little paws.

The general colour of the Douroucouli is a greyish-white, over which a silvery lustre plays in certain lights. The spine is marked with a brown line, and the breast, abdomen, and inside of the limbs, are marked with a very light chestnut, almost amounting to orange. The face is remarkable for three very distinct black lines, which radiate from each other, and which have earned for the animal the title of "Trivergatus," or "Three-striped." There are but very slight external indications of ears, and in order to expose the organs of hearing, it is necessary to draw aside the fur of the head. On account of this peculiarity, Humboldt separated the Douroucouli from its neighbours, and formed it into a distinct family, which he named "Aötes," or "Earless."

Guiana and Brazil are the countries where this curious little animal is found. Although by no means an uncommon species, it is not taken very plentifully, on account of its monogamous habits. The male and his mate may often be discovered sleeping snugly together in one bed, but never in greater numbers, unless there may be a little family at the time. Its cry is singularly loud, considering the small size of the animal which utters it, and bears some resemblance to the roar of the jaguar. Besides this deep-toned voice, it can hiss or spit like an angry cat, mew with something of a cat-like intonation, and utter a guttural, short, and rapidly repeated bark. The fur is used for the purpose of covering pouches and similar articles.

The beautiful little creature which is so well known by the name of the MARMOSET, or OUISTITI, is a native of the same country as the Douroucouli, and is even more attractive in its manners and appearance. The fur is long and exquisitely soft, diversified with bold stripes of black upon a ground of white and reddish-yellow. The tail is long and full; its colour is white, encircled with numerous rings of a hue so deep that it may almost be called black. A radiating tuft of white hairs springs from each side of the face, and contrasts well with the jetty hue of the head.

On account of the beauty of its fur, and the gentleness of its demeanour when rightly treated, it is frequently brought from its native land, and forced to lead a life of compelled civilization in foreign climes. It is peculiarly sensitive to cold, and always likes to have

H 2

its house well furnished with soft and warm bedding, which it piles up in a corner, and under which it delights to hide itself.

The Marmosets do not seem to be possessed of a very large share of intelligence, but yet are engaging little creatures if kindly treated. They are very fond of flies and other insects, and will often take a fly from the hand of the visitor. One of these animals with whom I struck up an acquaintance, took great pleasure in making me catch flies for its use, and taking them daintily out of my hand. When it saw my hand sweep over a doomed fly, the bright eyes sparkled with eager anticipation; and when I approached the cage, the little creature thrust its paw through the bars as far as the wires would permit, and opened and closed the tiny fingers with restless impatience. It then insinuated its hand among my closed fingers, and never failed to find and to capture the imprisoned fly.

When properly tamed, the Marmoset will come and sit on its owner's hand, its little paws clinging tightly to his fingers, and its tail coiled over his hand or wrist. Or it will clamber up his arm and sit on his shoulders, or if chilly, hide itself beneath his coat, or even creep into a convenient pocket.

The Marmoset has a strange liking for hair, and is fond of playing with the locks of its owner. One of these little creatures, which was the property of a gentleman adorned with a large bushy beard, was wont to creep to its master's face, and to nestle among the thick masses of beard which decorated his chin. Another Marmoset, which belonged to a lady, and which was liable to the little petulances of its race, used to vent its anger by nibbling the end of her ringlets. If the hair were bound round her head, the curious little animal would draw a tress down, and bite its extremity, as if it were trying to eat the hair by degrees. The same individual was possessed of an accomplishment which is almost unknown among these little monkeys, namely, standing on its head.

Generally, the Marmoset preserves silence; but if alarmed or irritated, it gives vent to a little sharp whistle, from which it has gained its name of Ouistiti. It is sufficiently active when in the enjoyment of good health, climbing and leaping about from bar to bar with an agile quickness that reminds the observer of a squirrel.

Its food is both animal and vegetable in character; the animal portion being chiefly composed of various insects, eggs, and it may be, an occasional young bird, and the vegetable diet ranging through most of the edible fruits. A tame Marmoset has been known to pounce upon a living gold fish, and to eat it. In consequence of this achievement, some young eels were given to the animal, and at first terrified it by their strange writhings, but in a short time they were mastered, and eaten.

Cockroaches are a favourite article of food with the Marmoset, who might be put to good service in many a house. In eating these troublesome insects, the Marmoset nips off the head, wings, and bristly legs, eviscerates the abdomen, and so prepares the insect before it is finally eaten. These precautions, however, are only taken when the cockroach is one of the larger specimens, the smaller insects being eaten up at once, without any preparation whatever.

Several instances of the birth of young Marmosets have taken place in Europe, but the young do not seem to thrive well in these climates. The colour of the young animal is a dusky grey, without the beautiful markings which distinguish them when adult, and the tail is destitute of hair.

The length of the full-grown Marmoset is from seven to eight inches, exclusive of the tail, which measures about a foot.

The two elegant little animals which are represented in the preceding page are members of the same genus as the Marmoset, inhabitants of nearly the same localities, and possessed of many similar qualities.

The PINCHE is remarkable for the tuft of white and long hair which it bears on its head, and which is so distinctly marked, that the little creature almost seems to be wearing an artificial head of hair. The throat, chest, abdomen, and arms, are also white, and the edges of the thighs are touched with the same tint. On each shoulder there is a patch of reddish-chestnut, fading imperceptibly into the white fur of the chest, and the greyish-brown hair that covers the remainder of the body. Its eyes are quite black.

The tail of the animal is long and moderately full; its colour slightly changes from

PINCHE.—*Jacchus Œdipus.*   MARIKINA.—*Jacchus Rosália.*

the russet-brown tint with which it commences, to a deeper shade of brownish-black. Its voice is soft and gentle, and has often been compared to the twittering of a bird.

The Pinche is quite as delicate in point of health as its slight form seems to indicate, and can with difficulty endure the privations of a voyage. When the animal is full-grown, the length of its head and body is about eight inches, and that of its tail rather exceeding a foot.

Among the various members of the monkey tribe, there is hardly any species that can compare with the exquisite little MARIKINA, either for grace of form, or soft beauty of colour.

The hair with which this creature is covered is of a bright and lustrous chestnut, with a golden sheen playing over its long glossy locks. To the touch, the fur of the Marikina is peculiarly smooth and silken ; and from this circumstance it is sometimes called the Silky Monkey.

Both for the texture and colour of the hair, the name is happily chosen, for the tint of the Marikina's fur is just that of the orange-coloured silk as it is wound from the cocoon, while in texture it almost vies with the fine fibres of the unwoven silk itself.

Another name for the same animal is the Lion Monkey, because its little face looks out of the mass of hair like a lion from out of his mane.

The colour of the hair is nearly uniform, but not quite so. On the paws it darkens considerably, and it is of a deeper tint on the forehead and the upper surface of the limbs than on the remainder of the body. Some specimens are wholly of a darker hue. In no place is the fur very short; but on the head, and about the shoulders, it is of very great length in proportion to the size of the animal.

The Marikina is rightly careful of its beautiful clothing, and is fastidious to a degree about preserving its glossy brightness free from stain. Whether when wild, it keeps its own house clean, or whether it has no house at all, is not as yet accurately ascertained ; but in captivity, it requires that all cleansing shall be performed by other hands. This slothfulness is the more peculiar, because the creature is so sensitive on the subject, that if it be in the least neglected, it loses its pretty gaiety, pines away, and dies.

It is fond of company, and can seldom be kept alone for any length of time. The food of the Marikina is chiefly composed of fruits and insects; but in captivity, it will eat biscuit and drink milk. It is a very timid animal, unable to fight a foe, but quick in escape, and adroit in concealment. Its voice is soft and gentle when the animal is pleased, but when it is excited by anger or fear, it utters a rather sharp hiss. The dimensions of the Marikina are much the same as those of the Pinche.

RUFFED LEMUR.—*Lemur Macáco.*

LEMURS.

THE form of the monkeys which are known by the name of Lemurs, is of itself sufficient to show that we are rapidly approaching the more quadrupedal mammalia, the which, however, we shall only reach through the wing-handed animals, or bats, and the strangely formed flying-monkey, which seems to span the gulf between the monkeys and bats.

The head of all the Lemurs is entirely unlike the usual monkey head, and even in the skull the distinction is as clearly marked as in the living being. Sharp, long, and pointed, the muzzle and jaws are singularly fox-like, while the general form of these animals, and the mode in which they walk, would lead a hasty observer to place them among the true quadrupeds. Yet, on a closer examination, the quadrumanous characteristics are seen so plainly, that the Lemurs can but be referred to their proper position among, or rather at the end of, the monkey tribe.

The word Lemur signifies a night-wandering ghost, and has been applied to this group of animals on account of their nocturnal habits, and their stealthy, noiseless step, which renders their progress almost as inaudible as that of the unearthly beings from whom they derive their name.

The RUFFED LEMUR is one of the handsomest of this family, challenging a rivalship even with the Ring-tailed Lemur in point of appearance.

The texture of the fur is extremely fine, and its colour presents bold contrasts between pure white and a jetty blackness, the line of demarcation being strongly defined. The face of the Ruffed Lemur is black, and a fringe of long white hairs stands out like a ruff round the face, giving to the creature its very appropriate title.

As is the case with all the Lemurs, it is a native of Madagascar and of the adjacent islands, and seems to take the place of the ordinary monkeys. Of all the Lemurs this species is the largest, its size equalling that of a moderately grown cat. Its voice is a sepulchral, deep roar, peculiarly loud, considering the size of the animal, and can be heard at a great distance in the stilly night. As the Lemurs delight in gathering together in large companies, the effect of their united voices is most deafening. The eyes are furnished with a transverse pupil, which dilates as darkness draws on, enabling the creature to see even in a dark night, and to make search after their daily, or rather their nightly food.

This species is timid at the presence of man, and hides itself at the sound of his footsteps. But if pursued and attacked, it takes instant courage from despair, and flinging itself boldly on its antagonist, wages fierce battle. In the conflict, its sharp teeth stand it in good stead, and inflict wounds of no trifling severity.

It is easily tamed, and although it is not a very intellectual animal, it displays much gentle affection, readily recognising its friends, and offering itself for their caresses, but avoiding the touch of those with whom it is not acquainted, or to whom it takes a dislike. It is very impatient of cold, and likes to sit before a fire, where it will perch itself for an hour at a time without moving, its attention solely taken up by the grateful warmth.

It is an active creature, being able to leap to some distance, and always attaining its mark with unfailing accuracy. While leaping or running rapidly, the tail is held in a peculiar and graceful attitude, following, indeed, Hogarth's line of beauty.

RING-TAILED LEMUR.—*Lemur Catta.*   WHITE-FRONTED LEMUR.—*Lemur albifrons.*   RED LEMUR.—*Lemur Ruber*

The RING-TAILED LEMUR, or MACACO, is at once recognisable by the peculiarity from which it derives its popular name.

It is not quite so large as the Ruffed Lemur, as it only measures a foot from nose to tail, the tail itself being some seven or eight inches in length. In captivity it soon becomes familiar, and when it chooses to exhibit its powers, is very amusing with its merry pranks. If several individuals are confined in the same cage, they are fond of huddling together, and involving themselves in such a strange entanglement of tails, limbs, and heads, that until they separate, it is almost impossible to decide upon the number of the animals that form the variegated mass.

It sometimes breeds in confinement, and then affords an interesting sight. The young Lemur is not so thickly clothed as its mother, but makes up deficiencies in its own covering by burying itself in the soft fur of its parent. Many a time have I seen the little creature sunk deeply in the soft fur of its mother's back, and so harmonizing with her, that the child could hardly be distinguished from the parent. Sometimes it would creep under the mother, and cling with arms and legs so firmly, that although she might move about her cage, the little one was not shaken off, but held as firmly as Ulysses to the Cyclops' ram.

There is a curious structure in the hand and arm of this Lemur, bearing consider- able analogy to the formation of the spider monkey's tail, which is mentioned on p. 84. By means of this construction of the limb, the fingers of the hand are closed when the arm is stretched out, so that the animal can suspend itself from a tree-branch, without incurring fatigue. It sometimes utters a sound which resembles the purring of a cat, and from that habit is derived the name of Cattus. The manner in which the dark spots and rings are distributed over the body and tail is well shown in the engraving, and need not be described.

The WHITE-FRONTED LEMUR derives its name from the patch of white hairs which appears on its forehead. Some naturalists suppose it to be the female of a similar animal on whose forehead a sable patch is substituted for the white, and is therefore called the Black-fronted Lemur. At present, however, the Black-fronted animal is considered to be a distinct species ; and the only difference between the sexes of the White-fronted Lemur seems to be, that in the male animal the forehead and some other portions of the fur are white, while in the female they are of a light grey. The general colour of the animal is a brownish chestnut, but in some examples a grey tint takes the place of the darker colour.

It is a gentle and engaging creature, and not at all shy, even to strangers, unless they alarm it by loud voices or hasty gestures. It is possessed of great agility, climbing trees, and running among the branches with perfect ease, and capable of springing through a space of several yards. So gently does it alight on the ground after it leaps, that the sound of its feet can hardly be heard as they touch the ground.

As will be seen from the figure on p. 103, the RED LEMUR possesses a fur which has somewhat of a woolly aspect, the hair separating into tufts, each of which is slightly curled It is a beautifully decorated animal, displaying considerable contrast of colour- ing. The body, head, and the greater portion of the limbs, are of a fine chestnut, with the exception of a large white patch covering the back of the head and nape of the neck, and a smaller one in the midst of each foot. The face, the tail, and paws, are black, as is all the under side of the body. This latter circumstance is most remarkable, as it is almost a general rule that the under parts of animals are lighter in tint than the upper. Around the sides of the face, the hair is of a paler chestnut than that which covers the body.

In habits it is similar to the Lemurs which have already been described. Being naturally a nocturnal animal, it passes the day in a drowsy somnolence, its head pushed between its legs, and the long, bushy tail wrapped round its body, as if to exclude the light and retain the heat. Should it be accustomed to be fed during the daytime, it shakes off its slumber for the purpose of satisfying the calls of hunger ; but even though urged by so strong an inducement, it awakes with lingering reluctance, and sinks to sleep again as soon as the demands of its appetite are satisfied. Its entire length is nearly three feet, of which the tail occupies about twenty inches. Its height is about a foot.

The curious animal which is known by the name of the DIADEM LEMUR, is generally thought to belong quite as much, if not more, to the Indris than to the Lemurs, and has, therefore, been placed by Mr. Bennett in a separate genus, which he names Propithecus.

The name of Diadem Lemur is given to this creature on account of the white semi- lunar stripe which runs across the forehead ; the curve being just the opposite to the crescent on the head of the Diana monkey, and therefore assuming the shape of a diadem. This white stripe is very conspicuous, and serves by its bold contrast with the black head and face, to distinguish the animal from any of its relatives. The shoulders and upper part of the back are of a sooty tint, not so black as the head, and fading almost imperceptibly into palest brown on the hinder quarters and the limbs. The under parts of the body are very light grey, nearly white. The paws are nearly black. The tail is tawny at its commencement, but gradually changes its colour by the admixture of lighter hairs, until at its tip it is nearly white, although with a slight golden tinge.

The hair of the tail is not so long as that of the body, which is long and rather silky in texture, with the exception of the fur about the lower end of the spine, which has a slight woolliness to the touch. As may be seen from the engraving, the thumbs of the

PROPITHECE, OR DIADEM LEMUR.—*Propithécus Diadéma.*

hinder paws are large in proportion, and suited for taking a firm grasp of any object to which the animal may cling; while the corresponding members of the fore-paws are not so largely developed, but yet can be used with some freedom. The face of the Propithece is not so long as that of the true Lemurs, and the round tipped ears are hidden in the bushy hair which surrounds the head. The length of the animal, exclusively of the tail, is about twenty-one inches, and the length of the tail is about four inches less.

Resembling the Lemurs in many respects, and given to similar customs, the animals which are known by the name of Loris are distinguished from the Lemurs by several peculiarities of structure.

SLENDER LORIS.—*Loris Gracilis.*

The first point which strikes the eye of the observer, is the want of that long and bushy tail which is possessed by the Lemurs, and which is only rudimentary in the Loris. The muzzle too, although sharp and pointed, is abruptly so, whereas that of the Lemur tapers gradually from the ears to the nose. The country which they inhabit is not the same as that which nurtures the Lemurs, for whereas the latter animals are found exclusively in Madagascar, the Loris is found in Ceylon, Java, Sumatra, and other neighbouring parts.

The SLENDER LORIS is a small animal, measuring only nine inches in length, and possessed of limbs so delicately slender, as to have earned for it the popular name by which it is distinguished from the Slow-paced Loris. Its colour is grey, with a slight rusty tinge, the under portions of the body fading into white. Round the eyes, the fur takes a darker hue, which is well contrasted by a white streak running along the nose.

Small though it be, and apparently without the power to harm, it is a terrible enemy to the birds and insects on which it feeds, and which it captures, "like Fabius, by delay."

Night, when the birds are resting with their heads snugly sheltered by their soft feathers, is the time when the Loris awakes from its daily slumbers, and stealthily sets forth on its search. Its large round eyes blaze in the dusky gloom like two balls of phosphorescent fire, and by the eyes alone can its presence be known. For the colour of its fur is such that the dark back is invisible in the obscurity, and the white breast and abdomen simulate the falling of a broken moonbeam on the bark of a branch. Its movements are so slow and silent, that not a sound falls on the ear to indicate the presence of a living animal.

Alas for the doomed bird that has attracted the fiery eyes of the Loris! No Indian on his war-path moves with stealthier step or more deadly purpose than the Loris on its progress towards its sleeping prey. With movements as imperceptible and as silent as the shadow on the dial, paw after paw is lifted from its hold, advanced a step and placed again on the bough, until the destroyer stands by the side of the unconscious victim. Then, the hand is raised with equal silence, until the fingers overhang the bird and nearly touch it. Suddenly, the slow caution is exchanged for lightning speed, and with a movement so rapid that the eye can hardly follow it, the bird is torn from its perch, and almost before its eyes are opened from slumber, they are closed for ever in death.

The SLOW-PACED LORIS, or KUKANG, is very similar in its habits to the animal just mentioned, but differs from it in size, colour, and several parts of its form.

The fur is of a texture rather more woolly than that of the Slender Loris, and its colour has something of a chestnut tinge running through it, although some specimens are nearly as grey as the Slender Loris. As may be seen from the engraving, a dark stripe surrounds the eyes, ears, and back of the head, reaching to the corners of the mouth. From thence it runs along the entire length of the spine. The colour of this dark band is a deep chestnut. It is rather larger than the preceding animal, being a little more than a foot in length.

In the formation of these creatures some very curious structures are found, among which is the singular grouping of arteries and veins in the limbs.

Instead of the usual tree-like mode in which the limbs of most animals are supplied with blood,—one large trunk-vessel entering the limb, and then branching off into numerous subdivisions,—the limbs of the Loris are furnished with blood upon a strangely modified system. The arteries and veins as they enter and leave the limb, are suddenly divided

KUKANG, OR SLOW-PACED LORIS.—*Nycticebus Javánicus.*

into a great number of cylindrical vessels, lying close to each other for some distance, and giving off their tubes to the different parts of the limb. It is possible that to this formation may be owing the power of silent movement and slow patience which has been mentioned as the property of these monkeys, for a very similar structure is found to exist in the sloth.

The tongue of the Loris is aided in its task by a plate of cartilage, by which it is supported, and which is, indeed, an enlargement of the tendinous band that is found under the root of the tongue. It is much thicker at its base than at the extremity, which is so deeply notched that it seems to have been slit with a knife. It is so conspicuous an organ, that it has been often described as a second tongue. The throat and vocal organs seem to be but little developed, as is consistent with the habits of an animal whose very subsistence depends upon its silence. Excepting when irritated, it seldom or never utters a sound ; and even then, its vocal powers seem to be limited to a little monotonous plaintive cry.

In captivity, this Loris appears to be tolerably omnivorous, eating both animal and vegetable food, preferring, however, the former. Living animals best please its taste, and the greatest dainty that can be offered to the creature is a small bird, which it instantly kills, plucks, and eats entirely, the bones included. Eggs are a favourite food with it, as are insects. It will take butcher's meat, if raw, but will not touch it if cooked in any way. Of vegetable substances, sugar appears to take its fancy the most, but it will eat fruits of various kinds, such as oranges and plantains, and has been known to suck gum arabic.

Another curious inhabitant of Madagascar is the INDRI, or AVAHI, a creature that has sometimes been considered as one of the lemurs, and placed among them by systematic naturalists. From the curled and woolly hair with which the body is covered it derives its name of "Laniger," or Wool-bearer. Just over the loins, and partly down the flanks, the soft wool-like hair takes a firmer curl than is found to be the case in any other part of the body or limbs. It is but a small animal, the length of its head and body being only a foot, and its tail nine inches. The general colour of the fur is a lightish brown, with a white stripe on the back of the thigh, and a tinge of chestnut in the tail. In

some individuals a rusty red, mingled with a yellow hue, takes the place of the brown; and in all the under parts are lighter than the upper. Its face is black, and the eyes are grey, with a greenish light playing through their large orbs.

The name Indri is a native word, signifying, it is said, "man of the woods." Its voice is not very powerful, but can be heard at some distance. It is of a melancholy, wailing character, and has been likened to the cry of a child.

The LITTLE GALAGO, which is represented in the lower figure of the accompanying engraving, is sometimes called by the name of the Madagascar Rat, on account of its rat-like form, and the colour of the fur. It is about the size of a small rat, and, as may be seen by a reference to the engraving, might easily be mistaken for one of those animals by a non-zoologist. The tint of its fur is a very light mouse-colour.

The ears of the Galago are large, and, during the life of the animal, are nearly transparent. The eyes are very large, and

AVAHI, OR INDRI.—*Indris Laniger.*

of that peculiar lustre which is always seen in the nocturnal animals. It is a native of Madagascar.

The MOHOLI GALAGO is a larger animal than the preceding, being nearly sixteen inches in length, inclusive of the tail. Its colour is grey, with irregular markings of a deeper hue. The under parts of the body are nearly white, and the limbs are slightly tinged with a golden lustre. The tail is not very bushy, excepting at the extremity, and its colour is a chestnut brown. The texture of the fur is very soft, and there is a slight woolliness in its setting.

Nocturnal in habits, it sleeps during the day, with its large ears folded over the head in such a manner as to give it the aspect of an earless animal. More active than the loris, the Moholi does not secure its prey by stealing on it with slow and silent movements, but leaps upon the flying insects on which it loves to feed, and seizes them in its slender paws. Besides insects, various fruits form part of the Moholi's food, more especially such as are of a pulpy nature, and it is said that the Moholi eats that vegetable exudation which is known by the name of Gum-Senegal. Its diurnal repose is taken in the curious nest which it builds in the forked branches of trees, using grass, leaves, and

LITTLE GALAGO.—*Gálago Minor.*   MOHOLI.—*Gálago Moholi.*

other soft substances for the purpose. In this lofty cradle the young are nurtured until they are of an age to provide for themselves.

The face is full of expression, in which it is aided by the large and prominent ears; and the creature is said to contract its countenance into strange grimaces, after the fashion of the ordinary monkeys. Like the monkeys, too, it can leap for some little distance, and springs from one branch to another, or from tree to tree with agility and precision. The Moholi Galago is an inhabitant of Southern Africa, having been found by Dr. Smith hopping about the branches of the trees that bordered the Limpopo river, in twenty-five degrees of south latitude.

At first sight, there is some external resemblance between the Galago and the little animal which is figured in the accompanying engraving. The ears, however, are not so large as those of the Galago, and the tail is less thickly covered with fur, being almost devoid of hair, except at its extremity, where it forms a small tuft. On reference to the figure, it will be seen that the hands are of extraordinary length, in proportion to the size of the creature. This peculiarity is caused by a considerable elongation of the bones composing the "Tarsus," or back of the hands and feet, and has earned for the animal the title of TARSIER. This peculiarity is more strongly developed in the hinder than in the fore-paws.

The colour of the Tarsier is a greyish-brown, with slight olive tint washed over the body. A stripe of deeper colour surrounds the back of the head, and the face and forehead are of a warmer brown than the body and limbs. It is a native of Borneo,

TARSIER.—*Tarsius Spectrum.*

Celebes, the Philippine Islands, and Banca. From the latter locality it is sometimes called the Banca Tarsier. Another of the titles by which it is known, is the Podji.

It is a tree-inhabiting animal, and skips among the branches with little quick leaps that have been likened to the hoppings of a frog. In order to give the little creature a firmer hold of the boughs about which it is constantly leaping, the palms of the hands are furnished with several cushions. The back of the hands are covered with soft downy fur, resembling the hair with which the tail is furnished. Excepting on the hands and tail, the fur is very thick and of a woolly character, but at the root of the tail, and at the wrists and ankles, it suddenly changes to the short downy covering.

The true position of that very rare animal the AYE-AYE, seems very doubtful, some naturalists placing it in the position which it occupies in this work, and others, such as Van der Hoeven, considering it to form a link between the monkeys and the rodent animals.

As will be seen by a reference to the figure, in its head and general shape it resembles the Galagos, but in the number and arrangement of its teeth it approaches the rodent type. There are no canine teeth, and the incisors are arranged in a manner similar to those of the rodents, the chief difference being, that instead of the chisel-like edge which distinguishes the incisor teeth of the gnawing animals, those of the Aye-aye are sharply pointed. These curious teeth are extremely powerful, and are very deeply set in the jaw-bones, their sockets extending nearly the entire depth of the bone.

The colour of the animal is a rusty brown on the upper portions of the body, the under parts, as well as the cheeks and throat, being of a light grey. The paws are nearly black. The fur of the body is thickly set, and is remarkable for an inner coating of downy hair of a golden tint, which sometimes shows itself through the outer coating. On the tail the hair is darker than on the body, greater in length, and in texture much coarser. The tail seems to be always trailed at length, and never to be set up over the body, like the well-known tail of the squirrel. The ears are large, and nearly destitute of hair.

It is probable that the natural food of the Aye-aye, like that of the preceding animals, is of a mixed character, and that it eats fruit and insects indiscriminately. In captivity it usually fed on boiled rice, which it picked up in minute portions, like Amine in the "Arabian Nights," using, however, its slender fingers in lieu of the celebrated bodkin with which she made her mock meal. But in its wild state it is said to search the trees for insects as well as fruits, and to drag their larvæ from their concealment by means of its delicate fingers. Buds and various fruits are also said to be eaten by this animal— possibly the buds may contain a hidden grub, and the entire flower be eaten for the sake of the living creature which it contains, as is the case with many a bud that is plucked by small birds in this country.

It is a nocturnal animal like the Galagos and Lemurs, and seeks its prey by night only, spending the day in sleep, curled up in the dark hollow of a tree, or in some similar spot, where it can retire from view and from light.

As is shown by the scientific name of the Aye-aye, it is a native of Madagascar, and even in that island is extremely scarce, appearing to be limited to the western portions of the country, and to escape even the quick eyes of the natives. Sonnerat, the naturalist, was the first to discover it, and when he showed his prize to the natives, they exhibited great astonishment at the sight of an unknown animal, and the exclamations of surprise are said to have given the name of Aye-aye to the creature. The name "Cheiromys," signifies "Handed Mouse," and is given to the animal because it bears some resemblance to a large mouse or rat which is furnished with hand-like paws instead of feet.

With the exception of the Aye-aye, all the Quadrumanous animals bear their mammæ upon the breast, and clasp their young to their bosoms with their arms. But in the Aye-aye, the milk-giving organs are placed on the lower portion of the abdomen, and thus a great distinction is at once made between this creature and the true quadrumana. Indeed, there are so many points of discrepancy in this strange being, that it is quite impossible to make it agree with the systematic laws which have hitherto been laid down, and naturalists place it in one order or another, according to the stress which they lay on different points of its organization.

The eyes are of a brownish-yellow colour, and very sensitive to light, as may be

AYE-AYE.—*Cheiromys Madagascariensis.*

expected in a creature so entirely nocturnal in its habits. The movements of the Aye-aye are slow and deliberate, though not so sluggish as those of the Loris. It is not a very small animal, measuring almost a yard in total length, of which the tail occupies one moiety.

On a review of this and the Lemurine monkeys, it can hardly fail to strike the observer that there must be something very strange in the climate or position of Madagascar—perhaps in both—that forbids the usual quadrumanous forms, and produces in their stead the Lemurs, the Indris, and the Aye-aye. So very little is known of this important island, that it may be the home of hitherto unknown forms of animal life, which, when brought under the observation of competent naturalists, would fill up sundry blanks that exist in the present list of known animals, and afford, in their own persons, the clue to many interesting subjects which are now buried in mystery.

The strange animal which is known by the name of the FLYING LEMUR, or COLUGO, presents a singular resemblance to the large bat which is popularly called the Flying Fox, and evidently affords an intermediate link of transition between the four-handed and the wing-handed mammals.

By means of the largely-developed membrane which connects the limbs with each other, and the hinder limbs with the tail, the Colugo is enabled to leap through very great distances, and to pass from one bough to another with ease, although they may be situated so far apart that no power of leaping could achieve the feat. This membrane is a prolongation of the natural skin, and is covered with hair on the upper side as thickly as any part of the body, but beneath it is almost naked. When the creature desires to make one of its long sweeping leaps, it spreads its limbs as widely as possible, and thus converts itself into a kind of living kite, as is shown in the figure. By thus presenting a large surface to the air, it can be supported in its passage between the branches, and is said to be able to vary its course slightly by the movement of its arms.

When the animal is walking or climbing about among the branches, the wide membrane is folded so closely to the body, that it might escape the observation of an

COLUGO.—*Galeopithécus volans.*

inexperienced eye. The membrane is not used in the manner of wings, but is merely
employed as a sustaining power in the progress through the air. It is evident, therefore,
that at every leap, the spot at which it aims must be lower than that from which it
starts, so that it is forced, after some few aerial voyages, to run up the trees and attain
a higher station. It is said that the Colugo will thus pass over nearly a hundred
yards.

Among other bat-like habits, the Colugo is accustomed to suspend itself by its hinder
paws from the branch of a tree, and in this pendant attitude it sleeps. Its slumbers are
mostly diurnal, for the Colugo is a night-loving animal, and is seldom seen in motion
until the shades of evening draw on. But on the approach of night, the Colugo awakes
from its drowsiness, and unhooking its claws from the branch on which it has hung
suspended during the hours of daylight, sets off on its travels in search of food.

The diet of this animal is said to consist of mixed animal and vegetable substances,
the former being eggs, insects, and small birds, while the latter is composed of various
soft fruits. Its paws are equally adapted for grasping the boughs of the trees among
which it passes its existence, and for seizing the prey on which it lives. The thumbs are
not capable of opposition to the fingers, and therefore cannot be used as are the thumbs
of the human hand.

It will be remembered that, in the Aye-aye, the structure of the mammæ is very
different from that of the true monkeys ; and in the Colugo, the same organs are marked
by a singular peculiarity of form and number. Instead of the usual supply of two
mammæ on the breast, the Colugo is furnished with four of those organs.

The female Colugo is motherly in her habits, and carries her young family with her
until they have attained a moderate size.

It is found in many of the islands that belong to the Indian Archipelago, and is tolerably common. As far as is known, there are several species of Galeopithecus; three, according to some naturalists, and four according to others.

The colour of the fur is very uncertain, even in the same species, some specimens being of a light brown, others of a grey tint, more or less deep; while many individuals have their fur diversified with irregular marblings or stripes, or spots of different shades and tints.

The teeth of the Galeopithecus are very curious in their shape, and present as great a contrast to the usual quadrumanous tooth as the entire form does to that of the true monkeys. The upper incisor teeth are separated from each other by a rather wide empty space, the lower incisors have their crowns deeply cut, as if they were being manufactured into combs, bearing, indeed, a very close resemblance to the rudely-manufactured wooden combs made by the inhabitants of the South Sea Islands.

None of the fingers of this animal are furnished with the broad flat nail which is found in the real monkeys, but each finger is armed with a sharp claw, decidedly hooked, and retractile. The thumbs are not opposable to the fingers. The hinder limbs are slightly larger than the arms. The Colugo is by no means a small animal, as, when it is full grown, it equals a large cat in size. The natives of the countries where this animal is found are in the habit of using it as an article of food. Strangers, however, find its flesh very unpleasant, on account of a strong odour with which it is pervaded.

As in this work it has been my endeavour to render the study of Animal Life as entertaining as possible, I have carefully avoided the use of scientific terms, which might give an air of pedantry to its pages, and deter the reader from venturing upon a subject so repellent. A greater stress has, therefore, been laid upon the disposition and habits of the various animals than on their purely physical form, and the descriptions have been rather of species than of genera. But if any reader should desire to learn the leading characteristics by which the genera are separated from each other and placed in their respective positions, he is referred to a "Compendium of Generic Distinctions," which will be found at the end of Vol. I., and by means of which, the reader will be enabled to assign almost any animal to its proper genus.

CHEIROPTERA;

OR, WING-HANDED ANIMALS.

FROM the earliest times in which the science of zoology attracted the attention of observant men, the discovery of a true systematic arrangement has been one of the great objects of those who studied animal life, and the forms in which it is outwardly manifested. In the writings of these pioneers of zoological science, from Aristotle, its father, even to the latest authors on this subject, we find that many animals, whether in groups or in single species, have long baffled investigation. Among the more conspicuous of these enigmatical beings are the strange and weird-like animals which are popularly known by the terse title of Bats, and, scientifically, by the more recondite name of Cheiroptera, a term derived from two Greek words signifying, the former, a hand, and the latter, a wing.

On a retrospect of the theories which have been broached on the subject of the Bats, we find that the singular diversity of opinion is quite on a par with the peculiar form of the animal which excited them.

Some authors place the Bats among the birds, because they are able to fly through the air, while others assign them a position among the quadrupeds, because they can walk on the earth. Some, again, who admitted the mammalian nature of the creatures, scattered them at intervals through the scale of animated beings, heedless of any distinction excepting the single characteristic on which they took their stand, and by which they judged every animal. These are but a few of the diverse opinions which ran riot among the naturalists of the former times, among which the most ingeniously quaint, is that which places the bat and the ostrich in the same order, because the Bat can fly, and the ostrich cannot.

By degrees the true mammalian character of the Bats became more clearly understood, and they were removed from the birds to take their rank among the higher forms. Even then, however, they were placed at the very end of the mammals, being considered as a connecting link which prevented a too abrupt change from the hairy to the feathered beings ; and it was left to the more recent investigators to discover, by careful anatomical research, the real position of the Bat tribe.

In general form the Bats are clearly separated from any other group of animals, and by most evident modifications of structure, can be recognised by the most cursory glance.

The first peculiarity in the Bat form which strikes the eye, is the wide and delicate membrane which stretches round the body, and which is used in the place of the wings with which birds are furnished. This membrane, thin and semi-transparent as it is, is double in structure, being a prolongation of the skin of the flanks and other portions of the animal, and, therefore, having its upper and under surface, in the same manner as the body of the creature itself. The two surfaces are so clearly marked, that with ordinary care, they can be separated from each other. Along the sides, this double membrane is rather stronger and thicker, but, as it extends from the body, it assumes greater tenuity, until at the margin it is so exquisitely thin, that the tiny blood-corpuscules, which roll along the minute vessels that supply the wing with nourishment, can be seen clearly through its integument, by the help of a good microscope.

In order to support this beautiful membrane, to extend it to its requisite width, and to strike the air with it for the purposes of flight, the bones of the fore-part of the body, and especially those of the arms and hands, undergo a singular modification.

As will be seen on reference to the accompanying engraving, which represents the skeleton of the Vampire, and which has been originally taken by the photographic process, the bones which thus constitute the arm and hand are marvellously elongated, becoming longer the farther they recede from the body.

The two bones of the fore-arm are extremely long, and the bone which is scientifically known by the name of the "ulna," is extremely small, and in many species almost wholly wanting. The reason for this arrangement is, that the great object of these two bones is, by the mode in which they are jointed to each other, to permit the arm to rotate with that movement which is easily shown by the simple process of turning the hand with its palm upwards. This latitude of motion would not only be useless to the Bats, but absolutely injurious, as the wing-membranes would not be able to beat the air with the steady strokes which are needful for maintaining flight. Therefore the arm is rendered incapable of rotation.

Passing onwards from the arms to the hands, the finger-bones are strangely disproportioned to the remainder of the body, the middle finger being considerably longer than the head and body together. The thumb is very much shorter than any of the fingers, and furnished with a sharp and curved claw. By means of this claw, the Bat is enabled to proceed along a level surface, and to attach itself to any object that may be convenient. In some of the Bats the thumb is much longer than that which is here figured.

The bones of the breast and the neighbouring parts are also formed in a peculiar manner, being intended to support the broad surface of the wing-membrane, and to enable

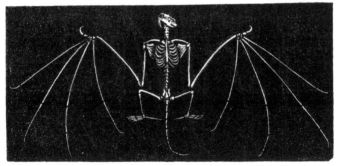

SKELETON OF VAMPIRE BAT.

it to beat the air with sufficient force. The collar-bones are long, considerably arched, and strongly jointed to the breast-bone and the shoulder-blades. In the insect-eating Bats, these bones are more developed than in the fruit-eaters; probably because the former need a better apparatus for the capture of their quick-winged prey, than the latter for seeking their vegetable food. Some species of Bat present a collar-bone which is half the length of the elongated upper arm.

The breast-bone is very long, and is widely expanded so as to form a strong point of attachment for the two collar-bones. There is also a bony crest running down its centre, which serves as a support for the enormous breast-muscles that work the wings. The ribs are long and well rounded, and, indeed, seem to be much greater in proportionate length than those of any other mammal.

The lower portions of the body and limbs are singularly small in proportion to the upper limbs. The legs are short and slender, and so arranged that the feet are rather turned outward, for the purpose of using their sharp claws freely. A kind of slender and spur-like bone is seen to proceed from the heel of each foot. When the skeleton is clothed with its softer textures, these curious bones run between the double membrane that joins the legs to the tail, and reach nearly half-way to the latter member. The exact history of this bony spur is not quite settled, some authors considering it to be a separate bone, while others think that it is merely a projection of the heel-bone, which remains permanently disunited. The feet are small, and each toe is furnished with a very sharp, strong, and curved claw, by means of which the animal is enabled to suspend itself from any object which affords a slight projection.

It will be seen on a glance at the formation of the bat, that the hands, or wings, if they may be so called, are the leading characteristics of the animal, and that, to those members, the whole of the body and the remaining limbs are subsidiary.

Having thus made a cursory review of the skeleton, we proceed to the outward form, and take for our first example the creature which has earned for itself a world-wide celebrity by the best means of obtaining mundane fame—the shedding of much blood.

The VAMPIRE BAT is a native of Southern America, and is spread over a large extent of country. It is not a very large animal, the length of its body and tail being only six inches, or perhaps seven in large specimens, and the spread of wing two feet, or rather more. The colour of the Vampire's fur is a mouse tint, with a shade of brown.

Many tales have been told of the Vampire Bat, and its fearful attacks upon sleeping men,—tales which, although founded on fact, were so sadly exaggerated as to cause a reaction in the opposite direction. It was reported to come silently by night, and to search for the exposed toes of a sound sleeper,—its instinct telling it whether the intended victim were thoroughly buried in sleep. Poising itself above the feet of its prey, and fanning them with its extended wings, it produced a cool atmosphere, which, in those hot climates, aided in soothing the slumberer into a still deeper repose. The Bat then applied its needle-pointed teeth to the upturned foot, and inserted them into the tip of a toe with such adroit dexterity, that no pain was caused by the tiny wound. The lips were then brought into action, and the blood was sucked until the Bat was satiated. It then disgorged the food which it had just taken, and began afresh, continuing its alternate feeding and disgorging, until the victim perished from sheer loss of blood.

For a time, this statement gained dominion, but, after a while, was less and less believed, until at last, naturalists repudiated the whole story as a "traveller's tale." However, as usual, the truth seems to have lain between the two extremes ; for it is satisfactorily ascertained, by more recent travellers, that the Vampires really do bite both men and cattle during the night, but that the wound is never known to be fatal, and, in most instances, causes but little inconvenience to the sufferer.

When they direct their attacks against mankind, the Vampires almost invariably select the foot as their point of operation, and their blood-loving propensities are the dread of both natives and Europeans. With singular audacity, the bats even creep into human habitations, and seek out the exposed feet of any sleeping inhabitant who has incautiously neglected to draw a coverlet over his limbs.

When they attack quadrupeds, they generally fix themselves on the shoulders and flanks of the animal, and inflict wounds sufficiently severe to cause damage unless properly attended to. It is quite a common occurrence that when the cattle are brought from the pastures wherein they have passed the night, their shoulders and flanks are covered with blood from the bites of these blood-loving bats. It might be said that the bleeding wounds might be accounted for by some other cause, but the matter was set at rest by a fortunate capture of a Vampire "red-handed" in the very act of wounding a horse.

Darwin, who narrates the circumstance, states that he was travelling in the neighbourhood of Coquimbo, in Chili, and had halted for the night. One of the horses became very restless, and the servant, who went to see what was the matter with the animal, fancied that he could see something strange on its withers. He put his hand quickly on the spot and secured a Vampire Bat. Next morning there was some inflammation and soreness on the spot where the bat had been captured, but the ill effects soon disappeared, and three days afterwards the horse was as well as ever.

It does not seem to be the severity of the wound which does the harm, but the irritation which is caused by pressure, whether of a saddle, in the case of a horse, or of clothing, in the case of a human being.

The Vampire seems to be very capricious in its tastes, for while one person may sleep in the open air with perfect impunity, another will be wounded almost nightly. Mr. Waterton, urged by his usual enthusiastic desire for personal investigation, slept for the space of eleven months in an open loft, where the Vampires came in and out every night. They were seen hovering over the hammock, and passing through the apertures that served for windows, but never made a single attack. Yet an Indian, who slept within a few yards, suffered frequently by the abstraction of blood from his toes. This distinction was not on account of colour, for a young lad about twelve years of age, the son of an English gentleman, was bitten on the forehead with such severity, that the wound bled freely on the following morning. The fowls of the same house suffered so terribly, that they died fast; and an unfortunate jackass was being killed by inches. He looked, to use Mr. Waterton's own language, "like misery steeped in vinegar."

Although these bats have so great a predilection for the blood of animals, they are not restricted to so sanguinary a diet, but live chiefly on insects which they capture on the wing. Indeed, they would have but a meagre diet were they to depend wholly on a

VAMPIRE BAT.—*Vampyrus Spectrum.*

supply of human or brute blood, for there are sufficient Vampires in existence to drain the life-blood from man and beast. Many other creatures have the same propensities— happy if they can gratify them; satisfied if they are withheld from so doing. The common leech is a familiar example of a similar mode of life; for it may be that not one leech out of a thousand ever tastes blood at all, although they are so ravenously eager after it when they have the opportunity for gratifying their sanguinary taste.

On reference to the figure of the Vampire Bat, it will be seen that the wide and flattened membrane which supports the body in the air, connects together the whole of the limbs and the tail, leaving free only the hinder feet, and the thumbs of the fore paws. This membrane is wondrously delicate, and is furnished not only with the minute blood-vessels, to which allusion has already been made, but with a system of nerves which possess the most exquisite power of sensation.

It has been long known that bats are able to thread their way among boughs of trees and other impediments with an ease that almost seems beyond the power of sight, especially when the dark hours of their flight are considered. Even utter darkness seems not to impede these curious animals in their aërial progress, and when shut up in a darkened place, in which strings had been stretched in various directions, the bats still pursued their course through the air, avoiding every obstacle with perfect precision. In order to ascertain beyond doubt whether this faculty were the result of a more than usually keen sight, or whether it were caused by some hitherto unknown structure, Spallanzani deprived a bat of its eyes, and discovered by this most cruel experiment, that the bat seemed as capable of directing its flight among the strings without its eyes as with them.

Whether this curious power were resident in any part of the animal's structure, or whether it were the result of a sixth and unknown sense, was long an enigma to naturalists. The difficulty, however, seems to have been solved by the investigations which have been made into the formation of the bat's wing, and it is now universally allowed, that to the exquisite nervous system of its wings the bat is indebted for the above-mentioned faculty.

The Vampires are said to unite in themselves the progressive power of quadrupeds and birds, and to run on the ground as swiftly as rats, while they fly through the air as easily as any bird. But this accomplishment of running is by no means general among the bats, whose mode of progress is awkward in the extreme, and when the animal is hurried or alarmed, positively ludicrous.

Bats are in general very much averse to the ground, and never, unless under compulsion, place themselves on a level surface. Their mode of walking is grotesque and awkward in the extreme ; and the arduous task of proceeding along the ground is achieved with such difficulty, that it seems almost to be painful to the animal which is condemned for the time to exchange its easy aërial course for the tardy and uncongenial crawl to which its earthly progress is limited. Quadrupedal in its form, although that form may be strangely modified, the bat will occasionally assume quadrupedal action, and walk on the ground by the aid of all its four feet. The method of advancing is as follows :

The bat thrusts forward one of the fore-legs or "wings," and either hooks the claw at its extremity over any convenient projection, or buries it in the ground. By means of this hold, which it thus gains, the animal draws itself forward, raises its body partly off the earth, and advances the hind leg, making at the same time a kind of tumble forward. The process is then repeated on the opposite side, and thus the creature proceeds in a strange and unearthly fashion, tumbling and staggering along as if its brain were reeling from the effects of disease. It steers a very deviating course, falling first to one side and then to the other, as it employs the limbs of either side.

None of the bats like to raise themselves into the air from a perfectly level surface, and therefore use all their endeavours to climb up some elevated spot, from whence they may launch themselves into the air.

They climb with great ease and rapidity, being able to hitch their sharp and curved claws into the least roughness that may present itself, and can thus ascend a perpendicular wall with perfect ease and security. In so doing they crawl backwards, raising their bodies against the tree or wall which they desire to scale, and drawing themselves up by the alternate use of the hinder feet. When they have attained a moderate height they are able to fling themselves easily into the air, and to take to immediate flight. They have the power of rising at once from the ground, but always prefer to let themselves fall from some elevated spot.

The reason is now evident why the bats take their repose in the singular attitude which has been already mentioned. When suspended by their hind feet, they are in the most favourable position for taking to the air, and when they desire to fly need only to spread their wings, and loosing their foothold, to launch themselves into the air.

There may be, and probably are, other reasons for the curious reversed attitude, but that which has already been given accounts in some measure for it. Even among the birds examples are found of a similar mode of repose ; members of the genus Colius, an African group of birds, sleep suspended like the bats, clinging with their feet, and hanging with their heads downwards. But these birds cannot assume this attitude for the purpose of taking to flight, as their wings are used as readily as those of most other feathered creatures, and therefore the reason which was given for the reversed position of the bats will not apply to the birds.

On the nose of the Vampire Bat may be observed a curious membrane of a leaf-like shape. This strange and not prepossessing appendage to the animal is found in some of the bats which inhabit our own country. Among the British bats which possess the leaf-decorated nose, the GREAT HORSESHOE BAT is the most conspicuous. Only the head of this animal is given in the engraving, as in its wings and body it differs but very little from other British bats.

The membrane which gives to this creature the title of Horseshoe Bat, is extremely large in proportion to the size of the animal, though not so large as in some of the foreign bats. It is double in form, that portion which is in front resembling a horseshoe in shape, and curving from the lips upwards, so as to embrace the nostrils. The second leafy membrane is placed on the forehead, and is sharply pointed.

HEAD OF HORSESHOE BAT.
Rhinólophus Ferrum-equinum.

The ears of this bat are large, pointed, and marked with a succession of ridges, which extend from the margins nearly half-way across the ears. The " tragus," or inner ear, is wanting in this bat, but its office seems to be fulfilled by a large rounded lobe at the base of the ear.

The colour of the fur is grey with a slight tinge of red above, while on the under portions of the animal the ruddy tint vanishes, and the hair is of a very pale grey. The membrane is of a dusky hue. The bat is not a very large one, the length of the head and body being only two inches and a half, while that of the extended wings is about thirteen inches. The ears are half an inch in breadth, when measured at their widest part, and are about three-quarters of an inch in length.

What may be the object of the wonderful nasal appendage seems to be quite unknown. The most obvious idea is, that it is given to the animals for the purpose of increasing the delicacy of their sense of smell in seeking food and avoiding foes. But even if such be the case, there seems to be no apparent reason why such a privilege should be granted to one species and denied to another—both animals being in the habit of seeking their nutriment and escaping pursuit in a similar manner. The generic term, Rhinolophus, which is applied to these bats, is derived from two Greek words, the former signifying a nose, and the latter a crest.

Another peculiarity of form which has been noticed in these animals, is the presence of two prominences on the groin, which have been taken for supplementary mammæ, and described as such. As, however, no mammary glands exist beneath these projections, they are evidently no true mammæ, and probably belong only to the skin.

The Great Horseshoe Bat seems to be less endurent of light than any of its British relatives, and takes up its abode in caverns so dark and gloomy that no other species of bat will bear it company. This instinct of concealment induces the bat to leave its home at a later and to return at an earlier hour than the other bats, and consequently it has only recently been found to exist in England. The first specimen which was captured had fixed its abode in rather a precarious situation, and was found in a building belonging to the Dartmouth powder mills. Since that time it has been discovered in many places, but always in some dark and retired situation.

There is another similar animal found in England, called the Lesser Horseshoe Bat (*Rhinólophus Hipposidéros*). This creature was for some time thought to be the young of the last mentioned animal, but is now known to be a distinct species. The name Hipposideros is Greek, and in that language signifies the same as Ferrum-equinum in Latin, *i.e.* Horseshoe.

The bats which we shall now examine are devoid of that strange nasal leafage which gives so unique an aspect to its wearer. The BARBASTELLE does not seem to be very plentiful in this country, although specimens have several times been taken in various parts of England. It is a singular coincidence that the first acknowledged British specimen was captured in a powder mill, as was the case with the Great Horseshoe Bat.

One of these animals which was for some weeks in the possession of Mr. Bell, was taken in Kent, at the bottom of a mine seventy feet in depth. It did not seem to be so active as some Long-Eared and other bats which were taken in the same locality, and preferred lying on the hearth-rug to using its wings. It fed readily on meat and would drink water, but never became so tame as its companions. Its captive life lasted only a few weeks, its death being apparently hastened by the attacks of the other bats, one of which was detected in the very act of inflicting a bite on the Barbastelle's neck.

The colour of the Barbastelle is extremely dark, so much so, indeed, that by depth of tint alone it can be distinguished from any other British bat. On the hinder quarters, a rusty brown takes the place of the brownish-black hue which characterizes the fore-part of the body. Underneath, the hair is nearly grey, being, however, much darker towards the neck.

The length of its head and body is just two inches, that of the ears half an inch, and

BARBASTELLE.—*Barbastellus communis.*

the expanse of wing measures between ten and eleven inches. The ears are tolerably large, and slightly wrinkled. The tragus is sharply pointed at its tip, and widened at its base. A full view of the face shows a rather deep notch in the outer margin and near the base of the ear.

The engraving represents the Barbastelle as walking on a level surface, and exhibits the strangely awkward mode by which these animals achieve terrestrial progression.

One of the most common, and at the same time the most elegant, of the British Cheiroptera, is the well-known LONG-EARED BAT.

This pretty little creature may be found in all parts of England; and on account of its singularly beautiful ears and gentle temper has frequently been tamed and domes-ticated. I have possessed several specimens of this bat, and in every case have been rewarded for the trouble by the curious little traits of temper and disposition which have been exhibited.

My last bat-favourite was captured under rather peculiar circumstances.

It had entered a grocer's shop, and to the consternation of the grocer and his assistant, had got among the sugar-loaves which were piled on an upper shelf. So terrible a foe as the bat (nearly two inches long) put to rout their united forces, and beyond poking at it with a broom as it cowered behind the sugar, no attempts were made to dislodge it.

At this juncture, my aid was invoked ; and I, accordingly, drew the bat from its hiding place. It did its little best to bite, but its tiny teeth could do no damage even to a sensitive skin.

The bat was then placed in an empty mouse-cage, and soon became sufficiently familiar to eat and drink under observation. It would never eat flies, although many of these insects were offered, and seemed to prefer small bits of raw beef to any other food. It was a troublesome animal to feed, for it would not touch the meat unless it were freshly cut and quite moist; forcing me to prepare morsels fit for its dainty maw, six or seven times daily.

It spent the day at the top or on the side of its cage, being suspended from its hinder claws, and would occasionally descend from its eminence in order to feed or to drink. While eating, it was accustomed to lower itself from the cage roof, and to crawl along the floor until it reached the piece of meat. The wings were then thrown forward so as to envelope the food, and under the shelter of its wings, the bat would droop its head over

LONG-EARED BAT.—*Pleiótus communis.*

the meat, and then consume it. On account of the sharp surface of its teeth, it could not eat its food quietly, but was forced to make a series of pecking bites, something like the action of a cat in similar circumstances.

It would drink in several ways, sometimes crawling up to the water vessel and putting its head into the water, but usually lowering itself down the side of the cage until its nose dipped in the liquid. When it had thus satisfied its thirst, it would re-ascend to the roof, fold its wings about itself, and betake itself to slumber once more.

I kept the little animal for some time, but it did not appear to thrive, having, in all probability, been hurt by the broom-handle which had been used so freely against it, and at last was found dead in its cage from no apparent cause. Although dead, it still hung suspended, and the only circumstance that appeared strange in its attitude was, that the wings drooped downwards instead of being wrapped tightly round the body.

In the attitude of repose, this bat presents a most singular figure. The wings are wrapped around and held firmly to the body ; the immense ears are folded back, and the pointed inner ear, or "tragus," stands boldly out, giving the creature a totally different aspect.

The enormous ears, from which the animal derives its name, are most beautiful organs. Their texture is exquisitely delicate, and the bat has the power of throwing them into graceful folds at every movement, thereby giving to its countenance a vast amount of expression. The figure on p. 121, exhibits the animal as it appears on the wing, and with its ears fully extended. But the present engraving shows the head of this bat, as it appears while the ears are disposed in slight folds and gently curved.

It sometimes happens that the Long-eared Bat has lived long in captivity, and even produced and nurtured its young under such conditions. For the following very interesting account of a maternal bat, I am indebted to the kindness of Mrs. S. C. Hall.

HEAD OF LONG-EARED BAT.

"While living in an old rambling country house in Ireland, without any companions of my own age,—an only solitary child left (after my 'lessons' were finished) to create my own amusements—I made friends, of course, with our own dogs and horses; and as all the servants loved 'little Miss,' and anxiously ministered to her desires, I became well acquainted with the habits and peculiarities of the wild creatures in our own grounds and neighbourhood. We were within a mile of the sea, and there was a beautiful walk from the dear old house, on to the cliff that sheltered our bathing cove, which I have traversed, accompanied by our Newfoundland dog, the old retriever-spaniels, and a fine deerhound, at nearly all hours of the day and night.

"A lovely ivy-covered cottage near the orchard, which, before I was born, was occupied by an old gardener, was at last given over to my menagerie, as the only way of keeping the 'big house' free from 'Miss Mary's pets.' My 'help' was a strong-bodied girl, one of the 'weeders,' who had the rare merit of not being afraid of anything 'barring a bull;' and she always intimated if I made a pet of a bull, she would 'wash her hands clean out of the menageeree for ever—Amin!'

"As I never did, poor Sally remained my assistant until the death of my dear grandmother broke up the establishment; and I came to England in the first blush of girlhood, to be civilised and educated, and made 'like other young ladies.'

"But those years were precious years to me; I grew, and fostered in those wild hours, an acquaintance with, and a love of Nature, which has refreshed my life with greenest memories. My dear young mother knew every bud and blossom of the parterre and the field, and though she disliked my seal, and obliged my young badger to be sent away (I was not very sorry for him, he bit so furiously, and would not be friends with the dogs, which the seal was), yet she tolerated my owl, my kites, and even a most prosperous colony of mice of many colours, and a black rat who was really an affectionate companion. My hare I was permitted to keep at the house, for he would hold no friendship with rabbits.

"Song birds I never attempted to cage, but robins and pigeons followed me (according to Sally), 'like their born mother.'

"The gable end of an old stable was covered by one of the finest myrtles I ever saw: it was twenty-two feet high and seventeen wide, and standing out here and there from the wall. Swallows and bats loved to shelter in the holes of the old building. I was just a small bit afraid of the 'leather-winged bat;' my nurse often told me how they sucked cows, and even scratched out children's eyes.

"But one cold spring morning I saw a boy tossing into the air and catching again what I fancied to be a large mouse: of course, my sympathy awoke at once, and I rushed to the rescue; it proved to be a half-dead bat, very large and fat, its beautiful broad ears were still erect, and when I took it in my hands I felt its heart beat. I placed it in a basket, covered it with cotton, and put it inside the high nursery fender. I peeped frequently under the lid, and at last had the pleasure of seeing it hanging bat-fashion on the side of the basket, its keen bright eyes watching every movement. When it was fully restored, I endeavoured to take it out, and then discovered that one of its hind feet had been crushed, and was hanging by a bit of skin. With trembling hands I removed the little foot, and applied some salve to the extremity.

" All this time the poor thing continued hooked on to the basket, and during the first day she would take no food, would not be tempted by meat or milk, by a fly or a spider. The next morning I saw her cowering in the cotton, and when I attempted to touch her she endeavoured to bite my finger, and made the least possible noise you can imagine. I then offered her a fly, and in a moment it was swallowed ; a bit of meat shared the same fate, and then she folded her wings round her, intimating, as I imagined, that she had had enough. All day she never moved, and at dusk, when I again tempted her with food, she took it. This continued for some days ; she became tamer, and seemed to anticipate 'feeding-time.'

" At last, to my astonishment, I saw a baby-bat covered with light brownish fur, but still looking as young mice look, under the folds of her wing (I do not know what else to call it). Doubtless Nature had taught her that for the sake of this little one she must take food. I believe it sucked, for, afterwards, when she again suspended herself against the side of the basket, the young bat was not in the cotton, and I fancied that it hung from the mother while imbibing nutriment.

" The old bat became furious if I attempted to touch the young one ; her soft hair stood up, and she would tremble all over, and utter little, short, sharp sounds. I wanted very much to see if the baby—like Chloe's puppies—was blind, but she would not allow an investigation. Certainly before a fortnight had passed, I saw its eyes, like little bright beads in the candle light.

" My bat and her baby excited great curiosity, and she was too frequently disturbed ; the young one lived for about a month, when, to my great grief, I found it dead in the cotton, the parent hanging, as usual, from the side of the basket. I am sorry to add, that the wee bat had what might have been a bruise, but which looked very much like a bite, at the back of the neck.

" The old bat became as tame as a mouse, would hang itself to any convenient portion of my dress, and devour whatever I gave it of animal food, and lick milk off my finger. It knew me well, would fly round my room in the evening, and go out at the window hawking for insects, and return in a couple of hours and hang to the window-sill, or to the sash, until admitted. At night, it would sometimes fasten in my hair, but never went near my mother or the servants. It did not seem to experience any inconvenience from the loss of its foot, and continued a great favourite for more than two years. I suppose the heat of my room prevented its becoming torpid in winter, though certainly it never prowled about as it used to do in spring and summer ; I do not think it ate in winter, but of this I cannot be certain. It disappeared altogether at last, falling a prey, I believe, to some white owls, who held time-honoured possession of an old belfry. I was very sorry for my bat, and should be glad to cultivate the intellect and affections of another, if I had the opportunity."

In the valuable work on " British Quadrupeds," by Mr. Bell, there is an account of a nursling bat, which presents many points of similarity with the foregoing description.

England possesses many species of these curious flying mammalia, nineteen of which are mentioned in " Bell's British Quadrupeds." Of these, the last which will be described in this work is nearly the largest of the British Cheiroptera, being only exceeded by the Large-eared Bat (*Myotis murinus*).

In length of head and body it is almost three inches, and the spread of its wings, from thirteen to fourteen inches. The tail is about an inch and three-quarters long, and is capable of considerable movement. The colour of the fur is a reddish brown, nearly uniform in tint over the whole body, and its texture is very soft. The ears are rather large, and the tragus is short, narrow at its root, and then expanding into a rounded head.

On account of the great height at which this bat loves to fly, it has been named " altivolans," or " high-flying," and seems to be among bats what the swift is among the swallow tribe.

It is curious, by the way, to mark the analogy that exists between the swallows and bats. Each of these groups loves the air, and is mostly seen on the wing. Their food consists of the flying insects, which they chase by their exquisite command of wing ; and it will be noticed that, as soon as the swallows retire to rest at dusk, after clearing the air of the diurnal insects, the bats issue from their homes, and take up the work, performing the same task with the insects of night, as the birds with those of day. Then, as the dawn breaks, out come the swallows again, and so they fulfil their alternate duties.

NOCTULE, OR GREAT BAT.—*Noctulinia altivolans.*

The NOCTULE is not so pleasant a companion as the Long-eared Bat, for it gives forth a most unpleasant odour. Its cry is sharp and piercing, thereby producing another analogy with the swifts, which are popularly known by the name of "Jacky-screamers."

The voice of all British bats is singularly acute, and can be tolerably imitated by the squeaking sound which is produced by scraping two keys against each other. There are many people whose ears are not sensible to the shrill cry of these animals—which, in some cases, is rather fortunate for them. I well remember being on Hampstead-heath, one summer's evening, when the air was crowded with bats hawking after flies, and their myriad screams were so oppressive, that I longed for temporary deafness. Yet my companion—an accomplished musician—was perfectly insensible to the shrill cries, which seemed to pierce into the brain like so many needles. It is also known that many ears are deaf to the stridulous call of the grasshoppers.

In order to show the sharply-pointed teeth of the insect-feeding bats, the skull of the common bat is here given.

One use of the tail is, evidently, that it should act as a rudder, in order to guide the flight while the creature is on the wing. There is, however, another purpose which it serves, and which would never have been discovered, had not the bat been watched. It seems that the female bat uses its tail, and the membrane which stretches on either side from the tail to the hind legs, as a cradle, in which to deposit its young when newly born and comparatively helpless.

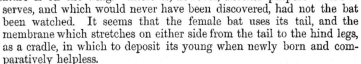

SKULL OF BAT.

Bats are generally found to assemble in great numbers wherever they find a convenient resting-place, and in such localities as church towers, rocky caverns, hollow trees, and the like, they may be found by the hundred together. These numerous assemblies are the cause of a large deposition of guano, which consists almost wholly of the refuse of insects, such as wings, legs, and the harder coverings. In this guano are found, by the aid of the microscope, very many curious infusorial objects, which may be separated from the guano by the usual modes of preparation.

The odour which arises from this substance is peculiarly sharp and pungent, and cannot easily be mistaken. The animals themselves are readily alarmed when disturbed in their home, they disengage themselves from their perches, and flap about in great dismay, knocking themselves against the intruder's face, much as the great nocturnal

FLYING FOX, OR ROUSSETTE.—*Pteropus Rubricollis.*

beetles are wont to do on summer's evenings. A visit to a bat-cave is, therefore, no pleasant affair.

The bats which have heretofore been mentioned feed on animal substances, insects appearing to afford the principal nutriment, and raw meat or fresh blood being their occasional luxuries. But the bats of which the accompanying engraving is an example, are chiefly vegetable feeders, and, in their own land, are most mischievous among the fruit-trees.

They are the largest of the present bat tribe, some of them measuring nearly five feet in expanse of wing. Their popular name is FLYING FOXES, a term which has been applied to them on account of the red, fox-like colour of the fur, and the very vulpine aspect of the head. Although so superior in size to the Vampires, the Flying Foxes are not to be dreaded as personal enemies, for, unless roughly handled, they are not given to biting animated beings.

But though their attacks are not made directly upon animal life, they are of considerable importance in an indirect point of view, for they are aimed against the fruits and other vegetable substances by which animal life is sustained. Figs and other soft fruits appear to be the principal food of these bats; and so pertinacious are the animals in their assaults on the crops, whether of field or tree, that they are held in no small dread by the agriculturist.

It is no easy matter to guard against such foes as these winged devourers, for as the air is an ever open path by which they can proceed on their destructive quest, and the darkness of night shields them from watchful eyes, the ordinary precautions which are taken against marauders would be useless.

There are but two alternatives for any one who desires to partake of the fruit which he has cherished—the one, to cover the whole tree with netting or similar fencing, and the other, to enclose each separate fruit-cluster with a sufficient protection. As the trees which the Kalongs, as these bats are often called, most affect, are of considerable size, the latter plan is that which is generally pursued. For this purpose, the natives weave from the split branches of the bamboo, certain basket-like armour, which is fastened round the fruit as it approaches maturity, and is an effectual guard even against the Kalong's teeth.

When the trees are small, they are sometimes covered entirely with netting, but not to such good purpose as when each fruit is separately protected. For these bats are so cunning, that they creep under the nets and render nugatory all the precautions which have been taken. One proprietor of a garden at Pernambuco was never able to secure a single fig from his trees, in spite of nets by night and guns by day. The bats are wise animals, and do not meddle with unripe fruit.

The flight of these creatures is unlike that of the more active insect-feeding Cheiroptera. The stroke of the wings is slow and steady, and instead of the devious course which characterizes the carnivorous bats as they flit about the air in chase of their insect prey, these frugivorous species fly in straight lines and to great distances.

The Kalongs do not seem to care much for dark and retired places of abode; and pass the day, which is their night, suspended from the trunks of large trees, preferring those which belong to the fig genus. On these boughs they hang in vast numbers, and by an inexperienced observer, might readily be taken for bunches of large fruits, so closely and quietly do they hang. If disturbed in their repose, they set up a chorus of sharp screams, and flutter about in a state of sad bewilderment, their night-loving eyes being dazzled by the hateful glare of the sun. They are apt to quarrel under such circumstances, and fight for their roosting, or rather their hanging places, much as birds do when retiring to rest for the night.

Bats do not seem to be very tempting additions to the *cuisine*, but man is an omnivorous animal, and eats everything, whether animal, vegetable, or mineral, the last-named diet being exemplified by the "stone-butter" of the German miners, and the clay balls of the Indian savage. Some nations there are which feed on their own kind. Many there are which live habitually on the Quadrumana that inhabit their country, and there are some who find a favourite article of diet in the Cheiroptera.

The species which is most generally eaten is the Edible Kalong (*Pteropus Edúlis*), a bat which is found in great quantities in the island of Timor and other places. It is a very large animal, the expanse of wing rather exceeding five feet, and the length of head and body being about a foot. The eye is a fine brown. The flesh of these bats is said by those who have ventured upon so strange a diet, to be very delicate in flavour, tender in substance, and white in colour.

It is probably to these animals that Bennett refers, in his "Whaling Voyage round the Globe."

"The only animals that came under our notice at Timor, were bats and foxes. The bats were of that large kind which sailors call Flying Foxes. When our woodcutters commenced their labours in the forest, the first blow of the axe caused a large flock of these creatures to mount in the air, and wing their way to a less precarious retreat. They flew in a body to the distance of more than two hundred yards, then returned as simultaneously to the vicinity of the spot which they had quitted, and ultimately settled in the depths of the jungle.

" Considering how little their vision is adapted for day duty, it was interesting to notice the systematic manner in which they directed their flight: one which arose some time after the others, taking immediately the right direction to follow and join the main body of fugitives."

In this latter passage is mentioned one distinguished peculiarity of these creatures, namely their habit of flying in long lines, somewhat after the manner of rooks returning to roost—

" The blackening train of crows to their repose."

One bat seems to take the lead, and the others follow at short and irregular intervals, pursuing the same course as their pioneer.

The bats which belong to this genus (Pteropus) are remarkable for the fact that they possess fewer vertebræ than any other known mammalian animal. In the entire spinal column, there are but twenty-four of these bones; this paucity of number being caused by the entire absence of a tail.

The hair with which the bat tribe is furnished, is of a very peculiar character, and although closely resembling the fur of a rat or mouse when seen by the unaided eye, is so unique in aspect when seen under a microscope, that a bat's hair can be detected almost at a glance. Each hair is covered with very minute scales, which are arranged in various modes around a central shaft.

The accompanying figure exhibits the central portion of a hair taken from one of the Indian bats, magnified five hundred diameters, or two hundred and fifty thousand times superficially. Near the root, the hair is almost devoid of these scales, and therefore appears much smaller than in the central and terminal portions. Some of these external scales bear a close resemblance to the scales which are placed on the surface of a butterfly's wing ; but can easily be distinguished from them by their smaller size, and the absence of the striated markings that are found on the scales of the butterfly's wing.

The strange similitude between the bat's hair, and the plant which is popularly known by the name of " Mare's-tail," cannot but strike any one who is in the least acquainted with botany. It may be, that so remarkable an outward resemblance would not exist unless there were some cause, at present hidden, which would account for it.

Before leaving the study of the bats, we must take a cursory view of the strange condition of life in which these animals pass the colder months of the year, which condition is known by the name of hibernation, because it takes place in the winter.

The insect tribes on which the bats chiefly subsist, and wholly so in this country, are either quiescent during the winter months, or are abroad in such limited numbers that they could not afford a subsistence to the bats or swallows. The latter creatures meet the difficulty by emigrating to more genial lands, and there finding the food which they would lose in these cold climes ; but the former are obliged by the laws of their being to remain in the country where they were born. It is evident, therefore, that unless some provision were made for them during the insectless time of year, every bat would perish of hunger.

HAIR OF INDIAN BAT.

Such a provision exists, and exerts its power by throwing the bats into a deep lethargy, during which they require no food and take no exercise, but just live throughout the winter in a state of existence that seems to partake more of the vegetable than the animal life.

During hibernation, the respiration ceases almost wholly, and if it takes place at all, is so slight as to defy investigation. The air in which these creatures pass the winter seems to undergo no change by the breath, as would be the case if only one inspiration were made ; and, strangest of all, the animal seems capable of existing for some time in gases that would be immediately fatal to it in the waking state, or even without any air at all. The temperature, too, sinks to that of the surrounding atmosphere, although as a general fact, the animal heat of these creatures is rather high, as is the case with most flying beings, whether mammals or birds.

Many curious and valuable trials have been made upon bats while in a state of torpidity, the subjects of experiment being placed in such a manner that the least act of respiration made itself clearly visible, by the movements of a delicate index. The wing was extended in such a manner, that the circulation of the blood was perceptible through its semi-transparent membrane, and a thermometer was arranged so as to register the temperature.

Very great care is requisite in conducting these experiments, because the least excitement, or the slightest raising of the temperature, suffices to rouse the somnolent animal, and to alter the conditions which are absolutely necessary for true hibernation. A hasty footfall, or an accidental tap given to the table on which the creature rested, would cause it to make several respirations, and to recover sufficient vitality to raise the temperature, and to consume some portion of oxygen from the air. The same animal which passed ten hours in a state of perfect somnolence, without producing any perceptible effect on the oxygen contained in the atmospheric air, consumed in a single hour more than four cubic inches of oxygen, when aroused and lively.

The curious subject of hibernation will be again noticed in connexion with the various animals, such as the marmot, dormouse, and others, which pass the cold months in a state of torpidity.

The analogy that exists between the bats and the birds is too evident to escape attention. But the most curious part of the analogy is the order in which the various portions of a mammalian animal are modified, so as to discharge the faculties which belong more properly to the feathered tribes.

The elongation of the fingers, and expansion of the membranous "wings," has already been mentioned, as well as the general development of the breast-bone. These two structures are in common with all bats, as are their corresponding portions in all birds. But there are some organizations which are found greatly developed in certain families of birds, and are repeated in certain of the bats.

The structure to which allusion is here made is that connexion of the lungs with the skin, or, rather, with the space between the skin and the body, that is found in many birds, especially those which pass a marine existence, and which enables the bird to inflate its skin with air, and so to increase its bulk largely without sensibly increasing its weight.

In the birds, this inflation is made by direct communication with the lungs; but in the bat, the air is conveyed into the membranous cells after a different fashion. A very small opening is found to exist at the bottom of the cheek-pouches of either side, and is furnished by an apparatus by means of which the air is prevented from escaping without the will of the animal. This opening affords a communication between the mouth of the bat and the space between the skin and flesh, which are only tied to each other by a few membranous threads at each side of the neck, and on the sides of the thumb.

When, therefore, the bat desires to inflate its body, it closes its mouth, and forces the air from its lungs through these cheek-passages into the empty space between the skin and flesh. The result of this operation is, that the skin is puffed out on all sides of the animal, so that the creature is immersed in a kind of atmospheric bath. So enormously is it distended by the amount of air which is introduced, that it loses all its shapely proportions, and looks like a little ball of fur, to which the head and limbs had been artificially attached.

The bats which are possessed of this wonderful faculty belong to the genus "Nycteris," and are found in Africa. What may be the object of so singular a power is not satisfactorily proved. That it gives very great buoyancy to the form is evident enough, but it also seems plain that it is intended for other designs than the obvious one of decreasing the proportionate weight of the animal.

LION, LIONESS AND HER YOUNG, AND LEOPARD.

FELIDÆ;

OR, THE CAT TRIBE.

THE beautiful and terrible animals which are known by the general name of the Cat Tribe, now engage our attention.

With the exception of one or two of the enigmatical creatures which are found in every group of beings, whether animal, vegetable, or mineral, the Cats, or FELIDÆ as they are more learnedly termed, are as distinct an order as the monkeys or the bats. Pre-eminently carnivorous in their diet, and destructive in their mode of obtaining food, their bodily form is most exquisitely adapted to carry out the instincts which are implanted in their nature.

All the members of the cat tribe are light, stealthy, and silent of foot, quick of ear and eye, and swift of attack. Most of them are possessed of the power of climbing trees or rocks, but some few species, such as the Lion, are devoid of this capability.

The teeth of the exclusively carnivorous animals are always of a form which permits them to seize and tear their prey, but does not give them the power of masticating their food after the manner of the vegetable feeders. We are all familiar with the mode in which the domestic cat consumes her food, whether it be a piece of butchers' meat which is given to her by the hand of man, or a mouse which she has captured by her own paws. Instead of the grinding process which is employed by monkeys and other creatures whose teeth are fitted for grinding their food, the cat tears the meat into conveniently sized morsels, and then eats the food by a series of pecking bites.

The annexed engraving of a Lion's teeth and jaws will explain the reason for this mode of action.

None of the teeth are furnished with the flat surfaces which are necessary for grinding the substances which may be placed between them; and this inability does not lie only in the teeth, but extends to the very framework of the jaws. As may be seen on reference to the engraving, the lower jaw is so largely developed at its base, and fits so deeply into its socket, that lateral motion is impossible.

In order to give a more perfect view of the lower jaw-bone, the bone immediately above it has been removed, and presents only its cut surface. This part of the structure is scientifically known as the "malar," or cheek-bone, and forms an arch, which has been termed the "zygomatic" arch. In the carnivorous, and more especially in the feline animals, this bone is extremely large in proportion, and is increased in

JAWS AND TEETH OF LION.

strength by its very decided curve. The great size, as well as the peculiar form of this bone, are required for the purpose of affording protection to the enormously powerful muscles by means of which these animals are enabled to tear their food, and also for the attachment of certain jaw-moving muscles. There is an upward as well as an outward curve in the malar bone, which gives strength precisely in the direction where it is most required.

On reference to the skeleton of the Lion, many curious structures will be seen. It would be impossible in the present volume to give a detailed history of even one portion of the bony framework around which the moving and vital organs of the Lion are arranged. Only a short description, therefore, will be here given; and in order to proceed methodically, we will start from the head.

The teeth and jaws have already been mentioned. On the top of the skull there runs a tolerably high bony crest, which reaches its greatest elevation at the very back of the head. This bone-ridge is intended for the attachment of the powerful muscles which raise the head, and enable the animal to perform its wonderful feats of strength.

Pausing awhile at this portion of the animal's form, and directing our view to the interior of the skull, a curious internal ridge of bone is seen, which arises to some little height, and separates the two great divisions of the brain from each other. In the cat tribe, this ridge arises entirely from that part of the skull which is known by the name of the "parietal bone;" but in other carnivorous animals, the "occipital bone" is the

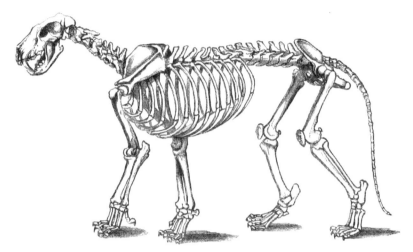

SKELETON OF LION.

principal source of this structure. The use of the bony ridge is not quite certain; but it seems likely that it may play an important part in guarding the brain from the severe shocks which must be occasioned by the movement of the animals when they leap upon their prey.

Reverting to the exterior form, and passing from the head to the neck, we find that the two first vertebræ partake of a similar enlargement to that which has already been observed on the back of the skull. The vertebra which is nearest to the head, and is called the "atlas," is broad and strong, and spreads laterally, while the second, or "axis," is long, and is developed upwards into a very powerful crest.

The ribs are beautifully formed, and placed rather widely apart, giving plenty of room for the heart and lungs to perform their duties effectually. The vertebræ that fill the space between the ribs and the hip-bones are very large, and are so exquisitely jointed together, that they unite a graceful flexibility of movement with great muscular power.

The limbs exhibit in their structure a beautiful unison of strength and lightness.

Powerful as are the bones which support the limbs, and heavily as they are framed, they are but just sufficient for the attachment of the enormous muscles which can carry the animal through the air for vast distances, and can strike down an ox with a single blow of the paw. So easy and so apparently gentle are the movements of the Lion's paw, that their power can only be judged by the effects. I have seen a Lion just wave his paw, and with that quiet movement send his mate rolling over and over on the ground, although her weight was but little inferior to his own.

The muscles in which such terrible power resides, move so easily that they hardly give any external indications of their true character. But when the skin, with its thick covering of hair, is removed, the iron muscles stand out in all their marvellous strength. Needs be, that the anatomist who undertakes the dissection of an adult Lion should be furnished with a large supply of the sharpest and most highly tempered knives; for the muscles are so hard and tough, that they make sad havoc with delicate instruments.

It will be seen, on reference to any member of the cat tribe, that the mode of walking employed by these creatures is different from that of man, monkeys, or bats. The weight of the body rests only on the toes, and not on the entire foot. This manner of walking is termed "digitigrade," from the Latin words "digitus," a finger, and "gradus," a step. As, however, this mode of progression would endanger the sharpness of the claws, if they were permitted to rest upon the ground, there is a beautiful structure by means of which the talons are kept from the earth, and preserved in their so-called sheaths until they are wanted for their legitimate use.

In the accompanying figure the mechanism of the claw is exhibited.

When the animal is at rest, the upper tendons draw the claw backwards, so that it is lifted entirely from the ground, and the weight of the body rests only on the soft pads which stud the under surface of the foot. But when the creature becomes excited, and thrusts out its paw for the purpose of striking a blow, or clutching at its prey, the upper tendons become relaxed, while the lower tendons are tightened, and the claw is thrown boldly forward, sharp and ready for either use.

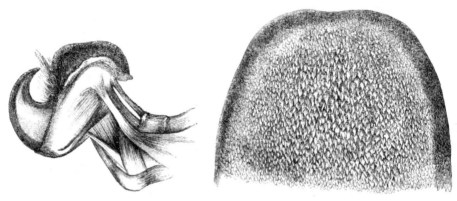

CLAW OF LION.   TONGUE OF LION.

The claw which is represented is that of the Lion, but the mechanism is common to all the true cats.

Another curious structure is common to the group of feline animals; and as the Lion seems to be their most perfect representative, the example has been taken from that animal.

Every one who cares for cats, and who in consequence is cared for by those graceful creatures, is familiar with the dry roughness of pussy's tongue, as she licks the hand of her human friend. This peculiarity of formation is the more conspicuous because it presents so great a contrast with the wet, smooth tongue of the dog; and, as a general rule, men are more accustomed to the lingual caresses of the dog than of the cat. The cause of the strangely dry tongue of the Felidæ is at once seen by reference to the tongue of a lion or tiger, or by applying a magnifying glass to the tongue of a domestic cat.

The entire surface of the tongue is covered with innumerable conical projections, which are so curved that their points lie towards the throat. On the central line of the tongue these projections are larger than at the side. Their chief, if not their only use, is to aid the cat in stripping the flesh from the bones of the animals which it has killed, and so to prevent the least avoidable waste of nutriment. Truly, in nature the economical system reigns supreme, and waste is an impossibility.

So strongly made are these armatures, that the constant licking of a cat's tongue will remove the living tissues from a delicate skin, while the tongue of the Lion can rapidly cause the outflow of blood. There is a well-authenticated anecdote of a tame Lion cub and its owner, which exhibits strongly the rasping power of the feline tongue.

THE LION.—*Leo Bárbarus.*

A gentleman had indulged himself in that rather dangerous pet, a tame Lion cub. The animal was very fond of its master, and would play with him as guilelessly as if it were a kitten.

One day the gentleman fell asleep, leaving one of his hands hanging over the side of his couch. His pet Lion came up to the couch when its master was slumbering, and by way of showing its affection, began to lick the exposed hand. In a very short time, the rough, file-like tongue cut through the delicate skin of the hand, and caused some little pain and a slight effusion of blood, which was eagerly licked off by the animal.

The pain which was caused by the too affectionate creature awoke its master, who naturally began to withdraw his hand from the caresses of the Lion. But at the first movement the Lion uttered a short, deep growl, which was repeated in a menacing manner at each attempt to remove the hand from its dangerous and painful position. Seeing that the Lion cub had become suddenly transformed from a domestic pet to a wild beast, which had for the first time lapped blood, and thirsted for more, its owner quietly slipped his other hand under his pillow, where he kept a ready-loaded pistol, and shot the poor Lion through the head. It was an act that went sorely against his will, but was the only course which he could have adopted in such an extremity, when there was no time for reflection, and when the hesitation of a moment might have cost a life.

Of the magnificent and noble creatures called Lions, several species are reported to exist, although it is thought by many experienced judges that there is really but one species of Lion, which is modified into permanent varieties according to the country in which it lives.

The best known of these species or varieties is the SOUTH AFRICAN LION, of whom so many anecdotes have been narrated. This noble animal is found in nearly all parts of Southern Africa, where the foot of civilized man has not stayed its wanderings. Before the tread of the white man, the Lion shrinks unwillingly, haunting each advanced post for a time, but driven surely and slowly backward as the human intellect gains opportunity for manifesting its supremacy over the lower animals. So entirely does man sweep the wild beasts from his presence, that even in the Cape colony, a living Lion is just as great a rarity as in England, and there are very few of the colonists who have ever beheld a living Lion except when pent in a cage.

The colour of the Lion is a tawny yellow, lighter on the under parts of the body, and darker above. The ears are blackish, and the tip of the tail is decorated with a tuft of black hair. This tuft serves to distinguish the Lion from any other member of the cat tribe. The male Lion, when full grown, is furnished with a thick and shaggy mane of very long hair, which falls from the neck, shoulders, and part of the throat and chin, varying in tint according to the age of the animal, and possibly according to the locality which it inhabits. The Lioness possesses no mane, and even in the male Lion it is not properly developed until the animal has completed his third year.

When fully grown, the male Lion measures some four feet in height at the shoulder, and about eleven feet in total length. These measurements are only applicable to the noble animals which have passed their lives in the free air of their native land, and have attained their majority with limbs unshackled and spirits unbroken.

The Lioness is a smaller animal than her mate, and the difference of size appears to be much greater than really is the case, because she is devoid of the thick mane which gives such grandeur and dignity to her spouse. Although smaller in size, she is quite as terrible in combat; and, indeed, the Lioness is ofttimes a foe much more to be dreaded than the Lion. When she has a little family to look after, Leæna is a truly fearful enemy to those who cross her path, assuming at once the offensive, and charging the intruders with a fierce courage that knows no fear and heeds no repulse.

Of the character of the Lion, opinions the most opposite have been promulgated.

Until later days the Lion was considered to be the very type of fiery courage and kingly generosity, indomitable in conflict with the strong, but merciful in sparing the weak and defenceless. Latterly, however, writers have passed to the opposite extreme, speaking of the Lion as a cowardly sneaking animal, and have even gone so far as to declare him to be no more formidable than a mastiff. It must be remarked that these opposite ideas have been put forth by men of practical experience, who have been personally acquainted with the king of beasts in his own domains.

Making due allowance for the "personal error," as astronomers would term the difference of idiosyncrasy in the narrators, we may safely conjecture that the truth lies somewhere between the two extremes, and that the Lion is not always so fierce an animal as is said to be the case by some, nor always so cowardly as it is said to be by others.

Even the same individual may be at one time ferocious and truculent, attacking a party of armed men, in spite of their fire-rampart, and carrying off one of their number from among them; or at another time it may be timid and cowardly, skulking out of sight if discovered, and flying in terror before the shouts and cries of a few savages.

Hunger seems to be the great cause of a Lion's defiance of danger; and it but seldom happens that a Lion which has had plenty to eat troubles itself to attack man or beast.

There seems to be a considerable spice of indolence in the Lion, which indeed is the case in most of the members of the cat tribe. It is capable of very great muscular efforts, and for a time will exert the most wary vigilance. But as soon as the existing cause is removed, the creature seems overcome with lethargy, and, seeking the cover of its lair, yields itself to repose.

Even when aroused by the calls of hunger, the Lion will not take more trouble than is necessary for the attainment of its end, and if it can strike down an antelope or jaguar with a blow of its paw, will be quite satisfied with its success, and will not trouble itself about such difficult game as a buffalo or a giraffe.

It is supposed by those who have had much experience of the leonine character, that the terrible "man-eating" Lions owe their propensity for human flesh to the indolence of their character or the infirmity of their frame, and not to their superior activity or courage. Unwilling, or unable, to expend strength and patience in the pursuit of the swift-footed antelope or powerful buffalo, the Lion prowls about the villages, thinking to find an easy prey in the man, woman, or child that may happen to stray from the protecting guardianship of the kraal and its dogs. Unarmed, man is weaker of limb, slower of foot, and less vigilant of senses than any of the wild animals, and therefore is a victim that can be slain without much trouble.

It is said that the taste for human flesh is often engendered by the thoughtless conduct of the very people who suffer from the "man-eaters." The Kaffirs are apt to leave their slain exposed in the bush, "a prey to dogs and all kinds of birds."

The Lion who passes near the spot where a dead Kaffir lies, is mightily pleased with the opportunity of obtaining a dinner on such charmingly easy terms; and being master of the situation, drives away hyænas, jackals, and vultures, until he has satisfied his lordly appetite. Having satiated himself, he retires to rest, and on awaking, repairs again to the site of his banquet in hope of making another such meal. He finds nothing but the fragments of bones, for the jackals and vultures have long ago consumed every morsel of flesh, and the hyænas have eaten the greater part of the bones. From that moment the Lion becomes a man-eater, and is a scourge to the neighbourhood. It beseemeth the whole armed population to rise and destroy this pest; for as long as the man-eater lives he will pay constant visits to the villages, and night after night, or even day after day, so great is his audacity, will he carry off his victims.

It is worthy of notice, that in all parts of the world where the larger felidæ live, certain individuals seem to isolate themselves from their kind by this propensity, and distinguish themselves for their predilection for human flesh.

As a general rule, the Lion is no open foe. He does not come boldly out on the plain and give chase to his prey, for he is by no means swift of foot, and, as has already been mentioned, has no idea of running into danger without adequate cause. He can make tremendous leaps, and with a single blow from his terrible paw can crush any of the smaller animals. So he creeps towards his intended prey, availing himself of every bush and tree as a cover, always taking care to advance against the wind, so that the pungent feline odour should give no alarm, and when he has arrived within the limits of his spring, leaps on the devoted animal and strikes it to the ground.

This mode of action gives a clue to the object of the fear-instilling roar which has made the Lion so famous.

As the Lion obtains his prey by stealth, and depends for nutrition on the success of his hunting, it seems strange that his voice should be of such a nature as to inspire with terror the heart of every animal which hears its reverberating thunders. Yet it will be seen, that the creature could find no aid so useful as that of his voice,

If the Lion has been prowling about during the evening hours, and has found no prey, he places his mouth close to the earth, and utters a terrific roar, which rolls along the ground on all sides, and frightens every animal which may chance to be crouching near. Not knowing from what direction the fearful sound has come, they leave their lairs, and rush frantically about, distracted with terror and bewildered with the sudden arousing from sleep. In their heedless career, one or two will probably pass within a convenient distance of the lurking foe.

These nocturnal alarms cause great trouble to those who travel into the interior of Africa. When night draws on, it is the custom to call a halt, and to release the draught oxen from their harness. A kind of camp is then made, a blazing fire is kept alight as a defence against the wild beasts, and the oxen are fastened either to the waggons or to the bushes by which the encampment is made.

The Lion comes and surveys the mingled mass of oxen, men, and waggons, but fears to

LION AND ZEBRAS.

approach too closely, for he dreads the blaze of a fire. In vain does he prowl around the encampment, for he can discover no stragglers from the protecting flame, and, moreover, finds that the watchful dogs are on the alert. So he retires to some little distance, and putting his mouth to the ground, pours forth his deepest roar. Struck with frantic terror, the stupid oxen break away from their halters, and quitting their sole protection, gallop madly away only to fall victims to the jaws and talons of the author of the panic.

It often happens that several Lions combine in their attacks, and bring their united forces to bear upon the common prey, each taking his appointed part in the matter. One of these joint attacks was witnessed by two English officers engaged in the late Kaffir war, with one of whom I am well acquainted.

A small herd of zebras were quietly feeding in a plain, all unconscious of the stealthy approach of several Lions, which were creeping towards them in regular order, under cover of a dense reed thicket. So quietly did the Lions make their advance, that their progress was unnoticed even by the zebra-sentinel. The Lions crept on, until they reached the sheltering thicket, when the sentinel took the alarm. It was too late—with a single bound, the leading Lion sprang over the reeds, felled one of the zebras, and set the others scampering in all directions so as to fall an easy prey to his companions.

It has happened that such alliances have come to a tragical end for the assailant as well as the victim.

"Early one morning," says Mr. Anderson, in his "Lake Ngami," "one of our herdsmen came running up to us in a great fright, and announced that a Lion was devouring a Lioness. We thought at first that the man must be mistaken, but his story was perfectly true, and only her skull, the larger bones, and the skin were left. On examining the ground more closely, the fresh remains of a young springbok were also discovered. We therefore conjectured that the Lion and Lioness, being very hungry, and the antelope not proving a sufficient meal for both, had quarrelled; and he, after killing his wife, had coolly eaten her also."

The same writer relates a curious instance of a wounded Lion being torn in pieces by a troop of his fellows.

In the attack of large animals, the Lion seldom attempts an unaided assault, but joins in the pursuit with several companions. Thus it seems to be that the stately giraffe is slain by the Lion, five of which have been seen engaged in the chase of one giraffe, two actually pulling down their prey, while the other three were waiting close at hand. The Lions were driven off, and the neck of the giraffe was found to be bitten through by the cruel teeth of the assailants.

When the Lion kills an eland, and does not happen to be very ravenously hungry, he feeds daintily on the heart and other viscera, not often touching the remainder of the flesh. In so doing, he rips open the abdomen with his powerful claws, and tearing out his favourite morsels, devours them. Sometimes, after satisfying his hunger, he will leave the eland lying on the ground apparently uninjured, the only visible wound being that which he has made by tearing the animal open.

Owing to the uniform tawny colour of the Lion's coat, he is hardly distinguishable from surrounding objects even in broad daylight, and by night he walks secure. Even the practised eyes of an accomplished hunter have been unable to detect the bodies of Lions which were lapping water at some twenty yards' distance, betraying their vicinity by the sound, but so blended in form with the landscape, that they afforded no mark for the rifle even at that short distance.

Under such circumstances, their glowing eyes afford the only means by which they can be discovered, and even with such assistance the position of the body cannot be made out. The felidæ tread so silently, that no footfall gives notice of their whereabouts; and aided by the beautiful mechanism of the "whiskers," they appear to be enabled to thread their stealthy way, almost without the aid of eyes.

Each whisker hair is, in fact, an organ endued with an exquisite sense of touch, and in connexion with a set of large nerves that convey to the brain the least touch. In the engraving is given a magnified representation of a single hair-bulb of one of the whiskers, together with the nerves by means of which the hair is converted into a tactile organ. It will be seen, on reference to the figure, that if the extremity of the hair is touched, a pressure will instantly be made on the nerves at its root. By means of these delicate feelers, the animals are able to guide themselves through the thickets, and to escape the risk of alarming their intended prey by too rude a contact with the branches.

Among the more inland settlers of Southern Africa, adventures with the Lion are of common occurrence. As may be expected, many of these rencontres are of a deeply tragic nature, while others are imbued with a decidedly comic element. A great number of original anecdotes of this nature have been most kindly placed at my disposal by Captain Drayson, who heard them from the lips of the actors themselves. In these narratives, the characters of both man and beast are well shown.

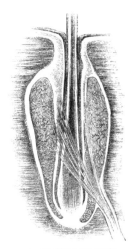

ROOT AND NERVES OF LION'S
WHISKER HAIR.

"ANY person who has mixed much with either Dutch, Hottentot, or Kaffir sportsmen, is sure to have heard many exciting and curious adventures connected with the chase of the Lion. From amongst a somewhat large stock I will now select one or two anecdotes which will serve to illustrate either the habits and character of the animal, or the method of hunting him.

A soldier, belonging to a line regiment, had heard that a great quantity of money might be obtained from amongst the Dutch Boers in the interior, by various simple processes with which he fancied himself acquainted.

Selecting a favourable opportunity, he deserted, taking care to well fill his haversack with meat, to serve him during his march across the wild uninhabited district which separated the Dutch locations from our frontier.

The soldier marched, during two days, some sixty miles or so, taking care when he slept to place the bag containing his meat under his head. On the third or fourth night, he lay down as usual to rest, with his head upon his pillow. It happened that in the country to which he belonged Lions were very common, and one of these unwelcome visitors happened to be prowling about in search of a supper, and dropped upon the military hero who was quietly snoring.

Whether the Lion were aware of the fierce calling of the sleeper, and therefore paid him some respect, is not mentioned; but, instead of carrying off the man, he merely clawed up the bag, and trotted away growling with his trophy. The only damage that he had inflicted on the soldier was the abstraction of a few inches of his scalp.

A Dutch Boer found the deserter wandering half starved on the plain, where he had been without food for a day and a night. The Boer fed and doctored him, but in return made him act as leader to the oxen and waggon, a position considered by the Dutch Boers to be the most degrading to man."

" NEAR the sources of the Mooi river there are several extensive plains on which large herds of elands and hartebeest were formerly found. Bordering on these plains are several ranges of hills, spurs from the Quathlomba mountains, and between these rocky spurs, kloofs or ravines exist, affording shelter for bush-buck, buffaloes, and many other animals which seek cover amongst either reeds or bushes.

At the time when the following scene occurred, there was scarcely an inhabitant in this locality besides a few Welshmen, who resided amongst the stony hills, and lived by the chase, and two or three Dutch Boers, the remnants of those who had accompanied Maritz in his migration from the old colony.

The Dutchmen had built themselves some wattle-and-daub huts, and were contented to remain where they were, as hunting and grazing-grounds were plentiful. A few thefts performed by their little neighbours, the Bushmen, had caused a commando to be raised, and, during the invasion of the hills that were then occupied by these little men, a boy had been captured by one of the Boers, and had been retained as a domestic. This individual will figure in the following scene with the Lion.

One evening, whilst one of these Boers was sitting with his son, a lad of about sixteen years of age, in front of his hut, smoking his stone pipe, and looking with pride upon his sleek herds which grazed about him, he noticed some object that moved slowly along the side of an old watercourse at a considerable distance from him. A telescope was an article of value which few of these residents possessed; it was therefore by patient watching only that the father and son at length discovered that the object was a Lion, which appeared to be carefully stalking a valuable black stallion grazing near the old watercourse. Instantly seizing their guns, which were as usual loaded and at hand, the two ran down towards the Lion, shouting as they went to the Hottentots who were engaged about the farm.

These individuals did not appear to be anxious about hurrying towards the scene of danger, and, consequently, the Dutchman and his son reached the stallion before any other aid arrived.

The course which they had followed caused them to lose sight of the Lion in consequence of intervening slopes of ground, so that, upon reaching the horse, which was grazing unconscious of danger, no Lion was to be seen. The young Boer, acting against the advice of his father, walked along the side of a ravine, in search of the grim monster. The old Boer repeatedly called to his incautious son to come back, and wait for the Hottentots and the dogs, which would soon come up; but, finding his advice disregarded, he left the horse, and walked towards his son, whom he found throwing stones into the long grass which fringed the edge of the ravine for the purpose of starting the Lion. When the Boer was about a hundred yards from the lad, he saw him stop, raise his gun, and fire suddenly, though apparently without aim, and then turn, running a few paces towards him. At the same instant, he saw the Lion make two prodigious bounds, and alight on his boy, whom he instantly dragged to the ground.

All this occurred in a very few seconds ; so that before the Boer, who ran to the rescue, arrived, the young Dutchman was mortally wounded. The Lion, crouching down among the long grass, retreated a few yards, then bounded over the rocks and reeds until out of sight, the shot which was fired by the old Boer being unheeded by him. When the father reached the fatal spot, he found his son senseless, and torn so fearfully as to preclude all possibility of recovery. He, however, had him conveyed home, but the lad never again spoke, and died during the night. Revenge was the first thought of the old Dutchman, who immediately sent round to his neighbours to warn them that a Lion was in their vicinity, and to beg their assistance on the following day in tracing the Lion to its den.

The night was passed by the Boer as usual ; for these men are very philosophic, and rarely allow any circumstance to interfere with their comfort. On the following morning, however, he was up very early, busily preparing for the great business of the day ; bullets were being cast and powder-horn filled, &c. &c., when he was suddenly interrupted by the entrance of his little Bushman, who had, since his capture by the Boer some years before, reached his full growth, and might be estimated at any age between sixteen and sixty.

'What do you here ?' asked the Dutchman.

The Bushman, who was armed with his tiny bow and arrows, answered by showing a small tuft of black hair like a shaving brush.

This was an intelligible answer to the Boer, who, with eagerness, demanded the particulars ; and the following is a translation of the Bushman's account.

When the Lion struck down the young Dutchman, the Bushman was sitting upon a rock which commanded a view of the scene. The little creature then watched the Lion in its retreat, and marked it down amongst some long grass and bushes at the distance of a mile or so. He then procured an old and nearly useless ox from the cattle kraal, and, arming himself with his bow and poisonous arrows, drove the beast close to the Lion's retreat, made it fast to a bush, and concealed himself in some long grass.

The Bushman, from his nocturnal habits, can see by night nearly as well as by day ; and so, when, shortly after dark, the Lion left his lair and walked on to the open plain outside, the Bushman was an attentive observer of his movements.

The ox soon attracted the attention of the Lion, which approached with caution upon its victim ; the Bushman at the same time holding his bow and arrows in readiness for an attack upon *his* victim. Soon the Lion sprang upon the ox, and, at the instant when he was engaged in the death struggle, the Bushman, with great rapidity, twice twanged his bow, and lodged two poisoned barbs in the Lion's flesh.

The ox was soon overcome, and was dragged amongst the reeds, whilst the Bushman sought shelter in the crannies of the rocks near the scene of his operations.

As soon as day began to dawn, the Bushman commenced his stealthy approach, through the grass and reeds, towards the Lion's lair, and was shortly sitting grinning on the carcass of the Lion, which, but a few hours before, was a terror to all the Hottentots on the farm, but now, overcome by the malignant poison with which the arrows had been prepared, was as harmless as one of the stones on which he lay.

Being anxious to proclaim his triumph, the Bushman merely cut off the tuft of hair from the tail of the Lion and returned with this trophy to the Dutchman, who was not, however, quite satisfied with the business, for he would have preferred to shoot the Lion himself ; moreover, he grudged the loss of the old ox, which he thought might have been spared to die the usual death of a draught ox, *i.e.* to work until it drops from fatigue, and to die where it falls. The Bushman, however, explained that, if he had wounded the Lion as it was walking along, it would have sprung upon him as soon as it felt the sharp arrow in its side ; but, when it was busily employed in killing the ox, it would only think that the ox had pricked it with its horns, and would neither see nor think of its human enemy.

Therefore it was safer and more certain to take the ox for a bait, and so, to save many a young and vigorous animal by the sacrifice of one old and worn-out beast."

"A Boer, a very humorous fellow, told me that he was returning to his waggons one evening, when he was far in the interior ; at the time, he had with him only the single charge of powder with which his gun was loaded, as he had been out buck-shooting all day.

Straight in his path he disturbed a Lion, which jumped up and turned to look at him. Very naturally, his first impulse was to fire, but remembering that he had but that one charge in his gun, he changed his tactics.

The Dutchmen usually wear large broad-brimmed felt hats, around which several ostrich feathers are fastened. The Boer jumped from his horse, and pulled off his hat, which he held with his teeth by the brim, so that the upper part only of his face could be seen above the conglomeration of feathers. He then dropped upon his hands and knees, and commenced crawling towards the Lion. Such a strange animal had never before been seen by the astonished *Leeuw*, which turned and fled without a moment's hesitation.

This method of alarming animals is not always successful ; for whilst I was on the frontier, a Hottentot, who had been told of a somewhat similar plan to frighten a savage ox, met with a severe accident.

The man had been instructed that to stoop down and look back at an animal from between the knees was a certain means of driving it away. So, being pursued by an infuriated ox, he stopped short, and doubled himself up for his peep ; but unfortunately without the desired result. For the animal charged home, ripped up the Totty's leather crackers, wounded him, and sent him sprawling into a bush."

"An old Dutch Boer, who lived under the shadow of the Draakensberg mountains, gave me the following account of an interview with a Lion. The man was a well-known sportsman, and lived principally by means of the dollars which he realized upon ivory and skins. He was accustomed to make a trip each year into the game country, and traded with the Kaffirs or other inhabitants, under very favourable auspices. His stock-in-trade consisted of his guns and ammunition, several spans of fine oxen, some horses, and about a dozen dogs.

A Lion, which appeared to have been roaming about the country, happened to pass near this Boer's location, and scenting the three coursers kept by the Boer, thought that the locality might suit him for a short period. A dense kloof, situated about a mile from the farm, afforded both shelter and water, and this spot the Lion selected as a favourable position for his head-quarters.

The Boer had not to wait for more than a day, before the suspicions which had been excited in his mind by some broad footmarks which he saw imprinted in the soil, were confirmed into a certainty that a full-grown Lion had passed near his residence.

It now became a question of policy, whether the Boer should attack the Lion, or wait for the Lion to attack him. He thought it quite possible, that *Leeuw*, having been warned off by the dogs, whose barking had been furious and continued during the night on which the Lion was supposed to have passed the farm, might think discretion to be the better part of valour, and consequently would move farther on, in search of a less carefully guarded locality upon which to quarter himself. He determined, therefore, to wait, but to use every precaution against a night surprise.

The Lion, however, was more than a match for the Boer. For during the second night, Roeberg, the stout after-ox of the pet span, was quietly carried off, and although there was some commotion amongst the dogs and cattle, it was supposed that the alarm had scared the Lion, which had then decamped.

The morning light, however, showed that the poacher had leapt the palisade which surrounded the kraal, and having killed the ox, had evidently endeavoured to scramble over it again, with the ox in his possession. The joint weight of the Lion and ox had caused the stakes to give way, and an exit had then been easily effected.

The spoor of the Lion was immediately followed by the Boer, who took with him a Hottentot and half-a-dozen of his best dogs. The traces were easily seen, and the hunters had no difficulty in deciding that the Lion was in the kloof. But this in itself was no great advance, for the kloof was about a mile in length, and three or four hundred yards in breadth; and the cover was composed of wait-a-bit thorns, creepers, and long grass, forming a jungle so thick and impenetrable, that for a man to enter appeared almost impossible.

It was therefore agreed that the Boer should station himself on one side, whilst the Hottentot went to the other side of the kloof, and that the dogs should be sent into the cover. This arrangement, it was hoped, would enable either the Dutchman or Hottentot to obtain a shot; for each concluded that the dogs, which were very courageous animals, would drive the Lion out of the kloof, and that it would, upon breaking cover, afford one or the other a good chance.

The excited barks of the dogs soon indicated that they had discovered the Lion, but they appeared to be unable to drive him from his stronghold: for although they would scamper away every now and again, as though the enraged monster were charging them, still they returned to bay at the same spot.

Both of the hunters fired several shots, upon the hope that a stray bullet might find its way through the underwood to the heart of the savage. But a great quantity of ammunition was expended, and no result achieved.

At length, as the dogs had almost ceased to bark, it was considered advisable to call them off. But all the whistling and shouting failed in recalling more than two out of the six, and one of these was fearfully maimed. The others, it was afterwards found, had been disposed of by the Lion in the most unceremonious manner; a blow from his paw had sufficed either to break the back or smash the skull of the nearest intruder.

It thus happens that the bravest dogs are not always the best adapted for Lion or buffalo hunting. A cur is, perhaps, the most suitable; for while a courageous dog will boldly face a Lion, and even venture within reach of his deadly stroke, and thus soon be "expended," a cur will continue to annoy and occupy the attention of the fierce game, but at the same time will take good care of its own safety. It is not expected that a dog is to struggle with either a Lion or a buffalo; its duty is merely to distract the animal, and prevent it from devoting too much of its time to the hunter. Well-bred dogs are nearly useless when employed against dangerous game.

This, the first attempt on the Lion, was a total failure, and the Boer returned home to lament the loss of his dogs, and to refresh himself after his exertions. During the night, he watched beside his kraal, but the Lion did not pay him a second visit.

Early on the following evening, he, accompanied by his Hottentot, started afresh for the kloof, and having marked the spot from which the Lion had on the former occasion quitted the dense thorny jungle, the two hunters ascended a tree, and watched during the whole night for a glimpse at their purposed victim. But whilst they were paying the residence of the Lion a visit, *he* favoured the farm with a call, and this time, by way of variety, carried away a very valuable horse, which he conveyed to the kloof, having been wise enough to walk out and return by a different path to that which he used on the former occasion. Consequently he had avoided the ambush which had been prepared for him.

When the Boer returned to his farm, he became furious at his new loss, abused the Totties and Kaffirs for their neglect and cowardice, but soon became reasonable, and determined on a plan which, although dangerous, was still the one which appeared the most likely to insure the destruction of this ravenous monster. This plan was to enter the dense kloof on foot, without dogs, and to endeavour by fair stalking to obtain his shot at the Lion.

Now, when we consider the difficulty of moving through any cover without making a noise, and also the watchful habits of every member of the feline race, we may be certain that to surprise the Lion was a matter of extreme difficulty, and that the probability was that the Dutchman would meet with a disaster.

At about ten o'clock on the morning after the horse slaughter, the Boer started for the kloof, armed with a double-barrelled smooth bore, and clothed in the most approved bush costume. He would not allow his faithful Hottentot to accompany him, because, as success mainly depended upon surprise, he considered that the highly flavoured Totty might be scented by the Lion; whereas he alone would be more likely to escape detection. By this arrangement the Boer demonstrated the truth of the proverb with reference to the pot and the kettle, for the Dutchmen are not fonder of lavations than their Hottentot servants, and it is probable that, although a wide-awake Lion might have scented the Totty at 600 yards down wind, he would have discovered the Boer under similar conditions at 400 yards. We must, however, take the Boer's reason as a just one, and conclude that to leave his Totty at home was a wise precaution.

On the first occasion, when the Lion was attacked by the Boer, it had been bayed by the dogs near some tall trees, far down in the kloof. If the animal had again selected the same location, the Boer would have had to creep through two or three hundred yards of thorny bush, and he would probably have alarmed the Lion long before he arrived within shot. He had thought over this, and had concluded that after dragging the carcass of the horse all the way from the farm, the Lion would not be disposed to drag it very far through the underwood in the kloof, and that, therefore, he should find the carcass of the horse at least at no great distance from the edge of the ravine, and probably the Lion close to it.

Now it is the nature of the Lion, when gorged, to sleep during the day; and if the animal has carried off any prey, it usually conceals itself near the remains to watch them until it is ready for another feast.

The Boer was aware of all this, and had laid his plans very judiciously. He approached the kloof slowly and silently, hit off the spoor of the Lion, and traced the spot where the horse had been allowed to remain on the ground for a short time.

Although he moved onwards very slowly and with great caution, he was soon surrounded by the bush; and the brightness of the plain was succeeded by the gloom of the kloof. Being a most experienced hand at bushcraft, he was enabled to walk or crawl without causing either a dried stick to crack or a leaf to rustle, and he was aware that his progress had been accomplished without noise; for the small birds, usually so watchful and so much on the alert, flew away only when he approached close to them, thus showing that their eyes and not their ears had made them conscious of the presence of man.

Birds and monkeys are the great obstacles in the bush to the success of a surprise, for the birds fly from tree to tree, and whistle or twitter, whilst the monkeys chatter and grimace, and express, by all sorts of harlequin movements, that some curious creature is approaching. When, therefore, the bushranger finds that birds and monkeys are unconscious of his presence until they see him, he may be satisfied that he has traversed the bush with tolerable silence, and has vanquished such formidable obstacles as sticks hidden by leaves, broken and dead branches, &c.

There is a vast difference between hearing or reading how any dangerous work has been accomplished, and doing that work itself. But we can, by imagining ourselves in the position of the performer, realize in a measure the sort of sensations which he must have experienced, and we can then weigh the effect which the circumstance would have produced upon our own moderately strong nerves. It is highly probable that those who sigh for new sensations, might possibly find them were they to enter a dense bush on foot, and expect momentarily to meet, within speaking distance, a Lion of capacious maw, or a long-tusked, heavy-footed elephant, or even such a moderate opponent as a bull buffalo.

The effect produced upon the system is much decreased when many individuals are together. To obtain the most satisfactory results, therefore, a person should undertake the journey alone, and he will soon learn to consider those only as epicures who thus conjointly enjoy solitude and excitement.

The Boer had penetrated scarcely fifty yards into the bush, when he had reason to suspect that he was close upon the lair of the Lion. He believed that such was a fact in consequence of the strong leonine scent, and from a part of the carcass of the horse being visible between the intervening branches. Instead, therefore, of advancing, as an incautious or inexperienced bushranger would have done, he crouched down behind a bush, and assumed a convenient attitude, so that he could remain still without inconvenience.

All the animal creation are aware of the advantages of a surprise, and the feline tribe especially practise the ambuscading system. The Boer therefore determined, if possible, to turn the tables on the Lion, and to surprise, rather than to be surprised. He concluded that the Lion, even when gorged with horse-flesh, would not be so neglectful of his safety as to sleep with more than one eye at a time, and that, although he had walked with great care through the bush, he had probably caused the Lion to be watchful; if, therefore, he should go up to the carcass of the horse, he might be pounced upon at once.

To sit down quietly within a few yards of a Lion, whose exact hiding-place was not known, required a certain amount of nerve; but the Boer knew what he was about, and had adopted the best and safest method to conquer his foe.

After remaining silent and watchful for several minutes, the Boer at length saw that an indistinctly outlined object was moving behind some large, broad-leafed plants, and at about twenty paces from him. This object proved to be the Lion, which was half-crouched behind some shrubs, and was attentively watching the bushes near the Boer. The head only was clearly visible, the body being concealed by the foliage.

It was evident that the Lion was aware that some person or thing had approached, but was not certain where this thing was now concealed. The Boer knew that this was a critical period for him, and therefore remained perfectly steady; he did not like to risk a shot at the forehead of the Lion, for it would require a very neat shot to insure a death wound, and the number of branches and twigs which were on the line of flight of the bullet would render a clear course almost impossible.

The Lion, after a careful inspection, appeared to be satisfied, and laid down behind the shrubs. The Boer then cocked both barrels of his heavy roer, and turned the muzzle slowly round, so that he covered the spot on which the Lion lay, and shifted his position so as to be well situated for a shot.

The slight noise which he made in moving attracted the attention of the Lion, who immediately rose to his feet. A broadside shot could not be obtained, so the Boer fired at a spot between the eyes; the bullet struck high, as is usually the case when the range is short and the charge of powder is heavy, but the Lion fell over on its back, rising, however, immediately, and uttering a fierce roar. As it regained its feet, it showed its side to the Boer, who sent his second bullet into its shoulder.

The Lion bounded off through the bush, much to the satisfaction of the Boer, who felt more calm as each snap of a branch showed that the animal was farther from him.

The Boer immediately started off home, and brought his Hottentots and dogs to assist in the search after the wounded animal, which the Boer concluded would be found dead, as the second wound, he thought, must be a mortal one.

Before sunset that evening, the skin of the Lion was pegged down outside the Boer's house, and the Hottentots were drunk with delight at the success of 'the master.'"

KOLBEN, a traveller who visited the Cape about the year 1705, described the appearance and character of the African Lion. He gives a rule by which all travellers may know to a certainty the state of mind in which *Leeuw* may be. He, however, does not mention whether he actually tested the truth of his assertions, but merely states as follows:—

" The Lions here are remarkable for their strength. When they come upon their prey they knock it down, and never bite till they have given the mortal blow, which is generally accompanied by a fearful roar. When the Lion is pinched with hunger, he shakes his mane and lashes his sides with his tail. When he is thus agitated it is almost certain death to come in his way, and as he generally lurks for his prey behind the bushes, travellers sometimes do not discover the motion of his tail till it is too late; but if a Lion shakes not his mane, nor lashes himself with his tail, a traveller may pass safely by him.

If we could drive a bargain with the Felis Leo that he should always thus signal to travellers, we might pass through the African wilderness with less risk than at the present time. But from the experience gained by more modern hunters, it appears that the Lion will frequently attack horses, oxen, &c., without any intimation from mane or tail.

The most formidable attacks are those which take place during a dark night, when it would be impossible to be prepared in consequence of not observing the shaking and lashing above referred to."

It has already been mentioned, that several naturalists accept the Lion of Western

GAMBIAN LION.—*Leo Gambianus.*

Africa as a species distinct from the Lion of Southern Africa, and have therefore given to the animal a different specific name, which is derived from the country in which it is found. Whatever may be said of the distinction between the Asiatic and African Lion, there seem to be scarcely sufficient grounds for considering the very slight differences which are found in Lions of Africa to be a sufficient warrant for constituting separate species. They may be permanent varieties, and even in that case are not nearly so different from each other as the mastiff from the spaniel.

From all accounts, however, it seems that the habits of all Lions are very similar, and that a Lion acts like a Lion, whether he resides in Africa or Asia.

We all are familiar with the self-gratulatory half-threatening mixture between a purr and a growl, which is emitted by the domestic cat when she has laid her paws on a mouse or a bird, and is divided in mind between the complacent consciousness of having won a prize by her own efforts, and the ever present fear that it should escape or be taken away. If we substitute a Lion for a cat, and suppose ourselves to be in the position of the victim, we may partly realize the feeling which must have filled the mind of a recent traveller and hunter in Southern Africa.

He had built for himself a "skärm," or slight rifle-pit, composed of stones, logs, and other convenient substances, and had watched during the night in hopes of finding game worthy the sacrifice of time and sleep. Nothing, however, had come within range of the concealed hunter excepting a white rhinoceros, which was shot, and fell dead on the spot. Wearied out with the prolonged vigil, the hunter dropped asleep, and lay for some time wrapped in unconsciousness.

But the active desert life requires that its votary should be ever prepared for any emergency, and even during sleep should be capable of instantaneous awaking ready for action. So it happened, that although the deep sleep of wearied nature had wrapped the hunter's senses in oblivion, a part of his being remained awake, ready to give the alarm to that portion which slept. Suddenly a sense of danger crept over the sleeper, and he awoke to a feeling that a monotonous rumbling sound, which reverberated in his ears, was in some way connected with imminent peril. A moment's reflection told him that none but a Lion could produce such sounds, and that one of those fearful animals was actually stooping over him, its breath playing on his face.

Taught by practical experience of the danger of alarming the Lion, the hunter quietly felt for his gun, which was lying ready loaded and cocked in front of him, and raised himself in order to get a glimpse at the foe. Slight as the movement was, it sufficed to alarm the lion, which uttered a sharp, menacing growl, speaking in a language well known to the intended victim. Knowing that not a moment could be lost, he pointed his weapon towards an indistinct mass, which loomed darkly through the mists of night, and fired.

The report of the gun was instantly mingled with the fierce roarings of the infuriated Lion, maddened with the pain of its wound, seeking to wreak its vengeance on its foe, and tearing up the ground in its fury, within a very few paces of the skärm. By degrees the fierce roars subsided into angry growls, and the growls into heavy moans, until the terrible voice was hushed, and silence reigned during the remainder of the night.

When the dawn broke, the hunter ventured from his place of concealment, and searched for the carcass of the Lion, which he found lying within fifty yards of the spot from whence the fatal shot had been fired. Even in that short space of time the hyænas and jackals had been busy over the body of their departed monarch, and had so torn his skin that it was entirely spoiled for any purpose except that of a memorial of a most fearful night.

The hero of this adventure was C. J. Andersson, who has recorded his valuable African experiences in his visit to " Lake Ngami."

The same author relates a curious anecdote of a half-starved, and entirely bewildered Lion, which contrived to get into the church at Richterfeldt. The unfortunate brute was so weakened by fasting, that the Damaras dragged him out of the edifice by his tail and ears, and speared him without trouble.

In the leonine character is no small craft, which displays itself in various modes. Keen of scent in perceiving the approach of an enemy, the Lion appears to be well aware of the likelihood that his own approach might be manifested by the powerful odour that issues from his body. He therefore keeps well to leeward of the animal which he pursues, and employs the direction of the wind to conceal him from the olfactory senses of his game, and the position of the rocks, trees, or reeds, to hide his approach from their organs of vision.

A curious property connected with the Lion's tooth is worthy of notice. It has happened that, when a man has been bitten by a Lion, and escaped from its fangs, he has long felt the after effects of the injury, and this in a singular manner. Although the wound has healed kindly, and to all appearance has left no evil result except the honourable scar, yet that wound has broken out afresh on the anniversary of the time when it was inflicted. There is probably some poisonous influence upon the Lion's tooth by which this effect is produced, for it has been recorded that two men have been attacked by the same Lion, one of whom, who was bitten upon his bare limb, suffered from the annual affliction, while the other, whose limb was protected by his coat, felt no after inconvenience of a similar nature from the bite of the same animal.

A similar effect, lasting for several years, has been produced by the bite of a rabid dog, where the poisonous effects of the envenomed tooth were not sufficiently powerful to produce the fearful disease of hydrophobia. In an instance with which I am acquainted, the wound continued to re-open annually at least for the space of six years, and possibly for some years longer. The bite of a venomous snake has sometimes been known to produce the same phenomenon.

The Lion is by no means so fastidious a feeder as is popularly supposed. It is true that he does very much like to strike down a living prey, and lap the hot blood as it wells from the lacerated victim. But he is very well satisfied with any dead animal that he may chance

I. L

to find, and indeed is in no way particular whether it be tainted or otherwise. So thoroughly is this the case, that Lion-hunters are in the habit of decoying their mighty game by means of dead antelopes or oxen, which they lay near some water-spring, knowing well that the Lions are sure to seize so excellent an opportunity of satisfying at the same time the kindred appetites of thirst and hunger.

In default of larger game, the Lion feels no hesitation in employing his mighty paw in the immolation of the small rodents, and frequently makes a meal on locusts, diversified with an occasional lizard or beetle. Led by implanted instinct, this animal will, when water is not to be found, quench its thirst by devouring the juicy water-melons that so marvellously store up the casual moistures of the desert, which would otherwise be exhaled in vapour before the fierce rays of the burning sun. Many other carnivorous animals, and one or two carnivorous birds, are known to possess the same instinctive knowledge. The scientific name of this water-melon is "Cucumis Caffer," and its native title "Kengwe," or "Kēme."

That a carnivorous animal should voluntarily take to vegetable food is a very curious fact, and seems to argue a high state of intellectual power. It is true that herbivorous animals, such as the rhinoceros and others, will resort to the same plant for the purpose of quenching their thirst ; but then it must be remembered that these latter creatures are but following their usual dietary system, while the Lion is acting in a manner directly opposed to his own flesh-loving nature.

The cautious habits which the Lion acquires when its domain has been invaded by man are most singular, and exhibit a considerable degree of reasoning power. The Lion which has never known man, knows no fear at the sight of man and his deadly weapons, attacking him with as much freedom as it would attack an antelope. But after it has had some experience of man and his wiles, it can only be induced by the calls of pressing hunger to venture upon an open attack, or to approach any object that looks as if it might be a trap.

Lions have been known to surround an escaped horse, and to prowl round it for two entire days, not daring to attack so apparently defenceless a prey, simply because its bridle was dangling from its neck, and made the creatures suspicious, even though the rein had accidentally been hitched over a stump. On another occasion, a Lion crept close to a haltered ox, saw the halter, and did not like it, crept away again until he reached a little hillock about three hundred yards away, and there stood and roared all night.

The hunters take advantage of this extreme caution to preserve the game which they have killed from any marauding Lion that may happen to pass in that direction. A simple white streamer tied to a stick, and waving over the dead beast, is amply sufficient to prevent the Lions from approaching so uncanny an object. Sometimes, when no streamer can be manufactured, a kind of clapper is substituted, which shakes in the wind, and by the unaccustomed sound, very much alarms the Lion. It does truly seem absurd, that so terrible a beast as the Lion should be frightened by the fluttering of a white handkerchief, or the clattering of two sticks—devices which would be laughed to scorn by a tomtit of ordinary capacity.

Nearly all the feline animals seize their prey by the back of the neck, but the Lion seems to prefer the flank or shoulders as his point of attack. It seldom happens that the Lion springs upon the back of his prey, as is the case with many of the felidæ, for in the chase of a large animal, he chooses rather to pull down the doomed creature by main strength, his hinder feet resting on the earth, and his fore-paws and fangs tearing deeply into the neck and shoulders of his victim. There are, of course, exceptional instances, but the general rule seems to be that the Lion either strikes down his prey with a furious blow of his paw, or drags it to the ground by hanging on its neck with teeth and claws.

The young of the Lion are various in number, sometimes amounting to three or four at a birth, thus entirely contraverting the well-known fable of the Lioness and Fox. For some time, the young Lion cubs present a curious appearance, their fur being faintly brindled in a manner very similar to that of the tiger, or, to give a more familiar illustration, resembling the coat of a tabby cat, very indistinctly marked upon a light tawny ground. These faint brindlings are retained for some months, when they gradually fade into the deeper brown which tinges the tawny fur, and after awhile become wholly merged in the darker hue. I have observed a similar absorption of the brindled markings in a kitten. In its earliest youth, it was of a lightish brown, marked with tolerably defined stripes ; but as it grew older, the dark streaks gradually became more faint, and, when the animal was about three months old, vanished entirely.

A cub-lion is just as playful an animal as a kitten, and is just as ready to romp with any one who may encourage its little wanton humours. Only it is hardly so safe a playfellow, for the very small Lion is as large as a very big cat, and sometimes becomes rather unpleasantly rough in its gamesomeness. It has no idea of the power of its stroke, and if it should deal a playful blow with its claws protruded, is apt to do damage which it never intended.

The weight of a Lion-cub is extraordinary in comparison with its size. I have personally tested the weight of several cubs, and was surprised at the massive build of the little creatures. Their bones are very large, and the muscular system very solid, so that a cub which about equals a large cat in actual measurement, far exceeds that animal in weight.

The development of the young Lion is very slow, three or four years elapsing before he can lay claim to the full honours of Lionhood, and shake his tawny mane in conscious strength.

At the tip of the Lion's tail is sometimes found a curious appendage, which was once thought to be a veritable claw, and to be used for the purpose of exciting the Lion to rage, when he lashed his sides with his tail. It is now, however, proved to be nothing but a piece of thickened skin, which is only slightly attached by its base to the member on which it rests, and falls off at a very gentle touch. A similar protuberance has been discovered on the tail of an Asiatic leopard.

Before bidding farewell to the African Lion, it is but right to refer to the species or variety which inhabits the more northern portion of this huge continent. According to the account of Jules Gerard, the French lion-hunter, the Northern Lion is far more formidable an antagonist than his Southern relative. But to an unprejudiced reader, the spirited narratives which are given in the name of that author seem rather to bear reference to the singular cowardice of the native Algerian mind when brought in contact with the Lion, than the absolute ferocity of the animal, or the courage of the hunter.

To take but one instance.

That a large party of warriors, each armed with loaded musket, should stand in a row with their backs against a rock, trembling in deadly fear, whilst a Lion walked coolly along the line, with tail erect, in calm defiance of the firelocks that waved their faltering muzzles before his gaze like ears of corn before the gale, speaks but little for the courage of the warriors, and, in consequence, for that of their impudent foe.

It is true, also, that the North African Lion is a terribly fearful opponent on a dark night, when he is met face to face, with but a few yards between his body and the rifle-muzzle of the hunter ; but so is the Lion of Southern Africa, in similar circumstances. All animals, like dogs, "bark best on their own threshold," and it behoves a man, who dares alone to make his nocturnal quest after the Lion, to bear a bold heart, a quick eye, and a ready hand. Yet these accomplishments are far more general than some writers would have us suppose, and there is many an unassuming hunter who sallies out at night and shoots a Lion or two without thinking that the beast was so inordinately ferocious, or himself so marvellously courageous.

There is really nothing in the character or history of the Lion of Algeria that could separate him from the Lion of Southern Africa.

As far as is known, the Lion which inhabits Asia is very similar in habits to that which is found in Africa, and therefore needs no detailed description. There is, however, one species, or variety, which ought to be noticed, on account of the peculiarity from which

MANELESS LION.—*Leo Goojrattensis.*

it derives its name. This is the "MANELESS LION" of Guzerat, so called from the very meagre mane with which its neck and throat are decorated.

When first this animal was brought before the notice of naturalists, it was supposed to be merely a young male, whose mane had not yet reached its full development. It is now, however, allowed to be either a distinct species, or, at all events, a permanent variety,—between which opinions there is such very trifling difference that one is nearly as decided as the other.

As may be seen from the engraving, the mane is not altogether absent, as the popular name might give cause to suppose, but is very trifling in comparison with the luxuriant mass of hair which droops over the shoulders of the African Lion. The limbs do not appear to be quite so long in proportion as those of the last-named animal, and the tail is shorter, with a more conspicuous tuft. This tuft, by the way, is the readiest point of distinction which separates the Lion from the other cats.

The natives term this animal the "Camel-tiger," because its uniform tawny fur bears some resemblance in tint to that of the camel.

That one animal should support its own life only by the destruction of another creature, appears to be rather a cruel disposition of nature, and repugnant to the beauty and kindness which prevails in the order of created things. Averse as are we, the created beings, to inflicting pain on any of our fellow-creatures, it cannot but seem strange that the Creator should have made so many animals to suffer a violent death, and apparently to endure torturing pangs, by the lacerations to which they are subjected by their destroyers. The reflection is a just one, and one which until late years has never received a worthy answer. Endeavours were made to reconcile the Divine love with this apparent cruelty, by asserting that the lower animals were endued with so low a sense of pain that an injury which would inflict severest torture on a man, would cause but a slight pang to the animal. Yet, as all animals are clearly sensitive to pain, and many of them are known to feel it acutely, this argument has but trifling weight. Moreover, the system which was insensible to pain would be equally dull to enjoyment, and thus we should reduce the animal creation to a level but little higher than that of the vegetables.

The true answer is, that by some merciful and most marvellous provision, the mode of whose working is at present hidden, the sense of pain is driven out from the victim as soon as it is seized or struck by its destroyer. The first person who seems to have taken this view of the case was Livingstone, the well-known traveller, who learned the lesson by personal experience. After describing an attack made upon a Lion he proceeds:—

" Starting and looking half round, I saw the Lion just in the act of springing on me. I was upon a little height; he caught my shoulder as he sprang, and we both came to the ground below together. Growling horribly close to my ear, he shook me as a terrier-dog does a rat. The shock produced a stupor similar to that which seems to be felt by a mouse after the first shake of the cat. It causes a sort of dreaminess, *in which there was no sense of pain or feeling of terror*, though I was quite conscious of all that was happening. It was like what patients, partially under the influence of chloroform, describe, who see all the operation, but feel not the knife. This singular condition was not the result of any mental process. The shake annihilated fear, and allowed no sense of horror in looking round at the beast. This peculiar state is probably produced in all animals killed by the carnivora ; and, if so, is a merciful provision by our benevolent Creator for lessening the pain of death."

This fearful experience is, although most valuable, not a solitary one, and is made more valuable by that very fact. I am acquainted with a similar story of an officer of the Indian army, a German nobleman by birth, who, while in Bengal, was seized and carried away by a tiger. He described the whole scene in much the same language as that of Livingstone, saying that as far as the bodily senses were concerned, the chief sensation was that of a pleasant drowsiness, rather admixed with curiosity as to the manner in which the brute was going to eat him. Only by his reasoning powers, which remained unshaken, could he feel that his position was one of almost hopeless danger, and that he ought to attempt escape. Perhaps, in so sudden and overwhelming a shock, the mind may be startled for a time from its hold upon the nerves, and be, so to speak, not at home to receive any impression from the nervous system. Many men have fallen into the jaws of these fearful beasts, but very few have survived to tell their tale. In the case of Livingstone, rescue came through the hands of a Hottentot servant, who fired upon the Lion, and who was himself attacked by the infuriated animal. In the latter instance, the intended victim owed his life to a sudden whim of the tiger, which, after carrying him for some distance, threw him down, and went off without him. The officer used playfully to attribute his escape to his meagre and fleshless condition, which, as he said, induced the epicurean tiger to reject a dinner on so lean and tough an animal as himself.

Those who have been in action are familiar with the indifference with which the severest wounds are received. There is one well-known instance of this apparent insensibility to pain, which occurred in the Crimean war. An officer was stooping to light his pipe at a camp-fire, when an enemy's shell plumped into the midst of the embers, and exploded, knocking the pipe out of his hands. He uttered an exclamation of annoyance at the loss of his pipe, unconscious that the fragments of shell had carried off several of his fingers and frightfully shattered other portions of his limbs. Even in cases of natural death a similar phenomenon occurs, and those who have expressed, in their last illness, the most utter terror of death, meet their dreaded fate with calm content, welcoming death as a friend instead of fearing him as a foe.

Upon the African continent, the lion reigns supreme, sole monarch over the feline race. But in Asia his claims to undivided royalty are disputed by the TIGER, an animal which equals the lion in size, strength, and activity, and certainly excels him in the elegance of its form, the grace of its movements, and the beauty of its fur. The range of the Tiger is not so widely spread as that of the lion, for it is never found in any portions of the New World, nor in Africa, and, except in certain districts, is but rarely seen even in the countries where it takes up its residence. Some portions of country there are, which are absolutely infested by this fierce animal, whose very appearance is sufficient to throw the natives into a state of abject terror.

In its colour the Tiger presents a most beautiful arrangement of markings and contrast of tints. On a bright tawny yellow ground, sundry dark stripes are placed, arranged, as may be seen by the engraving, nearly at right angles with the body or limbs. Some of these stripes are double, but the greater number are single dark streaks. The under parts of the body, the chest, throat, and the long hair which tufts each side of the face, are almost white, and upon these parts the stripes become very obscure, fading gradually into the light tint of the fur. The tail is of a whiter hue than the upper portions of the body, and is decorated in like manner with dark rings.

So brilliantly adorned an animal would appear to be very conspicuous among even the trees and bushes, and to thrust itself boldly upon the view. But there is no animal that can hide itself more thoroughly than the Tiger, or which can walk through the underwood with less betrayal of its presence.

The vertical stripes of the body harmonize so well with the dry, dusky jungle grass among which this creature loves to dwell, that the grass and fur are hardly distinguishable from each other except by a quick and experienced eye. A Tiger may thus lie concealed so cleverly, that even when crouching among low and scanty vegetation, it may be almost trodden on without being seen. The step too, is so quiet and stealthy, that it gives no audible indication of the creature's whereabouts, and the Tiger has, besides, a curious habit of drawing in its breath and flattening its fur, so as to reduce its bulk as far as possible. When a Tiger thus slinks away from the hunters or from any dreaded danger, it looks a most contemptible and cowardly creature, hardly to be recognised in the fiery beast, which, when driven to bay, rushes, regardless of danger, with fierce yells of rage and bristling hair, upon the foremost foe.

When seeking its prey, it never appears to employ openly that active strength which would seem so sure to attain its end, but creeps stealthily towards the object, availing itself of every cover, until it can spring upon the destined victim. Like the lion, it has often been known to stalk an unconscious animal, crawling after it as it moves along, and following its steps in hopes of gaining a nearer approach. It has even been known to stalk human beings in this fashion, the Tiger in question being one of those terrible animals called "Man-eaters," on account of their destructive propensities. It is said that there is an outward change caused in the Tiger by the indulgence of this man-slaying habit, and that a "Man-eater" can be distinguished from any other Tiger by the darker tint of the skin, and a redness in the cornea of the eyes. Not even the Man-eating Tiger dares an open assault, but crawls insidiously towards his prey, preferring, as does the lion, the defenceless women and children as the object of attack, and leaving alone the men, who are seldom without arms.

The Tiger is very clever in selecting spots from whence it can watch the approach of its intended prey, itself being couched under the shade of foliage or behind the screen of some friendly rock. It is fond of lying in wait by the side of moderately frequented roads, more particularly choosing those spots where the shade is the deepest, and where water may be found at hand wherewith to quench the thirst that it always feels when consuming its prey. From such a point of vantage it will leap with terrible effect, seldom making above a single spring, and, as a rule, always being felt before it is seen or heard.

It is a curious fact that the Tiger generally takes up his post on the side of the road which is opposite his lair, so that he has no need to turn and drag his prey across the road, but proceeds forward with his acquisition to his den. Should the Tiger miss his leap, he generally seems bewildered and ashamed of himself, and instead of returning to

TIGER.—*Tigris Regális.*

the spot, for a second attempt, sneaks off discomfited from the scene of his humiliation. The spots where there is most danger of meeting a Tiger, are the crossings of nullahs, or the deep ravines through which the water-courses run. In these localities the Tiger is sure to find his two essentials, cover and water. So apathetic are the natives, and so audacious are the Tigers, that at some of these crossings a man or a bullock may be carried off daily, and yet no steps will be taken to avert the danger, with the exception of a few amulets suspended about the person. Sometimes the Tigers seem to take a panic, and make a general emigration, leaving, without any apparent reason, the spots which they had long infested, and making a sudden appearance in some locality where they had but seldom before been seen.

In the districts where these terrible animals take up their abode, an unexpected meeting with a Tiger is by no means an uncommon event. While engaged in hog-spearing, the sportsmen have many times come suddenly upon a Tiger that was lying quite composedly in the heavy "rhur" grass from which the hog had started. In such cases, the terror of the native horses is excessive, for their dread of the Tiger is so great, that the very scent of a Tiger's presence, or the sight of a dried skin, is sufficient to set them plunging and kicking in their attempts to escape from the dreaded propinquity. One horse, which had been terrified by a Tiger, could not afterwards endure the sight of any brindled animal whatever, and was only restored to ordinary courage by the ingenious device of his master, who kept a brindled dog in the same stable with the horse until the poor beast became reconciled to the abhorred striped fur.

A very curious introduction to a Tiger occurred to a gentleman who was engaged in deer shooting.

He had crept up to a convenient spot, from whence he could command a clear view of the deer, which were lying asleep in the deep grass; had taken aim at a fine buck which was only at twelve yards' distance, and was just going to draw the trigger, when his attention was roused by a strange object which was waving above the grass, a few feet on the other side of the deer. It was the tail of a Tiger, which had approached the deer from the opposite direction, and had singled out the very animal which was threatened by the rifle. Not exactly knowing what kind of an object it was that stirred the grass, the sportsman re-adjusted his piece, and was again going to fire, when a Tiger sprang from the cover of the "moonje" grass, and leaped upon the very buck which had been marked out as his own. Under the circumstances, he did not choose to dispute the matter, but retreated as quietly as possible, leaving the Tiger in possession of the field.

The deer was an Axis, or Spotted Deer, animals which are very common in some parts of India, and are much appreciated by Tigers as well as men. Peacocks also abound in the same districts; in short, wherever spotted deer and peacocks may be found, Tigers are sure to be at no great distance from them. On one occasion, another sportsman had wounded a peacock, which fluttered about for a time, and then fell into a little open space in the bushes. As these birds, when winged, can run too fast to be overtaken by a man, the sportsman ran after the bird in order to catch it as it fell, and on entering the little area found himself in the presence of three Tigers, which had been evidently asleep, but were just roused by the report of the gun, and were looking about them in a dreamy and bewildered manner. The peacock lay dead close to the Tigers, who probably made a light repast on the game thus unexpectedly laid before them, for the sportsman took to his heels, and did not feel himself safe until he was fairly on board of his vessel.

The chief weapons of the Tiger are his enormous feet, with their sharp sickle-like talons, which cut like so many knives when the animal delivers a blow with his powerful limbs. Even were the talons retracted, the simple stroke of that sledge-hammer paw is sufficient to strike to the ground as large an animal as an ox; while, if the claws lend their trenchant aid to the heavy blow of the limb, the terrible effects may be imagined.

Besides the severity of the wound which may be inflicted by so fearful a weapon, there are other means of destruction that lie hid in the Tiger's claws. From some cause or other,—it may be presumed on account of some peculiar manner in which the claws affect the nervous system,—even a trivial wound has often been known to produce lockjaw, and to destroy the victim by the effects of that fearful disease. It may be, that the perturbation of mind caused by the attack of the Tiger, may have some hand in the matter. Captain Williamson, an officer of twenty years' experience in Bengal, states that he never knew a person to die from the wounds inflicted by a Tiger's claws without suffering from lockjaw previous to death; and he adds, that those cases which appeared the least alarming were the most suddenly carried off.

Many modes are adopted of killing so fearful a pest as the Tiger, and some of these plans are very ingenious.

There is the usual spring-bow, which is placed in the animal's path, the bow drawn to the arrow's head, and a string leading from the trigger across the path in such a manner that the creature presses against it with its breast, discharges the weapon, and so receives the arrow in its heart.

The bow is set by fastening it to two strong posts set by the side of the Tiger's path, the string of the bow being parallel with the path. The string is then drawn back to its utmost limits, and a stick placed between the bow and the string, thus keeping the weapon bent. A long wedge is inserted between the stick and the bow, and the liberating cord tied to its projecting end. Lastly, the arrow is laid on the string, and the engine is ready for action. Of necessity, as soon as the Tiger presses the cord, the wedge is drawn away, the guarding stick drops, and the bow hurls its deadly missile. So rapidly does this simple contrivance act, that the Tiger is generally hit near the shoulder. The arrow is usually poisoned by means of a thread dipped in some deadly mixture, and wrapped round the arrow-point.

There is another plan, in which human aid is requisite, namely, by building a strong bamboo enclosure, in which the hunter lies, armed with a spear. At nightfall the Tiger

TIGER AND DEER.

comes prowling along, and smelling the man, rears up on its hind legs, trying to claw down the bamboo bars. The hunter in the meanwhile takes his spear, and mortally wounds the brindled foe, by striking the spear-point between the bars of the edifice.

A still more ingenious mode of Tiger killing is that which is employed by the natives of Oude.

They gather a number of the broad leaves of the *prauss* tree, which much resembles the sycamore, and having well besmeared them with a kind of bird lime, they strew them in the animal's way, taking care to lay them with the prepared side uppermost. Let a Tiger but put his paw on one of these innocent looking leaves, and his fate is settled. Finding the leaf stick to his paw, he shakes it, in order to rid himself of the nuisance, and finding that plan unsuccessful, he endeavours to attain his object by rubbing it against his face, thereby smearing the ropy birdlime over his nose and eyes, and gluing the eyelids together. By this time he has probably trodden upon several more of the treacherous leaves, and is bewildered with the novel inconvenience; then he rolls on the ground, and rubs his head and face on the earth, in his efforts to get free. By so doing, he only adds fresh birdlime to his head, body, and limbs, agglutinates his sleek fur together in unsightly tufts, and finishes by hoodwinking himself so thoroughly with leaves and birdlime, that he lies floundering on the ground, tearing up the earth with his claws, uttering howls of rage and dismay, and exhausted by the impotent struggles in which he has been so long engaged. These cries are a signal to the authors of his misery, who run to the spot, armed with guns, bows, and spears, and find no difficulty in despatching their blind and wearied foe.

Another mode of destroying the Tiger is by means of a strongly constructed trap, made on the same principle as the ordinary mousetraps, which take their victim by dropping a door over the entrance. The Tiger trap is little more than the mousetrap, only made on a much larger scale, and of strong wooden bars instead of iron wires. The bait is generally a pariah dog, or a young goat, both of which animals give vent to their anxiety by loud wailings, and so attract the prowling foe. In order to secure the living bait from being drawn out of the trap by the Tiger's claws, it is protected by an inner cage, to which the animal cannot gain access without dropping the door against his egress. This plan, however, is not very generally followed, as it possesses hardly sufficient elements of success.

A more productive plan—productive, because the reward for killing a Tiger, together with the sum for which the skin, claws, and teeth sell, is sufficient to keep a native for nearly a twelvemonth,—is, by digging a hole in the ground near a Tiger's haunt, putting a goat in the hole, and tethering it to a stake which is firmly driven into the centre of the little pit. A stone is then tied in one of the goat's ears, which cruel contrivance causes the poor animal to cry piteously, and so to call the attention of the Tiger. On hearing the goat cry, the Tiger comes stealthily to the spot, and tries to hook up the goat with his paw. Not succeeding, on account of the depth of the pit, he walks round and round, trying every now and then to secure the terrified goat, and thus exposing himself fairly to the hunters, who, quietly perched on a neighbouring tree, and taking a deliberate aim with their heavy firelocks, lay him dead on the spot of his intended depredation.

A somewhat similar, but more venturous mode of proceeding, is that which is adopted by the Shikarries, as these native hunters are called.

When a Tiger has carried off a bullock, or some such valuable animal, the shikarrie proceeds to the spot, and after waiting sufficient time for the robber to gorge himself, and become drowsy, he sets off in search of the murdered bullock; a dangerous task, but one which is much lightened by the indications afforded by vultures, jackals, and other carrion-loving creatures, which never fail to assemble round a dead animal, of whatever race it may be.

Having found the half-eaten carcase, and ascertained that the Tiger is fast asleep, the hunter calls together as many assistants as possible, and with their aid, rapidly builds a bamboo scaffold, some twenty feet high, and four feet wide, which is planted close to the spot where the dead and mangled bullock lies. On the summit of the scaffold the shikarrie mounts; his gun and ammunition are handed up to him by his companions, his sharp "tulwar," or sword, is hung ready to his grasp, and after offering their best wishes for success, the assistants take their leave, each putting in a claim for some part of the spoils. The claws are the most coveted portion of the animal, for the natives construct from two of these weapons a charm, which, on the homœopathic principle, is supposed to render the wearer invulnerable to attacks from similar weapons.

After a while, the Tiger wakes from the drowsy lethargy which was caused by repletion, and after shaking himself, and uttering a few yawns, which draw the attention of the watchful hunter, proceeds to his temporary station, for the purpose of making another meal on the remains of the slaughtered animal.

The shikarrie takes advantage of the opportunity, and resting his gun on the platform, takes a deliberate aim, and lodges a bullet—often an iron one—in the body of the Tiger. Generally the aim is so true that the Tiger falls dead, but it sometimes happens that the wound, although a mortal one, is not instantaneously fatal, and the animal springs furiously upon the foe who dealt the blow. The Tiger is no climber, but rage will often supply temporary ability; and so fiercely does the animal launch itself against the scaffolding, that if made of a softer material, permitting the hold of the Tiger's claws, the creature might reach the hunter; or that if not firmly planted, the whole edifice would be brought to the ground. But the smooth, hard surface of the bamboo affords little hold for the sharp talons; and, even if the animal should succeed in approaching the platform where the hunter sits, a blow from the razor-edged tulwar strikes off a paw, and the tiger falls helplessly to earth, only to meet its fate by a second bullet from the deadly firelock.

Attracted by the report of the hunter's gun, the neighbours flock to the spot, each man armed according to his ability; and if the beast is killed outright, join in a chorus of laudation towards the successful hunter, and of anger towards his victim, which may now be insulted with perfect impunity. Besides the ordinary trophies, which consist of the skin, claws, teeth, and the ordinary reminiscences of success, other portions of the Tiger are eagerly sought by the natives, the tongue and liver bearing the highest value. These organs are appropriated to the medical art, and after being chopped into little dice-like cubes, are prepared after some Esculapian and mysterious fashion, and thenceforward hold rank as remedies of the first order.

Another, though less gallant, mode of killing Tigers is by setting certain enormous nets, supported on stakes, so as to form an inclosure, into which the animal is partly enticed and partly driven.

The height of the stakes to which the nets are suspended is about thirteen feet ; so that, allowing for the droop at the upper portion of the toils, the nets are about eleven feet in height at their lowest point. It is, however, rather a stupid, and withal hazardous, mode of Tiger-hunting, and is not very often employed. It requires the aid of a very large body of men, and besides there is always a risk of inclosing some large animal, such as the buffalo or elephant, which rushes madly forward, and with the irresistible impetus of its huge body bears to the ground nets, stakes, and sentinels, leaving a wide path free for the remainder of the inclosed game to follow.

In order to induce the Tiger to leave its lair and to enter the toils, all possible means are used. Fires are lighted, burning torches are waved, guns are fired, drums are beaten, and, lastly, fireworks are largely employed. The most effective kind of firework is one which is made on the rocket principle, the tube which holds the fiery composition being of iron, and the "tail," or shaft, of bamboo. The rocket is held in the hand like a spear, and the fuse lighted. When it begins to fling out its burning contents, and to pull against the hand of the thrower, it is launched by hand, as if it were a spear, in the direction of the concealed quarry. An extremely powerful impulse is given by the burning composition, and the missile rushes furiously onward, scattering on every side its burden of fiery sparkles, hissing and roaring with a terrible sound, and striking right and left with its long wooden tail.

No Tiger can endure this fiery dragon which comes on with such fury, and accordingly the terrified animal dashes out of cover, and makes for the nearest place of concealment. But so artfully managed is the whole business that his only path of escape takes him among the nets, and, once there, his doom is certain. He cannot leap over the toils, because they are too high, nor break them down, because they are so arranged that they would only fall on him, and inclose him in their treacherous folds. Should he endeavour to climb over the rope fence, he exposes himself as a target for bullets and arrows innumerable ; and, if he yields the point, and tries to conceal himself as best he may, he only delays his fate for a time, falling a victim to the watchful enemies who start him from his last fortress, and, from the safe eminence of an elephant's back, or the branches of a tree, pour their leaden hail on the devoted victim.

This mode of hunting, as well as the more legitimate custom of following the Tiger into the jungle, while mounted on elephants, requires the aid of many men, elephants, and horses, and cannot be undertaken every day. There is, however, another method of killing this terrible beast, which, when employed by hunters who understand each other's plans, and can place the fullest reliance on their mutual courage and tact, is more destructive to the fierce quarry than even the netting system, with its mob of beasts and men.

Two, or at the most three, hunters set out on their campaign, accompanied by their chosen "beaters" and other servants, and start with the determination of bearding the Tiger in his den, unaided by horse or elephant. It is a bold plan, yet, like many bold plans, succeeds through its very audacity.

The object of the beaters is by no means to give assistance when a Tiger is started, because they always run away as soon as the brute shows itself ; but to make so astounding a noise that the Tiger cannot remain in the vicinity. When they reach a likely, or as it is termed, a "Tigerish" spot, they shout, they yell, they fire pistols, they rattle stones in metal pans, they beat drums, they ring bells, they blow horns, and, by their united endeavours, produce such horrible discord, that not even a Tiger dare face such a mass of men and noise. This precaution is absolutely necessary, for the Tiger loves to hide itself in as close a covert as it can find, and, unless driven from its place of refuge by such frightful sounds as have been mentioned, would lie closely crouched upon the ground, and either permit the hunters to pass by, or leap on them with a sudden spring, and so obtain a preliminary revenge of its own death.

A few bold and active beaters are sent forward as scouts, whose business is to climb trees, and, from that elevated position, to keep watch over the country, and detect the Tiger if it attempt to steal quietly away.

There is a certain bushy shrub, called the korinda, which is specially affected by the Tigers on account of the admirable cover which its branches afford. It does not grow to any great height, but its branches are thickly leaved, and droop over in such a manner that they form a dark arch of foliage, under which the animal may creep, and so lie hidden from prying eyes, and guarded from the unwelcome light and heat of the noon-day sun. So fond are the Tigers of this mode of concealment that the hunters always direct their steps to the korinda-bush, knowing well that if a Tiger should be in the neighbourhood, it would be tolerably certain to be lying under the sombre shade of the korinda branches.

As it is necessary that pedestrian hunters should on a pinch be able to conceal themselves from the sharp eyes of the Tiger, the colour of their dress is a matter of some importance. Experience shows that there is no tint so admirably suited for the purpose as that warm reddish-brown which is assumed by dried leaves. Clothed in this dress, the hunter can so easily amalgamate his person with the surrounding objects, that not even the Tiger's eyes can distinguish his form. A hunter has actually thus lain on a piece of rock while a Tiger prowled along within fifteen feet of the unsuspected foe, and passed on without detecting his presence. Even when a Tiger does see a human being thus attired, it becomes suspicious, and, not knowing what to make of the strange object, moves slowly away from the cause of alarm. This costume is especially useful among rocky ground, with which it assimilates most perfectly.

If a Tiger be fairly traced to its ordinary lair, the sportsmen prefer to lie in wait at some convenient point, and either to await the voluntary egress of the quarry, or to send in the beaters, and cause the animal to be driven out in the proper direction. When this mode is adopted, it is found best to have, besides those which are held in hand, a whole battery of guns, eight or ten in number, which are laid on the ground, ready loaded and cocked, their muzzles all pointing towards the spot where the Tiger is expected to make its appearance. The object of this expedient is two-fold: firstly, to make sure of the animal in case the first shots fail to tell mortally; and, secondly, to be in readiness should a second or even a third Tiger be driven from the bush. It is so usual an occurrence for two Tigers to make their sudden appearance where only one was expected to lie, that the precaution is an absolutely necessary one.

Contrary to the habits of most animals, which take the utmost care of their young, and in their defence will expose themselves to the direst peril, the mother Tiger is in the habit of making her young family her pioneers, and, when she suspects anything wrong, of sending them forward to clear the way. Knowing this curious propensity, the experienced hunter will not fire upon a cub that shows itself, for the mother will, in most cases, be waiting to see the result of her child's venture. Therefore, they permit the cub or cubs to pass with impunity, and reserve their ammunition for the benefit of the mother as she follows her offspring.

Should the Tiger not fall to the shot, but bound away, the hunters know whether the wound is a mortal one by inspecting the marks made in the ground by the feet of the retreating animal. It is a curious fact, that however hard a Tiger may be hit, yet, if the wound be not a rapidly mortal one, the claws are kept retracted, and the foot-prints show no mark of the talons. But should the injury be one which will shortly cause death, the Tiger flings out its limbs with the paws spread to their utmost, and at every leap tears up the ground with the protruded talons.

A Tiger has many weak points where a bullet does its work with great rapidity. The brain and heart are of course instantaneously mortal spots, and the lungs come next in order. The liver is a very dangerous organ to touch, and a Tiger, when there struck, rarely lives for more than fifteen or twenty minutes.

Perhaps of all animals the Tiger is one of the easiest to kill, although the wound may not be an instantaneous cause of death. Whether the cause may lie in the habits or diet of the creature is not certain, but true it is, that a wound inflicted on a Tiger very soon assumes an angry appearance, becomes tainted, and affords a resting-place for the pestilent blow-flies, which take such a hold of the poor beast, that even a slightly wounded Tiger has been known to die, not from the immediate effects of the injury, but from the devouring maggots which swarmed in and about the wound.

In tracking the wounded Tiger, the blood-spots that are flung from the agitated animal are of vast service. They are easily distinguishable, even though they dry instantaneously on touching the ground. As it dries, each blood-patch is surrounded by innumerable tiny ants, which seem to crowd to the spot as if they had been created for that sole purpose, and from their numbers make the gory traces more apparent. But these bloody tracks are by no means a necessary consequence of wounding a Tiger, which very often receives a deadly hurt, and yet spills no single drop of blood. The cause of this strange fact is the loose manner in which the skin lies over the body. It may therefore happen that, when the Tiger is in energetic movement, a portion of the skin which, when the animal is at rest, would be over the shoulder-blades, is shifted to quite another spot. If at that moment a bullet passes into the body of the creature, and checks its active movement, the skin slips back again to its usual position, so that the hole in the skin and that in the body no longer coincide ; thus preventing the external outflow of blood.

When the Tiger is killed, it is necessary to guard it in some way from the direct beams of the sun, or even from actual contact with objects which have been heated by its burning rays.

Should the creature fall on a tolerably cool spot, all that is necessary is to cover it with bushy branches, grass, and other foliage; but if the locality should be a hot one, as is generally the case, further precautions must be taken, by dragging the dead animal under the shelter of some shady tree or bushes. The reason for this cautious proceeding is, that the Tiger's flesh rapidly yields to putrefaction, and thus loosens the hair from the skin. So, however fatigued the hunter may be after he has succeeded in killing his prey, he dares not give way to repose until he has taken all the necessary precautions. Even ten or fifteen minutes under a hot sun is sufficient to bring off the hair in large patches, leaving the hide in a state perfectly unfit for use. Should the animal lie on a hot rock, the result will be the same.

After taking the skin from the dead Tiger,—which in itself is no easy task,—the next business is to preserve it in such a manner that it will dry uniformly without contracting into unsightly folds, without putrefaction, and without suffering from the teeth of the swarming ants and other insects, which are the plague of all taxidermists in hot countries.

For the latter object, sundry preparations are used, arsenical soap being that which is most generally known. It is, however, an exceedingly dangerous substance, requiring very great care in manipulation. A more harmless preparation is composed of a very strong solution of salt, alum, and powdered "cutch," in which the hide is steeped before being dried. In order to insure regularity of drying, the skin is laid on the ground with the fur downward, and fastened to the earth by a great number of wooden pegs, which are driven through its edges, fringing its entire outline, inclusive of the head and limbs. The hot sunbeams soon draw away the moisture, and in a few hours the skin is dried, and may be packed for carriage. The size and value of the skins vary exceedingly, the latter according to the current prices of the day, and the former according to the age and growth of the animal. As a general rule, the finest skins are eleven feet six inches in length.

The colour, too, is more variable than might be supposed, some skins being much darker than others ; while occasionally, a specimen is discovered, the fur of which is so pale, as to earn for the animal the title of White Tiger. One of these animals is figured in the engraving ; the original was a well known specimen in London about the year 1820.

WHITE TIGER.

The colour of this animal was a creamy white, with the ordinary tigerine stripes so faintly marked that they were only visible in certain lights. It is probable that these White Tigers are only albinos, like the white pheasants, peacocks, crows, &c. which are so well known, and that they cannot even be ranked as permanent varieties. The markings are of that obscure neutral tint which is seen in the "eyes" of the albino peacock's tail, and on the feathers of the albino pheasant.

Not only is the Tiger skin considered as an article possessing a commercial value, but the fat commands an equally high price among the natives, who employ it as an infallible specific against rheumatic affections. It is prepared for use in rather a curious, and withal, a simple manner.

Were the fat to be exposed to the action of the atmosphere, it would soon become rancid, and then putrid; but by subjection to the native mode of treatment, it clarifies itself with no trouble to the preparer. As soon as removed from the animal, the fat is cut into long strips of a convenient size to enter the necks of sundry bottles, which are cleansed for the purpose. By the aid of a stick, as many as possible of these strips are pushed into the bottle, which is then corked, and set in the sunshine for a whole day. The heat of the sun's rays soon melts the fat, and liquifies it as if it were oil. In this state it is permitted to remain until the evening, when it cools down into a firm white mass, resembling lard. This prepared fat is as useful to Europeans as to natives, not so much to rub on their rheumatic joints, as to lubricate their guns and locks, on which may depend the life of the owner.

Those who have hunted the Tiger in a genuinely sportsmanlike manner, matching fairly man against beast, are unanimous in asserting it to be a very cunning animal, putting all the powers of the human intellect to the proof. As is the case with the fox,—our most familiar instance of astuteness among brutes,—each Tiger seems to have its peculiar individuality so strongly marked, that it must be separately matched by the hunter's skill.

Of the ordinary Tiger-hunt, or rather Tiger-mobbing, with its *posse comitatus* of elephants, horses, dogs, and men, no description will be given, as the subject has been rendered so familiar by many illustrated publications which have issued from the press in late years; and the space which would be required for a detailed narration of such scenes may be better employed in describing those portions of the Tiger's character which are not so popularly known.

When the Tiger strikes down and kills a large animal, such as an ox, he tears open the throat of his prey, and eagerly laps the blood as it streams hotly from the wound. Having solaced his appetite by this preliminary indulgence, he drags it to some place of concealment, where he watches over it until the evening, and makes up his mind for a prolonged banquet. Beginning at the hinder quarters, he eats his way gradually towards the head, occasionally moistening his sanguine feast by a draught of water from a neighbouring stream, but never ceasing from his gluttonous repast until he has so entirely gorged himself that he is incapable of taking another mouthful. He is in no way choice of palate, but eats everything as it comes, even to the skin and the very bones themselves. He now yields himself to sleep, and for three days lies in a semi-torpid state, never moving except to drink, and calmly enjoying the double happiness of a powerful appetite and a good digestion. After the three days have passed, he is ready for another feast, and returning to his prey, again gorges himself on the remains, caring little whether the taint of corruption has come upon them, and only desirous to assimilate as much animal matter as possible in a limited time.

Knowing the habits of the Tiger, the herdsman who has suffered the loss of one of his oxen takes his revenge by watching the marauder to his lair, waiting until the repleted animal has retired either to drink, or for his long sleep. He then rubs some arsenic into a few gashes which he cuts upon the hinder quarters of the stolen ox, and leaves the poison to do its work. In due time the Tiger returns to his prey, tears off and swallows the deadly food, and on feeling the burning agony caused by this most irritant of poisons, runs to the water-side, where he endeavours in vain, by repeated draughts of the cool stream, to quench the fire that consumes him. But a few hours now elapse before he lies dead by the water-side.

It would have been well for one cattle proprietor if he had adopted this safe expedient of destroying the animal that had robbed him.

Preferring the excitement of shooting the Tiger, he lay in wait for the beast as it returned to the dead ox for its second banquet, and fired at the marauder with uncertain aim, only frightening instead of destroying it. The Tiger was so alarmed at the report of the gun, that it would not run the risk of a similar danger, and yet was so fond of beef, that it could not refrain from attacking the herds. So it compromised the matter by making only a single meal on every ox which it killed, and was so fearful of exposing itself to peril, that it would only drink the blood of the slaughtered ox, and never return to it a second time. The consequence of this manœuvre was, that the Tiger used to kill two or three oxen at a time, merely for the purpose of drinking their blood.

The destruction of the Indian Tiger might be more complete were not the animal protected in various ways.

Religious principles take the chief ground, and, as is generally the case in India, choose the wrong side of the question. Many sects of that strange, polytheistic mythology, deem the Tiger to be a sacred animal, simply because it is so destructive and so dangerous, and will not suffer it to be killed unless it is one of the "man-eaters,"—whose propensities have already been mentioned.

Private predilections take second rank in this matter, and cause many a Tiger to roam unmolested, to the destruction of human and animal life, whose career might easily have been arrested, did not the will of an imperious ruler decree that the destructive animal should be at liberty to depopulate the country until such time as it pleased the self-willed autocrat to amuse a heavy hour by giving chase to the animal.

There are, in fact, some native chiefs—and, until later days, there were many more of them—who actually "preserve" the Tiger as carefully as English squires preserve foxes, and will permit none to be killed except those whose honourable fate calls them to die nominally by a royal hand. Nominally, because it only needs that the Oriental potentate should once discharge his weapon for the assembled crowd to give him the credit of killing the Tiger, even though the muzzle of his piece may at the time have been pointed to the zenith. Nowhere does courtiership reign so disguiseless as in the East.

The Tiger is a capital swimmer, and will take to the water with perfect readiness, either in search of prey, or to escape the pursuit of enemies.

It has been known to carry its aquatic audacity to such an extent, as to board a vessel, and by its unexpected advent, to cause an involuntary mutiny among the crew. Some jumped into a boat that was being towed astern, others leaped overboard, and sought safety in swimming, while others fled into the cabin, and barricaded the doors thereof. The Tiger, meanwhile, was left in possession of the vessel, but not comprehending the use of a rudder, he soon drove the vessel ashore, and springing to land, he indulged in a few growls at the occupants of the boat, and then disappeared in the jungle.

The Tiger swims rather high in the water, and therefore affords a good mark to those who are quick of aim. His natatory abilities are by no means small, and while swimming he can strike out with his paws most effectively, inflicting deep wounds wherever his outspread talons make good their aim. So cunning is the animal, that if there should be no cause for hurry, it will halt on the river's brink, and deliberately put its paw into the water, so as to ascertain the force of the stream. This point being made clear, it proceeds either up or down the river, as may best suit its purpose, and so makes allowance for the river stream, or the ocean tide.

The experienced natives say that there are more female than male Tigers, and that this discrepancy in numbers is caused by the unnatural behaviour of the adult males, who destroy every young one of their own sex upon whom they can lay a paw. All Tigers, when wild, seem to have a habit of rolling themselves in dusty spots, probably for the purpose of destroying the parasitic insects with which these animals are largely infested. This process is analogous to the dust-baths, of which birds are so fond.

In all barbarous ages, men have been accustomed to seek amusement by witnessing the combats which take place between various animals, among whom the armed human animal was the favourite. Next to the gladiatorial duels with which we are all familiar, stand the combats between various ferocious beasts, such as the lion, tiger, leopard, &c.

Even to the present day, and in the Christian world, vast multitudes of people derive ferocious gratification from the tortures of an irritated bull, and the imminent peril of human life and limb that forms the most exciting part of the spectacle. Therefore, it is no cause of wonder that in the heathen world, combats of a similar nature should minister largely to the savage joys of the inhabitants. Many animals are kept solely for the purpose of fighting each other, or of contending with certain enemies, whether human or bestial, which are brought to oppose them.

The cruel sports which delight the Oriental monarchs are familiar to all students of the Oriental character, who have found an admirable subject of contemplation in the last monarch who has ruled, or pretended to rule, the great kingdom of Oude. All kinds of animals were kept by this sensualist, simply for the purpose of fighting each other, and among the most celebrated of these warlike animals was the magnificent Tiger known by the name of JUNGLA.

This splendid animal has been brought to England, and I have been fortunate enough to procure a portrait, drawn from the living creature.

"Jungla" is one of the finest, if not the very finest Tiger that has ever set foot on English ground, and even when penned in the straight limits of a wooden cage that would not permit his noble head to be raised to its full height, and only gave room for a single short step backwards and forwards, his grand proportions were most striking. His present age is about five years.

In height he is about four feet, and the relative proportions can be judged from the illustration. The total length of the animal is said, by his keeper, to be thirteen feet six inches, and in girth he measures four feet eight inches. The principal peculiarity in the

JUNGLA.

appearance of this animal is, that nearly all the stripes are double, including those which partially surround the tail. Sometimes these dark streaks are very long, and sometimes comparatively short and very wide, leaving a broad interval of the golden-yellow fur between the outer and inner stripes. Between many of these streaks are placed a number of spots similar to those which appear on the leopard's skin, but the spots are small in size and not so distinctly outlined as the stripes. They are rather thickly scattered by the shoulders and flanks, occasionally making their appearance on the sides. Over the eyes some black lines are drawn, which closely resemble a stag's horn, and on the forehead runs a series of equally dark stripes, which remind the spectator of the figure of a bat with outstretched wings. The ears are black, with a solitary white spot upon the back of each ear.

His light yellow eyes are constantly changing their tint, at one moment becoming almost green, and at another time assuming a deep neutral tint. As is the case with all felines, the pupil of the eye varies rapidly in size, the passing of a hand near the front of the cage being sufficient to make them contract to half their previous diameter.

He has been matched against many antagonists, and always came off victorious in the fight, whether his opponent were a strong-horned and hard-headed buffalo, or a Tiger like himself. The last Tiger to which he was opposed was killed in fifteen minutes.

In India, many tales are told of the Tiger and its ferocious daring. It has often been known to leap on the roof of a native hut, tear up the slight covering with its claws, and leap into the room below. However, when a Tiger acts in this manner, the tables are generally turned, for the noise made by the scratchings and clawings on the roof give warning for the inhabitants to make their escape by the door, and bar the entrance behind them. It is not so easy to jump out of the house as into it, and in consequence, the

1. M

neighbours speedily change the course of events by getting on the roof in their turn, and shooting the burglarious quadruped through the opening which its own claws had made.

A rather ludicrous adventure occurred to an old woman who was on her way home. She had just arrived in sight of her doorway, when she perceived a large Tiger crawl up to the entrance, and allured, probably, by the scent of provisions, walk coolly into her house. With great presence of mind she closed the door on the intruder, and calling for aid from her friends, soon had the satisfaction of placing her hand upon the Tiger's carcass as he lay on her floor, pierced with the missiles hurled at him through the window.

Many of these beautiful animals have been brought to England, and through the medium of Zoological Gardens and travelling menageries are familiar to us all. When caught in its first infancy, or when born and bred in captivity, the Tiger is as tameable an animal as the lion or any of the feline race, displaying great attachment to its keeper, and learning many small accomplishments, such as jumping through hoops and over sticks, enacting the part of a couch to its keeper, letting him pull its huge jaws open, and all with perfect good humour. These exhibitions, however, are never quite safe, and ought not to be permitted.

On some occasions the animal may be in a bad temper, and not willing to go through its performances, and upon being urged strongly to act against its inclination, may turn upon its persecutor and inflict a fatal wound in a moment. The creature may not intend to commit murder, but its strength is so great that, having no mathematical knowledge of the theory of forces, it cannot calculate the effect of a blow from its paw, or a grip of its teeth. Such events have more than once occurred, one of which, the death of the well-known "Lion Queen," was singularly tragical. The Tiger was required by the "Lion Queen" to exhibit some part of his usual performances, and being in a sulky mood, refused to obey. The girl struck him with her whip, when he sprang upon her, forced her against the side of the cage, and seized her by the throat. She was almost immediately extricated from his grasp and removed from the cage, but although no apparently mortal injury had been inflicted, she died within a very few minutes.

If we look down the vista of long past centuries, we may recall the time when England was but one large hunting-field, where the wild beasts roved at their pleasure. That a human being should be struck down by a wild beast was in those days no uncommon event, but that a similar circumstance should occur at the present day and in the open streets of London, seems almost impossible. Yet it was but lately that a Tiger sprang upon a young lad, in Ratcliffe Highway, providentially without inflicting very serious damage.

The animal had escaped from his cage, and dashing down the street, more, perhaps, in terror than rage, leaped upon the boy, and bore him to the ground. Fortunately the keepers came up, and with their usual cool audacity rescued the sufferer, and recaptured the truant Tiger. The nonchalance with which these men treat the fiercest beasts, is most remarkable. They talk of a savage Tiger or leopard as if it were a kitten, and seem to handle the dangerous beast with as much composure as if it were, in truth, one of those animals.

The same Tiger is also celebrated for his battle with a lion, resulting in the death of the latter.

The two creatures had been put into one large cage, or box, which was divided by a partition in the centre, so as to separate the two animals. While the attendants were at their breakfast, the Tiger battered down the too frail barrier, and leaping into the lion's chamber, entered into fierce combat. Not even the keepers dared interfere to stop the battle, which raged until it was terminated by the slaughter of the lion. The poor beast never had a chance from the beginning, for it was weakened by three years' captivity, and had lost the swift activity of its wild nature. Its heavy mane defended its head and neck so well, that the Tiger could not inflict any severe injury on those portions, and the fatal wounds, under which it sank, were all upon the flanks and abdomen, which were torn open by the Tiger's claws.

It was a serious loss to the proprietor, for the lion had cost three hundred, and the Tiger, which, although the victor, did not escape unscathed, four hundred pounds. The lion was six or seven years of age at the time.

Dissimilar as are the lion and Tiger, there has been an example of a mixed offspring of these animals, the lion being the father and the Tigress the mother. The lion had been born and bred in captivity, and the Tigress had been captured at a very early age, so that the natural wildness of their character had been effaced by their captive life, in which they felt no need to roam after living prey, as their daily sustenance was always forthcoming.

It has already been mentioned, that the young of the lion are marked with faint stripes of the tigrine character. Similar streaks were observed on the fur of the Lion-Tiger cubs, but they were darker than those of the lion cub, and were permanent instead of vanishing as the creature increased in years. The shape of the head was like that of the lion, while the contour of the body resembled that of the Tiger. These curious little creatures were too valuable to be entrusted to the care of the mother, and therefore were removed immediately after birth, and placed under the fostering care of a goat and several dogs. Under this treatment they throve well, but did not reach maturity. This is not the only instance of a hybrid breed between the lion and Tiger. Although Europeans do not seem to succeed very perfectly in taming the Tiger, many native Indians meet with a better reward for their labours. Some of the fakirs or mendicant priests have so far subdued the savage nature of the Tiger, that they permit their fierce favourites to wander at large among the jungles and to enter or leave their rude dwellings at pleasure. They give these tame Tigers no raw animal food, but supply them with a mixture of boiled rice and ghee. One of these men was accustomed to walk in the streets of a neighbouring town accompanied by his favourite Tiger, which followed him like a dog, without requiring even the frail bondage of a cord and a collar. The inhabitants of the town were quite accustomed to the man and the beast, and felt no alarm when this strange pair promenaded their streets. English visitors, however, could not exhibit an equal stoicism, and were rather uneasy at the inquisitive air with which the Tiger walked round them. The fakir had wisely prohibited all people from touching his brindled favourite, to which caution it is probable that much of his success was owing. The natives were withheld from infringing this command by the dread of religious anathemas which were liberally threatened by the fakir, and the English visitors were respectfully begged to adhere to the same rule. However, in this case, such a precaution was totally unnecessary, for they felt in no way inclined to diminish the distance between themselves and the perilous-looking animal that regarded them with curious eyes.

Unlike the Tiger, which is confined to the Asiatic portion of the world, the LEOPARD is found in Africa as well as in Asia, and is represented in America by the Jaguar, or, perhaps, more rightly, by the Puma.

This animal is one of the most graceful of the graceful tribe of cats, and, although far less in dimensions than the tiger, challenges competition with that animal in the beautiful markings of its fur, and the easy elegance of its movements. It is possessed of an accomplishment which is not within the powers of the lion or tiger, being able to climb trees with singular agility, and even to chase the tree-loving animals among their familiar haunts. On account of this power, it is called by the natives of India "Lakree-baug," or Tree-tiger. Even in Africa it is occasionally called a "Tiger," a confusion of nomenclature which is quite bewildering to a non-zoologist, who may read in one book that there are no tigers in Africa, and in another, may peruse a narrative of a tiger-hunt at the Cape. Similar mistakes are made with regard to the American felidæ, not to mention the numerous examples of mis-called animals that are insulted by false titles in almost every part of the globe. For, in America, the Puma is popularly known by the name of the Lion, or the Panther, or "Painter," as the American forester prefers to call it, while the Jaguar is termed the "Tiger."

In Africa, the Leopard is well known and much dreaded, for it possesses a most crafty brain, as well as an agile body and sharp teeth and claws. It commits sad depredations on flocks and herds, and has sufficient foresight to lay up a little stock of provisions for a future day. A larder belonging to a Leopard was once discovered in the forked branches of a tree, some ten feet or so from the ground. Several pieces of meat were stowed

LEOPARD.—*Leopardus Varius.*

away in this novel receptacle, and hidden from sight by a mass of leaves piled upon them.

When attacked, it will generally endeavour to slink away, and to escape the observation of its pursuers ; but, if it is wounded, and finds no mode of eluding its foes, it becomes furious, and charges at them with such determinate rage, that, unless it falls a victim to a well-aimed shot, it may do fearful damage before it yields up its life. In consequence of the ferocity and courage of the Leopard, the native African races make much of those warriors who have been fortunate enough to kill one of these beasts.

The fortunate hunter is permitted to decorate his person with trophies of his skill and courage, and is looked on with envy by those who have not been able to earn such honourable distinctions. The teeth of the Leopard are curiously strung, with beads and wire, into a necklace, and hung about the throat of the warrior, where they contrast finely with their polished whiteness against the dusky hue of the native's brawny chest. The claws are put to similar uses, and the skin is reserved for the purpose of being dressed and made into a cloak, or " kaross," as this article of apparel is popularly termed. The tail is cut off, and, being hung to a string that passes round the waist, dangles therefrom in a most elegant and fashionable manner. If a Kaffir is able to procure some eight or ten tails, which he can thus suspend around his person, he is at the very summit of the aristocratic world, and needs no more attractions in the eyes of his comrades. Generally, these "tails" are formed from the skin of the monkey, which is cut into strips, and twisted so as to keep the hairy side of the fur outwards. But these are only sham tails, and are as nothing in comparison to the real tail which is taken from a veritable Leopard.

The natives seem in some way to connect the Leopard's skin with the idea of royalty, and to look upon it as part of the insignia of majesty, even when it is spread on the kingly throne, instead of hanging gracefully from the kingly shoulders. And, though the throne be but a mound of earth, and the shoulders be redolent with rancid grease, yet the native African monarch exercises a sway not less despotic than that of the former Turkish Sultans.

The Leopard, like most of the feline tribe, is very easily startled, and, if suddenly alarmed, will in most cases make off with the best speed possible. As the creature is so

THE SEVEN LEOPARDS.

formidable a foe, it may be imagined that to meet it on equal terms would be a proceeding fraught with the utmost danger. Yet this is not the case, for there are innumerable instances of such rencontres, where both parties seemed equally surprised by the meeting, and equally anxious to shorten its duration as much as possible. One of these adventures, which was told me by Captain Drayson, R.A., who had learned the tale from the hero—if so he may be called—of the narrative, was a most singular one, and one in which was more of peril than is usually the case.

A Dutch Boer,—one of the colonists of Southern Africa—was travelling across country, and, permitting the waggons to precede him at their slow uniform pace, amused himself by making a wide detour in search of game. Towards the end of his circuit, and just as he was coming in sight of the waggons in the far distance, he came upon a clump of scattered rocks, from which suddenly leaped no less than seven Leopards. In the hurry of the moment he acted in a very foolish manner, and fired his single-barrelled gun at the group. Fortunately for himself, the result of the adventure turned out better than he deserved; for, instead of springing upon the Boer, who was quite at the mercy of so formidable a party, the Leopards only started at the report of the gun, and one or two of them, leaping on their hind legs, clawed at the air as if they were trying to catch the ball as it sang by their ears.

The illustration is drawn from a sketch made by the narrator of the anecdote.

In its own country the Leopard is as crafty an animal as our British fox; and being aided by its active limbs and stealthy tread, gains quiet admission into many spots where no less cautious a creature could plant a step without giving the alarm. It is an inveterate chicken-stealer, creeping by night into the hen-roosts, in spite of the watchful dogs that are on their posts as sentinels, and destroying in one fell swoop the entire stock of poultry that happen to be collected under that roof. Even should they roost out of doors they are no less in danger, for the Leopard can clamber a pole or tree with marvellous rapidity, and with his ready paw strike down the poor bird before it is fairly awakened.

The following narratives of the Cape Leopard and its capture are taken from the anecdotes so kindly placed at my disposal by Captain Drayson.

THE LEOPARD acts in a very subtle manner, remaining in some unlikely spot near a village, and committing a great amount of havoc before its whereabouts is discovered. I knew that two Leopards were located in the bush at Natal within half a mile of the barracks, and yet they were never seen. The disappearance of a dog and a stray pig were the only indications that they gave to the non-observers of their being in the vicinity.

I became acquainted with their presence in rather a curious way. Being alone in the bush one day, as was my usual custom, I sat down under the shade of a dark Euphorbia, to watch the habits of a chameleon which I had caught. I set him upon a branch, and saw him try every change of colour of which he was capable. At first he was a dull green, then some spots of brown came over him, and he changed all over of a brownish tint; when I irritated him with my finger he opened his comical mouth and gave a gaping sort of hiss, whilst his swivel eyes pointed each in different directions at the same time.

Suddenly I heard the scream of a buck at a short distance from me; and concluding that the animal had been caught in a trap set by the Kaffirs, I grasped my gun, and pushed through the underwood towards the spot. Before I had gone far the noise ceased, and when I reached the place whence I conceived it had come, I saw nothing remarkable; there was no sign of a buck or of a trap. I therefore examined for spoor, and found that there had been a scuffle on the ground; and a few yards from the sign blood-spots lay on the leaves, together with small pieces of fur which I recognised as belonging to a Leopard.

I followed the trail for some distance, but at length lost it. On several successive occasions I went over this ground, and always found the spoor of one, and sometimes of two Leopards, either fresh or a day or two old.

It is a practice of this cunning animal to take up its position near a village, and then go to the farms of another village quite at a distance, so that its lair shall neither be suspected nor discovered.

THE LEOPARD when seen in its wild state is a most beautiful and graceful animal; its motions are easy and elastic, and its agility amazing. Although far inferior to the tiger in size, strength, and intrepidity, and though it shuns man, it is nevertheless, when wounded or driven to desperation, a most formidable antagonist. When hunted with dogs, the Leopard usually takes to a tree, if one should happen to be near. But to approach him here is a proceeding fraught with danger; for from this elevated position he will leap to the ground, and with one spring will be beside his pursuer, who will then fare badly unless he be sufficiently handy with his gun to kill (not wound) the animal in its advance. The Leopard usually selects some elevated position from which to bound upon his prey as it passes underneath.

I have been told by Hottentots and Kaffirs that this animal has the habit of lying on the ground half concealed by long grass or branches, and then twisting itself about so as to attract the attention of any antelope which may be near. The Leopard, being aware that curiosity is one of the failings of the antelope tribe, carries on its mysterious movements until its victim approaches to investigate what is going on, when it springs on and kills the weak-minded animal.

It is a well-known fact that the Leopard does a very good business when it devotes its attention to a herd of baboons. Success in this line speaks well for the Leopard; for he must be an adept in stalking who succeeds in surprising and capturing one of these wide-awake caricatures of humanity. I suspect, however, that the victims are either the old and infirm, or those reckless youngsters who have not paid sufficient attention to the instructions which their anxious parents have endeavoured to instil into them.

It may be said, and with some truth, that when hunting and shooting are made the regular business of life, and more important pursuits neglected, we are merely expending our abilities and sacrificing our energies upon a frivolous pleasure. These objections may certainly have some weight when they are directed against those who devote the whole of their time to mere sporting matters in such a place as England, where field sports should merely be taken up as a relaxation, and as a means of obtaining exercise and skill in those affairs which make an individual "more of a man." But these requirements cannot be employed against those who, having a great amount of leisure, occupy their time in hunting such animals as are to be found in India and Africa, and of ridding the country of man-eating tigers and lions, destructive Leopards, or other dangerous and formidable neighbours, —and even when engaged in the pursuit of less noble game. The African sportsman is either providing himself and his servants with venison, or is enabled to feed whole families of hungry Kaffirs, who have fasted from meat for many days.

To shoot or capture a Leopard is therefore useful as well as gratifying, and we shall be sure when we catch one of these beasts to have the opportunity of punishing either an old offender or one that is likely to become so.

When the Leopard has committed many deeds of rapine in one locality, he often appears to think it better to decamp and try some far-removed scene of operations.

A HOUSE some few miles from Natal had been frequently visited by a Leopard, which had carried off a dog, chickens innumerable, and a pig. To support a Leopard with so promiscuous and extravagant an appetite was rather unsatisfactory. So the combined intellect of three individuals plotted a trap for this robber, and an old hen was the bait. Scarcely had the night begun when a great cackling and various sounds of alarm were heard proceeding from the ancient fowl. She had been fastened on to the perch by some string, and it would be necessary for the Leopard to pull her off the perch before he let drop the door of the trap. The ordinary mouse-trap principle had been adopted, and the top of the cage secured by planks, on each end of which iron half-hundred weights were placed. The planks were also laid so close together that there was no room for a paw to be inserted, and the sides of the trap being made of stout stakes driven some feet into the ground, and lashed together at the top and bottom, made a very secure prison.

The Leopard was too cunning on the first occasion that he paid this trap a visit, and would not touch the hen; but a few nights afterwards he came again, seized the hen, and became a prisoner. I was told that when first trapped he was furious, and made the most frantic efforts to escape, trying, but vainly, to force the stakes asunder. Upon the appearance of a man, he became sullen and quiet, and slunk growling into the corner of his cage.

I visited him the morning after his capture, and was received with the most villainous grins and looks. He could not endure being stared at, and tried every plan to hide his eyes so that he need not see his persecutor. When every other plan failed, he would pretend to be looking at some distant object, as though he did not notice his enemy close to him. When I gazed steadily at him he could not keep up this acting for longer than a minute, when he would suddenly turn and rush at me until he dashed himself against the bars, and found that he was powerless to revenge himself.

Several Kaffirs who had suffered from his depredations visited him, and exhausted their abusive vocabulary by the epithets which they hurled at his devoted head. Even the civilized man finds it difficult to restrain his triumph over a fallen but dreaded foe, and the savage has no compunction about the matter. Around the cage, therefore, the Kaffirs are seated, and address the Leopard in the following terms:—

"You rascally cowardly dog! you miserable eater of chickens! so you are caught, are you? Do you remember the red and white calf you killed last moon? That calf was mine, you coward! Why didn't you wait until I came with my assagies and sticks? But we let you eat it that your skin might be more sleek when you were killed!" "Look at my assagy," says another; "I will strike it into your heart as I now do into the ground!"—(digs assagy deep into the ground several times)—"Ah, show your teeth, they will make me a necklace, and we will roast your heart!"

Suddenly, in the middle of their choice address, the Leopard would spring up, rush at the stakes, and away would fly all the boasting Kaffir warriors.

It was intended to have kept this animal, and shipped it in the bay for Cape Town, but during the second night it nearly escaped; and as some days would elapse before the regular cage could be made, it was found expedient to shoot it. It was a fine animal, in superb condition, and had committed an infinity of mischief in the neighbourhood.

I ASSISTED an old Kaffir in building a cage near the Umlass river soon after this, and one of the largest Leopards I ever saw was captured in a few days. This Leopard was caged and sent down to Natal, where it brought a good price.

Owing to the stealthy and watchful habits of this creature, it is not often seen, and although the spoor may show that Leopards are plentiful in a particular locality, it does not follow that the sportsman will obtain a shot at once. The marks of claws on the stems of trees, will frequently be seen in those parts of the bush which the Leopard frequents.

If the hunter expects an encounter with a Leopard, it is a very useful precaution to bind some leather and woollen stuff round the left hand and arm, so that if an accident should happen, and the Leopard come to close quarters where the gun could not be used, this shield would serve to protect the face or body, and a knife or revolver might then be used with greater chance of success than when the independent arm was being lacerated by the jaws of the monster.

Numberless instances are on record which prove the ferocity of the Cape Leopard, and those who purpose a campaign against Feræ of this description, would do well to remember that precaution is no indication of an absence of courage, for it frequently happens that men whose intellects are the densest are incapable of seeing danger, blunder into peril, and by great good luck blunder out again.

I once caught a man smoking his pipe on the front of a waggon which was loaded with some hundreds of pounds of powder. On another occasion a gunner of the artillery carried a burning portfire amongst two dozen or so of loaded shells, whilst he was looking for the nipper with which to cut off the end. And I also knew a man who went into a bush to attack a wounded Leopard with an old sword, and who was disabled for life in consequence of his temerity.

Now, none of these individuals ought to be called wonderfully brave for their proceedings, they were simply so thick-headed that they did not know how much risk they ran. A Dutch Boer who lived over the Draakens Berg mountains, and who planned an attack upon an old man-eater lion, which he followed on foot into a dense kloof, and which he there shot dead, deserved praise for his courage, as he must have been well aware of all the risks of the affair, and made his arrangements accordingly. In leopard, lion, elephant, and buffalo shooting, the accidents usually happen to those who pretend to despise, and who therefore do not take ordinary precautions.

The habits of the Indian Leopard are almost identical with those of its African relative. Equally cautious when caution is necessary, and equally bold when audacity is needed; the animal achieves exploits of a similar nature to those which have been narrated of the African Leopard. The following anecdote is a sample of the mixed cunning and insolence of this creature.

An ox had been killed, and the joints were hung up in a hut, which was close to a spot where a sentry was posted. In the evening the sentry gave an alarm that some large animal had entered the hut. A light was procured and a number of people searched the several rooms of which the hut was composed, without discovering the cause of the

BLACK LEOPARD.

alarm. They were just about to retire, when one of the party caught sight of a Leopard, which was clinging to the thatched roof immediately above the hooks on which the meat was suspended. No sooner did the animal discover that its presence was known, than it dropped to the floor, laid about it vigorously with its claws, and leaping through the doorway, made its escape, leaving several souvenirs of its visit in various scratches, one of which was inflicted on the sentry who gave the alarm, and kept him to his bed for several weeks.

The consternation caused by such an attack was very great, and many who escaped the Leopard's claws, suffered severely from bruises which they received in the general rush towards the door.

The usual colour of the Leopard's fur is a golden-yellow ground, which is thickly studded with dark rosette-shaped spots. The form of the rosettes and the colour of the fur are no means uniform.

There are some Leopards whose fur is so very dark as to earn for them the name of Black Leopard. This is probably only a variety, and not a distinct species. Although at first sight this Leopard appears to be almost uniformly black, yet on a closer inspection it is seen to be furnished with the usual pardine spots, which in certain lights are very evident. There have been often exhibited sundry Leopards of an exceedingly dark fur, and yet partaking largely of the distinct spottings of the ordinary Leopard. These were a mixed breed between the Black Leopard and the Leopard of Africa. The black variety of this animal is found in Java, and has by some authors been considered as a separate species under the title of "Felis (Leopardus) melas," the latter word being a Greek term, signifying "black."

The strength of the Leopard is marvellous when compared with its size. One of these animals crept by night into the very midst of a caravan, seized two wolf-greyhounds that were fastened to one of the tent pegs, tore up the peg to which they were tethered, and although both the dogs were linked together, and were of that powerful breed which is used for the pursuit of wolves and other fierce game, the Leopard dragged them clean out of the camp and carried them for some three hundred yards through dense thorny underwood. A pursuit was immediately set on foot, and the dogs rescued from the daring foe. To one of them aid came too late, for its skull was literally smashed by a blow from the Leopard's paw. The same animal had sprung upon and killed a goat which was picketed in the midst of the numerous servants that accompany an European.

Another Leopard committed an act of audacity which very much resembled the exploit of the roof-clinging Leopard mentioned on p. 168.

In a native hut some goats were kept, and as night had drawn on, the human inhabitants of the hut were beneath the shelter of their own roof. A Leopard which was prowling about, and was probably attracted either by the bleating or the scent of the goats, clambered up the low walls of the hut, and tearing away with his claws the fragile thatch, leaped into the middle of the room. In this case, the Leopard fared well enough, for the terrified inhabitants were without arms, and as soon as they saw the unexpected visitor come tumbling through the roof, they hid themselves like so many lean Falstaffs, in some wicker corn baskets that were standing in the hut, leaving the Leopard to his own devices and in full possession of the ground.

The Leopard has a curious and ingenious habit of obtaining a meal. He pays a visit to some village, and taking up a convenient post, at some little distance, sets up a loud and continuous growling.

The pariah dogs, which swarm in every village, present a curious contradiction of qualities. At the sound of a Leopard's voice they will rush furiously to the spot, uttering their yelling barks, as if they meant to eat up the enemy on the spot. But when they come to close quarters, self-preservation obtains the upper hand, and they run away as fast as they had appeared, turning again and baying at their foe as soon as they see that he is not pursuing them. These habits render them of invaluable assistance to the hunter, who employs the pariah dogs to point out the locality of his fierce quarry, and to distract its attention when found.

So at the sound of the angry growl, out rush the pariahs towards the spot from whence the sounds proceeded, yelping as if they would split their throats by the exertion. To draw the dogs away from the protecting vicinity of man is just the object of the concealed Leopard, who springs from his hiding place upon one of the foremost dogs, and bounds away into the woods with his spoil.

Fond as is the Leopard of well wooded districts, it appears to have a distaste for trees around which there is no underwood. The long grass jungle which is so favoured by the tiger, is in no way suited to the habits of the Leopard; so that if the hunter seeks for tigers, his best chance of success is by directing his steps to the grass jungles, while, if Leopards are the objects of his expedition, he is nearly sure to find them among wooded places where the trees are planted among underwood reaching some seven or eight feet in height.

When a Leopard is "treed," i.e. driven to take refuge in a tree, it displays great skill in selecting a spot where it shall be concealed so far as possible from the gazers below, and even when detected, covers its body so well behind the branches, that it is no easy matter to obtain a clear aim at a fatal spot. Its favourite arboreal resting places are at the junction of the larger limbs with the trunk, or where a large bough gives off several smaller branches. The Leopard does not take to water so readily as the tiger, and appears to avoid entering a stream unless pressed by hunger or driven into the water by his pursuers. When fairly in the water, however, the Leopard is a very tolerable swimmer, and can cross even a wide river without difficulty.

The Leopard has often been tamed, and, indeed, almost domesticated, being permitted to range the house at will, greatly to the consternation of strange visitors. This complete state of docility can, however, only take place in an animal which has either been born in captivity, or taken at so early an age that its savage propensities have never had time to expand. Even in this case, the disposition of the creature must be naturally good, or it remains proof against kindness and attention, never losing a surliness of temper that makes its liberation too perilous an experiment. The very same treatment by the same people will have a marvellously different effect on two different animals, though they be of the same species, or even the offspring of the same parents.

Some years ago, a couple of Leopards, which lived in England, afforded a strong proof of the innate individuality of these animals. One of them, a male, was always sulky and unamiable, and never would respond to offered kindnesses. The female, on the contrary, was most docile and affectionate, eagerly seeking for the kind words and caresses of her keeper. She was extremely playful, as is the wont of most Leopards, and was in the habit of indulging in an amusement which is generally supposed to be the speciality of the monkey tribe. Nothing pleased her so well as to lay her claws on some article of dress belonging to her visitors, to drag it through the bars of her cage and to tear it in pieces. Scarcely a day passed that this amusingly mischievous animal did not entirely destroy a hat, bonnet, or parasol, or perhaps protrude a rapid paw and claw off a large piece of a lady's dress.

The cubs of the Leopard are pretty, graceful little creatures, with short pointed tails, and spots of a fainter tint than those of the adult animal. Their number is from one to five. Even in captivity, the Leopard is a most playful animal, especially if in the society of companions of its own race. The beautiful spotted creatures sport with each other just like so many kittens, making, with their wild, graceful springs, sudden attacks upon one companion, or escaping from the assaults of another, rolling over on their backs, and striking playfully at each other, and every now and then uniting in a general skirmishing chase over their limited domains.

Even when they are caged together with lions and tigers, their playfulness does not desert them, and they treat their enormous companions with amusing coolness. I remember seeing rather a comical example of the sportful propensities which take possession of the Leopard. Several of the feline race, such as lions, tigers, and Leopards, were shut up in a rather large cage, and being docile animals, had been taught some of the usual tricks which are performed by tamed felidæ. They jumped through hoops, or over the keeper's whip, always taking advantage of the barred front of their den to afford a temporary support in their leaps ; they stood on their hind legs, they rolled on their backs, and opened their huge jaws at the word of command, and, in fine, went through the established feline accomplishments.

Among the inhabitants of the cage, two were specially conspicuous. One was a very fine lion, all-glorious in redundant mane and tufted tail, demure and dignified in movement, —as became the monarch of the predacious animals. The other was a slight, agile, malapert Leopard, who recked little of dignities, and, so that he could play a saucy trick, cared nothing for the personal stateliness of the object of his joke.

One day, the imprisoned animals had gone through their several performances with the usual accompaniments of growls and snarls, when the lion, as if to assert his dignity, —which had been somewhat chafed by his obedience to the commands of his keeper,— began to parade up and down the den in a solemn and stately manner, his nose thrown up, and his tail held perfectly erect, with the tufted tip bending to and fro in a majestic and condescending manner. The Leopard had, in the meantime, taken up his post on a little wooden bracket that was hitched over the upper bars of the cage, and formed a portion of the machinery that was employed in the exhibition. As this bracket was hooked over the bars of the cage, and the lion was parading in the very front of the den, it happened that the perpendicularly held tail, with its nodding tuft, passed immediately under the little bracket whereon the Leopard had poised himself in a compact and cat-like manner.

Every time the lion passed beneath, the Leopard protruded a ready paw, and hit the black tip of the lion's tail a rather hard pat. The owner of the aggrieved tail took no notice of this insult, so the Leopard improved his amusement by lying on the bracket in such a manner, that both its fore paws were at liberty. As the lion passed and repassed below, the Leopard struck the tail-tuft first to one side, and then to the other, so that it enjoyed two blows at the lion's tail instead of one. The lion, however, disdained to take the least notice, and the Leopard continued its amusement until the keeper put an end to the game by entering the cage, and commencing the performances afresh.

There are two titles for this animal ; namely, the Leopard, and Panther, both of which creatures are now acknowledged to be but slight varieties of the same species. The

OUNCE.—*Leopardus Uncia.*

OUNCE, however, which was once thought to be but a longer haired variety of the Leopard, is now known to be truly a separate species.

In general appearance it bears a very close resemblance to the leopard, but may be distinguished from that animal by the greater fulness and roughness of its fur, as well as by some variations in the markings with which it is decorated. From the thickness of its furry garment, it is supposed to be an inhabitant of more mountainous and colder districts than the leopard. The rosette-like spots which appear on its body are not so sharply defined as those of the leopard; there is a large black spot behind the ears. The spots exhibit a certain tendency to form stripes, and the tail is exceedingly bushy when compared with that of a leopard of equal size. The general colour of the body is rather paler than that of the leopard, being a greyish white, in which a slight yellow tinge is perceptible, and, as is usual with most animals, the upper parts of the body are darker than the lower. The Ounce is an inhabitant of some parts of Asia, and specimens of this fine animal have been brought from the shores of the Persian Gulf. In size, it equals the ordinary leopard of Asia or Africa.

The feline animals which have hitherto been described belong to the African and Asiatic continents, with their neighbouring islands. Passing to the New World, we find the feline races well represented by several most beautiful and graceful creatures, of which the JAGUAR is the largest and most magnificent example.

Closely resembling the leopard in external appearance, and in its arboreal habits, it seems to play the same part in America as the leopard in the transatlantic continents. It is a larger animal than the leopard, and may be distinguished from that animal by several characteristic differences.

JAGUAR.—*Leopardus Onca.*

In the first place, the tail is rather short in proportion to the size of its owner, and, when the animal stands upright, only just sweeps the ground with its tip. Across the breast of the Jaguar are drawn two or three bold black streaks, which are never seen in the leopard, and which alone serve as an easy guide to the species. The spots, too, with which its fur is so liberally studded, are readily distinguishable from those of the leopard by their shape and arrangement. The leopard spots are rosette-shaped, and their outlines are rounded, whereas those of the Jaguar are more angular in their form. But the chief point of distinction is found in a small mark that exists in the centre of the dark spots which cover the body and sides. In many instances, this central mark is double, and, in order to give room for it, the rosettes are very large in proportion to those of the leopard. Along the spine runs a line, or chain, of black spots and dashes, extending from the back of the head to the first foot, or eighteen inches, of the tail.

The colour is not quite the same in all specimens. Many Jaguar skins have an exceedingly rich depth of tinting, and are very highly valued, being worth rather more than three pounds. They are chiefly used for military purposes, such as the coverings of officers' saddles in certain cavalry regiments. Sometimes, a black variety of the Jaguar is found, its colour being precisely similar to that of the Black Leopard, mentioned on page 169.

The whole fur seems to take the tint of the dark spots, while the spots themselves are just marked by a still deeper hue. Probably, the cause of this curious difference in tint may be, that in the blood of the individual Jaguar there exists a larger quantity than usual of iron, which metal, as is well known, is found to form one of the constituents of blood. It can be extracted in the metallic form, and resembles very fine sand. In the human blood, late researches have discovered that the blood of the negro is peculiarly rich in iron, and it seems but reasonable that a similar cause will account for the very great variation in the leopard's and Jaguar's fur.

This beautiful animal is familiar to us through the medium of many illustrated works on natural history, and also on account of the numerous species which have been transmitted to this country. One of these creatures, which was brought to England by Captain Inglefield, and placed in the collection of the Zoological Gardens, was so gentle and docile, that it directly controverted the once popular notion that the Jaguar is an irreclaimable and untameable animal. It was a general pet on the voyage, and, from an account of its proceedings while on board ship, I am indebted to Captain Inglefield himself,

The Jaguar was named "Doctor," and was as well acquainted with its name as any dog. It was at times rather lazy, and loved to lie at full length on deck, and stretch its limbs to their full extent. It was so perfectly tame that Captain Inglefield was accustomed to lie down by the side of the spotted favourite, using its body as his pillow. When the vessel arrived in harbour, and people were anxious to view the Jaguar, the creature walked to the stable where it was to be exhibited, merely being led by its chain. It was a remarkable circumstance, that, although the animal was so entirely tame and gentle towards men, and would let them pull it about in their rough play, it could never be trusted in the presence of a little child, nor of a dog. In either case, the animal became excited, and used to stretch its chain to its utmost limit.

Uncooked meat was never permitted in its diet, and, except in one or two instances, when the animal contrived to obtain raw flesh, it was fed exclusively on meat that had been boiled. One of these exceptional cases was rather amusing.

At Monte Video, the admiral had signalled for the captains of H.M. ships to come on board and dine with him. His cook was, of course, very busy on the occasion, and more especially so, as there was at the time rather a scarcity of fresh provisions. The steward had been making the necessary arrangements for the entertainment, and came on board carrying a leg of mutton and some fowls. Just as he stepped on deck, the Jaguar bounced out of his hiding-place, and, clutching the meat and fowls out of the steward's hands, ran off with them. The fowls were rescued by the Captain, who got them away from the robber undamaged, with the exception of their heads, which had been bitten off and eaten, but the mutton was past reclaiming, and so, to the great disgust of the cook and steward, the bill of fare had to be altered.

When "Doctor" received his daily food, he used to clutch and growl over it like a cat over a mouse, but was sufficiently gentle to permit the meat to be abstracted. In order to take away the animal's food, two men were employed, armed with large sticks, one of whom took his place in front of the Jaguar, and the other in the rear. When all was arranged, the man in the rear poked "Doctor" behind, and, as he turned round to see what was the matter, the man in front hooked away the meat with his stick. However the animal might growl over its food, and snarl at any one who approached, it would become perfectly quiet and gentle as soon as the cause of anger was removed.

It was a very playful animal, and was as mischievous in its sport as any kitten, delighting to find any one who would join in a game of romps, and acting just as a kitten would under similar circumstances. As the animal increased in size and strength, its play began to be rather too rough to be agreeable, and was, moreover, productive of rather unpleasant consequences to its fellow voyagers. For, as is the custom with all the cat tribe, the Jaguar delighted in sticking its talons into the clothes of its human playfellows, and tearing them in a disastrous manner. The creature was so amusing that no one could resist the temptation of playing with it, and so the evil was remedied by docking the "Doctor's" claws of their sharp points.

This animal was about two years old when it was brought to England, and died but very lately. Two years after its arrival, Captain Inglefield went to see his old favourite, the "Doctor," and found that the Jaguar recognised him in spite of the long interval of time, and permitted him to pat its head and to open its mouth.

In its native land, the Jaguar ranges the dense and perfumed forests in search of the various creatures which fall victims to its powerful claws. The list of animals that compose its bill of fare is a large and comprehensive one, including horses, deer, monkeys, capybaras, tapirs, birds of various kinds, turtles, lizards, and fish; thus comprising examples of all the four orders of vertebrated animals. Nor does the Jaguar confine itself to the vertebrates. Various shell-fish, insects, and other creatures fall victims to the insatiate appetite of this ravenous animal.

It seems strange that such powerful creatures as horses should be reckoned among the prey of the Jaguar, for it would seem unlikely that the muscular force of the animal could be equal to the task of destroying and carrying away so large a quadruped as a horse. Yet such is truly the case; and the Jaguars commit infinite havoc among the horses that band together in large herds on the plains of Paraguay. A Jaguar has been known to swim across a wide river, to kill a horse, to drag it for some sixty yards to the water side, to plunge it into the stream, to swim across the river with its prey, to drag it out of the water after reaching the opposite bank, and, finally, to carry it off into a neighbouring wood. The natives of the country where the Jaguar lives assert that even when two horses have been fastened to each other, the Jaguar has been known to kill one of them, and to drag off the living and the dead horse in spite of the strength of the survivor.

These seem to be marvellous exploits, when the ordinary size of the Jaguar is taken into consideration. But Humboldt, than whom is no better or more trustworthy authority, says that he has seen a Jaguar, "which in length surpassed that of all the tigers of India which I had seen in the collections of Europe."

The favourite food of the Jaguar—when he can get it—is the flesh of the various monkeys. But to catch a monkey is not the easiest task in the world, and in general can only be achieved by leaping upon the prey from a place of concealment, or by surprising the monkeys while sleeping. Sometimes it is fortunate enough to get among a little band of monkeys before they are aware of the presence of the dreaded foe, and then seizes the opportunity of dealing a few fierce strokes of its terrible paw among the partly-awakened sleepers, thus dashing them to the ground, whither it descends to feast at leisure on the ample repast. The fierce hoarse roar of the Jaguar and the yells of terror that come from the frighted monkeys resound far and wide, and proclaim in unmistakeable language the deadly work that is going on among the trees.

Peccaries are also a favourite article of diet with the Jaguar, but he finds scarcely less difficulty in picking up a peccary than in knocking down a monkey. For the little, active, sharp-tusked peccary is even more swinishly dull than is usual even with its swinish relatives, and, being too thick-headed to understand danger, is a very terrible antagonist to man or beast. It seems to care nothing for size, weapons, or strength, but launches itself as fearlessly on a Jaguar or an armed man as on a rabbit or a child. So, unless the Jaguar can manage quietly to snap up a straggler, he has small chance with a herd of these warlike little pigs, which, if they caught a Jaguar among them, would cut him so severely with their lancet-like teeth, that he would ever repent his temerity, even if he escaped with his life.

One of the easiest animals to obtain is that huge and timid rodent, the capybara, which is not sufficiently swift of foot to escape by flight, nor agile of limb to bound out of reach of its enemy, nor furnished with natural arms with which to defend itself against his assaults. Should it take to the water, and so endeavour to elude pursuit, the Jaguar is in nowise disconcerted, for he is nearly as familiar with that element as the capybara itself, and thus seldom fails in securing his prey. When the Jaguar strikes down a large

JAGUAR AND MONKEYS.

animal, such as a horse or a deer, it performs its deadly task in a very curious manner. Leaping from some elevated spot upon the shoulders of the doomed animal, it places one paw on the back of the head and another on the muzzle, and then, with a single tremendous wrench, dislocates the neck. With smaller creatures, the Jaguar uses no such ceremony, but with a blow of the paw lays its prey dead at its feet.

With the exception of such animals as the long-tailed lizards, the food of the Jaguar is of a nature that human hunters would not disdain, and in many instances would meet the approbation of a professed epicure. Of turtles and their eggs the Jaguar is particularly fond, and displays great ingenuity and strength in the securing, killing, and eating such impracticable animals as turtles. Any one who has handled a common land tortoise would be wofully puzzled if he were ordered to kill that strong mailed creature without the aid of tools, and still more bewildered, were his only meal that day to consist of the flesh that was locked in so hard and impenetrable a covering. As to a huge turtle in the vigour of active health, scuttling over the sandy shores, throwing up showers of blinding dust with its flippers, and ready to snap at an intruder with its sharp-edged jaws; he must be a powerful man who would arrest the unwieldly creature in its onward progress, and a very clever one who would make a dinner upon the flesh of the reptile.

Yet the Jaguar contrives to catch, kill, and eat the turtle, displaying in this feat equal strength and ingenuity.

Watching a turtle as she—for it is generally the female turtles that are made the Jaguar's prey—walks riverwards, or seawards, as the case may be, after depositing her eggs under a slight covering of earth, there to be warmed into being by the genial rays of the sun, the Jaguar ... the creature as it is slowly making its way to its familiar element, and with a quick yet adroit movement of the paws, turns the turtle on its back. There the poor reptile lies, helpless, and waiting until its captor is pleased to consummate his work by killing and eating the animal which he has thus ingeniously intercepted. The Jaguar needs no saw to cut through the bony shell, nor lever to separate the upper from the lower portion, nor knife to sever the flesh from the bones, for his paw stands him in the stead of these artificial instruments, and serves his purpose right well. Tearing away as much as possible of the softer parts that lie by the tail, the Jaguar inserts his supple paw, armed with its sharp talons, and scoops out, as neatly as if cut by knives, the flesh together with the vital organs of the devoted chelonian. The difficulty of this task can only be rightly appreciated by those who have undertaken a similar task, and have achieved the feat of removing the interior of a tortoise or turtle without separating the upper and under shells.

The eggs of the turtle are nearly as important to the Jaguar as is the flesh of the mother turtle herself. After inverting the maternal turtle, the Jaguar will leave her in her impotent position, and going to the shore, coolly scoop out and devour the soft leather-covered eggs which she had deposited in the sandy beach in vain hopes of their seasonable development by the warm sunbeams.

Birds are simply struck down by a single blow of the Jaguar's ready paw; and so quick are his movements, that, even if a bird has risen upon the wing, he can often make one of his wonderful bounds, and with a light, quick stroke, arrest the winged prey before it has had time to soar beyond his reach. As to the fish, the Jaguar watches for them at the water side, and as soon as an unfortunate fish happens to swim within reach of the spotted foe, a nimble paw, with outstretched talons, is suddenly thrust forth, and the fish swept out of the water upon dry land.

The Jaguar is quite as suspicious and cautious an animal as any of the Old World felidæ, and never will make an open attack upon man or beast. Should a solitary animal pass within reach, the Jaguar hesitates not in pouncing upon it; but if a herd of animals, or a party of men, should be travelling together, the Jaguar becomes very cautious, and will dog their steps for many miles, in hopes of securing one of the party in the act of straggling. If the Jaguar should be very hungry indeed, and unable to wait patiently, it will yet temper audacity with caution, and though it will, under that urgent necessity seize one out of the number, it will always choose that individual which is hindermost, hoping to escape with its prey before the companions can come to the rescue. A Jaguar has been known to follow the track of travellers for days together, only daring to show itself at rare intervals.

In the countries where the Jaguar most abounds, many tales are rife respecting the strength, agility, and audacity of this fierce animal. When the earlier settlers fixed their rough wooden huts in the recesses of the American forests, the Jaguar was one of their most persistent and relentless foes. Did they set up a poultry-yard, the Jaguar tore open the hen-roosts, and ate the fowls. Did they fill their stables with horses, the Jaguar broke their necks, and did his best to carry the heavy carcasses to his forest home. Did they establish a piggery, the Jaguar snapped up sow and litter; and in fine, it was hardly possible to secure their live stock so effectually that it could not be reached by this ravenous beast. The only resource was to kill the Jaguar himself, and so to put an effectual stop to his depredations. But there are many Jaguars in a district; and for a term of years, the toil of ridding the country of these fierce marauders was a most arduous one. However, perseverance and indomitable courage gained the day at last, and the Jaguars were forced to retire from the habitations of men, and hide themselves in the thick uncultivated forest land.

The SERVAL, or "Bosch-katte," i.e. "Bush-cat," as it is appropriately termed by the Dutch colonists of the Cape, is an inhabitant of Southern Africa. It is a very pretty animal both with regard to the colour of its fur and the elegant contour of its body. The short, puffy tail, however, rather detracts from the general effect of the living animal. On

1.   N

SERVAL.—*Leopardus Serval.*

account of the bold variegations of the Serval's fur, its skin is in great request, and finds a ready sale among furriers, who know it by the name of the Tiger-cat.

The ground colour of the Serval's fur is of a bright golden tint, sobered with a wash of grey. The under portions of the body and the inside of the limbs are nearly white. Upon this ground are placed numerous dark spots, which occasionally coalesce and form stripes. In number and size they are very variable. The ears are black, with a broad white band across them, and from their width at the base, they give the animal a very quaint aspect when it stands with its head erect.

In disposition, the Serval appears to be singularly docile, and even more playful than the generality of the sportive tribe of cats. It is not a very large animal, measuring about eighteen inches in height, and two feet in length, exclusive of the tail, which is ten inches long, and covered with thick, bushy fur.

FEW animals have been known by such a variety of names as the PUMA of America. Travellers have indifferently entitled it the American Lion, the Panther, the Couguar the Carcajou (which is an entirely different animal), the Gouazouara, the Cuguacurana, and many other names besides. For the name of Lion, the Puma is indebted to its uniform tawny colour, so different from the conspicuous streaks and spots which decorate the fur o its congeners. It was entitled a Panther, on account of its pardine habits, which are almost identical with those of the spotted leopards of both continents. The word Couguar is a Gallican abbreviation of the Paraguay word Gouazouara ; and then the names Carcajou and Quinquajou are simply instances of mistaken identity. The Anglo-Americans compromise the matter by calling the creature a " painter."

It is rather a large animal, but, on account of its small head, appears to be a less powerful creature than really is the case. The total length of the Puma is about six feet and a half, of which the tail occupies rather more than two feet. The tip of the tail is black, but is destitute of the black tuft of long hair which is so characteristic of the true lion. Its limbs are extremely thick and muscular, as needs be for an animal whose life is spent almost entirely in climbing trees, and whose subsistence is gained only by the exercise of mingled activity and force.

GROUP OF PUMAS.

The colour of the Puma is an uniform light tawny tint, deeper in some individuals than in others, and fading into a beautiful greyish-white on the under parts. It is remarkable that the young Puma displays a gradual change in its fur, nearly in the same way as has been narrated of the lion cub. While the Puma cubs are yet in their first infancy, their coat is marked with several rows of dark streaks extending along the back and sides

PUMA.—*Leopardus Còncolor.*

and also bears upon the neck, sides, and shoulders many dark spots, resembling those of the ordinary leopard. But, as the animal increases in size, the spots fade away, and, when it has attained its perfect development, are altogether lost in the uniform tawny hue of the fur.

Until it has learned from painful experience a wholesome fear of man, the Puma is apt to be a dangerous neighbour. It is known to track human beings through long distances, awaiting an opportunity of springing unobservedly upon a heedless passer-by. A well-known traveller in American forest lands told me candidly, that he always ran away from "Grizzlys," *i.e.* grizzly bears, but that "Painters were of no account." He said that as long as a traveller could keep a Puma in sight, he need fear no danger from the animal, for that it would not leap upon him as long as its movements were watched.

Even in those rare instances where the Puma, urged by fierce hunger, issued boldly from the dark leafage of the woods, and ventured to track the very pathway that was trodden by the travellers, there was yet no real danger. The Puma would creep rapidly towards the party, and would, in a short time, approach sufficiently near to make its fatal spring. But if one of the travellers faced sharply on the crawling animal, and looked it full in the face, the beast was discomfited at once, and slowly retreated, moving its head from side to side, as if trying to shake off the influence of that calm steady gaze to which it had never been accustomed, and which was a positive terror to the rapacious animal. A caged leopard has displayed a similar uneasiness at a fixed gaze of a spectator, and has finally been so quelled that in its restless walk it dared not turn its face towards its persecutor.

Although it is not an object of personal dread to the civilized inhabitants of the forest lands, the Puma is a pestilent neighbour to the farmer, committing sad havoc among his flocks and herds, and acting with such consummate craft, that it can seldom be arrested in the act of destruction, or precluded from achieving it. No less than fifty sheep have fallen victims to the Puma in a single night. It is not, however, the lot of every Puma to reside in the neighbourhood of such easy prey as pigs, sheep, and poultry, and the greater number of these animals are forced to depend for their subsistence on their own success in chasing or surprising the various animals on which they feed. As is the case with the jaguar, the Puma is specially fond of the capybara and the peccary, and makes a meal on many smaller deer than even the latter animal.

YAGOUARONDI.—*Leopardus Yagouarondi.*

Such creatures as are unfortunate enough to please the taste of the Puma, are nearly always taken by surprise, and struck down before they are even aware of the vicinity of their tawny foe. The Puma loves to hide upon the branches of trees, and from that eminence to launch itself upon the doomed animal that may pass within reach of its active leap and its death-dealing paw.

While thus lying upon the branches, the creature is almost invisible from below, as its fur harmonizes so well with the brown bark which covers the boughs, that the one can scarcely be distinguished from the other. Even when imprisoned within the limits of a cage, where the eye has no great range of objects for inspection, the Puma will often lie so closely pressed against a shelf, or flattened upon the thick boughs which are placed in its cell, that the cage appears at first sight to be empty, even though the spectator may have come to it with the express object of inspecting the inhabitants. It may therefore be easily imagined how treacherous a foe the Puma may be when ranging at will among the countless trees of an American forest.

The flesh of this animal is said, by those who have made trial of it, to be a pleasant addition to the diet scale, being white, tender, and of good flavour. When taken young, the Puma is peculiarly susceptible of domestication, and has been known to follow its master just like a dog. The hunters of the Pampas are expert Puma slayers, and achieve their end either by catching the bewildered animal with a lasso, and then galloping off with the poor creature hanging at the end of the leather cord, or by flinging the celebrated bolas—metal balls or stones fastened to a rope—at the Puma, and laying it senseless on the ground with a blow from the heavy weapon.

The Puma is not the only example of a pardine animal which is destitute of the usual pardine spots and stripes.

The YAGOUARONDI possesses a fur of a nearly uniform colour, without either spots or streaks. Its colour is rather a variable brown, sometimes charged with a deep black tinge, and sometimes dashed with a slight freckling of white. When the animal is angry, the white grizzly tinge becomes more conspicuous than when its temper is undisturbed. The reason for this curious change of hue is, that each hair is alternately dark and white, the tips being all black. If, therefore, the Yagouarondi is in a placid humour, its fur lies closely to the body, and only presents its black surface to the eye. But if it is excited, and sets up its fur after the manner of an angry cat, the white markings of the hair immediately become visible. It is said to be a very savage animal when wild.

It is a native of Guiana, and several specimens have been brought to this country.

MARBLED CAT.—*Leopardus Marmoritus.*

The MARBLED CAT partakes more of the proverbial pardine spotted character than either of the two preceding animals, and although not so finely marked as the beautiful Ocelots, or Tiger Cats, possesses a fur prettily diversified with dark spots upon a light ground. The ground of the fur is generally of a greyish tawny, on which are scattered many spots, not so sharply defined as those of the Leopard, or the Tiger Cats. It is an inhabitant of Malacca.

MANY of the members of the large genus Leopardus, are classed together under the title of Ocelots, or, more popularly, of Tiger Cats. They are all most beautiful animals, their fur being diversified with brilliant contrasts of a dark spot, streak, or dash upon a lighter ground, and their actions filled with easy grace and elegance.

The common OCELOT is a native of the tropical regions of America, where it is found in some profusion. In length it rather exceeds four feet, of which the tail occupies a considerable portion. Its height averages eighteen inches. The ground colour of the fur is a very light greyish-fawn, on which are drawn partially broken bands of a very deep fawn-colour, edged with black, running along the line of the body. The band that extends along the spine is unbroken. On the head, neck, and the inside of the limbs, the bands are broken up into spots and dashes, which are entirely black, the fawn tint in their centre being totally merged in the deeper hue; the ears are black, with the exception of a conspicuous white spot upon the back and near the base of each ear. Owing to the beauty of the fur, the Ocelot skin is in great request for home use and exportation, and is extensively employed in the manufacture of various fancy articles of dress or luxury.

In its habits the Ocelot is quick, active, and powerful, proving itself at all points a true leopard, although but in miniature.

It is sufficiently fierce in its wild state to be an unchancy opponent if wounded or otherwise irritated. When in captivity, its temper seems rather capricious, depending, in all probability on the individuality of the animal, or the treatment of its keepers. Some of these creatures are always fierce and surly, setting up a savage growl when any one

OCELOT.—*Leopardus Párdalis.*

approaches their cage, spitting at the visitor like an angry cat, and striking sharp, quick blows with the paws. Others, again, are as quiet and well-behaved as the generality of domestic cats, like to be noticed, and, if they think that the visitor is about to pass by their cage without recognising them, call his attention by a gracious purr, and rubbing themselves against the bars. They will even offer themselves to be stroked and patted, and will bow their heads, just as a cat does on feeling the touch of a friendly hand.

The GREY OCELOT is so called on account of the comparatively light hue of the fur. The spots are not quite so numerous nor so bold as in the preceding animal, and the throat is remarkable for its whitish-grey tint, unbroken by spots or streaks. All these creatures are found in tropical America.

The Ocelot soon learns to distinguish friends from foes, and can easily be brought to a state of partial tameness.

Several of these animals, when I first made their acquaintance, were rather crabbed in disposition, snarled at the sound of a strange step, growled angrily at my approach, and behaved altogether in a very unsocial manner, in spite of many amicable overtures. After awhile, I saw that these creatures were continually and vainly attempting the capture of certain flies which buzzed about the cage. So I captured a few large blue-bottle flies, and poked them through a small aperture in the cage, so that the Ocelot's paw might not be able to reach my hand. At first, the Ocelots declined to make any advances in return for the gift, but they soon became bolder, and at last freely took the flies as fast as they were caught. The ice was now broken, and in a very short time we were excellent friends, the angry snarl being exchanged for a complacent purr, and the suspicious shrinking movements for a quiet and composed demeanour.

The climax to their change of character was reached by giving them a few leaves of grass, for which they were, as I thought they would be, more anxious than for the flies. They tore the green blades out of my hand, and retired to their sleeping-house for the purpose of eating the unaccustomed dainty undisturbed. After this they were quite at their ease, and came to the front of the cage whenever I passed.

Every one who has watched the habits of the domestic cat must have noticed how thankful she seems for a few leaves of grass. It is curious that a carnivorous animal should be so impelled by instinct as to turn for a time to vegetable food, and to become for the nonce, a herbivorous creature. Dogs, it is well known, will resort to the same plant, and appear to use it in a medicinal point of view.

GREY OCELOT.—*Leopardus Griscus.*

The eye of the Ocelot is a pale yellowish-brown, and tolerably full, with the linear pupil that is found in the smaller felidæ.

There are several species of these pretty and agile animals, among which the most conspicuous are the Common, the Grey, and Painted Ocelots, and the Margay, or Marjay, as it is sometimes called. The habits of these animals are very similar.

In its native woods, the Ocelot seeks its food chiefly among the smaller mammalia and birds, although it is sufficiently powerful to attack and destroy a moderately sized monkey. The monkeys it can chase into the tree branches, being nearly as expert a climber as themselves, but, as it cannot follow the birds into their airy region, it is forced to match its cunning against their wings. As is often done by the domestic cat, the Ocelot can spring among a flock of birds as they rise from the ground, and, leaping into the air, strike down one of them with its rapid paws. But its chief method of obtaining birds is by concealing itself among the branches of a tree, and suddenly knocking them over as they come and settle unsuspiciously within reach of the hidden foe.

PAINTED OCELOT.—*Leopardus Pictus.*

The PAINTED OCELOT resembles the preceding animal in the general aspect of its fur, but is marked in a richer manner.

The spots are more numerous, closer together, and more uniform than those of the common Ocelot. The black markings of the tail are of a very deep hue indeed, and occupy a large portion of that member. The throat is greyish white, with one or two very

MARGAY.—*Leopardus Tigrinus.*

bold black streaks drawn upon it, extending towards the shoulders. These streaks are branch-like in form, and are very clearly defined. The spots that run along the spine are solid, and of a deep velvety black.

When in captivity, the Ocelot seems to prefer birds and rabbits, or similar creatures, to any other food, and is able to strip the feathers from the bird before it begins its meal. The head appears to be its favourite morsel, and, with the head, the Ocelot generally commences its meal. The reader may remember that the jaguar, mentioned on page 174, had decapitated the fowls which it had snatched from the steward, and had eaten their heads before they could be reclaimed.

The MARGAY is a very handsome example of the Tiger Cats. The tail is rather more bushy towards the tip than those of the preceding animals, and the spottings are hardly so apt to run into hollow streaks or links. It will be observed that the spots are small and numerous towards the hind quarters.

It is, when caught young and properly treated, a very docile and affectionate animal, although it has been slanderously described as a wholly untameable and ferocious beast. Mr. Waterton mentions, in one of his essays on natural history, that when he was in Guiana he possessed a Margay which had been captured by a negro while still a kitten. It was nurtured with great care, and became so fond of its master that it would follow him about like a dog. Against the rats which inhabited the house, this Margay waged incessant war, creeping about the staircase in search of the destructive rodents, and pouncing with unerring aim on any rat that was unfortunate enough to make its appearance from out of its hiding-place behind the casements.

With an instinctive knowledge of rats and their habits, the Margay was accustomed to choose the closing hours of day as its best hunting time. The creature's assistance in rat-killing was most useful, for, during the owner's absence, the rats had gained entrance to his house, and, finding no one there to oppose their devices, took possession, and roamed about the rooms at their own will. Thirty-two doors had been gnawed through by the chisel-edged teeth of the rats, and many of the valuable window-frames had suffered irreparable damage from these long-tailed pests.

THE very handsome animal which is known by the name of Rimau-dahan, or more popularly as the Clouded or Tortoiseshell Tiger, was, until comparatively late years, a stranger to this country. One of the first specimens which visited England was exhibited for some time in a travelling menagerie, where it died. So indifferent or so ignorant were its proprietors, that after its death no trace was found of this unique animal, excepting a tradition that its hide had been cut up for the purpose of making caps for the keepers.

The spots and marks which cover the fur of the Rimau-dahan are so very irregular in shape and arrangement that a detailed description is almost impossible. Some of the patches are nearly oval, some are angular, some are particularly open, while others are enclosed within a well-defined dark edging. There are stripes like those of the tiger, solid spots like those of the leopard, hollow spots resembling those of the jaguar, and large black-edged spots like those of the ocelots. The black has a peculiarly rich and velvety appearance.

The ground colour of the fur is grey, tinged with brown, and however the other mark-ings may vary, there are always two bold uninterrupted bands of velvety-black running along the entire length of the animal, beginning at the back of the head, and only ending at the root of the tail. The tail itself is covered with dark rings, which contrast well with the very light ground of the fur. The hair is rather long, and beautifully fine in its tex-ture. Altogether, the Rimau-dahan, although so large an animal, bears a close resem-blance to the Marbled Cat, which has already been mentioned on page 182.

It seems to be a gentle animal, in despite of its size and strength, which are greater than those of the ocelots, and nearly approaches the tiger and leopard in those qualities. Two specimens, which were in the possession of Sir S. Raffles, were exceedingly well-behaved and playful animals, courting notice like petted cats, and rolling over on their backs the better to enjoy the caresses of those who would pat or stroke their beautiful soft fur. Nor did they confine their sportful propensities to human companions. One of them, while on board ship, struck up a great friendship for a little dog that was its co-voyager, and used to gambol with its diminutive playfellow in the most considerate manner, taking great care to do no damage through its superior strength and size. While on board, it was fed chiefly on fowls, and generally used to extract a little amusement out of its dinner before it proceeded to the meal. When it received the fowl, it was accustomed to pounce upon the dead bird just as if it had been a living one, and tear it to suck the blood. It would then toss the bird about for hours, just as a cat tosses a mouse, tumbling over it, and jumping about it, and, after it had thus amused itself for an hour or so, would at last condescend to eat its feathered toy.

The natives of Sumatra, where it is found, assert that it is by no means a savage animal, and that it generally restricts its depredations to the smaller deer and to birds. In the latter category are comprised the domesticated poultry, among which the Rimau-dahan is apt to make sad havoc. The curious name which is given to this animal is of native formation, and has been given to the creature on account of its arboreal propensities. It is said to spend much of its time upon the tree branches, and to lie in wait for its prey, crawling along a bough, with its head resting in the fork of the branches. The word " Dahan," or " Dayan," signifies the forked portion of a bough.

If the physiognomy of an animal is to be any test of its mental qualities, the Rimau-dahan is in truth—for a wild feline—a most gentle and forbearing creature. Its head is small in proportion to the body, and there is a very mild and pleasant expression in its countenance. It is not a very common animal, and even in its own land is not very often found. The southern portions of Sumatra are the localities which are most approved by this beautiful creature.

The tail of the Clouded Tiger is peculiarly capable of that curious expansion which is so familiar to us in the domestic cat when she is either very much pleased or very much irritated. Its limbs, although apparently rather short in proportion to the dimensions of the body, are very thick and powerful; and altogether, the Rimau-dahan presents the appearance of an animal which, if it chose to take up the offensive, might be a truely fearful foe.

An allied species, named popularly the TORTOISESHELL or SMALLER CLOUDED TIGER, and scientifically termed *Leopardus Macroceloïdes,* is found in the same locality as the

Rimau-dahan. It possesses many of the properties which belong to its larger relative, and is equally fond of climbing up, or resting on, the branches of trees.

A very fine and healthy specimen of this animal is at present in the Zoological Gardens, and is always attractive to visitors whenever it will venture from its straw

RIMAU-DAHAN.—*Leopardus Macrocelis.*

couch. On the thick branches which are placed in its cage this leopard loves to repose, and sometimes assumes the strangest, and apparently the most uncomfortable, attitudes. Lately, I saw the creature lying at full length on a nearly horizontal pole, its right cheek pillowed on the bar, and all its four legs hanging down at each side of the pole. It was, in fact, lying astride the bar with all four legs at once.

If the rimau-dahan be a gentle and quiet animal, it finds its contrast in the smaller, but more ferocious, creature, the COLOCOLO. The colour of this creature is almost wholly grey, with the exception of the under parts of the body, the throat, and inside of the limbs, which are white. Black streaks, occasionally diversified with a deep tawny hue, are drawn at intervals over the body and limbs; the legs are of a darker grey than the rest of the body, and the tail is covered with a series of partial black rings, which extend only half way round that member. These black stripes are almost invariably edged with a deep tawny hue, and, on the shoulders, flanks, and thighs, they are entirely tawny. The legs themselves are darker than the rest of the body, being of a very deep grey. In size, the Colocolo equals or surpasses the ocelots, and, to judge from collateral evidence, is a terrible enemy to the animals among which it lives.

A specimen of this creature was shot on the banks of a river in Guiana by an officer of rifles, who stuffed it, and placed the skin to dry on the awning of his boat. As the vessel

COLOCOLO.—*Leopardus ferox.*

dropped down the river, it passed under overhanging boughs of large trees, on which rested numerous monkeys. Generally, when a boat passes along a river, the monkeys which inhabit the trees that border its banks display great curiosity, and run along the boughs so as to obtain as close a view as possible of the strange visitant. Before the Colocolo had been killed, the passage of the boat had been attended as usual by the inquisitive monkeys, but when the stuffed skin was exhibited on the awning the monkeys were horribly alarmed, and, instead of approaching the vessel as they had before done, trooped off with prodigious yells of terror and rage. From this universal fear which the sight of the animal occasioned to the monkeys, it may be conjectured that the Colocolo, like the jaguar, the puma, and the ocelots, is in the habit of procuring its food at the cost of the monkey tribes.

The head of this animal is peculiarly flat and broad, and the ears are large and rounded. In its general aspect, it bears a slight resemblance to the contour of the serval.

THE pretty cat-like Leopard, which is known by the name of KUICHUA, is an inhabitant of Brazil, where it appears to be of very rare occurrence. It is chiefly remarkable for its beautiful tail, which, as will be seen from the engraving, is extremely long, very bushy, and boldly marked with black semi-rings upon an ochry-grey ground. The markings of the body are variable, as is the case with most of the Leopards, sometimes falling into solid or hollow spots, and sometimes coalescing until they form interrupted streaks. The face of the Kuichua is very short, and the neck long and thin. The very appropriate name, *Macrourus*, is a compound of two Greek works, signifying long-tailed.

ALTHOUGH so gentle in its demeanour when domesticated as to have earned for itself the name of "*Mitis*," or "placid," the Chati is, when wild, a sufficiently destructive animal. It is not quite so large as the ocelots, with which creatures it is a compatriot.

The colour of the Chati resembles that of the leopard, only is paler in general hue. The dark patches that diversify the body are very irregular—those which run along the back are solid, and of a deep black, while those which are placed along the sides have generally a deep fawn-coloured centre. Towards the extremity of the tail, the spots change

KUICHUA.—*Leopardus Macrourus.*

into partial rings, which nearly, but not quite, surround the tail. All specimens, however, are not precisely alike, either in the colour or the arrangement of the markings, but those leading characteristics which have just been mentioned may be found in almost every individual.

CHATI.—*Leopardus Mitis.*

When at large in its native woods, it wages incessant and destructive warfare against small quadrupeds and birds, the latter creatures being its favourite prey. The Chati is a vexatious and expensive neighbour to any one who may keep fowls, for it seems to like nothing so well as a plump fowl, and is unceasing in its visits to the henroost. It is so active and lithe an animal that it can climb over any palisade, and insinuate itself through a surprisingly small aperture ; and it is so wary and cautious in its nocturnal raids, that it generally gives no indication of its movements than that which is left next morning by the vacant perches, and a few scattered feathers flecked with blood-spots.

During the day it keeps itself closely hidden in the dark shades of the forests, sleeping away its time until the sun has set, and darkness reigns over its world. It then awakes from its slumber, and issues forth upon its destructive quest. On moonlight nights, however, it either stays at home, or confines its depredations to the limits of its native woods, never venturing near the habitations of man. Stormy and windy nights are the best adapted for its purpose, as it is sheltered from sight by the darkness, and from hearing by the rushing wind, which drowns the slight sounds of its stealthy footsteps. On such nights it behoves the farmer to keep a two-fold watch, and see well to his doors and windows, or he may chance to find an empty henroost in the morning.

In two years, no less than eighteen of these animals were caught by a land-owner within a space of five miles round his farm, so that their numbers must be truly great. They do not congregate together, but live in pairs, each pair seeming to appropriate its own hunting-ground.

In captivity it is a singularly gentle, and even affectionate, animal, possessed of most engaging habits, and full of pretty graceful tricks. One of these creatures, which was captured by the above-mentioned land-owner, became so entirely domesticated that it was permitted to range at liberty. But, although so gentle and tractable towards its owner that it would sleep on the skirts of its master's gown, its poultry-loving habits were too deeply implanted to be thoroughly eradicated, and it was quietly destructive among his neighbours' fowls. This propensity cost the creature its life, for the irritated farmers caught it in the very deed of robbing their henroosts, and killed it on the spot.

The native name for the Chati is Chibiguazu. It was found by experimenting on the captured Chatis, that the flesh of cats and of various reptiles was harmful to their constitution. Cat's-flesh gave them a kind of mange, which soon killed them, while that of snakes, vipers, and toads caused a continual and violent vomiting, under which they lost flesh and died. Fowls however, and most birds, were ravenously devoured, being caught by the head, and killed by a bite and a shake. The Chatis always stripped the feathers from the birds before beginning to eat them.

FROM the shorter heads, and other characteristics of the last few animals, it will be seen that we are rapidly approaching that type of the feline nature with which we are so familiar in the domestic cat. The PAMPAS CAT might easily be mistaken for a rather large domestic cat which had run at large for some time, and assumed the fierce, suspicious demeanour of the wild animal.

Its general colour is a yellowish grey, something like the tint which we call "sandy," when it belongs to the fur of a domestic cat or the scalp of a human being. The body is covered with numerous brown stripes, admixed with yellow, which run at a very small angle with the line of the body. On each side of the face two bold streaks are drawn from the eye over the cheeks, the lower stripe running round the neck, and uniting with the corresponding stripe of the opposite side. Two or three dark streaks appear across the upper portion of the legs. The depth of tint appears to be variable in different individuals, and the markings present slight discrepancies.

The fur of the Pampas Cat is extremely long, some of the hairs reaching a length of five inches. The tail is not very long, is well covered with bushy hair, and is devoid of the ring-like markings which are found in the same member in the ocelots.

The natives of Buenos Ayres and its vicinity name the Pampas Cat "Gato Pajero," the former word signifying a cat, and the latter being formed from the Spanish term "paja," or straw. It is so called because it frequents the jungles or reeds, and by the English residents is often termed the Jungle Cat. It is spread over a very large space of country, being found on the whole of the Pampas which are spread on the eastern side of South America, a range of some fourteen hundred miles. The food of the Pampas Cat consists

PAMPAS CAT.—*Leopardus Pajéros.*

chiefly of the moderately sized rodents which inhabit the same country in great profusion, and it is by no means so dangerous a foe to poultry as the ocelots or the chati.

The length of the animal, inclusive of the tail, is rather more than three feet, the tail occupying about eleven inches. Its height, when adult, is rather more than a foot.

Excepting for a certain upright and watchful carriage of the ears, the Egyptian Cat has a very domestic look about it.

This animal is supposed to be the species which was so honoured by the ancient Egyptians, that they refused to attack an invading army which bore a number of Cats in their front rank ; and even when their land was in possession of the hostile force, the people rose like one man, and demanded the life of a soldier who had killed one of these sacred animals. So deeply were these ideas implanted in their minds, and so determinately did they persist in their demand, that the invading general yielded to their religious enthusiasm, and actually delivered the unwitting offender into their hands.

The Egyptian Cat was not only honoured and protected during its lifetime, but even after death it received funeral honours such as only fall to the lot of distinguished or wealthy personages.

There were several methods of embalming in use among the Egyptians, by which the bodies of the dead were, for a time, withheld from the natural and beneficial process of decay, only to yield to its power a few hundred years later. Of these modes, only the most elaborate has left its records on the still existing bodies of the mighty dead. The carcass of the plebeian might be drenched and soaked in the antiseptic mixture, and so be preserved for a time. But it was the privilege for kings and rulers alone to have their bodies imbued with costly drugs and sweet spices, and to lie unchanged in their tombs for thousands of years, until their mummied remains were removed from their long repose, and exhibited to the public gaze of a people who, in their own royal time, were but a race of naked savages. The privilege which was denied to the workman was granted to his Cat, and we have in this country many specimens of mummied Cats, their bodies swathed, bandaged, and spiced in the most careful manner, partaking of this temporary immortality with a Rameses or a Pharaoh.

EGYPTIAN CAT.—*Felis Maniculata.*

The species of Cat which was thus glorified by these ghastly honours of the charnel-house, is the animal which is represented in the engraving. It is supposed to be the original stock from which descended the race of domestic Cats which found their home by the Egyptian's hearth, and were so piously cherished by that strange, intellectual, inexplicable people. It is indigenous to Nubia, and has been found on the western side of the Nile, inhabiting a district which was well furnished with brushwood, and broken up into rocky ground.

The general colour of this animal is something like that of the Pampas Cat, but not so clear or bright, as a brownish-grey tint is washed over the white portions. On the back, the colour is deeper than on the remainder of the body. The under portions of the body and inside of the limbs are a greyish-white, the grey disappearing under the throat and about the cheeks, leaving those parts of a pure white. Many streaks and dashes of black, or ochry-yellow, are spread over the body and limbs, two of the lighter stripes encircling the neck. Its eye is bright golden yellow.

The Egyptian Cat is about the size of an ordinary domestic cat, being nine or ten inches in height, and two feet five inches in length; the tail is about nine inches long.

Few of the Felidæ are so widely spread, or so generally known as the WILD CAT. It is found not only in this country, but over nearly the whole of Europe, and has been seen in Northern Asia, and Nepaul.

In England the Wild Cat is almost extinct, having been gradually exterminated by civilization and the conversion of forests and waste land into arable ground. It now very seldom occurs that a real Wild Cat is found even in an English forest, for the creature appears to be driven gradually northwards, finding its last fortress among the bleak and barren ranges of the Scottish hills. In Scotland it still lingers, but its numbers seem to diminish rapidly, and the time is not very far distant when the Wild Cat will be as entirely extinct as the wolf.

It is true that many so-called Wild Cats are found in the snares set by the game-keeper to protect the pheasants, hares, and partridges under his charge, but in ninety-nine cases out of every hundred, these captured robbers are nothing more than domesticated cats which have shaken off the trammels of their civilization, and have taken to a savage

WILD CAT.—*Felis Catus.*

life in the bush. Even tame and petted Cats have been known to take to poaching, and to bring to their owner a daily pheasant or partridge. There are few more dangerous foes to game than the domestic Cat, and the Wild Cat gets the credit of its misdeeds.

Whether the Wild Cat be the original progenitor of our domestic Cat is still a mooted point, and likely to remain so, for there is no small difficulty in bringing proofs to bear on such a subject. It is certain that if such be the case, the change from savage to domestic life must be of very long standing, for it is proved that certain distinctions between the Wild and domestic Cat are found in full force, even though the domestic Cat may have taken to a wild life for many a year. There are several points of distinction between the Wild and the domestic Cat; one of the most decided differences being found in the shape and comparative length of their tails.

As may be seen from the accompanying figure, the tails of the two animals are easily distinguished from each other. The upper figure represents the tail of the domestic Cat, which is long, slender, and tapering, while the lower represents the tail of the Wild Cat, which is much shorter and more bushy. Now it is proved that, even if several domestic Cats have escaped into the woods and there led a sylvan life, their long tapering tails have been transmitted to their posterity through many successive generations, in spite of their wild and marauding habits.

CATS' TAILS.

The colour of the Wild Cat is more uniform than that of the domestic animal, and is briefly as follows.

The ground tint of the fur is a yellowish, or sandy grey, diversified with dark streaks drawn over the body and limbs in a very tigrine manner. These stripes run, as do those of the tiger, nearly at right angles with the line of the body and limbs, so that the creature has been termed, with some justice, the British Tiger. A very dark chain of streaks and spots runs along the spine, and the tail is thick, short, and bushy, with a black tip, and many rings of a very dark hue. The stripes along the ribs and on the legs

1.   o

are not so dark nor so clearly defined as those of the spine. The tail is barely half the length of the head and body. The fur is tolerably long and thick, and when the animal is found in colder regions, such as some parts of Germany and Russia, the fur is peculiarly long and thick.

In the wilder and less cultivated parts of Scotland, the Wild Cat is still found, and is as dangerous an enemy to the game of Scotland as is the ocelot to that of tropical America.

The amount of havoc which is occasioned by these creatures is surprising. Mr. Thompson mentions, in his Notes on the Mammalia of Ireland, that a gamekeeper had frequently noticed certain grouse feathers and other *débris* lying about a "water-break" which lay in his beat, and had more than once come upon some of the birds lying without their heads, but otherwise in such excellent condition that they were taken home and served at table. Suspecting the Wild Cat to be the culprit, he set a trap, and captured two of these animals, an old and a young one.

Here, again, is exhibited the strange predilection which the Cat tribe seem to feel for the heads of the creatures on which they feed. No less than five grouse were discovered at the same time lying headless on the ground, and it is probable that their destroyers would have contented themselves with the heads only; and, like the blood-sucking tiger mentioned on p. 159, would have killed victim after victim for the sole purpose of feasting upon their heads. The keeper expected to secure one or two more of these feline marauders, for the young Wild Cats remain with their parents until they are full grown and able to take upon themselves the cares of wedded life.

The Wild Cat is said by some naturalists to be indigenous to Ireland, but is denied that honour by others. In Maxwell's "Wild Sports of the West" are several anecdotes of a fierce savage breed of Cats running wild, and depopulating the rabbit-warrens sadly. One of these animals, which was killed after a severe battle, was of a dirty-grey colour, double the size of the common house Cat, and its teeth and claws more than proportionately larger. This specimen was a female, which had been traced to a burrow under a rock, and caught in a rabbit-net. With her powerful teeth and claws she tore her way through the net, but was gallantly seized by the lad who set the toils. Upon him she turned her energies, and bit and scratched in a most savage style until she was despatched by a blow from a spade. The wounds which she inflicted were of so severe a character that lock-jaw was threatened, and the sufferer was sent to an hospital.

Besides these huge Wild Cats, which may, in all probability, be the true *Felis Catus*, there are many house Cats which run away from their rightful home, and, taking up their residence in the rabbit-warren, are as formidable enemies to rabbits and poultry as those of the larger kind. No less than five males were caught at one time in an outhouse, penned up until the morning, and then shot; after which execution the neighbouring warren largely increased its population.

The Wild Cat takes up its residence in rocky and wooded country, making its home in the cleft of a rock or the hollow of some aged tree, and issuing from thence upon its marauding excursions. It has even been known to make its domicile in the nest of some large bird. It is rather a prolific animal, and, were it not kept within due bounds by such potent enemies as the gun and the snare, would rapidly increase in numbers. As it is, however, the Wild Cat yields to these foes, and slowly, but surely, vanishes from the land. The number of its family is from three to five, or even six. The female is smaller than the male.

In total length, an adult male Wild Cat is about three feet, of which the tail occupies nearly a foot. This does not seem to be a very considerable length, as there are domestic Cats which equal or even exceed these dimensions; but it must be remembered that the tail of the Wild Cat is much shorter than that of the domestic animal.

Of the fiery energy which actuates this animal when attacked and roused to fury, the following extract from St. John's "Highland Sports" will give an excellent idea:—

"The true Wild Cat is gradually becoming extirpated, owing to the increasing preservation of game; and, though difficult to hold in a trap, in consequence of its great strength and agility, he is by no means difficult to deceive, taking any bait readily, and not seeming to be as cautious in avoiding danger as many other kinds of vermin. Inhabiting the most lonely and inaccessible ranges of rock and mountain, the Wild Cat is seldom seen during the daytime ; at night, like its domestic relative, he prowls far and wide, walking with the same deliberate step, making the same regular and even track, and hunting its game in the same tiger-like manner ; and yet the difference between the two animals is perfectly clear and visible to the commonest observer. The Wild Cat has a shorter and more bushy tail, stands higher on her legs in proportion to her size, and has a rounder and coarser look about the head.

The strength and ferocity of the Wild Cat, when hemmed in or hard pressed, are perfectly astonishing. The body when skinned presents quite a mass of sinew and cartilage.

I have occasionally, though rarely, fallen in with these animals in the forests and mountains of this country. Once, when grouse shooting, I came suddenly, in a rough and rocky part of the ground, upon a family of two old ones and three half-grown ones. In the hanging birch woods that border some of the Highland streams and rocks, the Wild Cat is still not uncommon ; and I have heard their wild and unearthly cry echo far in the quiet night, as they answer and call to each other. I do not know a more harsh and unpleasant cry than that of the Wild Cat, or one more likely to be the origin of superstitious fears in the mind of an ignorant Highlander.

These animals have great skill in finding their prey, and the damage they do to the game must be very great, owing to the quantity of food which they require. When caught in a trap, they fly, without hesitation, at any person who approaches them, not waiting to be assailed. I have heard many stories of their attacking and severely wounding a man, when their escape has been cut off. Indeed, a Wild Cat once flew at me in the most determined manner. I was fishing at a river in Sutherlandshire, and, in passing from one pool to another, had to climb over some rock and broken kind of ground. In doing so, I sank through some rotten heather and moss up to my knees, almost upon a Wild Cat, who was concealed under it.

I was quite as much startled as the animal herself could be, when I saw the wild-looking beast so unexpectedly rush out from between my feet, with every hair on her body standing on end, making her look twice as large as she really was. I had three small Skye terriers with me, who immediately gave chase, and pursued her till she took refuge in a corner of the rocks, where, perched in a kind of recess out of reach of her enemies, she stood with her hair bristled out, and spitting and growling like a common Cat. Having no weapon with me, I laid down my rod, cut a good-sized stick, and proceeded to dislodge her. As soon as I was within six or seven feet of the place, she sprang straight at my face, over the dogs' heads. Had I not struck her in mid air as she leaped at me, I should probably have got some severe wound. As it was, she fell with her back half broken amongst the dogs, who, with my assistance, despatched her. I never saw an animal fight so desperately, or one which was so difficult to kill. If a tame Cat has nine lives, a Wild Cat must have a dozen.

Sometimes one of these animals takes up his residence at no great distance from a house, and, entering the hen-houses and out-buildings, carries off fowls or even lambs in the most audacious manner. Like other vermin, the Wild Cat haunts the shores of the lakes and rivers, and it is, therefore, easy to know where to lay a trap for them. Having caught and killed one of the colony, the rest of them are sure to be taken, if the body of their slain relative is left in some place not far from their usual hunting ground, and surrounded with traps, as every Wild Cat who passes within a considerable distance of the place will to a certainty come to it. The same plan may be adopted successfully in trapping foxes, who also are sure to visit the dead body of any other fox which they scent during their nightly walk."

Although so scarce in these days of allotments and railways, the Wild Cat was once so common in England as to be an absolute pest, and was formerly numbered among the beasts of chase that contributed to the amusement of the dull unlearned leisure which fell to the lot of those olden aristocrats of our land whose only excitement was found in the

THE CAT.—*Felis Doméstica.*

act of destruction, either of men or beasts. As were almost all destructive beasts, it was protected by the great few who suffered no scath by its depredations, to the loss of the many small, whose little stock of poultry paid heavy toll to the licensed marauders. Even its fur was made a subject of legal enactment, being permitted to some orders of the people and forbidden to others.

In Ireland—if the large savage feline that ranges the waste lands be indeed the true *Felis Catus*—it goes by the appropriate name of the Hunting Cat.

WHEN ENGAGED in the study of an illustrated work on ethnology, with its portraits of the various forms which are assumed by the human race, a certain feeling of relief and repose takes possession of the mind when the reader turns from the savage races of mankind, with their selfish, restless, eager, bestialized expression, to the mild and intellectual countenances of the civilized nations. A similar sensation of repose is felt when we turn from the savage, hungry-looking Wild Cat to the placid face and tranquil expression of our favourite, the DOMESTIC CAT.

Although this country possesses an indigenous Cat, which would naturally be considered as the original progenitor of the Domestic Cat, which attaches herself so strongly to mankind, it is now generally admitted that for this useful and graceful animal we are indebted to another continent. In the description of the Wild Cat, it has been mentioned that the distinguishing marks which characterize the two species are so permanent as to defy eradication, and to mark decisively the "Felis Catus" from the "Felis Domestica." The comparative length of their tails is of itself a distinction, and one which seems never to be lost by either the wild or the domestic animal. Whether those two creatures have ever produced a mixed breed is a matter of much uncertainty, for although a wood or a warren may be infested with Cats living in a wild state, yet, in almost every case, they are only Domestic Cats in which the savage part of their nature has predominated, and conquered the assumed habits of domestication. They have acted as men sometimes act under similar temptation, and have voluntarily taken to a savage life. As far as is at present known, the Egyptian Cat, for which see p. 192, is the origin of our Domestic Cat.

In the long past times, when the Egyptian nation was at the head of the civilized world, the " Felis maniculata " was universally domesticated in their homes, while at the comparatively later days of English history the Domestic Cat was so scarce in England that royal edicts were issued for its preservation. Yet in those days, A.D. 948, the wild Cat was rife throughout the British Islands, and was reckoned as a noxious animal, which must be destroyed, and not a useful one which must be protected. It is conjectured that the Domestic Cat was imported from Egypt into Greece and Rome, and from thence to England.

In the eyes of any one who has really examined, and can support the character of the Domestic Cat, she must appear to be a sadly calumniated creature.

She is generally contrasted with the dog, much to her disfavour. His docility, affectionate disposition, and forgiveness of injuries ; his reliability of character, and his wonderful intellectual powers are spoken of, as truly they deserve, with great enthusiasm and respect. But these amiable traits of character are brought into violent contrast with sundry ill-conditioned qualities which are attributed to the Cat, and wrongly so. The Cat is held up to reprobation as a selfish animal, seeking her own comfort and disregardful of others ; attached only to localities, and bearing no real affection for her owners. She is said to be sly and treacherous, hiding her talons in her velvety paws as long as she is in a good temper, but ready to use them upon her best friends if she is crossed in her humours.

Whatever may have been the experience of those who gave so slanderous a character to the Cat, my own rather wide acquaintance with this animal has led me to very different conclusions. The Cats with which I have been most familiar have been as docile, tractable, and good-tempered as any dog could be, and displayed an amount of intellectual power which would be equalled by very few dogs, and surpassed by none.

With regard to the comparatively good and bad temper of the Cat and dog, there is as much to be said in favour of the former as of the latter animal, while, as to their mental capacities, the scale certainly does not preponderate so decidedly on the side of the dog as is generally imagined. Nor is my own experience a solitary one, for in almost every instance where my friends have possessed favourite Cats the result has been the same.

For example, the following lines are an extract from a letter, which was sent to me, narrating the habits of two of these animals :—

" I must now tell you something about our Mincing Lane Cats. Their home was the cellar, and their habits and surroundings, as you may imagine from the locality, were decidedly commercial. We had one cunning old black fellow, whose wisdom was acquired by sad experience. In early youth he must have been very careless ; he was then always getting in the way of the men and the wine cases, and frequent were the disasters he suffered from coming into collision with moving bodies. His ribs had been often fractured, and when Nature repaired them she must have handed them over to the care of her 'prentice hand,' for the work was done in rather a rough and knotty manner. This battered and suffering pussy was at last assisted by a younger hero, who, profiting by the teachings of his senior, managed to avoid the scrapes which had tortured the one who was self-educated.

These two Cats, senior and junior, appeared to swear (Cats will swear) eternal friendship at first sight. An interchange of good offices between them was at once established. 'Senior' taught 'junior' to avoid men's feet, and wine cases in motion, and pointed out the favourite hunting grounds, while 'junior' offered to his mentor the aid of his activity and physical prowess.

Senior had a cultivated and epicurean taste for mice, which he was too old to catch ;

he therefore entered into a solemn league and covenant with 'junior' to the following effect : It was agreed between these low contracting powers that 'junior' should devote his energies to catching mice for the benefit of 'senior,' who, in consideration of such feudal service, was daily to relinquish his claim to a certain allowance of cat's meat in favour of 'junior.'

This curious compact was actually and seriously carried out. It was an amusing and touching spectacle to behold young pussy gravely laying at the feet of his elder the contènts of his 'game bag;' on the other hand, 'senior,' true to his bargain, licked his jaws and watched 'junior' steadily consuming a double share of cat's meat.

'Senior' had the rare talent of being able to carry a bottle of champagne from one end of the cellar to the other, perhaps a distance of 150 feet. The performance was managed in this wise. You gently and lovingly approached the Cat, as if you did not mean to perpetrate anything wicked ; having gained its confidence by fondly stroking its back, you suddenly seized its tail, and by that member raised the animal bodily from the ground, its fore-feet sprawling in the air ready to catch hold of any object within reach. You then quickly bring the bottle of wine to the seizing point ; pussy clutches the object with a kind of despairing grip. By means of the aforesaid tail you carefully carry pussy, bottle and all, from one part of the cellar to another. Pussy, however, soon became disgusted with this manœuvre, and when he saw a friend with a bottle of champagne looming in the distance, he used to beat a precipitate retreat. So ends my tale."

In the course of this description of the Domestic Cat, I shall endeavour to introduce, as far as possible, entirely new anecdotes of this animal, which will bring forward certain traits of character that have never yet been laid before the public notice. Many of the incidents which will be recorded in the following pages are sufficiently wonderful to call forth an incredulous smile on the part of those who have no sympathy with this graceful and intelligent animal, and who have not given to its intellectual capacities the credit which they deserve. I therefore think it needful to state that every narrative of feline character which will be found in this work, either occurred within my own knowledge, or is substantiated by the authority of the correspondents who have favoured me with their narratives, many of whom enjoy a world-wide reputation in the realms of literature and science.

From putting forward some of these statements I have somewhat shrunk, knowing the incredulity which meets any controversion of a popular prejudice. But it seems a species of cowardice to withhold the truth through fear of opposition or ridicule, and, therefore, the following narratives are laid before the public simply because they are true, and not because they are credible.

The two anecdotes which have been just narrated will convey to the mind of any unprejudiced reader a certain respect for the amount of intellectual power possessed by both these animals, and for the exceeding good temper of the elder Cat while employed in his unwilling task of wine porterage.

As a general rule, a Cat that is well treated is as kindly an animal as a dog under similar circumstances, and towards young children still more so. There is, perhaps, no animal which is so full of trust as a Cat which is kindly treated, and none which, when subjected to harshness, is so nervously suspicious. Its very trustfulness of nature seems, when rebuffed, to react so forcibly upon its sensitive disposition as to cause an entire change of character, and fills it with a shy, timid suspicion. I have had many Cats, and never yet found one which would not permit almost any liberty to be taken with it. Indeed, there are few dogs which would suffer, without resentment, such unceremonious treatment as my Cats were called upon to meet daily.

One of these Cats, a huge, dignified, portly animal, would let me pick him up and carry him about in the most disrespectful manner. Any part of his body or limbs served as a handle, and he might be lifted by one or more of his legs, by a handful of his loose skin, by his tail, by his head, or by any portion of his person that happened to be most convenient, and would endure this ungracious manipulation with unruffled composure. Or he might be pitched into the air from one person to another, and used in the light of a quadrupedal ball without even uttering a sound of displeasure. Or he might be employed as a footstool, a " boa," or a pillow, and in either case would placidly go to sleep.

This kind of behaviour was the more extraordinary because his natural disposition was of a peculiarly dignified character, and no human being could feel a slight more keenly than " Purruts." Those whom he favoured with his confidence, and they were but few, might toss him about, make him jump over their hands, or leap on their shoulders and walk along their extended arms, and he would remain calm and complacent. But let any one laugh at him, and he immediately asserted his dignity by walking away very slowly, with his tail very upright and his whole person swaggering from side to side in a most self-asserting manner.

Only a short time ago, died one of the most accomplished and singular Cats that ever caught a mouse or sat on a hearth-rug.

Her name was " Pret," being an abbreviation of " Prettina," a title which was given to her on. account of the singular grace of her form and the beauty of her fur, which was soft as that of a chinchilla. Her colour was a very light grey tabby, and she was remarkable for an almost humanly expressive countenance, and an exceedingly long nose and tail. Her accomplishments were all self-taught, for she had never learned the usual routine of feline acquirements.

" Pret " was brought when quite a kitten from the Continent, being one of a rather peculiar breed of Cats, remarkable for the length of their tails and the softness of their fur. She accompanied her mistress in rather a lengthened journey, and finally settled down in England, not very far from the metropolis. Her mistress kindly sent me the following account of " Pret's " conduct during a severe illness :—

" Three years ago I had a lovely kitten presented to me. Her fur was of a beautiful blue-grey colour, marked with glossy black stripes, according to the most approved zebra or tiger fashion. She was so very pretty that she was named ' Pret,' and was, without exception, the wisest, most loving, and dainty pussy that ever crossed my path.

When Pret was very young, I fell ill with a nervous fever. She missed me immediately in my accustomed place, sought for me, and placed herself at my door until she found a chance of getting into the room, which she soon accomplished, and began at once to try her little best to amuse me with her little frisky kitten tricks and pussy-cat attentions. But soon finding that I was too ill to play with her, she placed herself beside me, and at once established herself as head nurse. In this capacity few human beings could have exceeded her in watchfulness, or manifested more affectionate regard. It was truly wonderful to note how soon she learned to know the different hours at which I ought to take medicine or nourishment ; and during the night, if my attendant were asleep, she would call her, and, if she could not awake her without such extreme measures, she would gently nibble the nose of the sleeper, which means never failed to produce the desired effect. Having thus achieved her purpose, Miss Pret would watch attentively the preparation of whatever was needed, and then come and with a gentle purr-purr announce its advent to me.

The most marvellous part of the matter was, her never being five minutes wrong in her calculations of the true time, even amid the stillness and darkness of night. But who shall say by what means this little being was enabled to measure the fleeting moments, and by the aid of what power did she connect the lapse of time with the needful attentions of a nurse and her charge ? Surely we have here something more than reason."

The never-failing accuracy of this wise little Cat was the more surprising, because she was equally infallible by night or day. There was no striking clock in the house, so that she could not have been assisted by its aid ; nor was it habit, for her assiduous attentions only began with the illness, and ceased with the recovery of the invalid. Instinct, popularly so called, will not account for this wonderful capability so suddenly coming into being, and so suddenly ceasing. Surely some spirit-guiding power must have animated this sympathetic little creature, and have directed her in her labour of love.

No animals seem to require human sympathy so much as Cats, or to be so capable of giving sympathy in return. "Pret" knew but one fear, and had but few hates. The booming sound of thunder smote her with terror, and she most cordially hated grinding-organs and singular costumes. At the sound of a thunder-clap poor Pret would fly to her mistress for succour, trembling in every limb. If the dreaded sound occurred in the night or the early morning, Pret would leap on the bed, and creep under the clothes as far as the very foot. If the thunder-storm came on by day, Pret would jump on her mistress' knees, put her paws round her neck, and hide her face between them.

She disliked music of all kinds, but bore a special antipathy to barrel organs; probably because the costume of the organ grinder was unpleasing to her eyes, as his doleful sounds to her ears. But her indignation reached its highest bounds at the sight of a Greenwich pensioner, accoutred in those grotesque habiliments with which the crippled defenders of the country are forced to invest their battered frames. It was the first time that so uncouth an apparition had presented itself to her eyes, and her anger seemed only equalled by her astonishment. She got on the window-sill, and there chafed and growled with a sound resembling the miniature roar of a small lion.

When thus excited, she used to present a strange appearance, owing to a crest, or ridge of hair, which used to erect itself on her back, and extend from the top of her head to the root of her tail, which latter member was marvellously expanded. Gentle as she was in her ordinary demeanour, Pret was a terrible Cat to fight when she saw cause, and seemed to be undaunted by size or number. She was amusingly jealous of her own territories, and if a strange Cat dared to come within range of her special domain would assault the intruder furiously, and drive it away.

She had a curious habit of catching mice by the very tips of their tails, and of carrying the poor little animals about the house, dangling miserably from her jaws. Apparently, her object in so doing was to enable her to present her prey uninjured to her mistress, who she evidently supposed would enjoy a game with a mouse as well as herself; for, like human beings, she judged the character of others by her own.

This strange custom of tail-bearing was carried into the privacy of her own family, and caused rather ludicrous results. When Pret became a mother, and desired to transport her kittens from one spot to another, she followed her acquired habits of porterage, and tried to carry her kittens about by the tips of their tails. As might be supposed, they objected to this mode of conveyance, and sticking their claws in the carpet, held firmly to the ground, mewing piteously, while their mother was tugging at their tails. It was absolutely necessary to release the kittens from their painful position, and to teach Pret how a kitten ought to be carried. After a while she seemed to comprehend the state of things, and ever afterwards carried her offspring by the nape of the neck.

At one time, while she was yet in her kittenhood, another kitten lived in the same house, and very much annoyed Pret by coming into the room and eating the meat which had been laid out for herself. However, Pret soon got over that difficulty by going to the plate as soon as it was placed in her accustomed spot, picking out all the large pieces of meat, and hiding them under a table. She then sat quietly, and placed herself as sentry over her hidden treasure, while the intruding Cat entered the room, walked up to the plate, and finished the little scraps of meat that Pret had thought fit to leave. After the obnoxious individual had left the room, Pret brought her concealed treasures from their hiding-place, and quietly consumed them. I never saw a more dainty Cat than Pret. She would not condescend to eat in the usual feline manner, but would hitch the talons of her right paw into the food that was given to her, carrying it to her mouth as delicately as if she had been accustomed to feed herself with a fork.

One curious little trait in her character is deserving of notice. She detested to see a pin, whether belonging to the hair or the dress, and devoted her energies to extracting the offending articles of costume, and laying them on the table.

In her friendships as well as her antipathies she was somewhat peculiar. She made acquaintance at one time with a puppy, a rabbit, and a game cock, and for the time was very affectionate in her conduct towards these strange allies.

She had curious tastes for a Cat, preferring well sweetened tea to milk, and bread crusts to meat. Moreover, she would not eat her meals unless the dish were placed near her mistress, and if this wish were not gratified, always sniffed contemptuously and turned away. She was an enthusiastic mouser, but her greatest talents were displayed in the capture of sparrows. She was accustomed to creep quietly into the garden, and to seek concealment under the thickest foliage that she could find. Being thus hidden from the watchful eyes of the little birds which flock in such numbers and with such easy impertinence to the suburban gardens, Pret would imitate the chirping of the sparrows with such wonderful success that she repeatedly decoyed a heedless sparrow within reach of her spring, leaped upon it, and carried it off in triumph to her mistress. While engaged in this singular vocal effort, she used to contort her mouth in the strangest manner, forcing her lower jaw so far from side to side, that it appeared every moment to be in danger of dislocation. On such occasions the distortion of the features was so great as to make her absolutely ugly.

She was one of the most playful Cats that I ever knew, and, even to the very last hours of her existence, would play as long as she had power to move a limb. Although the mother of several families, she was as gamesome as a kitten, and delighted in getting on some elevated spot, and dropping a piece of paper or a handkerchief for the purpose of seeing it fall. More than once she got on a chest of drawers, and insinuating her supple paw into a drawer that had been left slightly open, hooked out every article of apparel that it contained and let them drop on the floor.

When any one was writing, Pret was apt rather to disconcert the writer. She always must needs try her skill at anything that her mistress did, and no sooner was the pen in motion than Pret would jump on the table, and seizing the end of the pen in her mouth, try to direct its movements in her own way. That plan not answering her expectations, she would pat the fresh writing with her paw, and make sad havoc of the correspondence.

Clever as Pret was, she sometimes displayed a most unexpected simplicity of character. After the fashion of the Cat tribe, she delighted in covering up the remnants of her food with any substances that seemed most convenient. She was accustomed, after taking her meals, to fetch a piece of paper and lay it over the saucer, or to put her paw into her mistress' pocket, and extract her handkerchief for the same purpose. These little performances showed some depth of reasoning in the creature, but she would sometimes act in a manner totally opposed to rational action. Paper and handkerchiefs failing, she has been often seen, after partly finishing her meal, to fetch one of her kittens, and to lay it over the plate, for the purpose of covering up the remaining food. When kitten, paper, and handkerchief were all wanting, she did her best to scratch up the carpet and to lay the torn fragments upon the plate. She has been known, in her anxiety to find a covering for the superabundant food, to drag a table-cloth from its proper locality, and to cause a sad demolition of the superincumbent fragile ware.

Some of her offspring have partaken considerably of their mother's soft fur and gentle nature, but none of them are so handsome as their parent. One of her kittens, called "Minnie," was removed, and conveyed to another household, where was a young canary which I had bred. The Cat and the bird were formally introduced to each other, and for a time all went well. One day, however, the kitten, then three parts grown, was seen perched on the top of the wires, her paw being thrust into the cage. At first, the Cat seemed to be engaged in an attack upon the bird, but on a closer inspection it appeared that Minnie was simply playing with the little bird and was stroking its head with her soft paw, the canary seeming to comprehend the matter, and to be rather pleased with the caresses of the velvet paw than alarmed at the proximity of its natural enemy.

After a while, Minnie herself became a mother, and I conveyed herself and kitten to her former home. Although she had not seen the house since her early kittenhood, she recognised the locality at once, and pulling her kitten out of its basket, established it in her accustomed bed on the sofa. One of her offspring is now domiciled in my own house, and there was rather a quaint incident in connexion with its departure.

Minnie knew perfectly well that her kitten was going away from her, and after it had been placed in a little basket, she licked it affectionately, and seemed to take a formal farewell of her child.   When next I visited the house, Minnie would have nothing to do with me, and when her mistress brought her to me, she hid her face in her mistress' arms.   So I remonstrated with her, telling her that her little one would be better off with me than if it had gone to a stranger, but all to no purpose.   At last I said :—

"Minnie, I apologize, and I will not so offend again."

Whereupon, Minnie lifted up her head, looked me straight in the face, and voluntarily came on my knee.   Anything more humanly appreciative could not be imagined.   For many days after the abstraction of her offspring, Minnie would not approach the various spots which had been sanctified by the presence of her lost child, and would not even repose on a certain shawl, knitted from scarlet wool, which was the favourite resting place.

She is a compassionate pussy, and is mightily distressed at any illness that falls on any of the household.   When her mistress has been suffering from a severe cough, I have seen Minnie jump on the sofa, and put her paw sympathetically on the lips of the sufferer.   Sneezing seems to excite Minnie's compassion even more than coughing, and causes her to display even a greater amount of sympathy.

There are many varieties of the Domestic Cat, of which the most conspicuous are the MANX CAT and the ANGOLA.   In the accompanying engraving, the upper figure represents the former animal, and the lower the latter.   These two Cats present the strongest contrast to each other that can be imagined, the Angola Cat being gorgeous in its superb clothing of long silky hair and bushy tail, and the Manx Cat being covered with close-set fur, and possessing hardly a vestige of a tail.

A fine Angola Cat is as handsome an animal as can be imagined, and seems quite conscious of its own magnificence.   It is a very dignified animal, and moves about with a grave solemnity that bears a great resemblance to the stately march of a full-plumed peacock conscious of admiring spectators.   It is one of the largest of domestic Cats, and in its own superb manner will consume a considerable amount of food.   One of these animals, nearly the finest that I ever saw, made friends with me in a *café* at Paris, and used to sit on the table and eat my biscuits.   In order to test the creature's appetite, I once ordered two successive plates of almond biscuits, every crumb of which "Minette" consumed with a deliberate and refined air, and would probably have eaten as much more if it had been offered to her.   It must be considered, that she had plenty of friends who visited the same *café*, and that she was quietly levying contributions during the whole day and a considerable portion of the night, so that these two plates of biscuits were only taken in the usual course of events.

The Manx Cat is a curious variety, on account of the entire absence of tail, the place of which member is only indicated by a rather wide protuberance.   This want of the usual caudal appendage is most conspicuous when the animal, after the manner of domestic Cats, clambers on the tops of houses, and walks along the parapets.   How this singular variation of form came to be perpetuated is extremely doubtful, and at present is an enigma to which a correct answer has yet to be given.   It is by no means a pretty animal, for it has an unpleasant weird-like aspect about it, and by reason of its tailless condition is wanting in that undulating grace of movement which is so fascinating in the feline race.   A black Manx Cat, with its glaring eyes and its stump of a tail, is a most unearthly looking beast, which would find a more appropriate resting place at Kirk Alloway or the Blocksberg, than at the fireside of a respectable household.   Or it might fitly be the quadrupedal form in which the ancient sorcerers were wont to clothe themselves on their nocturnal excursions.

The prescience with which all animals seem to be in some measure gifted, has often excited the admiration of those who have witnessed its effects.   The Cat appears to possess an extremely large share of this gift, as has been frequently shown.   An instance of this previsional capacity occurred just before the burning of Peebles mill, in 1853.   A long account of this occurrence has been kindly sent to me, authenticated by the names of the various persons concerned in the matter, as well as by that of the writer.

MANX CAT.   ANGOLA CAT.

A family had resided for some time on the southern side of Cuddie Bridge, and had in their house a favourite Cat. Previous to the term of Michaelmas, 1852, the family changed their residence, and took a house on the opposite side of Eddlestone Water, leaving behind them the Cat, which refused to stir from her accustomed haunts. Pussy, however, took a dislike to the new inhabitants of the house, and finding her way across the bowling-green, entered into possession of the mill, where she doubtless found plenty of game. Here she remained for some eighteen months, in spite of several attempts made by her former owner to recover his lost favourite. Several times she had been captured and brought to his house, and on one occasion a kitten was retained as a hostage. But every endeavour was vain, and leaving her offspring in the hand of her detainers, and resisting all temptations, she set off again for her quarters at the mill; in her eagerness to get back to the mill even fording the river, "taking Cuddie at the broad side," as that action is popularly termed.

On the eighteenth of October, 1853, at ten o'clock in the evening, as the former owner of the Cat was standing by the church porch, his attention was caught by the fugitive Cat, which was purring and rubbing herself against his legs as affectionately as in the olden times. He took the Cat in his arms, and when he attempted to put her down, she clung tightly to his breast, and gave him to understand in her own feline language that she was going home with him. Six hours after this return of the wanderer the mill was discovered to be on fire, and in a short time was reduced to a heap of blackened and smouldering ruins.

Since that time the Cat has remained complacently with her former companions at Biggiesknowe, in spite of the ancient adage, which says that, "in Biggiesknowe, there is neither a bannock (*i.e.* oatmeal cake) to borrow nor lend." Reference will be made to this mill in a future portion of this work.

An objection may be made to the term "prescience" in this case, on the grounds that the fire might possibly have been smouldering when the Cat left the mill, and that the creature might have taken the alarm from seeing the fire in existence, and not from a prospective intimation of the future conflagration. But even supposing that this conjecture were true, it must be remembered that Cats are remarkable for their strong attachment to a fire, and that this animal would rather be attracted than alarmed by the grateful warmth of the burning wood. Moreover, from the time when the Cat found her former master to that when the fire was discovered, six hours had passed, and we may reasonably conclude that the animal had left the mill for some little time before renewing her broken acquaintance. It would be hardly probable that if the fire had been sufficiently powerful to make the Cat decamp from her residence, so many hours would have elapsed before the flames manifested themselves.

Among other differences between the habits of wild and domesticated animals, the effect which fire has upon them is very remarkable. We all know how the domestic Cat is always found near the fire, perched on the hearth-rug, or sometimes sitting inside the fender, to the imminent danger of her fur and whiskers. Yet there is nothing which so utterly terrifies the wild felidæ as the blaze of a glowing fire. Surrounded by a fiery circle the traveller sleeps secure, the waving flames being a stronger barrier between himself and the fierce hungry beasts than would be afforded by stone or wood of ten times the height.

Another Cat, also an inhabitant of Scotland, exhibited a mysterious intuitive power, which equalled, if not surpassed, that which has just been narrated.

She was the property of a newly-married couple, who resided towards the north of Scotland, where the country narrows considerably by reason of the deeply-cut inlets of the surrounding sea. Their cottage was at no great distance from the sea, and there they remained for some months. After a while the householders changed their locality, and took up their residence in a house near the opposite coast. As the intervening country was so hilly and rugged that there would have been much difficulty in transporting the household goods, the aid of a ship was called in, and after giving their Cat to a neighbour, the man and his wife proceeded by sea to their new home. After they had been settled for some weeks, they were surprised by the sudden appearance of their Cat, which presented itself at their door, weary, ragged, and half-starved. As might be expected, she was joyfully received, and soon recovered her good looks.

It is hardly possible to conceive whence the animal could have obtained her information. Even if the usual means of land transport had been taken, it would have been most wonderful that the Cat should have been able to trace the line of journey. But when, as in the present instance, the human travellers went by water, and the feline traveller went by land, there seems to be no clue to the guiding power which directed the animal in its course, and brought it safely to the desired goal.

A rather quaint use was lately made of the strange capacity which is possessed by Cats of finding their way home under difficulties which would cause almost every other animal to fail. Eighteen cats, belonging to different persons, were put in baskets, and carried by night to a distance of three miles, when they were set at liberty at a given moment. A wager was laid upon them, and the Cat that got home first was to be the winner. One of the animals arrived at its residence within an hour, and carried off the prize. Three only delayed their arrival until the next morning.

Although the natural disagreement of Cat and dog is so great that it has passed into a proverb, these two animals will generally become very friendly if they are inhabitants of the same house. In such a case the Cat usually behaves in a tyrannous manner towards her canine friend, and treats him in a most unceremonious manner. She will sit on his back and make him carry her about the room. She will take liberties with his tail or bite his ears, and if he resents this kind of treatment she deals him a pat on the nose, and either sets up her back at him defiantly, or leaps upon some elevated spot where he cannot reach her, and there waits until she supposes his ire to have subsided.

The attachment of the dog and the Cat is sometimes curiously manifested.

In a large metropolitan household there had been a change of servants, and the new cook begged as a favour to be permitted the company of her dog. Permission was granted, and the dog took up his quarters in the kitchen, to the infinite disgust of the Cat, who thought her dignity insulted by the introduction of a stranger into her own special domain. In process of time, however, she got over her dislike, and the two animals became fast friends. At last the cook left the family, and took away her dog with her.

After an absence of some length she determined on paying a visit to her former companions, her dog accompanying her as usual. Pussy was in the room when the dog entered, and flew forwards to greet him. She then ran out of the room, and shortly returned, bearing in her mouth her own dinner. This she laid before her old friend, and actually stood beside him while he ate the food with which she so hospitably entertained him. This anecdote was related to me by the owner of the Cat.

The extraordinary electrical character of the Cat is well known. On a cold, bright day, if a Cat be stroked, the hairs of the fur bristle up, and electrical sparks issue therefrom, accompanied with a slight crackling.

It appears, too, that the animal may be so surcharged with electricity that it will give a severe shock to the holder. In order to obtain this result, the Cat should be placed on the knees, and one hand applied to its breast while the other is employed in stroking its fur. Cracklings and sparkles soon make their appearance, and in a short time, if the party continues to stroke the animal, he will receive a sharp electrical shock that may be felt above the wrists. The Cat seems to suffer as much as the experimenter, for on giving forth the shock she springs to the ground in terror, and seldom will permit a repetition of the same process.

This electrical endowment may probably account for the powerful effects which are produced upon Cats by slight means. For example, if a hair from her mistress' head were laid upon "Pret," the Cat would writhe about on the floor and twist her body into violent contortions, and would endeavour with all her might to shake off the object of her fears. Even the mere pointing of a finger at her side was sufficient to make her fur bristle up and set her trembling, though the obnoxious finger were at six inches' distance from her body. On account of the superabundance of electricity which is developed in the Cat, this animal is found very useful to paralysed persons, who instinctively encourage the approach of a Cat, and derive a gentle benefit from its touch. Those who are afflicted with rheumatism often find their sufferings alleviated by the presence of one of these electrically gifted animals.

It is worthy of notice that Cats do not invariably display the same amount of electricity, but give out more or less of that marvellous power, according to the person who handles them. This phenomenon is evidently caused by the different amount of electricity which resides in different individuals.

There are some persons who are so highly electrical that whenever they take off an article which they have worn next the skin slight crackling is heard, accompanied with little electrical sparks. This outpouring of electricity becomes more powerful if the person drinks some exhilarating liquids, such as wine or spirits. Many delicate experiments have been made on this interesting subject, but as yet with few and unsatisfactory results. It has, however, been elucidated, that healthy men generally are positive in their electricity, while women are negative; in both cases there is an augmentation of power, electric or otherwise, towards and during the evening. Without warmth, the electrical phenomena are not shown, so that in winter a warm atmosphere is needed for conducting the experiments properly. Rheumatic affections seem to absorb or destroy the electricity, for during their presence the phenomena cannot be obtained.

Many instances are recorded of misplaced, or rather strangely placed, affection in Cats. They have been known to have taken compassion on all kinds of animals, and to have nourished them as their own. The well-known anecdote of the Cat and the leveret, which she brought up, is too familiar to be repeated in this work, but I have been lately favoured with an account of similar conduct on the part of a Domestic Cat.

A lady possessed a young rabbit, which fell ill and was carried by its mistress to be warmed before the fire. While it was lying on the hearth-rug the Cat entered the room, and seeing the sick rabbit went up to it, and began to lick and fondle it as if it had been one of her own kittens. After a while she took it by the neck, in the usual manner which the Cat adopts for the transportation of her young, and carrying it upstairs laid it in her own bed, which was snugly made up in a bandbox. However, her benevolent wishes were frustrated, for in spite of the attention which she lavished on her *protégée*, the poor little rabbit continued to pine away, and at last died.

Pussy's grief was so distressing that another young rabbit was substituted, and for a while the Cat bore it to her bed, and seemed as affectionate towards the little animal as towards its predecessor. As, however, with all her benevolent intentions she could not feed the rabbit, it was taken to its own mother for the purpose of receiving the nutriment which its foster mother was unable to give. Being thus separated from each other, the temporary link that bound the two creatures together appeared to be broken, and the Cat soon forgot her dead and living foster children.

A Cat has been known to take to a family of young squirrels, and to nurture them in the place of her own little ones which had been destroyed. This circumstance took place in the vicinity of the New Forest. The squirrels were three in number.

Cats are possessed of a large organ of love of approbation, and are never more delighted than when receiving the praises and caresses of those whom they favour with their friendship. To earn such praises puss will often perform many curious feats, that of catching various animals and bringing them to her owner being among the most common. My own Cat would bring mice to me quite unhurt, and permit me to take the terrified little creatures out of her mouth. She appeared not to care what happened to her mice, only looking for her reward of caresses and laudatory words.

It would be well if our favourite Cats would restrict themselves to such game as rats and mice, for they are rather indiscriminate in their zeal, and pay a tribute which may appear very valuable to themselves, but is by no means acceptable to the receiver. For example, when pussy jumps on one's knee, and deposits a cockroach, commonly called a "black beetle," in the hands or on the shoulder, it is impossible to resist a wish that she had tempered her zeal with discretion, and either left the long-legged nauseous insect to wander where it chose, or destroyed it at once with a blow of her paw. Birds, stoats, weasels, rats, rabbits, fish, and all kinds of animals, have been thus brought as a tit-bit of affection, and on more than one occasion the owner of a grateful Cat has been startled by the sudden gift of a living snake, which has been laid writhing and hissing in his hands.

The birds and mice that have been thus captured are seldom injured, although they often feign death as soon as they are within the resistless grip of their feline foe. So, after a bird has been laid on the floor or placed in the hands, it has often been known to awake as it were from a swoon, and to fly away. Perhaps the sudden grasp of the Cat's paws and teeth may have the same effect as has been already related of the lion's teeth and claws, and for a time produce insensibility to pain, and in some instances utter unconsciousness.

When Cats have been several times deprived of their kittens they become very cunning, and conceal their little ones so closely that they rear several successive families without detection. One of our own Cats was singularly ingenious in contriving a hiding-place for herself and family ; taking advantage of some defective laths in an outhouse roof, she squeezed herself through the aperture, and made her nest in a spot between the ceiling and the slates, where she could not be reached unless the slates were removed or the ceiling broken through. We could always hear the little maternal conversations that were carried on between the mother and her children, but could never get at one of the family until they chose to emerge on their own account.

One of them turned out a thorough vagabond, and after he had attained his full growth used to scratch and bite his mother shamefully, wresting from her by force the food which was intended for herself. He was such a savage animal, and so determined a robber, that as a last resource a death warrant was issued, and would have been carried into execution but for one preventing cause—the animal would not die. He was several times shot—I have seen him knocked off a wall by a charge of shot, and laid apparently lifeless on the ground ; yet, when he was approached, he jumped up, spat, snarled, and escaped. He had an arrow through him once, he was poisoned two or three times, and was once fairly pinned to the ground in his place of refuge among some hampers, by a long, sharp, steel spike, at the end of a pole. But he would not die, and did not die ; but continued to haunt the place with such cool pertinacity that we yielded the point.

A Cat of whom I lately heard chose a very curious spot in which to rear her little family. She made a nest on the summit of a pollard oak, and there brought up her kittens. Her spot of refuge was betrayed by the little animals in the tree, who were desirous to crawl down the stem, and, not daring to adventure on so perilous an undertaking, set up a loud and pitiful mewing.

Cats really seem to vary in their temperament as much as human beings. There are refined Cats, who find their proper sphere in the drawing-room ; there are boorish Cats, who are out of their element when removed from the kitchen or cellar ; there are robber Cats—of which the vagabond animal was an example—carrying on an open system of marauding ; and there are trickish Cats, who cheat their companions of their dinners. In fine, there is hardly a trait of human character which does not find its representation in one of these animals.

Some Cats appear to have a strong sense of honour, and will resist almost every temptation when they are placed in trust. Still, some temptations appear to be so powerful that the honourable feelings cannot resist them. For example, "Minnie" will resist every lure except a piece of fried sole ; and "Pret" could never withstand the allurements of a little jug of milk or bottled stout. She would have boldly averted her head from the very same liquids if they were placed in a basin or saucer ; but the little jug, into which she could just dip her paw and lick it, possessed irresistible fascinations for her.

That the palate of a Cat should be pleased with milk is natural enough, be the milk in jug or saucer ; but that bottled stout should delight the animal appears passing strange. Yet I have known several Cats who possessed a strong taste for fermented liquids, and I have seen one of these creatures eat a piece of bread soaked in pure brandy, and beg earnestly for a further supply. I conclude these remarks upon the Domestic Cat with an authorized account of some Normandy Cats.

In a château of Normandy lived a favourite Cat, which was plentifully supplied with food, and had grown fat and sleek on her luxurious fare. Indeed, so bounteously was her plate supplied, that she was unable to consume the entire amount of provision that was set before her. This superabundance of food seemed to weigh upon her mind ; and one day, before her dinner-time, she set off across the fields, and paid a visit to a little cottage near the roadside, where lived a very lean Cat. The two animals returned to the château in company, and after the feline hostess had eaten as much dinner as she desired, she relinquished the remainder in favour of her friend.

The kind-hearted proprietor of the château, seeing this curious act of hospitality, increased the daily allowance of meat, and afforded an ample meal for both Cats. The improved diet soon exerted its beneficial effects on the lean stranger, who speedily became nearly as comfortably sleek as her hostess.

In this improved state of matters, she could not eat as much as when she was half-starved and ravenous with hunger, and so after the two Cats had dined there was still an overplus. In order to avoid waste, and urged by the generosity of her feelings, the hospitable Cat set off on another journey, and fetched another lean Cat from a village at a league's distance. The owner of the château, being desirous to see how the matter would end, continued to increase the daily allowance, and had at last, as pensioners of his bounty, nearly twenty Cats, which had been brought from various houses in the surrounding country. Yet, however ravenous were these daily visitors, none of them tonched a morsel until their hostess had finished her own dinner. My informant heard this narrative from the owner of the château.

In the conduct of this hospitably minded Cat there seems to be none of the commercial spirit, which actuated the two Mincing Lane Cats, but an open-pawed liberality, as beseems a Cat of aristocratic birth and breeding. The creature had evidently a sense of economy as well as a spirit of generosity, and blending the two qualities together, became the general almoner of the neighbouring felines. There must have also been great powers of conversation between these various animals, for it is evident that they were able to communicate ideas to each other and to induce their companions to act upon the imparted information.

THE CHAUS.—*Chaus Lybicus.*

RETURNING once more to the savage tribe of animals, we come to a small, but clearly-marked group of Cats, which are distinguishable from their feline relations by the sharply pointed erect ears, decorated with a tuft of hair of varying dimensions. These animals are popularly known by the title of Lynxes. In all the species the tail is rather short, and in some, such as the Peeshoo, or Canada Lynx, it is extremely abbreviated.

The CHAUS, our first example of the Lyncine group, is not unlike the lion in the general tawny hue of its fur, but is extremely variable both in the depth of tint and in certain indistinct markings which prevail upon the body, limbs, and tail. The fur, however, is always more grizzled than that of the lion, and there seem to be in almost every individual certain faint stripes upon the legs and tail, together with a few obscure stripes or dashes of a darker colour upon the body.

Along the back, the hue is deeper than on the sides, and on the under parts of the body the fur is of a very pale tint. The extremity of the tail is black. The markings which are found on this animal are caused by the black extremities of some of the hairs. When these black-tipped hairs are scattered, they produce the grizzly aspect which has been mentioned as belonging to this animal, but when they occur in close proximity to each other, they produce either spots, streaks, or dashes, according to their number and arrangement. On the tail, however, they always seem to gather into rings, and on the legs into stripes. The cheeks are white, and below each eye is generally a white spot. There is an under coating of soft woolly hair, which is set next to the skin, and through this woolly coating the larger hairs protrude. It is this double set of hair which gives to the fur of the Chaus its rough fulness.

The Chaus, although it has been distinguished by the specific title *Lybicus,* is an Asiatic as well as an African animal, inhabiting the south of Africa, the shores of the Caspian Sea, Persia, and many parts of India. Of the specimens which are placed in our national collection, some have been taken at Madras, some in the Mahratta territories, some in Nepal, and some in Egypt. The localities where this creature is known to frequent are generally those spots where it finds marshy, boggy ground, and plenty of thick brushwood. It does not appear to care for wooded districts, where trees grow, for it is but a poor climber, and seeks its prey only on the ground. Its food consists chiefly of the smaller quadrupeds and birds, and it is also fond of fish, which it captures in the shallow waters by watching quietly for their approach, and then adroitly scooping them from their native element by a quick sweep of its paw. River banks, especially those where the vegetation grows dense and low, are favourite resorts of the Chaus, which can in those favoured localities find its two chief requisites :—a place of concealment, from whence to pounce upon any devoted bird or quadruped that may chance to come within reach of the deadly spring, and a convenient fishing place wherein to indulge its piscatorial propensities.

THE CAFFRE CAT.—*Chaus Caffer.*

ANOTHER species of the genus Chaus, is the animal which is generally known by the name of the Caffre Cat, but which properly belongs to the Lyncine group. In colour it is rather variable, some individuals being much paler than others, the general tint of the fur being a grey, here and there grizzled with black, and diversified with dark brindlings. On the legs the stripes become bolder and better defined. When young, the fur is paler than when the animal has attained its full growth. In size it rather surpasses a large domestic cat. As may be inferred from its name, it is an inhabitant of Southern Africa, being found at the Cape, and in those lands which are inhabited by the various native tribes which are popularly termed Caffres or Kaffirs.

AMONG the Lynxes, few species are better known, at all events by name, than the common CARACAL.

This animal is easily distinguishable from the other members of the Lyncine group by its very black ears. The name Caracal is given to the animal on account of this peculiarity, the word being a Turkish one, and literally signifying Black-eared. The Greek word *melanotis* bears a similar signification. The Persians have seized upon the same characteristic mark, and have termed the creature "Siagosh," which word bears an exactly similar import to the term Caracal. The colour of this creature is a pale brown, warmed with a tinge of red, varying slightly in different individuals. The under parts of the body are paler than the upper, and slightly besprinkled with spots. The colour of these spots is very variable, for in some individuals they are nearly black, while in others they are a reddish-chestnut. The lower lip, the tip of the upper lip, and the chin are quite white. The tail is very short. It is not a very large animal, being about equal to a rather large bull-terrier dog in size, and very much more active.

It is a peculiarly ferocious and surly animal, wearing a perpetual expression of malevolence, and always appearing to be, as it truly is, ready for a snarl and a bite.

In captivity it appears to be less pervious to the gentle power of kindness than almost any other feline animal, and very rarely can be induced to lay aside a suspicious and distrustful demeanour, which characterises its every movement. Even to its keeper it displays a sullen distrust, and when a stranger approaches its cage it resents the undesired visit as if an intentional insult had been offered, laying back its ears and uttering a malignant hiss and snarl, its eyes glaring with impotent rage. Although this repulsive demeanour has generally characterised the captive Caracal, there may be individuals of a very different disposition, ready to meet the advances of their keepers, if the keepers be endowed with a nature which is capable of drawing out the better feelings of the animals under their charge. More rests with the attendants upon captive animals than is supposed, and there is many a wild beast, such as the hyæna, the wolf, or the jaguar, which has been stigmatized as untameable, simply because its keeper did not know how to tame it. Therefore it may be that the Caracal, among other animals, is only waiting for the right man to appear, and that then it will become as docile as a dog under his firm, but gentle treatment.

There is one most valuable rule, learned by long experience among wild beasts, which ought to be engraven on the heart of any one who has to deal with these animals. Never cross the creature's disposition if there be any mode of avoiding it, but if it be necessary to do so, never yield on any pretext whatever. The animal ought to think that the will of its master is absolute, and that opposition is impossible. If the man should once yield to the beast he will have forfeited the entire prestige of his position, and will have lost an amount of influence which it will be almost impossible to recover.

The Caracal is essentially predaceous, feeding upon the various animals which fall victims to its active and muscular limbs. It is said to be able to destroy the smaller deer, and to display very great craft in the chase of the swifter quadrupeds and of birds. It is not particularly fleet of foot, nor, as far as is known, delicate of scent, so that it cannot fairly run down its prey by open chase like the long-winded wolf, nor follow it up by scent like the slow but sure stoat or weasel. But it is capable of making the most surprising springs, and of leaping on its prey with a marvellous accuracy of aim. It can also climb trees, and can chase its prey among the branches on which the doomed creatures had taken up their abode.

Like the hyæna, wolf, jackal, and many other flesh-eating animals, it does not content itself with the creatures which fall by the stroke of its own talons, or the grip of its own teeth, but will follow the lion or leopard in its nocturnal quest after prey, and thankfully partake of the feast which remains after the monarch of the woods has eaten as much as he can possibly contain. In truth, the lion seems oftentimes to carry out the ludicrously arrogant pretension of certain human rulers, and to proclaim, "I, the King of the Forest, have dined. Let the monarchs of earth take *their* dinner!" As is usual among quadrupedal and bipedal royalties, the lion-king has but little chance of making a second repast of any prey which his lordly paw may have immolated, for a band of hungry courtiers assemble round the victim, and after the royal appetite has been satiated, leave nothing but a few dry bones to tell of the animal that ranged freely through the forest but an hour or two ago.

No blame attaches to the black-eared Caracal for this dependent line of conduct, for, as has already been mentioned, the lion himself disdains not to avail himself of a ready killed prey, and to gorge himself thereon with as much satisfaction as if his own paw had dealt the lethal blow.

It is said that the Caracal will sometimes call in the aid of its fellows, and with their assistance will secure even a large animal. Some authors assert that they will unite, like hounds, in the chase of their prey, and will hunt it as regularly as a pack of wolves or wild dogs. But the general opinion seems to be that the Caracal, even when assisted by its companions, gives no open chase, but achieves its end by a few powerful bounds, a

THE CARACAL.—*Cáracal Melanótis.*

stroke with the paw, and a fierce grip with the fangs on the throat of its victim. Some authors assert that the Caracal is often tamed, and rendered useful in hunting; being trained to creep upon its prey and to spring from its place of concealment upon its unsuspecting quarry. When the trained Caracal seizes its prey it crouches to the earth, and lies motionless until its owner comes up and removes the slaughtered victim.

The strength of this animal is very great in comparison with its size. A captive Caracal has been known to leap upon a large dog and to tear it in pieces, although the dog defended itself to the best of its ability.

The Caracal is spread over a very wide range of country, being known to inhabit large portions of the Asiatic and African continents. Arabia, the Cape and its vicinity, Egypt, Nubia, and Barbary, are the habitations of this animal, which is also found spread over the greater part of India and Persia. The Arabs call this animal Anak-el-ard.

By name, if not by sight, the common LYNX of Europe is familiar to us, and is known as the type of a quick-sighted animal. The eyes of the Lynx, and the ears of the "Blind

EUROPEAN LYNX.—*Lyncus virgatus.*

Mole," are generally placed on a par with each other, as examples of especial acuteness of either sense.

The European Lynx is spread over a great portion of the Continent, being found in a range of country which extends from the Pyrenees to Scandinavia. It is also found in the more northern forests of Asia.

The colour of this animal is as variable as that of the caracal, or even more so, for the same individual will change the hue of its fur according to the season of the year. During the colder months the fur becomes larger, fuller, and more grizzled, the latter effect being produced by a change in the tips of the hairs, which assume a greyish-white. The usual colour of the Lynx is a rather dark grey, washed with red, on which are placed sundry dark patches, large and few upon the body, and many and small on the limbs. On the body the spots assume an oblong or oval shape, but upon the limbs they are nearly circular. The tail of the Lynx is short, being at the most only seven or eight inches in length, and sometimes extending only six inches. The length of the body and head is about three feet.

This animal resembles the caracal in its habits and mode of obtaining prey. Sheep often fall victims to the Lynx, but it finds its chief nourishment among hares, rabbits, and other small animals. Like the caracal it is an excellent climber of trees, and chases its prey among the branches with ease and success.

The fur of the Lynx is valuable for the purposes to which the feline skin is usually destined, and commands a fair price in the market. Those who hunt the Lynx for the purpose of obtaining its fur, choose the winter months for the time of their operations, as during the cold season the Lynx possesses a richer and a warmer fur than is found upon it during the warm summer months.

The SOUTHERN, OR PARDINE, LYNX is a peculiarly beautiful example of this group of Felidæ. It inhabits more southern districts than the last-mentioned animals, being found in Spain, Sardinia, Portugal, and other southern countries. From the leopard-like spots

SOUTHERN LYNX.—*Lyncus Pardinus.*

with which its ruddy chestnut fur is covered, it derives the name of Pardine **Lynx.** Its Spanish title is Gato-clavo.

THE New World possesses its examples of the Lyncine group as well as the Old World, and even in the cold regions of Northern America a representative of these animals may be found. This is the CANADA LYNX, commonly termed the "Peeshoo" by the French colonists, or even dignified with the title of "Le Chat."

The hair of this animal is longer than that of its southern relatives, and is generally of a dark grey, flecked or besprinkled with black. Large and indistinct patches of the fur are of a sensibly darker tint than the generality of its coat. Most of the hairs are white at their extremities, which will account for the apparent changes in colour which will be seen even in the same species at different times. Along the back and upon the elbow joint these dark mottlings become more apparent. In some specimens the fur takes a slight tinge of ruddy chestnut, the limbs are darker than the rest of the body, and the ears are slightly edged with white. It is probable that the same individual undergoes considerable changes, both in the colour and the length of its fur, according to the time of year.

The limbs of this Lynx are very powerful, and the thick heavily made feet are furnished with strong white claws that are not seen unless the fur be put aside. It is not a dangerous animal, and, as far as is known, feeds on the smaller quadrupeds, the American hare being its favourite article of diet.

While running at speed it presents a singular appearance, owing to its peculiar mode of leaping in successive bounds, with its back slightly arched, and all the feet coming to the ground nearly at the same time. It is a good swimmer, being able to cross the water for a distance of two miles or more. Powerful though it be, it is easily killed by a blow on the back, a slight stick being sufficient weapon wherewith to destroy the animal. The flesh of the Peeshoo is eaten by the natives, and is said, though devoid of flavour, to be agreeably tender. It is not so prolific as the generality of the feline tribe, as the number of its young seldom exceeds two, and it only breeds once in the year. The range of this animal

CANADA LYNX, OR PEESHOO.—*Lyncus Canadensis.*

is rather extensive, and in the wide district where it takes up its residence is found in sufficient plenty to render its fur an important article of commerce.

The length of this animal slightly exceeds three feet.

BOOTED LYNX.—*Lyncus Caligatus.*

The BOOTED LYNX derives its somewhat peculiar name from the deep black colouring with which its legs are partially stained. The side and the hinder portions of the legs are partially covered with black hair, which gives the animal, when seen from behind, a quaint aspect, as if it had been endued with a pair of short tight-fitting black buskins.

The fur of this animal is rather variable in its colouring, and it is found that the coat of the female is rather more yellow than that of the male. The tail is marked with several dark rings upon a whitish ground, the tip of the tail being black.

The general tint of the fur is a deep grey, sometimes varied by a reddish tawny hue, and sometimes plentifully besprinkled with black hairs. On the upper part of the legs there are some very faint stripes of a ruddy brown, and two similar bands may be observed on the sides of the face. When young, the fur is marked with dark stripes and blotches, which are found sparingly on almost every portion of the body, but are most conspicuous on the sides. It is spread over the two vast continents of Asia and Africa, being found in the southern parts of India and the greater part of Africa, from Egypt and Barbary to the Cape.

Its food consists of the smaller quadrupeds, and such birds as it can capture. It is by no means a large animal, being barely two feet in length exclusive of the tail, which measures rather more than a foot.

CHETAH.—*Gueparda jubata.*

THE beautifully marked and elegantly formed creature which is represented in the accompanying engraving is worthy the attention of all who are interested in the wondrous influence which can be exerted by the human mind upon the very being of the lower animals. The CHETAH, Youze, or Hunting Cat, as it is indifferently named, is, like the last-mentioned animal, an inhabitant of Asia and Africa. It is rather a large animal, exceeding an ordinary leopard in stature. This superiority in size appears to be greater

CHETAH AND SPRINGBOKS.

than it is, on account of the very long limbs of the Chetah, which give it the aspect of a very large animal. The head, however, is very small in proportion to its height, and the limbs, although very long, are slender, and devoid of that marvellous strength that lies latent in the true leopard's limb.

The title "jubata," or crested, is given to the Chetah on account of a short, mane-like crest of stiff long hairs which passes from the back of the head to the shoulders. Although the Chetah is popularly termed the "Hunting Leopard," it can lay but little claim to the pardine title, and has probably been placed among the true leopards more on account of its spotted hide than for its shape and structure. The claws of this animal are but partially retractile, nor are they so sharply curved, nor so beautifully pointed, as those of the leopard. The Chetah is unable to climb trees like the leopard, and in the general contour of its body evidently forms one of the connecting links between the feline and the canine races.

The Chetah is one of those animals which gain their living by mingled craft and agility. Its chief food is obtained from the various deer and antelopes which inhabit the same country, and in seizing and slaying its prey no little art is required. The speed of this animal is not very great, and it has but little endurance; so that an antelope or a stag could set the spotted foe at defiance, and in a short half-hour place themselves beyond his reach. But it is the business of the Chetah to hinder the active and swift-footed deer from obtaining that invaluable half-hour, and to strike them down before they are aware of his presence.

In order to obtain this end, the Chetah watches for a herd of deer or antelopes, or is content to address himself to the pursuit of a solitary individual, or a little band of two or three, should they be placed in a position favourable for his purpose. Crouching upon the ground so as to conceal himself as much as possible from the watchful eyes of the intended prey, the Chetah steals rapidly and silently upon them, never venturing to show himself until he is within reach of a single spring. Having singled out one individual from the herd, the Chetah leaps upon the devoted animal and dashes it to the ground. Fastening his strong grip in the throat of the dying animal, the Chetah laps the hot blood, and for the time seems forgetful of time or place.

Of these curious habits, the restless and all-adapting mind of man has taken advantage, and has diverted to his own service the wild destructive properties of the Chetah. In fact, man has established a kind of quadrupedal falconry, the Chetah taking the place of the hawk, and the chase being one of earth and not of air. The Asiatics have brought this curious chase to great perfection, and are able to train Chetahs for this purpose in a wonderfully perfect manner.

When a Chetah is taken out for the purpose of hunting game, he is hooded and placed in a light native car, in company with his keepers. When they perceive a herd of deer, or other desirable game, the keepers turn the Chetah's head in the proper direction, and remove the hood from his eyes. The sharp-sighted animal generally perceives the prey at once, but if he fails so to do the keepers assist him by quiet gestures.

No sooner does the Chetah fairly perceive the deer than his bands are loosened, and he gently slips from the car. Employing all his innate artifices, the quadrupedal hunter approaches the game, and with one powerful leap flings himself upon the animal which he has selected. The keepers now hurry up, and take his attention from the slaughtered animal by offering him a ladleful of its blood, or by placing before him some food of which he is especially fond, such as the head and neck of a fowl. The hood is then slipped over his head, and the blinded animal is conducted patient and unresisting to the car, where he is secured until another victim may be discovered.

It is a very curious fact, that although the Chetah is found in Africa as well as in Asia, it has not been subjected to the dominion of man by the African races, but is suffered to roam at large, unfettered and unblinded.

The natural disposition of this pretty creature seems to be gentle and placid, and it is peculiarly susceptible of domestication. It has been so completely trained as to be permitted to wander where it chooses like a domestic dog or cat, and is quite as familiar as that animal. Even in a state of semi-domestication it is sufficiently gentle. One sleek and well-conditioned specimen with which I made acquaintance behaved in a very friendly manner, permitting me to pat its soft sides, or stroke its face, and uttering short self-sufficient sounds, like the magnified purr of a gratified cat. Unfortunately, the acquaintance was rudely broken up by an ill-conditioned Frenchman, who came to the front of the cage, and with his stick dealt the poor animal a severe thrust in the side. The Chetah instantly lost its confident expression, and was so irritated by this rough treatment that it would not permit a repetition of the former caresses.

Certainly these caged animals have a wondrous perception of the intentions of those who visit them. I heard one curious instance of forbearance on the part of a caged tiger.

A little girl, about five or six years of age, was taken to see the lions and tigers in a travelling menagerie. They presented to her mind the idea that they were simply very large cats, only differing in size from her favourite cat at home. So she crept close to the cage, and getting on a stone, in order to lift her small person to a proper elevation, fearlessly thrust her arm through the bars, and began to stroke the nose of the tiger. The spectators, seeing the child thus engaged, very unwisely set up a general scream, which had the effect of startling the tiger, and of making it so suspicious, that a second attempt to stroke it now would have probably resulted in the loss of the arm.

The fur of the Chetah is rather rough, and is by no means so smooth as that of the African or Asiatic leopard. Its colour is very similar to that of the leopard, but the ground colour of the fur is of a deeper fawn. The spots which so profusely stud the body and limbs are nearly round in their form, and black in their tint. Excepting upon the face there seem to be no stripes like those of the tiger, but upon each side of the face there is a bold black streak which runs from the eye to the corner of the mouth. The hair about the throat, chest, and flanks is rather long, and gives a very determinate look to the animal.

The Chetah is known as an inhabitant of many parts of Asia, including India, Sumatra, and Persia, while in Africa it is found in Senegal, and at the Cape of Good Hope.

HYÆNAS AND BUFFALO.

HYÆNAS.

THE group of animals which are so well known by the title of HYÆNAS, are, although most repulsive to the view, and most disgusting in their habits, the very saviours of life and health in the countries where they live, and where there is necessity for their existence. In this land, and at the present day, there is no need of such large animals as the Hyænas to perform their necessary and useful task of clearing the earth from the decaying carcasses which cumber its surface and poison its air, for in our utilitarian age even the very hairs from a cow's hide are turned to account, and the driest bones are made to subserve many uses. We need not the Hyænas, with their strong teeth, their powerful jaws, their rapid digestion, and their insatiable appetite. For the animal substances which are cast out unburied on our land are generally either eaten or buried by certain of the insect tribes, who are of a verity visible providences to us, assimilating into their own being, or that of their progeny, the putrefying matter that, but for their providential interference, would pour out clouds of poisoned gases, rife with pestilence and disease.

In those countries, as well as in our own, there are carnivorous and flesh-burying insects, which consume the smaller animal substances ; but the rough work is left to those industrious scavengers the Hyænas, which content themselves with the remains of large animals.

In the semi-civilized countries of Africa and Asia, the Hyæna is a public benefactor, swallowing with his accommodating appetite almost every species of animal substance that can be found, and even crushing to splinters between his iron jaws the bones which would resist the attacks of all other carnivorous animals.

There are several species of Hyænas, which are found in Asia and Africa, such as the Striped Hyæna, sometimes called the Crested Hyæna, or Strand Wolf, the Brown Hyæna, and the Tiger Wolf, or Spotted Hyæna. The habits of all these animals are very similar. The animals comprising this group are remarkable for their slouching, shambling gait, which is caused by the disproportion that exists between their legs. The fore-legs, which are used for digging, are powerful and well developed, but the hinder pair are so short that the line of the back slopes suddenly downwards from the hips, and gives to the creature a most sneaking and cowardly look. There are only four toes on each foot.

Useful as is the Hyæna when it remains within its proper boundaries, and restricts itself to its proper food, it becomes a terrible pest when too numerous to find sufficient nourishment in dead carrion. Incited by hunger, it hangs on the skirts of villages and encampments, and loses few opportunities of making a meal at the expense of the inhabitants. It does not openly oppose even a domestic ox, but endeavours to startle its intended prey, and cause it to take to flight before it will venture upon an attack. In order to alarm the cattle it has a curious habit of creeping as closely as possible to them, and then springing up suddenly just under their eyes. Should the startled animals turn to flee, the Hyæna will attack and destroy them ; but if they should turn to bay, will stand still and venture no farther. It will not even attack a knee-haltered horse. So it often happens that the Hyæna destroys the healthy cattle which can run away, and is afraid to touch the sickly and maimed beasts which cannot flee, and are forced to stand at bay.

Among the warlike tribes that inhabit the greater part of Africa this cowardly disposition throws a sad discredit on the animal, and they lavish upon the Hyæna their copious vocabulary of abusive terms. Even a weapon which has been used for the purpose of killing a Hyæna is held by them as entirely defiled, and rendered unfit for the use of a warrior. Jules Gérard relates an incident of Hyæna hunting, which, although it reflects a little upon himself, he narrates with much humour.

He had left the encampment, and was proceeding hurriedly along the path, when he suddenly came upon a rough, hairy animal, which had been surprised by daybreak, and was shambling along towards its home with a limping, hobbling gait, and an air of blank astonishment. The animal, a Hyæna, made off as fast as it could, and the hunter, having left his gun with an attendant who was lingering behind, was fain to draw his sabre, and charge the retreating beast as he best could. The Hyæna was too quick for him, and plunging among the bushes disappeared into a cavity at the foot of a rock.

The hunter was determined to secure the animal if he could, so he tied his horse to a bush, and crawled into the little cavern. When fairly inside he found that he was within a deserted stone quarry, where he could stand erect and freely use his arms. The cavern was so dark, however, that he could not see the Hyæna, and the only indication of its presence was afforded by its teeth grinding upon the sword-blade, and endeavouring

STRIPED OR CRESTED HYÆNA, OR STRAND WOLF.—*Hyæna striata.*

to drag the weapon from his hand. In a few moments his eyes became accustomed to the obscurity, and he could perceive the Hyæna still holding on to the point of the sword. A sudden effort sufficed to free the weapon, and with a quick thrust, the blade was buried to the hilt in the creature's breast, laying the Hyæna dead on the floor of the cave.

Just as M. Gérard had withdrawn the dripping sword, and was about to drag the slain animal from the cave, his attendant arrived, accompanied by some negroes whom he had pressed into the service.

The hunter thought that he had deserved some credit for his hand-to-hand combat with so powerful an animal, and was unpleasantly disappointed when the Arab recommended him to return thanks that he had not used his gun, and advised him to discard the ensanguined sabre, as it would betray him. Indeed he found that he had committed a woeful blunder, and that it behoved him to achieve some specially daring deed in order to stop the slanderous tongues of the Arab tribes.

He afterwards found that the Arabs scorned to use a weapon against the Hyæna, which they killed in a most unique manner.

Taking a handful of wet mud, or similar substance, and presenting themselves at the mouth of the Hyæna's den, they extend their hand to the animal, and say mockingly, " See, how pretty I will make you with this henna !" They then dash the wet compost into the creature's eyes, drag him out by a paw, and gag him before he recovers from the sudden bewilderment. The poor beast is now handed over to the women and children, who stone it to death.

These Hyænas are very fond of dog-flesh, and employ a very ingenious mode of catching their favourite prey. The female Hyæna creeps quietly, and ensconces herself behind some bush or other concealment not far from a village or a temporary encampment. Her mate then plays his part by running boldly forwards, and making himself as conspicuous as possible, so as to draw the attention of some of the multitudinous dogs which prowl about human habitations. Out rush the dogs at the sight of the intruder, and the Hyæna runs off as fast as he can, taking care to pass near the spot where his mate is lying concealed. The result may be imagined.

It is not often the case that the Hyæna will commit itself to so bold an action, for it is never known to be venturesome unless compelled by dire hunger.

The STRIPED HYÆNA is easily to be distinguished from its relations by the peculiar streaks from which it derives its name. The general colour of the fur is a greyish-brown, diversified with blackish stripes, which run along the ribs, and upon the limbs. A large singular black patch extends over the front of the throat, and single black hairs are profusely scattered among the fur. When young, the stripes are more apparent than in adult age, and the little animal has something of a tigrine aspect about its face. The reason for this circumstance is twofold ; firstly, because the groundwork of the fur is lighter than in the adult Hyæna ; and secondly, because the stripes are proportionately much broader than in the full-grown animal, and therefore occupy more space.

Although the Hyæna is so cowardly an animal, yet, like all cowards, it becomes very bold when it finds that it can make its attack with impunity. Emboldened by numbers, and incited by fierce hunger, the Hyenas become the very pests of the native African towns ; roaming with impunity through the streets in search of the garbage that is plentifully flung from the houses, and conducting themselves with the greatest impudence. At nightfall the inhabitants are fain to close their doors firmly, for these dangerous brutes have been known to seize a sleeping man, and to kill him with the terrible grip of their powerful jaws.

In proportion to its size, the Hyæna possesses teeth and jaws of extraordinary strength, and between their tremendous fangs the thigh-bones of an ox fly in splinters with a savage crash that makes the spectator shudder. The skull of this animal is formed in a manner that at once points it out as belonging to a creature of enormous power. The "zygomatic" arches of bone that extend from the eyes to the ears are of exceeding strength and thickness ; and along the top of the head there runs a deep bony crest that projects beyond the brain-cavity, and serves for the attachment of the powerful muscles to which the animal owes its singular strength. So forcibly are these muscles exerted that the vertebræ of the neck are sometimes found to have united together— "anchylosed" according to the professional term, on account of the violent tension to which they were continually subjected.

The muzzle is but short, and the rough thorn-studded tongue is used, like that of the feline groups, for rasping every vestige of flesh from the bones of the prey.

The BROWN HYÆNA is so named on account of the colour of its fur, which is of a blackish-brown tint, diversified with a lighter hue upon the neck and throat, and a few indistinctly marked bands of a blackish-brown across the legs. The hair of this species is extremely long, and has a decided "set" backwards.

Sometimes the brown hue of the fur is washed with a warmer tint of chestnut, from which circumstance the animal has been termed "Crocuta rufa," the latter word signifying a ruddy hue, and being applied especially to hair.

THE last of the three acknowledged species of Hyæna is a larger and heavier built animal than either of the preceding species, from which it is easily distinguishable by the numerous and well-defined spots that are scattered over its body and limbs. The SPOTTED HYÆNA, or Tiger Wolf, as it is generally called, is, for a Hyæna, a fierce and dangerous animal, invading the sheep-folds and cattle-pens under the cover of darkness, and doing in one night more mischief than can be remedied in the course of years.

The spots, or rather the blotches, with which its fur is marked, are rather scanty upon the back and sides, but upon the legs are much more clearly marked, and are set closer

BROWN HYÆNA.—*Crocúta Brúnnea.*

together. The paws are nearly black. In the collection of the British Museum is a very young specimen, which, curiously enough, is devoid of the spots that mark its adult fur, thereby presenting a remarkable contrast to the animals which we have already mentioned. For example, the lion, which in mature age is of a uniform tawny hue, is covered when young with spots and stripes, which seem to partake equally of the tigrine and pardine character. The young puma, again, exhibits strongly marked spots of a deeper hue upon its pale tawny fur, and retains them for a considerable time. Indeed, even in the fur of an adult puma may be discerned the remnants of these maculations when the animal is placed in certain lights. The Striped Hyæna, again, exhibits more decisive markings while young than after it has attained its full growth, and there are many other similar instances. These examples would seem to justify the idea, that the young of these and similar animals were deeper in their colouring than their parents. Yet, in direct opposition to this seeming rule, we find the young of the Spotted Hyæna to possess a simple, ruddy, brown fur, similar in colour to that of the Brown Hyæna. It is worthy of notice, that whatever dark spots, stripes, or blotches exist upon an animal, whether in its young or its adult state, they may always be found either upon the back, following the line of the vertebræ, or upon the legs. And even in those numerous cases where, as in the leopard, tiger, ocelot, and other striped and spotted animals, the dark markings are persistent through the entire life of the creature, these dark spots and stripes are always found to be more powerfully developed upon the spine and on the legs. I would here offer a suggestion : that we may find a key to this curious enigma in the fact, that the darker fur seems, in these animals, to accompany the chief voluntary nerves, and therefore to become more conspicuous upon the

TIGER WOLF, OR SPOTTED HYÆNA.—*Crocuta maculata.*

line of the all-important nervous column that runs along the back, and of the great branch nerves which supply power and energy to the limbs. It will be borne in mind that the complicated ganglionic system of nerves that intertwines itself among the vital organs, and is woven into such manifold reticulations on the "epigastrium," is of a different character from the round cord-like nerves of motion, and is found mostly in those parts of the body where the fur is palest.

The Tiger Wolf is celebrated for the strange unearthly sounds which it utters when under the influence of strong excitement. The animal is often called the "Laughing Hyæna" on account of the maniacal, mirthless, hysterical laugh which it pours forth, accompanying these horrid sounds with the most absurd gestures of body and limbs. During the time that the creature is engaged in uttering these wild fearful peals of laughter it dances about in a state of ludicrously frantic excitement, running backwards and forwards, rising on its hind legs, and rapidly gyrating on those members, nodding its head repeatedly to the ground ; and, in fine, performing the most singular antics with wonderful rapidity.

The ancients, who had the vaguest possible ideas of the Hyæna, and considered it to be as fearful a foe to humanity as the lion, thought that the animal was accustomed to decoy stray travellers to its den by imitating the laughter of human revellers, and then to kill and devour those who had been deceived by the simulated revelry. Besides the ordinary teeth and claws with which a Hyæna was furnished, these ancient authors supplied the Hyæna with two additional rows of teeth in each jaw, and a supply of sharp darts at the extremity of the tail. The triple row of teeth was evidently borrowed from the shark, which is indeed a kind of marine Hyæna, and the caudal darts were clearly adopted from the skin of the porcupine.

The Hyæna is too vexatious a neighbour not to be persecuted, and frequently falls a victim to the treacherous spring-gun, in spite of the benefits which he confers on mankind by his unfailing energy in devouring every scrap of eatable food.

To set a gun for the purpose of Hyæna shooting is an easy matter, and is managed as follows. The loaded musket is fixed horizontally to a couple of posts, about the height of a Hyæna's head. A string is then fastened to the trigger, one end of which is passed behind the trigger guard, or through a ring placed for the purpose, and the other is firmly tied to a piece of meat, which is hung on the muzzle of the gun. When a passing Hyæna, prowling about in search of prey, is attracted by the meat, he seizes it between his teeth, and thus draws the trigger of the gun, lodging the bullet in his head. Tenacious of life as is the Hyæna, he falls dead on the spot.

In order to attract the notice of the Hyænas, a piece of putrid flesh is dragged along the ground so as to leave an odoriferous trail leading to the treacherous weapon.

Taught by experience, the Hyænas have become so suspicious of an object which they do not understand, and to which they are not accustomed, that the very sight of a piece of string alarms them, and guards them from self-immolation in many a trap. So the farmers, who chiefly set these explosive traps, match the creature's cunning by their own superior intellect, and substitute the stems of creeping plants for the hempen cord or leathern strings. These objects are regarded without suspicion, and by their assistance the outwitted Hyæna is laid low.

In chasing living animals the Hyæna employs the same caution that characterises his ordinary proceedings. When they seize their prey the Hyænas carefully avoid those spots where the affrighted animal might reach them with its hoofs, teeth, or horns. They never seem to spring on the animal's neck, but hang on to its flanks, dragging it to the ground by the mingled weight of their body and the pain of the wound. Many veteran oxen and horses are deeply scarred in the flanks by the teeth of the Hyæna, which has made its attack, but has been scared away or shaken off.

The eyes of the Hyænas are singularly repulsive in their expression, being round, dull, and almost meaningless.

There are man-eaters among the Hyænas, and these hominivorous animals are greatly dreaded, on account of the exceeding stealthiness and craft with which they achieve their object.

They very seldom endeavour to destroy the adult men and women, but limit their attacks to the young and defenceless children. On dark nights the Hyæna is greatly to be feared, for he can be guided to his prey by the light of the nocturnal fires which do not daunt an animal that is possessed by this fearful spirit of destructiveness, and at the same time can make his cautious approaches unseen. As the family are lying at night, buried in sleep, the Hyæna prowls round the inclosure, and on finding a weak spot the animal pushes aside the wattle bands of which the fence is made, and quietly creeps through the breach.

Between the human inhabitants and the fence, the cattle are picketed by night, and would fall an easy prey to the Hyæna if he chose to attack them. But he slips cautiously amid the sleeping beasts, and makes his way to the spot where lies a young child, wrapped in deep slumber. Employing the same silent caution, the Hyæna quietly withdraws the sleeping child from the protecting cloak of its mother, and makes its escape with its prey before it can be intercepted.

With such marvellous caution does this animal act, that it has often been known to remove an infant from the house without even giving the alarm.

It has already been mentioned that the Hyæna is in no wise fastidious in its diet, and that it will habitually consume the most indigestible of substances. Yet there seems to be something capricious about the function of assimilating food, which, even in the Hyæna, is subject to remarkable fluctuations. To one of these animals, after a fast of thirty-six hours, a dead rat was given, which, as might be expected, it immediately swallowed. In fifteen minutes, the creature rejected the skin and bones of the rat, though the same animal would have eaten with impunity the heavy bones or tough hide of a veteran ox, or even would have made a satisfactory meal on a few yards of leathern strap.

The following anecdotes of the Cape Hyæna and its habits are taken from the MS. of Captain Drayson, R.A., to which reference has already been made.

HYÆNAS QUARRELLING OVER THEIR PREY.

"THIS animal is very common in South Africa, and being cunning, and rarely venturing out by day, is likely to be longer a denizen of the inhabited districts than many other less formidable creatures. The height at the shoulder is about two feet six inches, and falls towards the rump; extreme length, about five feet ten inches. The head is short and very broad; muzzle and nose black; general colour, brown, irregularly

1.   Q

blotched with circular black spots. The tail sixteen inches; hairs on the back of the neck and withers long, forming a reversed mane.

The proper duty of this creature appears to be that of scavenger, and is, with regard to the beasts, what the vulture is to the birds; but owing to its great appetite, and naturally voracious disposition, it does not appear contented with merely the carrion which it might procure, but employs its strength and speed in destroying the flocks and herds of the colonists, or in killing such antelopes as it is enabled to capture.

If this animal possessed courage in proportion to its strength it would be a very formidable opponent to man, and, as it hunts frequently in packs, might test the skill and boldness of the hunter, but, fortunately, its principal characteristic is cowardice.

Owing to the custom prevalent amongst many of the South African tribes of exposing their dead to be devoured by beasts of prey, the Hyæna has acquired the taste for human flesh, and therefore cases are on record of the huts of Kaffirs having been entered by it, and the children carried off and devoured. Most ably does the Hyæna perform his functions in the economy of nature. Whilst the lion selects the choice parts of a slain animal, and the vulture those which he cannot eat, the Hyæna comes and finishes hide, bones, and other remnants which have been too tough for the digestion of the others.

It appears to be a law of nature that those animals which take the shortest time to fill their stomachs can go the longest time without eating. For example, the horse and the ox will take from half an hour to one hour and a half to feed, and they will both suffer if they are kept more than a day without food. The wolf and the dog can make a very satisfactory meal in about two minutes, and either can remain two or three days without suffering much for want of a meal. We may even remark that this instinctive mode of eating food is prevalent among human beings.

The rough ploughboy, whose meals are limited in number to one or two daily, and are composed of coarse bread and fat bacon, swallows in a few minutes these articles of food in great morsels which he can hardly force into his mouth, and which he scarcely takes the trouble to masticate. The food which is thus taken into the system will repel the feeling of faintness consequent on an empty stomach much more than if it were leisurely eaten and properly subjected to the action of the teeth. This result is only natural, for the better food is masticated, the sooner is it digested.

The Hyæna in the Zoological Gardens appears well acquainted with this fact, for on one occasion, being anxious to see how easily he crushed a huge bone of beef, I took my station in front of his cage, just before feeding time. After the usual laugh had been extracted from crowd and Hyæna, a leg of beef was forced under the bars, and was seized by the hysterical scavenger. A few strips of flesh were torn off and swallowed, and then there remained about nine inches of bone and sinew; instead of crushing these into little pieces, and then swallowing it, as I expected, the wise animal just turned the bone 'head on,' took it in his jaws, made a face, contorted his body, and that solid mass was deposited in the yawning sarcophagus. The crowd laughed and dispersed, but did not remark what experience had probably taught this prisoner, viz. that when he swallowed the bone whole he was not so famished by the next day's dinner-hour as when he ground it up into small pieces. This Hyæna, having but little variety of occupation for its mind, had probably devoted much patient thought to the adjustment of this fact.

The Hyæna usually lives in holes, or amongst rocks in retired localities, and when the sun has set he comes forth and searches for food. He then utters a long melancholy howl, which finishes with a sort of bark, and occasionally that fiend-like laugh which, when heard in the desert, amid scenes of the wildest description, calls up in the imagination of the solitary traveller the forms of some spectral ghouls searching for their unnatural feast.

The smell of the Hyæna is so rank and offensive that no animal, other than of its own species, will come near the carcass. Dogs, when they come across the scent of the Hyæna, at once show signs of fear; they will scarcely leave their master, and, with bristling manes and wild looks, examine every inch of ground over which they pass.

The spoor of the Hyæna is somewhat similar to, but larger than that of the dog; the nails not being retractile, usually leave an impression upon soft ground, which is not the case with the leopard. The inside toe of each foot is smaller than the outside, and the footmarks can be easily recognised and distinguished from those of dogs.

During one warm afternoon, whilst riding over the grassy slopes on the banks of the Umganie River, near Pietermaritzberg, and attended by a cunning old pointer, I saw the dog stand on the brink of an old watercourse, and bark fiercely at some object which appeared to be stationed below. I knew that the bark and the expression which accompanied it was the dog-language for 'there's something here,' so I dismounted, and walked towards the dyke. As I approached, the dog, with an aspect of alarm, sprang back, and then rushed forward again. From having had several unpleasant rencontres with poisonous snakes I had become very cautious, and advanced so slowly that I was only enabled to catch a glimpse of a Hyæna, which, upon seeing me, immediately retreated into an opening.

I descended the steep bank and found a large hole, which appeared to be the entrance to a subterraneous passage, by which the water obtained an exit. I collected a few sticks and some long grass, which I placed over the entrance, and then endeavoured to trace the course of this passage, to see if there were another opening.

About fifty yards from the first I found a second hole, which evidently led to the first; neither of these were large enough to admit me, and the dog could not have done much good even had he entered; but he appeared to have a great objection to approach too near to the den.

After some consideration, I determined to cut a quantity of the dry grass, to fill one opening with it, set it on fire, and then to watch near the other hole. This plan failed to unearth the creature, so I reversed the arrangement, but with no better success.

At length I fired several bullets into the opening, trusting that a stray shot might strike near the Hyæna, and that it would drive him into open ground. The sinuosities of the passage prevented the possibility of a fair shot.

Whilst thus engaged, the dog suddenly barked and dashed off. Upon reaching the top of the bank, I saw the Hyæna scrambling over the hills, closely followed by my dog. I mounted my pony, but the pace was too good for him. I however held the Hyæna in view for a considerable time as it passed over the successive ridges, but the pointer soon gave up his pursuit.

I think that when the ordinary game is driven away by sporting men, or killed by sportsmen, Hyæna-hunting with a pack of hounds would be found very good sport, and perhaps we should have Hyæna-hunters sneering at fox-hunters as much as some fox-hunters now do at 'thistle-whippers.'

The Hyæna is frequently caught in a trap of simple construction. Stakes are driven into the ground so as to form an inclosure, and a hanging door of stone, sustained by a cord, closes the aperture when it falls. A bait is placed at the farther end of the trap, and the whole contrivance is like a large mouse-trap. When caught, the Hyæna is despatched with spears and clubs, or is shot.

The traveller is frequently disturbed during the night by the daring Hyænas, who will sneak about his waggons in search of leather straps, trektows, and other savoury provender; and if a pair of shoes or some leather breeches happened to be left in an exposed situation during the dark hours, they may be considered lost without redemption, for such a supper would be an unlooked-for luxury by the gaunt brute."

One of these animals was discovered in a state of sad laceration. The two fore-paws were gone, and the legs themselves had been frightfully torn, evidently by some powerful beast of prey. The natives said that the Hyæna had been thus punished by the lion for interfering with his arrangements, and stated, moreover, that the lion frequently corrected the forward conduct of the Hyæna, by biting off every one of its paws. This statement, curious as it may seem, was corroborated by several experienced hunters.

Although in former days the Hyæna was supposed to be a wholly untameable animal, later experiments have shown that it is nearly as tractable and affectionate as a dog when it has the benefit of similar treatment. It has been known to accompany its master as familiarly as any dog, and to recognise him with airs of joy after a lengthened absence. The potency which some persons exert over animal natures is most remarkable. It may be that such persons pour much love upon all things, and therefore upon the animals with which they come in contact. So love, creating love,—which is the highest gift of God, and the sum of his divine attributes,—calls forth in animals the highest attributes of their nature, and, through this higher quality, develops their intellectual capacities.

AARD WOLF.—*Próteles cristátus.*

CIVETS.

It is generally the case with the greater divisions of animals that there exist certain intermediate forms of animal life, which seem to be rather higher than the one division, and lower than the other, being, in fact, transitional forms between the higher and the lower groups. Thus, the Colugo, or Flying Lemur, is an intermediate form between the monkey and bats, and the Aard Wolf is intermediate between the hyænas and the Civets, belonging, however, more to the latter than the former group of animals. It is much smaller than the hyænas, but larger than the Civets and genetts, and, indeed, has indifferently been called a hyæna, a jackal, or a Civet.

The form of the Aard Wolf much resembles that of the hyænas, the fore-quarters being powerful and well developed, and the hinder quarters low and sloping. The general aspect of the creature is very similar to that of the hyæna, for, in addition to the hyænine sloping back and weak hind legs, the fur is rough, coarse, and coloured in a manner not unlike that of the striped hyæna. The tail is very large in proportion to the size of the animal, and is thickly covered with long bushy hair, black at the extremity, and blackish-grey on the other portions of that member. The back of the neck and the shoulders are furnished with a thick bristling mane, which it can erect when excited, and it then resembles a miniature striped hyæna.

The claws of the fore-feet are sturdy, and firmly attached to the paws, so as to serve their proper use of digging. The Aard Wolf is an admirable excavator, and digs for itself a deep burrow, where it lies concealed during the day, buried in sleep at the bottom of its mine. From this habit of burrowing in the earth, the creature has derived its title of "Aard, or Earth Wolf."

A curious mode of domestic arrangement is carried out by these animals. Several individuals seem to unite in forming a common habitation. Several deep burrows are dug, having their common termination in a small chamber, where three or four Aard Wolves take up their residence. Whether each animal digs and uses its own burrow, or whether the tunnels, as well as the central chamber, are common to the inhabitants, is not known. It seems, however, to be probable that such a mode of procedure would be adopted, and that each member of the little community appropriated to itself the tunnel which its own paws had dug.

The colour of the Aard Wolf is grey, with a decided tinge of yellow. Several broad bands of darker fur are seen on the sides, and the paws are quite black. The hair of this animal is of two kinds,—a thick, short, woolly coating, which lies next to the skin, and a longer and coarser set of hairs, which protrude through the woolly coating, and hang downwards to some length. The adult Aard Wolf is about three feet six inches in total length, the tail being about a foot long.

The food of this animal is similar to that of the hyæna, and consists chiefly of carrion and small animals. It does not disdain to make an occasional meal on insects, for a number of ants were discovered in the stomach of an Aard Wolf that had been killed.

The CIVET, sometimes, but wrongly, called the Civet Cat, is a native of Northern Africa, and is found plentifully in Abyssinia, where it is eagerly sought on account of the peculiarly scented substance which is secreted in certain glandular pouches. This Civet perfume was formerly considered as a most valuable medicine, and could only be obtained at a very high price; but in the present day it has nearly gone out of fashion as a drug, and holds its place in commerce more as a simple perfume than as a costly panacea.

In this animal we may trace a decided resemblance to the Aard Wolf, both in the shape of the body and in the markings.

But the Civet bears itself in a very different manner, having more of the weasel than of the hyænine nature, and the colouring of the fur is of a much richer character than that of the previously mentioned animal.

It is nearly as large as the Aard Wolf, its total length being about three feet six inches, of which the tail occupies nearly one-third. Along the back, and even on part of the tail, runs a boldly marked crest or mane, which can be erected by the animal at pleasure, or can lie nearly, but not quite, evenly with the fur.

The substance which is so prized on account of its odoriferous qualities is secreted in a double pouch, which exists under the abdomen, close to the insertion of the tail. As this curious production is of some value in commerce, the animal which furnishes the precious secretion is too valuable to be killed for the sake of its scent-pouch, and is kept in a state of captivity, so as to afford a continual supply of the odoriferous material.

The mode by which the Civet perfume is removed from the animal is very ingenious. The animals which belong to this group are very quick and active in their movements, and, being furnished with sharp teeth and strong jaws, are dangerous beasts to handle. As may be imagined, the Civet resents the rough treatment that must be used in order to

CIVET.—*Viverra Civetta.*

effect the desired purpose, and snaps and twists about with such lithe and elastic vigour that no one could venture to lay a hand on it without sufficient precaution. So, when the time arrives for the removal of the perfume, the Civet is put into a long and very narrow cage, so that it cannot turn itself round. A bone or horn spoon is then introduced through an opening, and the odoriferous secretion is scraped from its pouch with perfect impunity. This end achieved, the plundered animal is released from its strait durance, and is permitted a respite until the supply of perfume shall be re-formed.

As the Civet might be inconvenienced by the continual secretion of this substance, Nature supplies a simple remedy, and the perfume falls from the pouch in pieces about the size of an ordinary nut. The interior of each half of the pouch is sufficiently capacious to hold a large almond. As the civet is formed, it is pressed through very small orifices into the pouch, so that if it is examined before it has merged itself into a uniform mass, it is something like fine vermicelli in appearance. The interior of the pouch is thickly coated with fine hairs, and entirely covered with the minute orifices or pores through which the perfume exudes. The creature is able to compress the pouch at will.

The Civet seems to be a very sleepy animal, especially during the daytime, and to be with difficulty aroused from its somnolence.

While it remains in the pouch, the "civet" is rather thick and unctuous, something like butter in texture.

The use which this curious secretion subserves in the economy of the creature is very dubious. It is not sufficiently liquid to be ejected against its pursuers, and so to repel them by its odour, as we know to be the case with the celebrated skunk of America, and other animals. It may be, that this substance can be re-absorbed into the system, and thus serve an important purpose; but, whatever its use may be, it is clear that it serves some worthy object, and that therefore the production of this secretion is deserving the attention of those who have the opportunity of making practical experiments.

The claws of the Civet are only partially retractile. The eyes are of a dull brown, very protuberant, and with a curiously changeable pupil, which by day exhibits a rather broad linear pupil, and glows at night with a brilliant emerald refulgence. The body is

ZIBETH.—*Viverra Zibetha.*

curiously shaped, being considerably flattened on the sides, as if the animal had been pressed between two boards.

Altogether, the Civet is a very handsome animal, the bold dashing of black and white upon its fur having a very rich effect. The face has a curious appearance, owing to the white fur which fringes the lips, and the long pure white whisker hairs of the lips, and eyes. When young, it is almost wholly black, with the exception of the white whisker hairs and the white fur of the lips. It seems to be an irritable animal, and, when angered, vents its indignation by fierce growls.

UPON the Asiatic continent, and its islands, the place of the civet is taken by several of the Viverrine tribe, one of which, the ZIBETH, bears a close resemblance to its African relative.

The Zibeth is a native of many parts of Asia, being found in China, India, the Philippines, Nepal, and other localities. It may be distinguished from the civet by the greater amount of white which is found in the fur, especially about the neck and throat, by the shorter hair, and by the greater number of dark rings upon the tail. The tail of the Zibeth is not so largely marked with black at its extremity as that of the civet. The mane or crest which runs along the back is comparatively small. The spots which mark the body are rather indistinctly outlined, and the general tint of the fur seems to be paler than that of the civet.

It is furnished with a musk-secreting pouch like that of the African civet. It is a lethargic animal in captivity, and even in a wild state passes the day in sleep, and only seeks its food after dark. Its usual diet is composed of birds and the smaller mammalia, but it will also eat various fruits, especially those of a sweet nature. In size it nearly equals the civet. In captivity it is a gentle creature, and is so completely tamed by the natives of the countries where it is found that it inhabits the house like a domestic cat and employs itself in similar useful pursuits.

THE animal which is known by the native name of TANGALUNG, bears some resemblance to the preceding animals. The black markings, however, are more distinct, and

TANGALUNG.—*Viverra Tangalunga.*

along the direction of the spine the fur is most deeply black. On the lower part of the throat and neck are three curiously shaped black bands, very wide in the middle and very narrow at each end, the central band being several times wider than the others.

The length of this animal is two feet six inches, the head measuring nearly seven inches in length, and the tail about eleven inches. The head is rather wide and rounded, and is suddenly contracted towards the nose, so as to form a rather short muzzle. The tail is nearly cylindrical, and does not taper so much as that of the zibeth, and the body is furnished with a close downy covering of soft hairs next the skin. It is partly to this woolly hair that the cylindrical outline of the tail is owing. The Tangalung is a native of Sumatra.

The RASSE is spread over a large extent of country, being found in Java, various parts of India, Singapore, Nepal, and other localities. The colour of its fur is a warm greyish-brown, upon which are placed eight parallel lines of elongated dark spots. The dark rings which mark the tail pass entirely round that member, while those which are found on the tail of the zibeth reach little more than half the circumference of the tail. The texture of the fur is rather coarse and stiff, and it is not very thickly set. The ears of this animal approach each other very closely at their base, being only separated by the space of an inch, whereas there is an interval of two inches between the ears of the zibeth.

In the Javanese language, the word "Rasa," from which the name Rasse is taken, signifies a sensation of the palate or the nostrils, so that it may be applied to the senses of smelling or tasting. It generally refers to odoriferous substances.

The perfume which is furnished by the Rasse is secreted in a double pouch, like that of the civet, and is removed from the animal in precisely the same manner. It is highly valued by the Javanese, who imbue their persons, their rooms, and their garments so strongly with this substance that a European nostril is grievously affected at the all-pervading odour. The substance itself is termed Dedes.

As far as is known of the disposition of this animal, it appears to be savage and irritable, bearing captivity very impatiently, and never losing its wild ferocious nature.

It is a very destructive creature among the animals on which it feeds, and on account of its long sharp teeth can inflict a severe bite when it is angry. In captivity it generally

RASSE.—*Viverra Malaccensis.*

feeds on eggs, various birds, and meat and fish, and a little rice. The natives say that salt is a poison to it.

DELUNDUNG.—*Linsang Grácilis.*

THE pretty animal which is represented in the accompanying engraving is remarkably rich in colouring, as well as graceful in form. The DELUNDUNG is a native of Java and

Malacca, and is destitute of the scent-pouches which are so curious a characteristic of the preceding Viverrine animals. It is not at all a common animal, and its habits are not very clearly known.

The general colour of the fur is a moderately deep grey, and upon the back are drawn four very large, saddle-shaped stripes of an exceedingly dark and rich brown, extremely broad on the spine, and becoming very narrow on the ribs. Along the sides run two rows or chains of similarly coloured markings, the upper band being occasionally merged in the broad stripes that cross the back. The lower band extends from the cheeks to the flanks. The legs are finely spotted, and the tail is covered with alternate rings of grey and dark brown, the rings becoming more distinct towards the point of the tail.

The creature has been termed Priónodon, or "Saw-tooth," on account of the curiously shaped teeth, which present a jagged, or saw-like appearance. Its limbs are very slender and delicately formed. Although a scarce animal in every part of Java, it is especially so in any part of the island except the eastern end, where it is found among the thick forests with which that locality is densely clothed.

BLOTCHED GENETT.—*Genetta Tigrina.*

GENETTS.

A SMALL, but rather important, group of the Viverrine animals, is that the members of which are known by the name of the GENETTS. These creatures are all nocturnal in their habits, as are the civets, and, like those animals, can live on a mixture of animal and vegetable food, or even on vegetable food alone. The Genetts possess the musk-secreting apparatus, which much resembles the pouch of the civet, although in size it is not so large, nor does it secrete so powerfully smelling a substance as that of the civets. The secreting organ, although it resembles a pouch, is not so in reality, being simply composed of two glands, united to each other by a strip of skin.

The best known of these animals is the COMMON, or BLOTCHED GENETT, an inhabitant of Southern Africa and of various other parts of the world, being found even in the south of France. It is a very beautiful and graceful animal, and never fails to attract attention from an observer. The general colour of the fur is grey, with a slight admixture of yellow. Upon this groundwork dark patches are lavishly scattered, and the full furry tail is covered with alternate bands of black and white. The muzzle would be entirely black but for a bold patch of white fur on the upper lip, and a less decidedly white mark by the nose. The feet are supplied with retractile claws, so that the animal can deal a severe blow with its outstretched talons, or climb trees with the same ease and rapidity which is found in the cat tribe.

Another pretty species of this genus is the PALE, or SENEGAL GENETT.

The fur of this animal is whiter than that of the Blotched Genett, and the markings are rather differently arranged. Along the spine a nearly unbroken dark stripe is drawn, and upon the neck and shoulders the spots have a tendency to merge into each other,

PALE GENETT.—*Genetta Senegalensis.*

and to form stripes, extending from the head along the neck and over the shoulders. On each side of the face is a bold black patch. The hinder legs are quite black at the ankle joint.

These animals are very susceptible of domestication, and in various Eastern districts are as familiar inhabitants of the house as the domestic cat. Like the house cat, the Genett signalizes itself in the destructive wars which it wages against rats and mice, being especially fitted for such a pursuit by its active limbs and lithe form. The Genetts seem, when wild, to prefer the low grounds in the vicinity of rivers to the higher forest lands, and are there captured.

They are not nearly so large as the civet, being only five inches in height at the shoulder, and about twenty inches in total length. The eye is of a light brown colour, and rather protuberant. The young of the Pale Genett has the spots of a light chestnut instead of the deep blackish-brown of the adult animal.

The AMER GENETT, an inhabitant of Abyssinia, is a boldly and handsomely marked creature. The general colour of its fur is a darkish yellow grey, on which are placed a number of well-defined dark spots. These markings run in fine regular lines, being

AMER.—*Genetta Amer.*

larger nearing the spine, and becoming smaller as they recede therefrom. The tail is boldly and equally covered with rings of the same dark fur as that of the spots on the body.

VERY different from the Genetts in its appearance is the CACOMIXLE, although it is closely allied to them.

CACOMIXLE.—*Bássaris Astúta.*

It is remarkable as being a Mexican representative of the Genett group of animals, although it can hardly be considered as a true Genett or a true Mungous. The colour of this animal is a light uniform dun, a dark bar being placed like a collar over the back of the neck. In some specimens this bar is double, and in all it is so narrow that when the animal throws its head backwards the dark line is lost in the lighter fur. Along the back runs a broad, singular, darkish stripe. The tail is ringed something like that of the Ringed Lemur, and is very full. The term Cacomixle is a Mexican word, and the animal is sometimes called by a still stranger name, "Tepemaxthalon." The scientific title "Bassaris" is from the Greek, and signifies a fox.

BANDED MUNGOUS.—*Mungos fasciátus.*   GARANGAN.—*Herpestes Javánicus.*

ICHNEUMONS.

THE two animals which are seen in this engraving are closely allied to each other, but are placed in different genera. The left-hand figure represents the creature which is known by the name of the BANDED MUNGOUS, and which is an inhabitant of Africa. It is a small animal, being about the size of a very large water-rat, and is peculiarly quick and energetic in its movements.

The colour of the Banded Mungous is a blackish grizzle, with a chestnut tinge pervading the hind quarters and the tail. Under the chin the fur is of a very light fawn colour. Across the back are drawn a row of darker lines, boldly marked towards the spine, but fading imperceptibly into the lighter tinted fur of the sides.

In habits it is singularly brisk and lively, ever restlessly in motion, and accompanying its movements with a curious and most unique sound, something like the croak of a raven. When excited it pours out a succession of quick chattering sounds, and when its feelings are extremely touched it utters sharp screams of rage. If its companions should cross its path in its temper it snaps and spits at them like an angry cat, and makes such very good use of its teeth that it leaves the marks of its passion for the remainder of the victim's life. Some of these animals, which have lived for a considerable time in the same cage, have lost a large portion of their tails by the teeth of their comrades. Still it is very playful, and sports with its companions in a curiously kitten-like manner.

It is extremely active with its fore-paws, armed as they are with their long claws, and scratches in a very absurd and amusing manner at anything that may take its attention. It is a very agile climber, running over the bars of its cage and up the tree-branches with great ease and rapidity, and can spring upon an object from some distance, and with admirable accuracy of aim. The eye of this animal is of a light brown, and very brilliant.

THE RIGHT-HAND figure upon the same engraving represents the GARANGAN, or Javanese Ichneumon. As is evident by the name, it is an inhabitant of Java. In size it equals the last mentioned animal. Its colour is nearly uniform, and consists of a bright rich chestnut on the body, and a lighter fawn colour on the head, throat, and under parts of the body.

This little animal is found in great numbers inhabiting the teak forests, where it finds ample subsistence in the snakes, birds, and small quadrupeds. The natives assert —whether truly or not—that when it attacks a snake it employs a ruse similar to that which is often used by a horse when it objects to being saddled. It is said to puff up its body, and to induce the snake to twine itself round its inflated person. It then suddenly contracts itself, slips from the reptile's coils, and darts upon its neck. There is some foundation for this assertion in the fact that the Garangan, in common with others of the same genus, does possess the power of inflating and contracting its body with great rapidity ; so much so, indeed, that during life it is not easy to measure the creature.

Although it is tolerably susceptible of education, it is rarely kept tame by the natives, because it is liable to occasional fits of rage, and when thus excited can inflict very painful wounds with its sharp teeth. Moreover, it is too fond of poultry to be trusted near the henroosts.

URVA, OR CRAB-EATING ICHNEUMON.—*Urva cancrivora.*

The URVA is easily distinguished from the preceding and the following animals by the narrow stripe of long white hairs that runs from the angle of the mouth to the shoulders, contrasting very decidedly with the greyish-brown tint of the rest of the fur. Some very faintly marked darker bars are drawn on the body, and the tail is marked with three or four faint transverse bars. This member is more bushy at the base than towards the extremity. The feet and legs are of a uniform dark tint.

THE ICHNEUMONS appear to be the very reptiles of the mammalian animals, in form, habits, and action, irresistibly reminding the spectator of the serpent. Their sharp and pointed snout, narrow body, short legs, and flexible form, permit them to insinuate themselves into marvellously small crevices, and to seek and destroy their prey in localities where it might well deem itself secure. There are many species of the genus Herpestes, or " creeper," one of which, the Garangan, has already been mentioned.

The common Ichneumon, or Pharaoh's Rat, as it is popularly but most improperly termed, is plentifully found in Egypt, where it plays a most useful part in keeping down the numbers of the destructive quadrupeds and the dangerous reptiles. Small and insignificant as this animal appears, it is a most dangerous foe to the huge crocodile, feeding largely upon its eggs, and thus preventing the too rapid increase of these fierce and fertile reptiles. Snakes, rats, lizards, mice, and various birds, fall a prey to this Ichneumon, which will painfully track its prey to its hiding-place, and wait patiently for hours until it makes its appearance, or will quietly creep up to the unsuspecting animal, and flinging itself boldly upon it destroy it by rapid bites with its long sharp teeth.

Taking advantage of these admirable qualities, the ancient Egyptians were wont to tame the Ichneumon, and permit it the free range of their houses, and on account of its

ICHNEUMON.—*Herpestes Ichneumon.*

habits paid it divine honours as an outward emblem of the Deity considered with regard to His sin-destroying mercy. There is much more in the symbolization of those old Egyptians than we deem, and they looked deeper into the character and the causes of outward forms than we generally suppose. Although the diminutive size of this creature renders it an impotent enemy to so large and well mailed a reptile as the crocodile, yet it causes the destruction of innumerable crocodiles annually by breaking and devouring their eggs. The egg of the crocodile is extremely small, when the size of the adult reptile is taken into consideration, so that the Ichneumon can devour several of them at a meal.

The colour of this animal is a brown, plentifully grizzled with grey, each hair being ringed alternately with grey and brown. The total length of the animal is about three feet three inches, the tail measuring about eighteen inches. The scent-gland of the Ichneumon is very large in proportion to the size of its bearer, but the substance which it secretes has not as yet been held of any commercial value. The claws are partially retractile.

The MOONGUS, sometimes called the INDIAN ICHNEUMON, is, in its Asiatic home, as useful an animal as the Egyptian Ichneumon in Africa. In that country it is an indefatigable destroyer of rats, mice, and the various reptiles, and is on that account highly valued and protected. Being, as are Ichneumons in general, extremely cleanly in manners, and very susceptible of domestication, it is kept tame in many families, and does good service in keeping the houses clear of the various animated pests that render an Indian town a disagreeable and sometimes a dangerous residence.

In its customs it very much resembles the cat, and is gifted with all the inquisitive nature of that animal. When first introduced into a new locality it runs about the place, insinuating itself into every hole and corner, and sniffing curiously at every object with which it comes in contact. Even in its wild state it exhibits the same qualities, and by a careful observer may be seen questing about in search of its food, exploring every little tuft of vegetation that comes in its way, running over every rocky projection, and thrusting its sharp snout into every hollow. Sometimes it buries itself entirely in some

MOONGUS.—*Herpestes Griseus.*

little hole, and when it returns to light drags with it a mole, a rat, or some such creature, which had vainly sought security in its narrow domicile.

While eating, the Ichneumon is very tetchy in its temper, and will very seldom endure an interruption of any kind. In order to secure perfect quiet while taking its meals, it generally carries the food into the most secluded hiding-place that it can find, and then commences its meal in solitude and darkness. The colour of the Moongus is a grey, liberally frecked with darker hairs, so as to produce a very pleasing mixture of tints. It is not so large an animal as its Egyptian relative.

The grizzled markings upon the fur of the Nyula are of a singularly beautiful character, and form a closely set zigzag pattern over the entire surface of the head, body, and limbs.

The pattern is very like that which is seen in some woven fabrics, or fine basket-work. Upon the back and body this pattern is tolerably large, but upon the head it becomes gradually smaller, and upon the upper portion of the nose is almost microscopically small, though as perfect and uniform as that upon the body, so that it is among the most elegantly coloured examples of the Ichneumons. The paws are dark, and devoid of that pretty variegation which extends over the upper surface of the animal.

NYULA.—*Herpestes Nyula.*

The word Ichneumon is Greek, and literally signifies a "tracker."

VERY CLOSELY allied to the Ichneumon, but differing from it in several points, the MEERKAT has been placed in the same genus with that animal by Cuvier and others, but has been separated by later naturalists, because there are only four toes on the hinder feet, and the number of the teeth is not the same. On account of the colour of its fur, it has been termed the Ruddy Ichneumon; and, from the brindlings in the tail, the Pencilled

MEERKAT.—*Cynictis Levaillantii.*

Ichneumon. It is rather a pretty animal, the tint of its coat being a light tawny brown, and the paws dark. The tail is rather bushy, and brindled with black hairs. It is a native of Southern Africa, and has received its specific title in compliment to the well-known African traveller, Le Vaillant.

1.   R

KUSIMANSE.—*Crossarchus Obscúrus.*

The curious animal which is known by the name of Kusimanse, or Mangue, is a native of Sierra Leone and Western Africa.

It is plantigrade in its walk, and has five toes on each foot. The teeth are of the same description as those of the succeeding animal. Its nose has something of the proboscis in its character, and its ears are small. The food of the Kusimanse consists of the smaller mammalia, of various insects, and some kinds of fruits. The general colour of the animal is a deep ruddy brown, but in certain lights, and when its coat is at all ruffled, the chocolate brown of its fur becomes plentifully grizzled with yellowish white. The reason for this change of tint is, that each hair is marked alternately with white and brown.

ZENICK.—*Suricáta Zenick.*

The Zenick, sometimes termed the Suricate, is a native of Southern Africa, but not very commonly found. It is not so exclusively carnivorous as the preceding animals, being fond of sweet fruits as well as of an animal diet. It is rather a small animal, measuring about eighteen inches in total length, its tail being six inches long. The feet are armed with long and stout claws, by means of which the creature can burrow with some rapidity. The colour is greyish brown, with a tinge of yellow, and the upper surface of the body is covered by several obscurely marked bars of a deeper brown hue. A silvery tint is washed over the limbs. The tail is brown, tinged with red, and black at the extremity. A few indistinct spots are sparsely scattered over the breast. The height of the animal is rather more than six inches.

The brain is large in proportion to the size of the animal, and, as may be expected, the creature is remarkably docile and intelligent. It is very sensitive to kindness, and equally so to harsh treatment, showing great affection towards those who behave well towards it, and biting savagely at any one who treats it unkindly. . When domesticated it ranges the house at will, and cannot be induced to leave its home for a life of freedom. Like the Ichneumon, it is an useful inmate of a house, extirpating rats, mice, and other living nuisances. It is offended by a brilliant light, and is best pleased when it can abide in comparative darkness. This nocturnal habit of eye renders it especially useful as a vermin exterminator, as it remains quiet during the hour while the rats, mice, and snakes lie still in their holes, and only issues from its hiding-place when the shades of night give the signal for the mammalian and reptilian vermin to sally forth on their own food-seeking quest. As its eyes are fitted for nocturnal sight, it becomes a terrible enemy to these creatures, creeping quietly upon them, and seizing them before they are aware of its proximity.

As far as is known, the sense of hearing is rather dull, and seems to assist the animal but little. The Zenick appears to bear some resemblance to our common polecat and ferret ; but it is altogether a curious animal, and stands nearly alone in the animal kingdom. Its walk is less gliding than that of the Ichneumons, and it is able to sit upon its hinder legs, and remain in the erect position for some time.

MAMPALON.—*Cynogale Bennettii.*

In Borneo, an allied animal is found, which is known in its native country by the title of Mampalon.

The so called "whisker hairs" which grow from the lips and behind the eyes are extremely long, and the feet are short, and furnished with five toes. When walking, the animal sets the entire sole of its foot on the ground, after the manner called "planti-grade." It is generally found in the neighbourhood of rivers. In total length it is about eighteen inches, the tail measuring nearly seven inches. The snout of this animal is rather long, but at its extremity is blunt and slightly depressed.

NANDINE.—*Nandinia binotata.*

Passing by several curious animals, we arrive at the pretty little creature which is known by the name of Nandine.

On account of the double row of spots which run along the body, the Nandine has been dignified with the title of "binotata," or "double-spotted," by almost every naturalist who has woven it into his system, even though the animal itself has been placed by some authors among the Civets, by some among the Ichneumons, and by others among the Paradoxures.

The general colour of the fur is a darkish and very rich brown, darker along the back, and lighter on the sides. The tail is covered with blackish rings which are but obscurely defined.

BINTURONG.—*Arctis Binturong.*

One of the largest examples of this group, is the dark, sullen, and sluggish Binturong.

This animal is a native of Malacca, from whence several living specimens and many skins have been brought to this country. The colour of the Binturong is a dead black, the hairs being long, coarse, and devoid of that gloss which is so often found upon black animals. The head is grey, and each ear is furnished with a long tuft of black hair. Round the edge of the ears runs a band of whitish grey.

The tail of the Binturong is thickly and heavily formed, longer than the body, and covered with exceedingly bushy hair. In some individuals, the black fur is mixed with white or grey hairs.

It seems to be a very indolent animal, passing the day in sleep, and being with difficulty aroused from its slumbers. When irritated, it utters a sharp fierce growl, shows its teeth, and curls itself up again to sleep. While sleeping, it lies partly on its side, curled round with its head snugly sheltered under its bushy tail. The muzzle of the Binturong is short and sharp, rather turned up at its extremity, and covered with long brown hairs which radiate around the face, and impart a very curious expression to the animal. The eyes are of a dull chestnut, unless the creature is excited, when they flash out with a momentary fire which dies away as soon as the cause is removed.

It is a good climber of trees, being assisted in this task by its tail, which is prehensile at the tip, and capable of grasping an object with some force. When in captivity it seems to prefer a vegetable to an animal diet, and feeds on rice, fruit, and other vegetable productions. But it is fond of eggs, birds, the heads of fowls, and other animal substances, and perhaps is best kept in health by a mixed diet. It enjoys a very excellent appetite, and whether its food be animal or vegetable, consumes an exceedingly large amount in comparison with the size of the consumer.

The length of the Binturong is about two feet six inches, exclusive of the tail, which always equals, and generally exceeds, the body in length. Its height varies from a foot to fifteen inches.

MASKED PAGUMA.—*Páguma Larváta.*

THE CURIOUS animal which is represented in the engraving, has, until lately, been placed among the weasels, under the title of Masked Glutton, and has only of late years been referred to its proper place in the scale of creation. The title of Larvatus, or Masked, is given to it on account of the white streak down the forehead and nose, and the white circle round the eyes, which gives the creature an aspect as if it was endued with an artificial mask. There is a pale olive-grey band extending from the back of each ear and meeting under the throat, and the general colour of the fur is an olive-brown, besprinkled and washed with grey. It has been found in China, from which country several specimens have been imported into England. There are many other species belonging to the same genus, such as the Nepal Paguma, the White Whiskered Paguma of Sumatra and Singapore, the Woolly Paguma from Nepal, and the Three-streaked Paguma of Malacca.

LUWACK, OR PARADOXURE.—*Paradoxurus Typus.*

THE ANIMALS which compose the little group of Paradoxures are very closely allied to the Pagumas and the Ichneumons, and appear to be confined to the Asiatic continent and its islands. The little group of animals to which the Luwack belongs was arranged by Cuvier under the generic title of Paradoxurus, literally, Puzzle-tail, because they have a curious habit of twisting their tails into a tight coil, and in their cat-like claws, and their civet-like teeth, present a strange mixture of characteristics.

The LUWACK, or common Paradoxure, is found plentifully in India, from whence many specimens have been brought to this country. As it has something of the viverrine look about it, Buffon and other naturalists placed it with the Genetts. It is a curious little creature, rather quick in its movements, and very inquisitive in its aspect, holding its head aside with an air of curiosity that is quite amusing. The eyes of this creature are very small and nearly black.

As the Luwack is tolerably widely spread, it is known by various names, according to the locality in which it lives. Its Malabar appellation is Pounougar-Pouné, a term which signifies "Civet Cat." The general tint of the fur is a yellowish black, but it assumes various hues, according to the light in which it is viewed. On each side of the spine run three rows of elongated spots, and upon the thighs and shoulders other spots are scattered. But if the animal is viewed in certain lights, the spots on the body seem to be merged into lines, while those on the breast disappear altogether. This change of appearance is caused by the mode in which the hairs are coloured, each hair being tipped with a darker hue, and some hairs being totally black. These latter hairs are very silken in texture, and much longer than the yellowish hairs of which the fur is mostly composed.

The Luwack, as are all the Paradoxures, is entirely plantigrade. Its feet are furnished with sharp claws, which are sufficiently retractile to be kept from the ground when the animal walks, and are preserved so sharp, that they can be used for tree-climbing with the greatest ease. Its tail is very remarkable on account of the tight spiral into which it is frequently rolled, and seems to be unlike the tail of any other animals. Although it can be so firmly curled, it is not prehensile, as might be supposed from its aspect when half unrolled.

One of these animals, which was kept in the Paris Museum, was accustomed to sleep during the day, coiled round upon its bed, and even by night appeared to feel a distaste for exertion. When evening came on, it would rouse itself from its slumbers, take food and drink, and again resign itself to sleep.

MUSANG.—*Paradoxurus Musanga.*

The MUSANG of Java is, although a destroyer of rats and mice, rather a pest to the coffee-plantations, which it ravages in such a manner as to have earned the title of the Coffee Rat. It feeds largely upon the berries of the coffee shrub, choosing only the ripest fruit, stripping them of their membranous covering, and so eating them. It is a remarkable fact that the berries thus eaten appear to undergo no change by the process of digestion, so that the natives, who are free from over scrupulous prejudices, collect the rejected berries, and are thus saved the trouble of picking and clearing them from the husk.

However, the injury which this creature does to the coffee berries is more than compensated by its very great usefulness as a coffee planter. For, as these berries are uninjured in their passage through the body of the animal, and are in their ripest state, they take root where they lie, and in due course of time spring up and form new coffee plantations, sometimes in localities where they are not expected. It may be that although the coffee seeds undergo no visible change in the interior of the Musang, they imbibe the animal principle, and thus become more fitted for the soil than if they had been planted without the intermediate agency of the creature.

The Musang is not content with coffee-berries and other vegetable food, although it seems to prefer a vegetable to an animal diet. When pressed by hunger, it seeks eagerly after various small quadrupeds and birds, and is often a pertinacious robber of the hen-roosts.

The habits of the Musang are well described by G. Bennett, in his "Wanderings in New South Wales:"—

" On the 14th of May, 1833, I purchased one of these animals from a native canoe, which came off to the ship on the coast of Java. It is commonly known among Europeans by the name of the 'Java Cat,' and is a native of Java, Sumatra, and perhaps other of the eastern Islands. This specimen was young and appeared very tame. The native from whom I procured it, had it enclosed in a bamboo cage, in which I also kept it for a short time. The colour of the back is blackish, intermingled with black; neck and abdomen of a yellowish colour; the eyes are full and large, of a yellowish brown colour; pupil perpendicular, becoming dilated at night. It resembles the cat in being more of a night than a day animal.

It feeds on plantains and other fruit, and also on fowls' bones. When busily engaged in picking the wing-bone of a fowl, it growls most savagely if disturbed in its repast, which well shows the nature of the beast. 'He eats only plantains,' said the Javanese from whom I purchased it, but could the animal have spoken for itself, it would probably have hinted that portions of the animals composing the feathered kingdom would also be acceptable by way of variety. It is tame and playful like a kitten, throwing itself on the back, playing with a bit of string, making at the same time a low whining noise.

It utters a sharp, quick, squeaking noise, as well as a low moaning, more particularly at night, or when in want of food, or of some water to quench its thirst. The specimen is a male; it is very playful, and climbed up my arm by aid of its claws, like a cat. When it drinks, it laps, like the dog or cat.

May 17th.—This morning, the animal had broken through and escaped from its cage during the night, and was about some part of the ship.

May 18th.—The whole of yesterday the creature was reported absent without leave; but early this morning it was found in the cabin of the second officer, asleep upon a jacket. It appears sufficiently tame to be left at liberty, so I did not immure it in a cage again, but kept a long piece of string attached to one of the hind legs, so as to limit its extent of range, when I find it necessary. Whilst writing in my cabin, the animal was either lying down quietly asleep, or else came to see what I was about, thrusting its little sharp snout among my papers, and amusing itself by playing with my pens and pencil.

This animal is called Mussong in Java, and I found it was also known by the same name among the natives on the north-east coast of Sumatra. It licks and cleans its furry coat with the tongue, like a cat, especially after it has been much handled; and seems almost to possess the caution and secretiveness of that tribe. It growls savagely when disturbed or teased. It lapped some coffee one morning, but became sick soon after.

It seems to be a fretful, impatient creature; and when it does not get its wants gratified, becomes terribly out of temper, or, rather, gets into a temper which is of a bad description. It then snaps ferociously at the fingers placed near it, but its young teeth can make but little impression: it is in downright earnest, however, for it bristles up, and advances its long whiskers, uttering a sort of peevish cries and growls.

It was lying on the pillow of my bed one morning, when I took the creature off, and placed it on as soft a place, which had been made up in the cabin on purpose for it. But this would not do; it did not like the removal, and there was no termination to its peevish, fretful cries, until it was removed back to the old place; where, being deposited, after licking itself about those parts of its furred coat that had been ruffled by handling, it stretched itself, and laid down quite contented.

The squeaking discontented noise of the creature during the night, when it is tied up, is very annoying. I suppose the desire of making nocturnal rambles, as is the nature of the tribe, was the cause to which the cries were to be attributed. At last I used to give it fowl-bones during the night to amuse itself, and, being occupied in crunching them, I was no more annoyed by its nocturnal cries.

When fighting, it uses the fore-paws, with extended claws, biting at the same time, retreating and advancing quickly, snapping, bristling up its large whiskers, and appearing a fierce object for one of the small animals of the creation. It does not spring at the object of attack like a cat, but jumps forward; it uses the claws of the fore-feet more than those of the hind, which, being longer and sharper, are more calculated for the purposes of defence, as well as in climbing. It well regards the object previous to attack, and then, with its little angular mouth expanded, it pounces upon and firmly grasps its prey.

The little beast has a very morose looking countenance, which some people skilled in physiognomy would call a sour, forbidding countenance; and, judging from what I have seen of this tamed and young specimen, it must be in the wild state a very savage animal.

Unlike the cat, when drinking, it does not care about wetting its feet, for it often places the fore-paws in the water at the time. It often plays with its long tail, as well as with anything that may be in the way, similar to what we observe in kittens; and often scratches against objects, growling at the same time, as if practising for future defence. It eats fowl readily, but not other kinds of meat so well; it ate some pine-apple with much avidity. It will carry away a bone given to it to a dark corner, growling and snapping at any one that may attempt to take it away.

Sometimes, when left to itself, it utters such loud, squeaking cries as to be heard all over the ship. One day, at dinner-time, when the animal was first on board, a noise was heard, but from whence it proceeded, or what it was, we could not tell, until the mystery was explained by the steward, who said that it was the foreign cat.

Like all animals, whether of the genus Homo, or lower in the creation, the Java Cat does not like to be disturbed at meals. This little, ill-humoured quadruped is particularly savage at that time, but, like the human race in all its numerous varieties, when feeding time is over, and it has had a sufficiency of provender, it will remain quiet, and be usually in a tolerably good humour; but when it is hungry, there is nothing but screeching, grumbling, and crying, until the appetite is satisfied.

I gave the animal one morning a dead cockroach, but after turning and twisting it about and licking it for some time, it would not eat it—perhaps it was not hungry. When the creature is excessively annoyed, it retires into some dark hole or corner, making a spitting noise, and is very furious against any one that may attempt to dislodge it from that place thus formed into a refuge for the ill-tempered.

When first set at liberty, it was missing for one or two days, having gone on a tour, by way of change of scene; he soon, however, returned to his old quarters, ran about the cabins, and, when sleeping during the day, would take the warmest and most comfort-able situations which the cabins afforded; it was as fully domesticated as a cat.

The Musang runs about quite domesticated, and climbs well, occasionally aiding itself by the tail having a prehensile power. He also runs about, particularly at night, and in the morning is usually found quietly asleep upon the softest bundle of clothes he can meet with in the cabin into which he has introduced himself. He dislikes much to be handled, or petted, or crammed, unless he crams himself, which he very often does.

At last I let the creature ramble about where it pleased in the after part of the ship; it reposed in the cabins or in other places it liked. It used to wander about and come at meal-times for food, until the 14th of June, when it was missing, and search being made about its usual haunts, the animal was discovered dead among some oakum in one of the cabins.

When at Pedir, on the south-east coast of Sumatra, I procured another young, but larger, specimen than the preceding; it was purchased for half-a-rupee. Although wild with strangers, with the native from whom the animal was purchased it was exceedingly domesticated. I have seen it follow him like a cat along the pathway for some distance, when he placed it out of his arms upon the ground. The natives gave it the same name here as at Java.

When I placed this animal in my cabin, it remained very quiet, not making so much noise as the last, but a few days afterwards it became so very vicious, not suffering any one to approach or touch it, without spitting, growling, and fighting so furiously, that I at last was obliged to destroy it."

The Musang, when in its native woods, constructs a nest not unlike that of the squirrel's, composed of leaves, dry grass, slender twigs, and other analogous substances. This habitation is usually fixed in the fork of a branch, and sometimes is placed in the hollow of a tree. Making this "nest" its head quarters, and sleeping there by day, it issues forth at night in search of food, making sad havoc with the hen-roosts when it can gain admission, and devouring every kind of ripe fruit which it can find. Pine-apples seem to be favourite articles of diet with this epicurean creature.

THE ANIMAL which is shown in the following engraving is remarkable for the singularity of its colouring, and the mode in which the fur is diversified with lighter and darker tints.

The colour of this animal's fur is a greyish-brown, on which are placed six or seven large and bold stripes, arranged saddle-wise upon the back, being very broad above, and

HEMIGALE.—*Hemigale Hardwickii.*

narrowing to a point towards the ribs. These bands are unconnected with each other. On the top of the head there is a narrow black line, and on each side of the face, a black line runs from the ear to the nose, surrounding the eye in its progress. The nose itself is black. Down the sides of the neck there are some obscure streaks, which are more conspicuous in a side light. The tail is marked with dark patches upon its upper surface, and latter half is black.

CRYPTOPROCTA.—*Cryptoprocta ferox.*

The name Hemigale is Greek, and signifies, "Semi-weasel"—and the specific title is given in honour of General Hardwick, who has done such good service to zoology.

The last of the great Viverrine group of animals is the CRYPTOPROCTA, a creature whose rabbit-like mildness of aspect entirely belies its nature.

It is a native of Madagascar, and has been brought from the southern portions of that wonderful island. It is much to be wished that the zoology of so prolific a country should be thoroughly explored, and that competent naturalists should devote much time and severe labour to the collection of specimens, and the careful investigation of animals while in their wild state.

Gentle and quiet as the animal appears, it is one of the fiercest little creatures known. Its limbs, though small, are very powerful, their muscles being extremely full and well knit together. Its appetite for blood seems to be insatiable as that of the tiger, and its activity is very great, so that it may well be imagined to be a terrible foe to any animals on whom it may choose to make an attack. For this savage nature it has received the name of "Ferox," or fierce. Its generic name of Cryptoprocta is given to it on account of the manner in which the hinder quarters suddenly taper down and merge themselves in the tail. The word itself is from the Greek, the former half of it signifying "hidden," and the latter half, "hind-quarters."

The colour of the Cryptoprocta is a light brown, tinged with red. The ears are very large and rounded, and the feet are furnished with strong claws. The toes are five in number on each foot.

In the foregoing description of the Viverrine animals, examples and figures are given of every remarkable genus which forms a portion of this curious group. Whether or not the Hyæna should be considered as belonging to the Viverrines is a question which is still mooted by many naturalists, who think that the Hyænines ought to be ranked as a divergent group of the Civet Cats.

With the exception of one or two species, these creatures are so little known that their habits in a wild state have yet to be fully described. This is the more to be regretted, because the native customs of an animal are more illustrative of its character, and give deeper insight into the part which it plays in the economy of nature, than can be gained by inspecting the same creature when shut up in the contracted space which its cage affords, or when a change in its nature has been wrought by the companionship of human beings. The habits of these agile and graceful animals are so interesting, when watched even in the limited degree which is afforded by our present means of observation, that they give promise of much curious information when noted in the wild freedom of their normal condition.

We lose much valuable knowledge of the habits of a new or scarce animal by the over-readiness of the discoverer to secure his prize. If one is fortunate enough to hit upon an animal which is new to science, or to meet with one which is rarely seen, he would do better service to Zoology by waiting awhile, and quietly watching the manner in which the animal conducted itself, than by hastily levelling his gun, and so giving to science nothing but a lifeless mass of dead matter, instead of a spirited history of a breathing and living being. For my own part, I would rather read in a library a good description of some strange animal, than see in a museum a stuffed skin about which nothing is known. There is always a greater probability of obtaining a dead specimen than its living history.

As is seen in the case of the Zenick, the Musang, and the Ichneumon, these creatures are quite as susceptible of human instruction as the feline or canine animals, and might be advantageously trained to human uses.

GROUP OF BRITISH DOGS.

DOGS.

THE large and important group of animals which is known by the general name of the Dog-Tribe, embraces the wild and domesticated Dogs, the Wolfs, Foxes, Jackals, and that curious South-African animal, the Hunting-Dog. Of these creatures, several have been brought under the authority of man, and by continual intermixtures have assumed that exceeding variety of form which is found in the different "breeds" of the domestic Dog.

The original parent of the Dog is very doubtful, some authors considering that it owes its parentage to the Dhole, or the Buansuah ; others thinking it to be an offspring of the Wolf ; and others attributing to the Fox the honour of being the progenitor of our canine friend and ally.   With the exception of a very few spots, the Dog is to be found spread over almost every portion of the habitable globe, and in all countries is the friend of man, aiding him either by the guardianship of his home and property, by its skill and endurance in the chase, or by affording him a means of transit over localities which no other animal could successfully encounter.

Before proceeding to the domesticated Dogs, we will examine the two species of Wild Dog which nearest approach them.

KHOLSUN, OR DHOLE.—*Cuon Dukhuensis.*

THE DHOLE, or KHOLSUN, as it is sometimes called, inhabits the western frontiers of British India, its range extending from Midnapore to Chamar, but does not appear to take up its residence in other parts of the same great country.   Even in the localities which are favoured by its presence, the Dhole seldom makes its appearance, and by many residents in India has been counted but as a myth of the natives.   It is a very shy animal, keeping aloof from man and his habitations, and abiding in the dense dark jungles, which extend for hundreds of miles, and afford little temptation for human beings to enter.

Among the peculiarities of the Dhole's character, its fondness for the chase is perhaps the most remarkable.   There is nothing peculiar in the fact that the Dhole unites in large packs and hunts down game, both large and small, because many of the canine race, such as the wolves and others, are known by many and tragical experiences to run down and destroy their prey in like manner. But the Dhole is apparently the only animal that, although individually so far the inferior of its fierce prey, in size, strength, and activity, has sufficient confidence in its united powers, to chase and kill the terrible tiger, maugre his fangs and claws.

From the observations which have been made, it seems that hardly any native Indian animal, with the exception of the elephant and the rhinoceros, can cope with the Dhole; that the fierce boar falls a victim, in spite of his sharp tusks, and that the swift-footed deer fails to escape these persevering animals. The leopard is tolerably safe, because the dogs cannot follow their spotted quarry among the tree branches, in which he fortifies himself from their attacks; but if he were deprived of his arboreal refuge, he would run but a poor chance of escaping with life from the foe. It is true that, in their attack upon so powerfully armed animals as the tiger and the boar, the pack is rapidly thinned by the swift blows of the tiger's paw, or the repeated stabs of the boar's tusks; but the courage of the survivors is so great, and they leap on their prey with such audacity, that it surely yields at last from sheer weariness and loss of blood.

It is probable that the sanguinary contests which often take place between the Dholes and their prey have a great effect in checking the increase of the former animals, and that, if such salutary influence were not at work, these bold and persevering hunters might increase to such an extent as to become a serious pest to the country.

In the chase, the Dhole is nearly silent, thus affording a strong contrast to the cheerful tongue of the foxhound in "full cry," or the appalling howl of the wolf when in pursuit of a flying prey. Only at intervals is the voice of the Dhole heard, and even then the animal only utters a low anxious whimper, like that of a Dog which has lost its master, or feels uneasy about its task. It is a swift animal in the chase, and Captain Williamson, who has seen it engaged in pursuit of its prey, thinks that no animal could lead the Dhole a long chase. The average number of individuals in the pack is about fifty or sixty.

The colour of the Dhole is a rich bay, darkening upon the feet, ears, muzzle, and tip of the tail. In height it equals a rather small greyhound. It does not assault human beings unless it be attacked, neither does it seem to fly from them, but, in case of a sudden meeting, pursues its avocations as if unconscious of the presence of an intruder. The countenance of this animal is very bright and intelligent, chiefly owing to the keen and brilliant eye with which it is favoured. The Greek word "Cuon" signifies a hound.

In the Wild Dog, which ranges Nepal and the whole of Northern India, the primitive type of the Dog was thought to be found. This animal, the BUANSUAH, presents many points of similarity to the Dhole, and is said to rival the latter creature in its tiger-killing propensities.

Like the Dhole, it is a shy animal, and never willingly permits itself to be seen, preferring to take up its residence in the thickest coverts which are afforded by the luxuriant vegetation of its native land. It hunts in packs, but, unlike the preceding animal, gives tongue continually as it runs, uttering a curious kind of bark, which is quite distinct from the voice of the domestic Dog, and yet has nothing in common with the prolonged howl of the wolf, the jackal, or the foxes.

The number of individuals in each pack is not very great, from eight to twelve being the usual average. They are possessed of exquisite powers of scent, and follow their game more by the nose than by the eye.

When captured young, the Buansuah readily attaches itself to its keeper, and, under his tuition, becomes a valuable assistant in the chase. Unfortunately, the Dog will too often refuse its confidence to any one except its keeper, and therefore is not so useful as it might otherwise be rendered. It is probable that the keeper himself has some hand in this conduct, and wilfully teaches his charge to repel the advances of any person save himself.

In the chase of the wild boar, the peculiar character of the Buansuah exhibits itself to great advantage, as its wolf-like attack of sudden snap is more destructive to its prey than the bite of an ordinary hound. For other game this creature is but an uncertain assistant,

BUANSUAH.—*Cuon Primævus.*

as it will often give up a chase just at the critical moment, and is too apt to turn aside from its legitimate quarry for the purpose of immolating a tame sheep or goat.

ALL the various Dogs which have been brought under the subjection of man are evidently members of one single species, *Canis familiaris*, being capable of mixture to an almost unlimited extent. By means of crossing one variety with another, and taking advantage of collateral circumstances, such as locality, climate, or diet, those who have interested themselves in the culture of this useful animal have obtained the varied forms which are so familiar to us. In general character, the groups into which domesticated Dogs naturally fall are tolerably similar, but the individual characters of Dogs are so varied, and so full of interest, that they would meet with scanty justice in ten times the space that can be afforded to them in these pages. It has been thought better, therefore, to occupy the space by figures and descriptions of the chief varieties of the domesticated Dog, rather than to fill the pages with anecdotes of individuals. Upwards of forty varieties of the Dog will be described in the following pages, and illustrated with figures which, in almost every instance, are portraits of well-known animals.

One of the most magnificent examples of the domesticated Dog is the THIBET DOG, an animal which, to his native owners, is as useful as he is handsome, but seems to entertain an invincible antipathy to strangers of all kinds, and especially towards the face of a white man. These enormous Dogs are employed by the inhabitants of Thibet for the purpose of guarding their houses and their flocks, for which avocation their great size and strength render them peculiarly fit. It often happens that the male inhabitants of a

THIBET DOG.—*Canis Familiáris.*

Thibetian village leave their homes for a time, and journey as far as Calcutta, for the purpose of selling their merchandise of borax, musk, and other articles of commerce. While thus engaged, they leave their Dogs at home, as guardians to the women and children, trusting to the watchfulness of their four-footed allies for the safety of their wives and families.

The courage of these huge Dogs is not so great as their size and strength would seem to indicate, for, excepting on their own special territories, they are little to be feared, and even then can be held at bay by a quiet, determined demeanour. Several of these handsome animals have been brought to England. Their colour is generally a deep black, with a slight clouding on the sides, and a patch of tawny over each eye. The hanging lips of the Thibet Dog give it a very curious aspect, which is heightened by the generally loose mode in which the skin seems to hang on the body.

The GREAT DANISH DOG is best known in England as the follower of horses and carriages upon roads; and, probably on account of being restricted to this monotonous mode of existence, is supposed to be rather a stupid animal. As, however, in its own country the Danish Dog is employed as a pointer, and does its work very creditably, we may suppose that the animal is possessed of abilities which might be developed by any one who would take pains to do so.

On account of its carriage-following habits, it is popularly called the Coach Dog, and,

DANISH DOG.—*Canis familiaris.*

on account of its spotted hide, receives the rather ignoble title of Plum-Pudding Dog. The height of the animal is rather more than two feet.

It is hardly possible to conceive an animal which is more entirely formed for speed and endurance than a well-bred GREYHOUND. Its long slender legs, with their whipcord-like muscles, denote extreme length of stride and rapidity of movement; its deep, broad chest, affording plenty of space for the play of large lungs, shows that it is capable of long-continued exertion; while its sharply pointed nose, snake-like neck, and slender, tapering tail, are so formed as to afford the least possible resistance to the air, through which the creature passes with such exceeding speed.

The chief use—if use it can be termed—of the Greyhound, is in coursing the hare, and exhibiting in this chase its marvellous swiftness, and its endurance of fatigue.

In actual speed, the Greyhound far surpasses the hare, so that, if the frightened chase were to run in a straight line, she would be soon snapped up by the swifter hounds. But the hare is a much smaller and lighter animal than her pursuer, and, being furnished with very short forelegs, is enabled to turn at an angle to her course without a check, while the heavier and longer limbed Greyhounds are carried far beyond their prey by their own impetus, before they can alter their course, and again make after the hare.

On this principle, the whole of coursing depends; the hare making short quick turns, and the Greyhounds making a large circuit every time that the hare changes her line. Two Greyhounds are sent after each hare, and matched against each other, for the purpose of trying their comparative strength and speed. Some hares are so crafty and

1.   s

GREYHOUND.—*Canis familiaris.*

so agile, that they baffle the best hounds, and get away fairly into cover, from whence the Greyhound, working only by sight, is unable to drive them.

Naturally, the Greyhound of pure blood is not possessed of a very determined character, and it is therefore found necessary to give these creatures the proper amount of endurance by crossing them with the bull-dog, one of the most determined and courageous animals in existence. As may be supposed, the immediate offspring of a bulldog and a Greyhound is a most ungainly animal, but by continually crossing with the pure Greyhound, the outward shape of the thick and sturdy bull-dog is entirely merged in the more graceful animal, while his stubborn pertinacity remains implanted in its nature.

The skeleton of the Greyhound is a curious one, and when viewed from behind, bears a marvellous resemblance to that of the ostrich.

The narrow head and sharp nose of the Greyhound, useful as they are for aiding the progress of the animal by removing every impediment to its passage through the atmosphere, yet deprive it of a most valuable faculty, that of chasing by scent. The muzzle is so narrow in proportion to its length, that the nasal nerves have no room for proper development, and hence the animal is very deficient in its powers of scent. The same circumstance may be noted in many other animals.

THE IRISH GREYHOUND is a remarkably fine animal, being four feet in length, and very firmly built. Its hair is of a pale fawn colour, and much rougher than that of the smooth English Greyhound.

IRISH GREYHOUND.—*Canis familiaris.*

Unless excited by the sight of its game, or by anger, it is a very peaceable animal ; but when roused, exhibits a most determined spirit. In former days, when wolves and wild boars infested the Irish forests, this Dog was used for the purpose of extirpating those animals ; but in these days their numbers are comparatively few. When fighting, it takes its antagonist by the back, and shakes the life out of its foe by main strength. One of these Dogs measured sixty-one inches in total length ; twenty-eight and a half inches from the toe to the top of the shoulder, and thirty-five inches in girth.

THE SCOTCH GREYHOUND is still rougher in its coat than its Irish relative, but hardly so large in its make : a very fine example of these Dogs, of the pure Glengarry breed, measures twenty-eight inches in height, and thirty-four inches in girth, being a little smaller than the Irish Dog which was mentioned above.

There seems to be but one breed of the Scotch Greyhound, although some families are termed Deerhounds, and others are only called Greyhounds. Each however, from being constantly employed in the chase of either deer or hare, becomes gradually fitted for the pursuit of its special quarry, and contracts certain habits which render it comparatively useless when set to chase the wrong animal. The Scotch Deerhound is possessed of better powers of scent than the Greyhound, and in chasing its game depends as much on its nose as on its eyes. And it is curious too, that although it makes use of its olfactory powers when running, it holds its head higher from the ground than the Greyhound, which only uses its eyes.

SCOTCH GREYHOUND.—*Canis familioris.*

THE RUSSIAN GREYHOUND is also gifted with the power of running by scent, and is employed at the present day for the same purposes which Irish Greyhounds subserved in former times.

Many Russian forests are infested with wild boars, wolves, and bears, and this powerful and swift Dog is found of great use in the destruction of these quadrupedal pests. In size it is about equal to the Scotch Greyhound. It is not exclusively used for the chase of the large and savage beasts, but is also employed in catching deer, hares, and other animals which come under the ordinary category of "game."

The fur of this Dog is thick, but does not run to any length.

THE NOBLE and graceful animal which is the representative of the Greyhound family in Persia, derives its origin from a source which is hidden in the mists of antiquity, and has been employed in the chase of swift-footed animals from time immemorial. Powerful of jaw, quick and supple of limb, the PERSIAN GREYHOUND is chosen to cope with that swift and daring animal, the wild ass, as well as with the no less rapid antelope, and the slower, but more dangerous, wild boar.

Of all these creatures, the wild ass gives the most trouble, for it instinctively keeps to rocky and mountainous neighbourhoods, which afford a refuge unassailable by the sure-footed Persian horse, and from which it can only be driven by such agile creatures as the native Greyhounds. So untiring is the wild ass, and so boldly does it traverse the rocky mountain spurs among which it loves to dwell, that a single ass will

RUSSIAN GREYHOUND.—*Canis familiaris.*

frequently escape, even though several relays of Greyhounds have been provided to take up the running at different parts of the course, as soon as their predecessors are fatigued.

For the antelope the Greyhound would be no match, and is therefore assisted by the falcon, which is trained to settle on the head of the flying animal, and by flapping its wings in the poor creature's eyes, to prevent it from following a direct course, and thus to make it an easier prey to the Greyhound which is following in the track. Of this curious mixture of falconry and hunting the Persian nobles are passionately fond, and peril their lives in ravines, and among rocks that would quail the spirit of our boldest foxhunters.

It is said that the Persian Greyhound is not the safest of allies, for if it should fail in its chase, it is reputed to turn its wasting energies upon its master, and to force him, Actæon like, to seek his safety in flight ; or, more fortunate than his cornuted prototype, to rid himself of his dependents by a blow from his ready scimitar. The Persian Greyhound is said to be especially addicted to this vice when it is imported into India.

This animal is rather slender in make, and its ears are "feathered" after the fashion of the Blenheim spaniel's ears. Nevertheless, it is a powerful and bold creature, and can hold its own among any assemblage of Dogs of its own weight.

A MORE UTTER contrast to the above-mentioned animal can hardly be imagined than that which is afforded by the ITALIAN GREYHOUND, a little creature whose merit consists in its diminutive proportions and its slender limbs. Hotspur, leaning all breathless on

PERSIAN GREYHOUND.—*Canis familiaris.*

his sword, and stiff with his wounds, was not more entirely the opposite of the carpet knight, with pouncet box to nose, and full of "parmaceti" babblings, than is the rough, fierce Greyhound of Persia, of the delicate, shivering, faint-hearted Italian Greyhound; sad type of the people from which it takes its name.

In truth, the Italian Greyhound is but a dwarfed example of the true smooth Greyhound, dwarfed after the same manner that delights our Celestial friends, when tried on vegetable instead of animal life. The weight of a really good Italian Greyhound ought not to exceed eight or ten pounds; and there are animals of good shape which only weigh six or seven pounds. One of the most perfect Dogs of the present day weighs eight and three-quarter pounds, and is fourteen and a quarter inches in height. His colour is uniformly black.

Attempts have been made to employ the Italian Greyhound in the chase of rabbits, but its power of jaw and endurance of character are so disproportioned to its speed, that all such endeavours have failed. A mixed breed, between the Italian Greyhound and the terrier, is useful enough, combining endurance with speed, and perfectly capable of chasing and holding a rabbit.

In this country, it is only used as a petted companion, and takes rank among the "toy-dogs," being subject to certain arbitrary rules of colour and form, which may render a Dog worthless for one year through the very same qualities which would make it a paragon of perfection in another. The Dutch tulip-mania afforded no more capricious versatility of criterion than is found in the "points" of toy Dogs of the present day. If the creature be of a uniform colour, it must be free from the least spot of white; and even a white stain on the breast is held to deteriorate from its perfection. The colour which is most in vogue is a golden fawn; and the white and red Dog takes the last place in the valuation of colour.

It is a pretty little creature, active and graceful to a degree, and affectionate to those who know how to win its affections. Even in the breed of our British smooth Grey-

ITALIAN GREYHOUND.—*Canis familiaris.*

hounds, this little animal has been successfully employed, and by a careful admixture with the larger Dog, takes away the heavy, clumsy aspect of the head which is caused by the bull-dog alliance, and restores to the offspring the elastic grace of the original Greyhound. It is generally bred in Spain and Italy, and from thence imported into this country, where the change of climate is so apt to affect its lungs, that its owners are forced to keep it closely swathed in warm clothing during the changeable months of the year.

THE LARGE and handsome animal which is called from its native country the NEW-FOUNDLAND DOG, belongs to the group of spaniels, all of which appear to be possessed of considerable mental powers, and to be capable of instruction to a degree that is rarely seen in animals.

In its native land the Newfoundland Dog is shamefully treated, being converted into a beast of burden, and forced to suffer even greater hardships than those which generally fall to the lot of animals which are used for the carriage of goods or the traction of vehicles. The life of a hewer of wood is proverbially one of privation, but the existence of the native Newfoundland Dog is still less to be envied, being that of a servant of the wood-hewer. In the winter, the chief employment of the inhabitants is to cut fuel, and the occupation of the Dogs is to draw it in carts. The poor animals are not only urged beyond their strength, but are meagrely fed with putrid salt fish, the produce of some preceding summer. Many of these noble Dogs sink under the joint effects of fatigue and starvation, and many of the survivors commit sad depredations on the neighbouring flocks as soon as the summer commences, and they are freed from their daily toils.

In this country, however, the Newfoundland Dog is raised to its proper position, and made the friend and companion of man. Many a time has it more than repaid its master for his friendship, by rescuing him from mortal peril.

Astrologically speaking, the Newfoundland Dog must have been originated under the influence of Aquarius, for it is never so happy as when dabbling in water, whether salt or fresh, and is marvellously endurant of long immersion. There are innumerable instances on record of human beings rescued from drowning by the timely succour brought by a Newfoundland Dog, which seems fully to comprehend the dire necessity of the sufferer, and the best mode of affording help. A Dog has been known to support a drowning man in a manner so admirably perfect, that if it had thoroughly studied the subject, it could not have applied its aiding powers in a more correct manner. The Dog seemed to be perfectly aware that the head of the drowning man ought to be kept above the water, and possibly for that purpose shifted its grasp from the shoulder to the back of the neck. It must be remembered, however, that all Dogs and cats carry their young by the nape of the neck, and that the Dog might have followed the usual instinct of these animals.

Not only have solitary lives been saved by this Dog, but a whole ship's crew have been delivered from certain destruction by the mingled sagacity and courage of a Newfoundland Dog, that took in its mouth a rope, and carried it from the ship to the shore.

Even for their own amusement, these Dogs may be seen disporting themselves in the sea, swimming boldly from the land in pursuit of some real or imaginary object, in spite of "rollers" and "breakers" that would baffle the attempts of any but an accomplished swimmer. Should a Newfoundland Dog be blessed with a master as amphibious as itself, its happiness is very great, and it may be seen splashing and snapping in luxuriant sport, ever keeping close to its beloved master, and challenging him to fresh efforts. It is very seldom that a good Newfoundland Dog permits its master to outdo it in aquatic gambols. The Dog owes much of its watery prowess to its broad feet and strong legs, which enable the creature to propel itself with great rapidity through the water.

As is the case with most of the large Dogs, the Newfoundland permits the lesser Dogs to take all kinds of liberties without showing the least resentment; and if it is worried or pestered by some forward puppy, looks down with calm contempt, and passes on its way. Sometimes the little conceited animal presumes upon the dignified composure of the Newfoundland Dog, and, in that case, is sure to receive some quaint punishment for its insolence. The story of the big Dog, that dropped the little Dog into the water and then rescued it from drowning, is so well known that its needs but a passing reference. But I know of a Dog, belonging to one of my friends, which behaved in a very similar manner. Being provoked beyond all endurance by the continued annoyance, it took the little tormentor in its mouth, swam well out to sea, dropped it in the water and swam back again.

Another of these animals, belonging to a workman, was attacked by a small and pugnacious bull-dog, which sprang upon the unoffending canine giant, and, after the manner of bull-dogs, "pinned" him by the nose, and there hung, in spite of all endeavours to shake it off. However, the big Dog happened to be a clever one, and spying a pailful of boiling tar, he bolted towards it, and deliberately lowered his foe into the hot and viscous material. The bull-dog had never calculated on such a reception, and made its escape as fast as it could run, bearing with it a scalding memento of the occasion.

The attachment which these magnificent Dogs feel towards mankind is almost unaccountable, for they have been often known to undergo the greatest hardships in order to bring succour to a person whom they had never seen before. A Newfoundland Dog has been known to discover a poor man perishing in the snow from cold and inanition, to dash off, procure assistance, telling by certain doggish language of its own of the need for help, and then to gallop back again to the sufferer, lying upon him as if to afford vital heat from his own body, and there to wait until the desired assistance arrived.

I might multiply anecdote upon anecdote of the wondrous powers of this spirited animal, but must pass on to make room for others.

There are two kinds of Newfoundland Dog; one, a very large animal, standing

NEWFOUNDLAND DOG.—*Canis familiaris.*

some thirty-two inches in height; and the other, a smaller Dog, measuring twenty-four or twenty-five inches high. The latter animal is sometimes called the Labrador Dog, and sometimes is termed the St. John's Dog. When crossed with the setter, the Labrador Dog gives birth to the Retriever. The large Newfoundland is generally crossed with the mastiff.

There are few Dogs which are more adapted for fetching and carrying than the Newfoundland. This Dog always likes to have something in its mouth, and seems to derive a kind of dignity from the conveyance of its master's property. It can be trained to seek for any object that has been left at a distance, and being gifted with a most persevering nature, will seldom yield the point until it has succeeded in its search.

A rather amusing example of this faculty in the Newfoundland Dog has lately come before my notice.

A gentleman was on a visit to one of his friends, taking with him a fine Newfoundland Dog. Being fond of reading, he was accustomed to take his book upon the downs, and to enjoy at the same time the pleasures of literature and the invigorating breezes that blew freshly over the hills. On one occasion, he was so deeply buried in his book, that he overstayed his time, and being recalled to a sense of his delinquency by a glance at his watch, hastily pocketed his book, and made for home with his best speed.

Just as he arrived at the house, he found that he had inadvertently left his gold-headed cane on the spot where he had been sitting, and as it was a piece of property which he valued extremely, he was much annoyed at his mischance.

He would have sent his Dog to look for it, had not the animal chosen to accompany a friend in a short walk. However, as soon as the Dog arrived, his master explained his loss to the animal, and begged him to find the lost cane. Just as he completed his explanations, dinner was announced, and he was obliged to take his seat at table. Soon after the second course was upon the table, a great uproar was heard in the hall ; sounds of pushing and scuffling were very audible, and angry voices forced themselves on the ear. Presently, the phalanx of servants gave way, and in rushed the Newfoundland Dog, bearing in his mouth the missing cane. He would not permit any hand but his master's to take the cane from his mouth, and it was his resistance to the attempts of the servants to dispossess him of his master's property that had led to the skirmish.

IT HAS BEEN mentioned that the Newfoundland Dog is employed during the winter months in dragging carts of hewn wood to their destination, and that it is unkindly treated by the very men who derive the most benefit from its exertions.

The ESQUIMAUX DOG, however, spends almost its entire life in drawing sledges, or in carrying heavy loads, being, in fact, the only beast of burden or traction in the northern parts of America and the neighbouring islands. Some, indeed, are turned loose at the beginning of the summer, and many get their living as they can, until winter summons them back again to scanty meals and perpetual toil. But many of the Esquimaux Dogs are retained in servitude for the entire year, and during the summer months are called upon to give their aid in draught and in carriage. Indeed, those Dogs which are thus kept to their work during the entire year are comparatively happy, for their work is not nearly so heavy as in the winter, and their food is much better.

The Esquimaux Dog is rather smaller than the Labrador, being only twenty-two or twenty-three inches in height. There is something very wolfish about the Dog, owing to its oblique eyes, bushy tail, and elongated muzzle. In its full face the Esquimaux Dog presents a ludicrously exact likeness of its master's countenance. The colour is almost invariably a deep dun, marked obscurely with dark bars and patches ; the muzzle is black.

When harnessed to the sledge, the Dogs obey the movements of their leader, who is always a faithful and experienced old Dog. There are no means of guiding the animals in their way, for each Dog is simply tied to the sledge by a leathern strap, and directed by the voice and whip of the driver. The whip is of very great importance to the charioteer, for by the sounds which he elicits from the lash, and by the ably-directed strokes which he aims at refractory Dogs, he guides the canine team without the aid of bit or bridle.

The old and experienced animal which leads the team knows the master's voice, and will dash forward, slacken speed, halt, or turn to right and left at command.

The actual stroke of the whip is used as little as possible, for when a Dog feels the sting of the biting lash, he turns round and attacks the Dog nearest to him. The others immediately join in the fight, and the whole team is thrown into admirable confusion, the traces being entangled with each other, and the sledge in all likelihood upset. When such a rupture occurs, the driver is generally forced to dismount, and to harness the Dogs afresh. Usually, the leading Dog is permitted to run his own course, for he is able to follow the right path with marvellous accuracy, and to scent it out, even when the thickly-falling snowflakes have covered the surface of the ground with an uniform white carpet, on whose glittering surface no impress is left of the subjacent earth.

These Dogs are able to travel for very great distances over the snow-clad regions of the north, and have been known to make daily journeys of sixty miles for several days in succession.

Captain Parry, in his well-known " Journal," remarks very happily, that " neither the Dog nor his master is half civilized or subdued," the former indeed being the necessary consequence of the latter. The Esquimaux bears no love towards his Dogs, and only looks upon them as animated machines, formed for the purpose of conveying him and his property from one place to another. He is a most exacting and cruel master, feeding scantily his Dogs on the merest offal, and then inflicting severest torture upon them if they

ESQUIMAUX DOG.—*Canis familiaris.*

break down in their work from want of nourishment, or if, incited by the pangs of hunger, they obey their natural instincts, and make a meal on the provisions which had been laid aside for his own use. The savage is ever ingenious in the art of torture, and the Esquimaux forms no exception to the rule.

The poor beasts have been known, when suffering from long-continued hunger, to devour their tough leather harness, and, as if excited by the imperfect meal, to fly upon the weaker members of the team, and to tear them to pieces. During this paroxysm of unrestrained fury, they would have made their masters their first victims, had they not been driven back by the sword and the bludgeon.

In consequence of the evil treatment to which they are subjected, the poor animals can have no affection for their cruel tormentors, and are afforded no opportunity for developing the mental qualities which they possess in very large degree. When placed under the care of a kind master, the Esquimaux Dog is a most affectionate animal, and displays considerable reasoning powers.

The Esquimaux Dog is rather larger than an English pointer Dog, although its true size appears to be less than it really is, on account of the comparative shortness of limb. Its fur is composed of a long outer covering of coarse hair, three or four inches in length, and an inner coating of short, woolly hair, that seems to defend the animal from the colds of winter. When the weather begins to wax warm, the wool falls off, and grows again as the winter draws near.

OF LATE years, a Dog which much resembles the last-mentioned animal has come into fashion as a house-dog, or as a companion. This is the POMERANIAN FOX DOG, commonly known as the " Loup-loup."

It is a great favourite with those who like a Dog for a companion, and not for mere use, as it is very intelligent in its character, and very handsome in aspect. Its long white fur, and bushy tail, give it quite a distinguished appearance, of which the animal

POMERANIAN DOG.—*Canis familiaris.*

seems to be thoroughly aware. Sometimes the coat of this animal is a cream colour, and very rarely is deep black. The pure white, however, seems to be the favourite. It is a lively little creature, and makes an excellent companion in a country walk.

OF THE Spaniel Dogs, there are several varieties, which may be classed under two general heads, namely, Sporting and Toy Spaniels; the former being used by the sportsman in finding game for him; and the latter being simply employed as companions.

The FIELD SPANIEL is remarkable for the intense love which it bears for hunting game, and the energetic manner in which it carries out the wishes of its master. There are two breeds of Field Spaniels, the one termed the "Springer," being used for heavy work among thick and thorny coverts, and the other being principally employed in woodcock shooting, and called in consequence the "Cocker." The Blenheim and King Charles Spaniels derive their origin from the Cocker. The three Dogs which are represented in the engraving are examples of the three most celebrated breeds of Springer Spaniels. The black Dog is a Sussex Spaniel; that which stands in the foreground is a Clumber; and the seated Dog is a Norfolk Spaniel. Some of these Dogs continually give tongue while engaged in the pursuit of game, and utter different sounds according to the description of game which they have reached; while others are perfectly mute in their quest. Each of these qualities is useful in its way, and the Dog is valued accordingly;

SPANIEL.—*Canis familiaris.*

only it is needful that if the Dog be one that gives tongue, it should not be too noisy in its quest, and should be musical in its note.

SPRINGER SPANIELS.

While hunting, the Spaniel sweeps its feathery tail rapidly from side to side, and is a very pretty object to any one who has an eye for beauty of movement. It is a rule, that however spirited a Spaniel may be, it must not raise its tail above the level of its back. For the purpose of sport, a Spaniel must be possessed of a thick coat, as it is subject to continual wetting from the dripping coverts through which it has to force its way. It should be also a tolerably large Dog, not weighing less than fourteen pounds if possible, and may with advantage weigh some thirty or forty pounds, as do the breed known by the name of the "Clumber" Spaniels. These last-mentioned animals work silently.

Examples are given in the accompanying illustration of three kinds of Cocker Spaniels. The dark Dog, that occupies the foreground, is a Welsh Cocker; and the other two Dogs are ordinary Cockers.

COCKER SPANIELS.

THE COCKER is altogether a smaller animal, seldom weighing above twenty pounds, and very often being only ten or twelve pounds in weight. It is an active and lively animal, dashing about its work with an air of gay enjoyment that assists materially in enlivening the spirits of its master. There are many breeds of this Dog, among which the English, Welsh, and Devonshire Cockers may be mentioned as well-known examples.

It is a courageous little creature, retaining its dashing boldness even when imported into the enervating Indian climate, which destroys the spirit of most Dogs, and even reduces the stubborn bull-dog to a mere poltroon. Captain Williamson, in his book of "Oriental Field Sports," records an instance of rash courage on the part of one of these little Dogs.

"I was shooting near some underwood, rather thinly scattered among reedy grass, growing on the edges of a large water-course, which took its rise near the foot of the large hill at Muckun Gunge, when suddenly one of a brace of fine cocking Spaniels I had with me ran round a large bush greatly agitated, and apparently on some game which I expected to put up.

I followed as fast as I could ; but Paris, which was the Dog's name, was too quick for me, and before I could well get round the bush, which was about ten yards from the brink of the ravine, had come to a stand, his ears pricked, his tail wagging like lightning, and his whole frame in a seeming state of ecstasy. I expected that he had got a hare under the bank, and, as the situation was in favour of a shot, I ran towards him with more speed than I should have done had I known that instead of a hare I should find, as I did, a tiger sitting on its rump, and staring Paris in the face. They were not above two yards asunder.

As soon as the Dog found me at his side, he barked, and giving a spring down, dashed at the tiger. What happened for some moments I really cannot say ; the surprise and danger which suddenly affected me banished at once that presence of mind which many boast to possess on all emergencies. I frankly confess that my senses were clouded, and that the tiger might have devoured me without my knowing a word of the matter. However, as soon as my fright had subsided, I began, like a person waking from a dream, to look about, and saw the tiger cantering away at about a hundred and fifty yards' distance, with his tail erect, and followed by Paris, who kept barking ; but when the tiger arrived at a thick cover, he disappeared.

I had begun in my mind to compose a requiem for my poor Dog, as I saw him chasing the tiger, which I expected every moment would turn about and let Paris know that he had caught a Tartar. Though Paris had certainly brought me to the gate of destruction, yet he as certainly saved me. I felt myself indebted to him for preservation, and consequently was not a little pleased to see him return safe."

This is not a solitary example of the achievement of so daring a feat. Another officer, belonging to the Bengal Artillery, was shooting near a jungle, and was attended by five or six Spaniels, for the purpose of putting up the bustards, floricans, peafowl, and other birds, when a tiger suddenly showed itself from a spot where it had lain concealed. Instead of retreating from the terrible animal, the Spaniels dashed boldly at the brindled foe, and although several of them were laid prostrate by the tiger's paw, the survivors remained staunch, and attracted the creature's attention so completely that their master was enabled to kill it without difficulty.

The report that the Dhole will attack the tiger is thus corroborated.

FROM its singular affection for the water, this Dog is termed the WATER SPANIEL, as a distinction from the Field Spaniel. In all weathers, and in all seasons, the Water Spaniel is ever ready to plunge into the loved element, and to luxuriate therein in sheer wantonness of enjoyment. It is an admirable diver, and a swift swimmer, in which arts it is assisted by the great comparative breadth of its paws. It is therefore largely used by sportsmen for the purpose of fetching out of the water the game which they have shot, or of swimming to the opposite bank of the river, or to an occasional island, and starting therefrom the various birds that love such moist localities.

Much of its endurance in the water is owing to the abundance of natural oil with which its coat is supplied, and which prevents it from becoming really wet. A real Water Spaniel gives himself a good shake as soon as he leaves the river, and is dry in a very short time. This oil, although useful to the Dog, gives forth an odour very unpleasant to human nostrils, and therefore debars the Water Spaniel from enjoying the fireside society of its human friends.

Some people fancy that the Water Spaniel possesses webbed feet, and that its aquatic prowess is due to this formation. Such, however, is not the case. All dogs have their toes connected with each other by a strong membrane, and when the foot is wide and the membrane rather loosely hung, as is the case with the Water Spaniel, a large surface is presented to the water.

The Water Spaniel is of moderate size, measuring about twenty-two inches in height at the shoulders, and proportionately stout in make. The ears are long, measuring from point to point rather more than the animal's height.

The KING CHARLES SPANIEL derives its name from the "airy monarch," Charles II., who took great delight in these little creatures, and petted them in a manner that verged on absurdity.

WATER SPANIEL. – *Canis familiaris.*

It is a very small animal, as a really fine specimen ought not to exceed six or seven pounds in weight. Some of the most valuable King Charles Spaniels weigh as little as five pounds, or even less. These little creatures have been trained to search for and put up game after the manner of their larger relatives, the springers and cockers, but they cannot endure severe exercise, or long-continued exertion, and ought only to be employed on very limited territory.

When rightly managed, it is a most amusing companion, and picks up accomplishments with great readiness. It can be trained to perform many pretty tricks, and sometimes is so appreciative of its human playfellows that it will join their games.

I knew one of these animals which would play at that popular boy's game, called "touch," as correctly as any of the boys who used to join in the game, and on account of its small size and great agility was a more formidable opponent than any of the human players. The same Dog carried on a perpetual playful feud with the cat, each seeking for an opportunity of dealing a blow and of getting away as fast as possible. It was most absurd to see the way in which the Dog would hide itself behind a door-step, a scraper, a large stone, or under a thick shrub, and panting with eager expectation, watch the cat walking unsuspiciously towards its ambush. As the cat passed, out shot the Dog, tumbled pussy over, and made off at the top of its speed, pursued by the cat in hot haste, all anxious to avenge herself of the defeat. In these chases, the cat always used to run on three legs, holding one paw from the ground as if to preserve its strength in readiness for a severe application to the Dog's ears.

"Prince," for that was the name of this clever little animal, was an accomplished bird's-nester, seldom permitting a too-confiding blackbird or thrush to build its hymeneal home in the neighbourhood without robbing it of its variegated contents. When the Dog first discovered how palatable an article of diet was a blackbird's egg, he used to push his nose into the nest and crush the eggs with his teeth, or would try to scrape them out with his paw. In both these methods, he wasted a considerable portion of the liquid contents of the eggs, and after a while invented a much better mode of action. Whenever he

KING CHARLES SPANIEL.—*Canis familiaris.*

discovered the newly-built nest of a thrush or blackbird, he would wait until there were some four or five eggs in the nest, and then would bite out the bottom of the nest, so as to let the eggs roll unbroken into his mouth.

One of these little animals, which belonged to a Gloucestershire family, was very clever and docile.

Every morning, he would voluntarily fetch his towel and brush, and stand patiently to be washed, combed, and brushed by the hands of his mistress. Generally, he was accustomed to take his meals with the family, but if his mistress were going to dine from home she used to say to him, "Prince, you must go and dine at the rectory to-day." The Dog would therefore set off for the rectory, rather a long and complicated walk, and after passing several bridges, and taking several turnings, would reach the rectory in time for dinner. There he would wait until he had taken his supper, and if no one came to fetch him, would return as he came.

The BLENHEIM SPANIEL is even smaller than the King Charles, and resembles it closely in its general characteristics. Both these animals ought to have very short muzzles, long silky hair without any curl, extremely long and silky ears, falling close to the head, and sweeping the ground. The legs should be covered with long silky hair to the very toes, and the tail should be well "feathered." The eyes of these little Dogs are extremely moist, having always a slight lachrymal rivulet trickling from the corner of each eye.

Although, from their diminutive size, these little Dogs are anything but formidable, they are terrible foes to the midnight thief, who cares little for the brute strength of a big yard-dog. Safely fortified behind a door, or under a sofa, the King Charles sets up such a clamorous yelling at the advent of a strange step, that it will disconcert the carefully arranged plans of professional burglars with much more effect than the deep bay and the fierce struggles of the mastiff or the bloodhound. It is easy enough to quiet a large Dog in the yard, but to silence a watchful and petulant King Charles Dog within doors, is

1.   T

quite a different matter. Many "toy" Dogs are equally useful in this respect, and the miniature terrier, which has lately become so fashionable, or the Skye terrier, are most admirable assistants in giving timely warning of a foe's approach, although they may not be able to repel him if he has once made good his entrance.

MALTESE DOG.—*Canis familiaris.*

A VERY celebrated, but extremely rare, "toy" Dog, is the MALTESE DOG, the prettiest and most loveable of all the little pet Dogs.

The hair of this tiny creature is very long, extremely silky, and almost unique in its glossy sheen, so beautifully fine as to resemble spun glass. In proportion to the size of the animal, the fur is so long that when it is in rapid movement, the real shape is altogether lost in the streaming mass of flossy hair. One of these animals, which barely exceeds three pounds in weight, measures no less than fifteen inches in length of hair across the shoulders. The tail of the Maltese Dog curls strongly over the back, and adds its wealth of silken fur to the already superfluous torrent of glistening tresses.

It is a lively and very good-tempered little creature, endearing itself by sundry curious little ways to those with whom it is brought in contact. The "toy" spaniels are subject to several unpleasant habits, such as snoring and offensive breath, but the Maltese Dog is free from these defects, and is therefore a more agreeable companion than the King Charles or the Blenheim Spaniels.

As the name implies, it was originally brought from Malta. It is a very scarce animal, and at one time was thought to be extinct; but there are still specimens to be obtained by those who have no objection to pay the price which is demanded for these pretty little creatures.

The LION DOG, so called on account of its fancied resemblance to the king of beasts, when it is shaven after the fashion of poodles, is a cross between the poodle and the Maltese Dog, possessing the tightly curled hair of the poodle without its elongated ears and determinate aspect.

A VERY decided contrast to the last-mentioned Dog is afforded by the ALPINE SPANIEL, more generally known by the title of the St. Bernard's Dog, on account of the celebrated

ST. BERNARD'S DOG.—*Canis familiaris.*

monastery where these magnificent animals are taught to exercise their wondrous powers, which have gained for them and their teachers a world-wide fame.

These splendid Dogs are among the largest of the canine race, being equal in size to a large mastiff. The good work which is done by these Dogs is so well known that it is only necessary to give a passing reference. Bred among the coldest regions of the Alps, and accustomed from its birth to the deep snows which everlastingly cover the mountain-top, the St. Bernard's Dog is a most useful animal in discovering any unfortunate traveller who has been overtaken by a sudden storm and lost the path, or who has fallen upon the cold ground, worn out by fatigue and hardship, and fallen into the death-sleep which is the result of severe cold.

Whenever a snow-storm occurs, the monks belonging to the monastery of St. Bernard send forth their Dogs on their errand of mercy. Taught by the wonderful instinct with which they are endowed, they traverse the dangerous paths, and seldom fail to discover the frozen sufferer, even though he be buried under a deep snow-drift. When the Dog has made such a discovery, it gives notice by its deep and powerful bay of the perilous state of the sufferer, and endeavours to clear away the snow that covers the 'ifeless form.

The monks, hearing the voice of the Dog, immediately set off to the aid of the perishing traveller, and in many cases have thus preserved lives that must have perished without their timely assistance. In order to afford every possible help to the sufferer, a small flask of spirits is generally tied to the Dog's neck.

The illustration which accompanies this notice of the Alpine Spaniel, is a portrait of the well-known Dog belonging to Mr. Albert Smith.

T 2

Of all the domesticated Dogs, the Poodle seems to be, take him all in all, the most obedient and the most intellectual. Accomplishments the most difficult are mastered by this clever animal, which displays an ease and intelligence in its performances that appear to be far beyond the ordinary canine capabilities.

A barbarous custom is prevalent of removing the greater portion of the Poodle's coat, leaving him but a ruff round the neck and legs, and a puff on the tip of the tail as the sole relic of his abundant fur.

POODLE.—*Canis familiaris.*

Such a deprivation is directly in opposition to the natural state of the Dog, which is furnished with a peculiarly luxuriant fur, hanging in long ringlets from every portion of the head, body, and limbs. The Poodle is not the only Dog that suffers a like tonsorial abridgment of coat; for under the dry arches of the many bridges that cross the Seine, in Paris, may be daily seen a mournful spectacle. Numerous Dogs of every imaginable and unimaginable breed, lie helpless in the shade of the arch, their legs tied together, and their eyes contemplating with woeful looks the struggles of their fellows, who are being shorn of their natural covering, and protesting with mournful cries against the operation.

There is a diminutive variety of the Poodle, which is termed the Barbet. This little Dog is possessed of all the intellectual powers of its larger relative, and on account of its comparatively small size, was formerly in great request as a lady's Dog. For this enviable post it is well fitted, as it is a cleanly little creature, very affectionate, and full of the oddest tricks and vagaries.

Some years since, I made acquaintance with a comical little Dog, named "Quiz," which I believe to have been a Barbet, though no one had ventured definitely to refer the strange little creature to any known variety.

He was very small, not larger than an ordinary rabbit, and was overwhelmed with such a torrent of corkscrew curls that his entire shape was concealed under their luxuriance; and, when he was lying asleep on the sofa, he reminded the spectator of a loose armful of mop thrums. While reposing, his head was quite undistinguishable from his tail; and when walking, his trailing curls collected such an ever-increasing mass of leaves, dry sticks, straws, and other impediments, that he was frequently obliged to halt, in order to be released from his encumbrances.

Casual passengers were constantly arrested in their walk by the singular animated mop that rolled along without any visible means of progression; and I have more than once been witness to a warm dispute respecting the position in nature which the strange animal might occupy. Some thought it *might* be a Dog, while others suggested that it was a young lion; but the prevailing idea referred little Quiz to a position among the bears.

He was a most amusing and clever little animal, readily picking up acquirements, and inventing new accomplishments of his own. He would sit at the piano, and sing a song to his own accompaniment, the manual, or rather the pedal, part of the performance being achieved by a dexterous patting of the keys, and the vocal efforts by a prolonged and modulated howl. He could also "talk," by uttering little yelps in rapid succession.

Like all pet Dogs, he was jealous of disposition, and could not bear that any one, not excepting his mistress, should be more noticed than himself.

When his mistress was ill, he was much aggrieved at the exclusive attention which was given to the invalid, and cast about in his doggish brain for some method of attracting the notice which he coveted. It is supposed that he must have watched the interview between medical man and patient, and have settled in his mind the attraction which exercised so powerful an influence upon the physician; for just as the well-known carriage drew up to the door, Quiz got on a chair, sat up on his hind legs, and began to put out his tongue, and hold forth his paw, as he had seen his mistress do, and evidently

MEXICAN LAPDOG.—*Canis familiaris.*

expected to be treated in a similar manner. His purpose was certainly gained, for he attracted universal attention by his *ruse*. He had not patience to keep his tongue out of his mouth, but rapidly thrust it out, and as rapidly withdrew it again.

Poor Quiz died very shortly after I made acquaintance with him, a victim to the cholera, which at that time was rife in Oxford.

THE VERY tiniest of the Dog family is the MEXICAN LAPDOG, a creature so very minute in its dimensions as to appear almost fabulous to those who have not seen the animal itself.

BLOODHOUND.—*Canis familiaris.*

One of these little canine pets is to be seen in the British Museum, and always attracts much attention from the visitors. Indeed, if it were not in so dignified a locality, it would be generally classed with the mermaid, the flying serpent, and the Tartar lamb, as an admirable example of clever workmanship. It is precisely like those white woollen toy Dogs which sit upon a pair of bellows, and when pressed give forth a nondescript sound, intended to do duty for the legitimate canine bark. To say that it is no larger than these toys would be hardly true, for I have seen in the shop windows many a toy Dog which exceeded in size the veritable Mexican Lapdog.

THE MAGNIFICENT animal which is termed the BLOODHOUND, on account of its peculiar facility for tracking a wounded animal through all the mazes of its devious course, is very scarce in England, as there is but little need for these Dogs for its chief employment.

In the "good old times" this animal was largely used by thief-takers, for the purpose of tracking and securing the robbers who in those days made the country unsafe, and laid the roads under a black mail. Sheep-stealers, who were much more common when the offence was visited with capital punishment, were frequently detected by the delicate nose of the BLOODHOUND, which would, when once laid on the scent, follow it up with unerring precision, unravelling the single trail from among a hundred crossing footsteps, and only to be baffled by water or blood. Water holds no scent, and if the hunted man is able to take a long leap into the water, and to get out again in some similar fashion, he may set at defiance the Bloodhound's nose. If blood be spilt upon the track, the delicate olfactories of the animal are blunted, and it is no longer able to follow the comparatively weak scent which is left by the retreating footsteps.

Both these methods have been successfully employed, but in either case great caution is needed. When the hound suspects that the quarry has taken to the water, it swims backward and forward, testing every inch of the bank on both sides, and applying its nose to every leaf, stick, or frothy scum that comes floating by.

In this country the Bloodhound is chiefly employed in deer-shooting, aiding the sportsman by singling out some animal, and keeping it ever before him, and by driving it in certain directions, giving to its master an opportunity for a shot from his rifle. Should the deer not fall to the shot, but be only wounded, it dashes off at a greatly increased pace, followed by the Bloodhound, which here displays his qualities. Being guided by the blood-drops that stud the path of the wounded animal, the hound has an easy task in keeping the trail, and by dint of persevering exertions is sure to come up with his prey at last.

The Bloodhound is generally irascible in temper, and therefore a rather dangerous animal to be meddled with by any one excepting its owner. So fierce is its desire for blood, and so utterly is it excited when it reaches its prey, that it will often keep its master at bay when he approaches, and receive his overtures with such unmistakeable indications of anger that he will not venture to approach until his Dog has satisfied its appetite on the carcase of the animal which it has brought to the ground. When fairly on the track of the deer, the Bloodhound utters a peculiar, long, loud, and deep bay, which, if once heard, will never be forgotten.

The modern Bloodhound is not the same animal as that which was known by the same title in the days of early English history, the breed of which is supposed to be extinct. The ancient Bloodhound was, from all accounts, an animal of extremely irritable temper, and therefore more dangerous as a companion than the modern hound.

The colour of a good Bloodhound ought to be nearly uniform, no white being permitted, except on the tip of the stern. The prevailing tints are a blackish-tan, or a deep fawn. The tail of this Dog is long and sweeping, and by certain expressive wavings and flourishings of that member, the animal indicates its success or failure.

CLOSELY allied with the bloodhound is the now rare STAGHOUND, a Dog which is supposed to derive its origin from the bloodhound and the greyhound, the latter animal being employed in order to add lightness and speed to the exquisite scent and powerful limbs of the former. Sometimes the foxhound is used to cross with this animal.

It is a large and powerful Dog, possessed of very great capabilities of scent, and able, like the bloodhound, to hold to the trail on which it is laid, and to distinguish it among the footprints of a crowd. Despite of the infusion of greyhound blood, the Staghound is hardly so swift an animal as might be conjectured from its proportions, and probably on account of its slow pace has fallen into comparative disrepute at the present day. Until the death of George III. the stag-chase was greatly in vogue; but since that time it has failed to attract the attention of the sporting world, and has gradually yielded to the greater charms of the foxhunt.

The real old English Staghound is now extremely rare, and is in danger of becoming entirely extinct. The Dog which is now used for the purpose of chasing the stag is simply a very large breed of the foxhound, which, on account of its superior length of limb, is more capable of matching itself against the swift-footed deer than the ordinary hound. These Dogs are very powerful when in a good state of health, and have been known to achieve very wonderful feats of speed and endurance. They have been known to run for a distance of fifty miles in pursuit of a stag; and one memorable run is recorded, where the stag, and the only two hounds which kept to its trail, were found dead close to each other. The stag had made one powerful effort, had leaped over a park wall, which the Dogs in their wearied state were unable to surmount, and had fallen dead just as it had gained a place of safety.

It is needful that the Staghound should be a courageous as well as a powerful animal; for when the stag is brought to bay it becomes a formidable antagonist, dashing

STAGHOUND.—*Canis ramiliaris.*

boldly at the nearest foe, whether man or Dog, and often inflicting by the stroke of its sharp antlers a mortal wound upon any Dog that may be within its reach. Some degree of cunning is also requisite, so that the Dog may not rush blindly upon its fate, but may craftily watch its opportunity, and seize its quarry without suffering for its boldness.

When the country was more open, and less broken up into fields and enclosures than is the case at the present, stag-hunting was a comparatively easy task, but in the present day, when a free Englishman can hardly walk half-a-mile without being checked by a wall or fence, or a warning notice, the stag has so much the advantage of the hounds and horses that the chase has gradually sunk into comparative disuse. With one or two exceptions, the royal Staghounds are now almost the only representatives of this once popular and exciting sport.

OF ALL the Dogs which are known by the common title of "hound," the FOX-HOUND is the best known. There are few animals which have received more attention than the Foxhound, and none perhaps which have so entirely fulfilled the wishes of its teachers. A well-known sporting author, who writes under the *nom de plume* of "Stone-henge," remarks, with pardonable enthusiasm, that "the modern Foxhound is one of the most wonderful animals in creation." The efforts which have been made, and the sums which have been spent, in the endeavour to make this animal as perfect as possible, are scarcely credible.

Without in the least disparaging any efforts to improve the nature and the character of any animal, we cannot but draw a sad comparison between the unwearying pains that are bestowed upon the condition of the Foxhound, and the neglected state of many a human being in the vicinity of the palatial dog-kennel and the magnificent stables. At one establishment, eight or ten thousand pounds per annum have been expended upon the Dogs and horses, and this for a series of many years. As might be expected, the command of such enormous sums of money, backed by great judgment on the parts of the owners and trainers of hounds, has produced a race of Dogs that for speed, endurance,

FOXHOUND.—*Canis familiaris.*

delicate scent, and high courage, approach as near to absolute perfection as can well be imagined.

By thus improving the condition of the domesticated Dog, the country has been benefited, for it is impossible to improve any inhabitant of a country without conferring a benefit on the land in which it is reared. Still, supposing that half the sums which are annually expended on training Dogs for the amusement of the upper classes had been employed in improving the condition of the uneducated and neglected poor, and had been backed by equal judgment, I cannot but fancy that the country would have received a greater benefit than is conferred upon it by the most admirable pack of hounds that can be conceived.

It is supposed that the modern Foxhound derives its origin from the old English hound, and its various points of perfection from judicious crosses with other breeds. For example, in order to increase its speed, the greyhound is made to take part in its pedigree, and the greyhound having already some admixture of the bull-dog blood, there is an infusion of stubbornness as well as of mere speed.

There are various breeds of Dogs which are remarkable for the very great development of some peculiar faculty, such as speed in the greyhound, courage in the bull-dog, delicacy of scent in the bloodhound, sagacity in the poodle, and so on. So that, when a breed of Dogs begins to fail in any of these characteristics, the fault is amended by the introduction of a Dog belonging to the breed which exhibits the needful quality in greatest perfection. It is remarkable that the mental character is transmitted through a longer series of descendants than the outward form. Even in the case of such widely different Dogs as the bull-dog and the greyhound, all vestige of the bull-dog form is lost in the fourth cross, while the determinate courage of the animal is persistent, and serves to invigorate the character of unnumbered successive progeny.

By using these means with the greatest care and judgment, the modern sportsmen have succeeded in obtaining an animal which is so accurate of scent, that it might almost challenge the bloodhound himself in its power of discovering it, and of adhering to it when found; so determined in character, that it has many a time been known to persevere in its chase until it has fallen dead on the track; and so swift of foot that few horses can keep pace with it in the hunting-field, if the scent be good, and ground easy. It is averred by competent authority, that no man can·undertake to remain in the same field with the hounds while they are running.

The speed which can be attained by Foxhounds may be estimated from the well-known match which took place upon the Beacon course at Newmarket. The length of the course is 4 miles 1 furlong and 132 yards, and this distance was run by the winning Dog, "Blue-cap," in eight minutes and a few seconds. The famous racehorse, "Flying Childers," in running over the same ground, was little more than half a minute ahead of the hound. Now, if we compare the dimensions of the horse and the hound, we shall form a tolerably accurate conception of the extraordinary swiftness to which the latter animal can attain. In that match, no less than sixty horses started together with the competitors, but of the sixty only twelve were with the Dogs at the end of this short run.

It must be remembered that, in addition to the severe and unceasing labour of the chase, in which the Dogs are always busily at work, either in searching for a lost scent, or following it up when found, the hounds are forced to undergo no small exertion in walking from their kennel to the "meet," which is frequently at some distance from their home; and then in walking back again when the chase is over.

That the animal should be enabled to perform these severe tasks, which often occur several times weekly, it is necessary that it should not be too large, lest it should fatigue itself with its own bulk, and go through considerable needless exertion in forcing its way through thickets where a lesser Dog would pass without difficulty; and it is equally necessary that it should not be too small, lest it should be unequal to the various impediments which cross its path, and by reason of its shorter limbs be unable to keep up properly with the rest of the pack.

According to the latest authorities, the best average height for Foxhounds is from twenty-one to twenty-five inches, the female being generally smaller than the male. However, the size of the Dog does not matter so much; but it is expected to match the rest of the pack in height as well as in general appearance.

It has been well remarked, by a writer to whom allusion has already been made, that a hound ought not to be looked upon as an individual, but as a component part of a pack, and, therefore, that a Dog which will be almost invaluable in one pack will be quite inadmissible into another. It is a great fault in a Dog to be slower than its companions, but it is a fault of hardly less magnitude to be too fast for them, and to run away at such a pace that it seems to be getting all the hunting to itself. To use an expressive, but conventional term, "suitiness" is one of the principal points in a pack of hounds, which ought to appear as if they all belonged to one family.

In its natural state, the head of the Foxhound has a different aspect from that which is presented by the trained Dog. This change of appearance is caused by the custom of cropping, or rather of trimming, the ears, so as to dock them of their full proportions, and to leave no more of the external organ than is necessary to protect the orifice. It is said that this process is necessary, in order to guard the animal's ears from being torn by the brambles and other thorny impediments which constantly come in its path, and through which the Dog is continually forced to thrust itself. But the custom does not seem to confer a corresponding benefit on the poor creature whose ears are subjected to the operator's steel, and it may be that the custom of cropping Dogs' ears will go out of fashion, as is happily the case with the equally cruel practice of cropping the ears of horses, and docking their tails.

This Dog is a sufficiently sagacious animal, and if it were subjected to the influence of man as frequently as the Terrier and other companions of the human race, would not lose by comparison with them. Even in the state of semi-civilization into which these Dogs are brought, their obedience to the voice and gestures of the huntsman is quite mar·vellous; and even when in their kennel they will come individually to be fed, no Dog venturing to leave its place until its name has been called.

As to the various sporting details connected with this animal, such as breeding, training, feeding, &c., they may be found in many sporting works, where they are elaborately discussed, but are not suitable for a work of the present character.

THE HARRIER, so called because it is chiefly employed in hunting the hare, is in the present day nothing more or less than a small foxhound, the description of the latter animal serving equally for that of the former, with the one exception of size. As has been mentioned in the account of the foxhound, the average height is about twenty-three inches, but the height of the Harrier ought not to exceed eighteen or nineteen inches.

Partly on account of its smaller size, and partly on account of the character of its work, the Harrier is not so swift an animal as the foxhound, and does not test so fully the speed and strength of the horses that follow in its track. It is a swifter animal in these days than was the case some few years back, because in the modern system of hare-hunting, poor "puss" is so rapidly followed by the hounds that she has no time to waste in those subtle contrivances for throwing the hounds off her track for which she is so justly famous, and which have often baffled the efforts of the best and strongest Harriers.

The points of a good Harrier are similar to those of the foxhound, and may be described as follows.

"There are necessary points in the shape of a hound which ought always to be attended to by a sportsman, for if he be not of a perfect symmetry he will neither run fast nor bear much work. He has much to undergo, and should have strength proportioned to it. Let his legs be straight as arrows, his feet round and not too large ; his shoulders back ; his breast rather wide than narrow ; his chest deep ; his back broad ; his head small; his neck thin; his tail thick and bushy ; if he carry it well, so much the better. Such hounds as are out at the elbows, and such as are weak from the knees to the foot, should never be taken into the pack.

"I find that I have mentioned a small head as one of the necessary requisites of a hound ; but you will observe that it is relative to beauty only, for as to goodness, I believe that large-headed hounds are in no wise inferior. The colour I think of little moment, and am of opinion with our fried Foote, respecting his negro friend, that a good Dog, like a good candidate, cannot be of bad colour."

These remarks were written by Beckford, in the year 1779, and are of such sterling value that they are accepted even in the present day as the criteria of a good hound. He proceeds to observe in the same letter from which the above description has been transcribed, that the shape of the Dog's head is as variable as the colour of his hide, and that some sportsmen prefer a sharp-nosed hound, while others care nothing for a Dog unless he have a large and roomy head. Each, however, in his opinion, is equally useful in its own way; for "speed and beauty are the chief excellences of the one, while stoutness and tenderness of nose in hunting are characteristic of the other." To these qualifications the modern huntsmen have added another, consisting of depth of the back ribs, in order to secure a stout build, and the capability of enduring daily work for a lengthened period.

Uniformity of size and colour is even more requisite in a pack of Harriers than of foxhounds. Such packs indeed are often composed of the latter variety of Dog, which are too small to be admitted into the regular foxhound pack. However, if a pack is composed of these dwarf foxhounds, the two best characters of the true Harrier are lost, namely, the musical tongue and the sensitive nose, and the only compensating quality that these animals possess is extreme speed. A pack of true Harriers is distinguished for the melodious tongues of its members, which can be heard at a distance of several miles, while the delicacy of their scent is so great that they can work out all the complicated doubles of the hare.

There are several breeds of the BEAGLE, which are distinguishable from each other by their size and general aspect.

The Medium-sized Beagle is not unlike the harrier, but is heavier about the throat than that animal, and has stouter limbs, and a comparatively larger body. The height of this Dog is from a foot to fourteen or fifteen inches.

The Rough Beagle is thought to be produced by crossing the original stock with the rough terrier, and possesses the squeaking bark of the terrier rather than the prolonged musical intonation of the Beagle. Some authorities, however, take the animal to be a distinct variety. The nose of this creature is furnished with the stiff whisker-hairs which are found on the muzzle of the rough terrier, and the fur is nearly as stiff and wiry as the terrier's.

The Dwarf Beagle, or Rabbit Beagle, as it is sometimes called, is the smallest of the three animals, delicate in form and aspect, but good of nose and swift of foot. So very small are some of these little creatures that a whole pack has been conveyed to and from the field in hampers slung over the back of a horse, or simply in the shooting pockets of the men. Their strength was thus preserved for the labours of the field, and they were

BEAGLE.—*Canis familiaris.*

saved from the fatiguing walk to the field and back again. Ten inches is the average height of a Rabbit Beagle.

These little Dogs are chiefly employed by those who hunt on foot, as they are not sufficiently swift to drive the hare from her doubles, and by patiently tracking her through all her wiles, " win like Fabius, by delay." Beagles used to be much in favour with the junior members of the universities, for the purpose of affording a pleasant afternoon's amusement. It is true that the legitimate object of chase, namely, the hare, is seldom forthcoming, but her place is readily supplied by a long-winded lad, who traverses the country at speed, trailing after him a rabbit-skin well rubbed with turpentine or aniseed. If the scent be good, and the course lie tolerably straight, the endurance of the hunter is severely tested, but if the miniature hounds come often to a check, any one of average powers can be in at the finish.

POINTERS.—*Canis familiaris.*

This little group of Dogs is representative of two breeds of the POINTER, the two foremost Dogs being examples of the modern English Pointer, and the third, of the Spanish Pointer. The latter of these Dogs is now seldom used in the field, as it is too slow and heavily built an animal for the present fast style of sporting, which makes the Dogs do all the ranging, and leaves to their master but a comparatively small amount of distance to pass over. The nose of this Dog is peculiarly delicate, as may be inferred from its exceedingly wide muzzle, and for those sportsmen who cannot walk fast or far, it is an useful assistant.

As may be seen from the engraving, the modern English Pointer is a very different animal, built on a much lighter model, and altogether with a more bold and dashing air about it. While it possesses a sufficiently wide muzzle to permit the development of the olfactory nerves, its limbs are so light and wiry that it can match almost any Dog in speed. Indeed, some of these animals are known to equal a slow greyhound in point of swiftness.

This quality is specially useful, because it permits the sportsman to walk forward, at a moderate pace, while his Dogs are beating over the field to his right and left. The sagacious animals are so obedient to the voice and gesture of their master, and are so well trained to act with each other, that at a wave of the hand they will separate, one going to the right and the other to the left, and so traverse the entire field in a series of "tacks," to speak nautically, crossing each other regularly in front of the sportsman as he walks forward.

When either of them scents a bird, he stops suddenly, arresting even his foot as it is raised in the air, his head thrust forward, his body and limbs fixed, and his tail stretched straight out behind him. This attitude is termed a "point," and on account of this peculiar mode of indicating game, the animal is termed the "Pointer." The Dogs are so trained that when one of them comes to a point he is backed by his companion, so as to avoid the disturbance of more game than is necessary for the purpose of the sportsman.

It is a matter of some difficulty to teach their lesson rightly, for the Dogs are quite as liable to error through their over-anxiety to please their master as through sluggishness or carelessness. Such Dogs are very provoking in the field, for they will come to a point at almost every strange odour that crosses their nostrils, and so will stand at pigs, sparrows, cats, or any other creature that may come in their way, and will hold so firmly to their "point" that they cannot be induced to move, except by compulsory means. This extreme excitability seems to be caused by too close adherence to the same stock in breeding, and is set right by a judicious admixture with another family.

According to "Stonehenge," the marks of a good Pointer are as follows. "A moderately large head, wide rather than long, with a high forehead and an intelligent eye, of medium size. Muzzle broad, with its outline square in front, not receding as in the hound. Flews (i. e. the overhanging lips) manifestly present, but not pendent. The head should be well set on the neck, with a peculiar form at the junction only seen in the Pointer. The neck itself should be long, convex in its upper outline, without any tendency to a dewlap or a ruff, as the loose skin covered with long hair round the neck is called. The body is of good length, with a strong loin, wide hips, and rather arched ribs, the chest being well let down, but not in a hatchet shape as in the greyhound, and the depth in the back ribs being proportionably greater than in that Dog. The tail, or 'stern,' as it is technically called, is strong at the root, but, suddenly diminishing, it becomes very fine, and then continues nearly of the same size to within two inches of the tip, where it goes off to a point, looking as sharp as the sting of a wasp, and giving the whole very much the appearance of that part of the insect, but magnified as a matter of course. This peculiar shape of the stern characterizes the breed, and its absence shows a cross with the hound or some other Dog."

The author then proceeds to recommend long, slanting, but muscular shoulder-blades, a long upper arm, a very low elbow, and a short fore-arm. The feet must be round and strong, and padded with a thick sole, the knee strong, and the ankle of full size. The colour is of comparatively small importance, but ought, if possible, to be white, so that the animal may be visible while beating among heather, clover, or turnips. Black or liver-coloured dogs are very handsome to the eye, but often cause much trouble to the sportsman, on account of the difficulty of distinguishing them among the herbage. White Dogs, with lemon-coloured heads, are the favourites of this author.

As the Pointer is seldom in contact with its master, except when in the field, its domestic qualities are rarely prized as they deserve to be. No Dog can be properly appreciated until it is a constant companion of man, and it is probable that many Dogs which are set down as stupid and untractable, are only so called because they have been deprived of the society of human beings, through whom alone their higher qualities can be developed, and have been confined to the kennel, the yard, or the field. The Pointer is but little known as a companion Dog, but when it is in the habit of living constantly with its owner speedily puts forth its intellectual powers, and becomes an amusing and interesting companion. One of my friends has kindly sent me the following account of a Pointer that belonged to him, and had been constantly with his master for a lengthened period of time. The animal was not an example of the thorough-bred Pointer, but was, nevertheless, a very respectable creature.

"I ONCE possessed a Dog whose nose, sight, and instinct were well developed; and as he was my companion for many a day, and my only friend for many months, some of his peculiarities may not be uninteresting.

The Dog could point a partridge, but he would eat it, too, if he had a chance; and often when I could not take a day's shooting I have observed my Dog doing a little

MODERN ENGLISH POINTER. — *Canis familiaris.*

amateur work on his own account. Very successful, also, was he in this occupation; and he frequently dined on a partridge or quail which. he had gained by means of his own skill. There was no concealing the fact that he was, however, an arrant coward; and he himself was perfectly conscious of this defect. As is usual amongst men, he endeavoured to conceal his weakness by the aid of a formidable exterior; and few who knew him not would ever venture even to insinuate that he was not as brave as a lion. If he happened to encounter any other Dog with which he was unacquainted he would immediately stand perfectly still, raise his tail, and keep it very firmly in one position; he would then elevate the hair on his back, and dragging up his jowls, would exhibit a formidable array of grinders. Thus exhibiting by no means a prepossessing appearance, he would merely growl whilst the other Dog walked round him, and he thus frequently prevented any liberties from being taken with him. No sooner had his visitor left him than his attitude would change; and with a glance, as much as to say, 'I did that very well,' he would jog along before me. In spite of his warlike positions, he was once terribly punished by a little terrier which resided in a butcher's shambles. Passing this locality, my Dog was set upon before even he had time to study attitudes or to assume a *pose*, so he made good use of his legs, and escaped with a few scratches. Now it happened that amongst his friends he had one which was a well-bred bull terrier, and after the mauling that he had received from the butcher's Dog I noticed that he was very much oftener with this friend than he had been before. The next time that I attempted to take him past the shambles he refused to come, and retreated home. I followed him, and, by dint of whistling, at length brought him out from his retreat, from which he was followed by his friend the bull terrier.

The two jogged along very pleasantly and cheerfully, my Dog evidently paying marked attention to his friend. When we approached the locality of the shambles my Dog ran along in front, whilst the bull terrier followed behind, and both looked as though 'up' to something. Opposite the shambles the terrier rushed out upon my Dog, which retreated with wonderful precipitancy behind his friend, who at once collared the assailant, and tumbled him over and over to the tune of the joyful barks of my old cur, which had evidently made the preliminary arrangements with his friend for this scene."

The same Dog was once taught a useful lesson in a singular manner. His master is an officer, and during the time when he possessed the Dog was annoyed by its constant intrusion into the mess-room when breakfast was on the table. Nothing could keep the Dog away from the tempting tables with their savoury viands, and as each member of the mess was liable to a fine every time that his Dog entered the room it was clear that these pertinacious intrusions must be stopped.

One morning the Dog crept into the room, after its custom, and fortunately there was no one at breakfast except its master. Attracted by the ham and fowls that lay so temptingly on the table, the Dog stealthily approached them, and stood pointing at the longed-for food, with watering mouth and eager eye. Seeing the Dog's attention thus occupied, his master slily tilted the teapot, so as to let a slender stream of the hot liquid trickle on the Dog's back. At first, its faculties were so absorbed in contemplation of the forbidden dainties, that it only acknowledged the hot liquid by a nervous twitching of the skin. As soon, however, as the fur was saturated, and the full effects of the boiling tea made themselves felt, the Dog sprang up with a yell of astonishment, and dashed howling through the door. Ever after its adventure with the teapot, no inducement could tempt the animal to enter that room, or come fairly within the threshold, and even if a chicken bone were held out as a bait the poor Dog would only lick its lips, and put on a plaintive and beseeching look as an appeal to the humanity of its tempter.

THE DALMATIAN DOG is even better known as a carriage or coach Dog than the Danish Dog, which has already been described and figured. Its shape is very like that of the pointer, but the artificially shortened ears give it a different aspect.

The ground colour of this animal's fur is nearly white, and is richly crossed with black spots, earning for it, in common with the Danish Dog, the title of "Plum-pudding." The height of this animal is about twenty-four or twenty-five inches. Some years ago, the Dalmatian Dog was very frequently seen in attendance upon the carriage of its owner, scampering along in high glee by the side of the vehicle, or running just in front of the horses, apparently in imminent danger of being knocked over every moment. Now, however, the creature has lost its hold on the fashionable world, and is but seldom seen.

This animal is seldom if ever permitted to be the constant companion of its master, and has therefore but little of that humanly intelligent look which marks the countenance of the companionable poodle or spaniel, and gives to the animal a certain semblance of its master.

We may see in every country a singular similitude between the human inhabitants of the land and the various animals which tread the same earth and breathe the same air. So we find that the countries which are the most productive of ferocious animals are most productive of ferocious men:—the Lion of Africa, the Tiger of India, the Grizzly Bear of America, the Polar Bear of the northern regions, being but lower types of the destructive humanity that prevails in those portions of the globe.

As this subtle bond of similar affections is found to pervade the wild animals and the human inhabitants of the same country, it is but natural that when the man and the brute are drawn closer together by domestication, and the higher Being enabled to pour its influence upon the lower, the similarity in their character should be still more apparent.

DALMATIAN OR COACH DOG.—*Canis familiaris.*

So we find that, whether in cats, Dogs, or horses, the animals which are most frequently made the companions of man, the disposition of the owner is reflected in the character of the beast. The large-hearted, kind-souled man will be surrounded with loving and gentle animals. His cat will sit and purr upon his shoulder fearless of repulse, his Dog will love and reverence his master with faithful worship, and his horse will follow him about the field in which it is freely grazing, and solicit the kind notice to which it is accustomed. On the other hand, the cross and snappish cat, the snarling Dog, and the crabbed-tempered horse are sure signs of corresponding qualities in the man that owns them, and will deter an observer of animal natures from placing his confidence in the man who could infuse such evil qualities into the creatures that surround him, and from whom they take their tone.

As the Dog is possessed of a disposition which is more easily assimilated with that of man than is the case with most animals, the affinity between itself and its master is constantly brought before our notice.

One man loves nothing so well as the largest Newfoundland or deerhound, while another is not satisfied unless his Dog be of the minutest proportions compatible with canine nature. One man places his faith in the terrier, another in the poodle ; one prefers the retriever, and another the spaniel. The man who pursues his sport at morning, in the face of the sun, is accompanied by the loud-tongued foxhound or beagle ; while the skulking nocturnal poacher is aided in his midnight thefts by the silent and crafty lurcher.

But of all the Dogs that are associated with man, and of all the men that make companionship with Dogs, the most repulsive, and most to be avoided by honest Dogs and men, are the bull-dog and his owner.

I may be accused of delivering too severe a judgment on Dog and man. Those who have been led by duty, curiosity, or chance through the unsavoury localities which are haunted by the members of the "Fancy," and have instinctively stepped aside from the fur-capped, beetle-browed, sleek-haired, suspicious ruffian, leading his sullen and scowling bull-dog at his heels, will hardly find terms too severe for the depraved human character that could encourage or cherish such an epitome of the most brutal features of the canine nature. Dog and man suit each other admirably ; and, had there been no human ruffian, there would have been no canine representation of his own ruffianism.

l.   U

That such a similarity should exist is an absolute necessity, inasmuch as the more powerful nature will inevitably expel the weaker, unless there is something in common between their characters, which will enable the higher being to convey its meaning to the lower, and the lower to receive obediently the mandates of the higher. As the two natures become more assimilated, they produce a corresponding effect in the outer form, and the resemblance extends to form and feature as well as to character. We notice the same effect to be produced among human beings when they are much thrown together, and a similar though not so evident a phenomenon takes place between the man and the brute.

The very form of the Dog tells its character as clearly as the human countenance betrays the disposition of the spirit which moulds its lines. It is most truly said by Bailey, in that mine of golden poetry, "Festus":—

> "All animals are living hieroglyphs,—
> The dashing Dog and stealthy-stepping cat,
> Hawk, bull, and all that breathe, mean something more
> To the true eye than their shapes show; for all
> Were made in love, and made to be beloved."

SETTER DOGS.—*Canis familiaris.*

As the pointers derive their name from their habits of standing still and pointing at any game which they may discover, so the SETTERS have earned their title from their custom of "setting" or crouching when they perceive their game. In the olden days of sporting, the Setter used always to drop as soon as it found the game, but at the present day the animal is in so far the imitator of the pointer, that it remains erect while marking down its game.

ENGLISH SETTER.—*Canis familiaris.*

There are several breeds of these animals, three of which are represented in the engraving. The Dog which occupies the foreground is an ordinary English Setter, the seated Dog towards the back is a Russian Setter, and the reclining Dog is an Irish Setter.

Each of these breeds possesses its particular excellences, which are combined in experienced and skilful hands by careful admixtures of one breed with another.

The Russian Setter is a curious animal in appearance, the fur being so long and woolly in texture, and so thoroughly matted together, that the form of the Dog is rendered quite indistinct. It is by no means a common animal, and is but seldom seen. It is an admirable worker, quartering its ground very closely, seldom starting game without first marking them; and possessed of a singularly delicate nose. In spite of its heavy coat, it bears heat as well as the lighter-clad pointer, and better than the ordinary English Setters, with their curly locks. When crossed with the English Setter it produces a mixed breed, which seems to be as near perfection as can be expected in a Dog, and which unites the good properties of both parents. A well-known sportsman, when trying these Dogs against his own animal, which he fondly thought to be unrivalled, found that the Russian animals obtained three points where his own Dog only made one. and that from their quiet way of getting over the ground they did not put up the birds out of gun-range, as was too often the case with his own swifter-footed Dogs.

The muzzle of this animal is bearded almost as much as that of the deerhound and the Scotch terrier, and the overhanging hair about the eyes gives it a look of self-relying intelligence that is very suggestive of the expression of a Skye terrier's countenance. The soles of the feet are well covered with hair, so that the Dog is able to bear plenty of hard work among heather or other rough substances.

The Irish Setter is very similar to the English animal, but has larger legs in proportion to the size of the body, and is distinguished from its English relative by a certain Hibernian air that characterises it, and which, although conspicuous enough to a practised eye, is not easy of description.

Taking as our authority the author above quoted, in the history of the pointer, the points of the Setter are shortly as follows :—" A moderately heavy head, but not so much so as in the pointer ; the muzzle not so broad nor so square in profile, the lower angle being rounded off, but the upper being still nearly a right angle. The eye is similar to that of the pointer, but not so soft, being more sparkling and full of spirit. The ear long, but thin, and covered with soft, silky hair, slightly waved. The neck is long, but straighter than that of the pointer, being also lighter and very flexible. The back and loins are hardly so strong as those of the pointer, the latter also being rather longer ; the hips also are more ragged, and the ribs not so round and barrel-like. The tail or ' flag' is usually set on a little lower, is furnished with a fan-like brush of long hair, and is slightly curled upwards towards the tip, but it should never be carried over the back or raised above the level of its root, excepting while standing, and then a slight elevation is admired, every hair standing down with a stiff and regular appearance. The elbow, when in perfection, is placed so low as to be fully an inch below the brisket, making the fore-arm appear very short. The hind-feet and legs are clothed with hair, or ' feathered,' as it is called, in the same way as the fore-legs, and the amount of this beautiful provision is taken into consideration in selecting the Dog for his points."

This description applies equally to the English and the Irish Setters.

While at work, the Setter has a strange predilection for water, and this fancy is carried so far in some Dogs that they will not go on with their work unless they can wet the whole of their coats once at least in every half-hour. If deprived of this luxury they pant and puff with heat and exertion, and are quite useless for the time.

It seems that the Setter is a less tractable pupil than the pointer, and even when taught is apt to forget its instructions and requires a second course of lessons before it will behave properly in the field. Owing to the rough coat and hair-defended feet of the Setter, it is able to go through more rough work than the pointer, and is therefore used in preference to that animal in the north of England and in Scotland,—where the heat is not so great as in the more southern countries,—where the rough stem of the heather would work much woe to a tender-footed Dog, and where the vicissitudes of the climate are so rapid and so fierce that they would injure the constitution of any but a most powerfully built animal.

This Dog, as well as the foxhound and harrier, is guided to its game by the odour that proceeds from the bird or beast which it is following ; but the scent reaches its nostrils in a different manner.

The foxhound, together with the harrier and beagle, follows up the odorous track which is left on the earth by the imprint of the hunted animal's feet, or the accidental contact of the under-side of its body with the ground. But the pointer, Setter, spaniel, and other Dogs that are employed in finding victims for the gun, are attracted at some distance by the scent that exhales from the body of its game, and are therefore said to hunt by " body-scent," in contradistinction to the hounds who hunt by "foot-scent." The direction in which the wind blows is, therefore, a matter of some consequence, and is duly taken advantage of by every good sportsman.

RETRIEVER Dogs, which are so called on account of their value in recovering or "retrieving" game that has fallen out of the reach of the sportsman, or on which he does not choose to expend the labour of fetching for himself, are of various kinds, and in every case are obtained by a crossing of two breeds. There are two principal breeds of Retrievers, the one being obtained by the mixture of a Newfoundland Dog and a setter, and the other by a cross between the water spaniel and the terrier.

The former of these breeds is the most generally known, and is the animal which is represented in the engraving. On inspection of this Dog, the characteristics of both parents are plainly perceptible in its form. For the larger kinds of game, such as hares

RETRIEVER.—*Canis familiaris.*

or pheasants, this Dog is preferable to the Terrier Retriever, as it is a more powerful animal, and therefore better able to carry its burden ; but, for the lesser description of game, the smaller Dog is preferable for many reasons.

The height of the large Retriever is from twenty-two to twenty-four inches ; its frame is powerfully built, and its limbs strong. A good nose is necessary, for the purpose of enabling the Dog to trace the devious and manifold windings of the wounded birds, which would baffle any animal not endowed with so exquisite a sense of smell. The fur of this Dog is curly and of moderate length, and is almost invariably black in colour. Indeed, many Dog-owners will repudiate a Retriever of any other colour but black.

To train a Retriever properly is rather a difficult task, demanding the greatest patience and perseverance on the part of the instructor. It is comparatively easy to teach a Dog to fetch and carry a load, but to teach him to retrieve in water is quite a different matter. On land the Dog can see the object from some little distance, but in the water his nose is so nearly on a level with the object for which he is searching, that he can only see a very little distance ahead, and must learn to guide his way by the voice and gesture of his master.

It is said that the greatest difficulty in the course of instruction is to keep the Dog from the water-rats, which are found so abundantly on the banks of rivers and ponds, and which afford such powerful temptations to a young and inexperienced animal.

Another obstacle in the tuition is the natural propensity of the Dog to bark when he is excited ; and as a young Dog is excited by almost everything that crosses his path, he generally tries his teacher's patience sorely before he learns to be silent and not to disturb the game by even a low whine. Again : the natural instinct of the Dog tells him to eat the animal which he has found, and it is not until he has been duly instructed that he learns to bring the game to his master without injuring it. July and August are the best months for teaching the Retriever, because the water is then comparatively warm, and there is no risk of disgusting the animals by forcing them into an icy bath, or of bringing on disease by overmuch exposure to a cold wind while their coats are wet and themselves wearied.

In order to keep the Dog from closing his teeth too firmly upon the game, he should always be made to lay down his spoil at his master's feet, or to loosen his hold as soon as his master touches the object which he is carrying. If the prey be snatched from his mouth, he instinctively bites sharply in order to retain it ; and when he gets into so bad a habit often damages the dead game so much that it is quite useless. Whenever a Dog is sent to fetch any object he must on no account be permitted to return without it, as, if he should once do so, he will ever afterwards be liable to give up the search as soon as he feels tired.

There are many other little difficulties in the training of the Retriever, some of them incidental to the Dog, simply because it is a Dog, and others belonging to the character of the individual animal. One great point to gain is, to make the Dog understand that the birds which he delights in fetching are killed by the gun and not by himself. Until he fully understands this lesson he is apt to dart off in chase of a bird as soon as he sees it, or perceives its scent, and to chase it until it is out of sight, just as we may see puppies chasing sparrows half over a field, barking at them as if they were to be caught as easily as if they were so many mice.

The smaller Retriever is produced by a cross of the terrier with the beagle, and in many points is superior to the large black Retriever. Should a larger animal be required, the pointer is employed in the cross instead of the beagle.

They are very quiet Dogs, and when on their quest do not make so much noise as the larger Retrievers, so that they are especially useful when the game is wild. The kind of terrier which is employed in the crossing depends on the caprice of the breeder, some persons preferring the smooth English Dog, and others the rough Scotch terrier. Being small Dogs they can be kept in the house, and become very companionable, so that when they go to their regular work they feel more love and respect for their master than would have been the case if they had been kept in a kennel, or sent to a cottage on board-wages.

Spaniels can be taught to retrieve, and will perform their task nearly as well as a Retriever itself. A thoroughly well-taught Dog is almost invaluable to the sportsman, and will command a large price. According to " Stonehenge," a well-instructed Clumber spaniel is worth thirty or forty guineas. If possible, the animal should in every case be taught by the person who intends to use him in the field, as neither the Dog nor its master can learn each other's ways without some experience, and without this knowledge neither can work well, or feel sure of the co-operation of the other.

These animals are also valuable for retrieving, because, like the smaller Retrievers, they are capable of sharing the house with their master, and are therefore more amenable to his authority, and more likely to follow out his wishes, than if their intercourse were restricted to the hunting-field. The peculiar and very unpleasant odour of the skin, which is found to exist in almost every kind of Dog, can be removed by careful and periodical washing—a practice which the animal soon learns to appreciate. There is, however, a drawback to the companionship of the Dog, in the parasitic insects with which it is generally infested, and which are too tenacious of life to be destroyed by immersion in water, or too strong to be dislodged by ordinary mechanical means.

The only method by which these disagreeable pests can be destroyed is by a rapidly acting poison, which kills them before they can retreat from its action. Such poisonous substances are too often dangerous to the Dog as well as to its parasites, and may seriously injure the animal instead of conferring any benefit upon it. Preparations of mercury are frequently used for this purpose, but are dangerous remedies for the reason above given, and are, moreover, rather tedious of application, requiring a careful rubbing in of the poison, and as careful a rubbing out again, together with the drawback of a muzzle on the poor Dog's mouth for three or four days, to prevent him from licking his irritated skin.

One very safe and very quick remedy is the "Persian Insect-destroying Powder," which has almost a magical effect, and is perfectly harmless to the Dog.

The best mode of applying this remedy is, first to dust the Dog well with the substance until every portion of him has received a few particles of the powder, and then to put him into a strong canvas bag, in which a small handful of the powder has been placed and shaken about well, so as to distribute it equally over the interior of the bag. Leave his head protruding from the bag, and put on his head and neck a linen cap, in which are holes for his nose and eyes, and let the interior of the cap be well treated with the powder. Lay him on the ground, and let him tumble about as much as he chooses, the more the better. In an hour or two let him out of the bag, and scrub his coat well the wrong way with a stiff brush.

If, during this operation, the Dog be placed on a sheet, or any white substance, it will be covered with dead and dying insects, and if the contents of the bag be emptied upon the white cloth, the number of moribund parasites will be rather astonishing. In a week or so the operation should be repeated, in order to destroy the creatures that have been produced from the unhatched eggs that always resist the powers of the destructive powder. I have personally tried the experiment, and have found the results to be invariably successful. The same substance is equally useful in freeing birds from their chief pest, the red mite, and is of deadly efficacy in the immolation of certain insects that are too often found in human houses.

THE MOST useful variety of the canine species is that sagacious creature on whose talent and energy depends the chief safety of the flock.

This animal seems to be, as far as can be judged from appearances, the original ancestor of the true British Dogs, and preserves its peculiar aspect in almost every country in Europe. It is a rather large Dog, as is necessary, in order to enable the animal to undergo the incessant labour which it is called on to perform, and is possessed of limbs sufficiently large and powerful to enable it to outrun the truant members of the flock, who, if bred on the mountain-side, are so swift and agile that they would readily baffle the efforts of any Dog less admirably fitted by nature for the task of keeping them together.

As the Sheep-dog is constantly exposed to the weather, it needs the protection of very thick and closely-set fur, which, in this Dog, is rather woolly in its character, and is especially heavy about the neck and breast. The tail of the Sheep-dog is naturally long and bushy, but is generally removed in early youth, on account of the now obsolete laws, which refused to acknowledge any Dog as a Sheep-dog, or to exempt it from the payment of a tax, unless it were deprived of its tail. This law, however, often defeated its own object, for many persons who liked the sport of coursing, and cared little for appearances, used to cut off the tails of their greyhounds, and evade the tax by describing them as Sheep-dogs.

The muzzle of this Dog is sharp, its head is of moderate size, its eyes are very bright and intelligent, as might be expected in an animal of so much sagacity and ready resource in time of need. Its feet are strongly made, and sufficiently well protected to endure severe work among the harsh stems of the heather on the hills, or the sharply-cutting stones of the highroad. Probably on account of its constant exercise in the open air, and the hardy manner in which it is brought up, the Sheep-dog is perhaps the most untiring of our domesticated animals.

There are many breeds of this animal, differing from each other in colour and aspect, and deriving their varied forms from the Dog with which the family has been crossed. Nearly all the sporting Dogs are used for this purpose, so that some Sheep-dogs have something of the pointer nature in them, others of the foxhound, and others of the setter. This last cross is the most common. Together with the outward form the

SHEPHERD'S DOG.—*Canis familiaris*

creature inherits much of the sporting predilections of its ancestry, and is capable of being trained into a capital sporting Dog.

Many of these animals are sad double-dealers in their characters, being by day most respectable Sheep-dogs, and by night most disreputable poachers. The mixed offspring of a Sheep-dog and a setter is as silently successful in discovering and marking game by night as he is openly useful in managing the flocks by day. As he spends the whole of his time in the society of his master, and learns from long companionship to comprehend the least gesture of hand or tone of voice, he is far better adapted for nocturnal poaching than the more legitimate setter or retriever, and causes far more deadly havoc among the furred and feathered game. Moreover, he often escapes the suspicion of the gamekeeper by his quiet and honourable demeanour during the daytime, and his devotion to his arduous task of guarding the fold, and reclaiming its wandering members. It seems hardly possible that an animal which works so hard during the day should be able to pass the night in beating for game.

Sometimes there is an infusion of the bull-dog blood into the Sheep-dog, but this mixture is thought to be unadvisable, as such Dogs are too apt to bite their charge, and so to alienate from themselves the confidence of the helpless creatures whom they are intended to protect, and not to injure. Unless the sheep can feel that the Dog is, next to the shepherd, their best friend, the chief value of the animal is lost.

It is well observed by Mr. Youatt, in his valuable work on these Dogs, that if the sheep do not crowd round the Dog when they are alarmed, and place themselves under his protection, there is something radically wrong in the management of the flock. He remarks, that the Dog will seldom, if ever, bite a sheep, unless incited to do so by its master, and suggests that the shepherd should be liable to a certain fine for every tooth-mark upon his flock. Very great injury is done to the weakly sheep and tender lambs by the crowding and racing that takes place when a cruel Dog begins to run among the flock. However, the fault always lies more with the shepherd than with his Dog, for as the man is, so will his Dog be. The reader must bear in mind that the barbarous treatment to which travelling flocks are so often subjected is caused by drovers and not shepherds, who, in almost every instance, know each sheep by its name, and are as careful of its wellbeing as if it were a member of their own family. The Dogs which so persecute the poor sheep in their bewilderments among cross-roads and the perplexity of crowded streets, are in their turn treated by their masters quite as cruelly as they treat the sheep. In this, as in other instances, it is "like man and like Dog."

As a general rule, the Sheep-dog cares little for any one but his master, and so far from courting the notice or caresses of a stranger will coldly withdraw from them, and keep his distance. Even with other Dogs he rarely makes companionship, contenting himself with the society of his master alone.

The SCOTCH SHEEP-DOG, more familiarly called the COLLEY, is not unlike the English Sheep-dog in character, though it rather differs from that animal in form. It is sharp of nose, bright and mild of eye, and most sagacious of aspect. Its body is heavily covered with long and woolly hair, which stands boldly out from its body, and forms a most effectual screen against the heat of the blazing sun, or the cold, sleety blasts of the winter winds. The tail is exceedingly bushy, and curves upwards towards the end, so as to carry the long hairs free from the ground. The colour of the fur is always dark, and is sometimes variegated with a very little white. The most approved tint is black and tan; but it sometimes happens that the entire coat is of one of these colours, and in that case the Dog is not so highly valued.

The "dew-claws" of the English and Scotch Sheep-dogs are generally double, and are not attached to the bone, as is the case with the other claws. At the present day it is the custom to remove these appendages, on the grounds that they are of no use to the Dog, and that they are apt to be rudely torn off by the various obstacles through which the animal is obliged to force its way, or by the many accidents to which it is liable in its laborious vocation. In the entire aspect of this creature there is a curious resemblance to the Dingo, as may be seen on reference to the account of that animal in a subsequent page.

It is hardly possible to overrate the marvellous intelligence of a well-taught Sheep-dog; for if the shepherd were deprived of the help of his Dog his office would be almost impracticable. It has been forcibly said by a competent authority that, if the work of the Dog were to be performed by men, their maintenance would more than swallow up the entire profits of the flock. They, indeed, could never direct the sheep so successfully as the Dog directs them; for the sheep understand the Dog better than they comprehend the shepherd. The Dog serves as a medium through which the instructions of the man are communicated to the flock; and being in intelligence the superior of his charge, and the inferior of his master, he is equally capable of communicating with either extreme.

One of these Dogs performed a feat which would have been, excusably, thought impos-sible, had it not been proved to be true. A large flock of lambs took a sudden alarm one night, as sheep are wont, unaccountably and most skittishly, to do, and dashed off among the hills in three different directions. The shepherd tried in vain to recall the fugitives; but finding all his endeavours useless, told his Dog that the lambs had all run away, and then set off himself in search of the lost flock. The remainder of the night was passed in fruitless search, and the shepherd was returning to his master to report his loss. However, as he was on the way, he saw a number of lambs standing at the bottom of a deep ravine, and his faithful Dog keeping watch over them. He immediately concluded that his Dog had discovered one of the three bands which had started off so inopportunely in the darkness; but on visiting the recovered truants he discovered, to his equal joy and wonder, that the entire flock was collected in the ravine, without the loss of a single lamb.

How that wonderful Dog had performed this task, not even his master could conceive. It may be that the sheep had been accustomed to place themselves under the guidance of the Dog, though they might have fled from the presence of the shepherd ; and that when they felt themselves bewildered in the darkness they were quite willing to entrust themselves to their well-known friend and guardian.

The memory of the Shepherd's Dog is singularly tenacious, as may appear from the fact that one of these Dogs, when assisting his master, for the first time, in conducting some sheep from Westmoreland to London, experienced very great difficulty in guiding his charge among the many cross-roads and bye ways that intersected their route. But on the next journey he found but little hindrance, as he was able to remember the points which had caused him so much trouble on his former expedition, and to profit by the experience which he had then gained.

The DROVER'S DOG is generally produced from the sheep-dog and the mastiff or fox-hound, and sometimes from the sheep-dog and the greyhound or pointer; the peculiar mixtures being employed to suit the different localities in which the Dog is intended to exercise its powers. In some places the Drover's Dog is comparatively small, because the sheep are small, docile, and not very active. But when the sheep are large, agile, and vigorous, and can run over a large extent of ground, a much larger and more powerful animal is needed, in order to cope with the extended powers of the sheep which are committed to its guardianship.

Although the Drover's Dog may be entrusted with the entire charge of the flock, its rightful vocation is the conveyance of the sheep from place to place. It will often learn its business so thoroughly, that it will conduct a flock of sheep or a herd of cattle to the destined point, and then deliver up its charge to the person who is appointed to receive them. Not the least extraordinary part of its performance is, that it will conduct its own flock through the midst of other sheep without permitting a single sheep under its charge to escape, or allowing a single stranger to mix with its own flock.

Such abilities as these can be applied to wrong purposes as well as to good ones, and there is a well-known story of a drover who was accustomed to steal sheep through the help of his Dog. His plan was to indicate, by some expressive gesture which the Dog well understood, the particular sheep which he wished to be added to his own flock, and then to send his flock forward under the guardianship of the Dog, while he remained with his companions at the public-house bar. The clever animal would then so craftily intermingle the two flocks that it contrived to entice the coveted sheep into its own flock, and then would drive them forwards, carrying off the stolen sheep among the number. If the stratagem were not discovered, the owner of the Dog speedily changed the marks on the sheep, and thus merged them with his own legitimate property. If the fraud were detected, it was set down as an excusable mistake of the Dog, the stolen animals were restored, and the real thief escaped punishment. However, detection came at last, as it always does, sooner or later.

The true CUR DOG is produced from the sheep-dog and the terrier, and is a most useful animal to the class of persons among whom it is generally found. It is rather apt to be petulant in its temper, and is singularly suspicious of strangers ; so that although it is rather an unpleasant neighbour by reason of its perpetually noisy tongue, it is of the greatest service to the person to whom it belongs. It is an admirable house-dog, and specially honest, being capable of restraining its natural instincts, and of guarding its owner's provisions, even though it may be almost perishing with hunger.

The Cur is the acknowledged pest of the passing traveller, especially if he be mounted, or is driving, as it rushes out of its house at the sound of the strange footstep, and follows the supposed intruder with yelps and snaps until it flatters itself that it has completely put the enemy to flight. About the house the Cur is as useful as is the colley among the hills, for it is as ready to comprehend and execute the wishes of its master at home as is the sheep-dog on the hills. Indeed, if the two Dogs were to change places for a day or two, the Cur would manage better with the sheep than the sheep-dog would manage the household tasks.

One principal reason of this distinction is, that a thorough-going sheep-dog is accustomed only to one line of action, and fails to comprehend anything that has no connexion with sheep, while the Cur has been constantly employed in all kinds of various tasks, and is, therefore, very quick at learning a new accomplishment. When the labourers are at their daily work they are often accustomed to take their dinners with them, in order to save themselves the trouble of returning home in the middle of the day. As, however, there are often lawless characters among the labourers, especially if many of them come from a distance, and are only hired for the work in hand, the services of the Cur Dog are brought into requisition. Mounting guard on his master's coat, and defending with the utmost honesty his master's little stock of provisions, he snarls defiance at every one who approaches the spot where he acts as sentinel, and refuses to deliver his charge into the hands of any but its owner. He then sits down, happy and proud of the caresses that await him, and perfectly contented to eat the fragments of that very meal which he might have consumed entirely had he not been restrained by his sense of honour.

Mr. Hogg, the "Ettrick Shepherd," says that he has known one of these Dogs to mount guard night and day over a dairy full of milk and cream, and never so much as break the cream with the tip of its tongue, nor permit a cat, or rat, or any other creature, to touch the milk pans.

The Cur Dog has—as all animals have—its little defects. It is sadly given to poaching on its own account, and is very destructive to the young game. It is too fond of provoking a combat with any strange Dog, and if its antagonist should move away, as is generally the case with high-bred Dogs, when they feel themselves intruding upon territories not their own, takes advantage of the supposed pusillanimity of the stranger, and annoys him to the best of its power; but if the stranger should not feel inclined to brook such treatment, and should turn upon its persecutor, the Cur is rather apt to invoke discretion instead of valour, and to seek the shelter of its own home, from whence it launches its angry yelpings as if it would tear its throat in pieces.

POSSESSING many of the elements of the sheep-dog, but employed for different purposes, the LURCHER has fallen into great disrepute, being seldom seen as the companion of respectable persons. It is bred from the greyhound and sheep-dog, and is supposed to be most valuable when its parents are the rough Scotch greyhound and the Scotch colley.

It is a matter of some regret that the Dog should bear so bad a character, as it is a remarkably handsome animal, combining the best attributes of both parents, and being equally eminent in speed, scent, and intelligence. As, however, it is usually the companion of poachers and other disreputable characters, the gamekeeper bears a deadly hatred towards the Lurcher, and is sure to shoot the poor animal at the earliest opportunity. For this conduct there is some pretext, as the creature is so admirably adapted for the pursuit and capture of game that a single poacher is enabled, by the aid of his four-legged assistant, to secure at least twice as much game as could be taken by any two men without the help of the Dog.

That punishment generally falls on the wrong shoulders is proverbially true, and holds good in the present instance. For the poor Dog is only doing his duty when he is engaged in marking or capturing game, and ought not to be subjected to the penalty of wounds or death for obeying the order which he has received. If any one is to be punished, the penalty ought to fall on the master, and not on his Dog, which is only acting under his orders, and carrying out his intentions.

The sagacity of this Dog is really wonderful. It learns to comprehend the unspoken commands of its master, and appreciates quite as fully as himself the necessity for lying concealed when foes are near, and, in every case, of moving as stealthily as possible. It is even trained to pioneer the way for its owner, and to give him timely warning of hidden enemies. Destructive to all game, whether winged or furred, the Lurcher is especially so in the rabbit warren, or in any locality where hares abound. Its delicate sense of smell

LURCHER.—*Canis familiaris.*

permits it to perceive its prey at a distance, and its very great speed enables it to pounce upon the hare or rabbit before it can shelter itself in the accustomed place of refuge. As soon as the Lurcher has caught its prey it brings it to its master, deposits it in his hands, and silently renews its search after another victim. Even pheasants and partridges are often caught by this crafty and agile animal.

Sometimes the game-destroying instincts of the Lurcher take a wrong turn, and lead the animal to hunt sheep, instead of confining itself to ordinary game. When it becomes thus perverted it is a most dangerous foe to the flocks, and commits sad havoc among them. One farmer, living in Cornwall, lost no less than fifteen sheep in one month, all of which were killed by Lurchers.

There are many breeds of the Lurcher, on account of the various Dogs of which the parentage is formed. The greyhound and sheep-dog are the original progenitors, but their offspring is crossed with various other Dogs, in order to obtain the desired qualifications. Thus, the greyhound is used on account of its speedy foot and silent tongue, and the sheep-dog on account of its hardiness, its sagacity, and its readiness in obeying its master. The spaniel is often made to take part in the pedigree, in order to give its well-known predilection for questing game, and the hound is employed for a similar purpose. But in all these crossings the greyhound must morally predominate, although its form is barely to be traced under the rough lineaments of the Lurcher.

As the Lurcher causes such suspicion in the minds of the gamekeeper or the landlord, the owners of these Dogs were accustomed to cut off their tails, in order to make them look like honourable sheep-dogs, and so to escape the tax which presses upon sporting Dogs, and to elude the suspicious glance of the game-preserving landlord and his emissaries. So swift is this animal that it has been frequently used for the purpose of coursing the hare, and is said to perform this task to the satisfaction of its owner. It can also be entrusted with the guardianship of the house, and watches over the property committed to its charge with vigilance and fidelity. Or it can take upon itself that character in reality which its cropped tail too often falsely indicates, and can watch a fold, keep the sheep in order, or conduct them from one place to another, nearly if not quite as well as the true sheep-dog from which it sprang.

OTTERHOUND.—*Canis familiaris.*

The OTTERHOUND is now almost exclusively employed for the chase of the animal from which it derives its name. Formerly it was largely used in Wales for the purpose of hunting the hare, and from that pursuit has derived the name of " Welsh Harrier."

It is a bold, hardy, and active animal, as is needful for any Dog which engages in the chase of so fierce and hard-biting a creature as the otter. As it is forced to take to the water in search or in chase of its prey, it is necessarily endowed with great powers of swimming, or it could never match that most amphibious of quadrupeds. Those who have seen an otter when disporting itself in its congenial element must have been struck with the exceeding rapidity and consummate ease of its movements, and can appreciate the great aquatic powers that must be possessed by any Dog which endeavours to compete with so lithe and agile an antagonist.

Great courage is needful on the part of the Dog, because the otter is, when irritated, a peculiarly fierce animal, and can inflict most painful wounds by the bite of its long sharp teeth. It is, moreover, so pliant of body that it can twist itself about almost like a snake, and, if grasped heedlessly, can writhe itself about as actively and slipperily as an eel, and unexpectedly plant its teeth in its antagonist's nose. Now, the nose is a very sensitive portion of all animal economy, and a wound or a bite in that region causes such exceeding pain that none but a well-bred Dog can endure the torture without flinching.

Such needful courage is found in the Otter Dog, but is sometimes rather prone to degenerate into needless ferocity. There are few animals, with the exception of the bull-dog, which fight so savagely as the Otterhound, or bite so fiercely and with such terrible results. The attack of the Otterhound is even more dangerous than that of the bull-dog, and its bite more to be dreaded. As is well known, where the bull-dog has once fixed his teeth there he hangs, and cannot be forced to loosen his hold without the greatest difficulty ; but when the Otterhound bites, it instantly tears its teeth away without relaxing its jaws, and immediately seizes its prey with a second gripe. The wounds which it inflicts by this ferocious mode of action are of the most terrible description, lacerating all the tissues, and tearing asunder the largest and most important vessels. The reason for this very savage mode of attack is evident enough. The otter is so quick and agile, that, if the Dog were to retain his hold, the otter would twist round and inflict a severe bite, so the Dog bites as fast and as often as he can, in order to give his antagonist the fewest possible chances of retaliation.

When a number of these Dogs are placed in the same kennel they are sadly apt to fight, and to inflict fatal injuries on each other from the sheer love of combat. If two of the Dogs begin to quarrel and to fight, the others are sure to join them ; so that, from the bad temper of a single Dog, half the pack may lose their lives.

As these Dogs are obliged to endure the most turbulent weather and the coldest streams, they are furnished with a very strong, rough, and wiry coat, which is capable of resisting the effects of cold and storm, and is also of much service in blunting the severity of the otter's bite. The face and muzzle are guarded with a profusion of longish and very rough " whisker" hairs.

Whether this animal is the production of a cross breed between two families of Dogs, or whether it forms a distinct family in itself, is a mooted point. According to the best authorities, the latter opinion seems to be the best founded. It is thought by those who consider the Dog to be of mixed breed, that it was originally the offspring of the deer-hound and terrier ; but as it retains the full melodious note of the hound, which is always injured or destroyed by an admixture with the sharp-voiced terrier, it appears to owe more of its parentage to that animal. Be this as it may, it is now treated as a separate breed, and may claim the honours of a pure lineage. In all probability it is a variety of the old southern hound, which was selected carefully for the work which it is intended to perform, and which in course of time has so settled down to its vocation as to have undergone that curious variation in form and aspect that is always found in animals or men which have long been employed in the same kind of work.

Any one of moderate experience among Dogs and their habits can, on seeing the animal, determine its avocation, just as any one who is conversant with men and their manners can, on seeing a man, at once announce his calling. There is something in the little peculiarities of the formation which tells its tale to the observing eye. There is a kind of moral and intellectual, as well as physical, atmosphere, that seems to surround every creature, and to tell of its essential nature, its education, and its habits. Animals appear to be peculiarly sensitive to this surrounding emanation, and to be attracted or repelled by an influence as powerful, though as invisible, as that which attracts or repels the different poles of a magnet. We feel it ourselves in the instinctive cordiality or repugnance which we perceive when brought in contact with a fresh acquaintance, and which very seldom misleads those who are content to follow their instincts. The nature of each being seems to pervade its every particle as it were—to overflow and shed its influence, consciously or otherwise, on every object with which it enters into communion. There are some men whose very presence warms and enlivens all whom they approach, and that not from any suavity of manner, for such men are often most abrupt and truth-telling in their demeanour ; and there are others who, however urbane may be their deportment, seem to cast from them a cold and freezing atmosphere that congeals all those around them, like the icebergs of the northern seas.

Although, on examining the form of the Otterhound, we should not be able to point out the description of game which it is accustomed to pursue, we should at once pronounce it to be a strong and hardy animal, a good swimmer, possessed of a delicate nose, and of stout courage. In each of these accomplishments the Otterhound excels, and needs them all when it ventures to cope with the fierce prey which it is taught to pursue.

The Otterhound is a tolerably large Dog, measuring nearly two feet in height at the shoulder. This is the height of the male, that of the female is an inch or two less.

THE FINE animal which is represented in the accompanying engraving can hardly be considered as belonging to a separate breed, but rather as a mixture between several families of domesticated Dogs.

According to competent judges, the BOARHOUND is derived from a mingling of the mastiff with the greyhound, crossed afterwards with the terrier. The reader will see why these three animals are employed for the purpose of obtaining a Dog which is capable of

BOARHOUND.—*Canis familiaris.*

successful attack on so dangerous and powerful a brute as the boar. The greyhound element is required in order to give the Dog sufficient speed for overtaking the boar, which is a much swifter animal than would be supposed from his apparently unwieldy and heavy frame. The admixture of the mastiff is needed to give it the requisite muscular power and dimensions of body, and the terrier element is introduced for the sake of obtaining a sensitive nose, and a quick, spirited action.

As might be imagined would be the case with an animal which derives its origin from these sources, the Boarhound varies very considerably in form and habits, according to the element which may preponderate in the individual. A Dog in which the greyhound nature is dominant will be remarkably long of limb and swift of foot; one in whose parentage the mastiff takes the greatest share will be proportionately large and powerful; while the Dog in whose blood is the strongest infusion of the terrier will not be so swift or so large as the other two, but will excel them in its power of scent and its brisk activity of movement.

To train the Dog rightly to his work is a matter of some difficulty, because a mistake is generally fatal, and puts an end to further instruction by the death of the pupil. It is comparatively easy to train a pointer or a retriever, because, if he fails in his task through over-eagerness or over-tardiness, the worst consequence is, that the sportsman loses his next shot or two, and the Dog is corrected for his behaviour. But if a Boarhound rushes too eagerly at the bristly quarry, he will in all probability be laid bleeding on the ground by a rapid stroke from the boar's tusks, and if he should hang back and decline the combat, he is just as likely to be struck by an infuriated boar as if he were boldly attacking it in front. A boar has been known to turn with such terrible effect upon a pack containing fifty Dogs, that only ten escaped scathless, and six or seven were killed on the spot.

Great tact is required on the part of the hound in getting into a proper position, so as to make his onset without exposing himself to the retaliating sweeps of the foam-flecked tusks, and at the same time to act in concert with his companions, so as to keep the animal busily engaged with their reiterated attacks, while their master delivers the death blow with a spear or rifle-bullet.

As we have no longer any wild boars ranging at will through the few forests which the advance of agriculture has suffered to remain as relics of a past age, the Boarhound is never seen in this country except as an object for the curious to gaze upon, or imported into this island through the caprice of some Dog-loving individual. But in many parts of Germany it is still employed in its legitimate avocation of chasing the wild boar, and is used in Denmark and Norway for the pursuit of that noble animal the elk. The latter creature is so large, so fleet, and so vigorous, that it would easily outrun or outfight any Dog less swift or less powerful than the Boarhound.

In the fur of the Boarhound the colour of the mastiff generally predominates, the coat being usually brown or brindled uniformly over the body and limbs, but in some animals the colour is rather more varied, with large brown patches upon a slate-coloured ground. The limbs are long and exceedingly powerful, and the head possesses the square muzzle of the mastiff, together with the sharp and somewhat pert air of the terrier. It is a very large animal, measuring from thirty to thirty-two inches in height at the shoulder.

The BULL-DOG is said, by all those who have had an opportunity of judging its capabilities, to be, with the exception of the game-cock, the most courageous animal in the world.

Its extraordinary courage is so well known as to have passed into a proverb, and to have so excited the admiration of the British nation that we have been pleased to symbolize our peculiar tenacity of purpose under the emblem of this small but most determined animal. In height the Bull-dog is but insignificant, but in strength and courage there is no Dog that can match him. Indeed, there is hardly any breed of sporting Dog which does not owe its high courage to an infusion of the Bull-dog blood ; and it is chiefly for this purpose that the pure breed is continued.

We have long ago abolished those cruel and cowardly combats between the bull and the Dog, which were a disgrace to our country even in the earlier part of the present century, and of which a few " bull-rings " still remaining in the ground are the sole relics. In these contests the Dog was trained to fly at the head of the bull, and to seize him by the muzzle as he stooped his head for the purpose of tossing his antagonists into the air. When he had once made good his hold it was almost impossible for the bull to shake off his pertinacious foe, who clung firmly to his antagonist, and suffered himself to be swung about as the bull might choose.

There seems, indeed, to be no animal which the Bull-dog will not attack without the least hesitation. The instinct of fight is strong within him, and manifests itself actively in the countenance and the entire formation of this creature.

It is generally assumed that the Bull-dog must be a very dull and brutish animal, because almost every specimen which has come before the notice of the public has held such a character. For this unpleasant disposition, a celebrated writer and zoologist attempts to account by observing that the brain of the Bull-dog is smaller in proportion to its body than that of any other Dog, and that therefore the animal must needs be of small sagacity. But " Stonehenge " well remarks, that although the Bull-dog's brain appears to the eye to be very small when compared with the body, the alleged discrepancy is only caused by the deceptive appearance of the skull. It is true that the brain appears to be small when compared with the heavy bony processes and ridges that serve to support the muscles of the head and neck, but if the brain be *weighed* against the remainder of the body, it will be found rather to exceed the average than to be below it.

BULL-DOG.—*Canis familiaris.*

The same writer is disposed to think the Bull-dog to be a sadly maligned animal, and that his sagacity and affections have been greatly underrated. He states that the pure Bull-dog is not naturally a quarrelsome creature, and that it would not bear so evil a character if it were better taught.

According to him, the Bull-dog is really a sufficiently intelligent animal, and its mental qualities capable of high cultivation. It is true that the animal is an unsafe companion even for its master, and that it is just as likely to attack its owner as a stranger, if it feels aggrieved. An accidental kick, or a tread on the toes, affords ample pretext for the animal to fasten on its supposed enemy; and when once it does fix its teeth, it is not to be removed except by the barbarous method which is considered to be legitimate for such a purpose, but which will not be mentioned in these pages. However, most of these short-comings in temper are said to be produced by the life which the poor Dog leads, being tied up to his kennel for the greater part of his time; and, when released from his bondage, only enjoying a limited freedom for the purpose of fighting a maddened bull, or engaging in deadly warfare with one of his own kind. Any animal would become morose under such treatment; and when the sufferer is a Bull-dog, the results of his training are often disastrous enough.

The shape of this remarkable animal is worthy of notice. The fore-quarters are particularly strong, massive, and muscular; the chest wide and roomy; and the neck singularly powerful. The hind-quarters, on the contrary, are very thin, and comparatively feeble; all the vigour of the animal seeming to settle in its fore-legs, chest, and head. Indeed, it gives the spectator an impression as if it were composed of two different Dogs; the one a large and powerful animal, and the other a weak and puny quadruped, which had been put together by mistake. The little fierce eyes that gleam savagely from the round, combative head, have a latent fire in them that gives cause for much suspicion on the part of a stranger who comes unwarily within reach of one of these Dogs. The underhung jaw, with its row of white glittering teeth, seems to be watering with desire to take a good bite at the stranger's leg; and the matter is not improved by the well-known custom of the Bull-dog to bite without giving the least vocal indication of his purpose.

1.   X

In all tasks where persevering courage is required, the Bull-dog is quietly eminent, and can conquer many a Dog in its own peculiar accomplishment. The idea of yielding does not seem to enter his imagination, and he steadily perseveres until he succeeds or falls. One of these animals was lately matched by his owner to swim a race against a large white Newfoundland Dog, and won the race by nearly a hundred yards. The owners of the competing quadrupeds threw them out of a boat at a given signal, and then rowed away as fast as they could pull. The two Dogs followed the boat at the best of their speed, and the race was finally won by the Bull-dog. It is rather remarkable that the Bull-dog swam with the whole of his head and the greater part of his neck out of the water, while the Newfoundland only showed the upper part of his head above the surface.

According to the authority which has already been quoted, a well-bred Bull-dog ought to present the following characteristics of form. "The head should be round, the skull high, the eye of moderate size, and the forehead well sunk between the eyes ; the ears semi-erect and small, well placed on the top of the head, and rather close together than otherwise ; the muzzle short, truncate, and well furnished with chop ; the back should be short, well arched towards the stern, which should be fine, and of moderate length. Many Bull-dogs have what is called a crooked stern, as though the vertebræ of the tail were dislocated or broken ; I am disposed to attribute this to in-breeding. The coat should be fine, though many superior strains are very woolly coated ; the chest should be deep and broad, the legs strong and muscular, and the foot narrow, and well split up like a hare's."

The MASTIFF, which is the largest and most powerful of the indigenous English Dogs, is of a singularly mild and placid temper, seeming to delight in employing its great powers in affording protection to the weak, whether they be men or Dogs. It is averse to inflicting an injury upon a smaller animal, even when it has been sorely provoked, and either looks down upon its puny tormentor with sovereign disdain, or inflicts just sufficient punishment to indicate the vast strength which it could employ, but which it would not condescend to waste upon so insignificant a foe.

Yet, with all this nobility of its gentle nature, it is a most determined and courageous animal in fight, and, when defending its master or his property, becomes a foe which few opponents would like to face. These qualifications of mingled courage and gentleness adapt it especially for the service of watch-dog, a task in which the animal is as likely to fail by overweening zeal as by neglect of its duty. It sometimes happens that a watch-dog is too hasty in its judgment, and attacks a harmless stranger, on the supposition that it is resisting the approach of an enemy. Sometimes the bull-dog strain is mixed with the Mastiff, in order to add a more stubborn courage to the animal ; but in the eyes of good judges this admixture is quite unnecessary.

It has already been mentioned that the Mastiff is fond of affording the benefit of its protection to those who need it. As, however, the Dog is but a Dog after all, it sometimes brings evil instead of good upon those who accept its guardianship.

During my school-boy days, a large Mastiff, called Nelson, struck up a great friendship with myself and some of my schoolfellows, and was accustomed to partake of our hebdomadal banquets at the pastrycook's shop, and to accompany us in our walks. One summer, as we were bathing in the Dove, a man pounced upon our clothes, and would have carried them off, had it not been for the opportune assistance of some older lads of the same school, who captured the offender after a smart chase, and tossed him into the river until he was fain to cry for mercy.

In order to prevent a repetition of a similar mischance, we determined to take Nelson with us, and put him in charge of our clothes. The old Dog was delighted at the walk, and mounted sentry over the pile of garments, while we recreated ourselves in the stream, and caught crayfish or tickled trout at our leisure. Unfortunately, a number of cows had lately been placed in the field, and, after the usually inquisitive custom of cows, they approached the spot where Nelson was lying, in order to ascertain the nature of the strange object on the river bank. Nelson permitted them to come quite close, merely uttering a few warning growls, but when one of the cows began to toss a jacket with her

MASTIFF.—*Canis familiaris.*

horns, his patience gave way and he flew at the offender. Off scampered all the cows, but soon returned to the charge. Nelson stood firm to his post, only retreating a few steps as the cows approached the garments which he was guarding, and then dashing at them again. However, the cows' hoofs and the Dog's feet began to wreak such dire mischief among the clothes, that we found ourselves compelled to drive away the assailants and carry our clothes to the opposite bank of the river, where no cows could interfere with us.

The head of the Mastiff bears a certain similitude to that of the bloodhound and the bull-dog, possessing the pendent lips and squared muzzle of the bloodhound, with the heavy muscular development of the bull-dog. The under-jaw sometimes protrudes a little, but the teeth are not left uncovered by the upper lip, as is the case with the latter animal. The fur of the Mastiff is always smooth, and its colour varies between a uniform reddish-fawn and different brindlings and patches of dark and white. The voice is peculiarly deep and mellow. The height of this animal is generally from twenty-five to twenty-eight inches, but sometimes exceeds these dimensions. One of these Dogs was no less than thirty-three inches in height at the shoulder, measured fifty inches round his body, and weighed a hundred and seventy-five pounds.

The CUBAN MASTIFF is supposed to be produced by a mixture of the true Mastiff with the bloodhound, and was used for the same purpose as the latter animal. It was not a native of the country where its services were brought into requisition, and from which it has consequently derived its name, but was imported there for the purposes of its owners, being taught to chase men instead of deer.

x 2

This Dog was employed with terrible success in the invasion of America by the Spaniards, and was, in the eyes of the simple natives, a veritably incarnated spirit of evil, of which they had never seen the like, and which was a fit companion to those fearful apparitions which could separate themselves into two distinct beings at will, one with four legs and the other with two, and destroy them at a distance with fiery missiles, against which they were as defenceless as against the lightning from above,

Even in more recent times, the services of these Dogs have been rendered available against the rebel forces of Jamaica, when they rose against the Government, and but for the able assistance of these fierce and sagacious animals, would apparently have swept off the European inhabitants of the island.

TERRIER.—*Canis familiaris.*

The TERRIER, with all its numerous variations of crossed and mongrel breeds, is more generally known in England than any other kind of Dog. Of the recognised breeds, four are generally acknowledged ; namely, the English and Scotch Terriers, the Skye, and the little Toy Terrier, which will be described in their order.

The ENGLISH TERRIER possesses a smooth coat, a tapering muzzle, a high forehead, a bright intelligent eye, and a strong muscular jaw. As its instinct leads it to dig in the ground, its shoulders and fore-legs are well developed, and it is able to make quite a deep burrow in a marvellously short time, throwing out the loose earth with its feet, and dragging away the stones and other large substances in its mouth. It is not a large Dog, seldom weighing more than ten pounds, and often hardly exceeding the moiety of that weight.

Although a light, quick, and lively creature, and fuming with anxiety at the sight or smell of the animals which are popularly termed "vermin," the pure English Terrier will seldom venture to attack a rat openly, although it will be of the greatest service in discovering and unharbouring that mischievous rodent. The sport which this Dog prefers is, that itself should startle the rats, while its master destroys them. If a rat should fasten upon this Dog, he will yelp and cry piteously, and, when relieved from his antagonist, will make the best of his way from the spot ; or if the rat should turn to bay, the Dog will usually scamper off and decline the combat. The celebrated rat-killing Terriers, of whose feats so much has been said, were all indebted for their valour to an infusion of the bull-dog blood, which gives the requisite courage without detracting from the shape of the Dog, or adding too much to its size. Of these bull-terrier Dogs, more will be said in their place.

The colour of the pure English Terrier is generally black and tan, the richness of the two tints determining much of the animal's value. The nose and the palate of the Dog ought to be always black, and over each eye is a small patch of tan colour. The tail ought to be rather long and very fine, and the legs as light as is consistent with strength.

The SCOTCH TERRIER is a rough-haired, quaint-looking animal, always ready for work or play, and always pleased to be at the service of its master. It is a capital Dog for those whose perverted taste leads them to hunt rats, or any kind of "vermin," and is equally good at chasing a fox to earth, and digging him out again when he fancies himself in safety. It was in former days largely employed in that most cruel and dastardly pursuit of badger-drawing, in which "sport" both the badger and the Dogs were so unmercifully wounded by the teeth of their antagonist, that even the winning Dog was often crippled, and the poor badger reduced to a state of suffering that would touch the heart of any but a hardened follower of these pursuits.

The colour of the Scotch Terrier is generally the same as that of the English Dog, saving that the black and tan tints are often besprinkled with grey, so as to give that peculiar modification of colouring which is popularly known by the name of "pepper-and-salt."

There is a peculiar breed of the Scotch Terrier which is called the Dandie Dinmont, in honour of the character of that name in Scott's "Guy Mannering." These Dogs are of two colours ; one a light brown with a reddish tinge, termed "mustard," and the other a bluish-grey on the body and tan on the legs, denominated "pepper." These little animals are very courageous ; although they often exhibit no proofs of their bold nature until they have passed the age of two years, appearing until that time to be rather cowardly than otherwise. This conduct is supposed to be occasioned by their gentle and affectionate disposition. The legs of this variety of Terrier are short in proportion to the length of the body, the hair is wiry and abundant, and the ears are large, hanging closely over the sides of the head.

The BULL-TERRIER unites in itself the best qualifications of the sporting Dogs, being very intelligent, apt at learning, delicate of nose, quick of eye, and of indomitable courage. In size it is extremely variable, some specimens being among the smallest of the canine tribes, while others measure as much as twenty inches in height. In this Dog it is quite unnecessary to have equal parts of the bull-dog and the Terrier ; for in that case the progeny is sure to be too heavily made about the head and jaws, and not sufficiently docile to pay instant and implicit obedience to the commands of its master. Until these points are removed, the Terrier cross should be continued, so as to restore the light, active form of the Terrier, together with its habit of ready obedience, while the courageous disposition remains. Indeed, the most ferocious Dogs, and the hardest fighters, are generally the immediate offspring of the bull-dog and Terrier, and are often erroneously described under the name of the former animal.

How entirely the external form of the bull-dog can be eradicated, while its dauntless courage remains intact, is shown in the graceful little Terriers which are used for rat-killing, and which are formed on the most delicate model.

The endurance and gallantry of these little creatures are so great that they will permit several rats, each nearly as large as themselves, to fix upon their lips without flinching in the least, or giving any indications of suffering. Yet the badly-bred Dog will yell with pain if even a mouse should inflict a bite upon this sensitive portion of its frame, and will refuse to face its little enemy a second time. One of these highly bred animals, which was celebrated in the sporting world under the title of "Tiny," weighed only five pounds and a half, and yet was known to destroy fifty rats in twenty-eight minutes and five seconds. It is estimated that this Dog must have killed more than five thousand rats, the aggregate weight of which nearly equals a ton and a half. He could not be daunted by size or numbers, and was repeatedly matched against the largest rats that could be procured.

He used to go about his work in the most systematic and business-like style, picking out all the largest and most powerful rats first, so as to take the most difficult part of the task while he was fresh. When fatigued with his exertions, he would lie down and permit his master to wash his mouth and refresh him by fanning him, and then would set to work with renewed vigour. He was a most excitable little creature during his younger days, running about the room with such preternatural activity that a gentleman to whom he was exhibited declared that he could not distinguish the Dog's head from his tail, or pronounce judgment on the colour of his fur.

As he grew older, however, he became more sedate in his demeanour, and used to sit in state every evening on a crimson velvet cushion edged with gold fringe, and flanked with a candle on each side, so that he might be inspected at leisure.

However quiet he might be in external demeanour, he was hardly less excitable in disposition, and actually died from the effects of over-excitement. He happened to hear or to smell a rat which was in a cage in another room ; and being chained in an adjoining apartment, and unable even to see the rat, he chafed and fretted himself into such feverish agitation that he died in a short time afterwards, although he was permitted to kill the rat. There are Dogs which have destroyed more rats in less time than this little creature ; but none which was nearly so successful in proportion to its size and weight.

A larger variety of the Bull-terrier was formerly in great request for dislodging foxes from their holes, or "earths," as their burrows are technically termed ; and one or two of these animals were invariably borne on the strength of each pack of foxhounds. There used to be a special strain of these Dogs, named Fox-terriers, which were bred and trained for this purpose alone.

The mental powers of this Dog are very considerable, and the animal is capable of performing self-taught feats which argue no small amount of intellect. There are several examples of Dogs which could in some degree appreciate the object of money, and which would take a coin to the proper shop and exchange it for food. A well-known black-and-tan Terrier, which lately resided at Margate, and was named Prince, was accustomed to make his own purchases of biscuit as often as he could obtain the gift of a halfpenny for that purpose. On several occasions the baker whom he honoured with his custom thought to put him off by giving him a burnt biscuit in exchange for his halfpenny. The Dog was very much aggrieved at this inequitable treatment, but at the time could find no opportunity of showing his resentment. However, when he next received an eleemosynary halfpenny, he wended his way to the baker's as usual, with the coin between his teeth, and waited to be served. As soon as the baker proffered him a biscuit, Prince drew up his lips, so as to exhibit the halfpenny, and then walked coolly out of the shop, transferring his custom to another member of the same trade who lived on the opposite side of the road.

Several instances of a similar nature have been recorded, but in no case does the animal appear to have comprehended the difference of value between the various coins of the realm. The elephant, for example, readily learns to take a coin from a visitor, and to exchange it for apples, cakes, or similar dainties, at a neighbouring stall. But he seems to be ignorant of the fact that he ought to receive twice as many cakes for a penny as for a halfpenny, and is quite contented so long as he gives a coin and receives cakes.

One of these Dogs, named Peter, an inhabitant of Dover, displays great ingenuity in adapting himself to the pressure of circumstances.

Several years since, he had the mishap to fall under the wheels of a carriage, and to be lamed in both his fore-legs. In consequence of this accident his limbs are so enfeebled that he cannot trust their powers in leaping, and therefore has taught himself to jump with his hind-legs alone, after the manner of a kangaroo. He can spring upon a chair or on a low wall without any difficulty, and does so after the usual manner of Dogs. But when he is forced to return again to earth he mistrusts his fore-limbs, and alights upon his hinder feet, making one or two small leaps upon those members before he ventures to place his fore-feet on the ground. When he is accompanying his master in the fields, and comes to a gate or a gap in a wall, he dares not leap through the aperture, as most Dogs would do, but hops up, and then down again, upon his hind-feet alone.

The real Bull-terrier of the first cross is a marvellously brave animal, falling but little short in courage from his bull-dog ancestor, and very far exceeding that animal in agility and intellectual quickness. Fear seems to make no part of a good Bull-terrier's character; and he dashes with brilliant audacity at any foe which his master may indicate to him, or which he thinks he ought to attack without orders. Mr. Andersson, in his valuable work entitled "Lake Ngami," gives an account of the courage and sagacity of one of these animals which accompanied him in his travels through South-western Africa. He had wounded a rhinoceros, which ran a few hundred yards, and then came to a stand.

"At break of day my men went on his trail. He had still strength enough to make a dash at them; and would probably have laid hold of some of them, had not a small bitch (half Terrier and half bull-dog, called Venus, in derision of her ugliness) caught the enraged animal by the lower lip, where she stuck with such tenacity that the rhinoceros, with all his fury, was unable to shake her off. She only relinquished her hold when her huge antagonist was fairly laid prostrate by a ball.

But the sagacity of this favourite Dog was as great as her courage. Being now in a game country, all sorts of beasts of prey abounded, more especially jackals, which might be seen running about by dozens. In order not to frighten the elephants, and other large animals, we were in the habit of encamping some little way from the water, to which Miss Venus regularly resorted to bathe and drink. On perceiving a jackal she instantly crouched, looking very timid. Reynard, mistaking her posture for an indication of fear, and probably thinking that from her diminutive size she would prove an easy conquest, boldly approached his supposed victim. But he had reckoned without his host, for the instant that the cunning Dog found her antagonist sufficiently near, she leaped like a cat at his throat, and, once there, the beast had no chance.

She then returned to camp, where her contented looks and bleeding jaws soon attracted the attention of the men, who immediately went on her track and brought the jackal, who was valued on account of his fur."

The quaint-looking SKYE TERRIER has of late years been much affected by all classes of Dog-owners, and for many reasons deserves the popularity which it has obtained.

When of pure breed the legs are very short, and the body extremely long in proportion to the length of limb; the neck is powerfully made, but of considerable length, and the head is also rather elongated, so that the total length of the animal is three times as great as its height. The "dew-claws" are wanting in this variety of domestic Dog. The hair is long and straight, falling heavily over the body and limbs, and hanging so thickly upon the face that the eyes and nose are hardly perceptible under their luxuriant covering. The quality of the hair is rather harsh and wiry in the pure-bred Skye Terrier; for the silky texture of the generality of "toy" Skyes is obtained by a cross with the spaniel. It is easy to detect the presence of this cross by the scanty appearance of the hair on the face.

The size of this animal is rather small, but it ought not to imitate the minute proportions of many "toy" Dogs. Its weight ought to range from ten to seventeen or eighteen pounds. Even amongst these animals there are at least two distinct breeds, while some Dog-fanciers establish a third.

It is an amusing and clever Dog, and admirably adapted for the companionship of mankind, being faithful and affectionate in disposition, and as brave as any of its congeners, except that epitome of courage, the bull-dog. Sometimes, though not frequently, it is employed for sporting purposes, and is said to pursue that avocation with great credit.

A HISTORY, however short, of the Dogs would be incomplete without some reference to that terrible disease called "Hydrophobia," which at times arises among the canine

SKYE TERRIER.—*Canis familiaris.*

race, and converts the trusted companion into an involuntary foe. From some cause, which at present is quite unexplained, the bite of a Dog which is affected with this terrible malady, or even the mere contact of his saliva with a broken skin, becomes endued with such deadly virulence, that the unfortunate person upon whom such an injury is inflicted is as certain to die as if he had been struck by the poison-fangs of the rattlesnake or cobra.

As far as is known, this dread malady appears to originate only in the canine tribe, being communicable to almost every other description of animal, man not excepted, and dooming them to a most painful illness and death. It is worthy of consideration, that the Dog does not perspire through the skin, and that the tongue and throat offer the only means by which the animal can avail itself of that needful exhalation. The symptoms of this malady are rather various in different individuals, but yet are of the same type in all.

There is an entire change of manner in the animal. The affectionate, caressing Dog becomes suddenly cross, shy, and snappish; retreating from the touch of the friendly hand as if it were the hand of a stranger. His appetite becomes depraved, and, forsaking his ordinary food, he eagerly swallows pieces of stick, straws, or any other innutritious substances that may lie in his way. He is strangely restless, seeming unable to remain in the same position for two seconds together, and continually snaps at imaginary objects which his disordered senses image in rapid succession before his eyes. Strange voices seem to fall upon his ears, and he ever and anon starts up and listens eagerly to the sounds which so powerfully affect him. Generally, he utters at intervals a wild howl, which tells its fearful tale even to unpractised ears, but in some cases the Dog remains perfectly silent during the whole of his illness, and is then said to be afflicted with the dumb madness. In most instances, the Dog is silent during the latter stages of the illness.

Before the disease has developed itself to any extent, the poor creature becomes thoughtful and anxious, and looks with wistful eyes upon his friends, as if beseeching them to aid him in the unknown evil that hangs so heavily upon him. He then retires to his usual resting-place, and sluggishly lies upon his bed, paying scarcely any attention to the voice of his master, but strangely uneasy, and ever and anon shifting his posture, as if endeavouring to discover some attitude that may bring ease and repose to his fevered limbs. Fortunately, the disposition to bite does not make its appearance until the disease has made considerable progress.

In these stages of the malady the Dog is often seen to fight with his paws at the corner of his mouth, as if endeavouring to rid himself of a bone that had become fixed among his teeth, and assumes much of the anxious aspect that is always seen in animals when their respiration is impeded. This symptom may, however, be readily distinguished by the fact that the Dog is able to close his mouth between the paroxysms of his ailment, which he is unable to do when he is affected by the presence of a bone or other extraneous substance in his throat.

There is, indeed, a mechanical hindrance to respiration, which, although not so outwardly apparent as the obstruction which is caused by a bone or similar substance, yet harasses the poor creature quite as painfully. As the poison, which has been infused into and taints the blood of the poor victim, works its dread mission through the frame, it infects some of the fluids that are secreted from the blood, and changes their external aspect as well as their inward essence. The saliva becomes thick and viscid in character, and is secreted in quantities so great that it obstructs the channels of respiration, and gives rise to those convulsive efforts on the part of the Dog which have already been mentioned.

Strangely enough, the infected Dog seems to partake of the serpent nature, and like the cobra or viper, to elaborate a deadly poison from harmless food. The snake feels but little inconvenience from the accumulation of venomous matter, as it is furnished with receptacles in which the lethal secretion may be lodged until it is needed. But the Dog has no such storehouse, and the poison is therefore diffused through the moisture of the throat and mouth, instead of being concentrated into one locality. There is another curious resemblance between the poison of serpents and that of rabid Dogs; namely, that while the venom of either creature produces such terrible effects when mixed with the blood, it may be swallowed with perfect safety, provided that the lips and mouth are free from sores.

I would offer a suggestion, that the instinct which induces the Dog to bite everything which may come within its reach, is intended to aid the creature in its cure, and that if it could only be induced to bite a succession of lifeless objects, it might rid itself of the venomous influence, and be restored to its normal state of health. So powerfully is this instinct developed, that the poor Dog will bite itself, and inflict the most fearful lacerations on its own flesh, rather than resist the furious impulse which fills its being. Horses and other animals which have been infected with this terrible disease have been known to feel the same necessity, and in default of other victims have torn the flesh from their own limbs.

An unquenchable thirst soon fastens upon the afflicted Dog, and drives him to the nearest spot where he can obtain any liquid that may cool his burning throat.

In the earlier stages of the complaint he laps without ceasing, but when the disease has destroyed the powers of his tongue and throat, he plunges his head into the water as far as the depth of the vessel will permit, in hope of bringing his throat in contact with the cooling fluid. It is generally supposed that a mad Dog will not touch water, and for this reason the malady was termed Hydrophobia, or "dread of water," but it is now ascertained that the animal is so anxious to drink, that he often spills the fluid in his eagerness, and so defeats his own object.

In the last stage of this terrible disease the Dog is seized with an uncontrollable propensity to *run*. He seems not to care where he goes, but runs for the most part in a straight line, seldom turning out of his way, and rarely attempting to bite unless he be obstructed in his course; and then he turns savagely upon his real or fancied assailant, and furiously snaps and bites without fear or reason. Not the least curious fact of this disease is, that it causes a singular insensibility to pain. A rabid Dog will endure terrible injuries without appearing to be conscious of them, and, in many cases, these poor creatures have been known to tear away portions of their own bodies as calmly as if they were lacerating the dead body of another Dog. A similar insensibility to pain is noticeable in human lunatics, who will often inflict the most terrible injuries on their own persons, with the most deliberate and unconcerned air imaginable. The nerves seem to be deprived of their powers, and to be insensible even to the contact of burning coals or red-hot metals. In anger, too, which is in truth a short-lived madness, pain is unfelt, and the severest wounds may be received unheeded.

It is possible that this locomotive instinct of the Dog may give a clue to the cure of this fearful malady, and that if a rabid Dog could be permitted to follow its instinct without molestation it might rid itself of its ailment by means of this unwonted exercise.

By this terrible malady the nerves are excited to the highest degree of tension, and it is not improbable that by violent and continual exercise the system might be enabled to throw off the "peccant humours" that infect every particle of the blood as it circulates through the veins, and envenom the natural moisture of the Dog's tongue.

There exists a curious parallel to this propensity for exertion in the celebrated Tarantula-dancing which was so famous in Naples during the sixteenth century. Those persons who were affected with this curious disease, which was for many years thought to be the effect of the bite of the Tarantula spider, were impelled to leap and dance continually in a kind of frenzy, until they sank from sheer fatigue. In many cases the dancing would continue for three or four days, and seemed to be cured best by the profuse perspirations which poured from the wearied frames of the dancers. In a similar manner the effects of a serpent's tooth may be driven from the system. When a person has suffered from the bite of a cobra, or other venomous snake, the most effectual treatment is to prevent him from falling into the lethargy which is produced by the poisonous infusion, and to keep him in constant and violent motion.

It is a remarkable fact that the Tarantismus, as this disease is termed, used in many cases to recur at regular annual intervals, as has already been related of the wounds caused by the lion's bite, and is the case with the healed wound which has been inflicted by the teeth of a rabid Dog. So subtle is this influence, and so thoroughly does it pervade the system, that where anger has risen in the mind of a person who has been bitten by a mad Dog, and by taking precaution has felt no evil results, the old sores have become flushed and swollen, and throbbed in unison with the angry feelings that occupied their mind.

How the nature of the Dog can be so utterly changed as to charge its bite with deadly venom, or how it is that the moist saliva of the rabid animal should communicate the disease to other beings, is at present but a mystery. There seems to be an actual infusion of the Dog nature into the animal which is bitten by a rabid Dog, or by one of the creatures which has been inoculated by the bite of one of these terrible beings. It is evident that the virus is resident in the saliva, because the malady has been communicated by the mere touch of the Dog's tongue upon a wound without the infliction of a bite from its teeth. Yet it is equally evident that the poisonous property belongs not to the saliva, but to the influence which is conducted by its means. In some strange fashion the spirit of the angry Dog seems to be infused into the victim of its bite, and it is well known that even where an angry Dog has in the heat of its passion inflicted a wound the result has been very similar to Hydrophobia, though the animal was not affected with that disease. Ordinarily, the bite of a Dog, such as the playful bite of a puppy, though sufficiently painful, carries no danger with it, but if the animal has only been touched with this malady its bite is but too frequently fatal. This death-dealing influence has been proved to remain in the saliva for four-and-twenty hours after the animal's death. Perhaps there may be something of electricity in the fatal influence, which requires a fluid conductor, for if the teeth of the animal have been wiped dry by passing through the clothing of its intended victim no evil results follow.

Not every one that is bitten by a rabid Dog is a sufferer from Hydrophobia, for it is needful that the constitution should be in a fit state to receive the poison, for its influence to produce any effect. We may notice a similar phenomenon among those who are vaccinated. Some persons appear to be almost proof against the vaccine virus, while others feel its effects so powerfully that they are thrown into a temporary fever, and the limb on which the vaccination is performed, swells to such a degree as to be extremely painful to the patient, and sometimes even alarming to the operator. In others, again, no visible effect is produced until they have undergone the operation two or three times, and then the disease develops itself fully and with great rapidity.

A rather remarkable circumstance connected with this subject took place within the last few years. A rabid Dog contrived to bite a large number of victims, including other Dogs, sheep, oxen, and human beings ; a surgeon attended the human sufferers, and treated the wounds by the severe application of nitrate of silver. All were treated in the same manner, but although the greater number escaped without further injury, several died from Hydrophobia ; and all those in whom the disease made itself manifest were light-haired persons, while those who escaped had dark hair.

The mode of treatment in such dire necessity is fortunately very simple, and can be applied by any one who is possessed of sufficient nerve and presence of mind. A piece of nitrate of silver, or lunar caustic, as it is popularly called, should be cut to a point like a common cedar-pencil, and applied to every part of the wound that can be reached. In default of the caustic, a hot iron, such as a steel fork, a knitting-needle, a skewer, or any similar household article, may be heated to a glowing redness, and applied in the same manner. The iron should be as hot as possible, for it is efficacious in proportion to its temperature, and is not nearly so painful in application if the heat is sufficiently powerful to destroy the nerves at once. A white-hot iron will not cause nearly so much suffering as if it were applied at a dull red heat.

Washing the injured part, applying cupping-glasses to the wound, and cutting away the surrounding portions, have been recommended by some writers, but are strongly condemned by men of large practical experience. They say that the water which is used for the purpose of washing away the poisonous substance will only dilute it, and render it more fluid for the blood to take up ; that the application of a cupping-glass will only draw blood into the wound, and so cause the mixture of the poison with the system ; and that in using the knife the blood which runs from the newly-made incision is apt to overflow into the poisoned locality, and so to convey the venom into the circulation by mixing with the fast-flowing blood as it bathes the enlarged wound.

There are few localities in England in which does not linger some old tradition of healing springs, or holy wells, whose waters are gifted with the blessed power of removing diseases, or of endowing the faithful applicant with mental or bodily accomplishments. A little below Gloucester is a ferry across the Severn, known by the name of the "Hock Crib," which is famous for its powers of healing men or animals which have been bitten by a rabid Dog. If one of these fearful animals has been detected among a flock of sheep, the whole flock is taken to be dipped in the "Hock Crib," even though none of them have been proved to have suffered from the bite of the Dog. Should man, ox, or sheep be bitten by a mad Dog, the sufferer has immediate recourse to the healing waters of this place as soon as possible after the infliction of the injury. After the ninth day, the charmed stream is said to lose its efficacy, and all sufferers from this evil are recommended to make trial of this aqueous remedy before the third day has elapsed.

When cattle of any kind are brought to be dipped, they are forced into the water until they are quite out of their depth, and then are pushed under water by means of a prong passed over their necks, until they are nearly drowned. This curious treatment is repeated until the poor animal is quite exhausted, and is said to be of unfailing efficacy. The ferrymen take upon themselves the task of dipping the patients, and it is probably on account of their unfailing presence, and the accommodation that is afforded for the object, that the Hock Crib is chosen for the purpose of dipping the afflicted animals.

There are one or two curious circumstances connected with this subject. It is said that the disease of Hydrophobia never originates with the female Dog ; and, moreover, that it is most commonly found in the fighting Dogs, and those animals which are kept for the illicit destruction of game. In Africa, and several other hot countries, the malady is unknown, although the animals swarm in very great numbers, and are exposed to the burning sun and the heated atmosphere, without the least assistance from human aid.

The time during which this disease may remain latent in the system is extremely variable. Sometimes it becomes manifest in a few days, while in other cases the virus has produced no tangible effects until the expiration of several months. In one case, however, the disease made its appearance after the seventh month. Mr. Youatt suggests that if every Dog could be kept in separate quarantine for the space of eight months, "the disease might be annihilated in this country, and could only appear in consequence of the importation of some infected animal." This opinion, however, will hardly hold its ground, for although all Dogs that are actually infected might be removed by this course of probation, there is no possibility of warranting that the disease might not again originate in some previously healthy individual, as it must have done in the first instance.

TURNSPIT.—*Canis familiaris.*

Just as the invention of the spinning-jenny abolished the use of distaff and wheel, which were formerly the occupants of every well-ordained English cottage, so the invention of automaton roasting-jacks has destroyed the occupation of the TURNSPIT DOG, and by degrees has almost annihilated its very existence. Here and there a solitary Turnspit may be seen, just as a spinning-wheel or a distaff may be seen in a few isolated cottages ; but both the Dog and the implement are exceptions to the general rule, and are only worthy of notice as being curious relics of a bygone time.

In former days, and even within the remembrance of the present generation, the task of roasting a joint of meat or a fowl was a comparatively serious one, and required the constant attendance of the cook, in order to prevent the meat from being spoiled by the unequal action of the fire. The smoke-jack, as it was rather improperly termed—inasmuch as it was turned, not by the smoke, but by the heated air that rushed up the chimney— was a great improvement, because the spit revolved at a rate that corresponded with the heat of the fire.

So complicated an apparatus, however, could not be applied to all chimneys, or in all localities, and therefore the services of the Turnspit Dog were brought into requisition.

At one extremity of the spit was fastened a large circular box, or hollow wheel, something like the wire wheels which are so often appended to squirrel-cages; and in this wheel the Dog was accustomed to perform its daily task, by keeping it continually working. As the labour would be too great for a single Dog, it was usual to keep at least two animals for the purpose, and to make them relieve each other at regular intervals. The Dogs were quite able to appreciate the lapse of time, and, if not relieved from their toils at the proper hour, would leap out of the wheel without orders, and force their companions to take their place, and complete their portion of the daily toil.

There are one or two varieties of this Dog, but the true Turnspit breed is now nearly extinct in this country. On the Continent, the spits are still turned by canine labour in some localities; but the owners of spit and Dog are not particular about the genealogy of the animal, and press into their service any kind of Dog, provided that it is adequately small, and sufficiently amenable to authority.

The PUG-DOG is an example of the fluctuating state of fashion and its votaries.

Many years ago the Pug was in very great request as a lapdog, or "toy" Dog, as these little animals are more correctly termed. The satirical publications of the last century are full of sarcastic remarks upon Pug-dogs and their owners, and delighted in the easy task of drawing a parallel between the black-visaged, dumpy-muzzled Dog, and the presumed personal attractions of its owner.

By degrees, however, this fashion passed away, as is the wont of fashions to do, and, as is equally their wont, has again returned in due course of time, and with renewed impetus. Although, in the interregnum that elapsed between the two periods of the Pug-dog's ascendancy, it was in very little request, yet in its recent popularity it has acquired so great a conventional value, that a thoroughly well-bred Dog will fetch as much as twenty or thirty pounds, or even more if it be a peculiarly fine specimen. The purity of the breed has been scrupulously preserved by one or two British Dog-fanciers, and to them the Pug-dog is indebted for its present position in the popular esteem.

It is a cheerful and amusing companion, and very affectionate in disposition. Sometimes it is apt to be rather snappish to strangers, but this is a fault which is common to all lap-dogs which are not kept in proper order by their possessors. For those who cannot spend much time in the open air it is a more suitable companion than any other Dog, because it can bear the confinement of the house better than any other of the canine species; and, indeed, seems to be as much at home on a carpet as is a canary on the perch of its cage. Moreover, it is almost wholly free from the unpleasant odour with which the canine race is affected.

The head of the Pug-dog ought to be round, and its forehead high, with a short, but not a turned-up, nose. The whole of the fore-front of the face, extending to the eyes, and technically termed the "mask," ought to be of a jetty black, marked clearly on the lighter ground of the face. The line which separates the two tints should be as sharply cut as possible. The tail should curl sharply and tightly round, lying on one side of the hinder quarters, and never standing upon the back. The height of the Pug-dog ought not to exceed fifteen inches, or its weight to be more than ten pounds.

The number of puppies which the Dog produces at a single litter is very large, varying from three or four to fifteen, or even a still greater number. They are born, as is the case with kittens and several other young animals, with closed eyes, and do not open their eyelids for the space of several days. As it is manifestly impossible for the mother to rear the whole of a very large family, their number must be reduced, either by destroying several of the little ones, which of course ought to be the weakest and smallest specimens, or by removing the supernumerary offspring and placing them under the care of another Dog which has lately taken upon herself the maternal duties. In this case it needs not that the wet nurse should be of the same kind with her charge, as it is found that health of constitution and a liberal supply of milk are the only necessary qualifications for that responsible office.

Sometimes the health of the mother will not permit her to rear her progeny; and in that case, if no worthy substitute can be found, the most humane mode of action is to remove the young puppies in succession, and so to avoid too severe a shock to the maternal feelings of their progenetrix. If they are all removed at the same time, the sudden deprivation is very likely to bring on a severe fever, and to endanger the already weakened life of the mother. If the process of removing and destroying the young ones has been repeated more than once, the mother becomes so watchful over her progeny that it is by no means easy to withdraw them without her cognizance. As an example of this maternal vigilance, I am enabled to give an anecdote which has been forwarded to me by Mrs. S. C. Hall, which exhibits not only the good memory of an often bereaved mother, but a most touching instance of maternal affection.

"In our large, rambling, country home, we had Dogs of high and low degree, from the silky and sleepy King Charles down (query, up?) to the stately Newfoundland, who disputed possession of the top step—or rather platform to which the steps led—of the lumbering hall-door with a magnificent Angora ram, who was as tame and almost as intelligent as Master Neptune himself. After sundry growls and butts the Dog and the ram generally compromised matters by dividing the step between them, much to the inconvenience of every other quadruped or biped who might desire to pass in or out of the hall.

The King Charles, named Chloe, was my dear grandmother's favourite; she was a meek, soft, fawning little creature, blind of one eye, and so gentle and faithful, refusing food except from the one dear hand that was liberal of kindness to her. Chloe's puppies were in great demand; and it must be confessed that her supply was very bountiful, too bountiful indeed, for out of the four which she considered the proper number at a birth, two were generally drowned. My grandmother thought that Chloe ought not to raise more than two; Chloe believed that she could educate four, and it was always difficult to abstract the doomed ones from the watchful little mother.

It so chanced that once, after the two pups had been drowned by one of the stablemen, poor Chloe discovered their little wet bodies in the stable-yard, and brought them to the live ones that remained in her basket. She licked them, cherished them, howled over them, but still they continued damp and cold. Gentle at all other times, she would not now permit even her dear mistress to remove them, and no stratagem could draw her from her basket. At last, we supposed, Chloe felt it was not good for the dead and the living to be together, so she took one of the poor things in her mouth, walked with it across the lawn to the spot where a lovely red thorn-tree made a shady place, dug a hole, laid the puppy in it, came back for the other, placed it with its little relative, scraped the earth over them, and returned sadly and slowly to her duties.

The story of the Dog burying her puppies was discredited by some of our neighbours; and the next time that Chloe became a mother the dead puppies were left in her way, for my grandmother was resolved that her friends should witness her Dog's sagacity. This time Chloe did not bring the dead to the living, but carried them at once to the same spot, dug their graves, and placed them quietly in it. It almost seemed as if she had ascertained what *death* was."

I am also indebted to the same lady for a short history of canine life, which corroborates the account of assistance requested by one Dog and given by another which may be found on p. 287.

"Neptune, the ram's antagonist, had a warm friendship for a very pretty retriever, Charger by name, who, in addition to very warm affections, possessed a very hot temper. In short, he was a decidedly quarrelsome Dog; but Neptune overlooked his friend's faults, and bore his ill-temper with the most dignified gravity, turning away his head, and not seeming to hear his snarls, or even to feel his snaps.

But all Dogs were not equally charitable, and Charger had a long-standing quarrel with a huge bull-dog, I believe it was, for it was ugly and ferocious enough to have been a bull-dog, belonging to a butcher,—the only butcher within a circle of five miles,—who lived at Carrick, and was called the Lad of Carrick. He was very nearly as authoritative as his bull-dog. It so chanced that Charger and the bull-dog met somewhere, and the result was that our beautiful retriever was brought home so fearfully mangled that it was a question whether it should not be shot at once, everything like recovery seeming impossible.

But I really think Neptune saved his life. The trusty friend applied himself so carefully to licking his wounds, hanging over him with such tenderness, and gazing at his master with such mute entreaty, that it was decided to leave the Dogs together for that night. The devotion of the great Dog knew no change ; he suffered any of the people to dress his friend's wounds, or feed him, but he growled if they attempted to remove him. Although after the lapse of ten or twelve days he could limp to the sunny spots of the lawn—always attended by Neptune—it was quite three months before Charger was himself again, and his recovery was entirely attributed to Neptune, who ever after was called Doctor Neptune,—a distinction which he received with his usual gravity.

Now here I must say that Neptune was never quarrelsome. He was a very large liver-coloured Dog, with huge, firm jaws, and those small cunning eyes which I always think detract from the nobility of the head of the Newfoundland ; his paws were pillows, and his chest broad and firm. He was a dignified, gentlemanly Dog, who looked down upon the general run of quarrels as quite beneath him. If grievously insulted, he would lift up the aggressor in his jaws, shake him, and let him go—*if he could go*—that was all. But in his heart of hearts he resented the treatment his friend had received.

So when Charger was fully recovered, the two Dogs set off together to the Hill of Carrick, a distance of more than a mile from their home, and then and there set upon the bull-dog. While we were at breakfast, the butler came in with the information that something had gone wrong, for both Neptune and Charger had come home covered with blood and wounds, and were licking each other in the little stable. This was quickly followed by a visit from the bristly Lad of Carrick, crying like a child—the great rough-looking bear of a man—because our Dogs had gone up the Hill and killed his pup ' Blue-nose ; ' ' The two fell on him,' he said, ' together, and now you could hardly tell his head from his tail.' It was a fearful retribution ; but even his master confessed that ' Blue-nose ' deserved his fate, and every cur in the country rejoiced that he was dead."

The DINGO, or Warragal, as it is called by the natives, is an inhabitant of Australia, where it is found in the greatest profusion, being, indeed, a pest of no ordinary character to those colonists who are employed in raising and maintaining large flocks of sheep.

The colour of this animal is a reddish-brown, sometimes plentifully sprinkled with black hairs over the back and ribs, the legs retaining the ordinary ruddy hue. Its muzzle is very sharp, as is generally the case with wild Dogs ; its ears are sharp, short, and erect ; its tail is pendent and rather bushy ; and its eyes small, cunning, and obliquely placed in the head. It was formerly thought to be an aboriginal inhabitant of Australia, but is now allowed to be an importation from some source which is at present uncertain.

Large packs of these wild Dogs ravage the localities in which they have taken up their residence, and have attained to so high a degree of organization that each pack will only hunt over its own district, and will neither intrude upon the territory which has been allotted to a neighbouring pack of Dingos, nor permit any intrusion upon its own soil. For this reason, their raids upon the flocks and herds are so dangerous that the colonists were obliged to call a meeting, in order to arrange proceedings against the common foe. Before the sheep-owners had learned to take effectual measures to check the inroads of these marauders, they lost their flocks in such numbers that they counted their missing sheep by the hundred. From one colony no less than twelve hundred sheep and lambs were stolen in three months.

The tenacity of life which is exhibited by the Dingo is almost incredible, and it appears to cling as firmly to existence as the opossum. Like the last-mentioned animal, the Dingo appears to feign death when it finds that escape is impracticable, and often manages to elude its opponents by the exercise of mingled craft and endurance. Mr. Bennett, in his well-known " Wanderings," mentions several instances of the wonderful tenacity of life exhibited by the Dingo, and the almost incredible fortitude with which it will submit to wounds of the most fearful description. One of these animals had been overtaken by its exasperated foes, and had been " beaten so severely that it was supposed

DINGO.—*Canis Dingo.*

that all the bones had been broken, and it was left for dead." After its supposed slayer had walked away from the apparently lifeless carcass, he was surprised to see the slain animal arise, shake itself, and slink away into the bush. Another apparently dead Dingo had been brought into the hut for the purpose of being skinned, and had actually suffered the operator to remove the skin from one side of its face before it permitted any symptoms of life or sensation to escape it.

Mr. Bennett further remarks, that this marvellous vitality of the Dingo accounts for the fact that the skeletons of these animals are not found in the places where they have been reported to lie dead. For, although the carrion-devouring beasts and birds will soon carry away every particle of the flesh of a dead animal, they always leave its larger bones as memorials of their ghoul-like repast. There are many similar accounts of the Dingo, and its fast hold of life.

As a general fact, the Dingo is not of a pugnacious character, and would at any time rather run away than fight. But when it is hard pressed by its foes, and finds that its legs are of no use, it turns to bay with savage ferocity, and dashes at its opponents with the furious energy of despair. It carries these uncivilized customs into domesticated life, and even when its restless limbs are subjected to the torpefying thraldom of chain and collar, and its wild, wolfish nature allayed by regular meals and restricted exercise, it is ever ready to make a sudden and unprovoked attack upon man or beast, provided always that its treacherous onset can be made unseen. After the attack, it always retreats into the farthest recesses of its habitation, and there crouches in fear and silence, whether it has failed or succeeded in its cowardly malice.

A Dingo which was kept for some years at the Zoological Gardens was accustomed to sit on its tail and bay the moon after the manner of dogs, making night hideous with its mournful monotone. Moreover, its voice was not silenced by the genial light of day, but rose continually in dolesome ululation, as if in perpetual lament for its captive lot.

In its native land it is a very crafty animal, rivalling the cunning fox in its ready wit when it feels itself endangered, and oftentimes outwitting even the intellectual power of its human foes. A litter of Dingo cubs was once discovered in a rocky crevice near the Yâs Plains, but as the mother was not with them the discoverer marked the locality, intending to return in a short time and to destroy the whole family at one fell swoop. After leaving the spot for such a length of time as he judged sufficient for the return of the mother, he came back to the den, and to his great discomfiture found it to be deserted. The maternal Dingo had probably seen the intruder, and had carried off her young family into a place of safety as soon as she found the coast clear. It is possible that she might not actually have witnessed the hasty visit which this unwelcome guest had paid to her family mansion, but on her return to her little ones had perceived by her sense of smell the late advent of a strange footstep.

JACKAL.—*Canis aureus.*

It is generally found that any large group of animals in one country will be represented in another land by creatures of similar character, and not very dissimilar form. In accordance with this general rule, we find that the part which the dingo plays in Australia is taken up in Asia and Africa by several animals belonging to the canine race, of which the most remarkable are the Jackals and certain wolves. From the former animals the continent of Europe is free ; and in these comparatively civilized times the wolves which still haunt several portions of Europe are simply looked upon as pests of which the country ought to be rid, and not as holding undisputed possession of the territory, and scouring at will over the land in nightly search after prey.

There are several species of the Jackal, two of which will be noticed and figured in this work.

1.   Y

The common JACKAL, or KHOLAH, as it is termed by the natives, is an inhabitant of India, Ceylon, and neighbouring countries, where it is found in very great numbers, forcing itself upon the notice of the traveller not only by its bodily presence, but by its noisy howling wherewith it vexes the ears of the wearied and sleepy wayfarer, as he endeavours in vain to find repose. Nocturnal in their habits, the Jackals are accustomed to conceal themselves as much as possible during the daytime, and to issue out on their hunting expeditions together with the advent of night. Sometimes, a Jackal will prefer a solitary life, and is then a most provoking neighbour to the habitations of civilized humanity; for it is so voracious in its appetite that it becomes a terribly destructive foe to domesticated animals, and so wily in its nature that it carries on its malpractices with impunity until it has worked dire mischief in home or fold. In these depredations, the audacity of the Jackal is as notable as his cunning. He will wait at the very door, biding his time patiently until it be opened and he may slink through the aperture. Pigs, lambs, kids, and poultry fall victims to his insatiate appetite, and he has been known to steal the sleeping puppies from the side of their mother without detection. The larder suffers as severely from his attacks as the henroost, for his accommodating palate is equally satisfied with cooked meat as with living prey.

Always ready to take advantage of every favourable opportunity, the Jackal is a sad parasite, and hangs on the skirts of the larger carnivora as they roam the country for prey, in the hope of securing some share of the creatures which they destroy or wound. On account of this companionship between the large and the small marauders, the Jackal has popularly gained the name of the Lion's Provider. But, in due justice, the title ought to be reversed, for the lion is in truth the Jackal's provider, and is often thereby deprived of the chance of making a second meal on an animal which he has slain. Sometimes, it is said, the Jackal does provide the lion with a meal, by becoming a victim to the hungry animal in default of better and more savoury prey.

There is a very unpleasant odour which arises from this creature, nearly as powerful and quite as offensive as that of the fox. In spite, however, of this drawback, the Jackal is often used as an article of food among the natives, and is said, by those who have tried it, to be pleasant to the palate, and very much superior to tough venison. A hungry lion, therefore, may be expected to find but little impediment in the rank odour of a slaughtered Jackal.

In India, the tiger is often followed during his nightly quests by a company of these animals, and in most cases by a single old Jackal, called in the native tongue, the Khole, or Kholah-balloo, whose expressive cries are well understood by the hunters, whether bipedal or quadrupedal. Many a tiger has been discovered and brought to his death by the yell of a Jackal, which led the pursuers on his track. When the tiger has killed some large animal, such as a buffalo, which he cannot consume at one time, the Jackals collect round the carcass at a respectful distance, and wait patiently until the tiger moves off and they can venture to approach.

As soon as the tiger moves away, the Jackals rush from all directions, carousing upon the slaughtered buffalo, and each anxious to eat as much as it can contain in the shortest time. So eager are they after their prey that they are jealous not only of their companions, but of the vultures that gather round every dead animal, and snap fiercely at them as they wheel round on their broad pinions; or try to push their beaks among the noses of the fighting and struggling Jackals. But although they may snap and snarl, they never seem to inflict any real injury. They are so audacious in their hunger that they will follow human hunters, and take possession of the dead game in a marvellously shameless manner.

They always keep a sharp watch for wounded animals, and pursue them with such relentless vigour that they are said never to permit their weakened prey to escape their fangs. One of these wild dogs, as they really seem to be, has been known to leap at the throat of a wounded Axis deer, and then to hang with such indomitable pertinacity that it resisted all the efforts of its wretched victim to free itself from so terrible a foe. When hanging by its teeth, it contracted its body into as small a compass as was compatible with its size.

Although not a brave animal individually, yet it will, when hard pressed, fight with great ferocity, and inflict extremely painful and dangerous wounds with its long and sharp teeth. It has a great dread of the civilized dog, but has more than once been known to turn the tables on its pursuers, and to call the help of its comrades to its aid. On one of these occasions two greyhounds had been sent in pursuit of a Jackal, which immediately made for a rising ground covered with grass and small bushes. Dogs and Jackal arrived at the spot almost simultaneously, when the Jackal gave a cry of distress, which was immediately answered by the appearance of a small pack of Jackals, which issued in every direction from the cover, and attacked the hounds. The owner of the dogs was at the time impounded in thick mud, and could not reach the spot in time to rescue his hounds from their furious enemies until they had been most severely mangled. One was quite unable to walk, and was carried home by bearers, and the other was so dreadfully bitten over his whole person that he appeared to have been fired at with buck-shot. Both dogs ultimately recovered, but not until the lapse of a long time.

On another occasion, when a pack of hounds was hunting a Jackal, a very much larger pack of Jackals came to the rescue, and in their turn attacked the hounds with such vehemence that they were unable to take the field for many weeks afterwards. So fierce were the assailants in their attack, that even when the hunters came to the aid of their hounds the Jackals flew upon the horses, and were so persevering in their onset that a rescue was not effected without considerable difficulty. If unmolested, the Jackal is harmless enough, and will permit a human being to pass quite closely without attempting to bite.

The Jackal is tolerably susceptible of human influence, and if taken when very young, or if born into captivity, can be brought to follow its master about like a dog, and to obey his orders. If it should be made captive when it has once tasted a free life, it behaves after the manner of the dingo, being shy, suspicious, and treacherous towards those who may come unexpectedly within reach of its teeth. It is rather remarkable that the animal loses its unpleasant odour in proportion to the length of its captivity. The name of "aureus," or golden, is derived from the yellowish tinge of the Jackal's fur. In size it rather exceeds a large fox, but its tail is not proportionately so long or so bushy as the well-known "brush" of the fox.

The BLACK-BACKED JACKAL is an inhabitant of Southern Africa, being especially abundant about the Cape of Good Hope, from which circumstance it is sometimes termed the Cape Jackal. In size it equals the common Jackal, but is easily distinguished from that animal by the black and white mottlings which are thickly spread over its back, and give a peculiar richness to the colouring of its fur. Its habits are precisely the same as those of the common Jackal, and need not be separately described.

It is a very cunning as well as audacious animal, and is extremely apt at extricating itself from any dangerous situation into which it has ventured in search of prey.

One of these animals had for several successive nights insinuated itself into a hen-roost, in Pietermaritzberg, and borne away its inmates without being detected or checked. The proprietor of the poultry finding that his fowls vanished nightly, and not knowing the mode of their departure, vowed vengeance against the robber, whoever he might be, and fixed a spring-gun across the only opening that gave access to the henhouse. In the course of the succeeding night the report of the gun gave notice that the thief had been at his usual work, and the bereaved owner ran out towards the discharged gun, hoping to find its charge lodged in the dead body of the marauder. However, the thief had made his escape, but had left behind him sure tokens of his punishment in the shape of several heavy spots of blood that lay along the ground for some little distance. Some hairs that were discovered in the cleft of a splintered bar, by which the animal had passed, announced that a Jackal was the delinquent.

In the morning the trail was followed up, but with little success, as it led across some roads where so many footsteps were constantly passing that the blood-spots were hopelessly destroyed, and the scent of the animal broken up by the trails of men and cattle. The road that led to the plains was carefully examined, but no traces of the

BLACK-BACKED JACKAL.—*Canis mesómelas.*

wounded animal could be discovered. Two days afterwards it was found, with a hind-leg broken, in a bundle of Tambookie grass, in the very middle of the village, and close to a butcher's shambles. The cunning animal evidently knew that if it went to the plains it must die of starvation, and might, moreover, be easily overtaken by its pursuers, so it concealed itself in the very spot where they would least think of looking for it, and where it was within easy reach of food.

The nightly shrieks with which the Black-backed Jackal fills the air are loud and piercing; but when heard at a distance are thought by some sportsmen to possess a certain melody to initiated ears.

The peculiar dark mottlings of the back form a band that extends from its neck and shoulders to the tail, is very broad in front, passing over the withers as far as the shoulders, and narrowing gradually towards the tail, where it becomes only two inches wide. The tail is of a fawn colour, and does not partake of this variable colouring, with the exception of the tip, which is black.

Lieutenant Burton remarks, that among the Somali the morning cry of the Jackal is used as an omen of good or evil, according to its direction and its tone. He also mentions that it is in the habit of attacking the peculiar fat-tailed sheep which inhabit that country, and carrying off their lambs. The fat-burdened tail forms an article of diet which seems to be greatly to the Jackal's taste, and which he procures by leaping suddenly upon the poor sheep, and then making a fierce bite at its tail. The terrified sheep starts off at best speed, and leaves a large mouthful of its tail between the Jackal's teeth. Kids and other small animals fall victims to this insatiate devourer.

In that country the Jackal, called by the natives "Duwas," dances nightly attendance upon the spotted hyæna.

GROUP OF WOLVES.

WOLVES.

FEW animals have earned so widely popular, or so little enviable, a fame as the WOLVES. Whether in the annals of history, in fiction, in poetry, or even in the less honoured, but hardly less important, literature of nursery fables, the Wolf holds a prominent position among animals.

There are several species of Wolf, each of which species is divided into three or four varieties, which seem to be tolerably permanent, and by many observers are thought to be sufficiently marked to be considered as separate species. However, as even the members of the same litter partake of several minor varieties in form and colour, it is very possible that the so-called species may be nothing more than very distinctly marked varieties. These voracious and dangerous animals are found in almost every quarter of the globe ; whether the country which they infest is heated by the beams of the tropical sun or frozen by the lengthened winter of the northern regions. Mountain and plain, forest and field, jungle and prairie, are equally infested with Wolves, which possess the power of finding nourishment for their united bands in localities where even a single predaceous animal might be perplexed to gain a livelihood.

The colour of the common WOLF is grey, mingled with a slight tinting of fawn, and diversified with many black hairs that are interspersed among the lighter coloured fur. In the older animals the grey appears to predominate over the fawn, while the fur of the younger Wolves is of a warmer fawn tint. The under parts of the animal, the lower jaw, and the edge of the upper lip, are nearly white, while the interior face of the limbs is of a grey tint. From this latter circumstance the Norwegians, with their usual superstitious dislike to calling an animal by its right name, dignify the Wolf by the title of " Graa-been," or Grey-legs. The equally superstitious Finns prefer the name of " Loajalg," or Broad-foot. Between the ears the head is almost entirely grey, and without the mixture of black hairs, which is found in greatest profusion along the line of the spine.

When hungry—and the Wolf is almost always hungry—it is a bold and dangerous animal, daring almost all things to reach its prey, and venturing to attack large and powerful animals,—such as the buffalo, the elk, or the wild horse. Sometimes it has been known to oppose itself to other carnivora, and to attack so unpromising a foe as the bear. Mr. Lloyd records an instance of this presumption on the part of the Wolves.

During a bear-hunt, when the hunting party was led by a dog that was following the footsteps of a bear, a small herd of Wolves, few in number, suddenly made their appearance, pounced on the dog, and devoured it. They then took up the trail, and when they came up with the bear entered into battle with him. The fight terminated in favour of the bear ; but not without much exertion and great danger to both parties, as was proved by the quantity of bear and Wolf fur that lay scattered about the scene of combat. So severely had the bear been treated that his fur was found to be quite useless when he was killed by the hunters a few days after the conflict.

This is not a solitary example of a fight between bears and Wolves, as the same author mentions a similar combat, which would apparently have had a different result. The bear had retreated to a large tree ; and, standing with his back against the trunk, boldly faced his antagonists, and for some time kept them at bay. At last, however, some of the Wolves crept round the tree, and seizing him unexpectedly in the flank, inflicted such severe wounds that he would soon have fallen a victim to their ferocity had not they been put to flight by the approach of some men.

It is by no means nice in its palate, and will eat almost any living animal,—from human beings down to frogs, lizards, and insects. Moreover, it is a sad cannibal, and is thought by several travellers who have noted its habits to be especially partial to the flesh of its own kind. A weak, sickly, or wounded Wolf is sure to fall under the cruel teeth of its companions ; who are said to be so fearfully ravenous that if one of their companions should chance to besmear himself with the blood of the prey which has just been hunted down, he is instantly attacked and devoured by the remainder of the pack.

In their hunting expeditions the Wolves usually unite in bands, larger or smaller in number, according to circumstances, and acting simultaneously for a settled purpose. If they are on the trail of a flying animal, as is represented in the large engraving on p. 325, the footsteps of their prey are followed up by one or two of the Wolves, while the remainder of the band take up their positions to the right and left of the leaders, so as to intercept the quarry if it should attempt to turn from its course. Woe be to any animal

WOLF.—*Canis lupus.*

that is unlucky enough to be chased by a pack of Wolves. No matter how swift it may be, it will most surely be overtaken at last by the long, slouching, tireless gallop of the Wolves; and no matter what may be its strength, it must at last fail under the repeated and constant attacks of the sharp teeth.

There is something remarkable about the bite of a Wolf. Instead of making its teeth meet in the flesh of its antagonist, and then maintaining its hold, as is done by most of the carnivora, the Wolf snaps sharply, fiercely, and repeatedly at its opponent or its quarry; delivering these attacks with such furious energy that when it misses its mark its jaws clash together with a sound that has been likened to the sudden closing of a steel-trap. These sharply snapping bites, so rapidly delivered, are of terrible efficacy in destroying an enemy, or bringing down the prey.

The skeleton of the Wolf which is here presented to the reader affords an instructive contrast with that of the lion on page 131, as exhibiting the bony framework around which is built the bodily organization of two distinct carnivorous types of animals.

Putting aside the differences that exist between the feline and the canine dentition, the general character of the whole form is worthy of notice, and points out the creature as belonging to the group of carnivorous animals which obtains its prey by running it down in a lengthy chase, rather than to those predaceous animals which destroy their prey by a single powerful spring. The limbs are larger in proportion than those of the lion, and the bones are more slenderly made. The head and neck are very differently formed. Those of the lion are intended to serve the purpose of an animal which leaps upon its prey, fixes its teeth in the flesh of its quarry, and there hangs until it has destroyed its prey; but the corresponding portions of the Wolf's anatomy belong evidently to an animal which is not intended by nature to exert the clinging hold of the cat tribe, but to overtake its prey by fair chase, to run, and to bite.

The sharp teeth with which the Wolf is furnished are strong enough to cut their way through substances which might be thought impervious to teeth. A hungry Wolf will devour a raw hide with enviable ease, and, when hard pressed by its unsatisfied appetite, has often been known to make a meal on thick leather traces that had been left unguarded for a few minutes.

Bold as is the Wolf in ordinary circumstances, it is one of the most suspicious animals in existence, and is infected with the most abject terror at the sight of any object to which its eyes, nose, or ears are unaccustomed.

Very fortunately for the hunters, this excess of caution on the part of the Wolf is the means of preserving their slaughtered game from the hungry maws of the Wolves that ever accompany a hunter, and hang on his steps in hope of obtaining the offal of such animals as he may slaughter, or of securing such creatures as he may wound and fail to kill on the spot. In order to preserve the carcass of a slain buffalo or deer, the

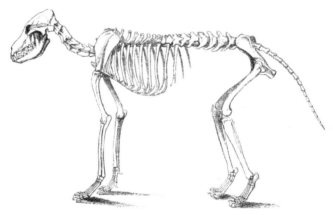

SKELETON OF WOLF.

hunter merely plants a stick by the side of the animal, and ties to the top of the stick a fluttering piece of linen, or any similar substance, and then goes his way, secure that the Wolves will not dare to approach such an object. In default of a strip of calico or linen, the inflated bladder of the dead animal is an approved "scare-wolf;" and, as a last resource, a strip of its hide is used for that purpose.

To this peculiarity have been owing, not only the preservation of game, but the lives of defenceless travellers. It has several times happened that a band of Wolves have been pressing closely upon the footsteps of their human quarry, and have been checked in their onward course by the judicious exhibition of certain articles of which the Wolves were suspicious, and from which they kept aloof until they had satisfied themselves of its harmlessness. As one article began to lose its efficacy, another was exhibited, so that the persecuted travellers were enabled to gain the refuge of some friendly village, and to baffle the furious animals by means which in themselves were utterly inadequate to their effects. A piece of rope trailed from a horse or carriage is always an object of much fear to the Wolves.

BLACK WOLF.—*Canis occidentalis.*

When the Wolf is once within a trap it becomes the most cowardly of animals, and will permit itself to be handled or wounded without displaying the least sign of animation, or attempting to resist the hand of its destroyer. The sensation of imprisonment appears to deprive it of all energy, and it sometimes happens that a trapped Wolf is so entirely destitute of self-control, that it has permitted the hunter to drag it from the trap, and to make it lie passively by his side while he reset the trap for the occupancy of another victim. On one occasion, a pitfall-trap contained two occupants, one a Wolf, and the other a poor old woman, who had unfortunately fallen into the pit when returning from her work. The Wolf was so cowed by finding itself entrapped, that it made no attempt to injure its fellow prisoner, but lay quietly at the bottom of the pit, and was shot in the morning by a peasant.

The BLACK WOLF of America was thought by some naturalists to be only a variety of the common Wolf, but it is now considered to be a distinct species. Not only does the colour of its fur vary from that of the common Wolf, but there are various differences of structure, in the position of the eye, the peculiar bushiness of the hair, and other peculiarities, which have entitled it to rank as a separate species.

The American Wolves partake of the general lupine character, being fierce, dangerous, and cowardly, like their European brethren. They are marvellously pusillanimous when they find themselves fairly inclosed ; and even if their prison-house be a large yard they crouch timidly in the corners, and do not venture to attack a human being if he enters the same inclosure. Audubon mentions a curious instance of this strange timidity in so fierce an animal, and of which he was an eye-witness.

PRAIRIE WOLF.—*Canis latrans.*

A farmer had suffered greatly from the Wolves, and had determined to take his revenge by means of pitfalls, of which he had dug several within easy reach of his residence. They were eight feet in depth, and wider at the bottom than at the top. Into one of these traps three fine Wolves had fallen; two of them being black, and the other a brindled animal. To the very great astonishment of M. Audubon, the farmer got into the pit, pulled out the hind-legs of the Wolves, as they lay trembling at the bottom, and with his knife severed the chief tendon of the hind-limbs, so as to prevent their escape. The farmer was thus repaying himself for the damage which he had suffered, for the skins of the captured Wolves were sufficiently valuable to reimburse him for his labour and previous losses.

Among the Esquimaux the Wolves are caught in traps made of large blocks of ice, and constructed in precisely the same manner as an ordinary mousetrap with a drop-door. The trap is made so narrow that the Wolf cannot turn himself, and when he is fairly inclosed by the treacherous door, he is put to death by spears, which are thrust through interstices left for that purpose.

There is a rather smaller species of Wolf, which is found in great numbers upon the American prairies, and named for that reason the PRAIRIE WOLF. These animals are always found hanging on the outskirts of the numerous herds of bisons that roam the prairies, and pick up a subsistence by assailing the weakly and wounded members of the herd. Small as is each individual Wolf, it becomes a terrible assailant when backed by numbers, and seldom fails to bring to the ground any animal which may be unfortunate enough to attract its attention.

When they have once brought their prey to the ground, they make marvellously short work. There is a scuffle of some two minutes in length, during which the Wolves are so eagerly plying their feet and jaws that nothing is visible except a cloud of dust and hair, in the midst of which is a mass of whisking tails. The dusty cloud then subsides, and the Wolves are seen moving slowly away from the scene of their late repast. They also are in the habit of accompanying the hunters through their long peregrinations over the prairies, always hanging behind at respectful distances, and at night encamping within easy range of the fire. They seem never to injure the hunter or his horse, preferring to make use of his superior powers in procuring them a daily supply of food. They are wise in so doing, as the hunter seldom requires more than the " hump," tongue, marrow-bones, and skin of the slaughtered bison, and leaves the remainder of the huge carcass for the Wolves.

ANOTHER well-known American Wolf is the COYOTE, or CAJOTE, in which there is something of the vulpine aspect. In habits it resembles the other Wolves. According to European ideas, the flesh of the Wolf would be thought a very strange, and decidedly repulsive, article of diet. But it is found by those who have had practical experience on this subject, that the Wolf, when properly dressed, affords a really excellent dinner, the tables being thus turned on him. The ribs are the portion which are most esteemed.

COYOTE.—*Canis ochropus.*

Like many other wild animals, the Wolf will feign death when it has fallen into the hands of its pursuers, and finds that escape is impossible. So admirably will it achieve this feat that it has often deceived the experienced eyes of the hunter, and, taking advantage of an unguarded moment on his part, has made good its escape. How perseveringly the animal will enact his part may be imagined from the description of a captive Wolf given by Captain Lyon, in his private journal.

The Wolf had been brought on board apparently dead, but as the eyes were observed to wink when an object was passed rapidly before them, a rope was fastened to his hind-legs, and he was suspended from the rigging, with his head downwards. Suddenly he threw off all disguise, and began to snap viciously in all directions; at one time aiming his attacks at the persons who surrounded him, and at another moment curling himself upwards and trying to bite the rope asunder. He was so very full of life that it required several heavy blows on his head, and the employment of a bayonet, to reduce him in reality to the state which he had previously been feigning.

It was formerly supposed that the Wolf was an untameable animal, but it is now known that there are few creatures which are more susceptible of affection than the Wolf, if it be captured when young, and treated rightly. It will follow its master like a dog, will obey his orders readily, will recognise him after a long term of absence, and in all things conduct itself with a propriety that is not always found in the domesticated dogs. Several instances of this tameable disposition of the Wolf are well known. One such example is afforded by the tame Wolf which belonged to Mr. F. Cuvier, and which recognised him after an absence of three years.

A Norwegian gentleman, named Grieff, " reared up two young Wolves until they were full-grown. They were male and female. The latter became so tame that she played with me, and licked my hands, and I had her often with me in the sledge in winter. Once when I was absent she got loose from the chain she was bound with, and was away for three days. When I returned home I went out on a hill, and called ' Where is my Tussa ?' as she was named, when she immediately came home, and fondled with me like the most friendly dog. She could not bear other people, but the male, on the contrary, was friendly with others but not with me, from the moment when he once seized a hen, and I whipped him with a carrier whip. As they were well treated, they got very large, and had fine skins."

When Wolves and dogs are domesticated in the same residence, a mutual attachment will often spring up between them, although they naturally bear the bitterest hatred to each other. A mixed offspring is sometimes the result of this curious friendship, and it is said that these half-bred animals are more powerful and courageous than the ordinary dog. Mr. Palliser possessed a remarkably fine animal of this kind, the father of which was a white Wolf, and the mother an ordinary Indian dog. Its fur was white, like that of its Wolf-parent.

When " Ishmah," as the dog was named, was first purchased from its Indian owners, he was so terrified at the white face of his new master, that he always ran away whenever he saw him, and could not be persuaded to come within two hundred yards. Ishmah was then tied up with a cord, but the moment that he was left to himself he held the cord to the ground with his paw, severed it in an instant with his sharp teeth, leaped out of the window, and dashed off to his former owners. After a while, however, he became reconciled to his white master, and proved to be a most faithful and useful ally ; dragging a small sledge that contained the heavier necessaries of a hunter's life, and partaking with his master all the pleasures and privations of a nomad existence. On account of his wolfish ancestry, he was rather apt to run off and play with the young Wolves instead of attending to his duty, but was never induced to throw off his allegiance. On one occasion the dog saved the life of his master by lying close to him on a bitterly freezing night, and with his long warm fur preserving him from the terrible death by frost.

In former days the British islands were infested with these savage brutes, and suffered greatly from their depredations, until the issuing of the famous edict which ordained that Wolves' heads should be accepted in the lieu of taxes, and which speedily caused their extinction. In Scotland and Ireland, however, they lingered for a much longer time, the last British Wolf being, as it is supposed, killed in Scotland in the seventeenth century.

The Wolf is rather a prolific animal, producing from three to nine young at a litter. In January the mother Wolf begins to prepare her habitation for the expected inmates, a task in which she is protected, and perhaps assisted, by her mate, who has won her in fair fight from his many rivals. He attaches himself solely to one single mate, and never leaves her until the young Wolves are able to shift for themselves. The nest in which the little family is nurtured is softly and warmly lined with dry moss and with the fur of the mother, which she pulls from her own body. March is the usual month for the appearance of the little family, and they remain under the maternal protection for seven or eight months. They begin to eat meat at four or five weeks of age, and are taught by their parents to join in the chase.

According to some systematic naturalists the FOXES are placed in the genus Canis, together with the dogs and the wolves. Those eminent zoologists, however, who have arranged the magnificent collections in the British Museum, have decided upon separating the Foxes from the dogs and wolves, and placing them in the genus Vulpes. To this decision they have come for several reasons, among which may be noted the shape of the pupil of the eye, which in the Foxes is elongated, but in the animals which compose the genus Canis is circular. The ears of the Foxes are triangular in shape, and pointed, and the tail is always exceedingly bushy.

A very powerful scent is poured forth from the Fox in consequence of some glands which are placed near the root of the tail, and furnish the odorous secretion. Glands of a similar nature, but not so well developed, are found in the wolves. The tenacity with which this scent clings to any object which it has touched is quite extraordinary. I remember an instance when a Fox was captured by an old labourer, in revenge for killing his fowls, and which he exhibited in an outhouse for a short time. The animal could not have been in the shed for more than twenty minutes, and yet the odour which it evolved was so pertinaciously adherent to everything which had been touched by the animal that the shed was not free from the tell-tale scent for many weeks.

At night, while walking over the Wiltshire Downs, and the various roads that intersect them, I have frequently been aware that a Fox had crossed the path, and could have followed up the scent for some distance.

It is by this scent that the hounds are able to follow the footsteps of a flying Fox, and to run it down by their superior speed and endurance. The Fox, indeed, seems to be aware that its pursuers are guided in their chase by this odour, and puts in practice every expedient that its fertile brain can produce in order to break the continuity of the scent,

FOX.—*Vulpes vulgaris.*

or to overpower it by the presence of other odours, which are more powerful, though not more agreeable. A hunted Fox will make the most extraordinary leaps in order to break the line of scent, and throw the hounds on a false track. It will run for a considerable distance in a straight line, return upon its own track, and then make a powerful spring to one side, so as to induce the dogs to run forward while it quietly steals away. It will take every opportunity of perfuming, or rather of scenting, itself with any odorous substance with which it can meet, in the hope of making the hounds believe that they have mistaken their quarry. In fine, there are a thousand wiles which this crafty animal employs and which are related by every one who has watched a Fox or hunted it.

Even when tamed it preserves its singular cunning. A tame Fox, that was kept in a stable-yard, had managed to strike up a friendship with several of the dogs, and would play with them, but could never induce the cats to approach him. Cats are very sensitive in their nostrils, and could not endure the vulpine odour. They would not even walk upon any spot where the Fox had been standing ; and kept as far aloof as possible from him.

The crafty animal soon perceived that the cats would not come near him, and made use of his knowledge to cheat them of their breakfast. As soon as the servant poured out the cats' allowance of milk, the Fox would run to the spot and walk about the saucer, well knowing that none of the rightful owners would approach the defiled locality. Day after day the cats lost their milk until the vulpine stratagem was discovered, and the milk was placed in a spot where it could not be reached by the Fox. There were three cats attached to the stables, and they all partook of the same detestation ; so that their abhorrence of the vulpine odour seems to belong to the general character of cats, and not to be the fastidious individuality of a single animal. He was also very successful in cheating the dogs of their food ; achieving his thefts by the force of superior intellect.

The same animal was cunning enough to procure a supply of milk, even after he had been prevented from robbing the cats. On one occasion, as the dairymaid was passing along with her pails, the Fox went up to her, and brushed himself against one of the milk-pails. In consequence of this contact, the milk became so tainted with the smell of the Fox that the dairymaid did not venture to bring it to the house, and rather thoughtlessly poured it out into a vessel, and gave it to the Fox. The crafty animal took advantage of the circumstance, and watched for the coming of the maid with her pails, in order to repeat the process. Several times he succeeded in his project, but when he found that the spoiled milk was given to the pigs, instead of being appropriated to his own use, he ceased his nefarious attempts.

He detested all ragged beggars, and was so energetic in his hostile demonstrations, that he realized the truth of the proverb, "Set a thief to catch a thief." The horses hated him with as thorough a detestation as that in which the cats held him. His presence in the stable would set the horses in confusion, and make them plunge about in a restless and uneasy manner.

The Fox resides in burrows, which it scoops out of the earth by the aid of its strong digging paws, taking advantage of every peculiarity of the ground, and contriving, whenever it is possible, to wind its subterranean way among the roots of large trees, or between heavy stones. In these "earths," as the burrows are called in the sportsman's phraseology, the female Fox produces and nurtures her young, which are odd little snub-nosed creatures, resembling almost any animal rather than a Fox. She watches over her offspring with great care, and teaches them by degrees to subsist on animal food, which she and her mate capture for that purpose.

The colour of the common Fox is a reddish-fawn, intermixed with black and white hairs. The hair is long and thick, being doubly thick during the colder months of the year, so that the fur of a Fox which is killed in the winter is more valuable than if the animal had been slain in the hot months. The tail, which is technically termed the "brush," is remarkably bushy, and partakes of the tints which predominate over the body, except at the tip, which is white. The height of this animal is about a foot, and its length about two feet and a half, exclusive of the tail.

There are several species of Foxes, which are found in various parts of the globe, some of which, such as the AMERICAN FOX, or MAKKEESHAW, sometimes called the Cross Fox, the Kit Fox, and the Arctic Fox, are tolerably familiar animals. The American Fox is very variable in the colour and markings of its fur, some specimens being of a pale yellow, some being blackish in their general tinting, and some of a reddish-fawn, while some specimens are remarkable for the manner in which the black, the white, the yellow, and the fawn are dispersed over the body and limbs. In almost every specimen there is a darkish transverse stripe over the shoulders, giving to the animal the title of Cross Fox.

This animal has its full share of the crafty spirit which is so notable in the nature of all Foxes. One of them, on whose track the hounds had been often laid, used always to baffle them at one particular point, the crest of a rather steep hill. Up to this spot the scent was perfectly good ; but at that particular spot the scent vanished, and so the Fox was lost. One of the disappointed hunters was so indignant at his repeated failures that he determined to lay aside the chase for a day, and to devote himself to the discovery of the means by which the creature could so invariably escape from the hounds and men. He therefore concealed himself near the charmed spot, and watched with much interest the proceedings of the hunted animal.

The Fox, after being driven from his cover, led the hounds a long chase through woods, ponds, and thickets, and at last came at full speed towards the crest of the hill. As soon as he had reached the spot, he laid himself down and pressed himself as closely as possible to the ground. Presently the hounds came along in full cry, and with a blazing scent, darting over the hill in hot pursuit, and never stopping until they reached the bottom of the hill. As soon as the last hound had passed, the Fox resumed his legs, crept quietly over the brow of the hill, and returned to his covert at leisure.

Another of these creatures made use of a very cunning device for the same purpose. In this instance, he always led his pursuers to the edge of a cliff that rose perpendicularly for several hundred feet, and then disappeared. The hunters had often examined the spot, and unsuccessfully, for it seemed that no wingless animal could venture to take such a fearful leap. The secret was, however, at last discovered by a concealed spy. The crafty Fox was seen coming quite at his leisure to the edge of the cliff, and then to look down. Some ten feet below the edge there was a kind of break in the strata of stone, forming a kind of step about a foot in width. By means of his claws the Fox let himself down upon this step, and then disappeared in a hollow which was invisible from above.

A man was lowered by ropes to the spot, and found that there was a wide fissure in the rock, to which the stony step formed an entrance. On searching the cavern it was found to have another and an easy outlet upon the level ground above. The Fox,

AMERICAN FOX.—*Vulpes fulvus.*

however, never used this entrance when the hounds were on his trail, but cut off the scent by scrambling over the cliff, and then emerged at the other outlet without danger of discovery.

Mr. C. W. Webber narrates an equally curious instance of the cunning of a Fox in escaping from his pursuers.

" There was a certain briary old field of great extent, near the middle of which we could, on any morning of the year, start a grey Fox. After a chase of an hour or so, just enough to blow the dogs and horses well, we invariably lost the Fox at the same spot, the fence-corner of a large plantation, which opened into a heavy forest on one side of this old field. The frequency and certainty of this event became the standing joke of the country. Fox-hunters from other neighbourhoods would bring their pack for miles, to have a run out of this mysterious Fox, in the hope of clearing up the mystery. But no. They were all baffled alike. We often examined the ground critically, to find out, if possible, the mode of escape, but could discover nothing that in any way accounted for it, or suggested any theory in regard to it. That it did not fly was very sure; that it must escape along the fence in some way was equally so. My first idea was, that the animal, as is very common, had climbed upon the top rail of the fence, and walked along it to such a distance, before leaping off, that the dogs were entirely thrown out. I accordingly followed the fence with the whole pack about me, clear round the plantation, but without striking the trail again, or making any discovery.

The affair now became quite serious. The reputation of our hounds was suffering; and, besides, I found they were really losing confidence in themselves, and would not run with half the staunch eagerness which had before characterized them. The joke of being regularly baffled had been so often repeated that they now came to consider it a settled thing that they were never to take another Fox again, and were disposed to give up in despair. Some of the neighbours had grown superstitious about it, and vowed that this must be a weir Fox, who could make himself invisible when he pleased.

At last I determined to watch at the fence-corner, and see what became of the Fox. Within about the usual time I heard him heading towards the mysterious corner, as the voices of the pack clearly indicated. I almost held my breath in my concealment, while I watched for the appearance of this extraordinary creature. In a little while the Fox made his appearance, coming on at quite a leisurely pace, a little in advance of the pack. When he reached the corner, he climbed in a most unhurried and deliberate way to the top rail of the fence, and then walked along it, balancing himself as carefully as a rope-dancer. He proceeded down the side of the fence next to the forest in which I was concealed.

I followed cautiously, so as to keep him in view. Before he had thus proceeded more than two hundred yards, the hounds came up to the corner, and he very deliberately paused and looked back for a moment, then he hurried along the fence some paces farther, and when he came opposite a dead but leaning tree which stood inside the fence, some twelve or sixteen feet distant, he stooped, made a high and long bound to a knot upon the side of its trunk, up which he ran, and entered a hollow in the top where it had been broken off, nearly thirty feet from the ground, in some storm. I respected the astuteness of the trick too much to betray its author, since I was now personally satisfied; and he continued for a long time, while I kept his secret, to be the wonder and the topic of neighbouring Fox-hunters, until at last one of them happened to take the same idea into his head, and found out the mystery. He avenged himself by cutting down the tree, and capturing the smart Fox.

The tree stood at such a distance from the fence that no one of us who had examined the ground ever dreamed of the possibility that the Fox would leap to it; it seemed a physical impossibility, but practice and the convenient knot had enabled cunning Reynard to overcome it with assured ease."

ONE of the most celebrated species of the Foxes is the ARCTIC FOX, called by the Russians PESZI, and by the Greenlanders TERRIENNIAK. This animal is in very great repute in the mercantile world on account of its beautifully silky fur, which in the cold winter months becomes perfectly white. During the summer the fur is generally of a grey, or dirty brown, but is frequently found of a leaden grey, or of a brown tint with a wash of blue. Towards the change of the seasons the fur becomes mottled; and by reason of this extreme variableness has caused the animal to be known by several different titles. Sometimes it is called the White Fox, sometimes the Blue Fox, sometimes the Sooty Fox, sometimes the Pied Fox, and sometimes the Stone Fox.

This animal is found in Lapland, Iceland, Siberia, Kamschatka, and North America, in all of which places it is eagerly sought by the hunters for the sake of its fur. The pure white coat of the winter season is the most valuable, and the bluish-grey fur of the summer months is next to the white the colour that is most in request. The soles of the feet are thickly coated with hair, from which circumstance it has derived its name of Lagopus, or hairy foot.

It is found that this animal possesses the power of imitating the cries of the birds on which it loves to feed, and it is probable that it employs this gift for the purpose of decoying its prey to their destruction. Although it is sufficiently cunning in obtaining its food, it seems to be remarkably destitute of the astute craft which aids the generality of the Foxes to avoid hidden dangers or to baffle their foes. It is easily induced to enter a trap, and will generally permit a hunter to approach within range of an easy shot. It

ARCTIC FOX.—*Vulpes Lagopus.*

is true that, when a human being approaches their burrows, the inmates retire into their homes ; but as they continually protrude their heads and yelp at their foe, the precaution is to very little purpose.

In size, the Arctic Fox is not the equal of the English species, weighing only eight pounds on an average, and its total length being about three feet. The eye is of a hazel tint, and very bright and intelligent. It lives in burrows, which it excavates in the earth during the summer months, and prefers to construct its simple dwellings in small groups of twenty or thirty.

THE LITTLE animal which is known by the name of the ASSE, or the CAAMA, is an inhabitant of Southern Africa, and is in great request for the sake of its skin, which furnishes a very valuable fur.

It is a terrible enemy to ostriches and other birds which lay their eggs in the ground, and is in consequence detested by the birds whose nests are devastated. The ingenuity of the Caama in procuring the contents of an ostrich's egg is rather remarkable. The shell of the egg is extremely thick and strong ; and as the Caama is but a small animal, its teeth are unable to make any impression on so large, smooth, hard, and rounded an object. In order, therefore, to obviate this difficulty, the cunning animal rolls the egg along by means of its fore-paws, and pushes it so violently against any hard substance that may lie conveniently in its path, or against another egg, that the shell is broken and the contents attainable.

The fur of this animal is highly esteemed by the natives for the purpose of making "karosses," or mantles. As the Asse is one of the smallest of the Foxes, a great number of skins are needed to form a single mantle, and the manufactured article is therefore held in high value by its possessor. Indeed, so valuable is its fur, that it tempts many of the Bechuana tribes to make its chase the business of their lives, and to expend their whole energies in capturing the animal from whose body the much-prized fur is taken.

The continual persecution to which the Caama is subjected, has almost exterminated it in the immediate vicinity of Cape Town, where it was formerly seen in tolerable

1.   z

ASSE, OR CAAMA.—*Vulpes Caama.*

plenty. Gradually, however, it retreats more and more northward before the tread of civilized man, and at the present day is but very rarely seen within the limits of the colony.

THE TWO animals which now claim our attention bear a considerable external resemblance to each other, albeit that similarity extends not to their formation. So different are they from each other, that they have been placed in a distinct genera by the almost unanimous voice of systematic naturalists.

The former of these animals, the OTOCYON, or Eared-dog, derives its name from the very great proportionate length of its ears. It is much smaller than the English Fox, and is of a tolerably uniform grey colour, except on the tail, which is covered with long black hair, and on the limbs, which are of a darker hue than the body. The ears are erect, well covered with fur, and nearly equal to the head in length. It is an inhabitant of Southern Africa. In several anatomical points, and especially in the arrangement and shape of its teeth, it is easily to be distinguished from the following animal.

The FENNEC, or ZERDA, is an inhabitant of Africa, being found in Nubia and Egypt. It is a very pretty and lively little creature, running about with much activity, and anon sitting upright and regarding the prospect with marvellous gravity. The colour of the Fennec is a very pale fawn, or "isabel" colour, sometimes being almost of a creamy whiteness. The tail is bushy, and partakes of the general colour of the fur, except at the upper part of the base and the extreme tip, which are boldly marked with black. The size of the adult animal is very inconsiderable, as it measures scarcely more than a foot in length, exclusive of the bushy tail, which is about eight inches long.

It is said that the Fennec, although it is evidently a carnivorous animal, delights to feed upon various fruits, especially preferring the date. Such a predilection is according to vulpine and canine analogies, for the common English Fox is remarkably fond of ripe fruits, such as grapes or strawberries, and the domestic dog is too often a depredator of those very gardens which he was enjoined to keep clear from robbers. But that the

OTOCYON.—*Otócyon Lalandii.*

animal should enjoy the power of procuring that food in which it so delights is a very extraordinary circumstance, and one which would hardly be expected from a creature

FENNEC.—*Vulpes Zaarensis.*

which partakes so largely of the vulpine form and characteristics. The date-palm is a tree of a very lofty growth, and the rich clusters of the fruit are placed at the very summit of the bare, branchless stem. Yet the Fennec is said to possess the capability of climbing the trunk of the date-palm, and of procuring for itself the coveted luxury.

This creature presents so strange a medley of characteristics that it has proved a sad stumbling-block to systematic zoologists, and has been so frequently transferred by them from one portion of the animal kingdom to another, that its position in their catalogues seems to vary as often as the different lists are published. One celebrated naturalist considers the Fennec to belong to the civets and genetts; another ranks it with the hyænas; while a third believes that its true position is among the Galagos. Now, however, it finds a resting-place in the genus Vulpes, being a congener with the various foxes of the Old and New Worlds.

It must here be remembered that the generic distinction of dogs and foxes can hardly be regarded as a settled matter, and that many practical naturalists favour the opinion that the foxes ought to be included in the genus "Canis." That the dog and the fox will produce a mixed offspring is now generally allowed. There are many authenticated accounts of such mixed breeds, dating from the earlier part of the present century up to the present time. Moreover, it has been found that the offspring of the dog and the fox is capable of reproduction when it is again crossed with the dog. Should this experiment be successfully conducted to a still farther extent, and the vulpo-canine offspring of both sexes be found capable of mutual reproduction, the difficult question to which we have referred will be finally solved.

Like the veritable foxes, the Fennec is accustomed to dwell in subterranean abodes, which it scoops in the light sandy soil of its native land. Bruce, who claims the honour of introducing this curious little animal to zoological science, avers that it builds its nest in trees. Rüppell, however, who may lay claim to more scientific knowledge than was possessed by Bruce, distinctly contradicts this statement, and asserts that it lives in "burrows" like other foxes.

This curious little animal is not entirely without its use to man; for its fur is of considerable value among the native tribes of the locality wherein it is found. The skin of the Fennec, called "motlose" in the native dialect, is said to furnish the warmest fur in Africa, and is highly prized for that quality. And as, on account of the diminutive size of the animal, a single skin forms but a very small portion of a garment, a mantle which is composed of "motlose" fur is valued very highly, and can with difficulty be purchased from its dark owner.

As is the case with the greater number of predaceous animals, the Fennec is but seldom seen during the daytime, preferring to issue forth upon its marauding expeditions under the friendly cover of night. Even when it has spent some time in captivity, it retains its restless nocturnal demeanour, and during the hours of daylight passes the greater portion of its time in semi-somnolence or in actual sleep. On a comparison with the Otocyon, the Fennec appears at first sight to bear so close a resemblance to that animal that either of the two creatures might easily be mistaken for the other. The slender body, the bushy tail, the sharply pointed snout, and the extraordinarily long ears, are so conspicuously notable that the two animals have frequently been confounded together, and actually figured under the same title. Yet, as has been already mentioned, when treating of the Otocyon, the distinguishing characteristics are so strongly marked as to justify their separation, not only into different species, but into different genera.

It is a quaint little creature in its aspect, and wears an air of precocious self-reliance that has quite a ludicrous effect in so small an animal. The colour of its eyes is a beautiful blue, and the "whisker" hairs which decorate its face are long and thick in their texture, and white in their colour. The honour of introducing the Fennec into Europe is claimed by two persons; the one being Bruce, the celebrated traveller, and the other being a Swedish gentleman of the name of Skioldebrand. The latter writer was certainly the first person who publicly brought the Fennec before the zoologists of Europe, but is supposed to have succeeded in his ambition by means which were hardly just or honourable.

The Fennec is identical with the fox-like animal that is named "Zerda" by Rüppell, and "Cerdo" by Illiger.

HUNTING-DOG.—*Lycaon venaticus.*

Just as the Aard wolf appears to form the link between the civets and the hyænas, being with some difficulty referred to either group of animals, so the Hunting-Dog seems to be the connecting link between the dogs and the hyænas. Its position, however, in the scale of animated nature is so very obscure that it has been placed by some zoologists among the dogs and by others among the hyænas. As, however, the leading characteristic of its formation appears to tend rather towards the canine than the hyænine type, the Hunting-Dog has been provisionally placed at the end of the dogs rather than at the end of the hyænas.

In its general aspect there is much of the hyænine character, and the creature has often been mistaken for a hyæna, and described under that name. There is, however, less of the hyænine type than is seen in the Aard wolf, for the peculiar ridge of hair that decorates the neck of the hyæna is absent in the Hunting-Dog, and the hinder quarters are not marked by that strange sloping form which is so characteristic of the hyæna and the Aard wolf itself. The teeth are almost precisely like those of the dogs, with the exception of a slight difference in the false molars, and therefore are quite distinct from those of the hyænas. But the feet are only furnished with four toes instead of five, which is a characteristic of the hyænas, and not of the dogs. Several other remarkable points of structure are found in this curious animal, some of them tending to give it a position among the dogs, and others appearing to refer it to the hyænas.

The general colour of the Hunting-Dog is a reddish or yellowish brown, marked at wide intervals with large patches of black and white. The nose and muzzle are black, and the central line of the head is marked with a well-defined black stripe, which reaches to the back of the head. The ears are extremely large, and are covered on both their faces with rather short black hairs. From their inside edge rises a large tuft of long white hair, which spreads over and nearly fills the cavity of the ear. The tail is covered with long bushy hair, which is for the greater part of a greyish-white hue, but is strongly tinged with black near its insertion. In nearly all specimens there is a whitish patch below each eye. These tints are somewhat variable in different individuals, but preserve the same general aspect in all.

There are many names by which this animal has been called; in the writings of some authors it is mentioned under the title of the Painted Hyæna, while by others it is termed the Hyæna-Dog. The Dutch colonists of the Cape of Good Hope, where this creature is generally found, speak of it by the name of Wilde Hund, or Wild Dog; and it is also known under the names of Simir and Melbia.

Its title of Hunting-Dog is earned by its habit of pursuing game by fair chase, and uniting in packs of considerable numbers for that purpose. As is the case with the generality of predaceous animals, it prefers the night for its season of attack, but will frequently undertake a chase in broad daylight. For the purpose of the chase it is well fitted, as it is gifted with long and agile limbs and with great endurance of fatigue.

The excellent nose and admirable hunting powers of the Hunting-Dog are really worthy of notice, when the performance of a pack of "Wilde-honden" is compared with those of a pack of foxhounds. How frequently the fox escapes from its pursuers is familiar to all who have paid the very least attention to the subject of field sports; yet we find that a pack of Hunting-Dogs will very seldom permit their prey to escape them, even though they are unassisted by the efforts of human allies.

A successful and practical sportsman, who has witnessed the performances of fox-hounds and Hunting-Dogs, is inclined to give the palm to the latter animals, for their almost invariable success in the chase. He suggests that to the ample nostrils and the wide forehead of the Hunting-Dog must be attributed much of the keen scent and the apt intelligence that renders these animals so successful in their united efforts. He also offers a further suggestion, that it appears as though freedom were a necessary adjunct to the hunting spirit, for we cannot train any animal to hunt with half the real zest which the same creature exhibits in its native or wild state.

This animal is not unfrequently found to prefer the easy task of attacking a sheepfold or a cattle-pen to the more laborious though more legitimate pursuit of prey in the open country. In such a case, it does terrible damage in a single night, and the owner of flocks and herds will sometimes find when he visits his cattle in the morning that many of them have grievously suffered from the inroads of these hungry animals.

The Hunting-Dogs are always very cautious in their approach when they are dealing with oxen, horses, or other powerful animals; but when they choose to make an onslaught upon a flock of sheep they use no precaution, and rush boldly to the hazardless enterprise. They are peculiarly addicted to biting off the tails of oxen, and causing thereby no small present suffering, and very great future inconvenience, for the climate of Southern Africa is so admirably adapted to the production and nourishment of certain predaceous flies, that a poor ox which has been deprived of his tail is in very bad case indeed, and suffers greatly from the gnats and other winged foes which congregate in clouds around any animal which is at all pervious to their attacks. As the Hunting-Dog is not very careful in using his teeth, and always takes as much as possible of his victim's tail, he sometimes makes such rude onslaughts that he inflicts mortal damage on his prey, especially on the colts and the calves.

When brought under human control, it is rather apt to retain its native ferocity, and to reject the companionship of mankind. Yet it has been known to enter into friendship with other animals, such as the hyæna and the lion, and was not more quarrelsome than is ordinarily the case among predaceous animals of different species. The experiment of its domestication has not as yet been fairly tried; and in all probability the creature will yield to the influence of man without any difficulty, whenever it may be subjected to the authority of a competent teacher. The innate treachery of its disposition may be traced to the suspicious wariness which is always found in those predaceous animals which are strong enough to obtain their food by the destruction of animal life, and which are not sufficiently powerful to feel themselves secure from the attacks of other animals which are larger or stronger than themselves.

GROUP OF BRITISH WEASELS.

WEASELS.

NEXT in order to the dogs, is placed the large and important family of the WEASELS, representatives of which are found in almost every portion of the earth. There is something marvellously serpentine in the aspect and structure of the members of this family,—the Mustélidæ, as they are called, from the Latin word *mustéla*, which signifies a Weasel. Their extremely long bodies and very short legs, together with the astonishing perfection of the muscular powers, give them the capability of winding their little bodies into the smallest possible crevices, and of waging successful battle with animals of twenty times their size and strength.

There are many species which are known to be ranged under the banners of this family, all of which are remarkable for their boldness and their ferocity, and many of which have gained much fame from their agility. Some of them, such as the badger and the ratels, are plantigrade in their manner of walking; *i. e.* they place the whole of the foot flat on the ground when they walk. Others, such as the stoats, polecats, and otters, are digitigrade, *i. e.* they only place the tips of the toes on the ground in walking. Nearly all the Weasels are excellent climbers, being able to run up the perpendicular trunks of trees with perfect ease, and to pervade the branches in search of their prey. They can also leap to a considerable distance; a circumstance which is worthy of notice, because their short legs would seem to be very little adapted for such an accomplishment.

In the engraving on page 343, the various British members of the Weasel tribe are depicted.

In the foreground is seen a group of common Weasels, one of them emerging from a cleft in the earth. Just above them a stoat is represented as having killed a hare, and engaged in drinking the blood as it flows from the fatal wound. Another stoat is ascending the tree. On the branches of the tree several martens are crouching. In the river a pair of otters are engaged in the capture of their finny prey, and a badger is standing on the river-bank.

The teeth of the Mustelidæ are eminently predaceous in their character: the canines being long, sharp, and slightly curved backwards, while the molars are studded with points and edges in a manner somewhat similar to those of the cat.

Still, the teeth are not so exclusively carnivorous as those of the cats, as in the molar teeth there is a slight approach to the structure which permits vegetable-feeding animals to masticate their food. Although their outer sides are furnished with the sharp-cutting edges that distinguish the carnivorous from the herbivorous teeth, their inner sides are supplied with more or less rounded ridges, resembling in some degree the corresponding portions in the teeth of the vegetable-feeding animals. This modification of the grinding teeth is most conspicuous in the last molar tooth of the upper jaw, which presents a tolerably wide and smooth surface to the action of the teeth that meet it in the lower jaw, and causes the substances that are subjected to their action to be crushed, as by the molar teeth of vegetable-feeding animals, and not only to be cut or lacerated, as would be the case if the teeth were wholly of the carnivorous nature.

The skull of every member of the Weasel tribe presents a peculiarity by which it may be distinguished, without much difficulty, from that of a feline or viverrine animal. The space between the orbit of the eye to the "occipital foramen," as the large aperture at the base of the skull is termed, is extremely large in proportion to the size of the skull itself, and gives to that portion of the structure a peculiar and characteristic aspect. The hinder portion of a Weasel's skull appears to be so elongated, when compared with the similar portion of a cat or a genett, that there is but little difficulty in distinguishing them.

In absolute size, they are not very formidable; but their nature is so fierce, their habits so sanguinary, and their muscular powers so highly developed, that they are most dangerous neighbours to a farm or a poultry-yard; and their courage is so determined, that when attacked they are no insignificant enemies even to man himself. They are nocturnal in their habits, passing the greater portion of the day in their darkened abodes, where they sleep away the unwelcome hours of light, and sally out in the evening for the purpose of procuring their nightly food. They are not, however, exclusively nocturnal, for it is a very common event to see a stoat or a Weasel engaged in the pursuit of its prey even during the brightest hours of the daylight; but as a general fact, they do not leave their homes until the shades of evening begin to fall.

The feet of the Weasels are furnished with five toes, and are armed with sharp and powerful claws. These claws, or talons, as they may be called, are in most of the species permanently protruded; but in some examples are very slightly retractile. The tongue of these animals is smooth to the touch, and partakes but very slightly of that dry roughness which is so conspicuous a characteristic of the feline tongue.

Injurious as are the generality of the Weasel tribe, and unpleasant neighbours as they may be to the poultry or rabbit fancier, they are of great consequence in the mercantile world, on account of the peculiarly beautiful fur with which their bodies are covered, and which is put to so many uses by mankind. Artists are indebted to the Weasels for the delicate elasticity of their best brushes, and the skins of many of the Weasel tribe are held in the highest esteem for the purpose of being formed into warm and costly clothing. The snowy ermine, that forms the mantles of kings, and lines their crowns, is a product of a very familiar member of the Mustelidæ; and the scarcely less coveted sable is taken from the spoils of another species of Mustelidæ.

PINE MARTEN.—*Martes Abietum.*

FIRST on the list of Weasels are placed the agile and lively MARTENS, or MARTEN-CATS, as they are sometimes termed. Two species of British Martens are generally admitted into our catalogues, although the distinction of the species is even as yet a mooted point. The chief distinction between the Pine and the Beech Martens is the different tint of the throat, which in the former animal is yellow, and in the latter is white. But it is said by many observers that this variation of tint is not of sufficient importance to warrant a separation of the species, and that the different sexes of the same species are marked by varying depth of colour in the throat, the male possessing a darker tinge of yellow than the female. There is also a slight difference of size between the two sexes. Taking, however, the arguments which have been adduced on both sides of the question, the balance of probabilities lies strongly on the side of those who consider the yellow-throated and the white-throated Martens to belong to different species.

The PINE MARTEN is so called because it is generally found in those localities where the pine-trees abound, and is in the habit of climbing the pines in search of prey. It is a shy and wary animal, withdrawing itself as far as possible from the sight of man; and although a fierce and dangerous antagonist when brought to bay, is naturally of a timid disposition, and shuns collision with an enemy.

It is a tree-loving animal, being accustomed to traverse the trunks and branches with wonderful address and activity, and being enabled by its rapid and silent movements to steal unnoticed on many an unfortunate bird, and to seize it in its deadly gripe before the startled victim can address itself to flight. It is a sad robber of nests, rifling them of eggs and young, and not unfrequently adding the parent birds to its list of victims.

Even the active and wary squirrel sometimes yields up its life to this agile and stealthy foe ; for in a hole which had been made the head-quarters of a Marten were found several of the bushy tails which are such familiar decorations of the squirrel's person. That the squirrels had been captured and eaten by a Marten is placed beyond doubt by the fact that the dead body of the destroyer was discovered within the hole, itself having fallen a victim to the venomous bite of a viper. Poetical justice was visible in this instance ; for there had evidently been a combat between the reptile and the Marten, both having succumbed to the deadly weapons of their adversary. It is probable that the snake was an intruder upon the Marten, and that the latter animal had, after receiving the fatal wound, retained sufficient strength to inflict such injuries upon its antagonist as to deprive it of the power of escape, and ultimately to cause its death.

The damage which a pair of Martens and their young will inflict upon a poultry-yard is almost incredible. If they can only gain an entrance into the fowl-house, they will spare but very few of the inhabitants. They will carry off an entire brood of young chickens, eat the eggs, and destroy the parents. Mr. William Thompson, in his admirable work on the Zoology of Ireland, relates an anecdote of the destructiveness of the Martens which exhibits in a very strong light the exceeding ferocity of these little animals.

A farmer, who had possessed twenty-one lambs, found one morning that fourteen of them had been killed by some destructive animal, and that the murderers had not eaten any of the flesh of their victims, but had contented themselves with sucking the blood. On the following night the remaining seven were treated in a similar manner, and the destroyers—a pair of Martens—were seen in the morning taking their departure from the scene of their sanguinary exploits. They were traced to their residences, and were found to have taken up their abode in a deserted magpie's nest in Tollymore Park.

It is found that the Martens of both species are very fond of usurping the nests of rooks, hawks, crows, magpies, and other birds, although they sometimes prefer the habitation of a squirrel, or the hole in a decaying tree. After a Marten has taken up its residence in the open nest of a crow, a hawk, or other similar locality, and is quietly sleeping in the daytime, its whereabouts is often manifested by the noisy crowd of small birds which surround the tree, and join in a unanimous outcry against their slumbering foe. In winter, it prefers the more genial retreats which are afforded by hollow trees, or the clefts of rocks, where it makes a warm bed with dry leaves or grass, and is securely sheltered from the cold atmosphere. But in the summer time, it migrates to the cooler and more airy domicile which is afforded by a deserted nest, and there takes up its abode.

A magpie's nest is a very favourite resort of the Marten, because its arched covering and small entrance afford additional security. A boy who was engaged in bird-nesting, and had climbed to the top of a lofty tree in order to plunder a magpie's nest, was made painfully sensible of an intruder's presence by a severe bite which was inflicted upon his fingers as soon as he inserted his hand into the narrow entrance. This adventure occurred in Belvoir Park, County Down, in Ireland.

The fur of the Pine Marten is rather valuable, especially if the animal be killed in the winter. A really fine skin is but little inferior to the celebrated sable, and can hardly be distinguished from it by inexperienced eyes. An ordinary skin, in good preservation, is worth about two shillings and sixpence, before it is dressed by the furrier, but its value is much enhanced by its quality. It is thought not to be so prolific an animal as the Beech Marten, seldom producing above three or four at a birth, while the latter animal has been known to nurture six or seven young at the same time. If this circumstance be generally true, it goes far towards proving that the Beech and the Pine Marten are really distinct animals. The head of this creature is smaller than that of the Beech Marten, and the legs are proportionately larger.

The length of the Pine Marten is about eighteen inches, exclusive of the tail, which measures about ten inches. The tail is covered with long and rather bushy hair, and is slightly darker than the rest of the body, which is covered with brown hair. The tint, however, is variable in different specimens, and even in the same individual undergoes considerable modifications, according to the time of year and the part of the world in which it is found. It has rather a wide range of locality, being a native of the northern parts of Europe and of a very large portion of Northern America.

THE BEECH MARTEN seems to be of rather more frequent occurrence than the Pine Marten, from which animal it may be distinguished by the white tint of the fur on its throat and the upper portion of its breast. On account of this circumstance, it is sometimes called the White-throated Marten. A slight yellow tinge is sometimes observed on its throat. There are several names by which this animal is known, such as the Marteron, the Martern, and the Stone Marten.

In its destructive habits and its thirst for blood, it resembles the animal which has already been described, and has earned for itself the title of "domestic," which was applied to it by Gesner, because it is in the habit of prowling about human habitations,

BEECH MARTEN.—*Mustes Foina.*

and of concealing itself in the barns and outhouses, for the purpose of gaining access to the poultry.

The Marten seems to be easily tamed to a certain degree, but beyond that point its wild instincts are too firmly rooted for speedy eradication. One of these creatures was procured when young by a shoemaker, and remained with him until it had reached maturity. It then escaped from its adopted home, and commenced a series of depredations among the fowls which were kept by the neighbours, returning every night, and concealing itself in the house. Its destructive energies became so troublesome that it was at last sentenced to death by the united voices of those who had suffered from its depredations, and paid the penalty of its many robberies.

Another Marten was captured in a rather curious manner. It had been driven from its home, and, in order to escape the dogs that were chasing it in hot pursuit, leaped over a precipice, and fell from a height of forty or fifty feet, without meeting with anything to break its fall. It lay on the ground as if dead, and one of the spectators descended the cliff, and captured the Marten. However, it soon gave indications of returning animation, by scratching and biting so fiercely that the captor was glad to put it into a bag. It soon became tame under kind treatment, and was led about a garden by a string which was tied round its waist.

After a while, as it was kept in a stable, it contracted a strong friendship for a horse, and was always to be found sitting upon the horse's back. If a visitor entered, it would growl and run about the horse in a very excited manner, sometimes sitting between his ears, and then running along his neck and lying on his back, and playing all kinds of similar antics. The horse seemed quite pleased with his curious little friend, and permitted the Marten to run over him as much as it pleased.

Unfortunately, this strange friendship was of very short standing, for the poor Marten contrived to get into a trap, and was found in the morning quite dead. When in captivity, it was fed on meat, birds, or even on bread and milk. It always preferred to receive the birds before they were killed, and could not endure any disturbance while engaged in its meal.

The Marten is a good swimmer, as well as an excellent leaper and climber, and has been often seen to swim across a tolerably wide river when it has been hard pressed in the chase.

Both the Pine and the Beech Martens are said to be very lively in a state of domestication, if they are rightly managed and carefully tended. They are among the most graceful of animals; and whether they are running, climbing, leaping, or moving in any way, all their actions are full of quick and easy grace. They are the more fitted for a state of domestication by being free from the very offensive scent which is given forth by many of the Mustelidæ, and which makes several of those otherwise pleasing creatures objects of utter dislike and abhorrence. There is, indeed, a very perceptible odour in these creatures, which is caused by the substance that is secreted in a kind of pouch which is placed near the insertion of the tail, but it is not offensive in its character, and to many persons is even agreeable. On this account, the Pine and Beech Martens are distinguished by the title of Sweet Martens, in contradistinction to the Polecat, which is termed the Foul-Marten, or Foumart, on account of the peculiarly unpleasant odour which is exhaled from its person.

Even in captivity, its agility is so great that, while it is engaged in its graceful antics, its shape can hardly be discerned. It is more watchful at night than in the daytime, but will often awake from its slumbers during the hours of light, and recreate itself with a little exercise. Although it is an essentially carnivorous animal, it will often eat various vegetable substances, when it is deprived of freedom, and very probably does so when it is at large in its native woods; a supposition which is strengthened by the partially herbivorous character of its molar teeth. It is said to be fond of nuts, which it strips of their shells while they are still hanging on the tree, leaving the shattered fragments adherent to the branches. So sharp are its teeth, and so powerful its jaws, that one of these animals has been known to gnaw its way through the wooden door of the room in which it was confined, and to make its escape through the orifice.

The Martens are nearly banished from the more cultivated English counties, but still linger in some numbers among the more rocky and wooded portions of Great Britain. In Carnarvon and Merionethshire they are still tolerably numerous, and are frequently hunted by hounds, as if they were foxes or other lawful game.

One of the most highly valued of the Weasels is the celebrated Sable, which produces the richly tinted fur that is in such great request. Several species of this animal are sought for the sake of their fur. They are very closely allied to the Martens that have already been described, and are supposed by some zoologists to belong to the same species. Besides the well-known Martes Zibellina, a North American species is known, together with another which is an inhabitant of Japan. These two creatures, although they are very similar to each other in general aspect, can be distinguished from each other by the different hue of their legs and feet: the American Sable being tinged with white upon those portions of its person, and the corresponding members of the Japanese Sable being marked with black.

The Sable is spread over a large extent of country, being found in Siberia,

Kamtschatka, and in Asiatic Russia. Its fur is in the greatest perfection during the coldest months of the year, and offers an inducement to the hunter to brave the fearful inclemency of a northern winter in order to obtain a higher price for his small but valuable commodities. A really perfect Sable skin is but seldom obtained, and will command an exceedingly high price. An ordinary skin is considered to be worth from one to six or seven pounds, but if it should be of the very best quality, is valued at twelve or fifteen pounds.

In order to obtain these much-prized skins, the Sable-hunters are forced to undergo the most terrible privations, and often lose their lives in the snow-covered wastes in which the Sable loves to dwell. A sudden and heavy snow-storm will obliterate in a single half-hour every trace by which the hunter had marked out his path, and, if it should be of long continuance, may overwhelm him in the mountain "drifts" which are heaped so strangely by the fierce tempests that sweep over those fearful regions. Should he not be an exceedingly experienced hunter, possessed of a spirit which is undaunted in the midst of dangers, and of a mind which is stored with the multitudinous precepts of hunters' lore, he is certain to sink under the accumulated terrors of his situation, and to

SABLE.—*Martes Zibellina.*

perish by cold and hunger in the midst of the snow-sea that rolls in huge white billows over the face of the country.

At the best, and when he meets with the greatest success, the privations which he is called upon to undergo are of the most fearful character, and he rarely escapes without bearing on his person the marks of the terrible labour which he has performed.

The Sables take up their abode chiefly near the banks of rivers and in the thickest parts of the forests that cover so vast an extent of territory in those uncultivated regions. Their homes are usually made in holes which the creatures burrow in the earth, and are generally made more secure by being dug among the roots of trees. Sometimes, however, they prefer to make their nests in the hollows of trees, and there they rear their young. Some authors, however, deny that the Sable inhabits subterranean burrows, and assert that its nest is always made in a hollow tree. Their nests are soft and warm, being composed chiefly of moss, dried leaves, and grass.

Their food is said to partake partially of a vegetable and partially of an animal character, according to the season of the year. In the summer time, when the hares and other animals are rambling about the plains and forests, the Sable takes advantage of their presence, and kills and eats them. But when the severity of the winter frosts has compelled these creatures to remain within their domiciles, the Sable is said to feed upon the wild berries that it finds on the branches. The hunters assert that the Sable is not content to feed only on the hares and such like animals, which constitute the usual prey of the larger Weasels, but that it is in the habit of killing and devouring the ermine and the smaller members of the Weasel tribe. Even birds fall victims to these agile and voracious animals, being often overtaken in their flight among the branches of trees by a well-aimed leap and a sharp stroke of the fore-paws.

Sometimes the ordinary supplies of food fail, and then the Sable enacts the part of parasite to some larger animal, such as a bear or a wolf, and, following on its track, endeavours to gain a subsistence by feeding on the remnants of the prey which may be taken by the superior powers of its unwitting ally.

The Sables are taken in various modes. Sometimes they are captured in traps, which are formed in order to secure the animal without damaging its fur. Sometimes they are fairly hunted down by means of the tracks which their little feet leave in the white snow, and are traced to their domicile. A net is then placed over the orifice, and by means of a certain pungent smoke which is thrown into the cavity, the inhabitant is forced to rush into the open air, and is captured in the net. The hunters are forced to support themselves on the soft yielding surface of the snow by wearing "snow-shoes," or they would be lost in the deep drifts which are perfectly capable of supporting so light and active an animal as the Sable, but would engulf a human being before he had made a second step.

It now and then happens that the Sable is forced to take refuge in the branches of a tree, and in that case it is made captive by means of a noose which is dexterously flung over its head.

On examining the fur of the Sable, it will be seen to be fixed to the skin in such a manner that it will turn with equal freedom in all directions, and lies smoothly in whatever direction it may be pressed. The fur is rather long in proportion to the size of the animal, and extends down the limbs to the claws. The colour is a rich brown, slightly mottled with white about the head, and taking a grey tinge on the neck.

Even in the localities where it is most usually found it is by no means a common animal, and is so cautious that it is not easily induced to enter a trap or to permit itself to be outwitted by its pursuers. Should the hunter prefer to catch the Sable in traps, he thinks himself fortunate if he secures a Sable in every eight or ten traps. Sometimes, on going the round of his traps, which is a task of great labour and difficulty, and involves a journey of many miles, he finds them all empty, and the baits gone, having been devoured by some crafty animal that has followed the hunter in his course for the sole purpose of robbing the traps of their baits.

When domesticated, the Sable exhibits no small amount of intelligence, and can be tamed with considerable success. One of these animals was an inhabitant of the palace belonging to the Archbishop of Tobolsk, and was so thoroughly domesticated, that it was accustomed to wander at will about the city and pay visits to the inhabitants. Two other specimens of the same creature were tamed, but not to so complete a degree. They used to sleep during the greater part of the day, but were peculiarly vigilant and restless during the night. After partaking of food, they always became exceedingly drowsy, and would sleep so soundly that they might be shaken, pinched, or even pricked with a sharp instrument without giving any signs of awakening. This curious somnolence would generally last from thirty minutes to one hour. They exhibited a great dislike to cats, and if they saw one of those animals, would rise on their hind-legs as if to fight it.

The mode of attack is the same in all the members of the Weasel tribe. They always endeavour, as far as possible, to steal unperceived upon their intended prey, and then to fasten suddenly upon the doomed animal by means of their sharp teeth, and tightly clinging paws. They always endeavour to seize their victim by the back of the head, and with a single bite drive their long canine teeth into the brain. They seem to be singularly and literally bloodthirsty in character, delighting to suck the blood of the animals which they have destroyed.

Unfortunately for the possessors of rabbits, poultry, or similar creatures, the whole of the Weasel tribe are sad epicures in their taste, and will wreak far more

JAPANESE SABLE.—*Martes Melánopus.*

destruction in a farmyard than if they were possessed of greater powers and smaller discrimination.

Oftentimes they are most wantonly destructive, killing great numbers of victims and contenting themselves with drinking the blood, without attempting even to tear the flesh in any other manner. This method of attack is well exemplified in the case of the stoat, and will be alluded to when that animal is described in its place among the Mustelidæ.

The size of the Sable is about equal to that of the marten, being about eighteen inches in length exclusive of the tail. It is not a very prolific animal, as it seldom produces more than five young at a birth, and is generally content with three. March and April are the months in which the young Sables are born, so that they are not likely to suffer from the want of proper nourishment until they have gained sufficient strength to search for food on their own account.

WOOD-SHOCK, OR PEKAN.—*Martes Canadensis.*

The PEKAN, more popularly termed the WOOD-SHOCK, is a native of Canada and other parts of America, and is of some value on account of its fur, which is nearly as useful, although not so valuable, as that of the sable, with which animal it is very closely allied. The colour of its fur is generally of a greyish-brown, the grey tint being found chiefly on its back, head, neck, and shoulders, and the legs, tail, and back of the neck marked with a much darker brown.

Its habitation is usually made in burrows, which it excavates on the banks of rivers, choosing that aqueous locality on account of the nature of its food, which consists of fish and various quadrupeds which live near the water. Hunting the Wood-shock is a diversion which is greatly in vogue, and is especially followed by the younger portions of the community, who find in this water-living, earth-burrowing, sharp-toothed animal, a creature which affords plenty of sport to themselves and their dogs, while it is not a sufficiently powerful antagonist to cause any great danger to its foes, if it should be driven to despair and assume the offensive, instead of yielding in sullen silence.

The POLECAT has earned for itself a most unenviable fame, having been long celebrated as one of the most noxious pests to which the farmyard is liable. Slightly smaller than the marten, and not quite so powerful, it is found to be a more deadly enemy to rabbits, game, and poultry, than any other animal of its size.

It is wonderfully bold when engaged upon its marauding expeditions, and maintains an impertinently audacious air even when it is intercepted in the act of destruction. Not only does it make victims of the smaller poultry, such as ducks and chickens, but attacks geese, turkeys, and other larger birds with perfect readiness. This ferocious little creature has a terrible habit of destroying the life of every animal that may be in the same chamber with itself, and if it should gain admission into a henhouse will kill every one of the inhabitants, although it may not be able to eat the twentieth part of its victims. It seems to be very fond of sucking the blood of the animals which it destroys, and appears to commence its repast by eating the brains. If several victims should come in its way, it will kill them all, suck their blood, and eat the brains, leaving the remainder of the body untouched.

There is a beautifully merciful provision in this apparently cruel habit of the Polecat, by which the creatures that are doomed to fall under its teeth and claws are spared from much suffering. The first bite which a Polecat delivers is generally sufficiently powerful to drive the long canine teeth into the brain, and to cause instantaneous insensibility, if not instantaneous death. Its habit of drawing the blood from the veins is another preservative against suffering, for the wounded animal is thus deprived of life while its senses are deadened by the injury to the brain, and the possibility of a lingering death prohibited. Nearly all the members of the Weasel tribe are remarkable for this development of a sanguinary nature, but in none of them is it more conspicuous than in the Polecat.

This animal is not only famous for its bloodthirsty disposition, but for the horrid odour which exhales from its body, and which seems to be partially under the control of the owner. When the Polecat is wounded or annoyed in any way, this disgusting odour becomes almost unbearable, and has the property of adhering for a long time to any substance with which it may come in contact. This odour is produced by a secretion which is found in a small pouch near the tail. The stoat and the weasel are more than sufficiently tainted with this powerfully rank odour, but in the Polecat it is insufferably fetid. From this circumstance the Polecat is termed the Foul-marten, or Foumart. Sometimes it is called the Fulimart, which is evidently a mere variation of the same word. It is also called the Fitchet, a name which is well known to artists as being the title of the animal from whose fur their best brushes are produced.

The hairs from which the brushes are manufactured are those long, sharp, and glistening hairs which protrude through the soft coating of woolly fur that lies next the skin and serves to preserve the animal from the effects of cold and moisture. The colour of these longer hairs is a rich shining brown, of a very dark hue, and that of the inner fur is a pale yellow. It follows, therefore, that the colour of the fur differs according to the comparative length of the two kinds of hair; that on the back being of a dark brown, because the long brown hairs are more numerously and closely set together; while the fur of the under portions are of a much lighter tint, because the brown hairs are shorter and fewer in number, and permit the soft yellow fur to appear. The outer skin, with its covering of fur, is of some value, and is used for the purpose of clothing as well as of the manufacture of brushes. Indeed, it is sometimes fraudulently employed in imitation of the true sable, and is sold under that title.

The Polecat does not restrict itself to terrestrial game, but also wages war against the inhabitants of rivers and ponds. Frogs, toads, newts, and fish are among the number of the creatures that fall victims to its rapacity. It has been known to take great numbers of frogs and toads, and to lay them up as a living store of food in a corner of its habitation, and to guard against their escape by a bite upon the brain of each victim, which produced a kind of perpetual drowsiness, and prohibited them from any active exertion. Large stores of eels have also been found in the larder of a Polecat,—a remarkable circumstance, when we consider the slippery agility of the eel, and its powers of swimming when immersed in its native element. Even the formidably defended nests of the wild bees are said to yield up their honeyed stores to the fearless attack of this rapacious creature.

As to rabbits, hares, and other small animals, the Polecat seems to catch and devour them almost at will. The hares it can capture either by stealing upon them as they lie asleep in their "forms," or by patiently tracking them through their meanderings, and hunting them down fairly by scent. The rabbits flee in vain for safety into their subterranean strongholds, for the Polecat is quite at home in such localities, and can traverse a burrow with greater agility than the rabbits themselves. Even the rats that are found so plentifully

POLECAT.—*Putorius fœtidus.*

about the water-side are occasionally pursued into their holes, and there captured. Pheasants, partridges, and all kinds of game are a favourite prey with the Polecat, which secures them by a happy admixture of agility and craft. So very destructive are these animals, that a single family is quite sufficient to depreciate the value of a warren or a covert to no small extent.

Although so injurious to the property of the farmer, the Polecat is not without its use. It certainly commits sad havoc among game, and if it can obtain admittance into a poultry-house, is sure to scatter destruction around it. But it is quite as deadly an enemy to the rats as to the poultry, although its rat-killing performances do not attract so much attention as its evil conduct towards game and poultry. In some parts of the world, the Polecat is taken under the protection of the farmers, who have an idea that the animal is penetrated with a sense of hospitality, and will do no damage to the property of the man whose farm-buildings afford it a shelter. It is true that the hen-roosts are frequently depopulated, but this mischance is laid on the shoulders of a Polecat which is the guest of some distant farmer, and which is not bound by any ties of gratitude.

The Polecat is a tolerably prolific animal, producing four or five young at a litter. The locality which the mother selects for the nursery of her future family is generally at the bottom of a burrow, which is scooped in light and dry soil, defended if possible by the roots of trees. In this subterranean abode a warm nest is constructed, composed of various dried leaves and of moss, laid with singular smoothness. The young Polecats

1.   A A

make their appearance towards the end of May or the beginning of June. Sometimes the nest is made in a deserted rabbit-burrow, sometimes in the deep and dark crannies which are often found in rocky localities, and sometimes in the interstices which exist between large and rough stones when they are thrown loosely into a heap. If a stone-heap should be covered with grass or brushwood, the Polecat seems to be better pleased than if the component parts were bleak and bare, and is more likely to take up its residence within its recesses. On one occasion, when a Polecat had entered a rabbit's burrow for the purpose of destroying the inmates, it was followed by a ferret, which had been sent into the burrow by its master for the same object. As the ferret remained in the burrow for a very long time, its master became impatient, and thinking that it might have settled itself to sleep, began to stamp on the ground and to shout into the hole, in order to awaken the slumbering animal.

After a while, to use the language of the narrator, "I heard a faint noise, which resembled the squeak of a ferret. I was all astonishment, and could not account for it ; I listened again, and the noise grew louder, as if proceeding from more than one animal. Presently, I observed the ferret's tail, and soon afterwards saw that it was struggling hard to bring something out of the burrow, which I thought must be a rabbit. The ferret continued to drag its prey to the mouth of the hole, when, to my astonishment, I found it closely interlocked with a male Polecat ; they had fastened like bull-dogs on each other's necks. The Polecat, when it saw me, did not quit its hold, but redoubled its efforts, and dragged the ferret back into the earth, when the squeaking was resumed.

I now feared for the ferret's life, but soon observed it again bringing its opponent to the mouth of the hole, so I drew back, hoping that it might be brought out for me to take a shoot. This never however happened, for the Polecat again rallied, kept its hold fast, and the belligerent parties once more disappeared.

I neither saw nor heard anything of them for some time, and again feared for my little champion's life. But a third time I saw it dragging out its antagonist with renewed vigour. The ferret brought its opponent to the mouth of the hole, when a desperate struggle ensued, and just as I expected to see the Polecat defeated, the ferret, quite exhausted, relinquished the combat and came hopping towards me, considerably mangled about the throat. Its enemy did not dare to follow, but stood deliberately snuffing up the air at the mouth of its burrow. I took aim at the Polecat, and, strange to say, my gun missed fire at least four or five times, when the little hero, turning quickly round, escaped into the earth, thus failing with my auxiliaries, my ferret and my gun.

I attribute the defeat of the ferret to the inevitable loss of vigour which every animal must suffer when confined under the dominion of man, and restrained from those habits of invariable instinct which in their natural state produces in them the great height of perfection."

In no manner is the dominion of mankind over the inferior animals more powerfully asserted than in his power of subjecting them to his influence, and bending their natural instincts to his service. There really appears to be hardly any group of animals, and, indeed, but very few families, which do not furnish their quota to the number of the servants of the human race.

We have already seen that several species of the monkey race are employed in the service of the human inhabitants of their native land. The bats appear to have escaped at present from the service of mankind, although it has been proved that they are possessed of a considerable amount of intellect, and can be tamed without difficulty. Among the cats, the chetah and the caracal are examples of wild animals whose destructive instinct has been employed in the service of mankind. Several species of the civet tribe have been taught to chase and destroy rats, mice, or other domestic pests, while the services which are rendered to mankind by the dog are too well known to need more than a passing reference.

Even those unpromising animals, the weasels, can be subjected to the wondrous super-eminence of the human intellect. The FERRET is well known as the constant companion of the rat-catcher and the rabbit-hunter, being employed for the purpose of following its prey into their deepest recesses, and of driving them from their strongholds into the open air, when the pursuit is taken up by its master. The mode in which the Ferret is employed will be presently related.

Some writers have thought the Ferret to be identical in species with the polecat, and have strengthened this opinion by the well-known fact that a mixed breed between these two animals is often employed by those who study the development and the powers of the Ferret.

However, the most generally received opinion of the present day considers the Ferret to be a distinct species. Mr. Bell, in his work on the British Quadrupeds, remarks that the different geographical range which is inhabited by these creatures is one of the most striking arguments in favour of the distinction of the species. The polecat is found in the northern parts of Europe, bearing the severest cold with impunity, and able to track its prey for many miles over the snow. But the Ferret is originally a native of Africa, and is most sensitive to cold, needing artificial means whereby it can be preserved from the cold air of our English climate, and perishing if it be exposed to the frosts of winter. When the Ferret is kept in a state of domestication, the box or hutch in which it resides must be amply supplied with hay, wool, or other warm substances, or the creature will soon pine away and die.

It sometimes happens that a Ferret escapes from its owners, and making its way into the nearest wood or warren, remains in its new quarters until the end of autumn, living quite at its ease, and killing rabbits and game at its leisure. But when the cold weather draws near, and the frosty nights of autumn begin to herald the frosty days of winter, the Ferret will do its best to return to its captivity and its warmer bed, or, failing in its attempt, will die. That a Ferret should escape is by no means an unlikely circumstance, for the creature is so active of limb and so serpentine of body that it can avail itself of the

FERRET.—*Mustela Furo.*

very smallest opening, and, when once at liberty, can conceal itself with such address that it is very rarely recovered.

Some years ago, an escaped Ferret was discovered in its usurped burrow, and most gallantly captured by a young lad who was at the time among the number of my pupils. He was prowling round a small, thickly-wooded copse, in search of birds' nests, when he saw a sharply-pointed snout protruding from a rabbit-hole in the bank which edged the copse, and a pair of fiery little eyes gleaming like two living gems in the semi-darkness of the burrow. Being a remarkably silent and reticent lad, he told no one of his discovery, but went into the village, and presently returned, bearing a little dead kitten which had just been drowned. He then crept to the foot of the bank which overhung the burrow, and holding the dead kitten by its tail, lowered it into the hole. The Ferret made an immediate spring at the prey which had made so opportune an arrival, and was jerked out of the burrow before it could loosen its hold.

The lad grasped the Ferret across the body, but as he was lying in such a manner that he could only use his left arm, the enraged animal began to bite his hand in the most furious manner. However, the young captor could not be induced to let the Ferret escape, and with great presence of mind whirled the creature round with such rapidity that it was soon rendered almost senseless by giddiness, and gave him an opportunity of grasping it with his right hand. The Ferret could not bite while thus held, and was borne triumphantly home, in spite of the wounds which had been inflicted on the hand. The bite of an enraged Ferret is of a very severe character, and, probably in consequence of the nature of its food, is difficult to heal and extremely painful.

It is a fierce little animal, and is too apt to turn upon its owner, and wound him severely before he suspects that the creature is actuated by any ill intentions. I once witnessed a rather curious example of the uncertainty of the Ferret's temper. A lad who possessed a beautiful white Ferret had partially tamed the creature, and thought that it was quite harmless. The Ferret was accustomed to crawl about his person, and would permit itself to be caressed almost as freely as a cat. But on one unfortunate morning, when its owner was vaunting the performances of his *protégée*—for it was a female —the creature made a quiet but rapid snap at his mouth, and drove its teeth through both his lips, making four cuts as sharply defined as if they had been made with a razor.

Still, the Ferret is really susceptible of kind feeling, and has been often known to be truly tamed. One of these animals was accustomed to accompany its master when he took a walk in the country, and was permitted to range at will. Round its neck a little bell was hung, so as to give indications of its presence, but it was so extremely tame that this precaution was hardly needed. It would follow its master like a dog, and if he ran away would hunt his footsteps, anxiously and eagerly seeking for his presence. This was a Polecat-ferret.

When Ferrets are used for the purpose of hunting rabbits, their mouths are securely muzzled before they are permitted to enter the burrows ; as, if their teeth were at liberty, they would in all probability kill the first rabbit which they met, and remain in the burrow for the purpose of sucking its blood. They are purposely kept without their ordinary meals before they are taken into the field, and are therefore especially anxious to secure their prey. Several modes of muzzling the Ferret are in vogue : some of them being as humane as is consistent with the act of fastening together the jaws of any animal, and others being most shamefully cruel. Not many years ago, it was the general custom to sew up the lips of the poor creature every time that it was used for hunting, and elaborate descriptions of this process are given in the sporting books of the period. Leathern muzzles are made especially for the purpose, and are the best that can be adopted ; but in their absence, the Ferret's mouth can be effectually closed by means of two pieces of string, one of which is placed round the neck and the other under the jaws, and the four ends tied together at the back of the neck.

Almost any Ferret will enter a rabbit-burrow and drive out the inmates, for the rabbits do not even think of resisting their pursuer, and flee before him with all their might. But there are comparatively few Ferrets that will venture to enter a rat-hole, especially after they have suffered once or twice from the sharp teeth of those voracious rodents. If the Ferret is accustomed to chase rabbits, it becomes totally useless for the purposes of the rat-catcher, for it will not venture even to face a well-grown and vicious old rat, and much less will it dare to enter the burrow. After suffering from the bite of a rat, the Ferret is seized with a very great respect for a rat's teeth, and will not willingly place itself within reach of those sharp-edged weapons. As has been graphically said by a practical rat-catcher, to force such a Ferret into a rat-hole is " like cramming a cat into a boot, and as for hunting, it is out of the question."

When a Ferret is possessed of sufficient skill and courage to face its long-tailed foes, and has been perfectly trained to the service, it can achieve wonders in open fight, and is a most valuable animal. As a general fact, a large grey old rat will beat off a Ferret, if it can only back itself into a corner, so as to prevent an attack from behind ; but when the Ferret is well trained to the business, it becomes a most destructive rat-slayer. There is a very graphic narrative in Mr. J. Rodwell's work on rats, which not only shows the wonderful powers of the Ferret, but gives a good description of the modes of attack and defence which are practised by both animals.

"One evening I called upon an acquaintance of mine, and found him just going to decide a wager respecting a large male Ferret of the polecat breed, which was to destroy fifty rats within the hour. It must be borne in mind that this Ferret was trained for the purpose.

The rats were placed in a large square measuring eight or ten feet from corner to corner. The Ferret was put in, and it was astonishing to see the systematic way in which he set about his work. Some of the larger rats were very great cowards, and surrendered with scarcely a struggle; while some of the smaller, or three-parts-grown ones, fought most desperately. One of these drew my particular attention. The Ferret, in making his attacks, was beaten off several times, to his great discomfiture; for the rat bit him most severely. At last the Ferret bustled the fight, and succeeded in getting the rat upon its back, with one of his feet upon the lower part of its belly. In this position they remained for some minutes, with their heads close to each other and their mouths wide open. The Ferret was rather exhausted with his former conflicts, and every move he made the rat bit him. At last he lost his temper, and making one desperate effort, he succeeded in getting the rat within his deadly grasp. He threw himself upon his side, and drawing the rat close to him, he fixed his teeth in its neck.

While thus engaged, a rat was running carelessly about. All at once, when near the Ferret, it threw up its head as if a new idea had struck it: it retreated till it met with another, and it was astonishing to see the instantaneous effect produced in the second. Off they ran together to the corner where the Ferret lay. The fact was, they scented the blood of either the rat or the Ferret, which in both was running in profusion. Without any further ceremony they seized the Ferret fast by the crown of the head, and drew themselves up for a comfortable suck of warm blood. The Ferret, feeling the smart, thought it was his old opponent that was struggling in his grasp, and bit his lifeless victim most furiously. Presently he let go the dead rat and seemed astounded at the audacity of the others. He began to struggle, and they seemed quite offended at being disturbed at their repast. He very soon, however, succeeded in catching hold of one of them, and the other ran away; but only for a few seconds. The Ferret demolished the whole fifty considerably under the hour."

POLECAT-FERRET.

Two kinds of Ferrets are employed for the purpose of hunting game; the one, a creamy-white creature, with bright pink eyes, and the other a much darker and fiercer-looking animal, which is the mixed offspring of the polecat and the Ferret. This is the animal which is called the Polecat-ferret in the above mentioned anecdote.

The same author mentions several curious instances of single combat between rats and Ferrets, in which the latter animals were successfully resisted. On one occasion, when he was walking in the fields, accompanied by the tame Ferret which has already been described, a sharp conflict took place between the Ferret and a female water-rat which was defending her young. Not seeing the first attack, the owner of the Ferret thought that his favourite had wounded its nose against a spike, for it was bleeding profusely, and seemed to be in great distress. Presently, however, the cause of its wounds became apparent, in the person of a large rat, which darted fiercely at him from the cover of a bunch of grass, and with the force of her spring fairly knocked him off his legs.

When the grass-tuft was removed, a litter of young rats was seen, over whom the mother was keeping such undaunted watch. She did not attempt to escape, but ever and anon, as the Ferret drew within a certain distance, she flew at him, and knocked him over, inflicting a fresh bite on every attack, so that the assailant was being worsted. At last, being encumbered with the weight of two little rats, which clung too firmly to their parent, she made a false leap, and was seized in the fatal embrace of the Ferret, who would soon have put an end to the valiant defender of her young had not the owner of the Ferret come to the rescue and disengaged the cruel teeth from their hold. But so furious was the mother rat, that when she was released from her foe she again flew at it, and inflicted several severe bites. Its owner then held the Ferret by its tail, and was carrying it away, when the rat, after making several ineffectual springs, actually leapt upon him, ran up his legs and body, and along his outstretched arm, so as to get at her hated enemy, on whom she inflicted another bite and fell to the ground. A second time she attempted this manœuvre, and when frustrated in her wishes, set up her back and bade defiance to man and beast.

To the honour of the human spectator, he took a great interest in the valiant little animal, and regularly supplied her with food until her offspring were able to shift for themselves.

The practice of muzzling the Ferret when it is wanted for the purpose of hunting seems not to be invariably necessary, for one practical sportsman asserts that, except in the case of very young Ferrets, the best plan is to leave the creature's mouth free, and to feed it moderately before it is permitted to enter the burrows. It sometimes happens that a rabbit is so stricken by fear that it cannot be induced to leave its home, and in that case the Ferret will waste its time in trying to urge the refractory animal to move. But if the mouth of the Ferret should remain free, it will then speedily kill the rabbit, and not being hungry, will leave the dead body in the burrow, and proceed in search of other victims.

In spite of all precautions, it does sometimes happen that the Ferret will not leave the burrow, and in that case it must either be laboriously dug out or relinquished to the certain fate which befalls all Ferrets when they are exposed to the frosty atmosphere of an English winter. To drive a Ferret out of a rabbit-burrow by means of fire and smoke is almost an impossibility, as the animal is capable of withstanding a considerable amount of asphyxiation with impunity, and the burrows are furnished with so many openings to the fresh air that the stifling vapour escapes before it can be sufficiently concentrated to have its proper effect on the delinquent.

It is evident from these few remarks that the nurture, training, and management of the Ferret is a work of some difficulty, and that a really good animal may be spoiled by the ignorance or carelessness of its owner. An excellent Ferret was once so cowed by the ill-result of a defeat in single combat with a rat, that it would never afterwards even face one of these animals. The rat had been caught in a box-trap by one of its hind legs, and the Ferret was put into the trap for the purpose of killing the rat. In a short time, however, when the trap was opened, the Ferret rushed out, bleeding greatly, and completely subdued. The fact was, that on account of the shape of the trap, the Ferret was unable to have recourse to its usual mode of attack, while the rat was placed in precisely the position which was best suited for defence.

I conclude this notice of the Ferret with a short anecdote which has been related by Mr. Jesse, in his "Gleanings in Natural History," and quoted by Mr. Bell in the British Quadrupeds, for the purpose of cautioning the owners of Ferrets from placing too strict a reliance on the tameness of so bloodthirsty an animal.

A poor woman ran into the house of a surgeon, screaming with terror, and bearing in her arms a poor child, which was terribly mangled, and had been attacked by a Ferret. It seemed that the mother had left her infant, a child of some few months of age, in its cradle, while she left her home on some household business. When she returned, she found the child in a fearfully wounded state, its face, neck, and arms being torn, some of the chief blood-vessels opened, and the eyes greatly injured.

After attending to the wounded child, the surgeon accompanied the mother to her house, when he was surprised by the savage conduct of the Ferret, which boldly advanced from a spot where it had hidden itself, and, as if roused to fury by the infant's cries, attempted to renew its sanguinary repast. It was met by a sharp kick, but undauntedly attacked the foot that dealt the blow, and attempted to run up the leg. Not until its back was broken by repeated kicks did the Ferret cease from its furious efforts; and even when struggling in death, it seemed to be powerfully excited by the child's cries. As the poor child had been heard by the neighbours to scream for more than half an hour before the return of its mother, it is probable that the savage creature had been employed for that time upon its sanguinary feast. The extraordinary boldness and ferocity of the Ferret are the more remarkable because it had been cited as a peculiarly shy animal.

On account of its water-loving propensities, the MINK is called by various names that bear relation to water. By some persons it is called the Smaller Otter, or sometimes the Musk Otter, while it is known to others under the title of the Water-Polecat. It also goes by the name of the NUREK VISON.

The Mink is spread over a very large extent of country, being found in the most northern parts of Europe, and also in North America. Its fur is usually brown, with some white about the jaws, but seems to be subject to considerable variations of tinting. Some specimens are of a much paler brown than others; in some individuals the fur is nearly black about the head, while the white patch that is found on the chin is extremely variable in dimensions. The size, too, is rather variable.

It frequents the banks of ponds, rivers, and marshes, seeming to prefer the stillest waters in the autumn, and the rapidly flowing currents in spring. As may be supposed from the nature of its haunts, its food consists almost wholly of fish, frogs, crawfish, aquatic insects, and other creatures that are to be found either in the waters or in their close vicinity. The general shape of its body is not quite the same as that of the

MINK.—*Vison Lutreola.*

marten or ferret; and assumes something of the otter aspect. The teeth, however, are nearer those of the polecat than of the otter; and its tail, although not so fully charged with hair as the corresponding member in the polecat, is devoid of that muscular power and tapering form which is so strongly characteristic of the otter. The feet are well adapted for swimming, on account of a slight webbing between the toes.

The fur of this animal is excellent in quality, and is by many persons valued very highly. By the furriers it passes under the name of "Mœnk," and it is known by two other names, "Tutucuri" and "Nœrs." As it bears a great resemblance to the fur of the sable, it is often fraudulently substituted for that article,—a deception which is the more to be regretted, as the fur of the Mink is a really excellent one, handsome in its appearance, and extremely warm in character. By some authors, the identity of the Mink with the water-polecat has been doubted, but, as it appears, without sufficient reason.

THERE is hardly any animal which, for its size, is so much to be dreaded by the creatures on which it preys as the common WEASEL. Although its diminutive proportions render a single Weasel an insignificant opponent to man or dog, yet it can wage a sharp battle even with such powerful foes, and refuses to yield except at the last necessity.

The proportions of the Weasel are extremely small, the male being rather larger than the opposite sex. In total length, a full-grown male does not much exceed ten inches, of which the tail occupies more than a fifth, while the female is rather more than an inch shorter than her mate. The colour of its fur is a bright reddish-brown on the upper parts of the body, and the under portions are of a pure white, the line of demarcation being tolerably well defined, but not very sharply cut. This contrast of red and white renders it an exceedingly pretty little animal. The tail is of a uniform tint with the body, and is not furnished with the tuft of jetty hairs that forms so conspicuous a decoration of the stoat.

The audacity of this little creature is really remarkable. It seems to hold every being except itself in the most sovereign contempt, and, to all appearance, is as ready to match itself against a man as against a mouse. Indeed, it carries its arrogant little pretensions so far, that, if elephants were inhabitants of this country, the Weasel would be quite willing to dispute the path with them. I remember being entirely baffled by the impertinence of one of these animals, although I was provided with a gun. While I was walking along a path that skirted a corn-field, a stir took place among some dried leaves by the hedge-side, and out ran something small and red along the bottom of the hedge. I instantly fired, but without success, at the moving object, which turned out to be a Weasel. The little creature, instead of running away, or appearing alarmed at the report and the shot, which tore up the ground around it, coolly ran into the middle of the path, and sitting up on its hind legs, with its paws crossed over its nose, leisurely contemplated me for a moment or two, and then quietly retired into the hedge.

It is a terrible foe to many of the smaller rodents, such as rats and mice, and performs a really good service to the farmer by destroying many of these farmyard pests. It follows them wherever they may be, and mercilessly destroys them, whether they have taken up their summer abode in the hedgerows and river-banks, or whether they have retired to winter quarters among the barns and ricks. Many farmers are in the habit of destroying the Weasels, which they look upon as "vermin," but it is now generally thought that, although the Weasel must plead guilty to the crime of destroying a chicken or duckling now and then, it may yet plead its great services in the destruction of mice as a cause of acquittal. The Weasel is specially dreaded by rats and mice, because there is no hole through which either of these animals can pass which will not quite as readily suffer the passage of the Weasel ; and as the Weasel is most determined and pertinacious in pursuit, it seldom happens that rats or mice escape when their little foe has set itself fairly on their track.

Not only does the Weasel pursue its prey through the ramifications of the burrows, but it possesses in a very large degree the faculty of hunting by scent, and is capable of following its prey through all its windings, even though it should not come within sight until the termination of the chase. It will even cross water in the chase of its prey. When it has at last reached its victim, it leaps upon the devoted creature, and endeavours to fix its teeth in the back of the neck, where it retains its deadly hold in spite of every struggle on the part of the wounded animal. If the attack be rightly made, and the animal be a small one, it can drive its teeth into the brain, and cause instantaneous insensibility. The gamekeeper has some reason for his dislike to the Weasel, as it is very fond of eggs and young birds of all kinds, and is too prone to rob the nests of eggs or young. It is said that an egg which has been broken by a Weasel can always be recognised by the peculiar mode which the little creature employs for the purpose. Instead of breaking the egg to pieces, or biting a large hole in the shell, the Weasel contents itself with making quite a small aperture at one end, through which it abstracts the liquid contents.

So determined a poacher is the Weasel that it has been seen to capture even full-grown birds. A Weasel has been seen to leap from the ground into the midst of a covey of partridges, just as they were rising on the wing, and to bring one of them to the earth. When the spectator of this curious occurrence reached the spot, he found the Weasel in the act of devouring the bird, which it had already killed. This adventure took place at Mansfield, at the end of the month of October. The birds were more than two feet from the ground when the attack was made upon them.

Another Weasel was seen to capture and kill a rook in a somewhat similar manner. The rooks had discovered the Weasel in a field, and, after their custom on such occasions, had gathered round it, and commenced mobbing it. Suddenly, just as one of the rooks made a lower stoop than usual, the Weasel leaped at its tormentor and dashed it to the ground. The dissonant cries of the rooks as they scolded the Weasel attracted the attention of a horseman who was passing by, who arrived at the spot just as the bird had been killed. It lay on the ground dead, from a wound in its neck; its murderer having taken shelter in a neighbouring hedge. As soon, however, as the horseman withdrew, the Weasel emerged from its hiding-place, and dragged the dead rook under the shelter of the bushes.

Although the Weasel proved the victor in this instance, it does not always meet with equal success, especially when it matches its mental powers against those of a superior kind. The predilection of this animal for eggs has already been mentioned, and the Weasel will take great pains in order to secure the coveted luxury. A gentleman, who had discovered a furtive nest made by one of his hens in a hedgerow, was witness to a curious scene. Just as the hen had laid an egg, she issued from her nest, cackling triumphantly, as is the manner of hens upon such occasions. A Weasel, which had been observed at a great distance stretching its neck as if watching for its prey, darted towards the spot, but just before it reached the nest it was anticipated by a crow, which seized the egg and bore it off in triumph. Desirous of investigating the matter further, the proprietor of the plundered fowl would not remove her nest, but

WEASEL.—*Mustéla vulgáris.*

took up his station on the succeeding day, in order to see whether crow or Weasel would return to the attack. No sooner had he arrived at his post than he saw the crow already perched on a neighbouring tree, and in a very short time the Weasel made its appearance also. By degrees the two animals drew nearer to the hen's nest, and as soon as her voice gave the signal, they simultaneously started for the spoil. As before, the wings were more than a match for the legs, and the crow again bore off the prize.

The Weasel has been seen to catch and to kill a bunting by creeping quietly towards a thistle on which the bird was perching and then to leap suddenly upon it before it could use its wings. When it seizes an animal that is likely to make its escape, the Weasel flings its body over that of its victim, as if to prevent it from struggling. In single combat with a large and powerful rat, the Weasel has but little hope of success unless it should be able to attack from behind, as the long chisel-edged teeth of the rat are terrible weapons against so small an animal as the Weasel. The modes of attack employed by the two animals are of a different character, the rat making a succession of single bites, while the Weasel is accustomed to fasten its teeth on the head or neck of its opponent, and there to retain its hold until it has drained the blood of its victim. The fore-legs of the Weasel are of very great service in such a contest, for when it has fixed its teeth, it embraces its opponent firmly in its fore-limbs, and rolling over on its side, holds its antagonist in its unyielding grasp, which is never relaxed as long as a spark of life is left.

In these mortal contests, the Weasel has a considerable advantage in its long and powerful neck, which can be twisted with a most snake-like ease, and which gives the possessor a very serpentine aspect on occasions.

Like the polecat, and others of the same group of animals, the Weasel is most destructive in its nature, killing many more animals than it can devour, simply for the mere pleasure of killing. It is curious to notice how the savage mind, whether it belong to man or beast, actually revels in destruction, is maddened to absolute frenzy by the sight of blood, and is urged by a kind of fiery delirium to kill and to pour out the vital fluid. Soldiers in the heat of action have often declared that everything which they saw was charged with a blood-red hue, but that the details of the conflict had entirely passed from their minds. A single Weasel, urged by some such destructive spirit, has been known to make its way into a cage full of freshly-caught song-birds, and to destroy every single bird. The little assassin was discovered lying quite at its ease in a corner of the cage, surrounded with the dead bodies of its victims. The angry bird-catchers sought at once for a stone wherewith to avenge themselves of the destroyer, but before they could procure a weapon, the Weasel glided through one of the little holes through which the birds obtain access to the water, and was speedily concealed in a hedge beyond hope of discovery.

Even such large animals as hares have been said to fall victims to the Weasel. But it must be borne in mind that in many parts of Great Britain the stoat goes by the title of Weasel, and under that name obtains the credit for many of the achievements which ought to have been attributed to the rightful perpetrator. It is said to kill and eat moles, and this idea is strengthened by the fact that Weasels have more than once been captured in mole-traps. These unfortunate animals were evidently snared in the act of traversing the same passages as the mole, but whether their object was the slaughter of the original excavators is not clearly ascertained.

The exceeding audacity of the Weasel has been already mentioned, and for proofs of this disposition the following anecdotes are cited.

Two gentlemen were riding in the vicinity of Cheltenham, one of whom dismounted in order to inspect some cattle in a field, leaving his horse in the charge of his companion. Presently, a Weasel came out of the neighbouring hedge, and fastened on the fetlock of one of the horses, grasping so firmly that it would not loosen its hold until it had been crushed under foot by the owner of the horse. Some little while ago, a party of Weasels were seized with an idea that they must prevent any one from passing near their habitation. A boy, who was obliged in his way homewards to pass very close to the prohibited spot, was actually chased away several times by the "futterits," as he called them, and dared not oppose the fierce little creatures. A carrier happening to come in the direction, accompanied the boy to the spot, and was immediately attacked by the Weasels. A few sharp blows from his whip laid the principal assailants dead at his feet, and the others, seeing the fate of their comrades, left the field to their conquerors.

There are many similar anecdotes extant, which are easily believed by those who have seen the consummate assurance with which a party of Weasels will run from their habitations and inspect a passing traveller. In a certain hedge, near Ashborn, in Derbyshire, was a well-known spot whence Weasels were accustomed to emerge in some numbers, and to run across the path with entire indifference to the passengers.

At all times the Weasels are sufficiently precarious in their temper, and extremely apt to take offence; but when a mother Weasel imagines that her little ones are likely to be endangered by man or beast, she becomes a really dangerous opponent. Even so small an animal is capable of inflicting a very severe bite, and when she is urged by the desperate courage which is implanted in the breast of every mother, is not unlikely to succeed in her object before she is repelled. Moreover, she does not trust to her sole efforts, but summons to her assistance the inhabitants of the same little community, and with their aid will drive away an unarmed man from the neighbourhood of their habitations. Several such instances are on record, in one of which a powerful man was so fatigued with his exertions in keeping off his assailants, that he would soon have sunk under their united attacks had he not been rescued by the timely assistance of a horseman who happened to pass near the spot, and who came to the rescue with his whip. Urged by their bloodthirsty instinct, the Weasels all directed their efforts to the throat, and made their attacks in such rapid succession that their opponent was solely occupied in tearing away the active little creatures and flinging them on the ground, without being permitted the necessary leisure for killing or maiming his pertinacious antagonists.

It seems that the Weasels will unite their forces for the purposes of sport as well as for those of attack, and will hunt down their game in regular form. Not long ago, as a gentleman was walking in the fields, he saw a number of small objects moving in a line, which he at first took for partridges, but which turned out to be Weasels, which were evidently following the track of some animal by its scent. Having his gun with him, he fired, and found that he had shot no less than six Weasels which had brought themselves into a line with the gun.

A most curious example of mingled courage and presence of mind displayed by this animal is related by Mr. Bell in his History of British Quadrupeds.

As a gentleman was riding over his grounds, he saw a kite pounce upon some object and carry it from the ground. In a short time the kite showed symptoms of uneasiness, trying to free itself from some annoying object by means of its talons, and flapping about in a very bewildered manner. In a few minutes the kite fell dead to the earth, and when the spectator of the aerial combat approached, a Weasel ran away from the dead body of the bird, itself being apparently uninjured. On examination of the kite's body, it was found that the Weasel, which had been marked out for the kite's repast, had in its turn become the assailant, and had attacked the unprotected parts which lie beneath the wings. A considerable wound had been made in that spot, and the large blood-vessels torn through.

The same writer relates a curious anecdote of the conduct of a Weasel towards a snake which was placed in the same box. The snake did not attempt to attack the Weasel, nor the Weasel the snake, both animals appearing equally unwilling to become the assailant. After a while, the Weasel bit the snake once or twice near the nose, but not with any degree of violence, and as the two creatures appeared to be indifferent to each other, the snake was removed. That this peaceable demeanour on the part of the Weasel was not owing to any sluggishness on its own part, was made sufficiently evident by the fact that when a mouse was introduced into the same box, the Weasel immediately issued from its corner, and with a single bite laid the mouse dead. The experiment was made for the purpose of ascertaining whether the Weasel would kill and eat a snake, which had been asserted to be the case.

The fondness of the Weasel for frogs has already been recorded. A curious instance of the nonchalant manner in which the Weasel will sometimes gratify this predilection, occurred at a church near Oxford where I for some time officiated. One morning, during service, a Weasel was seen to creep into the chancel through a small door which led into the churchyard, and to walk gently into the middle of the floor. It sat up and reconnoitred the locality for a few moments, and then retired. But in a very short time it returned with a frog in its mouth, carried its prey into the middle of the floor, and there ate it, undisturbed by the presence of the congregation or the sound of many voices.

That the Weasel, when its numbers are not very great, is a valuable ally to the farmer and the poultry-fancier, is now generally acknowledged. But there are instances where it has played the part which is generally attributed to a powerful ally, and has, after successfully extirpating the foes against whom it was summoned, taken possession of the country which it came to save. There was a certain fish-pond which was suddenly invaded by a large body of rats, which bored the banks in every direction, caught and ate the fish, and were so insolent in the confidence of their numbers and strength that they would sit openly at the mouth of their burrows, and boldly challenge any one who approached too near them. The nuisance increased with great rapidity, when it was unexpectedly checked by the advent of a party of Weasels, which in their turn took possession of the burrows, and in a short time had driven away or killed every one of the rats. The fish were thus saved, and their owner felt a very warm gratitude towards the Weasels for their timely interference on his behalf. However, the Weasels, having eaten all the rats, began to extend their operations farther afield, and invaded the neighbouring premises in search of more game. Chickens, eggs, and young rabbits were continually carried off, and the owner of the pond was soon as anxious to rid himself of the Weasels as he had been desirous of destroying the rats. The Weasels, however, were not so easily driven from their usurped burrows. and continued to hold their ground.

The Weasel affords another example of the hasty manner in which so many animals are calumniated. It is said by Buffon to be wholly untameable, sullen, and savage, and to be insensible to every kindness that could be lavished upon it. Yet we find that the true disposition of the Weasel is of a very different character, and that there is hardly any of our British animals which is more keenly susceptible of kindness, or which will more thoroughly repay the kind treatment of a loving hand. A lady who had taken a fancy to a Weasel, and had succeeded in gaining its affections, wrote a most charming account of the habits of the little creature which she had taken under her protection. She writes as follows :—

" If I pour some milk into my hand," says this lady, " it will drink a good deal, but if I do not pay it this compliment it will scarcely take a drop. When satisfied, it generally goes to sleep. My chamber is the place of its residence ; and I have found a method of dispelling its strong smell by perfumes. By day, it sleeps in a quilt, into which it gets by an unsewn place which it has discovered on the edge ; during the night, it is kept in a wired box or cage, which it always enters with reluctance, and leaves with pleasure. If it be set at liberty before my time of rising, after a thousand little playful tricks, it gets into my bed, and goes to sleep in my hand or on my bosom. If I am up first, it spends a full half-hour in caressing me ; playing with my fingers like a little dog, jumping on my head and on my neck, and running round on my arms and body with a lightness and elegance which I have never found in any other animal. If I present my hands at the distance of three feet, it jumps into them without ever missing. It exhibits great address and cunning to compass its ends, and seems to disobey certain prohibitions merely through caprice.

During all its actions it seems solicitous to divert and to be noticed ; looking at every jump and at every turn to see whether it be observed or not. If no notice be taken of its gambols, it ceases them immediately, and betakes itself to sleep, and even when awakened from the soundest sleep it instantly resumes its gaiety, and frolics about in as sprightly a manner as before. It never shows any ill-humour, unless when confined, or teased too much ; in which case it expresses its displeasure by a sort of murmur, very different from that which it utters when pleased.

In the midst of twenty people this little animal distinguishes my voice, seeks me out, and springs over everybody to come at me. His play with me is the most lively and caressing imaginable. With his two little paws he pats me on the chin, with an air and manner expressive of delight. This, and a thousand other preferences, show that his attachment to me is real. When he sees me dressed for going out, he will not leave me, and it is not without some trouble that I can disengage myself from him ; he then hides himself behind a cabinet near the door, and jumps upon me as I pass, with so much celerity that I often can scarcely perceive him.

He seems to resemble a squirrel in vivacity, agility, voice, and his manner of murmuring. During the summer he squeaks and runs about the house all the night long ; but since the commencement of the cold weather I have not observed this. Sometimes, when the sun shines while he is playing on the bed, he turns and tumbles about and murmurs for a while.

From his delight in drinking milk out of my hand, into which I pour a very little at a time, and his custom of sipping the little drops and edges of the fluid, it seems probable that he drinks dew in the same manner. He seldom drinks water, and then only for want of milk, and with great caution, seeming only to refresh his tongue once or twice, and even to be afraid of that fluid. During the hot weather it rained a good deal ; I presented to him some rain-water in a dish, and endeavoured to make him go into it, but could not succeed. I then wetted a piece of linen cloth in it, and put it near him, and he rolled upon it with extreme delight.

One singularity in this charming animal is his curiosity. It is impossible to open a drawer or a box, or even to look at a paper, but he will examine it also. If he get into any place where I am afraid of permitting him to stay, I take a paper or a book, and look attentively at it, on which he immediately runs upon my hand, and surveys with an inquisitive air whatever I happen to hold. I must further observe, that he plays with a young cat and dog, both of considerable size, getting about their necks, backs, and paws, without their doing him the slightest injury."

This amusing little creature was fed chiefly with small pieces of fresh meat, which it preferred to receive from the hand of its mistress.

This is not a solitary instance of a Weasel being effectually tamed, for M. Giely has recorded his success in taming a Weasel, which he had trained so perfectly that it would follow him wherever he went. Indeed, it seems but reasonable to suppose, that as the ferret has been rendered subservient to man, and has been domesticated to a considerable extent, the Weasel might be equally susceptible of the same influence, and be employed for the same purposes.

Indeed, it is very unlikely that a totally untameable animal should exist, for, as far as has yet been known, the very creatures which gave the most unpromising indications of ferocity or obstinacy have been the most remarkable for their docility under the treatment of certain individuals. We should not be overpassing the bounds of credibility were we to assert, that no creature in which is the breath of life is capable of withstanding the potent influence which is given to mankind for that very purpose, always provided that it be used with gentleness, firmness, and much patient love.

The number of young which the Weasel generally produces at each birth is four or five, and there are said to be usually two or even three litters in each year. The nest is generally placed in the warm cover which is afforded by a hollow tree, in the crevices that exist in rocky ground, or in burrows which are made in dry sandy soil. The nest is composed of dry moss and leaves.

The fur of the Weasel is sometimes powerfully influenced by the effects of the severe cold, and has been known to become nearly white during a sharp and protracted frost. It is worthy of notice that, in such cases, the tip of the tail does not partake of the general change of tint, but retains its bright red hue, precisely as the tail of the ermine retains its jetty blackness while the remainder of the fur is either white or cream-coloured. Mr. Bell remarks that he has seen a Weasel which had retained its wintry whiteness in two spots on each side of the nose, although the remainder of the fur had returned to its usual reddish hue during the summer months. This specimen was captured in the extreme north of Scotland. While clad in the white garments of winter, in which state it is frequently found in Siberia and Northern Europe, it is the animal which was called Mustela nivalis, or snowy Weasel, by Linnæus. Even in England it is rather variable in tint, independently of the influence of climate ; some individuals being less brightly tinged with red than others, while occasional specimens are found in which the fur is of an exceedingly dark brown.

To persons who have had but little experience in the habits of wild animals, it is generally a matter of some surprise that the celebrated Ermine fur, which is in such general favour, should be produced by one of those very animals which we are popularly accustomed to rank among "vermin," and to exterminate in every possible way. Yet so it is. The highly-prized ERMINE and the much-detested STOAT are, in fact, one and the same animal, the difference in the colour of their coats being solely caused by the larger or smaller proportion of heat to which they have been subjected.

In the summer time, the fur of the Stoat—by which name the animal will be designated, whether it be wearing its winter or summer dress,—is not unlike that of the weasel, although the dark parts of the fur are not so ruddy, nor the light portions of so pure a white, as in that animal. The toes and the edges of the ears are also white.

The change of colour which takes place during the colder months of the year is now ascertained, with tolerable accuracy, to be caused by an actual whitening of the fur, and not by the gradual substitution of white for dark hairs, as was for some time supposed to be the case.

The hairs are not entirely white, even in their most completely blanched state, but partake of a very delicate cream-yellow, especially upon the under portions, while the slightly bushy tip of the tail remains in its original black tinting, and presents a singular contrast to the remainder of the fur. In these comparatively temperate latitudes, the Stoat is never sufficiently blanched to render its fur of any commercial value, and the hair appears to be longer, thicker, and whiter in proportion to the degree of latitude in which the animal has been taken. As may be supposed, from the extreme delicacy of the skin in its wintry whiteness, the capture of the Stoat for the purpose of obtaining its fur is a matter of no small difficulty. The traps which are used for the purpose of destroying the Stoat are formed so as to kill the animal by a sudden blow, without wounding the skin ; and many of the beautiful little creatures are taken in ordinary snares.

The object of the whitened fur of the Stoat is popularly supposed to be for the purpose of enabling the animal to elude its enemies by its similarity to the snow-covered ground on which it walks, or to permit it to creep unseen upon its prey. It seems, however, that many animals partake of the same tinting, some of which, such as the polar bear, are so powerful, that they need no such defence against enemies, and so active in the pursuit of the animals on which they feed, that their success in obtaining food seems to depend but little upon colour. The arctic fox, which has already been mentioned on page 336, and the lemming, which will be recorded in a future page, are examples of this curious mutation of colour.

Putting aside for the present the mode in which the fur changes its colour, the real object of the change appears to be for the purpose of defending the wearer against the intense colds which reign in those northern regions, and which, by a beautiful provision, are obliged to work the very change of colour which is the best defence against their powers. It is well known that black substances radiate heat more effectually than objects which are bright and polished. This fact is popularly shown in the bright teapots with which we are so familiar, and which are known, by practical experience, to retain the heat for a much longer period than if their surface had been roughened or blackened.

The reader will not fail to remark a certain coincidence between the snowy hairs that deck the frosty brows of old age with a reverend crown and the white fur that adds such beauty to the frost-beset Stoat. It may be that the energies of the animal are forced, by the necessity which exists for resisting the extremely low temperature of those icy regions, to concentrate themselves upon the vital organs, and are unable to spare a sufficiency of blood to form the colouring matter that tinges the hair. There is evidently an analogy between the chilly feeling that always accompanies old age and the frosty climate that causes the Stoat's fur to whiten.

It is well known that examples of albinos occur in almost every kind of quadruped and bird, and it seems probable that the deprivation of colour is in very many cases owing to the weak constitution of the individual. One of these albinos was a bird, which was caught and tamed, and although it was of a cream colour when it was captured, yet assumed the usual dark plumage of the species at the first moulting season that occurred after its capture. As the bird also appeared to be much more healthy and lively than when it was clad in white feathers, it seems likely that the albino state may have been caused by weakness of constitution.

It is clear that, whatever may be the immediate cause of the whitening of the hair, the change of tint is caused by the loss of the colouring matter which tinges the hair, and that there must be some connexion between the frost-whitened Stoat, the age-whitened human hair, and the abnormal whiteness of various albinos. I would also mention, in connexion with this subject, the curious instances where the hair of human beings has been suddenly blanched by powerful emotion. This fact has been disputed by several physiologists, but is now acknowledged to be true. Besides the various well-attested examples which are on record, I am enabled to give my own personal testimony to the truth of this singular phenomenon, as I have frequently seen a person whose hair was changed in a single night from dark to grey by sudden grief and terror, and the whole system fatally deranged at the same time.

In this country, where the lowest temperature is considerably above that of the ordinary wintry degrees, the Stoat is very uncertain in its change of fur, and seems to yield to or to resist the effects of the cold weather according to the individuality of the particular animal.

In the autumn, when the Stoat is beginning to assume its wintry dress, and in the spring, when it is beginning to lose the snowy mantle of the wintry months, the fur is generally found to be marked with irregular patches of dark and white spots, the sides of the face appearing to be especially variable in this respect. Sometimes the animal resists the coldest winters, and retains its dark fur throughout the severest weather, and it sometimes happens that a Stoat will change its fur even though the winter should be particularly mild. Mr. Thompson records, in his work on the Natural History of Ireland, that he saw a Stoat which was captured on the 27th of January, 1846, which was

STOAT, OR ERMINE (Summer Coat).—*Mustela Erminea.*

wholly white, with the exception of a brown patch on each side of its face. Yet the winter had been remarkably mild, without any frost or snow, although there had been abundance of rain and storms. Two white Stoats were killed in Ayrshire, in 1839, which were almost entirely white, though the frosts had been extremely mild, and the snow had altogether been absent.

As, in the former of these examples, the weather is said to have been extremely wet, t may be presumed that the moisture of the atmosphere and ground may have some connexion with the whitening of the hair. On account of the better radiating powers of dark substances, the dew and general moisture is always found to be deposited in greater quantity on dark or dull, than on white or polished substances. Any one may easily prove this fact, by watching the effects of the dew on a white and a red rose growing in close proximity to each other.

The Stoat is considerably larger than the weasel, measuring rather more than fourteen inches in total length, of which the tail occupies rather more than four inches. There is, however, considerable difference in the size of various individuals.

It is a most determined hunter, pursuing its game with such pertinacious skill that it very seldom permits its intended prey to escape.

Although tolerably swift of foot, it is entirely unable to cope with the great speed of the hare, an animal which frequently falls a victim to the Stoat. Yet it is enabled, by its great delicacy of scent and the singular endurance of its frame, to run down any hare on whose track it may have set itself, in spite of the long legs and wonderful speed of its prey. When pursued by a Stoat, the hare does not seem to put forward its strength as it does when it is followed by dogs, but as soon as it discovers the nature of its pursuer, seems to lose all energy, and hops lazily along as if its faculties were benumbed by some powerful agency. This strange lassitude, in whatever manner it may be produced, is of great service to the Stoat, in enabling it to secure an animal which might in a very few minutes place itself beyond the reach of danger, by running in a straight line.

In this curious phenomenon, there are one or two points worthy of notice.

Although the Stoat is physically less powerful than the hare, it yet is endowed with, and is conscious of, a moral superiority, which will at length attain its aim. The hare, on the other hand, is sensible of its weakness, and its instincts of conservation are much weaker than the destructive instinct of its pursuer. It must be conscious of its inferiority, or it would not run, but boldly face its enemy, for the hare is a fierce and determined fighter when it is matched against animals that are possessed of twenty times the muscular powers of the Stoat. But as soon as it has caught a glimpse of the fiery eyes of its persecutor, its faculties fail, and its senses become oppressed with that strange lethargy which is felt by many creatures when they meet the fixed gaze of the serpent's eye. A gentleman who once met with a dangerous adventure with a cobra, told me that the creature moved its head gently from side to side in front of his face, and that a strange and soothing influence began to creep over his senses, depriving him of the power of motion, but at the same time removing all sense of fear. So the hare seems to be influenced by a similar feeling, and to be enticed as it were to its fate, the senses of fear and pain benumbed, and the mere animal faculties surviving to be destroyed by the single bite.

I have no doubt but that this phenomenon is nearly connected with the curious benumbing of the nerves, and the deprivation of fear which is recorded by Livingstone in his well-known account of his adventure with a lion, which is mentioned on page 149 of this work. The preservative faculties of the hare are excited by the loud noisy dogs that make so violent an attack upon the hare, and which consequently makes use of all her muscular and intellectual powers to escape from them. But the silent, soft-footed, gliding Stoat steals quietly on its victim without alarming it by violent demonstrations, soothes it to its death and kills it daintily.

Be it noticed that there are human types of the Stoat, or rather that the visible animal is but an outward emblem of the inward nature.

If in the course of the chase, the hunted animal should cross a stream, the Stoat will do the same, although, when it is engaged in the pursuit of water-voles, it seldom ventures to follow them into an element where they are more at their ease than their pursuer. Still, although it may not choose to match itself against so accomplished a swimmer and diver as the water-vole, it is no mean proficient in the natatory art.

Mr. Thompson relates a curious instance of the prowess which is displayed by the Stoat in crossing a tolerably wide expanse of water. " A respectable farmer, when crossing in his boat over an arm of the sea, about one mile in breadth, which separates a portion of Islandmagee (a peninsula near Larne, county Antrim,) from the mainland, observed a ripple proceeding from some animal in the water, and on rowing up, found that it was a ' weasel '—Stoats are called weasels in Ireland—which he had no doubt was swimming for Islandmagee, as he had seen it going in a direct line from the shore, and it had reached the distance by a quarter of a mile when taken. The poor animal was cruelly killed, although its gallant swimming might have pleaded in favour of its life."

As to the food of the Stoat, the animal seems to be very easily contented in this respect, killing and eating almost any description of wild quadrupeds, birds and reptiles. Of rabbits it is very fond, and kills great numbers of them, especially when they are young.

A curious scene between a Stoat and rabbit was once witnessed in Epping Forest. A piercing cry was heard among some underwood, from which issued a poor rabbit, bearing with it a Stoat, clinging to its neck. The Stoat, on finding that its actions were observed, quitted its prey and ran up a tree. One of the keepers, who witnessed the scene, had not his gun with him, and sent his companion to fetch it. Just as he arrived, bearing the weapon, the Stoat descended the tree, and running to the rabbit, which had lain as if paralyzed on the ground, tried to drag it away, but was stopped by the contents of the gun, which involved the Stoat and its victim in a common fate.

Birds' nests of all kinds are plundered by this incorrigible poacher, for its quick eye and keen nose enable it to discover a nest, be it ever so carefully hidden; its agile limbs and sharp claws give it the power of climbing any tree-trunk, and of clinging to any branch which will bear the weight of a nest and eggs; while its lithe and serpent-like body enables it to insinuate itself into any crevice that is sufficiently large to afford ingress and egress to the parent birds. The pheasant and partridge are said to be sad sufferers from the Stoat, which is mercilessly slain by the keeper with the aid of traps or gun, the former being the preferable mode of destroying "vermin." The traps in which Stoats are to be caught are most ingeniously placed in certain tempting "runs" to which the Stoat, being a dark-loving animal, is sure to be attracted. For several days the baits are laid on the traps, which are left unset, so that the Stoats find out the locality, and think that they have fallen upon a most hospitable ground. When they have accustomed themselves to eat the baits with impunity, the keeper sets the traps, and immolates the hapless visitants.

When the female Stoat is providing for the wants of a young family, she forages far and wide for her offspring, and lays up the produce of her chase in certain cunningly contrived larders. In a wood belonging to Lord Bagot, a Stoat nursery was discovered, having within it no less than six inhabitants, a mother and her five young. Their larder was supplied

STOAT (Winter Dress).

with five hares and four rabbits, neither of which had been in the least mangled, with the exception of the little wound that had caused their death. In another nest of Stoats were found a number of small animals, such as field-mice, birds, and frogs, all packed away in a very methodical manner. In two nests which were found in Tollymore Park, the Stoats had laid up an abundance of provision. In one of them, there were six or seven mice, besides other small animals, all laid with their heads in the same direction. In the other nest was a more extensive assortment of dead animals. A dozen mice, a young rabbit, and a young hare were laid in the storehouse, together with the feathers and tail of a woodcock, showing that even that wary bird had fallen a victim to the Stoat.

Although the Stoat is so formidable an enemy to rats and mice, and destroys annually such numbers of these destructive animals, it sometimes happens that the predaceous animal finds its intended prey to be more than its match, and is forced ignominiously to yield the contest. One of these animals was seen in chase of a rat, which it was following by scent, and at a great pace. After a while, the Stoat overtook the rat, and would have sprung upon her, had not its purpose been anticipated by a sudden attack from the rat, which turned to bay, and fiercely flung herself with open jaws on her pursuer. The Stoat was so startled at so unexpected a proceeding, that it fairly turned tail and ran away.

1.

B B

The rat now took up the pursuit, and chased the Stoat with such furious energy that she drove her enemy far from the place. It is probable that the rat had a young family at hand, and was urged to this curious display of courage by the force of her maternal feelings.

On account of its agile limbs, sharp teeth, and ferocious disposition, even a single Stoat would be an unchancy opponent for an unarmed man. But if several Stoats should unite to attack a single man, he would find himself in bad case, armed or not. Such a circumstance has been lately communicated to me, my informant having heard it from the lips of the principal actor in the scene.

A gentleman was walking along a road near Cricklade, when he saw two Stoats sitting in the path. He idly picked up a stone, and flung it at the animals, one of which was struck, and was knocked over by the force of the blow. The other Stoat immediately uttered a loud and peculiar cry, which was answered by a number of its companions, who issued from a neighbouring hedge, and sprang upon their assailant, running up his body with surprising rapidity, and striving to reach his neck. As soon as he saw the Stoats coming to the attack, he picked up a handful of stones, thinking that he should be able to repel his little enemies, but they came boldly on, in spite of the stones and of his stick. Most providentially a sharp wind happened to be blowing on that day, and he had wound a thick woollen comforter round his neck, so that he was partially protected.

Finding that he had no chance of beating off the pertinacious animals, he flung his stick down, fixed his hat firmly over his temples, and pressing his hands to his neck, so as to guard that perilous spot as much as possible from the sharp teeth of the Stoats, set off homewards as fast as he could run. By degrees, several of the animals dropped off, but others clung so determinately to their opponent, that when he arrived at his stables, no less than five Stoats were killed by his servants as they hung on his person. His hands, face, and part of his neck were covered with wounds ; but owing to the presence of mind with which he had defended his neck, the large blood-vessels had escaped without injury. The distance from the spot where he had been attacked to his own house was nearly four miles.

He always declared that when he struck the Stoat with the stone, its companion called out "Murder !"

The Stoat is, like the weasel, possessed of a powerful and exceedingly unpleasant odour ; yet even this disagreeable accompaniment does not always suffice to preserve it from being killed and eaten by predaceous animals more powerful than itself. Even so fastidious an animal as the domestic cat has been known to capture a Stoat, to eat part of it herself, and to distribute the remainder to her kittens, who partook of the powerfully scented food without manifesting any reluctance.

Although so wild an animal, it has been tamed with as great success as the weasel and the ferret, displaying the same gentle and active playfulness as has been already mentioned as belonging to the weasel when in a state of domestication. The animal was suffered to roam at will about the house, and never gave any intimation that it wished to make its escape. It was an amusingly playful little creature, delighting to leap upon the members of the family, and run up their backs. But its greatest pleasure seemed to be in attacking a couple of old stuffed magpies that stood upon a shelf. It used to jump upon them, twist its serpentine body round their necks, drag out their feathers between its teeth, and would not unfrequently, in the exuberance of its spirits, knock the bird off the shelf, when magpie and Stoat would come to the ground together.

THE lively little animal which is known by the name of the TAYRA is an inhabitant of tropical America, where it is found in moderate numbers, though not in very great profusion.

The colour of the Tayra is a uniform black, slightly tinged with brown, with the exception of a large white patch which covers the throat and upper portion of the chest. It is said to take up its residence in burrows, which it scoops for itself in the ground. In captivity it is extremely lively and amusing, performing every movement in a sharp, quick manner, and accompanying its actions with an odd little chuckle, something like

TAYRA.—*Gálera bárbara.*

that of a hen calling to her chicks. The eye of the Tayra is small, bright, and brown in colour. It is sometimes known by the title of the Great Weasel, under which name it has been described by Azara, who has, unfortunately, not left any account of its habits in a wild state. Its nature, manners, and customs, are, however, said to resemble those of the following animal.

The size of the Tayra is nearly equal to that of the common Marten.

The GRISON, or HURON, is a native of the Brazils, and is very common about the vicinity of Paraguay.

In its natural disposition it is exceedingly fierce, and is a terrible foe to almost every animal that it chooses to attack. Even in a state of domestication the savage instinct cannot be eradicated, for even when the Grison is rendered sufficiently tame to suffer the touch of the human hand, and to return the caresses of those to whose presence it was accustomed, it has been known to break loose from its confinement, and to slaughter some unfortunate animal that happened to be within its reach.

A Grison that belonged to Mr. Bell contrived to get out of a cage in which it had been placed, and to attack a young alligator that had been brought into the same chamber. The alligators were, as the above-mentioned author quaintly remarks, " stupidly tame, and had, on a certain evening, been laid before a fire in order to rejoice in the welcome heat. In the morning, when their owner entered the room, he found that the Grison had made its escape, and had attacked one of the alligators with such savage fury that it had torn a considerable hole under one of the fore-legs, just where the large nerves and blood-vessels run, and had inflicted so terrible an injury that the poor creature died from the effects of its wound. The other alligator, although unhurt, was in a strongly excited state, snapping angrily at every one who approached it."

Another Grison, that was domesticated by M. F. Cuvier, committed a similarly fatal assault on a rather valuable animal. Although it was always well supplied with food, it became so excited at the presence of a lemur, that it broke the bars of its cage, and inflicted a mortal injury on the poor animal which had so unexpectedly called forth the innate ferocity of its character.

Yet this animal was remarkable for its docility and gentle playfulness, and was always ready for a game with any one who would spend a few minutes in the mock combats in which it delighted. The play of all wild-natured animals is a mock fight, and is often rather prone to become a real battle, if their combative nature be too much excited. The Grison would, when challenged to play, turn on its back, seize the fingers of its human playfellow between its jaws, hold them to its mouth, and press them gently with its teeth. It never bit with sufficient force to cause pain, so that its ferocious onslaught on the lemur would not have been expected from an animal of so gentle a nature. It was possessed of a very retentive memory, and could recognise its friends by the touch of their fingers, without needing to see their owners.

It is a peculiarly impertinent creature in its demeanour, and has a curious habit of rearing its long neck, and bearing its head in a very snake-like fashion. When it assumes this attitude, its bright little black eyes have a curiously pert air, as they look out from under the white, hood-like, hairy covering with which the head is furnished. All its movements are brisk and cheerful, and while running about its cage it continually utters a faint, grasshopper-like chirp.

The colour of the Grison is very peculiar, and is remarkable as being of lighter colour on the back than on the under portions of the body. This divergence from the usual rule is very uncommon, and is only seen in one or two animals. The muzzle, the under part of the neck, the abdomen, and legs, are of a dullish black colour; while the entire upper surface of the body, from the space between the eyes to the tail, is covered with a pale

GRISON.—*Grisonia vittata.*

grey fur, each hair being diversified with black and white. The tinting of this lighter fur is rather variable; in some individuals it is nearly white, while in others it has a decided tinge of yellow.

The ears of this species are very small, and the tongue is rough. The hairs which give the distinctive colouring to the upper parts of the body are longer than those which cover the remaining portions of the body and the limbs. In total length it measures about two feet, the tail being rather more than six inches in length.

The odour which proceeds from the scent-glands of the Grison is peculiarly disgusting, and offends human nostrils even more than that of the stoat and polecat.

In the clumsy-looking animal which is called the RATEL, a beautiful adaptation of nature is manifested. Covered from the tip of the nose to the insertion of the claws with thick, coarse, and rough fur, and provided, moreover, with a skin that lies very loosely on the body, the Ratel is marvellously adapted to the peculiar life which it leads.

Although the Ratel is in all probability indebted for its food to various sources, the diet which it best loves is composed of the combs and young of the honey-bee. So celebrated is the animal for its predilection for this sweet dainty, that it has earned for itself the title of Honey Ratel, or Honey Weasel. The reason for its extremely thick coating of fur is now evident. The animal is necessarily exposed to the attacks of the infuriated bees when it lays siege to their fastnesses, and if it were not defended by a coating which is impenetrable to their stings, it would soon fall a victim to the poisoned weapons of its myriad foes.

In every way, the Ratel is well adapted to the circumstances in which it is placed. Not being a swift animal, it cannot escape from foes by its speed ; but if it can gain but a few minutes' respite, it can sink itself into the ground by the vigorous action of its powerful paws, and thus can avoid the attacks of almost any antagonist. Should it be overtaken before it can reach its accustomed home, or dig a new one, it throws itself on its back, and uses its teeth and claws with such force that it will beat off any ordinary antagonist. The extreme looseness of its skin renders it a very formidable combatant, for when it is seized by any part of its body, it can turn round, as it were, in its skin, and fix its teeth most unexpectedly in the body of its foe.

Partly for this reason, and partly from the singular endurance of its nature, the

RATEL.—*Mellivora Ratel.*

Ratel is most tenacious of life, and will be comparatively unhurt by attacks that would suffice to kill many an animal of ten times its size.

During the daytime, the Ratel remains in its burrow ; but as evening begins to draw near, it emerges from its place of repose, and sets off on its bee-hunting expeditions. As the animal is unable to climb trees, a bee's nest that is made in a hollow tree-limb is safe from its attacks. But the greater number of wild bees make their nests in the deserted mansions of the termite, or the forsaken burrows of various animals. It is said that the Ratel finds its way towards the bees' nests by watching the direction in which the bees return towards their homes.

The movements of the Ratel are not at all graceful, but the animal is lively enough in captivity, and always affords much amusement to the spectator by the grotesque character of its recreation. One of these creatures, which is familiar to every visitor of the Zoological Gardens, and is in possession of a tolerably large house, is in the habit of constantly going through the most extraordinary performances, and thereby attracting the attention of a numerous body of spectators.

In the enclosure that has been allotted to this animal, the Ratel has, by dint of constantly running in the same direction, made for itself an oval path among the straw that is laid upon the ground. It proceeds over the course which it has worked out, in a quick active trot, and every time that it reaches either end of the course, it puts its head on the ground, turns a complete summersault, and resumes its course. At intervals, it walks into its bath, rolls about in the water for a second or two, and then addresses itself with renewed vigour to its curious antics.

The colour of the Ratel is black upon the muzzle, the limbs, and the whole of the under portions of the body ; but upon the upper part of the head, neck, back, ribs, and tail, the animal is furnished with a thick covering of long hairs, which are of an ashy-grey colour. A bright grey stripe, about an inch in width, runs along each side and serves as a line of demarcation between the light and the dark portions of the fur. The ears of the Ratel are extremely short. The lighter fur of the back is variously tinted in different individuals, some being of the whitish-grey which has been already mentioned, and others remarkable for a decided tinge of red. The length of the Cape Ratel is rather more than three feet, inclusive of the tail, which measures eight or nine inches in length. In its walk it is plantigrade, and has so much of the ursine character in its movements that it has been called the Indian or Honey Bear. It is sometimes known under the title of " Bharsiah."

The animal which has just been described is an inhabitant of Southern Africa, being found in great profusion at the Cape of Good Hope. There is, however, an Indian species of Ratel, which very closely resembles the African animal, and in the opinion of some writers is identical with it.

The Indian Ratel is said to be an extremely voracious animal, prowling about the vicinity of human habitations, and not unfrequently paying a visit to the burial-grounds in search of newly interred corpses. It is necessary for the friends of the deceased person to barricade the grave with thorny bushes, in order to defend it from the sharp and powerful claws of the Ratel, which can work their way through the earth with singular rapidity. It is very commonly found along the course of the Ganges and Jumna, especially frequenting the lofty banks for which those rivers are noted.

It is so expert a burrower that it is said to be able to bury itself beneath the surface in ten minutes, even though working in hard and stiff soil ; while digging, it plies its limbs with such exceeding good-will that it flings the loosened soil to a distance of some yards. When taken young, it is easily tamed, and becomes a very amusing animal, diverting the spectators by the singular antics which it plays. But if an adult specimen should be captured, it cannot reconcile itself to the loss of its liberty, and struggles vainly to make its escape, until it dies from the mingled effects of hunger and excitement.

Flesh of all kind is acceptable to the Indian Ratel, and it seems to have a great predilection for rats, mice, and birds in a living state. It is generally drowsy by day, and only rouses itself from its slumbers at the approach of evening. The natives speak of it under the name of " Beejoo."

The WOLVERENE, more popularly known by the name of the GLUTTON, has earned for itself a world-wide reputation for ferocity, and has given occasion to some of the older writers on natural history to indulge in the most unshackled liberty of description.

Voracious it certainly is, having been known to consume thirteen pounds of meat in a single day, and it is probable that if the animal had been living in a wild state it could have eaten even a larger amount of food. It was said by the older naturalists to prey upon deer, which it killed by cunningly dropping on the ground a heap of the moss on which the deer feeds, and then climbing upon a branch which overhung the spot. As soon as the deer passed beneath the tree, the Glutton was said to leap upon its shoulders, and to cling there until it had brought the deer to the ground. This and similar tales, however, rest on no good foundation.

It is known that the Glutton feeds largely on the smaller quadrupeds, and that it is a most determined foe to the beaver in the summer months. During the winter it has little chance of catching a beaver, for the animals are quietly ensconced in their home, and their houses are rendered so strong by the intense cold that the Glutton is unable to break through their ice-hardened walls.

The Wolverene is an inhabitant of Northern America, Siberia, and of a great part of

WOLVERENE.—*Gulo Luscus.*

Northern Europe. It was once thought that the Glutton and the Wolverene were distinct animals, but it is now ascertained that they both belong to the same species.

The general aspect of this animal is not unlike that of a young bear, and probably on that account it was placed by Linnæus among the bears under the title of Ursus Luscus. The general colour of the Wolverene is a brownish-black; the muzzle is black as far as the eyebrows, and the space between the eyes of a browner hue. In some specimens, a few white spots are scattered upon the under jaw. The sides of the body are washed with a tint of a warmer hue. The paws are quite black, and the contrast between the jetty fur of the feet and the almost ivory whiteness of the claws is extremely curious. These white claws are much esteemed among the natives for the purpose of being manufacted into certain feminine adornments.

The paws are very large in proportion to the size of the animal, and it is supposed that this modification of structure is intended to enable the Wolverene to pass in safety over the surface of the snow. Indeed, the feet are so large, that the marks which they leave on the snow are often mistaken for the footprints of a bear. As the tracks of the Wolverene are often mixed with those of the bear, it is evident that the latter animal must often fall a prey to the former during the winter months. When the animal which it kills is too large to form a single meal, the Wolverene is in the habit of carrying away the remains, and of concealing them in some secure hiding-place, in readiness for a second repast.

The eyes of the Wolverene are small, and of a dark brown, and are not remarkable for their brilliancy,

A fine specimen of this animal is at present in the Zoological Gardens, where its form and habits may be well studied. Except when it opens its mouth, and displays the double row of glittering teeth, it does not give the spectator the idea of being a particularly savage or voracious animal, but has rather a good-humoured aspect. Although not very quick in its movements, it is rather restless, and is seldom still except when sleeping. It climbs about the branches of a tree with great ease, and seems to luxuriate in its own curious way among the boughs, rolling itself upon them, and patting the branches with its paws in quite a playful manner. Its perfect command over itself while thus recreating itself appears very curious, because it has but little of the look of a climbing animal.

It can leap from a tolerable height without seeming to take any precaution, or to consider that it had achieved any great feat. When it descends from its tree, it will not long remain on the ground, but climbs about the bars of its cage with great ease and activity, always, however, seeming to ascend with greater readiness than it descends. Sometimes it runs several times in succession round the enclosure, keeping up a kind of canter or short gallop, and ever and anon pausing to see if a piece of cake or other delicacy has been pushed through the bars.

In its native country, the animal is detested by the hunters, whether they belong to Europe or America. For the Wolverene is in the habit of following the sable-hunters on their rounds, and of detaching the baits from the traps, thereby rendering the whole circuit useless. If a sable or marten should happen to be entrapped, the Wolverene does not eat the dead animal, but tears it out of the trap and carries it away. In America, it is specially obnoxious to the hunters, because its fine sense of smell enables it to discover the storehouses of provisions—"caches" as they are technically termed—which the provident hunters lay by in order to fall back upon in case of bad success. If it should unfortunately discover one of these repositories, it sets itself determinately to work, tears away all obstacles, and does extreme damage to the provisions, by eating all the meat, and scattering on every side all the vegetable food.

In captivity, its greatest dainty is said to be the body of a cat, for which strange diet it will leave every other kind of food.

The Wolverene is not a very prolific animal, as it seldom produces more than two at a birth. The maternal residence is generally placed in the crevice of a rock, or in some secluded situation, and the young Wolverenes make their appearance about May.

The SKUNK has obtained the unenviable reputation of being literally in worse odour than any other known animal. All the weasels are notable for a certain odour which emanates from their persons, but the Skunk is pre-eminent in the utter noisomeness of the stench which it exhales when annoyed or alarmed. To the animal itself, the possession of this horrid effluvium is a most valuable means of defence, for there is no enemy that will dare to attack a creature that has the power of overwhelming its foes with so offensive an odour that they are unable to shake off the pollution for many hours.

There seems to be no animal that can withstand the influence of this abominable odour. Dogs are trained to hunt this creature, but until they have learned the right mode of attacking the fetid game, they are liable to be driven off in consternation. Dogs that have learned the proper mode of attacking the Skunk, do so by leaping suddenly upon the creature, and despatching it before it can emit the fetid secretion. The scent proceeds from a liquid secretion which is formed in some glands near the insertion of the tail, and which can be retained or ejected at will. When the Skunk is alarmed, it raises its bushy tail into a perpendicular attitude, turns its back on its enemy, and ejects the nauseous liquid with some force.

Should a single drop of this horrid secretion fall on the dress or the skin, it is hardly possible to relieve the tainted object of its disgusting influence. A dog, whose coat had suffered from a discharge of a Skunk's battery, retained the stench for so long a time that even after a week had elapsed it rendered a table useless by rubbing itself against one of the legs, although its fur had been repeatedly washed. The odour of this substance is so penetrating that it taints everything that may be near the spot on which it has fallen, and renders them quite useless. Provisions rapidly become uneatable, and clothes are so saturated with the vapour that they will retain the smell for several weeks, even though they are repeatedly washed and dried. It is said that if a drop of the odorous fluid should fall upon the eyes, it will deprive them of sight. Several Indians were seen by Mr. Gresham who had lost the use of their eyes from this cause.

On one occasion, a coach full of passengers was passing along the road, when a Skunk ran across the path and tried to push its way through a fence. Not succeeding in so doing, it evidently seemed to think that the coach was the cause of its failure, and ceasing its attempt to escape, deliberately sent a shower of its vile effluence among the passengers. Secure in its means of defence, the Skunk is remarkably quiet and gentle of demeanour, and has more than once enticed an unwary passenger to approach it, and to attempt to seize so playful and attractive an animal.

Mr. Audubon has recorded a curious adventure which befel him in his younger days. In one of his accustomed rambles, he suddenly came upon a curious little animal,

SKUNK. - *Mephitis varians.*

decorated with a parti-coloured coat and bushy tail, and so apparently gentle in demeanour that he was irresistibly impelled to seek a nearer acquaintance. As he approached, the creature did not attempt to run away, but awaited his coming with perfect equanimity. Deceived by its gentle aspect, he eagerly ran towards the tempting prize, and grasped it by its bushy tail, which it had raised perpendicularly as if for the purpose of tempting him to make the assault. He soon repented of his temerity, for he had hardly seized the animal when he was overwhelmed with so horrible a substance, that his eyes, mouth, and nostrils were equally offended, and he was fain to fling away the treacherous foe. After this adventure he became very cautious with respect to pretty little playful animals with white backs and bushy tails.

There is nothing in nature that is wholly evil, and even this terrible fluid is proved to be possessed of medicinal virtues, being sometimes used for the purpose of giving relief to asthmatic patients. There is rather a curious story respecting a clergyman who had been accustomed to use the scent-glands of the Skunk for this purpose, and to keep them in a closely-stopped bottle. It unfortunately happened, one Sunday, that, having been attacked with a fit of asthma, he took his bottle into the pulpit, and when his breathing became troublesome, he opened the bottle, and applied it to his nostrils. Whether he obtained the required relief or not is not recorded, but he was entirely spared the trouble of going on with his sermon, as the congregation made a hasty retreat, and left him nearly alone in the church.

The chief drawback to the medicinal use of this mephitic substance is, that after it has been in use for some time, the whole frame of the patient becomes so saturated with the vile odour that he is not only unpleasant to his neighbours, but almost unbearable to himself. It would be a curious experiment if any one could force one Skunk to cast its ill-smelling secretion upon another, in order to discover whether the scent is as nauseous to the animal that secretes it as it is to all other animals.

There is a curious analogy between the mode of defence which is employed by the Skunk and that which is used by the cuttle-fish, and in both cases it seems to be the result of various emotions, of which fear and combativeness are the chief.

In its fur, the Skunk is extremely variable, but the general markings of its coat are as follows. The fur is of a brown tint, washed with black, and variegated by white streaks along its back. The tail is long and extremely bushy, being covered with long hairs of a creamy-white hue. Its habitation is commonly in burrows, which it scratches in the ground by means of its powerful claws. The creature is about the size of a cat, being about eighteen inches in length from the nose to the root of the tail, which measures fourteen or fifteen inches. The legs are short, and the animal is not endowed with any great activity by nature. It is an American animal, and is found towards the northern parts of that continent.

SCARCELY less remarkable for its ill-odour than the skunk, the TELEDU is not brought so prominently before the public eye as the animal which has just been described.

It is a native of Java, and seems to be confined to those portions of the country that are not less than seven thousand feet above the level of the sea. On certain portions of these elevated spots, the Teledu, or Stinkard, as it is popularly called, can always be found. The earth is lighter on these spots than in the valleys, and is better suited to the habits of the Teledu, which roots in the earth after the manner of hogs, in search of the worms and insects which constitute its chief food. This habit of turning up the soil renders it very obnoxious to the native agriculturists, as it pursues the worms in their subterraneous meanderings, and makes sad havoc among the freshly-planted seeds. It is also in the habit of doing much damage to the sprouting plants by eating off their roots.

We are indebted to Mr. Horsfield for an elaborate and interesting account of the Teledu, an animal which he contrived to tame and to watch with singular success. The following passages are selected from his memoir.

" The Mydaus forms its dwelling at a slight depth beneath the surface, in the black mould, with considerable ingenuity. Having selected a spot defended above by the roots of a large tree, it constructs a cell or chamber of a globular form, having a diameter of several feet, the sides of which it makes perfectly smooth and regular ; this it provides with a subterraneous conduit or avenue, about six feet in length, the external entrance to which it conceals with twigs and dry leaves. During the day it remains concealed, like a badger in its hole ; at night it proceeds in search of its food, which consists of insects and other larvæ, and of worms of every kind. It is particularly fond of the common lumbrici, or earth-worms, which abound in the fertile mould. These animals, agreeably to the information of the natives, live in pairs, and the female produces two or three young at a birth.

The motions of the Mydaus are slow, and it is easily taken by the natives, who by no means fear it. During my abode on the Mountain Prahu, I engaged them to procure me individuals for preparation ; and as they received a desirable reward, they brought them to me daily in greater numbers than I could employ. Whenever the natives surprise them suddenly, they prepare them for food ; the flesh is then scarcely impregnated with the offensive odour, and is described as very delicious. The animals are generally in excellent condition, as their food abounds in fertile mould.

On the Mountain Prahu, the natives, who were most active in supplying me with specimens of the Mydaus, assured me that it could only propel the fluid to the distance of about two feet. The fetid matter itself is of a viscid nature : its effects depend on its great volatility, and they spread through a great extent. The entire neighbourhood of a village is infected by the odour of an irritated Teledu, and in the immediate vicinity of the discharge it is so violent as in some persons to produce syncope. The various species of Mephitis in America differ from the Mydaus in the capacity of projecting the fetid matter to a greater distance.

The Mydaus is not ferocious in its manners, and, taken young, like the badger, it might be easily tamed. An individual which I kept some time in confinement afforded me an opportunity of observing its disposition. It soon became gentle and reconciled to its situation, and did not at any time emit the offensive fluid. I carried it with me from Mountain Prahu to Bladeran, a village on the declivity of that mountain, where the temperature was more moderate. While a drawing was made, the animal was tied to a small stake. It moved about quietly, burrowing the ground with its snout and feet, as if in search of food, without taking notice of the bystanders, or making violent efforts to disengage itself ; on earth worms (lumbrici) being brought, it ate voraciously ; holding one extremity of a worm with its claws, its teeth were employed in tearing the other. Having consumed about ten or twelve, it became drowsy, and making a small groove in the earth, in which it placed its snout, it composed itself deliberately, and was soon sound asleep."

TELEDU.—*Mydaus méliceps.*

The colour of the Teledu is a blackish brown, with the exception of the fur upon the top of the head, a stripe along the back, and the tip of the short tail, which is a yellowish-white. The under surface of the body is of a lighter hue. The fur is long and of a silken texture at the base, and closely set together, so as to afford to the animal the warm covering which is needed in the elevated spots where it dwells. The hair is especially long on the sides of the neck, and curls slightly upwards and backwards, and on the top of the head there is a small transverse crest. The feet are large, and the claws of the fore limbs are nearly twice as long as those of the hinder paws.

In the whole aspect of the Teledu there is a great resemblance to the badger, and, indeed, the animal looks very like a miniature badger, of rather eccentric colours.

THE curious animal whose portrait is presented to the reader is known under several titles, among which the SAND-BEAR is that by which it will be designated in these pages. It is also called the Indian Badger, and sometimes the Balisaur, a name which is corrupted from the Hindostanee word Balloo-soor, signifying Sand-Hog. There is a very great resemblance between this animal and the well-known English badger, from which creature, however, it may easily be distinguished by the greater comparative length of its legs, and the more hog-like snout.

The general colour of the fur of the Sand-Bear is a yellowish-white, diversified by two black bands that run on each side of the head, and unite by the muzzle. The upper of these bands includes the ear and eye in its course, and curves downwards at the shoulder, where it is nearly met by the dark hue of the fore-limbs. The claws are

SAND-BEAR, OR BALISAUR.—*Arctonyx Collaris.*

slightly curved, extremely powerful, and well suited for digging in the ground, as the toes are united for their entire length. The tail is extremely short.

In its wild state the Sand-Bear is said to be fierce in disposition, and sufficiently powerful to beat off a dog that would not hesitate to attack a wolf or a hyæna. When attacked or irritated, the Sand-Bear raises itself on its hind legs, after the manner of the bears, and threatens its antagonist with its fore-limbs, in which it seems fully to trust. Its food is of a mixed character, but appears to be more of a vegetable than an animal nature. It is not a very common animal, and is generally found in the hill country.

ALTHOUGH one of the most quiet and inoffensive of our indigenous animals, the BADGER has been subjected to such cruel persecutions as could not be justified even if the creature were as destructive and noisome as it is harmless or innocuous. For the purposes of so-called "sport," the Badger was captured and kept in a cage ready to be tormented at the cruel will of every ruffian who might choose to risk his dog against the sharp teeth of the captive animal.

Although the Badger is naturally as harmless an animal as can be imagined, it is a terrible antagonist when provoked to use the means of defence with which it is so well provided. Not only are the teeth long and sharp, but the jaws are so formed, that when the animal closes its mouth the jaws "lock" together by a peculiar structure of their junction with the skull, and retain their hold without the need of any special effort on the part of the animal. The subject is by no means a pleasant one, and will not be further noticed.

Unlike the generality of the weasel tribe, the Badger is slow and clumsy in its actions, and rolls along so awkwardly in its gait that it may easily be mistaken for a young pig in the dark of the evening, at which time it first issues from its burrow. The digging capacities of the Badger are very great, the animal being able to sink itself into the ground with marvellous rapidity. For this power the Badger is indebted to the long curved claws with which the fore-feet are armed, and to the great development of the muscles that work the fore-limbs.

BADGER.—*Meles Taxus.*

When the Badger is employed in digging a burrow, it makes use of its nose in order to push aside the earth, which is then scraped away by the fore-paws and flung as far back as possible. In a very short time, the accumulation of earth becomes so considerable that it impedes the animal's movements, and if permitted to remain would soon choke up the tunnel which the miner is so industriously excavating. The hinder paws are now brought into play, and the earth is flung farther back by their action. As the excavation proceeds, the accumulated earth becomes so inconvenient that the Badger is forced to remove it entirely out of the burrow, by retrograding from its position and pushing the loose earth away in its progress. Having thus cleared the tunnel from the impediment, the Badger proceeds to fling the earth as far away as possible, and until it has done so will not resume its labours.

In this burrow the female Badger makes her nest and rears her young, which are generally three or four in number. The nest is made of well-dried grass, and stored with provisions in the shape of grass-balls, which are firmly rolled together, and laid up in a kind of supplementary chamber that acts the part of a larder. There are also several ingeniously contrived sinks, wherein are deposited the remnants of the food and other offensive substances.

The food of the Badger is of a mixed character, being partially vegetable and partly animal. Snails and worms are greedily devoured by this creature, and the wild bees, wasps, and other fossorial hymenoptera find a most destructive foe in the Badger, which scrapes away the protecting earth and devours honey, cells, and grubs together, without being deterred from its meal by the stings of the angry bees. The skin of the Badger is so tough, and lies so loosely on the body, that even if a bee or a wasp could find a bare spot wherein to plant its sting, the Badger would in all probability care little for the wound; and as the covering of hair is so dense that no bee-sting can force its way through the furry mantle, the Badger is able to feast at its ease, undisturbed by the attacks of its winged antagonists.

As is the case with the generality of weasels, the Badger is furnished with an apparatus which secretes a substance of an exceedingly offensive odour, to which circumstance is probably owing much of the popular prejudice against the "stinking brock."

The Badger is very susceptible of human influence, and can be effectually tamed with but little trouble. It is generally set down as a stupid animal, but in reality is possessed of considerable powers of reasoning. One of these animals has been known to set at defiance all the traps that were intended for its capture, and to devour the baits without suffering for its temerity. On one occasion, the animal was watched out of its burrow, and a number of traps set round the orifice, so that its capture appeared to be tolerably certain. But when the Badger returned to its domicile, it set at nought all the devices of the enemy, and by dint of jumping over some of the traps and rolling over others, gained its home in safety.

The colours of the Badger are grey, black, and white, which are rather curiously distributed. The head is white, with the exception of a rather broad and very definitely marked black line on each side, commencing near the snout and ending at the neck, including the eye and the ear in its course. The body is of a reddish-grey, changing to a white-grey on the ribs and tail. The throat, chest, abdomen, legs and feet are of deep blackish-brown. The average length of the Badger is two feet six inches, and its height at the shoulder eleven inches.

ALTHOUGH by no means a large animal, the OTTER has attained a universal reputation as a terrible and persevering foe to fish. Being possessed of a very discriminating palate, and invariably choosing the finest fish that can be found in the locality, the Otter is the object of the profoundest hate to the proprietors of streams and by all human fishermen. It is so dainty an animal that it will frequently kill several fish, devouring only those portions which best please its palate, and leaving the remainder on the banks to become the prey of rats, birds, or other fish-loving creatures.

When the Otter is engaged in eating the fish which it has captured, it holds the slippery prey between its fore-paws, and, beginning with the back of the neck, eats away the flesh from the neck towards the tail, rejecting the head, tail, and other portions. In well-stocked rivers, the Otter is so extremely fastidious that it will catch and kill four or five good fish in a single day, and eat nothing but -the fine flaky meat which is found on the shoulders. The neighbouring rustics take advantage of this epicurean propensity, and make many a meal upon the fish which have been discarded by the dainty Otter. Sometimes, as in the very dry or the very cold seasons, the Otter is forced to lay aside its fastidious notions, and is glad to find an opportunity of appeasing its hunger with any kind of animal food. Driven by hunger, the Otter has been known to travel overland for five or six miles, and is sometimes so hardly pressed that it will have recourse to vegetable substances in default of its usual animal food. In such trying seasons, the Otter is too apt to turn its attention to the farmyard, and to become very destructive to poultry of all kinds, to young pigs, and lambs. One of these animals was captured in a rabbit-warren, whither it had evidently wandered with the intention of feeding on the rabbits.

For the pursuit of its finny prey the Otter is admirably adapted by nature. The body is lithe and serpentine ; the feet are furnished with a broad web that connects the toes, and is of infinite service in propelling the animal through the water ; the tail is long, broad, and flat, proving a powerful and effectual rudder by which its movements are directed ; and the short, powerful legs are so loosely jointed that the animal can turn them in almost any direction. The hair which covers the body and limbs is of two kinds, the one a close, fine, and soft fur, which lies next the skin and serves to protect the animal from the extremes of heat and cold, and the other composed of long, shining, and coarser hairs, which permit the animal to glide easily through the water. The teeth are sharp and strong, and of great service in preventing the slippery prey from escaping.

The colour of the Otter varies slightly according to the light in which it is viewed, but is generally of a rich brown tint, intermixed with whitish-grey. This colour is lighter along the back and the outside of the legs than on the other parts of the body, which are of a paler greyish hue. Its habitation is made in the bank of the river which it frequents, and is rather inartificial in its character, as the creature is fonder of occupying some natural crevice or deserted excavation than of digging a burrow for itself. The nest of the Otter is composed of dry rushes, flags, or other aquatic plants, and is purposely placed as

OTTER.—*Lutra vulgáris.*

near the water as possible, so that in case of a sudden alarm the mother Otter may plunge into the stream together with her young family, and find a refuge among the vegetation that skirts the river banks. The number of the young is from three to five, and they make their appearance about March or April.

Although at the present day the custom of Otter-hunting has necessarily fallen into disuse, it sometimes occurs that a stray Otter is discovered in some stream, and is in consequence the subject of continual annoyances until it finally falls under the hands of its persecutors. When attacked, the Otter is a fierce and desperate fighter, biting and snapping with the most deadly energy, and never yielding as long as life remains within the body. The bite of an angry Otter is extremely severe; for the creature has a habit of biting most savagely, and then shaking its head violently, as if it were trying to kill a rat. There are few dogs which can conquer an Otter in fair fight, and the combat is generally ended by the spear of one of the hunters. Even when transfixed with the deadly weapon, the Otter gives no sign of yielding, but furiously bites the staff, sullen and silent to the last.

The track which the Otter makes upon the bank is easily distinguishable from that of any other animal, on account of the "seal," or impression, which is made by a certain round ball on the sole of the foot. On account of the powerfully-scented secretion with which the Otter is furnished by nature, it is readily followed by dogs, who are always eager after the sport, although they may not be very willing to engage in single fight with so redoubtable an opponent. An Otter has been known to turn savagely upon a dog that was urged to attack it, to drag it into the water, and to drown it. The best dogs for the purpose are said to be the Otterhounds, which have already been mentioned on page 301. Even human foes are resisted with equal violence. On one occasion, an Otter was hard pressed in the water, and endeavoured to escape into an open drain, when it was prevented from carrying out its purpose by one of the hunters, who grasped it by the tail, and tried to force it into the water. The aggrieved animal twisted itself sharply round, and made so savage a snap at its antagonist's hand that it severed the end of his thumb at a single bite. When the Otter has once fairly fixed its teeth, it cannot be forced to relinquish its grasp without the greatest difficulty ; and even when it is dead its jaws are said to retain their hold with unremitting firmness. When the animal is hunted, it swims and dives with such singular agility that the only mode of effecting its capture is by watching its progress below the surface by means of the train of air-bubbles which mark its course, and by forcing it to dive again before it has recovered its breath. By a repetition of this manœuvre the poor creature is wearied, and at last falls an unwilling prey.

The fur of the Otter is so warm and handsome that it is in some request for commercial purposes. The entire length of the animal is rather under three feet and a half, of which the tail occupies about fourteen or fifteen inches. On the average, it weighs about twenty-three pounds ; but there are examples which have far surpassed that weight. Mr. Bell records an instance of a gigantic Otter that was captured in the river Lea, between Hertford and Ware, which weighed forty pounds.

ALTHOUGH so fierce and savage an animal when attacked, the Otter is singularly susceptible of human influence, and can be taught to catch fish for the service of its masters rather than for the gratification of its own palate. The CHINESE or INDIAN OTTER affords an excellent instance of this capability ; for in every part of India the trained Otters are almost as common as trained dogs in England. It seems odd that the proprietors of streams should not press the Otter into their service instead of destroying it, and should not convert into a faithful friend the animal which at present is considered but as a ruthless enemy.

Even in England, the Otter has frequently been tamed and trained for the purposes of sport. A well-known sporting gentleman, an inhabitant of Carstairs, was possessed of one of these animals, which had been trained with singular success. " When called, the Otter immediately answered to the appropriate name of Neptune. The animal, it appeared, was caught two years ago, being then only a few weeks old. It was actually suckled by a pointer, and, showing early signs of docility, was made over to the gamekeeper. In process of time, the animal increased in aptitude and sagacity, and was soon enabled to undertake the duty of an economical fisherman, frequently procuring a dish of excellent burn trout at such seasons when the angler's art, from adverse winds or foul streams, was in vain.

In the morning after these fishing exploits, which sometimes occupied the greater part of the night, Neptune was always found at his post, and the stranger might be astonished to see him among several brace of pointers and greyhounds. No one understood better how to keep at his own side of the house. In fact, according to the gamekeeper, he was ' the best cur that ever ran.'

Neptune was an amiable creature. He would allow himself to be gently lifted by his tail, but invariably objected to any interference with his snout. As an angler, his reputation is advancing rapidly, and one or two of Mr. M——'s neighbours intend to borrow him for a day or two in the spring, for the purpose of ascertaining the quality and size of the larger trout in the pools on their estates."

Another of these animals was accustomed to go to work in a very systematic manner. It always plunged into the water very quietly, and, keeping close by the bank, took its course up the stream, disturbing the fish by smart blows with its tail. If a fish remained by the bank, the Otter passed by and did not seem to notice it, but if the fish should dart in front of its pursuer, it was instantly seized and brought near the surface of the water—probably in order to lessen the force of its struggles. When the Otter had brought its prey to shore, it always discovered some reluctance in parting with the fish which it had caught, and signified its disapprobation by a plaintive whine.

Mr. Richardson gives a very interesting account of an Otter which he tamed, and which was accustomed to follow him in his walks like a dog, sporting by his side with graceful playfulness, and swimming at perfect liberty in the stream. This animal, however, could never be induced to yield her prey to her master, but when she saw him approaching would quickly swim to the opposite bank of the river, lay down her fish, and eat it in peace. The animal was accustomed to wander at her own will in the house and garden, and would eat all kinds of garden pests, such as snails, worms, and grubs, detaching the snails from their shells with great dexterity. She would also leap upon the chairs as they stood by the windows and catch and eat flies as they fluttered on the window-panes. She struck up a warm friendship with an Angora cat, and on one occasion when her friend was attacked by a dog, she flew at the assailant, seized him by the jaw, and was so excited that her master was obliged to separate the combatants and to send the dog out of the room.

The mode of instruction which is followed in the education of the Otter is sufficiently simple. The creature is by degrees weaned from its usual fish diet, and taught to live almost wholly on bread and milk ; the only fish-like article which it is permitted to see being a leathern caricature of the finny race, with which the young Otter is habituated to play, as a kitten plays with a crumpled paper or a cork, which does temporary duty for a mouse. When the animal has accustomed itself to chase and catch the artificial fish, and to give it into the hand of its master, the teacher extends his instructions by drawing the leathern image smartly into the water by means of a string, and encouraging his pupil to plunge into the stream after the lure and bring it ashore. As soon as the young Otter yields the leathern prey, it is rewarded by some dainty morsel which its teacher is careful to keep at hand, and soon learns to connect the two circumstances together.

CHINESE OTTER.—*Lutra Chinensis.*

Having become proficient in the preliminary instructions, the pupil is further tested by the substitution of a veritable, but a dead fish, in lieu of the manufactured article, and is taught to chase, capture, and yield the fish at the command of its master. A living fish is then affixed to a line in order to be brought by the Otter from the water in which it is permitted to swim ; and lastly, the pupil is taught to pursue and capture living fish, which are thrown into the water before its eyes. The remaining point of instruction is to take the so-far trained animal to the water-side, and induce it to chase and bring to shore the inhabitants of the stream, as they rove free and unconstrained in their native element.

In many parts of the world the Otter is admirably trained for this purpose, and is taught to aid its master, not only by capturing single fish, but by driving whole shoals of fishes into the ready nets.

When in pursuit of its finny prey, the Otter displays a grace and power which cannot be appreciated without ocular investigation. The animal glides through the watery element with such consummate ease and swiftness, and bends its pliant body with such flexible undulations, that the quick and wary fish are worsted in their own art, and fall easy victims to the Otter's superior aquatic powers. So easily does it glide into the water, that no sound is heard, and scarcely a ripple seen to mark the time or place of its entrance ; and when it emerges upon the shore, it withdraws its body from the stream with the same noiseless ease that characterizes its entrance. The Otter is a playful creature, and is very fond of engaging in mock aquatic combats, which display the extraordinary powers of the creature to the very best advantage. When on shore, the Otter can proceed at a considerable pace, and when in haste, employs a curious "loping" gallop as its means of progression.

UPON the northern shores of the Pacific Ocean, and especially in those parts where the Asiatic and American continents approach nearest to each other, an extremely large species of Otter is found, which has the peculiarity of preferring the sea-coast to the fresh-water lakes and rivers for the greater part of the year.

SEA OTTER, OR KALAN.—*Enhydra Lutris.*

The KALAN, or SEA OTTER, is very much larger than its fresh-water relations, being rather more than twice the size of the common Otter, and weighing as much as seventy or eighty pounds. During the colder months of the year, the Kalan dwells by the sea-shores, and can be found upon the icy coasts of the Northern Pacific, where it is extremely active in the capture of marine fish. When the warmer months begin to loosen the icy bonds of winter, the Sea Otter leaves the coasts, and in company with its mate proceeds up the rivers until it reaches the fresh-water lakes of the interior. There it remains until the lessening warmth gives warning for it to make its retreat seawards before the fierce frosts of those northern regions seal up the lakes and deprive it of its means of subsistence.

It is rather a scarce animal, and is not so prolific as many of its relations. The fur of the Kalan is extremely beautiful, shining with a glossy velvet-like sheen, and very warm in character. It is in consequence valued at a very high price. The colour of the fur is rather variable, but its general hue is a rich black, slightly tinged with brown on the upper portions of the body, while the under portions of the body and the limbs are of a lighter hue. In some specimens the head is nearly white, and in one or two instances the white tinge extends as far as the neck. Indeed, the proportions of dark and white fur differ in almost every individual.

All the Otters are long of body and short of limb, but in the Kalan this peculiarity is more apparent than in the ordinary Otters, on account of the curious setting on of the hinder limbs and the comparative shortness of the tail, which is barely more than seven inches long, while the head and body measure three feet in length. The food of the Sea Otter is not restricted to fish, but is composed of various animal productions, such as crustacea and molluscs. Some writers assert that, in default of its more legitimate food, it varies its diet by sea-weeds and other vegetable substances.

DURING the progress of this work, several allusions have been made to the destructive principle, as illustrated in the character of certain animals, and a few suggestions have been offered as to its origin, its manifestation, and its object. The subject is too deep in its purport and too wide in its bearings to be comprehended within the limits of a single article, and it must therefore be resumed from time to time, as its various phases are exemplified by the nature of the various creatures which draw the breath of life.

As in the animals which have already been mentioned the principle of terrestrial destruction has been manifested, so we find a further development of the same idea in the Otter, the destroyer of the waters. In order that we may rightly appreciate the part which the Otter plays in the great and ever-changing drama of Nature, it needs that we should as far as possible place ourselves in the position of the creatures among whom its destructive mission is fulfilled.

A shoal of fish is swimming quietly through the clear stream, thinking of nothing but themselves, their food, and their physical enjoyment of existence. Suddenly, from some unknown sphere, of which they can form no true conception, comes flashing among them a strange and wondrous being, from whose presence they flee in instinctive terror. Flight is in vain from the dread pursuer, which seizes one of their companions in its deadly grasp, and in spite of the resistance of the struggling prey, bears it away into an unknown realm, whose wonders their dim sight cannot penetrate, and whose atmosphere is too etherial for their imperfect frames to breathe and live. Ever and anon the terrible pursuer is mysteriously among them, like the destroying angel among the Egyptians, and, as often as it is seen, snatches away one of their number in its fatal grasp, and vanishes together with its victim into the unseen realms above.

To the fish, the Otter must appear as a supernatural being, for it comes from a world which is above their comprehension, and returns thereto at will, a visible and incarnate Death. All animals, creations, and existences, have some idea of a being that is superior to themselves, and that being, which to their minds conveys the highest idea, is to them the Divinity. So that to the fish, the Otter may stand in the light of deity—a remarkable type of the heathen ideas of the Divine nature.

As various races and individuals of mankind are endowed with greater or smaller capacities, they must form an idea of a deity which is consonant with their own natures, and it therefore follows that the loftiest natures will worship the highest God. Therefore, we find in the history of the Israelitish nation, that the narrow-minded Jews copied the surrounding heathens in paying their fearful worship to the fiery Moloch, the cruel and murderous deity of wrath; while the poets and prophets prostrated their spirits in loving adoration before Jehovah, the great Source of all, from whom, through whom, and by whom all things, beings, and essences came into existence.

At the present day, and even in this country, the same contracted ideas are too evident, for there are many narrow-minded persons who are incapable of receiving a deity that is more loving than themselves, and can only appreciate one that is more powerful. Their form of praise is expressed by fear and trembling, and the amount of their reverence is measured by the amount of punishment which they think he can inflict upon them. So with the savage natives of the Southern seas, who consistently honour the representations of their deity by piteous deprecations of his anger, and lie trembling before him in slavish fear. Servile terror is the form of respect which they pay towards those whom they honour, and which they unscrupulously exact from those by whom they desire to be honoured.

Still, there is a great truth in this power-worship of the savage and undeveloped nature, for it is a step in the improvement of the human race when they learn to acknowledge any being as superior to themselves, even though the ground on which they base that superiority may not be of the most elevated description. For all power, of whatever kind, is in its essence spiritual, however material and even revolting its outer manifestations may appear, and is therefore an attribute of the Supreme, although misunderstood and misapplied.

In reality, the attribute which we call Destruction, ought to be termed Conservation and Progression, for without its beneficent influence all things would be limited in their number and manifestation as soon as they first came into existence, and there would be no improvement in physical, moral, or spiritual natures. In such sad case, it would be possible to find a centre and circumference to creation, whereas it is truly as unlimited as the very being of its Creator.

Suppose, for example, that the huge Saurians of the geological eras had been permitted to retain their place upon the earth, and that the land and water were overrun with megatheria, iguanodons, and other creatures of like nature. Suppose, to take our own island as a limited example, that the land was peopled with the naked and painted savages of its ancient times, unchanged in numbers, in habits, and in customs. It is evident that in either case the country would be unable to retain the higher animals and the loftier humanity of the present day, and that in order to escape absolute stagnation it is a necessity that old things should pass away and that the new should take their place. How limited would not the human race be were it not subject to physical death! But a very few years and the earth would be over-peopled, setting aside the question of bodily nourishment, which requires the destruction of other beings, either animal or vegetable. The same rule holds good with regard to moral as well as physical improvement, for it is necessary that all mental progress should be caused by a continual destruction, a death of erroneous ideas, before the corresponding truths can obtain entrance into the mind.

Apply the same principle to the entire creation, and it will become evident that the destructive attribute is essentially the preserver and the improver. Death, so-called, is the best guardian of the human race, and its preserver from the most terrible selfishness, and the direst immorality. If men were unable to form any conception of a future state, and were forced to continue in the present phase of existence to all eternity, they would naturally turn their endeavours to collecting as much as possible of the things which afford sensual pleasure, and each would lead an individual and selfish life, with no future for which to hope, and no aim to which to aspire.

The popular error respecting the destructive principle is, that it is supposed to be identical with annihilation, than which notion nothing can be more false in itself, or more libellous to the Supreme Creator of all things. Death is to every man a terror, an abasement, or an exaltation, as the case may be; but, in truth, to those who are capable of grasping this most beautiful subject, destruction is shown as transmutation, and death becomes birth. Nothing that is once brought into existence can ever be annihilated, for the simple reason that it is an emanation of the Deity, who is life itself, essential, eternal, and universal. The form is constantly liable to mutation, but the substance always remains.

In every pebble that lies unheeded on the ground are pent sundry gaseous substances, which only await the delivering hand of the analyzer to be liberated and expanded; possessing in their free and etherealized existence, many powers and properties which they were debarred from exercising while imprisoned in their condensed and materialized form. To the ordinary observer, the stone thus transmuted in its form appears to be destroyed, but its apparent death is in reality the beginning of a new life, with extended powers and more ethereal substance. Thus it is that physical death acts upon mankind, and in that light is it regarded by the true and brave spirit, with whom to live is toil, and death is a new birth into life, of which he is conscious even here. Death is to such minds the greatest boon that could be conferred upon them, for just as the destruction or death of the pebble etherealizes and expands the elements of its being, so by the death or destruction of the body, the spirit is liberated from its material prison, and humanity is divinized through death.

GROUP OF BEARS (ISABELLA).

BEARS.

THE BEARS and their allies form a family which is small in point of numbers, but is a very conspicuous one on account of the large size of the greater number of its members, and the curious habits of the entire family of the Ursidæ, as these creatures are learnedly named, from the Latin word *ursus*, which signifies a Bear.

These animals are found in almost every portion of the earth's surface, and are fitted by nature to inhabit the hottest and the coldest parts of the world. India, Borneo, and other burning lands are the homes of sundry members of this family, such as the Bruang and the Aswail, while the snowy regions of Northern Europe and the icebound coasts of the Arctic Ocean are inhabited by the Brown Bear and the Nennook or Polar Bear. The diet of the Ursidæ is of a mixed character, and the creatures appear to be capable of sustaining existence upon a purely animal or purely vegetable diet, or to be carnivorous or vegetarian at will. Indeed, it is found that when Bears are kept in captivity, they may be restricted to vegetable food with the best result, both to themselves and their owners. With a few rare exceptions, the Bears are singularly harmless animals when undisturbed, contenting themselves with fruit, honey, nuts, snails, roots, and other similar articles of diet, and rarely attacking the higher animals, except when driven by necessity.

In their gait the Bears are all plantigrade, and on account of the large surface which is placed on the ground when they walk, they are capable of erecting themselves on their hinder limbs, and of supporting themselves in an erect position with the greatest ease. When attacked in close combat, they have a habit of rearing themselves upon their hinder feet, and of striking terrific blows with their fore-paws, which, if they take effect upon their object, cause the most dreadful injuries.

The paws of the Bears are armed with long and sharp talons, which are not capable of retraction, but which are most efficient weapons of offence when urged by the powerful muscles which give force to the Bear's limbs. Should the adversary contrive to elude the quick and heavy blows of the paw, the Bear endeavours to seize the foe round the body, and by dint of sheer pressure to overcome its enemy. In guarding itself from the blows which are aimed at it by its adversary the Bear is singularly adroit, warding off the fiercest strokes with a dexterity that might be envied by many a pretender to the pugilistic art.

Few antagonists are so formidable to the experienced hunter as the Bear, whether it be the Brown Bear of Northern Europe, the Black or Grizzly Bear of America, the Aswail of India, or the Polar Bear of the Arctic regions; and although there are a few instances where a man has conquered a Bear in fair hand-to-hand combat, there are few animals whom a hunter would not rather oppose than the Bear, provided that he were deprived of fire-arms, and furnished only with a knife or hatchet. On one or two occasions, a foolhardy and ignorant person has ventured to attack and to kill a Bear in single combat, but in such instances the victory has almost always been attributable to some accident which never could have been foreseen, and on which no real hunter would have calculated. In fact, the more experienced the hunter, the less will he venture himself against the beast, which, according to Scandinavian aphorism, "has the strength of ten men and the sense of twelve."

With fearful ingenuity, the Bear, when engaged with a human foe, directs its attacks upon the head of its antagonist, and if one of its powerful strokes should take effect, has been known to strike the entire scalp from off the head at a single blow. Mr. Lloyd, who had the great misfortune to be struck down by a Bear, and the singular good fortune to escape from its fangs, says that when he was lying on the ground at the mercy of the angry beast, the animal, after biting him upon the arms and legs, deliberately settled itself upon his head, and began to scarify it in the most business-like manner, leaving wounds of eight and nine inches in length. The experience of this practised Bear-hunter goes to show that the Bear does not make use of its claws when its opponent has been once struck down, but inflicts the subsequent injuries wholly with its teeth. It does not appear from Mr. Lloyd's account that the senses of a person who is seized by a Bear are blunted in the manner which takes place when a lion or tiger is the assailant.

All the Bears are the more terrible antagonists from their extreme tenacity of life, and the fearful energy which they compress into the last moment of existence when they are suffering from a mortal wound. Unless struck in the heart or brain, the mortally wounded Bear is more to be feared than if it had received no injury whatever, and contrives to wreak more harm in the few minutes that immediately precede its decease

BROWN BEAR.—*Ursus Arctos.*

than it had achieved while still uninjured. Many a hunter has received mortal wounds by incautiously approaching a Bear which lay quiescent in apparent death, but was really only stunned for the moment by the shock of the injury which it had received, and which in a very few minutes would have deprived it of life.

SEVERAL species of Bears are now recognised by systematic naturalists, the principal examples of which will be noticed in the following pages.

The Bear which is most popularly known in this country is the BROWN BEAR; a creature which is found rather plentifully in forests and the mountainous districts of many portions of Europe and Asia. As may be supposed from its title, the colour of its fur is brown, slightly variable in tint in different individuals, and often in the same individual at various ages. In many specimens it is found that the neck is encircled with a white band when the animal is young, but that this curious mark is soon merged into the general brown tint of the fur as the animal increases in years and dimensions. This white neck-band was once supposed to be the mark of a male cub, but it is now ascertained that it belongs equally to the male and female sex. In general it is merged into the brown fur after the second or third year, but in some instances it remains throughout the entire life of the animal, which is on that account termed a " Ring Bear."

The size to which a well-fed and undisturbed Brown Bear will grow is really surprising, for although it loses its growing properties after its twentieth year, it seems permanently to retain the capability of enlargement, and when in a favourable situation will live to a very great age. The weight of an adult Brown Bear in good condition is very great, being sometimes from seven to eight hundred pounds when the creature is remarkably fine, and from five to six hundred pounds in ordinary cases. Mr. Falk remarks, that a Bear which he killed was so enormously heavy, that when slung on a pole it was a weighty burden for ten bearers.

The Brown Bear is not so formidable a foe to cattle and flocks as might be supposed from the strength, courage, and voracity of the animal, as it has been often known to live for years in the near vicinity of farms without making any inroads upon the live stock. Fortunately for the farmers and cattle owners of Northern Europe, the Brown Bear is chiefly indebted for its food to roots and vegetable substances, or the sheds and folds would soon be depopulated. As a general fact, the Bear does not trouble itself to pursue the cattle, and in many cases owes its taste for blood to the absurd conduct of the cattle, which are apt to bellow and charge at the Bear as soon as it makes its appearance. The Bear is then provoked to retaliation, and in so doing, learns a táste for blood which never afterwards deserts it. When a Bear has once taken up the business of cattle-stealing, there is no peace in the neighbourhood until the country is freed from the presence of the marauder. It is said that the Bear is more virulent in the destruction of cattle when the weather is wet and cloudy than when it is dry and clear.

Ants form a favourite article of diet with the Bear, which scrapes their nests out of the earth with its powerful talons, and laps up the ants and their so-called " eggs " with its ready tongue. Bees and their sweet produce are greatly to the taste of the Bear, which is said to make occasional raids upon the bee-hives, and to plunder their contents.

Vegetables of various kinds are favourite articles of diet with the Bear, and in the selection of these dainties the animal evinces considerable taste. According to Mr. Lloyd, "the Bear feeds on roots, and the leaves and small limbs of the aspen, mountain-ash, and other trees : he is also fond of succulent plants, such as angelica, mountain-thistle, &c. To berries he is likewise very partial, and during the autumnal months, when they are ripe, he devours vast quantities of cranberries, blueberries, raspberries, strawberries, cloudberries, and other berries common to the Scandinavian forests. Ripe corn he also eats, and sometimes commits no small havoc amongst it ; for seating himself, as it is said, on his haunches in a field of it, he collects with his outstretched arms nearly a sheaf at a time, the ears of which he then devours."

Even in captivity the Bear retains this fruit-loving propensity. One of these animals, which was being maltreated by a cruel owner, was benevolently purchased by one of my friends, an officer in the Guards, who had no sooner concluded the bargain than he repented of his kindness, for the Bear was so demonstrative in its expressions of gratitude that he began to be rather uneasy, and having no possible locality wherein to lodge his new acquisition, he felt himself in some perplexity as to its lodging. However, he got the Bear into a post-chaise, and having taken the precaution to purchase a great many pottles of strawberries, he urged the post-boy to drive at his best speed, and set himself to propitiate his new acquaintance. The Bear took the strawberries in a very polished manner, and ate them deliberately, rejecting the green calices as fastidiously as if it had been accustomed to good society all its life. However, the fruit vanished so fast, that the unfortunate proprietor became alarmed for his own safety, and was not fairly relieved from his fears until he was deposited at the door of the barracks in which the head-quarters of his regiment were at that time established. The Bear, on seeing so many red-coated strangers, became alarmed in its turn, and fled for protection to the only person with whom it was acquainted.

It so happened that the mess-dinner was just served, and that the proprietor of the Bear had but time to make a hasty toilet, and gain the mess-room. On this occasion the commanding officer was delayed for a few minutes, and while the assembled guests were awaiting his arrival, the Bear walked into the room, having sniffed its way after its master. The unexpected intruder advanced to the table, and, mounting upon the colonel's chair, began to inspect the festive arrangements. Just as the Bear had lifted a dish-cover off the joint at the head of the table—a feat which it performed as dexterously as if it had been accustomed to wait at table all its life—the colonel entered the room, and when he saw the strange intruder who had taken such unceremonious possession of his seat, he demanded, with some irritation, "who brought the animal there?" and was told he was only a friend of H——'s, whom he had forgotten to introduce.

The Bear speedily became a favourite in the regiment, and was promoted to the office of sentinel over the property contained in a baggage-waggon. Unfortunately, the poor animal's sense of justice was so acute that it executed its responsible office with too much zeal. On one occasion, a soldier had gone to the waggon with the intention of robbing it of some of the property contained therein, and quietly inserted his arm under the coverings. His intended depredation was, however, soon checked by the teeth of the watchful Bear, which bit his arm with such severity that the limb was rendered useless for the rest of the man's life. Some little time after this occurrence, a child belonging to the regiment made a similar attempt upon the waggon, and was killed by the Bear in its anxiety to fulfil the trust that had been committed to its charge. As the animal was manifestly an unsafe one, and it was feared that the creature might gain a thirst for blood, it was condemned to be shot, although not without much regret on the part of judge and executioners.

The various military adventures of this Bear are very curious, but would occupy too large a space for the present work.

During the autumn, the Bear becomes extremely fat, in consequence of the ample feasts which it is able to enjoy, and makes its preparations for passing the cold and inhospitable months of winter. About the end of October the Bear has completed its winter house, and ceases feeding for the year. The saccharine-loving instinct of the Bear which leads it to discover a bee's nest, however carefully it may be concealed, and to undergo much toil and trouble for the sake of the sweet banquet, seems to be given to the animal for the purpose of enabling it to lay up within its own body a supply of fat which shall serve the double purpose of sustaining the creature in proper condition during its long fast, and of loading the body with carbon for the purpose of producing the state of lethargy in which the animal passes the winter. It is well known that sugar has the property of producing fat to a very great extent, and as it possesses more of the saccharine property than any other natural substance, the Bear is led by its instinct to search for and to devour this valuable food with untiring assiduity.

Again, the excess of carbon, whether it be diffused in the atmosphere or concentrated in the body, is always productive of sleep, or rather of lethargy, as is seen by the constant drowsiness of human beings when overloaded with this condensed carbon, or when they are placed in a room which is charged with the carbonic acid gas that has been exhaled from the lungs of its inhabitants.

A curious phenomenon now takes place in the animal's digestive organs, which gives it the capability of remaining through the entire winter in a state of lethargy, without food, and yet without losing condition. As the stomach is no longer supplied with nourishment, it soon becomes quite empty, and, together with the intestines, is contracted into a very small space. No food can now pass through the system, for a mechanical obstruction—technically called the "tappen"—blocks up the passage, and remains in its position until the spring. The "tappen" is almost entirely composed of pine-leaves, and the various substances which the Bear scratches out of the ants' nests.

From the end of October to the middle of April the Bear remains in his den, in a dull, lethargic state of existence; and it is a curious fact that if a hybernating Bear be discovered and killed in its den it is quite as fat as if it had been slain before it retired to its resting-place. Experienced hunters say that even at the end of its five months' sleep, the Bear is as fat as at its beginning. Sometimes it is said that the Bear loses the "tappen" too soon, and in that case it immediately loses its sleek condition, and becomes extremely thin. During the winter, the Bear gains a new skin on the balls of the feet, and Mr. Lloyd suggests that the curious habit of sucking the paws, to which Bears are so prone, is in order to facilitate the growth of the new integument. The den in which the Bear passes a long period of its life is mostly found under the sheltering defence of rocks or tree-roots, but is sometimes composed of moss which the Bear gathers into a hillock, and into which it creeps. These moss-houses are not so easily discovered as might be supposed, for the habitation bears a very close resemblance to an ordinary hillock, and when the ground is covered with a uniform carpet of snow, might easily be passed without detection. Bears are nearly as careful of their comfort as cats, and take the greatest pains to prepare a soft and warm bed, in which they lie at ease during their long sleep. The flooring of their winter-house is thickly covered with dried leaves and all kinds of similar substances, the smaller branches of the pine-tree being in great request for this purpose. In the Swedish language this moss-house is known by the name of " Korg."

Heavy and unwieldy as the Bear may seem to be, it is possessed of marvellous activity, and when disturbed in its den rushes out with such astonishing rapidity that it will baffle the aim of any but a cool and experienced hunter. One writer, who witnessed the sudden issuing of a Bear from its den and its escape from its pursuers, compares the animal to those children's toys that are popularly called " skip-jacks," and which execute somersaults by means of a twisted string, a wooden lever, and a little shoemakers' wax.

If captured when young, the Brown Bear is readily tamed, and is capable of mastering many accomplishments. It is a very playful animal, and seems to have a keen sense of the ludicrous, which sometimes causes it to overpass the bounds of good breeding. To its owner it displays a great affection, and can be trained to follow him about like a dog. Two of these animals belonging to Mr. Lloyd, and which he had tamed, were very game-some in their disposition, although, as they increased in size and strength, their frolicsome disposition became rather annoying. They were extremely fond of their master, and would seek him on every occasion. If he fastened the door of his room against his troublesome pets, they would clamber up the side of the house, and gain access by the window. It is said that if domesticated Bears be permitted to remain in a secluded place they will pass the winter in a torpid state.

The affectionate nature of the Brown Bear is not only exercised towards human friends, but towards each other. Two of these animals which were born in England were exceedingly attached to each other. One of the two was sold and removed from its companion, which immediately became uneasy at the protracted absence of its playfellow. So deeply was its affectionate heart wounded by the separation, that it became nearly mad, and at last contrived to make its escape from its place of confinement, evidently with the intention of searching after its lost friend. It was captured and replaced in its cage, but its health became so seriously affected that its owners were obliged to repurchase its companion and restore it to its disconsolate relation.

Savage as is the Bear when attacked, it is naturally of a kind and playful disposition, seldom inflicting injury except when urged by fear or hunger. Mr. Atkinson, in his valuable work on Siberia, relates a curious and interesting anecdote of the gentleness which naturally actuates the Brown Bear.

Two children, of four and six years of age, had wandered away from their home, and were after a little time missed by their parents, who set out in search of their offspring. To their horror and astonishment they found their children engaged in play with a large Bear, which responded to their infantine advances in a most affectionate manner. One of the children was feeding its shaggy playfellow with fruit, while the other had mounted on its back and was seated on its strange steed strong in the fearlessness of childish ignorance. The parents gave a terrified scream on seeing the danger to which their children were exposed, and the Bear, on seeing their approach, quietly turned away from the children and went into the forest.

The same writer records a curious adventure with a Bear, which partakes largely of the ludicrous. A woman had lost her donkey, and after a long and fatiguing search she at last came on the missing animal. Being very much irritated with the truant for his misconduct, she fell to scolding and beating him with the handle of a broom which she happened to be carrying. Her vituperation and castigation were however suddenly checked by the discovery that the animal which she was beating so unceremoniously was not her donkey, but a great Brown Bear. The astonishment of the two seems to have been mutual, for the Bear was evidently as much confused by the unwarranted assault as was the woman by the sight of her antagonist ; so that after looking at each other for a few moments, the Bear turned tail and ran away as fast as his legs would carry him.

It is but seldom that the Bear will make an unprovoked attack on a human being, and when he does so, it is generally because he is rendered desperate by the pangs of hunger. In such a case, the Bear is greatly to be dreaded by the benighted traveller, especially if he happen to be journeying alone and has no companion who may share his watch.

That wild beasts of all kinds are scared away by fire is a well-known fact, but the hungry Bear is of so cunning a nature that it even sets at defiance the flaming circle which would at other times afford a secure protection to the sleeping traveller. It is true that the Bear does not venture to cross the fiery barrier, but it contrives to avoid the difficulty in a most ingenious manner. Going to the nearest stream, it immerses itself into the water so as to saturate its fur with moisture, and then, returning to the spot where the intended prey lies asleep, the animal rolls over the flaming embers, quenching the glowing brands, and then makes its attack upon the sleeper. This curious fact is well known among the natives of Siberia, so that they have good grounds for the respect in which they hold the Bear's intellectual powers.

The Bear is possessed of several valuable accomplishments, being a wonderful climber of trees and rocks, an excellent swimmer, and a good digger.

During the time when it is engaged in feeding, the Bear is constantly in the habit of climbing up all kinds of elevated spots, for the purpose of obtaining food, either vegetable or animal. Leaves of various trees are a favourite article of diet with the animal, as are also the nests of the wild bees and ants. Trusting to its powers of swimming, the Bear does not hesitate to cross considerable rivers in search of food or in order to escape from its enemies, and it is in the habit of taking frequent baths during the hotter months of the year for the sake of cooling its heated frame. Its digging capabilities are brought into use on many occasions, such as the demolition of an ant's nest previous to swallowing the inhabitants, or in scraping for itself a comfortable habitation for the winter.

The number of cubs which the female Bear produces is from one to four, and they are very small during the first few days of their existence. They make their appearance at the end of January or the beginning of February, and it is a curious fact that, although the mother has at the time been deprived of food for nearly three months, and does not take any more food until the spring, she is able to afford ample nourishment to her young without suffering any apparent diminution in her condition. It is said by those who have had personal experience of the habits of the Bear, that the mother takes the greatest care of her offspring during the summer, but that when winter approaches, she does not suffer them to partake of her residence, but prepares winter quarters for them in her immediate neighbourhood. During the winter, another little family is born, and when they issue forth from their home, they are joined by the elder cubs, and the two families pass the next winter in the mother's den.

The SYRIAN BEAR, which is otherwise known by the names of DUBB, or RITCK, is doubly interesting to us, not only on account of its peculiarly gentle character, but from the fact that it is the animal which is so often mentioned in the Scriptural writings under the title of the Bear. The animals which are represented as issuing from the wood and avenging the insults offered to Elisha, and the Bear which David attacked and killed in defence of his flock, belonged to the species which is now known by the name of the Syrian Bear.

Even at the present day, the precise number of species into which the members of the Bear tribe are resolvable, is not very satisfactorily ascertained. It seems evident, however, that the Ritck, Isabella Bear, or Syrian Bear, may fairly be considered as a separate species.

The colour of this animal is rather peculiar, and varies extremely during the different periods of its life. While it is in its earliest years, the colour of its fur is a greyish-brown, but as the animal increases in years, the fur becomes gradually lighter in tint, and when the Bear has attained maturity, is nearly white. The hair is long and slightly curled, and beneath the longer hair is a thick and warm covering of closely-set woolly fur, which seems to defend the animal from the extremes of heat or cold. Along the

SYRIAN BEAR, OR DUBB.—*Ursus Isabellinus.*

shoulders and front of the neck, the hair is so perpendicularly set, and projects so firmly, that it gives the appearance of a mane, somewhat resembling that of the hyæna.

At the present day, the Syrian Bear may be found in the mountainous parts of Palestine, and has been frequently seen upon the higher Lebanon mountains. The summit of the mountain itself is composed of two snow-clad peaks, and it is remarkable that the Bear has only been found on one of these peaks, "Makmel" as it is called, while the other—Gebel Sanin—is apparently free from these animals. The Bear appears to remain upon the upper portions of the mountains during the hours of daylight, but as soon as the evening draws near it descends from its rocky fastness in search of food, and often causes considerable alarm to the traveller.

The food of the Syrian Bear is mostly of a vegetable nature, although the creature is perfectly capable of feeding on animal substances, and frequently does so. In consequence of its vegetarian tastes, it often inflicts considerable damage on the cultivated lands that may happen to lie within the boundary of its range. It is especially fond of a species of chick-pea which is largely cultivated in those regions, and in its endeavours to appease its enormous appetite does incalculable damage to the ripening crops.

To this species belonged an animal which enjoyed a high reputation at Oxford and elsewhere on account of his singularly gentle and amusing manners. The Bear, which was generally known by the name of "Tig," being an abbreviation of the somewhat lengthy name of Tiglath Pileser, was for some time a noted celebrity in Oxford, whither he was brought in his early boyhood. High-spirited and rather tetchy in temper, he was very affectionate to those who treated him with consideration, and was perfectly amenable to proper discipline.

Like my dog Rory, he was accustomed to indue a regulation cap and gown, and under this learned shade to perambulate the college, and partake of the hospitality of its members. He would sometimes repel with some asperity the familiarity with which he was greeted by a strange dog, but was in general so quiet in his demeanour that he caused no alarm among those who knew him, even when indulging in some strange freak of humour.

On one occasion he had been treated to sweetmeats at the house of a village dealer in such commodities, and entertained so affectionate a reminiscence of the spot, that he contrived to escape from bondage, and made at once for the coveted dainties. The owner of the shop took to flight at his entrance, and when his pursuers entered the shop they found Mr. Tig seated upon the counter, helping himself to brown sugar with a liberal paw, and displaying such an appreciation of his good fortune that it was not without much trouble that he was removed from the scene of his repast. He was rather peculiar in his tastes, and had attained to a highly civilized state of epicureanism, for his chief delicacies were not, as might be supposed, the produce of the garden or the field, but the more sophisticated dainties of hot muffins and cold ices. He was a most social animal, and if left alone, even for a short time, would cry and lament in the most pitiful of tones.

This gregarious disposition was so excessively developed that when the poor animal was abruptly deprived of his accustomed intercourse with human friends his health speedily gave way under the horrors of solitude ; he refused to eat, ran continually about his den, in the hope of making his escape and rejoining his collegiate acquaintances, and was one day found lying dead in his cage.

The fur of this Bear is rather valuable on account of its warmth and beauty, and the fat and the gall are also held in much esteem for various purposes, chiefly medicinal.

AMERICA furnishes several species of the Bear tribe, two of which, the Grizzly Bear and the MUSQUAW, or BLACK BEAR, are the most conspicuous.

The Black Bear is found in many parts of Northern America, and was formerly seen in great plenty. But as the fur and the fat are articles of great commercial and social value, the hunters have exercised their craft with such determination that the Black Bears are sensibly diminishing in number. The fur of the Black Bear is not so roughly shaggy as that of the European or the Syrian Bear, but is smooth and glossy in its appearance, so that it presents a very handsome aspect to the eye, while its texture is as thick and warm as that of its rougher-furred relations.

This creature is but little given to animal food, and will restrict itself to a vegetable diet unless pressed by hunger. It is, however, very fond of the little snails which come up to feed on the sweet prairie-grass as soon as it is sufficiently moistened by showers or dew to suit the locomotive capabilities of those wet-loving molluscs, and is extremely fond of honey, in search of which dainty it displays great acuteness and perseverance.

Few trees afford so unstable a footing, that the Black Bear will not surmount them in order to reach a nest of wild bees, and there are few obstacles which his ready claws and teeth will not remove in order to enable him to reach the subjacent dainty. Even if the honey and comb be deeply concealed in the hollow of a tree, and the entrance by which the bees find ingress and egress to and from their habitation be too small for the insertion of a paw, the Bear will set steadily to work with his teeth, and deliberately gnaw his way through the solid wood until he has made a breach sufficiently wide to answer his purpose. When once he has succeeded in bringing the combs to light, he scrapes them together with his fore-paws, and devours comb, honey, and young, without troubling himself about the stings of the surviving bees.

The hunters, who are equally fond of honey, find that if it is eaten in too great plenty it produces very unpleasant symptoms, which may be counteracted by mixing it with the oil which they extract from the fat of the Bear. This custom of eating mingled oil and honey affords a partial explanation of the prophecy, " Butter and honey shall he eat," which was necessarily put forth in language which was in accordance with the popular ideas of the period.

The flesh of the Bear is held in high esteem among the colonists and native hunters, and when properly prepared is considered a great delicacy by the denizens of civilized localities. The hams, when cured after the approved recipe, are greatly esteemed by epicures. The Brown Bear of Europe is also famed for the excellent quality of the meat which it furnishes.

The fat of the Bear is, as is well known, considered as an infallible specific for increasing the growth of the hair and promoting its gloss, and is therefore a valuable article of commerce. The only portion of the fat that is legitimately employed for this purpose is the hard white fat which is found in the interior of the body. As might be expected from the enormous amount of titular "Bear's-grease" which is annually consumed, even in England, but a very small proportion of the substance which is called by that name has ever formed part of a Bear's person. The pig steps in to make good the deficiency, and the greater portion of the material which is sold under the name of Bear's-grease, is in reality nothing more or less than hog's-lard, coloured and scented in order to charm the eye and nostrils of the purchaser. There is yet another use to which the fat of the Bear is put, which will be presently mentioned.

The chase of this Bear is an extremely dangerous one, and there are but very few Bear-hunters, however dexterous they may be, who do not in the end succumb to the claws and teeth of one of these powerful animals. Although it is naturally a very quiet and retiring creature, keeping itself aloof from mankind, and never venturing near his haunts except when incited by the pangs of fierce hunger, it is a truly furious beast when hemmed in by its antagonists, and all hope of escape cut off. Seated erect, with its eyeballs darting fury, its ears laid closely upon its head, its tongue lolling out of its mouth, and every gesture glowing with fierce energy, it presents a sight that is sufficient to unnerve any but an experienced hunter, who has learned by long practice to preserve a cool demeanour under the most exciting circumstances. Horses are almost useless at such a juncture, for unless they have been most carefully trained to the task, they are seized with such mental terror at the sight and scent of the infuriate animal that they give way to their frantic fears, and become wholly unmanageable by their rider. As the Bear stands, or rather sits at bay, it deals such terrible and rapid blows with its ready paws that it strikes down the attacking dogs as if they were so many rabbits, and ever and anon makes a furious charge at its enemies. Nothing but a rifle-ball seems to check the creature when it is wrought up to this pitch of fury, for even the severest wounds from a knife, seem, unless they reach the heart, to have only the effect of exciting the animal to more furious rage.

The Musquaw has a curious habit of treading frequently in the same path, so that after a little time it makes out for itself certain roads, which are easily detected by the practised eye of the hunter, and often lead to the destruction of the animal which trod them.

During the month of June the Bears are very thin, and their flesh is considered to be of no value whatever; so that they enjoy a short period of unmolested ease. As they are especially fierce at this time of the year, the hunters have a double reason for keeping aloof from the animals which they persecute with a deadly pertinacity throughout the other portions of the year. Their peculiar ferocity at this time is attributable to the fact that the male Bears are engaged in seeking their mates, and when it happens, as is often the case among wild animals, that two or more males take a fancy to the same female, they fight for the desired prize with unrelenting fury.

Although the white hunters chase and kill the Bear without any remorse of conscience, the copper-coloured races are so impressed with the intellectual powers of this cunning and dangerous animal, that they endeavour to appease the manes of a slaughtered Bear with various singular and time-honoured ceremonies. The head of the slain animal is decorated with every procurable trinket, and is then laid ceremoniously upon a new blanket. Tobacco-smoke is then solemnly blown into the nostrils of the severed head by the successful hunter, and a deprecatory speech is made, in which the orator extols the courage of the defeated animal, pays a few supplementary compliments to its still living relations, regrets the necessity for its destruction, and expresses his

MUSQUAW, OR AMERICAN BLACK BEAR.—*Ursus Americanus.*

hopes that his conduct has been, on the whole, satisfactory to the dead Musquaw and its relations.

This curious custom is the more remarkable, as it bears a close analogy to the belief of the Scandinavians, who are little less fastidious in their conduct towards the Bear. No true Norwegian will ever speak of a Bear as a Bear, but prefers to mention it as "the old man with the fur cloak;" or, more tersely and poetically, the "Disturber."

As is the case with the Bears which have already been mentioned, the Black Bear is in the habit of passing the cold months of winter in some comfortable residence which it has prepared in the course of the summer. Practical hunters, however, remark that unless the Bear is exceedingly fat at the commencement of the cold season, it does not venture to betake itself to its winter home, but gets through the winter without hybernation. When they can be detected in their dens, the hybernating Bears are often so oppressed with irresistible sleep, that they can hardly be induced to move sufficiently to enable their discoverer to plant a fatal wound. One old Nimrod told a companion who had newly entered on the sport of Bear-hunting, that he had often been forced to push the sleeping Bear with the muzzle of his rifle, in order to make the somnolent animal raise its head.

This species of Bear is remarkably prolific, the number of cubs which are produced at a birth being from one to four. When newly born they are very small, being only six or eight inches in length, and covered with grey hair. The month in which they make their entry into the world is either January or February, and they remain under strict maternal control until they are six months of age. For the first year of their existence the fur continues to retain the grey hue, but when they reach their second year the light-hued hair gives place to the glossy black coat which distinguishes the Musquaw. They shed their coat twice in the year; namely, in spring and autumn; so that when the winter arrives, they are defended from its rigours by a new and warm covering of thickly planted hair. On account of this change in the colour of the fur, the juvenile Musquaw has been considered as a separate species, and admitted into systematic catalogues under the name of Yellow, or Cinnamon Bear.

THERE are few animals which are so widely and deservedly dreaded as the GRIZZLY BEAR. This terrible animal is an inhabitant of many portions of Northern America, and is the acknowledged superior of every animal that ranges over the same country.

The other members of the ursine family are not given to attacking human beings, unless they are alarmed or wounded, but the Grizzly, or "Ephraim," as the creature is familiarly termed by the hunters, displays a most unpleasant readiness to assume the offensive as soon as it perceives a man, be he mounted or on foot, armed or otherwise.

Yet the Bear is not entirely without the innate dread of humanity which is instinctively implanted in every known animal, for, although it will attack a man without hesitation, it will not venture to follow up his track, and even if it should come across the air which is tainted by his presence, the Grizzly Bear will escape as fast as he can run. To this curious instinct the hunters have more than once owed their lives.

One man, who was engaged in duck-shooting, and whose gun was only loaded with shot, was suddenly alarmed at seeing a Grizzly Bear cantering towards him, having clearly already made up his mind to attack him. For the moment, the old man was in despair, but his presence of mind soon returned, and he made his escape in a very ingenious manner. Plucking some of the light fibres from his rough coat, he threw them in the air, in order to ascertain the direction of the wind, and then moved to one side, so as to cause the wind to blow from himself towards the advancing foe. As soon as the Bear perceived the strange scent, it stopped, sat upon its hind legs, wavered, and finally made off, leaving its intended prey master of the field.

If, however, the anger of this terrible animal should be aroused by the pain of a wound, it cares little for men or their scent, but rushes furiously upon them, dealing the most fearful blows with its huge paws, armed with their array of trenchant talons, and holding its powerful teeth in readiness for a close combat. So tenacious of life is the Grizzly Bear, that unless it receives a wound in the head or heart it will continue its furious struggles, even though it be riddled with bullets and its body pierced with many a gaping wound. These warlike capacities render the creature respected by the natives and colonists, and the slaughter of a Grizzly Bear in fair fight is considered an extremely high honour. Among the native tribes that dwell in the northern portions of America, the possession of a necklace formed from the claws of the Grizzly Bear is considered as enviable a mark of distinction as a blue ribbon among ourselves. No one is permitted to wear such an ornament unless the Bear had fallen under his hand; consequently, the value of the decoration is almost incalculable. So largely is this mark of distinction prized, that the Indian who has achieved such dignity can hardly be induced to part with his valued ornament by any remuneration that can be offered.

Mr. Palliser, who was fortunate enough to kill five of these terrible creatures, without suffering from their teeth or claws, bears ample testimony to the fury with which they make their assaults, and the need of a cool determination in the hunter who matches himself against such a foe. Just as the Bear approaches within a few yards of its adversary, it sits up on its hind legs for a moment, and then rushes forward with almost inconceivable velocity. But the moment when the Bear remains quiescent affords sufficient time for a determined hunter to take a steady aim, and to lodge a bullet in the heart or brain of the savage foe.

When the hunter is sufficiently confident in his powers of nerve to match himself against the Bear, he can generally come upon his game by searching among the lower lying grounds, which are filled with rugged timber and scrub fruit-trees.

GRIZZLY BEAR.—*Ursus ferox.*

It is generally supposed that the Grizzly Bear is unable to ascend trees, but it is now ascertained that the animal is quite an adept in tree-climbing, and makes use of the scandent art for the purpose of supplying itself with a bountiful and leisurely repast. As the Bear is very fond of acorns, and does not choose to gather them separately from the branches on which they grow, it ascends the trees, and with its powerful fore-limbs administers such severe blows and shakings to the boughs that the ripe acorns shower down like hail to the earth, whither the ingenious animal speedily descends in order to reap the benefit of its exertions. Yet it is frequently found that a man who has been chased by a Grizzly Bear has succeeded in saving his life by ascending a tree which the Bear has made repeated but ineffectual efforts to climb. The two accounts may be reconciled by the supposition that while the Bear is young, and comparatively agile in proportion to its weight, it is capable of ascending a perpendicular tree-trunk; but that when it becomes large and unwieldy, its limbs are not sufficiently powerful. to raise so great a weight from the earth by so slight a hold as that which is afforded by the claws as they affix themselves to the rough bark.

The colour of the Grizzly Bear is extremely variable, so much so, indeed, that some zoologists have suggested the existence of two distinct species. Sometimes the colour of the fur is a dullish brown, plentifully flecked with grizzled hairs, and in other specimens the entire fur is of a beautiful steely grey. In every case, however, these grizzled hairs are very conspicuous, so that there appears to be a certain tendency to whiteness in the surface of the fur. From this peculiarity, the specific title of "candescens," or whitish, has been affixed to the Grizzly Bear by Major Smith, and the creature has, in one or two hunters' narratives, been erroneously described as the White Bear.

In its earlier years, the young Grizzly Bear may boast of a really beautiful fur, which, although very long, thick, and shaggy, is not of that coarse, wiry texture which is notable as belonging to the coat of the adult animal. The fur of the juvenile Bear is of a brown colour, with a dark stripe along the spine, and is so enormously thick and long, that as the animal shuffles along, it shakes up and down with every step. The gait of this creature is rather peculiar, as it swings its body in a curious and exceedingly awkward manner, and rolls its head from side to side in unison with the movement of its body.

The fore-limbs of this animal are enormously powerful, and the feet of a full-grown adult are eighteen inches in length, and armed with claws of five inches long. These claws are extremely sharp, and when the animal delivers a blow with its paw, the sharp-edged talons cut the adversary's frame as if they were so many chisels. A singular peculiarity is found in these claws. The animal possesses the power of using them separately, and has been repeatedly seen to grasp a dry clod of earth in its foot, and to crumble it to pieces by the mere movement of the claws upon each other. The head is extremely large in proportion to the body, and the tail is so short that it is entirely hidden beneath the heavy fur that covers the hinder quarters. The native Indians are in the habit of amusing themselves with the perplexity of persons who are not aware of this circumstance, and whom they persuade that the carcass of a dead Grizzly Bear is easily lifted if seized by the tail,—a proceeding which bears a strong analogy to the method of capturing a bird by covering its tail with salt.

All animals stand in great fear of this formidable beast, and display the greatest terror even at the sight or the scent of a Bear-skin that has been stripped from the body. Even the powerful bison falls a victim to the Grizzly Bear, which has been seen to spring upon the foremost bull of a herd, dash it to the ground, and destroy it by a succession of tremendous blows with its armed paws. Another of these animals contrived to carry off a bison that had been shot by a hunter, and, after dragging it to some distance from the spot where it fell, to bury it in a pit which it had dug for the reception of its prey. It is said that the other predaceous animals hold the Grizzly Bear in such respect that they will not venture to touch a deer which has been killed by this powerful creature, and that the very imprint of the Bear's feet upon the soil is a warning which not even a hungry wolf will disregard.

As might be expected, this disinclination to meddle with the Grizzly Bear extends to the dead animal itself, and to its skin and carcass. One of these creatures had been shot, and its skin taken from the body, but as the hunter was not strong enough to carry the weighty hide, he was forced to leave it unguarded for fifteen hours, exposed to the attacks of the myriad nocturnal prowlers that swarm in those regions. Yet, when he came at daybreak next morning to secure his prize, he found that neither the skin nor the carcass had sustained the least damage from the teeth of the wolves, although any other animal would have been totally devoured in a very short time. Horses evince such terror at the sight and smell of the Grizzly Bear that they will not permit the skin to be laid on their backs until they have been carefully trained to the unwelcome task.

They are not very easily tamed, except when captured at an extremely early age; but even in that case, they are rather rough in their manners, and are but dangerous play-fellows. They are extremely playful creatures when young, and are very amusing in their habits. One of these animals, which was captured by Mr. Palliser, behaved in a very amusing manner during the voyage homewards, and caused much mirth by its absurd pranks. "Indeed," as the writer observed, "the Bear proved to be the most entertaining member of the whole ship's company. He ate, drank, and played with the sailors, and proved such a source of amusement to them, that the captain, whom I have since had the pleasure of meeting, told me that he would gladly engage always to take a Bear with him when he went to sea in future."

On board of a passage-boat, a sudden shower of rain drove all the passengers, including the Bear, below deck, and Mr. Palliser's attention was roused by peals of laughter over the dining cabin. "On going above, to discover the cause of the merriment, I saw that the Bear was gone and his chain broken. The pilot, who had been relieved a few minutes before, now led me forward to inspect his caboose, which was surrounded by the passengers and deck hands, all in fits of laughter.

I could not make out the reason of it at first, until one of the bystanders pulled a corner of the blanket of the pilot's bed, when, to my surprise, the jerk was answered by an indolent growl. My friend Bruin having got drenched by the shower, had broken his chain in disgust, and actually found his way to the pilot's bed, clambered into it, and rolled himself carefully between the blankets. The good-humoured pilot was not in the least angry, but, on the contrary, highly amused, replying to my apologies as I kicked out his strange bed-fellow, 'Oh! never mind, mister; why, what's the hindrance to the blankets being dried again?'"

The same animal had contracted a strong friendship with a little antelope which was a fellow-voyager with himself; and on one occasion performed a most chivalrous service in behalf of its defenceless little friend.

As the antelope was being led through the streets, towards the vessel, a large mastiff flew at it, and was with difficulty kept at bay by the voice and stick of the person who was leading the terrified little creature. Mr. Palliser, who was following with the Bear, rushed to the rescue, but was outrun by the Bear, who dashed boldly forward and closed with the assailant in a moment. A fierce combat ensued, in which the Bear refrained at first from using his teeth or claws, and contented himself with seizing the mastiff in his powerful arms, and flinging him on his back with such violence that it rolled over and over on the ground. The dog, cheered by the voice of its master, succeeded at last in giving the Bear a tolerably sharp grip between its teeth. Incensed by the pain, Bruin lost his temper, and seizing the dog in his arms, squeezed the breath nearly out of its body, and was preparing to use its teeth, when the dog, which was rapidly choking under the terrible pressure of the Bear's arms, contrived to extricate itself by a sudden struggle, and ran away with piteous howls, leaving the Bear master of the field.

The length of a well-grown adult male is rather more than eight feet six inches, and the girth round the body is equal to the length. The weight of such an animal is rather more than eight hundred pounds. Specimens still larger are sometimes killed, but the average weight and dimensions are as given above.

The powerful claws of this animal are employed not only for combat, but in digging up the earth for various purposes, such as the search after various roots and bulbs, and the interment of some large animal which they have killed. The instinct for burying their prey is so largely developed in these creatures that they have more than once been deceived by the craft of a hunter, who, when resistance or escape was impracticable, has simulated death in order to disarm the wrath of the terrible animal. Thinking the man to be dead, and not being irritated by wounds, the Bear proceeds to scratch a pit in the earth, and to drag the unresisting prey into the hollow, and to cover him carefully with grass and leaves, pressing them well down, so as to conceal him effectually. Satisfied with its precaution, the Bear betakes itself to rest, and the buried hunter seizes the opportunity of slipping quietly away while the animal is engaged in repose.

Several Grizzly Bears have been brought to this country, and have attracted great attention by their amusing and playful habits. A further interest attaches to them from the fact that two of them underwent a surgical operation while under the influence of chloroform. Bears are subject to ophthalmia, especially when in confinement, and are often totally deprived of sight by this disease. Until the discovery of the anæsthetic powers of chloroform, the poor animals were doomed to hopeless blindness, but at the present day, the Bear is rendered as quiet and harmless as a guinea-pig under the influence of this potent vapour.

In order to place the sponge that contained the chloroform fairly under the animal's nostrils, it was necessary to bring its head close to the bars of the cage, an operation which was with difficulty effected by the united efforts of four strong men. The sponge was then affixed to its snout, and in a very short time the animal was lying on the floor of its cage, without sense or motion. The door was then opened, and the Bear's head being laid on a plank outside the cage, the operator speedily removed the obstacle. The animal was

THIBETIAN SUN-BEAR.—*Helarcto Tibetanus.*

then replaced in the cage, where it lay for five or six minutes without motion, and at last, contrived to get on its legs, and walk very unsteadily into its den. The next morning saw the Bear sitting at its ease, restored to the blessings of sight, and feeling no apparent inconvenience from the contrast between the brilliant morning's light and the thick dulness that had for so long a time oppressed its vision.

THE animal which is represented in the accompanying engraving is an example of a group of Bears which have received the title of Sun-Bears, from their habit of basking in the rays of the burning sun, instead of withdrawing to their dens, as is the custom with the generality of Bears, as long as the sun pours its meridian beams on the earth. The name Helarctos, by which the genus is designated, is composed of two Greek words, the former signifying the sun, and the latter a Bear.

The country in which the Thibetian Bear resides is manifest by its name. It has also been discovered in the Nepal range of mountains. The fur of this creature is tolerably thick and smooth, and is generally of a black colour, with the exception of the lower lip and a large patch of white hairs on the breast, which is narrow at the lower part, and, widening as it approaches the chin, separates into two short horns, which partially extend towards the shoulders. The entire spot bears, therefore, some resemblance to the letter Y rather imperfectly delineated, and with its upright stem rather shortened.

The body of the Thibetian Bear is heavily, but strongly made, and the limbs seem to be rather less agile than those of the American or Scandinavian Bears. The claws are not so powerful as those of the generality of Bears, the ears are comparatively large, and the neck is peculiarly thick. It seems chiefly to rely for its subsistence on fruit, roots, and various vegetable productions. It is not quite so large as the Bears which have already been mentioned.

A VERY curious example of the Sun-Bears is found in the species which is known by the name of the BRUANG, or MALAYAN SUN-BEAR, and has been rendered famous by the spirited description of its appearance and habits which has been given by Sir Stamford Raffles.

MALAYAN SUN-BEAR.—*Helarctos Malayanus*.

The fur of this animal is particularly fine and glossy, and the hair is shorter than in the generality of the Bear tribe. The colour of its fur is a very deep black, with the exception of a large semi-lunar shaped patch of white on the breast, and a yellowish-white patch on the snout and upper jaw, which afford a striking and curious contrast to the uniformly black colour of the fur. The lips and tongue of this Bear are extremely flexible, and are capable of being prolonged to an almost incredible extent. It is supposed that the great length of its tongue, and the exceedingly flexible power of that organ, are intended for the purpose of enabling the animal to obtain the honey from the nest of the wild bee, by insinuating its lithe tongue into the apertures of the hive, and licking the sweet food from the waxen treasuries.

The head of the Bruang is rather thick, and the neck is singularly powerful in comparison with the size of the head. The eyes are very small, and the iris is of a rather pale lilac colour, and tolerably lively in its appearance. It is not a large animal, measuring when adult only four feet six inches in length, but it is extremely powerful in proportion to its size, being able to grasp and tear from the ground the strongly-rooted plantains of Borneo, which are so large that the Bear is hardly able to embrace them in its grasp. The claws of the Bruang are extremely long.

When in its wild state, it is almost entirely a vegetable eater, preferring fruit before most articles of diet, and making great havoc among the tender shoots of the cocoa-nut trees. In some parts of Sumatra, where the villages have been deserted, the cocoa-nut groves have been entirely destroyed by the insatiate appetite of the Bruang.

As it is easily tamed, it is frequently seen in a state of domestication, and is a very amusing and gentle creature, associating freely with children, and earning by its uniformly quiet conduct the privilege of unrestricted liberty. Sir Stamford Raffles, who possessed one of these Bears, permitted it to live in the nursery, and never was obliged to chain, chastise, or otherwise punish the good-tempered animal. Being something of an epicure, and often admitted to his master's table, the Bruang would refuse to eat any fruit except mangosteens, or to drink any wine except champagne. It may seem remarkable that a Bear should display any predilection for fermented liquids, and more so that it should be so fastidious as to select champagne as the wine which it honoured with its preference.

BORNEAN SUN-BEAR.—*Helarctos Euryspilus.*

Such, however, was the case, and the animal was so fondly attached to the champagne-bottle, that the absence of his favourite liquid was the only circumstance that would make him lose his temper. His affectionate disposition led him to extend his friendship to various of his acquaintances, and he was on such excellent terms with the entire household, that he would meet on equal footing the cat, the dog, and a small Lory, or Blue-mountain bird, and amicably feed with them from the same dish.

One of these Bears that was successfully domesticated was able to eat animal as well as vegetable food, but was fed exclusively on bread and milk, of which it consumed rather more than ten pounds per diem. It is possessed of much flexibility of body, and is very fond of sitting on its hind legs, thrusting out its long tongue to an extraordinary distance, and ever and anon withdrawing it into the mouth with a peculiar snapping sound. While thus engaged, it makes the most grotesque and singular gestures with the fore-limbs, and rolls its body from side to side with unceasing assiduity. It seldom remains in one position for any length of time, and, although its movements are not characterized by much energy or rapidity, it is evidently possessed of much power over its limbs, and if it were disposed to enter into strife would probably use its long talons to good purpose.

RESEMBLING the Malayan Bruang in general habits and disposition, the BORNEAN BRUANG, or SUN-BEAR, is acknowledged to belong to a different species from the animal which has just been described. The colour of its fur is nearly as black as that of the Bruang; but the patch upon its breast is of an orange hue, instead of the greyish white which is so conspicuous in the Malayan Bruang.

Like that animal, it can sit or stand on its hinder limbs with the greatest ease, and possesses nearly as much flexile power of lip and tongue as the Bruang. It has a curious habit of placing its superabundant food upon its hinder paws, as if to guard it from the defilement of sand or dust, and feeds itself by slow degrees with dainty carefulness.

It is extremely fond of fruits and various vegetables, and is, in its native country, a dreaded foe to the cocoa-nut trees and their fruit. The animal is so excellent a climber

ASWAIL, OR SLOTH BEAR.—*Melursus Lybius.*

that it cannot be baffled by loftiness of trunk or smoothness of bark ; and when the creature has attained the summit of the tree it frequently destroys the life of the tree by devouring the topmost shoot for the sake of its delicate succulence. The cocoa-nuts themselves are objects of interest to the Bornean Bruang, who is extremely fond of the peculiarly-flavoured liquid that is found in the interior of the nut ; and when he has reached a cluster of ripe nuts will tear them from the tree and fling them on the ground.

In captivity it is gentle, playful, and amusing, and possesses very curious and almost ludicrous habits, which render it an object of interest to its visitors. Profoundly sensible of human sympathies, and almost as fond of notice as a cat, the Bornean Bruang will accept with evident delight the caresses of its visitors, and is pleased to be patted or stroked by kindly hands, provided that it does not happen to be in a bad humour at the time. Should the animal consider itself to be insulted—a matter of rather frequent occurrence —it will contumeliously reject all advances, and will not consent to receive any mark of attention until the offender is fairly out of sight.

UNWIELDY in its movements, and grotesque in its form, the ASWAIL, or SLOTH BEAR is one of the most curious members of this group of animals. It is found in the mountainous parts of India, and is equally dreaded and admired by the natives of the same country.

Although a sufficiently harmless creature if permitted to roam unmolested among its congenial scenery of mountain and precipice, it is at the same time an extremely dangerous foe if its slumbering passions are aroused by wounds or bodily pain of any kind. It needs, however, that the wound be tolerably severe to induce the animal to turn upon the person that inflicted the injury ; for should it only be slightly wounded, it runs forward in a straight line, as if it were actuated by the one idea of getting as far as possible away from the object which had caused it so much bodily suffering, and can but seldom be finally captured.

As a general rule, the Aswail remains within its sheltered den during the hot hours of the day, as its feet seem to be extremely sensitive to heat, and suffer greatly from the bare rocks and stones which have been subjected to the burning rays of that glowing Indian sun. On one or two occasions, however, where the wounded Bear had been successfully tracked and killed, the soles of the poor animal's feet were found to be horribly scorched and blistered by the effects of the heated rocks over which the creature had recklessly passed in its haste to escape from its enemies. On account of this extreme sensitiveness of the Aswail's foot, it is very seldom seen by daylight, and is generally captured or killed by hunters who track it to its sleeping place, and then attack their drowsy prey.

The Aswail is said never to eat vertebrate animals except on very rare occasions, when it is severely pressed by hunger. Its usual diet consists of various roots, bees'-nests, together with their honey and young bees, grubs, snails, slugs, and ants, of which insects it is extremely fond, and which it eats in very great numbers. Probably on account of its mode of feeding, its flesh is in much favour as an article of diet, and though rather coarse in texture, is said by those who have had practical experience of its qualities to be extremely good.

The fat of this Bear is very highly valued among the natives and the European residents, being used chiefly for the lubrication of the delicate steel work that is employed in the interior of gun-locks. For this purpose the fat is prepared in a similar manner to that of the tiger, being cut into long strips, forced into closely stoppered bottles, and placed during the entire day in the blazing rays of the sun. The powerful sunbeams soon melt the fat into a homogeneous mass, and when the evening begins to draw on, the contents of the bottle are found to settle into a firm and white substance, which has the property of remaining untainted even in that heated climate, where, if no such precaution were taken, it would in a very few hours become a mass of putrescent abomination. The prepared fat is especially valuable for gun-locks, as it preserves the bright steel from rust, and does not clog by constant service, as is the case with almost every other animal oil.

In connexion with this subject it may be as well to mention that the ordinary "trotter oil," or "neats'-foot oil," may be prepared for the most delicate work in a similar manner. If a bottle of this oil be placed in the sun's rays, and a few strips of lead dropped into the vessel, an extraordinarily heavy deposit begins to take place, and fills the lower part of the bottle. The upper part, however, remains bright and limpid as crystal, and by a repetition of the same process may be so effectually purified that it will never be liable to that annoying viscidity which detracts so much from the value of animal oils that have been for some time in use. It is in this manner that watch makers purify the oil for the lubrication of the delicate machinery of their trade.

Very little is known of the habits of this Bear while in its wild state, but it would appear from the conduct of two young animals that inhabited the same cage in the Zoological Gardens, that it must be a gentle and affectionate creature.

It is, at all events, known that the maternal Aswail is in the habit of carrying on her back those of her offspring that are not able to make full use of their own means of progression. The two animals that were kept in the Zoological Gardens were accustomed to lie close to each other, and while in that easy position used to suck their paws after the usual ursine fashion, uttering at the same time a kind of bearish purr, as an expression of contentment. This sound, although it partakes of the nature of a whine, admixed with the purr, is not without a musical intonation, and may be heard at some little distance. Indeed, it has not unfrequently happened that the Bear has been betrayed to its pursuers by the continuous sound which it utters while lying half asleep within its den.

The hair which covers the body and limbs is of singular length, especially upon the back of the neck and the head, imparting a strange and grotesque appearance to the animal. The colour of the fur is of a deep black, interspersed here and there with hairs of a brownish hue. Upon the breast, a forked patch of whitish hairs is distinctly visible.

When it walks, its fore-feet cross over each other, like those of an accomplished skater when accomplishing the "cross-roll," but when it remains in a standing attitude its feet are planted at some distance from each other.

These Bears seem to be very liable to the loss of their incisor teeth, and even in the skulls of very young animals the teeth have been so long missing that their sockets have been filled up by nature as if no teeth had ever grown there. On account of this curious deficiency, the first specimen which was brought to England was thought to be a gigantic sloth, and was classed among those animals under the name of *Bradypus Ursinus*, or Ursine Sloth. In one work it was candidly described as the Anonymous Animal. Other names by which it is known are the Jungle Bear, and the Labiated or Lipped Bear.

This last-mentioned title has been given to the animal in consequence of the extreme mobility of its long and flexible lips, which it can protrude or retract in a very singular manner, and with which it contorts its countenance into the strangest imaginable grimaces; especially when excited by the exhibition of a piece of bun, an apple, or other similar dainty. It is fond of sitting in a semi-erect position, and of twisting its nose and lips about in a peculiarly rapid manner in order to attract the attention of the bystanders, and ever and anon, when it fails to attract the eyes of its visitors, it slaps the lips smartly together, in hopes to strike their sense of hearing.

When captured young, it is easily tamed, and can be taught to perform many curious antics at the bid of its master. For this purpose it is often caught by the native mounte-banks, who earn an easy subsistence by leading their shaggy pupil through the country, and demanding small sums of money for the exhibition of its qualities. On account of its association with these wandering exhibitors, it has been called by the French naturalist "Ours Jongleur." Whether owing to the natural docility of the animal, or to the superior powers of its instructor, it performs feats which are more curious and remarkable than the ordinary run of performances that are achieved by the Learned Bears of our streets.

In either case, it is always a saddening sight, for, however ingenious may be the instructor, or however docile the pupil, the unnatural performances of the poor animal always seem to be out of place. We have no right to attempt to humanize a Bear or any other animal; for in so doing we are preventing it from working the task which it was placed in the world to fulfil. The Bear—as may be said of every animal—is the result of a divine idea in the mind of the Creator, and it ought to be our business to aid the creature in developing that idea as far as possible, and not to check its development by substituting some other idea of our own, which, with all we can do, must necessarily be a false one. Even the imprisoned Bears which mount a tall pole for the purpose of obtaining cakes and fruit from their visitors, are performing their mission much more truly than the most accomplished Bear that ever traversed the country, and are, in consequence, much more agreeable to the eye of any one who values the animal creation on account of the moral qualities which are implanted in them from their birth, for us to develop to their highest extent, and in which we may read an ever living word proceeding from the ever-creating hand of God.

Moreover, all those who in studying natural history desire to look deeper than the surface, and to direct their attention rather to the inward being of the various animals than to their outward forms, will find that every creature in which is the breath of life has a physical, a moral, and sometimes a spiritual analogy with the more expanded organisms of humanity, and owes its position among created beings to that very analogy. In every human being are comprised all the mental characteristics that are outwardly embodied in the various members of the animal kingdom, and it is impossible to mark any attribute of the lower animals which does not find a further and a higher development in the human existence in one or other of its manifestations.

This subject is too wide to receive even a cursory notice in the present article, but will be again taken up on a future page.

There is generally an aquatic member of each group of animals throughout the vertebrate kingdom, and among the Bears this part is filled by the NENNOOK, or POLAR BEAR, sometimes called, on account of its beautifully silvery fur, the WHITE BEAR. As has already been mentioned, the Bears are good swimmers, and are able to cross channels of considerable width, but we have, in the person of the Nennook, an animal that is especially formed for traversing the waters and for passing its existence among the ice-mountains of the northern regions.

Probably in consequence of the extreme cold which prevails in the high latitudes where this creature is found, its food is almost entirely of an animal nature, and consists of seals and fish of various kinds. In order to capture the fish in their own element, or to make prey of the active and wary seals, it is necessary that the Nennook should be endowed with no ordinary powers of body and sense. Its capabilities of scent are extraordinarily fine, for it will perceive, by the exercise of that sense alone, the little breathing-holes which the seals have made through the ice, even though the icy plain and the breathing-holes are covered with a uniform coating of snow. Even the Esquimaux dog, which is specially trained for this very purpose, is sometimes baffled by the extreme difficulty of discovering so small an aperture under such difficult circumstances.

So active is this Bear, and so admirable are its powers of aquatic locomotion, that it has been seen to plunge into the water in chase of a salmon, and to return to the surface with the captured fish in its mouth. And when it is engaged in the pursuit of seals, as they are lying sleeping on a rock or an ice-raft, it is said to employ a very ingenious mode of approach. Marking the position in which its intended prey lies, it quietly slips into the water, and diving below the surface, swims in the intended direction, until it is forced to return to the surface in order to breathe. As soon as it has filled its lungs with fresh air, it again submerges itself, and resumes its course, timing its submarine journeys so well, that when it ascends to the surface for the last time it is in close proximity to the slumbering seal. The fate of the unfortunate victim is now settled, for it cannot take refuge in the water without falling into the clutches of its pursuer, and if it endeavours to escape by land it is speedily overtaken and destroyed by the swifter-footed Bear.

The endurance of the Bear while engaged in swimming is very great, for it has been seen swimming steadily across a strait of some forty miles in width. Even the large and powerful walrus is said to fall a victim to the superior prowess of the Polar Bear. Although its appetite is of so decidedly carnivorous a nature in the northern regions, it assumes a milder character in southern climes, and contents itself with vegetable aliment. In England, it has been fed for a considerable time on bread alone, of which it consumes about six pounds per diem, and its fondness for cakes and buns is well known to every frequenter of the Zoological Gardens. Even in its wild state, it is in the habit of varying its food by sundry roots and berries, and is often found engaged in searching for these dainties at some distance from the sea-shore.

So powerful an animal as the Polar Bear must necessarily be very dangerous when considered in the light of a foe, and as it is rather tetchy and very uncertain in its temper, it often affords ample scope by which its pursuers may test their prowess. Sometimes it runs away as soon as it sees or smells a human being, but at others it is extremely malicious, and will attack a man without any apparent reason. As is the case with nearly all the Bears, it is very tenacious of life, and even when pierced with many wounds will fight in the most desperate manner, employing both teeth and claws in the combat, and only yielding the struggle with its life.

The colour of the Nennook's fur is a silvery white, tinged with a slight yellow hue, rather variable in different individuals. Even in the specimens that were confined in the Zoological Gardens there was a perceptible difference in the tint of their fur, the coat of one of them being of a purer white than that of the other. The yellowish tinge which has been just mentioned is very similar to the creamy-yellow hue which edges the Ermine's fur. The feet are armed with strong claws of no very great length, and but slightly curved. Their colour is black, so that they form a very bold contrast with the white fur that falls over the feet. Even at a considerable distance, and by means of its mere outline, the Polar Bear may be distinguished from every other member of the Bear tribe by its peculiar shape. The neck is, although extremely powerful, very long in proportion to the remainder of the body, and the head is so small and sharp that there is a very snake-like aspect about that portion of the animal's person.

POLAR BEAR.—*Thalarctos maritimus.*

The shape of the head is rather remarkable, for whereas in the Brown and other Bears the muzzle is separated from the forehead by a well-marked depression, in the Polar Bear the line from the forehead to the nose is almost continuous. The foot of the Nennook is of surprising comparative length, for it is equivalent in length to one-sixth of the entire length of the body, whereas in the Brown Bear it is but one-tenth of that measurement. The sole of the foot is covered with a thick coating of warm fur, which is in all probability intended for the double purpose of protecting the extremities from the intense cold of the substance which it is formed to traverse, and of enabling the creature to tread firmly on the hard and slippery ice.

From these and other peculiarities of form it is now acknowledged as a separate species of Bear, and even removed into a different genus by many naturalists, although the earlier writers on this subject supposed that it was merely a permanent variety of the Brown Bear, which had obtained a white coat by constant exposure to the terrible cold of these wintry regions, and whose form had been slightly modified by the ever-repeated habits of its strange life.

The skeleton which is here presented to the reader is that of the Polar Bear, and has been selected because it affords an excellent example of the peculiar bony formation around which the body of the Bear is built, and at the same time exhibits some of the characteristics which distinguish the Polar from the other Bears. The reader will especially notice the length of the neck, the peculiar flatness of the skull, and the very great comparative length of the feet.

Although so powerful an animal, and furnished by nature with such dreadful arms of

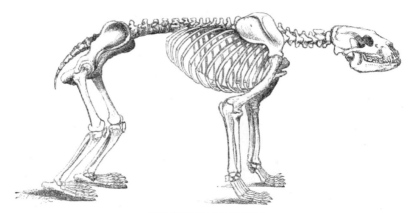

SKELETON OF POLAR BEAR.

offence, it is considered by the experienced Bear-hunters of Norway to be inferior in combat to the common Brown Bear, and is less dreaded by them as an antagonist.

Its powers of endurance are necessarily great, for its means of subsistence are always precarious, and in many cases are extremely small indeed. As the Bear is in the habit of passing so much time upon the ice, and generally devours upon its frozen surface the prey which has been captured, it is liable to be affected by the sudden and extraordinary changes that are constantly taking place in the vast ice-fields of these cold regions. Pieces of ice on which the Bears are quietly sleeping after their repast, become noiselessly dissevered from the main body, and are carried off to sea for a very great distance before the Bear is aware of its enforced voyage. Scoresby records such an instance, where he met with a Polar Bear upon a piece of drift ice that was floating at sea some two hundred miles distant from the land. As the ice nourishes no animals that could afford nutriment to the white-coated resident, the Bear is forced to depend for its entire subsistence upon the fish that it may be able to capture. Out at sea, however, the fishy tribe are not so easily procured as near the shore, and the hunger-endurent powers of the Bear are thoroughly tested before it can again place its shaggy foot on the welcome soil.

Owing to these marine excursions the Polar Bear is forced to pay unwilling visits to civilized shores which it loves not, and where it is obliged to fall upon the sheep and cattle of the residents in order to appease its hunger. The ire of the owners is greatly excited by the loss of their cattle, and the unfortunate Bear—a thief in spite of itself—is soon destroyed by the bereaved proprietors. Sometimes a whole party of Polar Bears is thus carried off, and for a while they inflict infinite damage on the country where they land.

As the Nennook passes its life among the wintry regions of the north, its hybernation has been often discredited, and it has been said to make a partial migration southwards, so soon as the terrible frosts of the Arctic winter close up the pools whereto the seals and other animals which constitute its prey are in the habit of resorting. Other writers, again, assert that the Polar Bear ceases feeding in the winter, as do the other members of the same group, and that the young Nennooks are produced while the mother is safely housed in her den. There is a truth in both these opinions, for it is now ascertained that the female Polar Bear is in the habit of hybernating, but the male Nennook passes his winter in the active exercise of his faculties.

The winter home of the Polar Bear is always made in some sheltered situation, such as the cleft of a rock, or the foot of a precipitous bank. In a very short time after the animal has taken up her residence in her new abode, she is effectually concealed from observation by the heavy snow-drifts, which cover the whole country with such strangely-shaped hills and valleys that the Bear's den is entirely undiscoverable by the eye. Sometimes the Bear will wait until a heavy fall of snow has taken place, and then will dig away the snow so as to form a cavern of the requisite size. In all cases, the snow appears to be a necessary element in the wellbeing of the animal during its long winter's repose. If the female Bear should not be about to take upon herself the cares of maternity, she does not think herself bound to lie hidden during the winter, but traverses the ice-fields together with the male, and becomes very fat during the cold months of the year. These nomad individuals do not confine their peregrinations to the sea-shore, but extend their journeys inland to a considerable distance, being sometimes found as far as thirty miles from the sea-coast.

The young of the Nennook are generally two in number; and when they make their first appearance outside the snow-built nursery in which their few months of existence have been passed, are about the size of shepherds' dogs, and in excellent condition. Their mother, however, is sadly reduced by her long fast and the calls which have been made upon her by her offspring; so that she re-enters the world in a very poor condition of aspect and temper, as might be expected of so ravenous and hungry an animal. Watchful over the safety of her cubs, and unburdened by any superfluous flesh, she is a very dangerous personage to be casually met with; for she is so savage with hunger that her temper is in a constant state of irritation; and she is so jealous of the safety of her offspring that she suspects every moving object to be an enemy.

The flesh of the Polar Bear is eatable, and is highly esteemed by the Arctic voyagers, who eagerly welcome a supply of fresh and wholesome meat such as is furnished by the animal in question. It is said, however, that the liver ought to be avoided as an article of diet, as it is apt to cause painful and even dangerous symptoms to those who have partaken of it. Yet the liver of the American Black Bear is said to be a peculiar luxury when dressed on skewers, kabob fashion, with alternate slices of fat.

It will be observed, after the perusal of the foregoing pages, that the Bears are found in almost every part of the world, with two notable exceptions, viz. Africa and Australia. With regard to the latter of these countries, it may be remembered that the entire creation, whether animal or vegetable, is of so strange a nature that it cannot be subjected to the rules which govern the rest of the world. There is, it is true, a tree-climbing creature in Australia, of a somewhat clumsy and ursine aspect, which is popularly called the Australian Bear, but which is in reality no Bear at all, but a member of the curious family of the Macropidæ, which contains the kangaroos, bandicoots, and opossums, and will be shortly described in its proper place under the title of the Koala. With regard to the African continent, the existence or non-existence of Bears is by no means decided.

Many of the ancient historians make constant mention of African Bears. Juvenal, for instance, speaks of Numidian Bears, Virgil and Martial of Libyan Bears, while it is recorded in the annals of the Roman empire, that in the year B.C. 61, a hundred Numidian Bears were exhibited in the circus, each Bear led by a negro-hunter. None, however, of the later African travellers have clearly seen Bears in that country, and it is certain that from the days of Pliny up to the present time no true Bears have been found in Africa. Still, it is very possible that these animals may be yet discovered in that vast continent; for there seems to be no reason why Bears should be unable to exist in some parts of so large a country, although they might not be able to find subsistence in those portions which have already been investigated.

PRESERVING somewhat of the ursine aspect and much of the ursine habits, the RACOON, or MAPACH, as it is sometimes named, is an active, spirited, and amusing animal. As it is readily tamed, although rather subject to occasional infirmity of temper, and is inquisitive, quaint, and lively withal, it is a great favourite with such persons as have kept it in captivity.

The colour of this animal is rather peculiar, and not very easy to describe.

The general tint of the body and limbs is an undecided blackish grey, the grey and black predominating according to the position of the observer and the arrangement of the fur. The hairs that form the coat of the Racoon are of two kinds, the one of a soft and woolly character, lying next to the skin, and the other composed of long and rather stiff hairs that project through the wool for some distance. The woolly fur is of a uniform grey, while the longer hairs are alternately marked with black and greyish white. Upon the top of the head and across the eyes the fur is of a very dark blackish brown ; and upon the knee-joint of each leg the fur is of a darker tint than on the rest of the body. The tail is rather short and bushy in character, and is marked with five, or sometimes six blackish rings upon a ground of dark grey. In the British Museum is preserved a specimen of the Racoon, which is entirely white, its fur being of as pure a whiteness as that of the ermine itself.

In its gait and general carriage is visible an admixture of the plantigrade with the digitigrade ; for when it stands or sits it plants the entire sole of its foot upon the ground, but when it runs in haste it only touches the earth with the tips of its toes. Generally, it is nocturnal in its habits, passing the whole of the daytime in sleep, snugly curled up in the warm blanket of its own rich fur, and slumbering heavily with its head sunk between its hinder limbs.

As is indicated by the peculiar nature of its teeth, the Racoon is capable of feeding on animal or vegetable food, but seems to prefer the latter. Indeed, there seem to be few things which the Racoon will not eat. One of these animals ate a piece of cedar pencil which it snatched out of my hand, and tried very hard to eat the envelope of a letter on which I was making notes. Not succeeding in the attempt, it consoled itself by tearing the paper into minute morsels, employing teeth and paws in the attempt. It did its best to get a ring off my finger, by hitching one of its crooked claws into the ring and pulling with all its strength, which was very considerable in proportion to the size of the animal. Its brown eyes lighted up with animation when engaged in play, and it was very fond of pushing its paw through the bars of its cage, in order to attract attention.

A Racoon that was kept in a barrack-yard in Canada, in company with a bear, an owl, and various finned and feathered creatures, was considered to be the most interesting of all the little menagerie. It was extremely tame, but could not be trusted near poultry, as it had a bad habit of pouncing suddenly upon them, grasping them in its hand-like paws, and biting off their heads in a moment. It would then devour the head and afterwards the body in a leisurely manner. There were many bats in the neighbourhood, and the soldiers were in the habit of capturing those nocturnal depredators, and throwing them on the ground within reach of the Racoon's chain. Before the bat could flap its wings, the Racoon would leap upon it, roll it rapidly in its paws for a while, and then despatch it with a single bite.

It was rather a vengeful animal, and possessed of a tenacious memory for an insult. The great owl that was partaker of the same residence had one day been irritated with the Racoon and had pecked it on the back. The Racoon treasured the insult in its heart, and waited a favourable time for revenge. The opportunity was not long delayed, for on the first occasion that the owl ventured within reach of the Racoon's chain, the aggrieved animal crept slily towards its foe, and adroitly snatched out all the feathers of the owl's tail.

In its native state it is a great devourer of oysters, crabs, and other similar animals, displaying singular ingenuity in opening the stubborn shells of the oysters, or in dispatching the crabs without suffering from their ready claws. Sometimes it is said to fall a victim to the oyster, and to be held so firmly by the closing shells that it cannot extricate itself, and perishes miserably by the rising tide. Its oyster-eating propensities have been questioned, but are now clearly proven. The sand and soil that fringe the oyster-beds are frequently seen to be covered with the footmarks of this animal.

It is always fond of water, drinking largely, and immersing its food, so as to moisten it as much as possible. When engaged in this curious custom it grasps the food in both

RACOON.—*Prócyon Lotor.*

its forepaws, and shakes it violently backward and forward in the water. On account of this remarkable habit it has been dignified with the title of Lotor, a washer. The German naturalists term it Wasch-Bär, or Washing Bear. In captivity it is anything but abstemious, and rejects plain water, provided that it can be furnished with fermented liquids, strong and sweet. Referring to this propensity, Lawson, who was Surveyor-General of Carolina in the year 1714, says of the Racoon that, "if taken young, it is easily made tame, but is the drunkenest creature living if he can get any liquor that is sweet and strong." He furthermore relates that this animal is in the habit of catching crabs by putting its foot into their holes, and dragging out the crab as soon as it seizes the intruding limb.

Roving at night through the woods, and being gifted with singular subtlety as well as agility, it is frequently chased by the residents, who think a 'Coon hunt to be one of the most exciting of sports. Certainly, to judge from the animated descriptions of such scenes, the whole affair must be marvellously picturesque to the eye as well as exciting to the mind. The usual plan of hunting the 'Coon, is to set an experienced dog on its trail, and to chase it until it takes refuge in a tree. A blazing fire of pine-chips is then built under the tree, which illuminates its branches and renders the smallest leaf perceptible. A good climber then ascends the tree, and speedily dislodges the concealed animal. Audubon gives a very lively account of a Racoon hunt, ending as follows :—

"Off we start again. The boys had got up with the dogs, which were baying at a Racoon in a small puddle. We soon joined them with a light. 'Now, stranger! watch and see!' The Racoon was all but swimming, and yet had hold of the bottom of the pool with his feet. The glare of the lighted torch was doubtless distressing to him ; his coat was ruffled, and his rounded tail seemed thrice its ordinary size, his eyes shone like emeralds ; with foaming jaws he watched the dogs, ready to seize each by the snout if it came within reach. They kept him busy for several minutes ; the water became thick with mud ; his coat now hung dripping and his draggled tail lay floating on the surface. His guttural growlings, in place of intimidating his assailants, excited them the more ;

and they very unceremoniously closed upon him, curs as they were, and without the breeding of gentle dogs. One seized him by the rump, and tugged, but was soon forced to let go ; another stuck to his side, but soon taking a better directed bite of his muzzle than another dog had just of his tail, 'Coon made him yelp ; and pitiful were the cries of luckless tyke.

The Racoon would not let go, but in the meantime the other dogs seized him fast, and worried him to death, yet to the last he held by his antagonist's snout. Knocked on the head by an axe, he lay gasping his last breath, and the heaving of his chest was painful to see. The hunters stood gazing at him in the pool, while all around was by the flare of the torch rendered trebly dark and dismal. It was a good scene for a skilful painter."

In size, the Racoon equals a small fox, to which animal it bears a slight external resemblance. The number of its young is usually two or three, and they make their appearance in the month of May.

THE AGOUARA, OR CRAB-EATER.—*Prócyon Cancrivorus.*

THE AGOUARA, or CRAB-EATING RACOON, is a native of a warmer portion of America than the common Racoon, and has derived its name of Crab-eater from its habit of feeding on all kinds of crustacea and molluscs, whether marine or terrestrial, although perhaps it is not more addicted to cancricide than the animal which has just been described. In size it is larger than the common Racoon, and its colour is not quite the same.

The tail is short in proportion to the body, and is marked with six black rings upon a greyish or blackish-yellow ground. The fur of the body is rather variable in different individuals, but is generally composed of a blackish-grey washed with a tinge of yellow, the darker and the lighter tints predominating on different parts of the body and limbs. On the head, neck, and along the spine, the black tinge prevails, while the rest of the body and the sides of the neck are almost wholly of the yellowish-grey hue. A very dark brown patch incloses each eye, and, passing backwards almost to the ears, is merged into a dark spot on the crown of the head. The Racoon has been mentioned by several travellers under different names, such as Raton, Mapach, Agouarapopè, Yllanraton, Maxile, Wasch-Bär, and Cioutlamacasque.

THE animals which compose the curious genus that is known by the name of Narica, are easily recognised on account of the singular length of the nose, which is prolonged so

COAITI, OR COAITI-MONDI.—*Nasua Rufa.*

as to form a miniature and mobile proboscis. In their general habits and diet, they very strongly resemble the racoons, and are as admirable climbers of trees as can be found in the animal kingdom.

The extraordinary snout with which the Coaitis are gifted is very useful to the possessor, being employed for the purpose of rooting in the ground in search of worms and insects, together with other important uses. When they drink, the Coaitis lap the water after the manner of dogs, and when so engaged, turn up their flexible snouts, so as to keep that useful member from being wetted more than is necessary. They are inhabitants of Southern America, and are found in small companies upon the trees among which they reside, and on the thin branches of which they find the greater part of their food. Two examples of the Coaitis will be briefly described.

The COAITI-MONDI, or RED COAITI, derives its name from the reddish-chesnut hue which prevails over the greater portion of the fur, and is only broken by the black ears and legs, the maroon-coloured bands upon the tail, and the white hairs which edge the upper jaw, and entirely cover the lower. The texture of the fur is rather harsh and wiry, and of no very great importance in commerce. Upon the paws are certain curious tubercles, which alone would serve to identify the animal were it entirely destroyed with the exception of a single foot. It is extremely active in the ascent and descent of trees, and pursues its prey among the limbs with great certainty. Its food consists of sundry vegetable and animal substances, but the creature seems to prefer the latter to the former.

When the Coaiti descends a tree, it does so with its head downward, securing itself from falling by hitching the claws of the hinder feet into the inequalities of the bark, and displaying by the act no small amount of flexibility in the jointing of the hinder limbs. It is a nocturnal animal, and does not display its true liveliness until the shades of evening begin to draw on, but lies curled up in a curious but comfortable attitude, its long and bushy tail serving for blanket and pillow. Towards evening, however, the Coaiti arouses itself from its lethargy, and becomes full of life and vigour, careering about the branches with extraordinary rapidity of movement and certainty of hold, and agitating its mobile nose with unceasing energy, as if for the purpose of discovering by the snout the presence of some welcome food. It is a merciless robber of birds' nests, and will eat parent, eggs, or young, with equal appetite.

Although possessed of a very irritable temper, the Coaiti is tamed without difficulty to a certain extent, but is always capricious in its affections, and cannot be trusted without

1.   E E

danger. When attacked by men or dogs, the Coaiti fights desperately, and can inflict such dangerous wounds with its double-edged canine teeth, that it is, although so small an animal, no despicable antagonist.

ANOTHER species of Coaiti inhabits the same regions as the last-mentioned animal. This is the NARICA, or QUASJE, which is sometimes called the BROWN COAITI, in order to distinguish it from the red species. Sometimes the name is spelled as Quaschi.

It is a very lively and amusing animal, and possessed of singular powers of nose and limb. Distrustful by nature, it will very seldom venture to approach a strange object until it has endeavoured to ascertain the nature of the unknown, by means of its sense of smell, which is marvellously acute. It seems to be as inquisitive as it is distrustful, and will not be satisfied until it has by gradual degrees approached and examined anything which it does not quite understand.

One of these animals, which was kept in confinement for some time, was extremely tame to those who understood the peculiarity of its temper, but was irresistibly morose and sulky with those who would not respect its customs. Any stranger who ventured to

NARICA, OR QUASJE.—*Nàsua Nàrica.*

approach the animal was repelled with open mouth and threatening cries, unless he propitiated the creature by offering it some delicacy of which it was fond. It would then lay aside its suspicious demeanour, and become suddenly confidential, returning the caresses of its newly-found friend, and searching eagerly for a further supply of food.

It proved to be quite a useful inhabitant of the house when it was domesticated, for it was accustomed to roam over the premises in chase of mice and rats, which it pursued unrelentingly through house, hay-loft, and stables. It was also accustomed to pay visits into the garden, where it spent much of its time in catching snails and slugs, and in digging after worms,—a task for which its powerful claws are eminently calculated to adapt it. When it was supplied with meat, it was accustomed to tear its food to pieces with its claws before carrying it to the mouth; and in the act of feeding, it always supplied itself by hitching one of its claws in the morsel which it was about to carry to its mouth. It struck up a friendship with a little dog, and would permit its four-footed friend to occupy the same bed, but would never endure the society of any other animal.

The colour of this creature is extremely variable, as it seldom or never happens that two specimens are marked in precisely the same manner. In some individuals the dark portion of the fur is brown, mottled with black; but the general hue of the fur is a brown, tinted more or less with chesnut, and occasionally being so pale as to be of a warm fawn colour. The under surface of the body and the internal face of the limbs are of a grey hue, tinged with yellow or orange, according to the individual, and extending, in some cases, to the sides of the neck and the lower jaw. The coat of the Narica is rather thick, and the texture of the fur is harsh; it does not lie closely to the body, but presents a rather shaggy and rough aspect.

THE singular creature which is known under the title of KINKAJOU, or POTTO, has been the means of perplexing systematic naturalists in their laudable attempts to place each animal in its proper position.

On account of its external aspect and its general habits, it has been considered as one of the Lemurine family, and was termed in consequence the Yellow Macauco, or the Yellow Lemur (*Lemur flavus*). As, however, the structure of its teeth and limbs is entirely different from that of the lemurs, and very closely approaches the carnivorous type, it has been placed among the flesh-eating animals, under the name of Mexican Weasel (*Viverra caudivolvula*). But the flat surfaces of its under teeth, and its curiously prehensile

KINKAJOU, OR POTTO.—*Cercoleptes caudivólvulus.*

tail, are characteristics of sufficient importance to remove it from the pure carnivora, and place it among the animals which are capable of eating both animal and vegetable food, so that it has at present found a resting-place at the end of the ursine animals.

It is an inhabitant of Southern America, and is spread over a very large extent of country, so that it is known in different places under different appellations, such as Honey Bear, Manaviri, or Guchumbi. When full grown, the Kinkajou is equal to a large cat in size, but is very much stronger in proportion to the dimensions of its body. The colour of the animal is a very light dun, obscurely traversed by narrow darker bands, that run over the back towards the ribs, and partly follow their course. Another darker band is observable round the neck, but all these marks are so very indistinct, that they can only be seen in a favourable light.

The most remarkable point in this animal is the extreme length and flexibility of the tongue, which the creature is able to protrude to a marvellous extent, and which it can insinuate into the smallest crevices in search of the insects which have taken shelter therein. It is said that the animal employs its long tongue for the purpose of thrusting that organ into the bee-cells, and licking out the sweet contents of the waxen treasury. With its tongue it can perform many of the offices of an elephant's trunk, and will

PANDA, OR WAH.—*Ailurus Fulgens.*

frequently seize and draw towards its mouth the articles of food which may be beyond the reach of its lips. It has also been seen to use its tail for the same purpose.

Assisted by its prehensile tail, the Kinkajou is an admirable and fearless climber, possessing the capability of suspending its body by the hinder feet and the tail, and remaining in this inverted attitude for a considerable space of time.

It is eminently nocturnal in its habits, being sadly distressed by the effect of daylight upon its eyes. The pupils of the eyes are capable of great dilatation, and when the unwelcome light of day falls upon them, they contract to a singularly small size, and their owner testifies by its uneasy movements the inconvenience which it feels. Not even the owl appears to be more disconcerted by the glare of the noontide light than does the Kinkajou. During the day, the animal lies buried in profound repose, from which it can with difficulty be aroused; but when the unwelcome light has passed away, the Kinkajou becomes extremely lively, and exhibits considerable activity of limb and playfulness of character, and runs up and down the branches with great skill, uttering at intervals a low, bleating kind of sound, and descending every now and then to drink. In descending, it makes use of its hinder claws in the same manner as has been related of the coaitis. In its native state, its food is of a mixed nature, consisting of fruits, insects, honey, small birds, eggs, and other similar substances.

It is easily tamed, and when domesticated is of a sportful nature, delighting to play with those persons whom it knows and trusts, and making pretence to bite, after the manner of puppies and kittens. It is very susceptible to kindness, and is fond of the caresses which are offered by its friends. In its wild state, however, it is a rather fierce animal, and when assaulted, offers such a spirited resistance even to human foes, that it will beat off any but a determined man, supposing him to be unarmed and unassisted.

THERE are few of the Mammalia which are decorated with such refulgently beautiful fur as that which decks the body of the WAH, or PANDA, as it is also called.

This beautiful creature is a native of Nepal, where it is known under the different names of Panda, Chitwa, and Wah,—the last-mentioned name being given to it on account of its peculiar cry. The fur of the Panda is of a bright rich chestnut-brown, which rapidly darkens into a peculiarly rich black upon the ribs and the outside of the legs. The head is of a whitish-fawn colour, with a ruddy chesnut spot under each eye. The tail is of the same chesnut hue as the body, and is marked with a series of dark rings. The head is very short, and thick muzzled, presenting a curious contrast to the coaitis and racoons.

It is generally found among the trees that grow near rivers and mountain-torrents, but does not seem to occur in sufficient numbers to render its beautiful fur an object of commercial value. This is the more to be regretted, as the coat of the Panda is not only handsome in appearance, but is very thick, fine, and warm in texture, being composed of a double set of hairs, the one forming a thick woolly covering to the skin, and the other composed of long glistening hairs that pierce through the wool and give the exquisitely rich colouring to the surface of the fur. The soles of the feet are not merely defended by nailed and thickened cuticles, but are furnished with a heavy covering of woolly hair, which in some species is of a light grey colour, and in others of a snowy white, that contrasts strangely with the deep rich black of the legs and paws.

The food of the Panda is usually of an animal character, and consists chiefly of birds, their eggs, and the smaller Mammalia and insects, many of which it discovers on the trees whereon it is generally found.

GROUP OF BRITISH SHREWS.

INSECTIVORA.

THE animals which are comprised in the Insect-eating group are well represented in England, in which country we find the Mole, the various Shrews, and the Hedgehog, as examples of the TALPIDÆ, or the family of the Moles.

As the food of these creatures is almost exclusively composed of insects, snails, worms, and similar animals, it is necessary that their teeth should be formed in a manner suitable to seizing and retaining their prey. Accordingly, on opening the mouth of a mole, a shrew, or a hedgehog, we find that none of the teeth are provided with flattened surfaces for the purpose of grinding the food, but that even the molar teeth are covered with sharp points, which are admirably suited for piercing and retaining their active prey, or for tearing it to pieces when it has been killed. All the insectivorous animals are plantigrade in their walk.

Some of these creatures, such as the shrew, present so close an external resemblance to the common mice, that they are popularly supposed to belong to the same class, and are called by the same general name. Many species live beneath the surface of the earth, and seek in that dark hunting-ground the prey which cannot be enticed to the surface in sufficient numbers to supply adequate nourishment for the ever-hungry worm-devourers.

MOLE.—*Talpa Europœa.*

OF all the insect-eating animals there is none which is better known by name than the common MOLE, and very few which are less known by their true character.

On inspecting a living Mole that has been captured on the surface of the earth, and comparing it with the multitudinous creatures that find their subsistence on the earth's surface, rejoicing in the full light of day, and free to wander as they please, we cannot but feel some emotions of surprise at the sight of a creature which is naturally debarred from all these sources of gratification, and which passes its life in darkness below the surface of the ground.

Yet this pity, natural though it be, will be entirely thrown away, for there is scarcely any creature that lives which is better fitted for enjoyment, or which is urged by more fiery passions. Dull and harmless as it may appear to be, it is in reality one of the most ferocious animals in existence, and will engage in the fiercest combats upon very slight provocation. While thus employed, its whole faculties are so entirely absorbed in its thirst for revenge, that it will leave the subterraneous shafts which it has been so busily excavating, and join battle with its foe in the full light of day. Should one of the combatants overpower and kill the other, the victorious Mole springs upon the vanquished enemy, tears its body open, and eagerly plunging its nose into the wound, drinks the blood of its slaughtered enemy, and feasts richly on the sanguine banquet.

Such a combat was lately witnessed by one of my friends, who kindly wrote the account of the proceeding, and of the fate of one of the combatants.

" During a fine autumn afternoon, I was taking a walk in some woods near Shooter's Hill, and had reached a very retired part of the path, when I suddenly heard a considerable scrambling amongst the dried leaves and underwood. Upon stooping to obtain a view of whatever might be causing the disturbance, I caught sight of two little black creatures engaged in fierce combat. They tumbled over one another, and were so much concealed by the leaves that they could not be easily distinguished. Scarcely had I obtained this view of the combatants than one of them began to retreat, and was quickly followed by his opponent. Seeing the direction in which they were running, I made way through the briars as quickly as possible, and placed myself directly in the path of the creatures. They came on quite unconscious of my presence, and dodged about within a few feet of me. I could now perceive that the creatures were either Moles or rats, and determined, whichever they might be, to capture one or both.

Disregarding the thorns and thistles, I jumped through the underwood, and was then close to the animals, which immediately ceased their personal quarrel, and began sinking in a most ghostlike manner into the soil. This proceeding showed me that they were Moles. Not being particularly acquainted with the arrangement of the Mole's teeth, or with its disposition, I did not venture to take hold of either creature with my hand, but being anxious to effect a capture, I caught hold of the tail of one of the little fellows, and with the help of a sharp tug I pulled him out of the ground, and gave him an aërial voyage.

He came to the ground without any apparent injury, and again began busily sinking a shaft. This time the back looked so very inviting that I seized him by the short fur upon that portion of his body, and then found that I had him quite secure. He appeared very philosophical under the circumstances, and did not perform any unnecessary contortions, having very soon discovered that he was a safe prisoner. The next proceeding was to place him in my handkerchief, and to discover whether he could, either by his teeth or claws, make his way through the silk. No sooner was he suspended in this impromptu prison than he tried his utmost to work his way through the walls, but the silk yielded to him and would not open. A minute's exertion appeared to satisfy him, after which he laid himself calmly down. Having been occupied for some little time with this creature, I then sought after his companion, which had, however, effected his subterranean retreat, and was probably two feet beneath the surface.

My prisoner was conveyed in safety to my domicile, and was placed in a large tub, over which for security a board was placed, and in which was some earth. The little gentleman was quite at home in his tub, and enjoyed himself amazingly with a 'Diet of Worms.' To supply him, however, with this luxury was not an easy task; half an hour's digging in a yielding soil brought to light scarcely a sufficient quantity of food for one day's consumption. Small worms of about two inches in length were quickly disposed of; whilst fine long ones were put out of sight in two meals. After feasting upon half a dozen or so, the Mole would look very much like a boy full of pudding, and appeared to take a little doze. No sooner, however, did a worm give him a flap on the nose with its tail than he awoke, and, with renewed vigour and appetite, demolished half a dozen more victims, when he would again subside into a quiet slumber.

He lived in great ease and luxury during three days, at the end of which time he met an untimely end. The board upon his tub was accidentally knocked off by an awkward man, who forgot or neglected to replace it. A sly old tom-cat saw the Mole, and impelled by his own organ of destructiveness, killed, but would not eat our hero."

The cause of this curious combat was probably at no great distance, in the shape of a female Mole, for whose favour the two animals were so fiercely fighting. The Mole pursues its intended mate with extraordinary zeal and perseverance; and as the number of males is considerably greater than that of females, it seldom happens that a Mole succeeds in winning his bride until he has gained her in battle. So furious are all the passions of the Mole, that we may think ourselves fortunate that the creature is so small; for if it were as large as a tiger it would be by far the more formidable animal.

Even with its limited capability, it has more than once proved itself a dangerous creature ; for on one occasion, a Mole that had been made prisoner turned fiercely on its captor, and fixed its teeth into his hand with such pertinacious courage that it would not loosen its hold until it had been squeezed nearly to death between the teeth of its antagonist, who was obliged to have resort to that unpleasant mode of defence in order to free himself from the infuriated little animal. Another of these creatures flung itself upon a young lady's neck, and inflicted a severe wound before its purpose could be comprehended or its movements arrested.

With the exception of sight, the senses of the Mole seem to be remarkably developed.

The sense of scent is singularly acute, and enables the animal to discover the presence of the earthworms on which it feeds, and to chase them successfully through their subterranean meanderings,—a kind of terrestrial otter. So acute is the sense of smell, that the experienced Mole-catchers are in the habit of keeping a dead Mole at hand when they are engaged in their destructive avocation ; and after setting their traps, draw the dead Mole over every part of the trap or adjoining soil which their hands have touched, so as to overpower the taint of human contact. This is an effectual precaution, as the Mole is endowed with a powerful, very peculiar, and very unpleasant scent, that adheres pertinaciously to the hand, and cannot be entirely removed without repeated lavation.

The hearing of the Mole is proverbially excellent ; and it is probable that the animal is aided in its pursuit of worms by the sense of hearing as well as that of smell. Much of the Mole's safety is probably owing to its exquisite hearing, which gives it timely notice of the approach of any living being, and enables it to secure itself by rapidly sinking below the surface of the earth. To tread so softly that the blind Mole may not hear a footfall, is an expression which has become a household word.

The sense of touch is peculiarly delicate, and seems to be chiefly resident in the long and flexible nose, which is employed by the Mole for other purposes than those of scent. When the creature is placed upon the surface of the ground, and is about to sink one of its far-famed tunnels, it employs its nose for that purpose almost as effectually as its armed fore-paws. I have often seen the animal engaged in the process of sinking a pit, and have observed that it always begins by running about very rapidly, wriggling its snout backwards and forwards upon the ground, as if to discover a soft spot. When it has fixed upon a suitable locality, it begins its excavation by rooting in the ground with its nose, and making a shallow groove in the earth by way of a commencement. Not until it has formed this preliminary trench does it bring its digging claws into action ; and even while employing its natural spades in the actual digging and casting up of loose earth, it still makes use of its nose as a pioneer, so to speak, and seems to learn, by means of the delicate sense of touch with which that organ is gifted, the nature of the soil through which the animal intends to make its way.

It seldom happens that all the senses of an animal are developed to an equal extent, so that where one or two are singularly acute it is generally at the expense of the others. Such is the case with the Mole ; for although the scent, touch, and hearing are remarkable for their excellence, the sight is so extremely defective that it may almost be considered as a nullity. It is true that the Mole possesses eyes ; but those organs of vision are so small, and so deeply hidden in the fur, that they can be but of little use to their owner, except to mark the distinctions between light and darkness. The eyes are so exceedingly small that their very existence has been denied, and it is only by a careful search that they can be seen at all.

The simplest mode of observing the Mole's eyes in perfection is to immerse the living animal in water. It fancies that it is in danger of drowning, and therefore exerts its power of protruding the eyes from the furry coat, in order to employ every means for escaping from the deadly peril. Its eyes are then perceptible, as little, black, beady objects that glitter through the fur, but do not appear to enjoy any great powers of vision. This power of protruding and withdrawing the eyes is rendered necessary by the subterranean habits of the animal, which require that it should be able to protect its eyes from the loose mould through which the creature is constantly passing.

In order to adapt the Mole to the peculiar life which it leads, the entire framework of its body is very wonderfully constructed. As its chief employment consists in digging, the entire strength of the animal seems to be concentrated in the fore-quarters, where the bones and muscles are developed to a marvellous extent. If the Mole be stripped of its skin, the muscles of the fore-limbs will be found to be so powerful that they roll between the fingers, hard, slippery, and almost incompressible. These enormous muscles give power and motion to the very efficient digging apparatus with which the Mole is endowed.

The fore-paws are extremely large, and furnished with strong and flattened nails. They are turned rather obliquely, as seen in the figure on p. 423, in order to give free scope to their exertions. The bones of the fore-arm are of very great thickness, and bowed in that peculiar manner which always indicates enormous strength in the possessor. But the most striking and curious peculiarity in the structure of the Mole is the singularly long shoulder-blade, which, by its great length and strength, affords attachment to the powerful muscles which alone could give the requisite force to the broad, spade-like paws. The paws are devoid of the soft fur that shields the rest of the body, and are covered with a thick but naked skin. It is chiefly to these paws that any mould is found adherent when the Mole is captured, for the soft and velvet-like fur permits no earthy stain to defile its glossy smoothness.

The Mole's fur is remarkably fine in its texture, and is affixed to the skin in such a manner that it has no particular "grain," and lies smoothly in every direction. Were it not for this peculiarity, the Mole would find great difficulty in proceeding along its galleries with the necessary celerity. The skin of the Mole is remarkably tough and thick, and is often used by the peasantry for the purpose of making purses. The process of manufacture is simple enough, consisting merely in cutting the animal across, just behind the shoulders, stripping the skin from the hinder portions, drying it carefully, and closing it by means of a string run round the edge.

The Mole is said to be an excellent swimmer, and to be able to cross rivers, when led to such an act by any adequately powerful motive. How far true this assertion may be, I cannot prove by personal experience; but I think that it is likely to be possible, for I have seen a Mole swim across the bend of a brook—a distance of some few yards—and perform its natatory achievement with great ease. I was not near enough to ascertain the mode of its progression, but it seemed to use its fore-paws as the principal instruments of locomotion. This circumstance took place in Wiltshire.

From all accounts, the Mole seems to be a thirsty animal, and to stand in constant need of water, drinking every few hours in the course of the day. In order to supply this want it is in the habit of sinking well-like pits in different parts of its "runs," so that it may never be without the means of quenching its thirst. Everything that the Mole does is marked with that air of desperate energy which is so characteristic of the animal. The labourers in different parts of England all unite in the same story, that the Mole works for three hours "like a horse," and then rests for three hours, labouring and resting alternately through the day, and with admirable perception of time.

The well-known "mole-hills" which stud certain lands, and which disfigure them so sadly, however much their unsightliness may be compensated by their real usefulness, are of various kinds, according to the sex and age of the miner. The small hillocks which follow each other in rapid succession are generally made by the female Mole before she has produced her little family, and when she is not able to undergo the great labour of digging in the harder soil. Sometimes the "run" is so shallow as to permit the superincumbent earth to fall in, so that the course which the Mole has followed is little more than a trench. This is said to be produced by the little coquetries that take place between the Mole and its future mate, when the one flies in simulated terror, and the other follows with undisguised determination. Deeper in the soil is often found a very large burrow, sufficiently wide to permit two Moles to pass each other. This is one of the high-roads which lead from one feeding-ground to another, and from which the different shafts radiate.

But the finest efforts of talpine architecture are to be found in the central fortress, from which the various roads diverge, and the nest which the maternal Mole forms for the security of her young.

The fortress is of a very peculiar construction, and is calculated to permit the ingress or egress of the Mole from almost any direction, so that when its acute senses give notice of the approach of an enemy, it can make its retreat without difficulty.

The first operation is to build a tolerably large hill of compact and well-trodden earth. Near the summit of this mound the excavator runs a circular gallery, and another near the bottom, connecting the two galleries with five short passages. It then burrows into the centre of the mound, and digs a moderately large spherical hole, which it connects with the lower gallery by three passages. A very large passage, which is a continuation of the high-road, is then driven into the spherical chamber by dipping under the lower gallery, and is connected with the circular chamber from below. Lastly, the Mole drives a great number of runs, which radiate from the rest in all directions, and which all open into the lower circular gallery. It will be seen from this short description, that if a Mole should be surprised in its nest it can withdraw through its central chamber and so reach the high-road at once, or can slip through either of the short connecting galleries and escape into any of the numerous radiatory runs.

In the central or middle chamber of the edifice the Mole places a quantity of dried grass or leaves, upon which it sleeps during its hours of repose. This complicated room is seldom used during the summer months, as at that time the Mole prefers to live in one of the ordinary hillocks.

The nest which the female contrives is not so complicated as the fortress, but is well adapted for its purpose. The hillock in which the nest is made is always a very large one, and is generally placed at some distance from the fortress. Its interior is very large, and is generally filled with dried grass, moss, or other similar substances, and it is said that in some of these nests have been found certain roots on which the young Moles can feed during the first weeks of their existence. The young are usually born about April, but their appearance in the world is not so determinately settled as that of many animals, as young Moles are found continually from March until August. The average of their number is four or five, although as many as seven young have been found in one nest. There is but one brood in a year.

The colour of the Mole is usually of a blackish-grey, but it is extremely variable in the tinting of its fur, and it is not uncommon to find in a single locality specimens of every hue from brown to white. There are specimens in the British Museum of almost every tint, and I have long had in my possession a cream-coloured Mole-skin, which was obtained I believe in Wiltshire, as it was furnished by a Mole-catcher that resided in that county. The fur is so beautifully smooth and soft that it has sometimes, though rarely, been employed as an article of wearing apparel, or used as a light and delicate coverlet. The fur, or "felt," is best and most glossy if the animal is taken in the winter.

Whether the Mole is more useful or hurtful to the agricultural interest is at present a mooted point, and seems likely to continue so. It cannot be denied, however, that the subterraneous passages of the Mole, added to those of the earthworms, form a very complete system of subsoil drainage, and that the creature is continually enriching the surface of the ground by bringing fresh earth from a considerable depth, and thus involuntarily performing the office of a plough or a spade.

ONE of the animals which forms a link between the Moles and the Shrews, and seems to possess some of the peculiar habits of each, is the curiously formed SCALOPS, or SHREW-MOLE.

This creature seems to be peculiar to Northern America, and is generally found near the banks of rivers, being very water-loving in its habits. Like the ordinary mole, the Scalops passes the greater portion of its existence below the surface of the ground, and finds a subsistence among the worms and other creatures which it captures during its subterraneous meanderings. The muzzle of the Scalops is even more remarkable than that of the common mole, being much longer in proportion to the size of the animal, and is cartilaginous at its extremity. The claws of the fore-feet are very long and flattened, and are arranged in such a manner as to present a sharp point to the earth when the creature exercises them in the act for which they were intended. The hinder feet and legs are extremely small, and the tail is but short. There is no apparent outward vestige

SCALOPS, OR SHREW-MOLE.—*Scalops aquáticus.*

of an ear, and the eyes are almost invisible. In size it equals the common European Mole, being about seven inches in total length.

Another similar animal, the Desman of Russia, has been frequently mistaken for the Shrew-Mole, but can be easily distinguished at a casual glance by the greater length of its tail, and its superior size ; its total length being sixteen or seventeen inches.

CAPE CHRYSOCHLORE, OR CHANGEABLE MOLE.—*Chrysochlóris holosericea.*

The CHRYSOCHLORE, or SHINING MOLE, or CHANGEABLE MOLE, has derived its various names from the very peculiar character of its fur.

The colour of the Chrysochlore's coat is of a character that resembles "shot" silk, or the peculiar changeable metallic radiance which is thrown from the feathers of many birds. According to the light in which the animal is viewed, the fur is in some parts of a golden or bronzed green, and in others of a bronzed red, these brilliant hues interchanging among themselves as the animal moves, or as the light falls at different angles upon the fur. The name Chrysochlore is derived from two Greek words, signifying gold-green, and is happily applicable to the very singular colouring of the animal. The term *holosericea* is also Greek, and signifying "wholly silken," in reference to the lustrous surface of the hairs.

Even putting aside the strange chromatism of the fur, the creature is a very remarkable one in many respects, and especially deserving of notice on account of its teeth and its feet.

The teeth of the Changeable Mole are arranged in a very peculiar mode, being separated from each other by an interval that is equal to their thickness, so that when the jaws are closed, the teeth of either jaw fit exactly into the interstices that are left between those of the opposite jaw, like the iron serrations of a steel-trap. It has been well remarked that "the Chrysochlore affords, it is believed, the only example in the animal world of teeth being opposed by their anterior and posterior faces. The skeleton is altogether a singular one, for there are no less than nineteen pairs of ribs, and in one species twenty pairs have been made out. The first rib is thick and broad in proportion to the others.

The fore-feet are furnished with four toes, the fourth toe being very small, and tipped with a nail of ordinary size. The other three toes are armed with most formidable claws, by means of which the animal is enabled to dig into the earth. The middle toe carries a claw of surprising dimensions, as may be seen on reference to the engraving. The hinder feet are five-toed, and of no very great proportional size. The eye is externally invisible, being covered with skin, so that the animal appears to be practically blind. There are no ears, and no tail. The size of the creature is rather less than that of the common European Mole.

As may be perceived from its title, it is a native of the Cape of Good Hope, although it was formerly thought to inhabit Siberia. It is also known by the name of the Hottentot Chrysochlore; and the French name is Taupe dorée, or Gilded Mole. Its food consists, like that of the other Moles, of worms and various insects.

RADIATED MOLE, OR STAR-NOSED MOLE.—*Astromyctes cristatus.*

Even in a stuffed specimen, or in an uncoloured engraving, the aspect of the RADIATED MOLE is a most grotesque and singular one; but its quaint uniqueness is much more striking when the animal is alive and in full health.

The most remarkable point in this animal is the muzzle, which is produced into a long, slender proboscis, round the extremity of which are arranged a number of soft, fleshy rays, of a bright rose-colour, radiating like the petals of a daisy, or the tentacles of a sea-anemone. These curious rays, or caruncles, as they are more scientifically termed, can be spread or closed at pleasure, and present a strange spectacle when in movement. Their probable object is that they may serve as a delicate organ of touch, to aid the animal in procuring the worms and insects on which it feeds. The openings of the nostrils are situated in the centre of the radiated disc. The number of the caruncles is about twenty.

On account of the proportionately lengthened tail, the animal is sometimes called the Long-tailed Mole; for the tail is two inches and a half in length, while the head and body only measure four inches and a half. Another name by which it is known is the Condylure, or "knotty tail," an epithet which has been applied to it because, when a specimen is dried, the skin of the tail contracts so firmly over the vertebræ that the separate bones exhibit their form through the skin, and give to the tail a knotted aspect. The colour of the fur is much like that of the common Mole, being a velvety blackish-grey on the upper portions of the body, and paler on the under parts. The eyes are extremely small, and there is no external indication of ears. It is an inhabitant of Canada and the United States.

TUPAIA-TANA.—*Tupaia Tana.*

THE insect-eating animals which have already been described are in the habit of searching for their prey under the surface of the earth, and are furnished with extremely imperfect means of sight. But the curious examples of Insectivora which are collected into the single genus Tupaia are of a very different nature, living in the full light of day, and seeking their insect prey among the branches of the trees on which they dwell. It needs, therefore, that animals which obtain their food in such a manner should be endowed with excellent powers of vision; and we find accordingly that the Tupaias—which animals will be represented by two examples—are furnished with good eyes and quick sight. Indeed, the entire aspect of these creatures reminds the observer more of the squirrels than of the moles. The Tupaias are inhabitants of Sumatra and parts of India.

The head of this animal is very singular in its shape, which is well represented in the engraving. The upper jaw is slightly longer than the lower, and the muzzle considerably produced, so that the head has a strangely dragon-like aspect, which is heightened by the position of the ears, which are set very far back, and by the long sharp rows of teeth which arm each jaw. The long bushy tail of the Tupaia gives it a kind of resemblance to the squirrel, a resemblance which is appreciated by the native

Sumatrans, who call the squirrels and the Tupaias by the same name. The feet are planti-grade, and terminated by five toes on each foot, armed with small, but sharp nails, which assist the animal in climbing, and are sufficiently elevated to be spared from friction against the ground. In the hinder feet the fourth toe is the longest. The hair is of a silky texture, and tinged with brown and yellow by reason of the alternate coloured rings with which each hair is marked. On the tail the hair is long and bushy, and hangs equally on each side, after the manner that is learnedly termed " distichous."

These animals are variously named by different zoological writers, and the genus in which they are placed is by some termed Tupaia, after the native name, by some Hylogale, and by others Cladobates. The last mentioned term seems to be in greatest favour, and is a very appropriate one, signifying " branch-traverser."

ANOTHER example of this curious genus is the PRESS, or FERRUGINEOUS TUPAIA. This pretty creature is so exceedingly like a squirrel, as it runs about the branches of the trees, that it can hardly be distinguished from that animal except by the elongated outline of its head, as it is defined sharply against the sky. It is a small animal, measuring only some thirteen or fourteen inches in total length, of which the tail occupies rather more than five inches. The length of the head is two inches, and the height of the animal, as it stands, is rather more than three inches.

PRESS.—*Tupaia Ferruginea.*

The colouring of its fur is very elegant. The prevailing tint is a brownish-maroon, which in some parts, such as along the spine, is deepened into a rich brownish-black, and in others, such as on the ribs and flanks, is warmed into a reddish tint. On account of this peculiar hue, which resembles the red rust of iron, the epithet of " ferruginea " has been applied to the animal. This change of colour is caused by the mode in which the hairs are marked in alternate rings of black and maroon. Those which run along the back are black, with a fawn-coloured ring in the middle, but those which grow upon the ribs are fawn, with a black ring in the middle. The ears are black. Upon the under surface of the body the fur is of a whitish-yellow, which on the abdomen and the internal face of the limbs fades into grey. The long and bushy hairs which decorate the tail are so dotted with white that their aggregation upon the tail gives to that member a greyish-brown effect.

Although the teeth of all the Tupaias are evidently of an insectivorous description, the Press, as well as its congeners, is said to feed chiefly on coleopterous insects, but to vary its diet with certain fruits. It is said that the Press partakes so far of the carni-vorous propensities of the mole, that it will sometimes pounce upon small birds as they are hopping among the branches, and make a meal upon their bodies. One of these animals that was tamed, and accustomed to roam about the house at will, was very fond of milk and fruits, and used to attend at every meal for the purpose of obtaining these coveted luxuries.

ELEPHANT SHREW.—*Macroscélides Proboscídeus.*

THE elongation of the nose, which has already been noticed in the Tupaias of Sumatra, seems to have reached its utmost limit in those curious inhabitants of the Cape that are called, from their elephantine elongation of nose, the ELEPHANT SHREWS. Several species of Elephant Shrews are known to exist, all of which, with one exception, are inhabitants of Southern Africa. The solitary exception, Macroscelides Roretti, is found in Algeria.

The peculiarly long nose of the Elephant Shrew is perforated at its extremity by the nostrils, which are rather obliquely placed, and is supposed to aid the animal in its search after the insects and other creatures on which it feeds. The eyes are rather large in proportion to the size of the animal.

The tail is long and slender, much resembling the same organ in the common mouse, and in some specimens, probably males, is furnished at the base with glandular follicles, or little sacs. The legs are nearly of equal size, but the hinder limbs are much longer than the fore-legs, on account of the very great length of the feet, which are capable of affording support to the creature as it sits in an upright position. As might be presumed from the great length of the hinder limbs, the Elephant Shrew is possessed of great locomotive powers, and when alarmed, can skim over the ground with such celerity that its form becomes quite obscured by the rapidity of its movement through the air. Its food consists of insects, which it captures in open day.

Although the Elephant Shrew is a diurnal animal, seeking its prey in broad daylight, its habitation is made below the surface of the ground, and consists of a deep and tortuous burrow, the entrance to which is a perpendicularly-sunk shaft of some little depth. To this place of refuge the creature always flies when alarmed, and as it is so exceedingly swift in its movements, it is not readily captured or intercepted.

The colour of the fur is a dark and rather cloudy brown, which is warmed with a reddish tinge upon the sides and flanks, and fades on the abdomen and inner portions of the limbs into a greyish-white. The generic name, Macroscelides, is of Greek origin, in allusion to the great length of its hinder limbs, and signifies "long-legged." It is but a small animal, as the length of the head and body is not quite four inches in measurement, and the tail is about three inches and a quarter.

Passing in a regular gradation from the moles to the shrews and hedgehogs, we pause for a while at the powerfully scented animal that is called, by virtue of its perfumed person, the Musk-Rat of India, and is also known by the titles of Mondjourou, and Sondeli.

This animal is a native of various parts of India, and is very well known on account of the extremely powerful scent which exudes from certain glands that are situated in the under parts of the body and on the flanks.

The odoriferous substance, which is secreted by the above-mentioned glands, is of a musky nature, and possesses the property of penetrating and adhering to every substance over which the Musk-Rat has passed. The musky odour clings so pertinaciously to the objects which are impregnated with its tainting contact, that in many cases they become entirely useless. Provisions of all kinds are frequently spoiled by the evil odour with which they are saturated; and of so penetrating a nature is the musky scent, that the combined powers of glass and cork are unable to preserve the contents of bottles from its unpleasant influence. Let but a Sondeli run over a bottle of wine, and the contained liquid will be so powerfully scented with a musky savour that it will be rendered unfit for civilized palates, and must be removed from the neighbourhood of other wines, lest the contaminating influence should extend to them also.

In colour it is not unlike the common shrew of England; having a slight chestnut, or reddish tinge, upon a mouse-coloured ground, fading into grey on the under parts of the body. In size, however, it is much the

SONDELI.—*Sorex murinus.*

superior of that animal; being nearly as large as the common brown or "Hanoverian" rat. The hair is very short, and the peculiar reddish-brown hue of the fur is caused by the different tintings of the upper and under fur.

During the autumnal months of the year, the country roads and by-paths are frequently rendered remarkable by the presence of little mouse-like animals, with long snouts and peculiarly squared tails, that lie dead upon the ground, without mark of external injury to account for the manner of their decease.

There are probably many other such corpses upon the wide and grassy meadow lands, but, owing to the nature of the ground, they are not so conspicuous as those upon the smoothly trodden paths. The presence of these deceased creatures is the more remarkable, because there are so many predatory animals and birds, such as cats, weasels, stoats, owls, and hawks, which would be very likely to kill such small prey, but, having slain them, would be almost sure to eat them. These unsepultured remains are the bodies of the Shrew-mouse of England, otherwise known by the name of Erd Shrew. Another title by which this little animal is known, in some parts of England, is the Fetid Shrew; a name which has been given to it on account of the powerful scent which it exudes; and the creature is called in Scotland, the Ranny, a name which is evidently modified from the Latin term, *araneus*, or spider-like, which has been applied to this animal by several writers, because it was said to bite poisonously like a spider.

The teeth of the true Shrew are very peculiar, so much so, indeed, that they

1.   F F

cannot be mistaken for those of any other animal. Their peculiarities are mostly remarkable in the incisor teeth, which are extremely long; those of the upper jaw being curved and notched at their base, while those of the lower jaw project almost horizontally. There are no canines, and the molars differ slightly in arrangement, according to the species. In the Erd Shrew the tips of the teeth are tinged with a blood-coloured brown.

The head of the Shrew is rather long, and its apparent length is increased by the long and flexible nose which gives so peculiar an aspect to the animal, and serves to distinguish it at a glance from the common mouse, which it so nearly resembles in general shape and colour. The object of this elongated nose is supposed to be for the purpose of enabling the animal to root in the ground after the various creatures on which it feeds, or to thrust its head among the densest and closest herbage. Many insects and their larvæ are found in such localities, and it is upon such food that the Shrew chiefly subsists. Worms are also captured and eaten by the Shrew, which in many of its habits is not unlike the mole.

ERD SHREW, OR SHREW-MOUSE.—*Corsira vulgáris.*

The habitation of the Shrew is in certain little subterraneous tunnels, which it excavates in the soil, and which serve as a hunting-ground as well as a home. Like the mole, the Shrew is very impatient of hunger, and cannot endure a protracted fast, although it may not be so inordinately voracious as that velvet-coated animal, which it is said will die of hunger if it be kept without food for six hours. It has been suggested, that the many dead Shrews which are found in the autumn owe their deaths to starvation, the worms having descended too deeply into the ground for them to follow, and the insects, being pinched with the cold, having concealed themselves in their wintry hiding-places.

If this be the case, the curious phenomenon of dead Shrews lying uninjured on the ground will be readily cleared up, although it will not account for the singular fact that the dead animals are not carried off by cat, weasel, or owl. For this portion of the phenomenon another reason must be found; which probably exists in the rank and powerful scent which saturates the body of the Shrew, and which is sufficiently unpleasant to deter cats and other animals from eating its flesh. Owls, however, will eat the Shrew, as has been found by examination of the pellets which are ejected by owls and other birds of prey, and which contain the skin, feathers, bones, and other indigestible portions of the creatures on which they prey. Twenty such pellets, or casts, as they are technically termed, were examined for the purpose of ascertaining their component parts, and no less than seven Shrew skeletons were discovered in the *débris*. Moles are said to be among the number of the Shrew's enemies, and to make occasional havoc among the pretty little creatures.

Sometimes the Shrews mutually kill each other, for they are most pugnacious little beings, and on small ground of quarrel enter into persevering and deadly combats; which, if they took place between larger animals, would be terrifically grand, but in such little creatures appear almost ludicrous. They hold with their rows of bristling teeth with the pertinacity of bull-dogs, and, heedless of everything but the paroxysm of their blind fury, roll over each other on the ground, locked in spiteful embrace, and uttering a rapid succession of shrill cries, which pierce the ears like needles of sound. It is a most fortunate circumstance that the larger animals are not so vindictively pugnacious as the moles and the Shrews; for it would be a very hard case if we were unable to put two horses or two cows in the same field without the certainty of immediate fight, and the probability that one of the combatants would lose its life in the struggle. Such, however, is the case with the Shrews; for if two of these little quadrupeds be confined in the same box, they are sure to fight to the death, and the consummation of the combat is, generally, that the vanquished foe is eaten by the victor.

However great may be the damage which the bite of such tiny teeth may inflict upon each other, yet the bite of a Shrew is so insignificant as to make hardly any impress even on the delicate skin of the human hand. Popular prejudice, however, here steps in, and attributes to the bite of the Shrew such venomous properties that in many districts of England the viper is less feared than the little harmless Shrew. The very touch of the Shrew's foot is considered as a certain herald of evil, and animals or men which had been "Shrew-struck" were supposed to labour under a malady which was incurable except by a rather singular remedy, which partakes somewhat of the homœopathic principle, that "similia similibus curantur."

The curative power which alone could heal the Shrew-stroke lay in the branches of a Shrew-ash, or an ash-tree which had been imbued with the shrewish nature by a very simple process. A living Shrew was captured and carried to the ash-tree which was intended to receive the healing virtues. An auger-hole was made into the trunk, the poor Shrew was introduced into the cavity, and the auger-hole closed by a wooden plug. Fortunately for the wretched little prisoner, the entire want of air would almost immediately cause its death. But were its little life to linger for ever so long a time in the ash-trunk, its incarceration would still have taken place, for where superstition raises its cruel head, humanity is banished.

The popular ideas respecting the Shrew's bite, which once reigned even over the scientific world, and are still in full force throughout many portions of the rural districts, may be gathered from the following extract from a curious old zoological author named Topsel, in his "History of Four-footed Beasts and Serpents," published in London in the year A.D. 1658, p. 406:—

"It is a ravening beast, feigning itself gentle and tame, but, being touched, it biteth deep, and poysoneth deadly. It beareth a cruel minde, desiring to hurt anything, neither is there any creature that it loveth, or it loveth him, because it is feared of all. The cats, as we have said, do hunt it, and kill it, but they eat not them, for if they do, they consume away and die. They annoy vines, and are seldom taken, except in cold; they frequent ox-dung, and in the winter time repair to houses, gardens, and stables, where they are taken and killed.

If they fall into a cart-road, they die, and cannot get forth again, as *Marcellus, Nicander,* and *Pliny* affirm. And the reason is given by *Philes,* for being in the same, it is so amazed, and trembleth, as if it were in bands. And for this cause some of the ancients have prescribed the earth of a cart-road to be laid to the biting of this mouse as a remedy thereof. They go very slowly; they are fraudulent, and take their prey by deceit. Many times they gnaw the oxes hoofs in the stable.

They love the rotten flesh of ravens; and therefore in *France,* when they have killed a raven, they keep it till it stinketh, and then cast it in the places where the Shrew-mice haunt, whereunto they gather in so great a number, that you may kill them with shovels. The *Egyptians,* upon the former opinion of holiness, do bury them when they die. And thus much for the description of this beast. The succeeding discourse toucheth the medecines arising out of this beast; also the cure of her venomous bitings.

The Shrew, which by falling by chance into a cart-rode or track, doth die upon the same, being burned, and afterwards beaten, or dissolved into dust, and mingled with goose-grease, being rubbed or anointed upon those which are troubled with the swelling coming by the cause of some inflammation, doth bring into them a wonderful and most admirable cure and remedy. The Shrew being slain or killed, hanging so that neither then nor afterwards she may touch the ground, doth help those which are grieved and pained in their bodies, with sores called fellons or biles, which doth pain them with a great inflammation, so that it be three times environed or compassed about the party so troubled. The Shrew which dyeth in the furrow of a cart-wheel, being found and rowled in potter's clay or a linnen cloth, or in crimson, or in scarlet woollen cloth, and three times marked about the impostrumes, which will suddenly swell in any man's body, will very speedily and effectually help and cure the same.

The tail of a Shrew being cut off and burned, and afterwards beaten into dust, and applyed or anointed upon the sore of any man, which came by the bite of a greedy and ravenous dog, will in very short space make them both whole and sound, so that the tail be cut from the Shrew when she is alive, not when she is dead, for then it hath neither good operation, nor efficacy in it."

It is probable that this virulent hatred of the Shrew, and this groundless terror of its bite, was caused by the rank scent which exudes from the creature, and the acknowledged fact that the Shrew is frequently seen in the close vicinity of reposing cattle. But as the Shrew is an insectivorous animal, it has been well suggested that its habit of frequenting the neighbourhood of cattle may be in consequence of the flies and other insects which are always found in such localities, and on which the Shrew hopes to make a meal.

It has already been mentioned that the Shrew will eat one of its own species if slain in battle, and it is therefore evident that its food does not wholly consist of insects and worms, but is occasionally varied by other and more generous diet. One of these little creatures has been discovered and killed while grasping a frog by the hind-leg; and so firmly did it maintain its grasp, that even after its death the sharp teeth still clung to the limb of the frog. Whether the creature intended to eat the frog, or whether it was urged to this act by revenge or other motive, is uncertain.

The nest of the Shrew is not made in the burrow, as might be supposed, but is built in a suitable depression in the ground, or in a hole in a bank. It is made of leaves and other similar substances, and is entered through a hole at the side. In this nest are produced the young Shrews, from five to seven in number, and, as may be imagined, extremely diminutive in size. They are generally born in the spring.

The total length of the adult Shrew is not quite four inches, of which the tail occupies very nearly the moiety. The tail is remarkable for being square in form, instead of cylindrical, and on account of that circumstance it has received from some authors the specific name of *tetragonúrus*, or Square-tail.

SIMILAR to the erd Shrew in general aspect, but easily to be distinguished from that animal by its colour and other peculiarities, the WATER SHREW stands next on our list.

This little creature was for many years supposed to be identical with the erd Shrew, and its aquatic propensities thought to be the ebullition of joyous existence, which was not content with disporting itself upon the earth, but must needs seek a further vent for its happiness among the waters. However, the Water Shrew is now acknowledged to be a separate species, and may be distinguished from the erd Shrew by the following characteristics.

The fur of the Water Shrew is nearly black upon the upper portions of the body, instead of the reddish-brown colour which tints the fur of the erd Shrew. The under parts of the body are beautifully white, and the line of demarcation between the two colours is very distinctly drawn. The fur is very soft and silken in texture, and, when the animal is submerged under the surface of the water, possesses the useful property of repelling moisture, and preserving the body of the animal from the injurious effects of the water. When the Water Shrew is engaged in swimming, those parts of the fur which are submerged below the surface appear to be studded with an infinite number of tiny silvern beadlets, that give to the whole animal a very singular aspect. This phenomenon is produced by the minute air bubbles that cling to the fur, and which exude from the space that is left between the hairs. This curious appearance is well shown in the large engraving of British Shrews on page 422.

A further distinction, and one which is more valuable than that which is furnished by the colour of the fur, is the fringe of stiff white hairs which edges the tail and the toes, and which is evidently of great use in the natatory movements of the animal.

The Water Shrew finds its food in various ways. Sometimes it burrows in the muddy river banks, rooting in the soft earth with its elongated nose, and dislodging the larvæ of certain insects that pass that stage of their existence in the mud. It also chases and captures various aquatic insects as they move through the water, and will not disdain to feed upon moths and other similar creatures which have fallen or have been blown into the water and then drowned.

In all its movements, the Water Shrew is extremely graceful and active, displaying equal agility, whether its movements be terrestrial or aquatic. As the sphere of its vision does not appear to be very extended, it can easily be approached while it is engaged in its little gambols, and can be watched without much difficulty.

I have repeatedly observed the proceedings of a little colony of these creatures, and was able to sit within a yard or two of their haunts without their cognizance of my person. They are most sportive little creatures, and seem to enjoy a game of play with thorough appreciation, chasing each other over the ground and through the water, running up the stems of aquatic plants, and tumbling off the leaves into the water, scrambling

WATER SHREW.—*Cróssopus Fódiens*

hastily over the stones around which the stream ripples, and playing a thousand little pranks with the most evident enjoyment. Then they will suddenly cease their play, and begin to search after insects with the utmost gravity, rooting in the banks, and picking up stray flies, as if they never had any other business in view.

As it is in the habit of repeatedly passing over the same ground in these mutual chases, it soon treads a kind of path or road upon the land, which, although very obscurely marked, is yet sufficiently well defined to attract the attention of any one who is conversant with the habits of these little creatures.

Being an excellent diver, and fond of submerging itself wholly beneath the surface, the Water Shrew would suffer great inconvenience were its ears to be constantly filled with the liquid element in which it moves; and in order to avert such an inconvenience, a special provision of nature is needed. For this purpose the ears are peculiarly formed, so that as soon as the animal is wholly submerged, the pressure of the water acts upon three small valves, which fold together and effectually prevent the entrance of a single drop of water into the cavity of the ear. As soon as the animal rises to the surface, the pressure is removed, and the ears unfold like the petals of a flower, when the sun shines warmly on them.

From repeated observations, it seems that the Water Shrew is not entirely confined to the neighbourhood of water, neither is it totally dependent for its subsistence on aquatic insects, for it has been frequently seen at some distance from any stream or pond. It must be remarked, however, that a very small rivulet is amply sufficient for the purposes of the Water Shrew, which will take up its residence for several years in succession on the banks of a little artificial channel that is only used for the purpose of carrying water for the irrigation of low-lying fields.

It is believed that the Water Shrew is a more prolific animal than the erd Shrew, for whereas the latter creature produces from five to seven young at a litter, the former is blessed with a family of seven, eight, or nine in number, six or seven being the ordinary average. The total length of the Water Shrew is not quite four inches and a half, the length of the head and body being a little more than three inches, and that of the tail being about two inches. Its snout, although long, is not quite so narrow and pointed as that of the erd Shrew, and its ears are remarkably small. When it swims, it has a curious habit of spreading out its sides, so as to flatten the body as it floats upon the water.

THE largest of the British Shrews is that species which is called the OARED SHREW, on account of the oar-like formation of the feet and tail; which are edged with even longer and stiffer hairs than those which decorate the same parts in the water Shrew.

OARED SHREW, OR BLACK WATER SHREW.—*Cróssopus ciliátus.*

As may be imagined from this structure, the habits of the animal are aquatic in their nature, and its manners are so closely similar to those of the preceding species, that it may easily be mistaken for that animal, when seen at a little distance, so as to render the difference in size less conspicuous, and the colour of the under portions of the body less apparent.

It has already been stated that the back of the water Shrew is of a velvety black, and the abdomen and under portions of the body of a beautiful and clearly defined white. In the Oared Shrew, however, the back is profusely sprinkled with white hairs, and the fur of the abdomen and flanks is blackish-grey instead of pure white. The middle of the abdomen, however, together with that of the throat, is strongly tinged with yellow; the throat being more of an ashy yellow than the abdomen.

Although not so common as the erd and the water Shrew, it is of more frequent occurrence than is generally supposed, and has been found in many parts of England where it was formerly supposed to be wanting. The total length of the Oared Shrew is about five inches and a quarter, the head and body measuring rather more than three inches, and the tail being about two inches in length. Its nose is not quite so sharp or narrow as that of the water Shrew, and the ears are decorated with a slight fringe of white hair. The latter third of the tail is flattened, as if for swimming, while the remaining two-thirds are nearly cylindrical, but are slightly squared, as has been already mentioned of the common Shrew.

On account of the general darkness of its fur, it is sometimes called the Black Water Shrew, and is catalogued in the British Museum under that title. The generic name, Crossopus, is of Greek origin, and signifies "fringed-feet."

There is another British Shrew, called the Rustic Shrew (*Corsira Rústica*), which is very common throughout Ireland, and is also found in many parts of England. Among the Shrews is found the smallest known mammalian animal of the present day; being even smaller than the tiny harvest-mouse of England, which has been made so famous by Mr. White's elegant description of itself and its habits in his "Natural History of Selborne." This most minute quadruped is only one inch and a half in length, exclusive of the tail, which measures about an inch. The name of this minikin among mammals is the Etruscan Shrew, and its habitation is in Italy. Specimens are said to have been discovered in Algeria.

THE specific title of Paradoxus, or puzzling, has very appropriately been given to the AGOUTA; a little animal which is peculiar to Hayti, and which combines in its own person several characteristics that properly belong to different families of animals.

Were the observer to pay regard only to the external peculiarity of fur, ears, and tail, he would be inclined to place it among the opossums; but if he were to lay the greatest stress upon the teeth, he would probably assign it a place among the shrews.

AGOUTA, OR SOLENODON.—*Solénodon Paradoxus.*

It seems, however, that it is really allied to the latter group of animals, and it is therefore placed in the position which it at present holds in zoological catalogues. The fur of the Agouta is long, harsh, and coarse in texture, and its colour is an undecided red, tinged with yellow. The nose is extremely elongated, like that of the shrews, and strengthened at its base by a slender bone, so that it appears to be intended for the purpose of digging in the earth like those animals. The nostrils are placed at the extremity of the snout, and are divided from each other by a distinct furrow. The cheeks and lips are decorated with whisker-hairs of very great length; the eyes are very small; the ears are moderate in size, and rounded, and almost devoid of hairy covering. All the feet are terminated with five toes, and the long claws are curved, rather compressed, and evidently fitted for the purpose of scraping at the soil.

The tail is moderately long, measuring about nine inches in length, and is rounded throughout its length, remainder of the head and body being rather more than a foot long. It is not covered with hair, but is rather naked, and for the greater part of its length is scaly. The lower jaw is rather shorter than the upper.

The teeth of the Agouta are very remarkable, both for their arrangement and their form, but are very difficult to describe. The two middle incisors of the upper jaw are extremely large, almost triangular in form, and are separated from the small lateral incisors by a considerable interval. The most singular part of the dentition is, however, found in the incisors of the lower jaw, of which Van der Hoeven speaks as follows:—" The two middle incisors of the lower jaw are small, narrow, placed between two long conical, *hollowed onthe inside by a deep groove;* the second grooved incisor of the lower jaw distinguishes this genus from all the others of which the dental system is known hitherto." —*Handbook of Zoology,* vol. ii. p. 727.

The dentition of the Agouta would seem to indicate that the creature was insectivorous in its diet, but Mr. Hearne, who possessed one of these animals in a living state, remarks that its food is chiefly grain, although it is also capable of eating animal food. In general appearance the Agouta somewhat resembles the barn-rat of England, and might easily be mistaken for that animal if seen while in motion, and for a short time only. There is supposed to be but one species of this curious genus. The generic name, Solenodon, is of Greek origin, and signifies channel-toothed.

ALTHOUGH the water shrew has earned for itself its aquatic title, it is not nearly so constant an inhabitant of the water as the DAESMAN or DESMAN, an animal whose very form is sufficient to stamp it as a creature that lives almost exclusively in the water. A casual glance at the external formation of the Daesman will at once pronounce the animal to be made for swimming and diving, and its admirable adaptation for aquatic evolutions is more evident as the structure of the creature is more closely examined.

The legs and feet, which in the aquatic shrews are provided with rows of stiff bristles, in order to assist the animal in its progress through the water, are in the Daesman entirely modified into oars; the powers of terrestrial movements being subservient to those of aquatic locomotion. The toes are connected with each other by well defined webs, and the greater portion of the legs are concealed under the skin. The tail is modified for the same purpose, and is evidently used as a rudder by which the creature may direct its course.

The most remarkable point in the appearance of the Daesman is its extraordinarily elongated nose, which bears no slight resemblance to the proboscis of an elephant, and, indeed, is quite as valuable to its possessor. This prolonged nose is extremely mobile, and can be applied to various purposes: one object of the elongated nose is extremely singular, and deserves special notice.

The habitation in which the Daesman lives is a most complicated house, the entrance to which is under the surface of the water, so that the creature may escape into its stronghold whenever it has cause to fear danger. The subterraneous tunnel in which the creature lives extends for a considerable distance around the starting point, and often embraces an extent of more than twenty feet in its various windings. As the animal does not become torpid during the winter, it needs a plentiful supply of food and air. The former necessary can be obtained easily enough, but as the inclement frost of its native country covers the surface of the water with a thick coat of ice, and at the same time binds the earth in an icy chain, the poor Daesman is often sadly harassed for want of air, as it cannot find exit from its burrow, and there is no other mode of getting into the fresh atmosphere.

In this strait the long and flexible nose of the Daesman stands it in good stead; for it runs about its burrow in search of any little fissures which may communicate with the open air, and by thrusting the mobile organ into any such fortunate crevice, is enabled to obtain sufficient air to sustain the vital powers. Should the winter be a particularly severe one, many Daesmans are killed by the insufficiency of ventilation in their houses, and are found in the spring lying dead in their burrows.

It is very seldom that a Daesman is seen upon dry ground, and even at the hymeneal season, which makes all animals courageous, it is never seen at any distance from the water, and contents itself with running along the extreme edge of the water, or making temporary resting-places in the heavy leafage of aquatic plants. Even these trifling

THE DAESMAN, OR DESMAN.

aberrations from the usual tenor of its way are only caused by its eagerness in seeking its intended mate, and are limited to the short season of matrimonial enterprise. During the remainder of the year the Daesman never voluntarily quits the water; and even if it makes little journeys from one pond or stream to another, it is generally found to make use of ditches or wet channels as the roads by which it proceeds; or, in default of such aqueous paths, to traverse the distance by means of a subterraneous tunnel.

The scent which exudes from the Daesman is of a musky character, and so extremely powerful that it is locally known by the name of the Musk-Rat. It is of a most penetrating character, and so thoroughly saturates every substance that may have come in contact with it, that the odour is with the greatest difficulty removed. The pike and other predaceous fish which inhabit the same waters are accustomed to eat the Daesman, whenever they can succeed in capturing it, and, by the odour of their prey, their flesh becomes so tainted that it is unfit for human consumption. The glands which produce this powerful scent are placed near the tail, and arranged in a double row.

DAESMAN, OR DESMAN.—*Gálemys Pyrenáica.*

The food of the Daesman is chiefly of an animal character, as might be imagined from the arrangement and shape of the teeth, and the general habits of the animal. In the stomachs of several of these creatures that have been dissected, were found the remains of larvæ of various kinds, and of earthworms, but nothing of a vegetable nature. Yet it has been asserted by several writers that the animal diet is sometimes mixed with vegetable food; and that the Daesman will on occasions make a meal of aquatic roots and of acorns, of which latter articles it lays up a store in the recesses of its burrow. Small fish and frogs are known to form part of its food.

The habits which have just been mentioned are common to the entire genus Galemys; two species of which are known to exist, the one being the Russian Daesman, and the other the animal which is depicted in the engraving.

The Russian Daesman is about seventeen inches in total length, the head and body being ten inches long, and the tail seven inches. On account of its aquatic propensities, and the peculiar aspect of its incisor teeth, the Daesman was formerly thought to be a rodent animal, and allied to the beavers, among which creatures it was classed, under the name of *Castor moschatus*, or Musky Beaver. Its fur is much esteemed on account of its rich colour, long silky texture, and warm character. The colour of the Russian Daesman is brown on the upper portions of the body, becoming darker on the flanks, and fading suddenly into silvery white on the abdomen. The peculiar warmth of the fur is owing to a thick and woolly felt of fine hairs, which lies beneath the long silken hairs that form the apparent fur of the animal, and which affords an effectual defence against the liquid element in which the creature passes so much of its time.

The tail of this animal is shorter than the body, and very remarkable in its shape, for at its base it is compressed, but rapidly becomes rounded, and swells with such abruptness, that it may almost deserve the term of bulbous. It then decreases in size as rapidly as it had increased, and, in proportion as it becomes smaller, it becomes vertically compressed. The entire member is, like that of the beaver, thickly set with scales, through the intervals of which protrude a number of short and bristly isolated hairs.

Both by dimensions and colour, the French Daesman is easily to be distinguished from its Russian congener, for it is barely more than half the size of the Russian animal; the tail is differently formed, and the colour is of a distinct character. The tail of the French Daesman is devoid of the peculiar swelling that characterises that member in the Russian Daesman, and tapers gradually to a point. For three-fourths of its length the tail is nearly cylindrical, but becomes vertically compressed for the remaining fourth. It is, moreover, as long as the body. The colour of the fur is a very warm brown, almost amounting to maroon, the flanks are a greyish-brown, and the abdomen is a greyish-white. There is also a slight difference in the webbing of the feet, for the toes of the fore-feet are only half enveloped in the skin, and the external toe of the hinder feet is unconnected with the others.

BULAU, OR TIKUS.—*Gymnura Rafflesii.*

A VERY remarkable animal now comes before us, the BULAU, TIKUS, or GYMNURA, as it is indifferently termed.

This creature, which is an inhabitant of Malacca and Sumatra, bears no slight external resemblance to the opossum of America, the similarity being increased by its long and harsh hair, and the long scaly tail, sparely furnished with very short hairs. The generic name, Gymnura, is derived from two Greek words signifying naked tail, and is, therefore very appropriately applied to this animal. All the feet are terminated by five toes, the three middle toes being longer than the others. The muzzle is much lengthened, but is cut off rather abruptly at its termination. The eyes are small in proportion to the dimensions of their owner, and the ears are small, rounded, and devoid of hairy covering.

One distinguishing peculiarity of the animal is, that the fur which covers the body and head is pierced by a number of very long bristling hairs, that project for a considerable distance from the body, and are much longer on the neck and shoulders than on any other portion of the body. The colour of the creature is a rather peculiar mixture of black and white, which are arranged as follows:—the greater part of the body, the upper portion of the legs, and the beginning of the tail, are black; while the head, the neck, and flanks, and the remainder of the tail, are white. There is also a black stripe over each eye, which forms a bold contrast with the white fur of the head.

Like the preceding animals, the Bulau is possessed of glands which secrete a substance of a powerful musky smell. For the introduction of this animal to science we are indebted to Sir Stamford Raffles, who brought it from Sumatra, and, taking it for one of the viverrine animals, described it under the name of Viverra Gymnura.

THE extraordinary animal which has been recently brought before the notice of zoologists, under the characteristic name of PEN-TAIL, is a native of Borneo, from which country it was brought by Mr. Hugh Low.

It is about the size of a small rat, but appears to be of greater dimensions on account of its extremely long tail with the remarkable appendage at its extremity. As may be seen from the engraving, the tail is of extraordinary length when compared with the size of the body, and is devoid of hair except at its extremity, where it is furnished with a double row of stiff hairs on each side, which stand boldly out, like the barbs of a quill pen, or the feathers of an arrow. The remainder of the tail is covered with scales, which are square in their form, like those of the long-tailed rats, and of considerable size. The colour of the tail is black, and the bristly barbs are white, so that this member presents a peculiarly quaint aspect.

The fur which covers the body of the Pen-tail is extremely soft in texture, and is of a blackish-brown tint above, fading into a yellowish-grey beneath. As the tips of the hairs are tinged with a yellow hue, the precise tint of the fur is rather indeterminate, and is changeable, according to the position of the hairs which are exposed to view. The specimen which is preserved in the British Museum was captured by Mr. Low in the house of Sir James Brooke, the celebrated Rajah of Sarāwak.

PEN-TAIL.—*Ptilocercus Lowii.*

It is presumed that the long tail of the Pen-tail is used for the purpose of balancing itself in its progress among the branches of trees; but this conjecture is only problematical, as the habits of the animal are not yet known. By the arrangement and form of the teeth, it is supposed to be allied to the Tupaias, which are described on page 430, and with which animals it would have been placed but for an unavoidable omission.

The generic name, Ptilocercus, is compounded of two Greek words, signifying "feather-tailed," and is therefore very appropriately given to this singular creature.

The HEDGEHOG finds representatives in many parts of the world, which seem to be possessed of the same propensities whether they are found in England, in India, or in Africa. There are several species of this curious animal, which are remarkable for two or three peculiarities of form and habit.

The external characteristic which immediately strikes the attention of the beholder is the formidable array of bristling spines with which the back is more or less covered, and which offers a *chevaux-de-frise* of sharp spikes towards any animal that may present itself as an enemy. Another peculiarity, is the power possessed by these creatures of rolling themselves into a round ball, by placing the head on the breast, drawing up the legs, and curling the body firmly round these members. By this posture, the Hedgehogs render themselves invulnerable to almost every animal that may attack them, and defend the legs, abdomen, and other portions of the body that are left unprotected by nature. When in this curious attitude, the Hedgehog cannot be unrolled by main force, as long as any life remains in the body, for there is an enormously developed muscle, with a very thick

LONG-EARED HEDGEHOG.—*Erinaceus auritus.*

margin, which spreads over the back and round the sides, and which, when contracted, holds the creature in so firm an embrace that it will be torn in pieces rather than yield its point.

The technical name of this muscle is *panniculus carnósus*, and it is by means of this muscle that bears and other animals are able to shake their skins when they are irritated by any substance that clings to the hair, and which they cannot reach with their teeth.

The Hedgehogs are plantigrade in their gait, and, like the generality of plantigrade animals, are not particularly active or rapid in their movements. Although they generally prefer a deliberate pace when they are not alarmed or hurried, they can get over the ground with no small speed when they feel themselves called upon to make such an exertion.

The feet of the Hedgehog are furnished with five toes, those of the fore-feet terminated with tolerably strong claws, which, although not so evidently fossorial as those of the moles and other insectivorous animals, are yet very capable of digging, and are used effectually for that purpose. The soles of the feet are naked. The limbs and the entire under surface of the body are undefended by the stiff prickles which are so thickly set upon the back, and are clothed with hair of a more or less dense character, according to the particular species. In every species, however, the hair is of a peculiar character, and is intermixed with a goodly number of tolerably stiff hairs of a bristly character.

The food of the Hedgehog consists chiefly of insects, worms, snails, and similar creatures, but it is of essentially carnivorous taste, and is in no wise particular what the kind of food which it eats, provided that it be of an animal character.

These details of form and habit are common to all the Hedgehogs; and the other peculiarities of the Erinacea, as these animals are learnedly named, will be mentioned in connexion with the two species that will be figured and described in these pages.

The LONG-EARED HEDGEHOG derives its name from the exceeding dimensions of its ears, which project from its head in such a manner as to give to the animal a very porcine aspect. This species is found in Siberia and in all the eastern regions of Asiatic Russia, and has also been captured in Egypt. It is a smaller animal than the common Hedgehog of Europe, but is very variable in its dimensions, according to the locality in which it is found. The limbs are comparatively long and slender, and the long hair that clothes the lower portions of the body is extremely fine in its texture. The array of prickly spines that guard its back does not extend so far as in the European species, and are of a rather peculiar colouring. At the base, each spine is marked with a whitish ring, the centre is brown, and the tip is tinted with yellow. The colour of the eye is bluish-grey.

The common HEDGEHOG, HEDGE-PIG, or URCHIN, is one of the most familiar of our indigenous mammalia, being found in every part of Great Britain which is capable of affording food and shelter.

The hard round spines which cover the upper part of its body are about an inch in length, and of a rather peculiar shape, which is well represented in the accompanying sketch. This form is wonderfully adapted to meet the peculiar objects which the spine is intended to fulfil, as will be seen in the following account.

The spine, which is here given, is supposed to be lying nearly horizontally upon the back of the animal, a position which it assumes whenever the Hedgehog chooses to relax the peculiar muscle which governs the spines, and which seems to retain the creature in its coiled attitude. The point of the quill or spine is directed towards the tail. It will be seen that the quill is not unlike a large pin, being sharply pointed at one extremity, and furnished at the other with a round, bead-like head, and rather abruptly bent near the head. If the skin be removed from the Hedgehog, the quills are seen to be pinned, as it were, through the skin, being retained by their round heads, which are acted upon by the peculiar muscle which has already been mentioned.

SPINE OF HEDGEHOG.

It is evident, therefore, that whenever the head of the quill is drawn backward by the contraction of the muscle, the point of the quill is erected in proportion to the force which is exerted upon the head, so that when the animal is rolled up, and the greatest tension is employed, the quills stand boldly out from the body, and present the bayonet-like array of points in every direction.

These curiously formed spines are useful to the Hedgehog for other purposes than the very obvious use of protecting the creature from the attacks of its foe. They are extremely elastic, as is found to be the case with hairs and quills of all descriptions, and the natural elasticity is increased by the sharp curve into which they are bent at their insertion into the skin. Protected by this defence, the Hedgehog is enabled to throw itself from considerable heights, to curl itself into a ball as it descends, and to reach the ground without suffering any harm from its fall. A Hedgehog has been seen repeatedly to throw itself from a wall some twelve or fourteen feet in height, and to fall upon the hard ground without appearing even to be inconvenienced by its tumble. On reaching the ground, it would unroll itself, and trot off with perfect unconcern.

The thorn-studded skin of this animal is not without its use even to mankind, and is still employed for various useful purposes.

In some parts of the country it is used in weaning calves, and is an infallible mode of effecting that object. When the farmer desires to wean the young calf, he fixes a

Hedgehog's skin upon the calf's muzzle, so that when it goes to suckle its mother it causes such irritation that she will not permit her offspring to approach, and drives it away as often as it attempts to effect its purpose. It is also used in order to cure carriage-horses of the troublesome habit of "boring" to one side while being driven, for when fixed on the pole or the traces it gives the animal such effectual reminders whenever it begins to "bore," that it soon learns to pull straight, and thus to avoid the unpleasant aids to memory that bristle at its side. Even to scientific pursuits the Hedgehog's quills are made to render its services, being used as pins whereby certain anatomical preparations are displayed in spirits of wine, and which are not liable to that provoking rust which is so apt to attack metallic pins when immersed in spirits, and which often render the most elaborate dissections perfectly useless.

Another purpose to which the Hedgehog's skin was formerly applied was the hackling of hemp before it was made up into coarse cloth. This custom was followed by the ancient Romans, but is now obsolete, being superseded by artificial instead of natural combs.

The under surface of the body, together with the limbs, is covered with long bristles and undulating soft hair, which passes rather abruptly into the stiff quills that defend the back, and is so long that it almost conceals the limbs when the animal is walking on level ground. In the adult animal the quills are hard and shining, they thickly cover the entire back and top of the head, and are of a greyish-white colour, diversified with a blackish-brown ring near the middle. In the young animal, however, the spines are comparatively few in number, very soft in texture, and nearly white in colour, so that for the first few days of their life the little creatures look like balls of white hair.

The tail of the adult Hedgehog is scarcely visible, being hidden by the bristling quills, which exceed its length by nearly one-fourth. In the young animal, however, the tail is apparent enough, as there are, as yet, no quills to conceal it, and it is carried nearly in a line with the length of the body. The total length of a full-grown Hedgehog is rather more than ten inches, the length of the tail being only three-quarters of an inch, and that of the head three inches. The ears are moderately long in their dimensions, being about an inch in length.

The young of the Hedgehog are born about May, and are so unlike the parents that they have been mistaken for young birds by inexperienced observers. It is a very singular fact, and one which is almost if not entirely unique, that not only are they born with their eyes closed, as is the case with kittens, puppies, and many other animals, but with their ears closed also. The soft white quills, which present so curious an appearance as they lie upon the transparent pink skin, very soon begin to deepen in their colour, and to increase in number, so that about the end of August the little animals resemble their parents in everything but size. The number of young which are produced at a birth is from three to four.

The nest in which the little Hedgehogs are produced and nurtured is most ingenious in its structure, being so admirably woven of moss and similar substances, and so well thatched with leaves that it will resist the effects of the violent showers that generally fall during the spring, remaining perfectly dry in the midst of the sharpest rain.

Marching securely under the guardianship of its thorn-spiked armour, the Hedgehog recks little of any foe save man. For, with this single exception, there are, in our land at least, no enemies that need be dreaded by so well-protected an animal. Dogs, foxes, and cats are the only creatures which possess the capability of killing and eating the Hedgehog, and of these foes it is very little afraid. For dogs are but seldom abroad at night while the Hedgehog is engaged in its nocturnal quests after food ; and the fox would not be foolish enough to waste its time and prick its nose in weary endeavours to force its intended prey out of its defences. Cats, too, are even less adapted to such a proceeding than dogs and foxes.

It is indeed said that the native cunning of the fox enables it to overreach the Hedgehog, and to induce it to unroll itself by an ingenious, but, I fear, apocryphal process. Reynard is said, whenever he finds a coiled-up Hedgehog, to roll it over and over with his paw towards some runnel, pond, or puddle, and then to souse it unexpectedly into the

HEDGEHOG.—*Erinaceus Europæus.*

water. The Hedgehog, fearing that it is going to be drowned, straightway unrolls itself, and is immediately pounced on by the cunning fox, which crushes its head with a single bite, and eats it afterwards at leisure. In America, the puma is said to eat the Hedgehog in a very curious manner. Seizing the animal by the head, it gradually draws the animal through its teeth, swallowing the body and stripping off the skin.

Man, however, troubles himself very little about the Hedgehog's prickles, and when disposed to such a diet, kills, cooks, and eats it without hesitation.

The legitimate mode of proceeding is to kill the animal by a blow on the head, and then to envelop it, without removing the skin, in a thick layer of well-kneaded clay. The enwrapped Hedgehog is then placed on the fire, being carefully turned by the cook at proper intervals, and there remains until the clay is perfectly dry and begins to crack. When this event has taken place, the cooking is considered to be complete, and the animal is removed from the fire. The clay covering is then broken off, and carries away with it the whole of the skin, which is adherent by means of the prickles. By this mode of cookery the juices are preserved, and the result is pronounced to be extremely excellent.

This primitive but admirable form of cookery is almost entirely confined to gipsies and other wanderers, as in these days there are few civilized persons who would condescend to partake of such a diet. Utilitarians, however, can render the creature subservient to their purposes by using it as a guardian to their kitchens. Its insect-devouring powers are of such a nature that it can be made a most useful inhabitant of the house, and set in charge of the "black beetles."

It is domesticated without the least difficulty, and speedily makes itself at home, if it be only supplied with a warm bed of rags or hay in some dark crevice. The rapidity with which it extirpates the cockroaches is most marvellous, for their speed and wariness are so great that the Hedgehog must possess no small amount of both qualities in order to destroy them so easily. A Hedgehog which resided for some years in our house was accustomed to pass a somewhat nomad existence, for as soon as it had eaten all the cockroaches in our kitchen it used to be lent to a friend, to whom it performed the same valuable service. In a few months those tiresome insects had again multiplied, and the Hedgehog was restored to its former habitation.

The creature was marvellously tame, and would come at any time to a saucer of milk in broad daylight. Sometimes it took a fancy to promenading the garden, when it would trot along in its own quaint style, poking its sharp nose into every crevice, and turning over every fallen leaf that lay in its path. If it heard a strange step, it would immediately curl itself into a ball, and lie in that posture for a few minutes until its alarm had passed away, when it would cautiously unroll itself, peer about with its little bead-like eyes for a moment or two, and then resume its progress.

From all appearances, it might have lived for many years had it not come by its death in a rather singular manner. There was a wood-shed in the kitchen-garden, where the bean and pea-sticks were laid up in ordinary during the greater part of the year, and it seemed, for some unknown reason, to afford a marvellous attraction to the Hedgehog. So partial to this locality was the creature that whenever it was missing we were nearly sure to find it among the bean-sticks in the wood-shed. One morning, however, on searching for the animal, in consequence of having missed its presence for some days, we found it hanging by its neck in the fork of a stick, and quite dead. The poor creature had probably slipped while climbing among the sticks, and had been caught by the neck in the bifurcation

It has just been mentioned that the Hedgehog was in the habit of drinking milk from a saucer, and this fact leads to the prevalent idea that the Hedgehogs are accustomed to suck cows while they are lying on the ground. Naturalists have generally denied this statement, saying, as is true enough, that the little mouth of the Hedgehog is so small that it would not be capable of sucking the cow, and that, even if it could do so, its needle-pointed teeth would be so painful to the cow that she would drive away the robber as soon as she felt its teeth. So far they are quite correct, for both their propositions are undoubtedly true. But, nathless, there is great truth in the assertion that the Hedgehog drinks the milk of cows. I have received several communications on this subject, where my correspondents assert that they have seen the creature engaged in that pursuit, and I have been told by several credible witnesses that they have been spectators of the same circumstance. But in neither case was it asserted that the animal was really sucking the cow, but that it was lying on the ground, lapping up the milk as it oozed from the over-filled udder of the animal before the hour of milking had arrived. Granting this to be a fact, the creature can yet do no real injury to the farmer or the dairyman, as the amount of milk which it thus consumes is very small, and would have been wasted had it not been lapped up by the Hedgehog's greedy tongue.

The Hedgehog is also accused of stealing and breaking eggs, to which indictment it can but plead guilty.

It is very ingenious in its method of opening and eating eggs ; a feat which it performs without losing any of the golden contents. Instead of breaking the shell, and running the chance of permitting the contents to roll out, the clever animal lays the egg on the ground, holds it firmly between its fore-feet, bites a hole in the upper portion of the shell, and, inserting its tongue into the orifice, licks out the contents daintily.

Not contenting itself with such comparatively meagre diet as eggs, the Hedgehog is a great destroyer of snakes, frogs, and other animals, crunching them together with their bones as easily as a horse will eat a carrot. Even the thick bone of a mutton-chop, or the big bone of a fish, is splintered by the Hedgehog's teeth with marvellous ease. On one account it is rather a valuable animal, for it will attack a viper as readily as a grass-snake, being apparently proof against the venom of the serpent's fangs. Experiments have been tried in order to prove the poison-resisting power of this strange animal, which seems to be invulnerable to every kind of poison, whether taken internally or mixed with the blood by insertion into a wound.

On one occasion, a Hedgehog was placed in a box together with a viper, and, after a while, began to attack it. The snake, being irritated, rose up, and bit its assailant smartly on the lip. The Hedgehog took but little notice of the incident, but, after licking the wounded spot once or twice, returned to the charge. At last it succeeded in killing the viper, and, after having done so, ate its vanquished enemy, beginning at the tail, and so working upwards. The animal always seems to eat a snake in this fashion, and on one occasion was known to proceed with its banquet while the poor snake was still living.

Poisons of all kinds have been tried upon the Hedgehog without the least effect. Prussic acid, arsenic, and other deadly substances have been unsuccessfully administered, and the animal has been known to make a very satisfactory meal on cantharides without experiencing any ill effects from these cauterising insects. How it is that the constitution of the creature can resist the effects of such powerful substances is not, as yet, known. It is, however, a subject of much interest, and, if it could be elucidated, would probably be of incalculable service to mankind.

On one occasion, when a Hedgehog was employed in the demolition of a snake, it proceeded in a remarkably cautious manner, as if it had been a practised combatant, and had learned how to inflict injury on its foe without suffering in return. On being roused by the touch of the snake, the Hedgehog—which had been coiled up—unrolled itself, bit the snake sharply, and immediately resumed its coiled attitude. Three times it repeated this proceeding, and when after the third bite the snake's back was bitten through, the Hedgehog stood by the side of its victim, and deliberately crushed the snake's body throughout its entire length by biting it at intervals of about half an inch. Having thus placed itself beyond the reach of retaliation, it took the tip of the snake's tail in its mouth, began to eat it, and finished the reptile in the course of twenty-four hours.

The exploits of the Hedgehog in serpent-killing are useful enough in their way, but it too often happens that the carnivorous propensities of the animal are exercised upon less harmful creatures than vipers or other "vermin." Indeed, the poultry-fancier and the game-preserver have too much reason for ranking the Hedgehog itself under that expressive and somewhat comprehensive epithet. Many are the instances on record where the creature has been detected in the act of destroying rabbits, poultry, and various kinds of game, and has been unexpectedly discovered to have been the per-petrator of sundry acts of robbery which had been laid upon the shoulders of the fox, the weasel, or the polecat.

On one occasion, the proprietor of a fine bantam cock was roused by a great disturbance in the place where the fowl was kept, and on going down to see what might be the matter, found his feathered favourite struggling in the jaws of a Hedgehog, which had caught it by the leg and would speedily have devoured it had not its owner come, happily, to the rescue. Again, no less than fifteen turkey poults had been destroyed in the course of a single night, three having been abstracted and the others killed. A number of steel-traps were laid around the scene of devastation, and on the following morning three male Hedgehogs were found in the traps, having evidently returned for the purpose of bringing away the victims of their previous raid.

All kinds of game fall occasional victims to the Hedgehog's appetite, and the partridge, the hare, and the pheasant seem to suffer equally from the voracity of this strange animal. A Hedgehog has been seen in the act of destroying a hare, and had inflicted such injuries that the poor creature died in a very short time after it had been rescued from the jaws of its assailant. This circumstance occurred in Cumberland. Rabbits, too, are frequently eaten by this animal, and Hedgehogs have several times been taken in traps that have been set for other "vermin," and baited with portions of dead rabbits.

That hares, rabbits, and other terrestrial animals should be captured by so apparently clumsy an animal as the Hedgehog is sufficiently remarkable, but that the wary pheasant and the well-winged partridge should fall victims to the creature is more than singular. Yet there are many accredited instances where the Hedgehog has been captured in the very act of killing and eating partridges, and has even been killed while the head of a young partridge still protruded from its mouth. One of these creatures has been detected in the act of eating a hen-pheasant which had been placed in a cage to which it had gained access by squeezing itself through a marvellously small aperture. Another pheasant had

1.   G G

been killed on the previous day, but its death had been laid at the door of the stoat. Earth and air thus seem to furnish their quota of nourishment for the Hedgehog, which extends its depredations to the aqueous element, and displays a cultivated taste for fish. So fond is this carnivorous creature of the finny tribe, that it has been frequently caught in traps which have been baited with fish for the express purpose of decoying the Hedgehog into their treacherous jaws.

Whether in its wild state it is able to capture the little birds, is not accurately known, but in captivity it eats finches and other little birds with great voracity. One of these animals, that was kept in a state of domestication, ate no less than seven sparrows in the course of a single night, and another of these creatures crushed and ate in the course of twenty-four hours more than as many sparrow-heads, eating bones, bill, and neck with equal ease.

Its legitimate prey is found among the insect tribe, of which it consumes vast numbers, being able, not only to chase and capture those which run upon the ground, but even to dig in the earth and feed upon the grubs, worms, and various larvæ which pass their lives beneath the surface of the ground. A Hedgehog has been seen to exhume the nest of the humblebee, which had been placed in a sloping bank, as is often the case with the habitation of these insects, and to eat bees, grubs, and honey, unmindful of the anger of the survivors, who, however, appeared to be but little affected by the inroads which the Hedgehog was making upon their offspring and their stores.

According to the generality of writers, among whom we may reckon Mr. White, the immortalizer of Selborne, the food of the Hedgehog is not entirely animal, but is varied with sundry vegetable substances, such as roots, haws, crabs, and other wild fruits. Others, however, deny the vegetable diet of the Hedgehog. In the "Natural History of Selborne," however, we find a very interesting account of the manner in which the Hedgehog devours the roots of the plantain without injuring the leaves, by grubbing with its snout, and biting off the stems so delicately that the leaves fall untouched. The roots of grasses are also said to form part of the Hedgehog's food.

As might be supposed from the destructive tendency which is, on certain occasions, so strongly developed in the Hedgehog, the animal is a determined fighter whenever it engages in battle, and is capable of inflicting severe wounds with its sharp teeth and powerful jaws. Should several Hedgehogs be confined in one spot, and a stranger be admitted among them, the new-comer will assuredly be forced to fight for his position, and, in all probability, will either kill one of his opponents, or will fall by the teeth of his adversary. In either case the victor becomes a quadrupedal cannibal, and, not satisfied with having destroyed his foe, proceeds to eat him. In such a case, the slain combatant is totally devoured, with the exception of the skin and its prickles, which remain as a token of battle and a trophy of victory.

All Hedgehogs are, however, not endowed with an equal amount of combativeness, but are extremely different in their dispositions. Some are most gentle and retiring in their habits, while others are savage and ferocious to a degree, and seem to be totally devoid of fear, so that they will attack boldly any object which annoys them, perfectly regardless of its character or its size.

The Hedgehog has generally been considered as a dull and stupid animal, incapable of being tamed, and mindful only of its own comfort. Such, however, is really not the case, for when the animal meets with a kind and thoughtful owner, who will try to develop the best feelings of the creature, it proves to be quite affectionate in its character, and will display no small amount of fearless attachment to its master. It would, in all probability, have been better appreciated had it not been, unfortunately, the object of terror or detestation to those who are unacquainted with its habits, and who are either alarmed at its prickly array of quills, or have imbibed certain prejudicial notions concerning its harmful qualities.

It has already been mentioned that the Hedgehog is fond of milk, but it would hardly be imagined that the animal would condescend to partake of strong drink, and that to such a degree that it would be reduced to a state of helpless intoxication. Such, however, is the case, as has been recorded by Dr. Ball of a Hedgehog which he possessed, and to which he administered a strong potation of sweetened whisky. The experiment was not made with any intention of injuring the animal, but for the purpose of testing the popular assertion that the creature would thereby be rendered tame. After saying that the intoxicating draught soon showed its power on the animal, Dr. Ball proceeds as follows : —

"Like the beasts that so indulge, he was anything but himself, and his lack-lustre, leaden eye, was rendered still less pleasing by its inane, drunken expression. He staggered towards us in a ridiculous, get-out-of-my-way sort of manner ; however, he had not gone far before his potation produced all its effects — he tottered, then fell on his side ; he was drunk in the full sense of the word, for he could not even hold by the ground. We could then pull him about, open his mouth, twitch his whiskers, &c.—he was unresisting. There was a strange expression in his face of that self-confidence which we see in cowards when inspired by drinking.

We put him away, and in some twelve hours afterwards found him running about, and, as was predicted, quite tame, his spines lying so smoothly and regularly that he could be stroked down the back and handled freely. We turned him into the kitchen to kill cockroaches, and know nothing further of him."

The home of the Hedgehog is made in some retired and well-protected spot, such as a crevice in rocky ground, or under the stones of some old ruin. It greatly affects hollow trees, wherever the decayed wood permits it to find an easy entrance, and not unfrequently is found coiled up in a warm nest which it has made under the large gnarled roots of some old tree, where the rains have washed away the earth and left the roots projecting occasionally from the ground. Beside these legitimate habitations, the Hedgehog is frequently found to intrude itself upon the homes of other animals, and has been often captured within rabbit burrows. Perhaps it may be led to these localities by the double motive of obtaining shelter from weather and enemies and of making prey of an occasional young rabbit.

In its retreat the Hedgehog usually passes the winter in that semi-animate condition which is known by the name of hibernation.

The hibernation of the Hedgehog is more complete than that of the dormouse or any other of our indigenous hibernating quadrupeds, for they always have a stock of food on which they can rely, and of which they sparingly partake during the cold months of the year. The Hedgehog, however, lays up no such stores, nor, indeed, could it do so, for, as has already been mentioned, its food is almost entirely of an animal nature.

The hibernation of the Hedgehog has lately been denied, because Hedgehogs are occasionally found at large during the winter months. Yet this is no proof to the contrary, for it has already been noticed that the bears are occasionally in the habit of roaming about during the winter, instead of lying motionless in their dens, as is the general custom, yet no one denies the hibernation of the bear in consequence of that well-known circumstance. The subject of hibernation has been most elaborately worked out by Dr. Marshall Hall, who has published the result of his experiments in "Todd's Cyclopædia of Anatomy," and has made many curious observations on the hibernating qualities of the animal which is now under consideration.

In this able dissertation, Dr. Hall warns observers against confounding together the torpor which is produced by excessive cold and that peculiar torpid state which is called hibernation. Indeed, it is always found that although a Hedgehog, or other hibernating animal, will pass into its semi-animate condition at a moderately low temperature, it will be roused at once by severe cold, and will not again resume its lethargy until the temperature be somewhat moderated. "All hibernating animals," he observes, "avoid exposure to extreme cold. They seek some secure retreat, make themselves nests or houses, or congregate in clusters, and if the season prove unusually severe, or if their retreat be not well chosen, and they be exposed in consequence to excessive cold, many become benumbed, stiff, and die."

Those who experiment upon so delicate a subject as hibernation must bear this in mind, and remember also that the least disquieting of the animal will injure the condition under which it sustains its torpidity, even though it should be of so slight a nature as touching the table on which it is placed, or walking with a heavy step across the room. One experimenter, who thought that intense cold was the cause of the torpidity, surrounded a hibernating Hedgehog with a freezing mixture, in the hope of plunging the animal into a more profound sleep. The result, however, was entirely different from his expectation, for the excess of cold first awoke the sleeping animal and afterwards froze it to death.

If the sleeping Hedgehog be touched, or otherwise disturbed, it rouses itself from its lethargy, walks about a little, takes some food, if there should be any at hand, and soon returns to its somnolent condition.

It is a very curious fact that if a hibernating animal be suddenly decapitated, before it has time to awake from its sleep, the action of the heart continues to last for a considerable time, as if it were endowed with a kind of independent life. In one experiment, not only was the brain removed, but the entire spinal cord removed; yet the heart continued to pulsate for two hours, and for more than twelve hours would contract if touched with the point of a penknife. The animal on which this experiment was made had been continually torpid for rather more than six days.

As might be gathered from the result of this wonderful experiment, the respiration of hibernating animals is extremely slight, so slight, indeed, as to be almost imperceptible. Long and delicate rods have been so fixed to the slumbering animal that the smallest movement was apparent, and yet they never moved perceptibly unless the animal were roused by a touch or the sudden shock of an incautious footstep. It is a curious fact that whenever the hibernating Hedgehog is thus roused it utters a deep sonorous respiration, which is a test of its being truly hibernating. Should it be only in the ordinary sleep, the creature only stirs uneasily, and silently coils itself more firmly than before.

The sight of the Hedgehog does not appear to be so excellent as its powers of scent, which are admirably developed, as may be seen by opening the side of a Hedgehog's face.

One of these animals has been seen to chase a partridge across a road, following her through the hedge with perfect precision; and another was observed to discover the presence of mankind by means of its powers of scent, as it was in a position from whence it could not see its fancied enemies. The Hedgehog had already passed the observers, who remained perfectly quiet in order to watch its proceedings, but after it had run for a few paces, it suddenly stopped, seemed suspicious of some danger, stretched its nose in the air, and stood on its guard. In a few moments it seemed to have set itself at ease, and resumed its course. The spectators then slightly shifted their position, so as to bring the animal again within the range of their "wind," when the creature repeated the same process, and did not appear entirely at its ease for some little time.

ALTHOUGH unable to contract itself into a ball, after the manner of the true Hedgehogs, the TANREC, or MADAGASCAR HEDGEHOG, as it is sometimes called, is closely allied to these animals, and in many respects bears some resemblance to them.

In size, this animal is about the equal of the European Hedgehog, but is rather more elongated in its form, and furnished with longer legs, so that when it walks it does not carry its abdomen so close to the ground, as is the case with the preceding animal. The muzzle of the Tanrec, or Tenrec, as the name is sometimes written, is extremely elongated, rather sharply pointed, and brown in colour; the ears are small and rounded, and the tail is absent, a peculiarity which has earned for the animal its specific title of ecaudátus, or tail-less. The generic name, Centétes, or more correctly, Kentétes, is of Greek origin, and signifies "thorny," in allusion to the short and thorn-like spines with which the body is covered.

The colour of the Tanrec is rather variable at different times, on account of the variegated tints which bedeck the array of quills that adorn and defend its back. These quills are black towards their tips, and yellowish towards their bases, so that either tint predominates, according to the arrangement of the quills. In length they are inferior to those of the Hedgehog, the largest not exceeding an inch. The throat, abdomen, and inside faces of the limbs are covered with rather coarse yellowish hairs, and the sides and flanks are decorated with long silken hairs of the same colour as the spines.

Like the Hedgehog, the Tanrec is a hibernating animal, sleeping for at least three months of the year, secure in the burrow which it has excavated by means of the powerful and crooked claws which are attached to its feet.

Some writers assert that its period of torpidity is during the heat of summer, while others, who have had practical knowledge of the animal and its habits, say that its periodic somnolence takes place during the cold and wintry months. These contradictory accounts can be reconciled by the fact, that the Mauritian winter is from June to November, and that the months which in that island are reckoned as summer months, are winter months with ourselves.

It is not very commonly seen, even in the localities which it most frequents, as it is a nocturnal animal, and, except when under the protection of the shades of night, very seldom leaves the burrow in which it has taken up its residence. The locality which it chooses for its subterranean residence is generally well chosen for the purpose of security, being usually among the old roots of clumps of bamboos, which defend and conceal the entrance, and offer an almost insurmountable obstacle to any foe that might desire to dig the animal out of its den.

The natural food of the Tanrec consists of worms, insects, snails, reptiles, and various similar substances, but the creature will condescend to feed for a time on more sophisticated

TANREC.—*Centétes ecaudátus.*

dainties, such as boiled rice. It is supposed that an unmixed vegetable diet would be very hurtful to the animal's well-being.

Possessed of a most overpowering and unpleasant smell of musk, the Tanrec is not an animal which would be supposed to furnish an agreeable article of diet to any one, except to a starving man in the last extremity of hunger. Yet the natives of Madagascar esteem it among their rarest luxuries, and are so tenacious of this very powerful food, that they can hardly be induced to part with a specimen which they have captured, and which they have already dedicated, in anticipation, to the composition of some wonderful specimen of the cook's art.

The Tanrec is an inhabitant of Madagascar, as may be deduced from its popular title of Madagascar Hedgehog, but has been taken to the Mauritius and there naturalized.

THERE are other species of the Madagascar Hedgehog, besides the tanrec, among which are recognised the TENDRAC, or SPINY TENREC (*Centétes spinosus*), and the BANDED TENREC (*Centétes Madagascarensis*).

The former of these animals is inferior in size to the tanrec, being only five or six inches in length. The colour of this animal is rather rich and varied, owing to the deep tinting of the quills and the soft hues of the long and flexible hairs which stud the body intermixed with the quills. The hair is of pale yellow, and the quills are of a deep red or mahogany tint towards their points, and white towards their bases. The long coarse hairs which cover the abdomen and the legs are annulated. This animal is said to be generally found in the neighbourhood of water, whether fresh or salt, and to make deep burrows near the bank. The natives esteem it highly as an article of food.

THE BANDED TENREC, or VARIED TENREC, as the name is sometimes given, is also a native of Madagascar, and has derived its title of Banded, or Varied, from the bold colouring of the quills and hair.

The general colour of the back is a blackish-brown, diversified with three bold stripes of yellowish-white, that afford a strong contrast with the dark ground-hues of the back. The centre one of these stripes extends along the entire length of the animal, and the two others commence by the ear and terminate by the flank. The hair that covers the under portions of the body is of a yellowish-white colour.

GROUP OF MACROPIDÆ.

MACRÓPIDÆ.

THE EXTRAORDINARY animals which are grouped together under the title of Macropidæ are, with the exception of the well-known opossum of Virginia, inhabitants of Australasia and the islands of the Indian Archipelago.

Many of these creatures, such as the kangaroo, some of the opossums, and the petauristes, are of such singular formation, and so remarkable in their habits of life, that if they had not been made familiar to us through the mediumship of menageries, museums, and the writings of accredited travellers, we should feel rather inclined to consider them and their habits to be but emanations from the fertile brain of some imaginative voyager, who was taking full advantage of the proverbial traveller's licence. Even at the present day, our familiarity with these animals in no way derogates from our wonder at their strange conformation; and the structure of many of them is so complicated, and involves so many considerations, that the study of the Macropidæ and their habits is as yet but little advanced. Anatomists such as Owen, Meckel, John Hunter, and scientific travellers such as Gould, have done much towards clearing up many dubious points in the history of these animals, but the subject is yet comparatively in obscurity, and much remains to be achieved by future zoologists.

Many acknowledged species are known but as "specimens," no accounts of their mode of life, the localities which they most frequent, their food, or their habits, having as yet been given to the world; while it is more than suspected that in many of the vast unexplored portions of Australasia may yet be found numerous species of these animals which are as yet unknown to science, and which will supply many of the links which are needed to complete the system of nature.

There is hardly any practical writer on zoology who does not lament the very incomplete state of our knowledge on this subject; and those who have thrown themselves most zealously into the work, and have achieved the greatest success, have been the most ready to acknowledge the enormous gap that has yet to be filled, and to urge others to prosecute their researches in regions which have as yet been untraversed by the foot of civilized man, and which are the most likely to be the dwelling-places of creatures on which, as yet, an educated white man has never set his eye. Several genera are known to be extinct, and there are interesting accounts of fossil discoveries in Australia, which bring to light the remains of gigantic animals of the same kind as those which now inhabit that country.

So distinct are many of the animals of Australia from those of the Old World, that more than one zoologist has confessed that they seem to be the result of another and a later creation than that by which the animals of the northern hemisphere received their being.

The peculiarity which gives the greatest interest to this group of animals, is that wonderful modification of the nutritient organs, which has gained for them the title of MARSUPIALIA, or pouched animals — a name which is derived from the Latin word *marsupium*, which signifies a purse or pouch. This singular structure is only found in the female Marsupials, and in them is variously developed according to the character of the animal and the mode of life for which it is intended.

The more minute details concerning the marsupium, or pouch, will be found in the course of the work in connexion with the particular species to which it belongs, but the general idea of that structure is much as follows :—

The lower part of the abdomen is furnished with a tolerably large pouch, in the interior of which the mammæ, or teats, are placed. When the young, even of so large an animal as the kangaroo, make their appearance in the world, they are exceedingly minute — the young kangaroo being only an inch in length — and entirely unable to endure the rough treatment which they would meet with were they to be nurtured according to the manner in which the young of all other animals are nourished. Accordingly, as soon as they are born, they are transferred by the mother into the pouch, when they instinctively attach themselves to the teats, and there hang until they have attained considerable dimensions. By degrees, as they grow older and stronger, they loosen their hold, and put their little heads out of the living cradle, in order to survey the world at leisure. In a few weeks more they gain sufficient strength to leave the pouch entirely, and to frisk about under the guardianship of their mother, who, however, is always ready to receive them again into their cradle if there is any rumour of danger; and if any necessity for flight should present itself, flies from the dangerous locality, carrying her young with her.

In some of the Marsupials the pouch is hardly deserving of the name, being modified into two folds of skin, so that the mother is obliged to find other means of carrying her young from place to place. In the structure of the animal there is an admirable provision for sustaining the pouch and its contents, and preventing it from exerting too painful a " drag " upon the skin and walls of the abdomen. Two supplementary bones, called, from their position in the pouch, the marsupial bones, issue from the pelvis, and are directed forward almost parallel to the spine. On account, however, of the method in which certain muscles wind round the marsupial bones, and taking into consideration the fact that these structures are found in both sexes, Mr. Owen considers that their chief aim is not so much in affording support to the pouch as in compressing the numerous glands, so as to aid the feeble young in gaining nourishment.

We will now leave their general consideration, and proceed t. ......mine some of the principal species which are contained in this wonderful group of ...

At the head of the Macropidæ are placed a small but inter.. .... .su,ial animals, which are called Phalangistines, on account of the curiou. ... . i: .vuich two of the toes belonging to the hinder feet are joined together as far as the ..al.ng.. The feet are all formed with great powers of grasp, and their structure is intended to fit them for procuring their food among the branches of the trees, on which they pass the greater portion of their existence.

These creatures fall naturally into three subdivisions—namely, the Petaurists, or those which are furnished with a parachute-like expansion of the skin along the flanks, much resembling a similar structure in the colugo, or flying lemur, which has been already described in page 88 ; the Phalangists, or those which are devoid of the parachute, and are furnished with a long prehensile tail ; and the Koalas, or those which are devoid of both parachute and tail. According to many excellent authorities, these three subdivisions are, in fact, three genera, which comprise the whole of the Phalangistines, and which render any further separation into genera entirely unnecessary.

OPOSSUM MOUSE.—*Acróbates pygmæus.*

First, and least of the Phalangistines, is the beautiful little animal which is called the Opossum Mouse in some parts of the country, and the Flying Mouse in others.

This pretty little creature is about the size of our common mouse, and when it is resting upon a branch, with its parachute, or umbrella of skin, drawn close to the body by its own elasticity, it looks very like the common mouse of Europe, and at a little distance might easily be taken for that animal. In total length it rather exceeds six inches, the length of its head and body being about three inches and a half, and that of the tail not quite three inches. On account of its minute size, this animal is also called the Pigmy Petaurist.

In the colour of the upper portions of the body the Opossum Mouse is of the well-known mouse tint, slightly sprinkled with a reddish hue ; but on the abdomen, and under portions of the skin-parachute, the fur is beautifully white. The line of demarcation between the hair is very well defined, and there is a narrow stripe of darker brown that marks out the line of juncture. When the animal is at rest, the parachute closes by its own elasticity, and gathers itself into folds, which have a very pretty effect, on account of the delicate white fur which becomes exposed by the action, and which undulates in rich and graceful folds, alternating with the dark fur of the back and the still darker stripe that forms the line of demarcation.

The tail of the Opossum Mouse is nearly as long as the body, very slender, and remarkable for the manner in which the hairs are affixed to it. The hairs that fringe the greater part of the tail are about one-sixth of an inch in length, reddish-grey in colour, rather stiff, and are set on the tail in a double row, like the barbs of a feather. A similar formation has already been described in the history of the pen-tail of Saráwak, on page 443. This mode of arrangement is called " distichous."

The food of the Petaurists is generally of a vegetable character, consisting of leaves, fruits, and buds, but the sharply pointed molars of the Opossum Mouse approach so closely to the insectivorous type that the creature is probably able to vary a vegetable diet by occasional admixture with animal food.

The parachute-like expansion of the skin is of very great service to the animal when it wishes to pass from one branch, or from one tree, to another without the trouble of descending and the laborious climbing up again. Trusting to the powers of its parachute, the little creature will boldly launch itself into the air, stretching out all its limbs, and expanding the skin to the utmost. Upborne by this membrane, the Opossum Mouse can sweep through very great intervals of space, and possesses no small power of altering its course at will. It cannot, however, support itself in the air by moving its limbs, like the bats, nor can it make any aerial progress when the original impetus of its leap has expired.

This little creature is very common at Port Jackson.

HEPOONA ROO.—*Petaurus Australis.*

The HEPOONA ROO, or GREAT FLYING PHALANGER, is rather a remarkable animal in appearance. It is an inhabitant of New Holland, and is found in tolerable plenty about Port Jackson and Botany Bay.

The colour of the Hepoona Roo is rather variable, but is generally as follows. The upper part of the body is brown, tinged with grey, and a much darker brown stripe runs along the course of the spine. The head is darker than the general hue of the body, and on the top of the head the brown tint is warmed by the admixture of hairs of a fawn colour. The under portions of the abdomen and the parachute are white, very perceptibly washed with yellow, a peculiarity which has earned for the animal the title of flaviventer, which has been applied to it by some naturalists. The feet are blackish-brown, and the toes of the hinder limbs thickly supplied with hair. The skin is brown.

The tail of the Hepoona Roo is almost as long as the body, and is heavily covered with long and soft fur of a general brown tint, warming to a reddish-rust near its insertion, and darkening into a blackish-brown near its tip.

Sometimes the fur of this animal varies so widely from the colour which has just been described, that it can hardly be recognised as the same animal, except by a very careful inspection. In some specimens the back is ashy-grey, and the under portions of a dirty greyish-yellow, while in others the coat is variegated with brown, grey, and white, the only dark spot being the tip of the tail, which still retains its deep brown hue. A similar phenomenon takes place with the weasels, when their hair becomes white during a very sharp winter.

In one or two instances, the fur is totally white, and in such cases it is evident that the animal can only be considered as an albino.

The head of the Hepoona Roo is small, and its large and expressive ears are covered with hair. It is not a very small animal, as the total length is rather more than three feet, the head and body occupying one foot eight inches, and the tail rather exceeding eighteen inches in length.

SUGAR SQUIRREL, OR SQUIRREL PETAURUS.—*Petaurus Sciureus.*

On account of the wonderful resemblance which exists between the members of the genus Petaurus and the flying squirrels that belong to the family of rodents, the Petaurists have, ever since their discovery, been popularly known by the same title. There seems to be little doubt but that the Petaurists are the representatives of these flying rodents, and that the strange animal creation of Australasia is a kind of repetition of the ideas which formed the animal creation of the older world, but carried out in a different manner and for different purposes.

The animal which is represented in the accompanying engraving is known by several popular names, the most common of which is the SUGAR SQUIRREL. It is also called the NORFOLK ISLAND FLYING SQUIRREL, and the SQUIRREL PETAURUS.

It is not nearly so large an animal as the hepoona roo, being only sixteen inches in total length, of which measurement the tail occupies one moiety.

The fur of the Sugar Squirrel is very beautiful, being of a nearly uniform brownish-grey, of a peculiarly delicate hue, and remarkably soft in its texture. The parachute membrane is grey above, but is edged with a rich brown band, and a bold stripe of blackish-brown is drawn along the curve of the spine, reaching from the point of the nose to the root of the tail. The head is somewhat darker than the rest of the body. The under parts of the body are nearly white.

Its long and bushy tail is covered with a profusion of very long, full, soft hair, greyish-brown above, and of a beautiful white underneath. The extremely long tail with which these animals are furnished appears to be of exceeding service to them in balancing their bodies as they make their desperate leap through space, and may also be useful in aiding them to modify the original direction of their sweep through the air.

This supposition is strengthened by the fact, that many long-tailed animals employ that member for the same purpose when they are perched in any critical position where an accurate balance is needful. I have seen a large spider-monkey—the same animal whose exploits have already been recorded on page 112—employ her long prehensile tail for the same purpose. She was seated upon a loose horizontal cord, holding as usual by her hands and tail. But when I gave her an apple, she removed both her hands from the cord, grasping it firmly with her hinder feet, and then permitted her tail to hang its full length, so that she could balance herself by swinging it from side to side, according to the necessity of the moment.

This was the more remarkable, as the animal is noted for the pertinacity with which it grasps any neighbouring object with its tail, and never likes to move without securing itself by its tail to the various objects as it goes along, or even to the string by which it is led.

The Sugar Squirrel, like the other Petaurists, is a nocturnal animal, and is seldom seen in the daytime. During the hours of daylight it remains concealed in one of the hollow branches of the enormous trees that grow in its native country, and can only be detected in its retreat by the marvellous organs of vision with which the native Australians are gifted. As soon as evening comes on, the Sugar Squirrels issue from their darksome caverns, and immediately become very frolicsome, darting from tree to tree, and going through the most extraordinary and daring evolutions with admirable ease.

It seems to be a gamesome little animal, and fond of the society of its own species, although it does not appear to respond very readily to the caresses or advances of human playfellows. Being fond of society, the Sugar Squirrels associate in small companies as soon as they emerge from their retreats, and thus are enabled to enjoy their graceful pastime to their hearts' content. Any cage, however, must be most annoying to these active little creatures, who are accustomed to sweep through very considerable spaces in their leap. Mr. Bennet remarks, that the Sugar Squirrel has been known to leap fairly across a river forty yards in width, starting from an elevation of only thirty feet.

Even in captivity they retain their playfulness, and as soon as night brings their expected day, they awake from the heavy lethargy which oppresses them during the hours of light, and uncoiling themselves from the very comfortable attitude in which they sleep, they begin to be very lively, and to traverse their cage with great agility, chasing one another about their residence, and leaping as far as the confined space will permit them.

In climbing and leaping, as well as in grasping the branches towards which they aim their flight, the creatures are greatly aided by the manner in which the thumb of the hinder feet is set on the foot, so as to be opposable to the others, thus enabling the creature to clasp the branches in the same manner as the quadrumana.

THE beautiful little animal which has been called by the expressive name of ARIEL, is about the size of a small rat, and in the hue of the upper portions of the body is not unlike that animal.

The colour of the fur upon the upper portions of the body is a light brown, which darkens considerably upon the parachute membrane. On the under surface it is white, the white fur just turning over the edge of the parachute, and presenting a pretty contrast with the dark brown colour of its upper surface. The tail is nearly of the same colour as the body, with the exception of the tip, which is dark. On account of its graceful movements, and the easy undulating sweep of its passage through the air, it has earned

ARIEL PETAURUS.—*Petaurus Ariel.*

for itself the appropriate name of Ariel, in remembrance of the exquisite and tricksy sprite that animates the world-celebrated drama of the "Tempest."

It is not an uncommon animal, and is frequently seen at Port Essington.

The TAGUAN, or PETAURIST, is the largest of the Petaurists, and is supposed to be the only species that belongs to the genus Petaurista. The peculiarity of its teeth and other portions of its structure will be found in the table of generic differences at the end of the volume.

This animal is a native of New Holland, where it breeds in great abundance, although it is but seldom seen in a living state by any but the natives. It is, like the rest of its tribe, a nocturnal animal, taking up its residence in the hollows of large decaying trees, and remaining buried in sleep until the evening has set in, and the shades of night extend their welcome veil over its actions. While it is lying buried in sleep in the depths of its arboreal retreat, it is safe from almost any foe except the ever hungry and ever watchful native of New South Wales, whose keen eye is capable of detecting almost anything eatable, however deeply it may be hidden from sight.

A slight scratch on the bark of a tree, or a chance hair that has adhered to the side of the aperture into which the animal has entered, tells its tale as clearly to the black man as if he had seen the creature ascend the tree and enter its domicile. He is even able to gather from the appearance of the scratch and the aspect of the hairs how many hours have elapsed since the animal left the traces behind it, and can conjecture very accurately whether the intended prey is still within its residence, or whether it be away from home. Should the indications prove favourable, the native proceeds to cut little holes in the tree, in which he thrusts his toes and fingers, and ascends the huge trunk as easily as a brick-layer walks up a ladder. Having reached the aperture, he strikes the tree sharply once or twice with the back of the hatchet, so as to learn, by the echo which is returned to the blow, the position of the animal within the hollow. He then rapidly cuts a hole through the tree into the cavity, seizes the concealed animal by its tail, jerks it out before it has time to use its claws or teeth, dashes it against the tree, and drops it on the ground dead.

It is rather remarkable, that the creature will not emerge from its concealment when awakened by the sound of the axe so near its presence, and is not even induced by the quick jarring of the wooden walls of its habitation to attempt escape from imminent danger. The precaution of jerking the creature quickly from its domicile is most necessary, for the strong, sharp, and curved claws of the animal are formidable weapons when the creature is disposed to use them for combat, and, together with its sharp teeth,

PETAURIST, OR TAGUAN.—*Petaurista Taguanöides.*

can inflict terrible laceration upon its foe. It is of a sufficiently pugnacious disposition, and when it is enraged is a desperate fighter with teeth and claws.

The flesh of the Taguan is said to be very good, and as the animal is a tolerably large one, it is a favourite article of diet among the white and black inhabitants of the country. It is, however, so extremely difficult of capture, that, without the assistance of native aid, the white men would seldom be able to make a dinner on this creature. But as travellers or hunters are generally accompanied by one or more "black fellows," they are well supplied with Taguans by the quick eye and ready hand of their sable allies.

In colour the Taguan is extremely variable, but the general arrangement of its colour is as follows.

The back is of a rather deep blackish-brown, darker or lighter in different individuals, the feet and muzzle are nearly black, and the under surface of the body and membrane is white. The upper surface of the parachute membrane is rather grizzled, on account of the variegated tints of black and grey with which the hairs are annulated. Many varieties, however, of colour exist in the animal, and there are hardly any two specimens in which the tints are precisely alike. The brown hue of the fur is in some examples deepened into a rich black-brown; others are almost entirely grey on the upper surface of the body and parachute membrane; while specimens of a beautiful white are not of very unfrequent occurrence. In all cases, however, the fur of the under portions, and inner faces of the limbs, preserves its white hue.

The whole of the fur is extremely long, being no less than two inches in length on the back. It is very soft and silken in texture, and is remarkably loose and glossy, so that it waves in the air at every movement of the animal, or at the touch of every breath of wind that may stir the atmosphere. On the tail the hair is remarkably long and bushy, and gradually deepens in colour from a pale brown at the base to a dark, blackish-brown at the tip.

The animal is found inhabiting the vast forest ranges that run from Port Phillip to Moreton Bay, and is seldom, if ever, found in any part of the country except in the eastern or south-eastern districts of New South Wales.

The food of the Taguan consists of leaves, buds, and the young shoots of trees, chiefly of the eucalypti, which it eats only during the hours of night. It seldom troubles itself to descend to the ground, for it can easily pass from one tree to another by means of the wonderful apparatus with which it is gifted, but when it does come to earth, prowls about in search of some vegetation that may afford an agreeable variety to the too uniform diet of leaves and buds.

The animals which form the genus Cuscus, and of which the SPOTTED CUSCUS is a good example, have been separated from their neighbours on account of the structure of the tail, which, instead of being covered with hair, is naked except at its base, and is thickly studded with minute tubercles. They are inhabitants of the Molucca Islands, Amboyna and New Guinea, and have never been found in New South Wales nor in Van Diemen's Land. The name Cuscus is Latinized from the native term couscous, or coëscoës; and the specific term, maculatus, or spotted, refers to the peculiar markings which decorate the fur of the species which is represented in the engraving.

In size the Cuscus is equal to a tolerably large cat, as a specimen of average size will measure about three feet in total length, the tail being fifteen or sixteen inches long, and the head and body about eighteen or nineteen inches. There are, however, several examples where the animal has attained to a considerably greater dimension. It is a tree-loving animal, and is very seldom seen away from the congenial haunts among which it loves to dwell, and for traversing which it is so admirably adapted by nature.

The tail of this creature is remarkably prehensile, and the animal never seems to be content unless this member be twisted round some supporting object. Whenever the Cuscus thinks that it is in danger, or that it may be seen by an enemy, it immediately suspends itself by its tail from a branch, and there hangs, swaying about in the wind among the leaves as if it were some lifeless fruit.

It is said that this curious propensity is turned to good account by any one who wishes to capture a Cuscus without any trouble on his own part except a large amount of patient waiting. When the Cuscus is conscious of the human gaze, and has suspended itself by its tail from a branch, it hangs in counterfeited death until it fancies that the peril is overpast. Nothing will induce the animal to give the least signs of life as long as the eye is not taken from it. According to popular report, for the absolute truth of which I do not vouch, it is said that if the man will steadily keep his eye on the suspended animal, it will hang until its wearied muscles refuse to support the weight of its body, and it drops helplessly to the ground.

The movements of the Cuscus among the branches are not characterised by the dashing elegance which characterises the arboreal feats of the petaurists, but are slow and cautious, the creature never venturing to put itself in a perilous position without having secured itself firmly by its tail. On this account it is thought, with some reason, to bear analogy to the slow-moving lemurs, to which it bears some sort of external resemblance. The food of this animal generally is of a vegetable nature, and consists of fruits, leaves, buds, young twigs, and other similar substances; but the creature is capable of eating animal food also, and seems to be in the habit of eating various insects and the eggs of birds. In some of its relations the carnivorous power is developed to a still greater degree.

The fur of the Cuscus is beautifully soft and silken in its texture, and is of some value for conversion into articles of human attire or luxury, such as cloaks and mantles. The colour of the fur is singularly variable, even if the Spotted Cuscus be really a separate species, and still more so if, according to many skilful zoologists, it can only be considered as a single variety.

The ground tint of the Spotted Cuscus is a whitish-grey. Upon this pale tint are scattered very large and bold spots of deep brown, covered with a reddish-chestnut. Sometimes it is almost wholly white, with only one or two small spots scattered sparingly over the body. The tail is yellowish-white. Another specimen will be almost entirely of the darker colour, and marked as follows : — The shoulders and head of a curious grey grizzle, and the remainder of the body to the tail greyish-white. A number of large angular black spots or patches are so placed upon this pale field, that they communicate with each other, and form a kind of indistinct black pattern on the

SPOTTED CUSCUS.—*Cuscus maculatus.*

creature's back. The colour of these dark patches is nearly black, and would be so entirely but from a number of white hairs which are seen among the black. These descriptions are taken from actual specimens. Another species, called the Ursine Cuscus, is of a uniform deep brown.

These animals are in some request among the white and the native population of the country which they inhabit, for they not only furnish valuable fur or " peltry," as the skin of these and similar creatures is popularly termed, but also afford nourishment to their captors. The flesh of the Cuscus is thought to be remarkably good by those who have partaken of it, and is said to be quite equal to that of the kangaroo. There is a certain rather powerful and not very agreeable scent that issues from the Cuscus and most of its relations, which does not, however, disqualify the creature from forming a most valued portion of the hunter's dietary. This scent proceeds from some small glands which are situated near the insertion of the tail.

In captivity it is not a particularly interesting animal, being dull and slow in its movements, and seldom exhibiting any energy, except, perhaps, when it ought rather to keep itself quiet. One of these creatures, which had been for some time partially domesticated, was very sluggish and unimpressible in its manner until a companion was placed in the same cage. The two animals immediately became violently excited,

SOOTY PHALANGIST.—*Phalangista fuliginósa.*

attacked each other fiercely, and growled, and scratched, and bit, with infinitely more energy than would have been expected from creatures of such apparently apathetic natures.

These specimens were great water-drinkers, and would eat bread, although they evidently gave the preference to meat, thus confirming the opinion that their diet is naturally of a mixed character.

Passing by the curious little dormouse-like animals which are classed under the genus Dromicia, we arrive at the true Phalangists, the first of which is the TAPOA, or SOOTY PHALANGIST, an animal which has been gifted with its rather dismal title in consequence of the uniform smoky-black colour of its fur.

The Sooty Phalangist is tolerably common in Van Diemen's Land, where it is much sought after on account of its skin, which is highly valued by white and black men for the purpose of being manufactured into a soft, warm, and beautiful fur. As with the preceding animal, there is considerable variation in the tint of the coat, some specimens being entirely clothed with a uniform dark, dull, blackish-brown, while the fur of others

1.   H H

is warmly tinged with a chestnut hue. The tail of this animal is extremely full, the hair being thick, long, and very bushy, more so than that of the body and limbs. One of the most remarkable points in the colouring of this animal is the fact that the abdomen and the under portions of its body retain the brown hue of the upper portions instead of being covered with the beautiful white or yellowish fur which is found in nearly all the preceding animals. The ears of the Tapoa are rather elongated, and triangular in form, thickly covered with hair on the outside, but naked on their inner faces.

In the structure of this creature a rather peculiar formation is well defined, and as it is one of the distinctive marks by which the genus Phalangista is separated from its neighbours, it is well worthy of notice. The tail is, to all appearance, entirely covered with a heavy coating of thick, long, and loose hair, but if that member be lifted up, so as to expose the under surface, and carefully examined, it will be seen that at the extremity the tail is bare of fur, and that a naked stripe runs for some little distance from the tip towards the base. During the lifetime of the animal, this naked stripe, together with the nose and the soles of the feet, are of a light flesh colour.

Fox-like in nature as well as in form, the Vulpine Phalangist has well earned the name which has been given to it by common consent. It has also been entitled the Vulpine Opossum, and in its native country is popularly called by the latter of these names.

It is an extremely common animal, and is the widest diffused of all the Australian opossum-like animals. Like the preceding animals, it is a nocturnal being, residing during the day in the hollows of decaying trees, and only venturing from its retreat as evening draws on. The nature of its food is of a mixed character, for the creature is capable of feeding on vegetable food, like the Petaurists, and also displays a considerable taste for animal food of all kinds. If a small bird be given to a Vulpine Phalangist, the creature seizes it in his paws, manipulates it adroitly for a while, and then tears it to pieces and eats it. It is rather a remarkable fact, that the animal is peculiarly fond of the brain, and always commences its feast by crushing the head between its teeth and devouring the brain.

In all probability, therefore, the creature makes no small portion of its meals on various animal substances, such as insects, reptiles, and eggs. As to the birds on which it so loves to feed, it may very probably, although so slow an animal, capture them in the same manner as has been related of the lemurs, viz. by creeping slowly and cautiously upon them as they sleep, and swiftly seizing them before they can awaken to a sense of their danger. It is a tolerably large animal, equalling a large cat in dimensions, and is, therefore, able to make dire havoc among such prey whenever it chooses to issue forth with the intention of making a meal upon some small bird that may chance to be sleeping in fancied security.

The fore-paws of the Vulpine Phalangist are well adapted for such proceedings, as they are possessed of great strength and mobility, so that the animal is able to take up any small object in its paws, and to hold it after the manner of the common squirrel. When feeding, it generally takes its food in its fore-paws, and so conveys it to its mouth. In captivity it does not seem to be a very intelligent animal, even when night brings forth its time of energy, and it but little responds to the advances of its owner, however kind he may be. It will feed on bread and milk, or fruits, or leaves, or buds, or any substance of a similar nature, but always seems best pleased when it is supplied with some small birds or animals, and devours them with evident glee.

The flesh of the Vulpine Phalangist is considered to be very good, and the natives are so fond of it that, notwithstanding the laziness that is engrained in their very beings, except when they are under the influence of some potent excitement, they can seldom refrain from chasing an "opossum," even though they have been well fed by the white settlers. When the fresh body of a Vulpine Phalangist is opened, a kind of camphorated odour is diffused from it, which is probably occasioned by the foliage of the camphor-perfumed trees in which it dwells, and the leaves of which it eats.

The fur of this animal is not valued so highly as that of the Tapoa, probably because it is of more common occurrence, for the colour of the hair is much more elegant, and its quality seems to be really excellent. Some few experiments have been made upon the capabilities of this fur, and, as far as has yet been accomplished, with very great success. Good judges have declared that articles which had been made from this fur presented a great resemblance to those which had been made from Angola wool, but appeared to be of superior quality. The hat-makers have already discovered the value of the fur, and are in the habit of employing it in their trade.

The natives employ the skin of the "opossum" in the manufacture of their scanty mantles, as well as for sundry other purposes, and prepare the skins in a rather ingenious manner. As soon as the skin is stripped from the animal's body, it is laid on the ground, with the hairy side downwards, and secured from shrinking by a number of little pegs which are fixed around its edges. The inner side is then continually scraped with a shell, and by degrees the skin becomes perfectly clean and pliable. When a sufficient number of skins are prepared, they are ingeniously sewn together with thread that is made from the tendons of the kangaroo, which, when dried, can be separated into innumerable

VULPINE PHALANGIST.—*Phalangista vulpina.*

filaments. A sharpened piece of bone stands the sable tailor in place of a needle. From the skin of the same animal is also formed the "kumeel," or badge of manhood, a slight belt, which no one is permitted to wear until he has been solemnly admitted among the assembly of men.

In its colour, the Vulpine Phalangist is rather variable, but the general hue of its fur is a greyish-brown, sometimes tinted with a ruddy hue. The tail is long, thick, and woolly in its character, and in colour it resembles that of the body, with the exception of the tip, which is nearly black. The dimensions of an old male are given by Mr. Bennett as follows: Total length, two feet seven inches; the head being four inches in length, and the tail nearly a foot.

THE QUAINT-LOOKING animal which is popularly known by the native name of KOALA, or the AUSTRALIAN BEAR, is of some importance in the zoological world, as it serves to fill up the gulf that exists between the phalangistines and the kangaroos.

It has been well remarked that this creature, arboreal in its habits, and really ursine in its general aspect, is the representative of the sun-bears of the Indian Archipelago, or of the sloths of America. The Koala is nocturnal in its habits, and is not very frequently found, even in the localities which it most affects. It is not nearly so widely spread as most of the preceding animals, as it is never known to exist in a wild state except in the south-eastern regions of Australia.

Although well adapted by nature for climbing among the branches of trees, the Koala is by no means an active animal, proceeding on its way with very great deliberation, and making sure of its hold as it goes along. Its feet are peculiarly adapted for the slow but sure mode in which the animal progresses among the branches by the structure of the toes of the fore-feet or paws, which are divided into two sets, the one composed of the two inner toes, and the other of the three outer, in a manner which reminds the observer of the feet of the scansorial birds and the chameleon. This formation, although well calculated to serve the animal when it is moving among the branches, is but of little use when it is upon the ground, so that the terrestrial progress of the Koala is especially slow, and the creature seems to crawl rather than walk.

As far as is yet known, its food is of a vegetable nature, and consists chiefly of the young leaves, buds, and twigs of the eucalypti, or gum-trees, as they are more popularly called. When it drinks, it laps like a dog.

It seems to be a very gentle creature, and will often suffer itself to be captured without offering much resistance, or seeming to trouble itself about its captivity. But it

KOALA, OR AUSTRALIAN BEAR.—*Phascolarctos cinéreus.*

is liable, as are many gentle animals, to sudden and unexpected gusts of passion, and when it is excited by rage it puts on a very fierce look, and utters sharp and shrill yells in a very threatening manner. Its usual voice is a peculiar soft bark.

The head of this animal has a very unique aspect, on account of the tufts of long hairs which decorate the ears. The muzzle is devoid of hair, but has the curious property of feeling like cotton velvet when gently stroked with the fingers. There is a naked patch of skin that begins at the muzzle and extends for a small space towards the head, and over the whole of this bare patch the peculiar velvety feeling is exhibited. The upper jaw projects slightly over the lower. The generic name, Phascolarctos, is of Greek origin, signifying "pouched bear," and is very appropriate to the animal. As soon as the young Koala is able to leave the pouch, the mother transposes it to her back, where it clings with its hand-like paws, and remains there for some considerable time.

It is said by those who have seen the animal in its wild state, that it is truly deserving of the name of Australian Sloth, which has been applied to it because it is able to cling with its feet to the branches after the manner of the sloths, and to suspend itself from the boughs much after the same fashion.

TREE KANGAROO.—*Dendr…ogos ursinus.*

This animal is rather prettily coloured, the body being furnished with fur of a fine grey colour, warmed with a slight reddish tinge in the adult animal, and fading to a whitish-grey in the young. The claws are considerably curved, and black ; and the ears are tufted with long white hairs. In size it equals a small bull-terrier dog, being, when adult, rather more than two feet in length, and about ten inches in height, when standing. The circumference of the body is about eighteen inches, including the fur.

On account of the tree-climbing habits of the Koala, it is sometimes called the Australian Monkey as well as the Australian Bear.

THE animals which come next under consideration are truly worthy of the title of Macropidæ, or long-footed, as their hinder feet are most remarkable for their comparative length, and in almost every instance are many times longer than the fore-feet. This structure adapts them admirably for leaping, an exercise in which the Kangaroos, as these creatures are familiarly termed, are pre-eminently excellent.

FIRST on the list appears the singular animal which is well represented in the engraving, and which, on account of its peculiar habit, is known by the name of the TREE KANGAROO. In general form, this animal is sufficiently Kangaroo-like to be enrolled at once among the members of that group of Macropods, but the comparative shortness of the hinder feet and the length of the fore-feet, together with some peculiarity in the dentition, have induced the later zoologists to place it in a separate genus from the true Kangaroo.

The fur of the Tree Kangaroo is so remarkably dark that its deep tinting serves as an infallible mark of distinction, by means of which it may be recognised even at some distance. It is on account of the dark, glossy blackness of the fur, that the creature is called ursinus, or bear-like, as the hairs of its fur are thought to bear some resemblance to those which form the coat of the American black bear.

The colouring of its fur is generally as follows : the whole of the back and the upper parts of the body are a deep, glossy black, the hairs being rather coarser, and running to some length. These hairs are only of one kind, for in the fur of the Tree Kangaroo there is none of that inner coat of fine, close, woolly hair which is found in the other Kangaroos, and which lies next to the skin. The whole of the fur is, therefore, composed solely of the long and stiff hairs that are usually found to penetrate through the interior covering of woolly fur, and to lie upon its surface. The under parts of the body are of a yellowish hue, and the breast is washed with a richer and deeper tint of chestnut. The tail is of the same colour as the body, and is of very great length, probably to aid the animal in balancing itself as it climbs among the branches of the trees on which it loves to disport itself.

To see a Kangaroo on a tree is really a most remarkable sight, and one which might well have been deemed a mere invention had it not often been attested by credible witnesses. I have repeatedly seen one of these creatures clambering about a tree-trunk with perfect ease, and ascending or descending with the security of a squirrel. The animal looks so entirely in its wrong place, that when the black-haired, long-legged creature hops unexpectedly upon a tree and hooks itself among the branches, with its long black tail dangling below it, the entire aspect of the animal is absolutely startling, and suggestive of the super—or, perhaps, the infer—natural to the mind of the spectator. This species is not, however, the only one that can ascend trees, an art which is practised with some success by the Rock Kangaroo.

The food of this species consists of vegetable substances, such as the young bark, twigs, berries, and leaves of the trees upon which it lives, but very little is known of its habits in a wild state. It is an inhabitant of New Guinea.

AMONG the largest of the Macropidæ is the celebrated KANGAROO, an animal which is found spread tolerably widely over its native land.

This species has also been called by the name of giganteus, on account of its very great size, which, however, is sometimes exceeded by the woolly Kangaroo. The average dimensions of an adult male are generally as follows : the total length of the animal is about seven feet six inches, counting from the nose to the tip of the tail; the head and body exceed four feet, and the tail is rather more than three feet in length; the circumference of the tail at its base is about a foot. When it sits erect after its curious tripedal fashion, supported by its hind-quarters and tail, its height is rather more than fifty inches ; but when it wishes to survey the country, and stands erect upon its toes, it surpasses in height many a well-grown man. The female is very much smaller than her mate, being under six feet in total length, and the difference in size is so great that the two sexes might well be taken for different species.

The weight of a full-grown male, or " boomer," as it is more familiarly called, is very considerable, one hundred and sixty pounds having often been attained, and even greater weight being on record. The colour of the animal is brown, mingled with grey, the grey predominating on the under portions of the body and the under-faces of the limbs. The fore-feet are black, as is also the tip of the tail.

Without being truly gregarious, the Kangaroo is seldom seen entirely alone, but in scattered groups of seven or eight in number, and even the members of these little bands are not closely united, but are seen singly disposed at some distance from each other. There are certainly instances on record where very large numbers of Kangaroos have been seen in true flocks, herding closely together, and being under the superintendence of one leader. These animals, however, belong to another species.

As the Kangaroo is a valuable animal, not only for the sake of its skin, but on account of its flesh, which is in some estimation among the human inhabitants of the same land, it is eagerly sought after by hunters, both white and black, and affords good sport to both on account of its speed, its vigour, and its wariness. The native hunter, who trusts chiefly to his own cunning and address for stealing unobserved upon the animal and lodging a spear in its body before it is able to elude its subtle enemy, finds the Kangaroo an animal which will test all his powers before he can attain his object, and lay the Kangaroo dead upon the ground.

There is also another but not so sportsmanlike a method of killing the Kangaroo, which is often in use among the aborigines, and which partakes of the nature of a battue in England, or a bear " skal " in Norway.

A number of armed men associate themselves together, and, having laid deep counsel about the plan of the hunt, proceed cautiously forward until they come upon a number

KANGAROO.—*Macropus major.*

of Kangaroos. They then silently arrange themselves so as to surround the unconscious animals which are feeding carelessly in the plain. At a preconcerted signal a portion of the hunters issue from their concealment and shower their deadly missiles upon the Kangaroos. The poor alarmed creatures flee from the danger, and are met by another party of the same band, who also ply their spears and clubs with deadly effect. Backwards and forwards run the bewildered animals, assailed on all sides by sharp and heavy missiles hurled by the strong arm and directed by the keen eye of the native hunters; and so well are the plans laid, and with such accurate aim are the deadly weapons thrown, that it seldom happens that a single Kangaroo escapes from the scene of massacre.

A time of feasting then follows, for these wild children of nature have no conception of thrift, and would think themselves very hardly used were they not allowed to eat every particle of food which they could obtain, even though they would be forced to endure the pangs of hunger for many a day afterwards. The quantity of meat that a native Australian will eat at a single meal, and the gallons of water that he will drink, are so astounding as almost to surpass belief.

Besides these modes of hunting, the native makes use of pitfalls, snares, nets, and other devices, by means of which he contrives to entrap the animal without putting himself to the trouble of hunting it.

The white hunters, however, go to work in a very different manner, looking more to the sport than to the number of Kangaroos killed. They are in the habit of breeding and training a certain valuable and peculiar strain of hounds, called, from their quarry, "Kangaroo dogs," and which hunt by sight like the greyhound. These animals are long, large, and powerful; but, even with all these advantages, are no match for a full-grown

Boomer or Forester, as the animal is indifferently called, whenever he chooses to turn to bay and bid defiance to his pursuers.

A very graphic account of a Kangaroo hunt was sent to Mr. Gould, and is published by him in his very valuable monograph on the Macropidæ of Australia. A portion of the letter is extracted, and runs as follows :—

"The 'Boomer' is the only Kangaroo which shows good sport, for the strongest Brush Kangaroo cannot live above twenty minutes before the hounds. But as the two kinds are always found in perfectly different situations, we were never at a loss to find a 'Boomer,' and I must say that they seldom failed to show us good sport.

We generally 'found' in a high cover of young wattles, but sometimes in the open forests, and then it was really pretty to see the style in which a good Kangaroo would go away. I recollect one day in particular, when a very fine Boomer jumped up in the very midst of the hounds in the 'open;' he at first took a few jumps with his head up, in order to look about him, to see on which side the coast was clearest, and then, without a moment's hesitation, he started forward and shot away from the hounds, apparently without an effort, and gave us the longest run I ever saw after a Kangaroo.

He ran fourteen miles by the map, from point to point, and if he had had fair play, I have very little doubt but that he would then have beaten us; but he had taken along a tongue of land which ran into the sea, so that, being pressed, he was forced to try to swim across the arm of the sea, which, at the place where he took the water, cannot have been less than two miles broad. In spite of a fresh breeze and a hard sea against him, he got fully half-way over, but he could not make head against the waves any farther, and was obliged to turn back, when, being faint and exhausted, he was soon killed.

The distance he ran, taking the different bends in the line, cannot have been less than eighteen miles, and he certainly swam two. I can give no idea of the length of time it took him to run this distance, but it took us something more than two hours, and it was evident by the way the hounds were running that he was a long way before us ; it is also plain that he was still fresh, as quite at the end of the run he went on the top of a long, high hill, which a tired Kangaroo will never attempt to do, as dogs gain so much on them in going up-hill. His hind-quarters weighed within a pound or two of seventy pounds, which is large for the Van Diemen's Land Kangaroo, though I have seen larger.

We did not measure the length of the hop of this Kangaroo, but on another occasion, when the Boomer had taken along the beach and left its prints in the sand, the length of each jump was found to be just fifteen feet, and as regular as if they had been stepped by a sergeant."

The Boomer is a dangerous antagonist to man and dog, and unless destroyed by missile weapons will often prove more than a match for the combined efforts of man and beast.

When the animal finds that it is overpowered in endeavour by the swift and powerful Kangaroo dogs, which are bred for the express purpose of chasing this one kind of prey, it turns suddenly to bay, and placing its back against a tree-trunk, so that it cannot be attacked from behind, patiently awaits the onset of its adversaries. Should an unwary dog approach within too close a distance of the Kangaroo, the animal launches so terrible a blow with its hinder feet, that the long and pointed claw with which the hinder foot is armed cuts like a knife, and has often laid open the entire body of the dog with a single blow. Experienced dogs, therefore, never attempt to close with so terrible an antagonist until they are reinforced by the presence of their master, who generally ends the struggle with a bullet. Sometimes, however, the Kangaroo is so startled by the apparition of the hunter that it permits its attention to wander from the dogs, and is immediately pulled down by them.

If the hunter should be on foot, he needs beware of the Kangaroo at bay, for the creature is rather apt to dash through the dogs and attack its human opponent, who is likely to fare badly in the struggle unless he succeeds in launching a fatal missile at the advancing animal.

Sometimes the Kangaroo comes to bay near water, and then takes a singular advantage of the situation. If any dog should be bold enough to come within reach, the Kangaroo picks up its foe in its fore-paws, and leaping to the water, holds the dog under the surface until it is dead. On one occasion, a Boomer had come to bay in some shallow water, and was already engaged in drowning a dog, when it was assailed by the remainder of the hounds, which had just arrived. Nothing daunted by their onset, the Kangaroo kept its dying foe under water by holding it down with one of its hind-feet, and held itself prepared to repeat the process upon the next dog that should attack.

But the Kangaroo is wise enough to postpone an actual combat until it is absolutely forced to fight, and uses every stratagem in its endeavours to escape. When pressed very hardly by the hounds, the Boomer has often been known to make a sudden leap at right angles to its former course, and to make good its escape before the dogs could recover themselves. This mode of proceeding is, however, rather a dangerous one, as the animal has more than once broken one of its legs by the sudden strain that is thrown upon the right or left leg, as the case may be.

When running, the creature has a curious habit of looking back every now and then, and has sometimes unconsciously committed suicide by leaping against one of the tree-stumps which are so plentifully found in the districts inhabited by the Kangaroo.

The doe Kangaroo displays very little of these running or fighting capabilities, and has been known, when chased for a very short distance, to lie down and die of fear. Sometimes when pursued, it contrives to elude the dogs by rushing into some brushwood, and then making a very powerful leap to one side, so as to throw the dogs off the scent. She lies perfectly still as the dogs rush past her place of concealment, and when they have fairly passed her, she quietly makes good her escape in another direction. When young, and before she has borne young, the female Kangaroo affords good sport, and is called, from her extraordinary speed, the "Flying Doe."

The extraordinary pouch in which the young of the Kangaroo and other marsupiated animals are nourished has already been casually mentioned, and as it is highly developed in the Kangaroo, it will be described in connexion with this animal.

The young animal when first born is of extremely minute dimensions, hardly exceeding an inch in total length, soft, helpless, and semi-transparent as an earth-worm. After birth it is instantly conveyed into the pouch, and instinctively attaches itself to one of the nipples, which are very curiously formed, being re-tractile, like the finger of a glove when not in use, and capable of being drawn out to a considerable degree when they are needed by the young animal. In the accompanying engraving this structure is very well delineated.

In this internal cradle the young Kangaroo passes the whole of its earlier stages of development, and when it has attained some little bodily powers occa-sionally loosens its hold, and pokes its head out of the pouch, as if to see how large the world really is. By

YOUNG KANGAROO IN ITS MOTHER'S POUCH.

degrees it gains sufficient strength to crop the more delicate herbage, and, in course of time, it leaves the pouch altogether, and skips about the plains under the ever watchful protection of its mother. No sooner, however, is the little animal tired, or does the mother see cause of danger, than it scrambles back again into the pouch, and does not emerge until it is refreshed by repose, or until all danger has passed away.

WOOLLY KANGAROO.—*Macropus Laniger.*

Nearly eight months elapse between the time when the young Kangaroo is first placed in the pouch and the period of its life when it is able to leave the pouch and seek subsistence for itself. Even after it has become too large to continue its residence in its former cradle, it is in the habit of pushing its head into the pouch and refreshing itself with a draught of warm milk, even though a younger brother or sister should be occupant of the living cradle. The little animal weighs about ten pounds when it becomes too heavy for its mother to carry.

This Kangaroo is a very hardy animal, and thrives well in England, where it might probably be domesticated to a large extent if necessary, and where it would enjoy a more genial climate than it finds in many districts of its native land. One of the favoured localities of this species is the bleak, wet, and snow-capped summit of Mount Wellington.

At different times of the year the coat of the Kangaroo varies somewhat in its colouring and density. During the summer the fur is light and comparatively scanty, but when the colder months of the year render a warmer covering needful, the animal is clothed with very thick and woolly fur, that is admirably calculated to resist the effects of the damp, cold climate. It is a very singular fact that those specimens which inhabit the forests are much darker in their colour than those which live in the plains. The young Kangaroos are lighter in their colouring than their parents, but up to the age of two years their fur deepens so rapidly that they are darker than the old animals. After that age, however, the fur fades gradually, until it finally settles into the greyish-brown of the adult animal.

The eye of the Kangaroo is very beautiful, large, round, and soft, and gives to the animal a gentle, gazelle-like expression that compensates for the savage aspect of the teeth, as they gleam whitely between the cleft lips.

THE largest of the Macropidæ, of which there are already known upwards of eighty species, is the WOOLLY KANGAROO, or RED KANGAROO, as it is more popularly called, on account of its peculiarly tinted fur.

WHALLABEE.—*Halmatúrus Ualabatus.*

The character of the fur is rather singular, for it does not lie so closely to the body as that of the common Kangaroo, and is of a peculiar texture, which somewhat resembles cotton wool. The hairs are not very long, and their woolly, matted appearance, makes them seem shorter than they really are. The size of this animal is very great, for an adult male measures rather more than eight feet in total length, the head and body being five feet long, and the tail a little short of thirty-eight inches.

By the colour of the fur alone the Woolly Kangaroo can be distinguished from its long-legged relatives, independently of other minute differences. The general tint of the fur is of a rusty yellow, changing to grey upon the head and shoulders, the head being washed with a slight brown tint. The sides of the mouth are white, through which protrude a few long, stiff, black hairs, and which are planted in greater numbers over the angle of the mouth, forming an indistinct black patch. The female is distinguished by a broad white mark which runs from the angle of the mouth to the eye. The toes are covered with black hairs.

An ashy-grey tint is seen upon the under portions of the body in the male sex, but in the female these parts are beautifully white. The limbs are greyish-white, washed with rust, and the tail is of the same colour as the limbs.

The tail is uncommonly large and powerful, and of vast service to the animal in supporting the heavy frame while the creature is standing erect upon the tripod formed by its hinder feet and its tail. The hairs of the tail are comparatively short and scanty, so that they do not give to the tail that peculiar woolliness which is so distinguishing a characteristic of the creature's fur. It may as well be mentioned in this place that the Kangaroo does not employ the tail in leaping from the ground, but seems to use it partly as a kind of third leg, by which it supports itself when at rest, and partly as a kind of balance, by which it maintains its equilibrium as it leaps through the air.

The muzzle of the Woolly Kangaroo is not so thickly covered with hair as that of the preceding animal. This species is an inhabitant of Southern Australia.

Passing by the Nail-tailed Kangaroos, so called from the strange nail-like appendage that is found at the extremity of their tails, and which is concealed by the tuft of long black hair which terminates that member, we arrive at the WHALLABEE, or WALLABY, as the word is sometimes spelled.

The genus to which this animal belongs is easily distinguished from the genus Macropus, by reason of the muzzle being devoid of hair. This creature is not nearly so large as the common or the woolly Kangaroo, being only four feet six inches in total length, of which measurement the tail occupies two feet.

The fur of the Whallabee is rather long and coarse in texture, being decidedly harsh to the touch. The colour is rather curious, being a darkish-brown washed with a warm rusty hue, and obscurely pencilled with whitish-grey. The whole of the under portions of the body are of a yellowish tint, and the feet and the wrists are quite black. The tail is also rather singular in its colouring, by which it is divided into three nearly equal portions. The dorsal third of the tail is of the same colour as the back, but the remaining two-thirds change abruptly from brown to black.

The animal is an inhabitant of New South Wales, and is of tolerably frequent occurrence in the neighbourhood of Port Jackson. It is sometimes known by the name of the Aroë Kangaroo. The singular word Ualabatus has no particular meaning, being only the harsh Latinized form of the native name Whallabee. The genus embraces a considerable number of species, some twelve or thirteen being acknowledged to belong to it.

ONE of the most singular of this singular group of animals is the ROCK KANGAROO, which has derived its popular name from its rock-loving habits.

In the tree Kangaroo we have already seen a remarkable instance of unexpected powers, and the Rock Kangaroo will shortly be seen to be possessed of equal, if not of superior bodily prowess. The agility with which this animal traverses the dangerous precipices among which it lives is so very great, that when the creature is engaged in skipping about the craggy rocks that shroud its dwelling-place from too vigilant eyes, it bears so close a resemblance to a monkey in its movements, that it has, on many occasions, been mistaken for that active animal. Not only does it resemble the quadrumana in its marvellously easy manner of ascending rocks, but it also emulates those creatures in the art of tree-climbing, being able to ascend a tree-trunk with ease, provided that it be a little divergent from the perpendicular.

By means of its great scansorial capabilities, the Rock Kangaroo is enabled to baffle the efforts of its worst foes, the dingo and the native black man. In vain does the voracious and hungry dingo set off in chase of the Rock Kangaroo, for as soon as the creature has gained the shelter of its congenial rocks, it bounds from point to point with an agility which the dingo can by no means emulate, and very soon places itself in safety, leaving its baffled pursuer to vent its disappointment in cries of rage.

The only method in which the dingo is likely to catch one of these animals, is by creeping unsuspectedly into its den, and seizing it before it can make its escape. Both the dingo and the Rock Kangaroo are in the habit of making their resting-place in some rocky crevice, and it might happen that the Kangaroo might choose too low a domicile, and perchance make choice of the very same crevice that a dingo was about to appropriate to himself. In order, however, to escape such dangers, the habitation of the Rock Kangaroo is generally furnished with two or more outlets, so that its chances of escape are proportionably multiplied.

It is by means of this precaution that the creature baffles the best efforts of the natives. Should a native be fortunate enough to spear a Rock Kangaroo, but not fortunate enough to kill it on the spot, the animal dives at once into its rocky abode, and there awaits its death; for the rocks are too hard to be destroyed by the tools of the aborigines, and if the sable hunter has recourse to fire, and tries to smoke out his intended prey, the smoke rolls harmlessly through the rocky burrow and makes its exit through the various entries, without causing very much inconvenience to the concealed inmate.

It is found that when the Rock Kangaroo comes from its cavernous home, it is in the habit of taking the same route along the rocks, so that by continually passing over the

ROCK KANGAROO.—*Petrógale penicilláta.*

same ground, its sharp and powerful claws make a very visible track over the stones, and afford an infallible guide to the acute sense of the black hunter, who is enabled to follow up the trail and to ascertain the precise crevice in which the animal has taken up its abode.

Generally nocturnal in its habits, the Rock Kangaroo is not seen so often as might be expected, considering the frequency of its occurrence. Now and then, however, it ventures from its dark home and braves the light of day, skipping daintily over the rocky prominences, or lying in the full blaze of the sunlight, and enjoying the genial warmth of the noontide beams. The native and colonial hunters watch eagerly for a basking Kangaroo, for when thus engaged, it is so fully taken up with appreciation of the warm sunbeams, that it can be approached and shot without difficulty. As its flesh is thought to be remarkably excellent, the animal is eagerly sought after by the hunters. It is rather gregarious in its habits, being generally found in little parties of two or three in number.

The colour of this animal is rather varied, but is generally of a purplish or vinous grey, which warms into a rich rusty red upon the hind-quarters and the base of the tail. The chest is purplish-grey, pencilled with white, the chin is white, and a very conspicuous white band runs along the throat to the chest. The fur is not in very great repute, as, although long, it is rough and harsh to the touch. The total length of an adult male is about four feet, the tail being about twenty-three inches in length. The tail is furnished with a moderately sized tuft of dark hairs, each hair being about three inches in length, a peculiarity which has earned for the animal the name of Brush-tailed Kangaroo. The body is strong and robust in its form, and the claws of the hinder feet are powerful in their make, as might be expected in an animal of such habits. The feet are so densely covered with fur that the claws are nearly hidden in the thick hairy coat. It is a tolerably hardy animal, and thrives well in England.

The habitation of the Rock Kangaroo is in the south-eastern portions of Australia, and on account of its peculiar habits it is a very local animal, being restricted to those districts which are furnished with rocks or mountain ranges.

The Brush-tailed Bettong, or Jerboa Kangaroo, as it is sometimes called, affords an excellent example of the genus Bettongia, in which are collected a small group of Kangaroos that are easily distinguished by their peculiarly short and broad heads. In size it equals a common hare, the head and body being about fourteen inches in length, and the tail about eleven inches, without including the tuft which decorates its extremity. The general colour of the animal is a palish brown liberally pencilled with white, and the under parts are of a pale greyish-white. The "brush" is black, and the under side of the tail is brownish-white.

It is a nocturnal animal, and lies curled up during the entire day, issuing forth from its nest as the shades of evening begin to draw on. The nest of the Brush-tailed Bettong is a very ingenious specimen of architecture, and is so admirably constructed, that it can hardly be detected by a European eye, even when it is pointed out to him. The native, however, whose watchful eye notes even the bending of a leaf in the wrong place, or the touch of a claw upon the tree trunk, seldom passes in the vicinity of one of these nests without discovering it and killing its inmates, by dashing his tomahawk at random into the mass of leaves and grass.

BRUSH-TAILED BETTONG.—*Bettongia Penicillata.*

As this animal resides chiefly on grassy hills and dry ridges, it is no easy matter to make a nest that shall be sufficiently large to contain the female and her young, and yet so inconspicuous as not to attract attention. The manner in which the nest is made is briefly as follows.

The animal searches for some suitable depression in the earth, enlarging it till it is sufficiently capacious, and builds a curious edifice of leaves and grass over the cavity, so that when she has completed her task, the roof of the nest is on a level with the growing grass. For additional safety, the nest is usually placed under the shelter of a large grass tuft or a convenient bush.

The manner in which the animal conveys the materials of its nest to the spot where they are required is most remarkable. After selecting a proper supply of dried grass, the creature makes it up into a sheaf, and twisting her prehensile tail round the bundle, hops away merrily with her burden. It is almost impossible to comprehend the extreme quaintness of the aspect which is presented by a Jerboa Kangaroo engaged in this manner without actual experience, or the aid of a very admirable and spirited drawing. When the animal has completed its nest, and the young are lying snugly in its warm recesses, the young family is effectually concealed from sight by the address of the mother, who invariably drags a tuft of grass over the entrance whenever she leaves or enters her grassy home.

It is an active little creature, and not easily caught even by fair speed, and has a habit of leaping aside when it is hard pressed and jumping into some crevice where it effectually conceals itself. It is extremely common over the whole of New South Wales. The colour of the fur is a grey-brown above, and the under parts of the body are of a greyish-white.

The KANGAROO RAT, called by the natives the POTOROO, is a native of New South Wales, where it is found in very great numbers.

It is but a diminutive animal, the head and body being only fifteen inches long, and the tail between ten and eleven inches. The colour of the fur is brownish-black, pencilled along the back with a grey-white. The under parts of the body are white, and the fore-feet are brown. The tail is equal to the body in length, and is covered with scales, through the intervals of which sundry short, stiff, and black hairs protrude.

This little animal frequents the less open districts, and is very quick and lively in its movements, whether it be indulging in its native gamesomeness or engaged in the search for food. Roots of various kinds are the favourite diet of the Kangaroo Rat, and in order

KANGAROO RAT.—*Hypsiprymnus minor.*

to obtain these dainties the animal scratches them from the ground with the powerful claws of the fore-feet. It is specially fond of potatoes, and often commits considerable havoc in a kitchen-garden by exhuming and carrying away the seed-potatoes. In retaliation for these injuries the owner of the garden sets traps about his potato-grounds, and by means of baiting them with the coveted roots entices numbers of Potoroos into the treacherous snare.

The movements of the Kangaroo Rats do not in the least resemble those of the Kangaroos themselves, for although they can sit erect upon their hind-legs, they cannot make those vigorous leaps which are so characteristic of the Kangaroos, nor can they manipulate their food with their fore-paws and carry it to their mouth by means of those limbs. Their gait, especially when chased, is a curious kind of gallop, very unequal, but tolerably swift. They are very timid and harmless animals, and when captured or attacked do not kick or make any violent resistance, contenting themselves with expressing their indignation by an angry hiss.

They are not so exclusively nocturnal as many of the preceding animals, and seem to be equally lively by day as by night. When the animal is sitting upon its hinder portions, the tail receives part of the weight of the body, but is not used in the same manner as

KANGAROO HARE.—*Lagorchestes Leporoïdes.*

the tail of the true Kangaroos, which, when they are moving slowly and leisurely along, are accustomed to support the body on the tail, and to swing the hinder legs forward like a man swinging himself upon crutches.

CONSIDERABLY larger than the preceding animal, the KANGAROO HARE may at once be distinguished from it by the hair-covered muzzle which is a distinguishing mark of the genus Lagorchestes.

The colour of the coat is very like that of the common hare, but the fur is short, rather hard, and slightly curled. The upper parts of the body are a mixture of black and cream, the sides are tinged with a yellow hue, and the under parts are a greyish-white. The skin is white. There is much variety in the tinting of different specimens of this animal, some being of a much redder hue than others. The fore-legs are black, and the fore-feet are variegated with black and white, the hinder feet being of a brownish-white. A buff-coloured ring surrounds the eye, and the back of the neck is washed with yellow. It sometimes happens that a light rust-colour takes the place of the buff. The tail is of a very pale brownish-grey.

The Kangaroo Hare inhabits the Liverpool Plains and the greater part of the interior of Australia, to which region it seems to be limited, seldom, if ever, being seen nearer the sea. It has many hare-like traits of character, such as squatting closely to the ground in a "form," and then sitting, in hopes of eluding notice, until it is roused to active exertion by actual contact. When it once takes to flight, it runs with amazing celerity, and doubles before the hounds in admirable style, not unfrequently making good its escape in the opposite direction by a well-executed "double."

Mr. Gould relates a curious incident that occurred to him while he was engaged in the pursuit of a Kangaroo Hare, attended by two dogs. The hounds had pressed the animal closely, when it doubled before them, retraced its course at full speed, making directly for Mr. Gould, who was following up his dogs. The animal came within twenty yards without seeing him, and then, instead of turning aside, leaped clear over his head.

The total length of this animal is about two feet, the tail occupying about thirteen inches. The Kangaroo Hare is not able to dig after the manner of many of the preceding animals.

WOMBAT.—*Phascólomys ursínus.*

THE WOMBAT, or AUSTRALIAN BADGER, as it is popularly called by the colonists, is so singularly unlike the preceding and succeeding animals in its aspect and habits, that it might well be supposed to belong to quite a different order ; indeed, in all its exterior character with the exception of its pouch, it is a rodent animal, and in its internal anatomy it approaches very closely to the beaver.

As might be imagined from its heavy body and short legs, the Wombat is by no means an active animal, but trudges along at its own pace, with a heavy rolling waddle or hobble, like the gait of a very fat bear. It is found in almost all parts of Australia, and is rather sought after for the sake of its flesh, which is said to be tolerably good, although rather tough, and flavoured with more than a slight taint of musk. The fur of the Wombat is warm, long, and very harsh to the touch, and its colour is grey, mottled with black and white. The under parts of the body are greyish-white, and the feet are black. The muzzle is very broad and thick. The length of the animal is about three feet, the head measuring seven inches.

In its temper the Wombat is tolerably placid, and will permit itself to be captured without venting any display of indignation. Sometimes, however, it is liable to violent gusts of rage, and then becomes rather a dangerous antagonist, as it can scratch most fiercely with its heavy claws, and can inflict tolerably severe wounds with its chisel-like teeth. Easily tamed, it displays some amount of affection for those who treat it kindly, and will come voluntarily to its friends in hopes of receiving the accustomed caress. It will even stand on its hind legs, in token of its desire to be taken on the knee, and when placed in the coveted spot will settle itself comfortably to sleep.

Generally, however, the Wombat is not a very intelligent animal, and exhibits but little emotion of any kind, seeming to be one of the most apathetic animals in existence. When in captivity it is easily reconciled to its fate, and will feed on almost any vegetable substance, evincing considerable partiality for lettuce-leaves and cabbage-stalks ; milk also is a favourite article of diet, and one of these animals was said by Mr. Bennett to be in the habit of searching after the milk vessels when set out to cool in the night air, to push off the covers, and to bathe in the milk as well as drink it.

1. I I

In its wild state it is nocturnal in its habits, living during the day in the depths of a capacious burrow, which it excavates in the earth to such a depth that even the persevering natives will seldom attempt to dig a Wombat out of its tunnel. Owing to this habit of burrowing, it is very destructive if left in an unpaved yard, for it soon excavates several subterraneous passages, and puts the stability of houses and walls into sad jeopardy.

The creature seems to be remarkably sensitive to cold, considering the severe weather which often reigns in its native country. It is fond of hay, which it chops into short pieces with its knife-edged teeth. The natives say that if a Wombat is making a journey, and happens to come across a river, it is not in the least discomfited, but walks deliberately into the river, across the bed of the stream, and, emerging on the opposite bank, continues its course as calmly as if no impediment had been placed in its way.

The teeth of the Wombat present a curious resemblance to those of the rodent animals, and are endowed with the same powers of reproduction as those of the beaver and other animals of the same order. The feet of the Wombat are broad, and the fore-feet are provided with very strong claws, that are formed for digging in the earth. There are five toes to each foot, but the thumb of the hinder feet is extremely small, and devoid of a claw. This animal is remarkable for possessing fifteen pairs of ribs—in one case sixteen pairs of ribs were found—only six pairs of which reach the breast-bone. Remains of a fossil species of Wombat have been discovered in New Holland, together with the relics of an allied and gigantic species, which, when living, must nearly have equalled the hippopotamus in dimensions.

BANDED BANDICOOT.—*Perámeles fasciáta.*

The BANDICOOT, two examples of which will be described in the following pages, form a little group of animals that are easily recognisable by means of their rat-like aspect, and a certain peculiar, but indescribable mode of carrying themselves. The gait of the Bandicoot is very singular, being a kind of mixture between jumping and running, which is the result of the formation of the legs and feet. During progression, the back of the creature is considerably arched. The snout is much lengthened and rather sharply pointed, and the second and third toes of the hinder feet are conjoined as far as the claws. The pouch opens backwards.

The BANDED BANDICOOT, or STRIPED-BACKED BANDICOOT, derives its name from the peculiar marking of its fur.

The general colour of its coat is a blackish-yellow, as if produced by alternate hairs, the black tint predominating on the back and the yellow on the sides. Over the hinder quarters are drawn some boldly marked black lines, which, when viewed from behind, form a singular and rather pleasing pattern, the dark stripes being made more conspicuous by bands of whitish-yellow. These marks continue as far as the root of the tail, and a single, narrow dark line runs along the whole upper side of the tail, which is of the same colour as the body. The fur is rather light upon the head, and the under parts of the body, together with the feet, are white, slightly tinged with grey.

This animal is very widely spread over the eastern and south-eastern parts of Australia, but is mostly found in the interior. It specially loves the stony ridges that are so common in its native land, and although not very often seen by casual travellers, is of very frequent occurrence. Its pace is very swift, and its gait is said to bear some resemblance to that of the pig. Its food is of both kinds, and consists of insects and their larvæ, and of various roots and seeds. Its flesh is held in some repute by natives and colonists.

It is but a small animal, measuring only eighteen inches in total length. When the animal is killed, it is not easily flayed, as the skin adheres so tightly to the flesh that its removal is a matter of some difficulty, when there is need for preserving the skin in its integrity.

LONG-NOSED BANDICOOT.—*Perameles nasuta.*

The LONG-NOSED BANDICOOT is not unlike the preceding animal in form, but differs from it in the colouring of its fur, and the greater length of its snout.

The face, head, and body, are of a brown tint, pencilled with black on the upper portions, and the sides are of a pale brown, sometimes warmed with a rich purplish hue. The edge of the upper lip is white, as are also the under portions of the body, and the fore-legs and feet. This fur is very harsh to the touch. The total length of this animal is about twenty-one inches, the tail being five inches in length.

The food of the Long-nosed Bandicoot is said to be of a purely vegetable nature, and the animal is reported to occasion some havoc among the gardens and granaries of the colonists. Its long and powerful claws aid it in obtaining roots, and it is not at all unlikely that it may, at the same time that it unearths and eats a root, seize and devour the terrestrial larvæ which are found in almost every square inch of ground. The lengthened nose and sharp teeth which present so great a resemblance to the same organs in insectivorous shrews, afford good reasons for conjecturing that they may be employed in much the same manner.

The dentition of the Bandicoot is rather interesting, and will be found detailed at some length in the table of generic distinctions at the end of the volume.

THE large-eared, woolly-furred little animal which is here represented, is closely allied to the bandicoots, but at once distinguishable from them by the peculiarity of structure which has earned for it the generic title of Chæropus, or " swine-footed."

Upon the fore-feet there are only two toes, which are of equal length, and armed with sharp and powerful hoof-like claws, that bear no small resemblance to the foot of a pig, and are not only porcine in their external aspects, but in the track which they leave upon the ground when the creature walks on soft soil. Slenderly and gracefully swinish, it is true, but still piggish in appearance though not in character.

CHÆROPUS.—*Chæropus castanótis.*

The CHÆROPUS was formerly designated by the specific title of ecaudatus, or tailless, because the first specimen that had been captured was devoid of caudal appendage, and therefore its discoverers naturally concluded that all its kindred were equally curtailed of their fair proportions. But as new specimens came before the notice of the zoological world, it was found that the Chæropus was rightly possessed of a moderately long and somewhat rat-like tail, and that the taillessness of the original specimen was only the result of accident to the individual, and not the normal condition of the species. The size of the Chæropus is about equal to that of a small rabbit, and the soft, woolly fur is much of the same colour as that of the common wild rabbit.

It is an inhabitant of New South Wales, and was first discovered by Sir Thomas Mitchell on the banks of the Murray River, equally to the astonishment of white men and natives, the latter declaring that they had never before seen such a creature. The speed of the Chæropus is considerable, and its usual haunts are among the masses of dense scrub foliage that cover so vast an extent of ground in its native country. Its nest is similar to that of the bandicoot, being made of dried grass and leaves rather artistically put together, the grass, however, predominating over the leaves. The locality of the nest is generally at the foot of a dense bush, or of a heavy tuft of grass, and it is so carefully veiled from view by the mode of its construction that it can scarcely be discovered by the eyes of any but an experienced hunter.

The head of the Chæropus is rather peculiar, being considerably lengthened, cylindrically tapering towards the nose, so that its form has been rather happily compared to the neck and shoulders of a champagne bottle. The hinder feet are like those of the bandicoots, and there is a small swelling at the base of the toes of the fore-feet, which is probably the representative of the missing joints, more especially as the outermost toes are always extremely small in the bandicoots, to which the Chæropus is nearly allied. The ears are very large in proportion to the size of the animal. The pouch opens backwards. The food of the Chæropus is said to be of a mixed character, and to consist of various vegetable substances and of insects.

THE teeth of the Dasyurines, sharp-edged and pointed, indicate the carnivorous character of those animals to which they belong. At the head of these creatures is placed the TASMANIAN WOLF, or DOG-HEADED THYLACINUS, as it has often been named on account of the curious aspect of its thick head, and powerful, truncated muzzle.

TASMANIAN WOLF.—*Parácyon Cynocéphalus.*

Although not perhaps the fiercest of the Dasyurines, it is the largest and the most powerful, well deserving the lupine title with which it has been by common consent designated, and representing in Tasmania the true wolves of other countries. It is not a very large animal, as needs must be from the nature of the country in which it lives, for there would be but small subsistence in its native land for herds of veritable wolves, and the natural consequence would be that the famished animals would soon take to eating each other in default of more legitimate food, and by mutual extirpation thin down the race or destroy it altogether.

The natural subsistence of the Tasmanian, or Zebra Wolf, as it is sometimes called by virtue of the zebra-like stripes which decorate its back, consists of the smaller animals, molluscs, insects, and similar substances. The animal is also in the habit of prowling along the sea-shore in restless search of food among the heterogeneous mass of animal and vegetable substances that the waves constantly fling upon the beach, and which are renewed with every succeeding tide. The mussels and other molluscs which are found so profusely attached to the sea-edged rocks form a favourite article of diet with the Tasmanian Wolf, who is sometimes fortunate enough to discover upon the beach the remains of dead seals and fish, and can easily make a meal on the shore crabs which are found so plentifully studding the beach as the tide goes out.

Though hardly to be considered a swift, or even a quick animal, the Tasmanian Wolf contrives to kill such agile prey as the bush kangaroo, and secures the duck mole, or duck bill, in spite of its natatory powers and its subterranean burrow. When the animal is hungry, it seems to become a very camel in its capability of devouring hard and thorny substances, for it has been known to kill—no easy matter—and to swallow—an apparent impossibility—the echidna itself, undismayed by its panoply of bayonet-like prickles. The deed seems so incredible that it would hardly have been believed, had it not been proved beyond doubt by the slaughter and subsequent dissection of a Tasmanian Wolf, in whose stomach were found the remains of a half-digested echidna.

As soon as civilized inhabitants took up their abode in Tasmania, this animal made great capital out of the sheep flocks and henroosts, and for some time committed sad ravages among them, greatly to the detriment of the colonists. By degrees, however, the weapons of the white man prevailed, and the Tasmanian Wolf was driven back from its former haunts where it once reigned supreme. Still continuing to prowl round the habitation of mankind, many individuals of this species were fain to pick up what loose and uncertain subsistence they could contrive to appropriate, and, being forced to live in copses and jungles, became the representatives of the hyæna as well as of the Wolf.

In the earlier days of the colony, the Tasmanian Wolf was of very frequent occurrence, but is now seldom seen except in the cold and dreary localities where it takes up its residence. These animals are found in considerable numbers on the summits of the western mountains, at an elevation of nearly four thousand feet above the level of the sea, and there thrive, even though their lofty domains are plentifully covered with snow.

The home of the Tasmanian Wolf is always made in some deep recess of the rocks, away from the reach of ordinary foes, and so deeply buried in the rocky crevices that it is impenetrable to the light of day. In this murky recess the female produces her young, which are generally three or four in number, and in its dark cavern the animal spends the whole of its day, only venturing from home at night, except under the pressure of some extraordinary circumstances.

As may be seen from the engraving, the feet of the Tasmanian Wolf are so dog-like in their nature, that they cannot enable the animal to ascend trees, and as the tail is not in the least degree prehensile, it is evident that the creature is not capable of chasing its prey among the branches, as is the case with many of the allied animals.

In size it is about equal to the jackal, being generally about four feet in total length, of which measurement the tail occupies some sixteen inches. Some few specimens, however, are said to attain a very great size, and to measure nearly six feet in total length. Its height at the shoulders is about eighteen or nineteen inches. It is a fierce and most determined animal, and if attacked will fight in the most desperate manner. One of these animals has been seen standing at bay, surrounded by a number of dogs, and bidding them all defiance. Not a single dog dared venture within reach of the teeth of so redoubtable a foe.

As it is a nocturnal animal, it seems little at its ease when in the uncongenial glare of daylight, and, probably on account of its eyes being formed for the purpose of nocturnal light, is very slow in its movements by day. It always seems to be greatly annoyed by too strong a light, and constantly endeavours to relieve itself from the unwelcome glare by drawing the nictitating membrane over its eyeballs, after the manner of owls when they venture forth by daylight.

The animal is a very conspicuous one, on account of the peculiar colouring of its fur, and the brightly defined stripes which decorate its back.

The general tint of the fur is a greyish-brown, washed with yellow, each hair being brown at its base and yellow towards the point. Along the back runs a series of boldly defined stripes, nearly black in their colour, beginning just behind the shoulders and ending upon the base of the tail. The number of these stripes is various, being from fourteen to seventeen on an average. At the spot where they commence they are very short, but lengthen rapidly as they approach the tail, reaching their greatest length over the haunches, over which they are drawn to some extent. In many specimens the stripes are forked upon the haunches. Towards the tail the stripes again become short, and upon the base of the tail are so short that they only cover its upper surface. The under parts of the body are grey. The tail is slightly compressed, and gradually tapers to its extremity. The eyes are large and full, and their colour is black. The edge of the upper lip is white.

In this animal the marsupial bones are absent, their places being indicated by some fibrous cartilages that are found in the locality which these bones might be expected to fill. The character of the fur is not very fine, but it is short, rather woolly, and closely set upon the animal's skin. In front of the eye there is a small black patch, which runs round the eye, and surrounds it with a dark line.

As may be imagined, from the very expressive name which has been appropriated to the animal which is represented in the engraving, its character is not of the most amiable, nor its appearance the most inviting.

Few animals have deserved their popular titles better than the creature to whom the first colonists of Van Diemen's Land unanimously gave the name of NATIVE DEVIL. The innate and apparently ineradicable ferocity of the creature can hardly be conceived except by those who have had personal experience of its demeanour. Even in captivity its sullen and purposeless anger is continually excited, and the animal appears to be more obtuse to kindness than any other creature of whom we have practical knowledge. Generally, a caged animal soon learns to recognise its keeper, and to welcome the hand that supplies

TASMANIAN DEVIL.—*Diabolus ursinus.*

it with food; but the Tasmanian Devil seems to be diabolically devoid of gratitude, and attacks indiscriminately every being that approaches it.

I have frequently had opportunities of testing the character of this curious animal, and have always found it to be equally savage and intractable. Without the least cause it would fly at the bars of its cage, and endeavour by dint of teeth and claws to wreak its vengeance on me, while it gave vent to its passionate feelings in short, hoarse screams of rage. There was no reason for these outbursts of anger, for the animal behaved in precisely the same manner whenever any visitor happened to pause in front of its domicile.

It is a very conspicuous animal, and not easily to be mistaken for any other species. The coat of the Tasmanian Devil is very appropriately black, dashed here and there with spots, patches, or stripes of a pure white, which afford a bold and singular contrast to each other. In different individuals there is considerable variety in the distribution of these two colours, but the character of the markings is similar in all.

The general hue of the fur is a deep dead black, the fur being devoid of that rich silky glossiness which gives to the coats of many black animals so pleasing an effect. Across the breast there is nearly always a very conspicuous white mark, which in some individuals takes the form of a semilunar band, and in others is contracted to a mere spot. Generally, another white mark is found to extend saddlewise across the end of the spine, just before the insertion of the tail. This mark is also susceptible of great variation, being of considerable dimensions in some specimens, and extremely small in others. Now and then a white streak or patch is seen upon the shoulders, but in many individuals the shoulders are of equal blackness with the remainder of the body. Behind the eyes is a tuft of very long hairs, and another similar tuft is placed immediately above them.

As might be presumed from the heavy make of its body, and the thickset shortness of its limbs, the animal is not at all brisk or lively in its movements, and seldom displays much energy except when under the influence of the easily-excited irascibility for which it is so widely renowned. The head is short and thickly made, the muzzle very blunt, and the mouth wide. The gait of the animal is plantigrade, and its movements are in general dull and sluggish.

The length of this animal is about twenty-one inches, exclusive of the tail, which measures about seven inches in length, and is moderately well covered with fur.

Despite of its comparatively small size, this creature is hardly less destructive than the Tasmanian wolf, and in the earlier days of the colonists wrought sad havoc among the sheep and poultry, especially among the latter. In those days it swarmed in great numbers, but it is now nearly extirpated out of some districts, and is so persecuted by the righteous vengeance of the farmers, that a solitary specimen can scarcely now be seen in the locality where its nightly visits used to be of continual occurrence. Many of these depredators were shot, caught in traps, or otherwise destroyed, and suffered a poetical justice in furnishing a meal for those at whose expense they had often feasted. The flesh of this rather ungainly animal is said to be far from unpleasant, and to have some resemblance to veal.

The traps in which these nocturnal robbers are caught are baited with flesh of some kind, generally with butcher's offal, for the animal is a very voracious one, and is always sensitive to such attractions. Like the Tasmanian wolf—to which animal it is closely allied—it is in the habit of prowling along the sea-shore in search of the ordinary coast-loving molluscs and crustaceans, or in hopes of making a more generous feast on the dead carcases which the tides will sometimes leave upon the beach. In captivity it will eat almost any kind of food, and is found to thrive well upon bread and milk, with an occasional addition of meat. When it is indulged in the latter delicacy it speedily tears in pieces the meat with which it is furnished, and is in nowise baffled by the presence of moderately sized bones, which it can crack with wonderful ease by means of its strong teeth and powerful jaws.

The great power of its jaws, backed by its unreasoning ferocity, which seems to be literally incapable of comprehending the feeling of fear, renders it extremely formidable when attacked. Indeed, there are hardly any dogs, however strong and well-trained they may be, which can boast of a victory gained over a Tasmanian Devil in single fight.

It is rather a productive animal, the number of its family being from four to five at a birth. The habitations of this species is ascertained to be made in the depths of the forests, concealed as far as possible from the light of day, which grievously affects the eyes of this, as of all other strictly nocturnal animals.

The pain which is caused to the creature by the unwelcome brilliancy of ordinary daylight is constantly indicated by the ceaseless movements of the nictitating membrane over the eyeball, even when the animal is shrouded in the comparative dimness of a straw-filled den, and shades itself from the glare by crouching in the darkest corner of its cage.

Aided by the strong fossorial claws of the fore-feet, the Tasmanian Devil digs for itself a deep burrow in the ground, or, taking advantage of some natural hollow or crevice, shapes the interior to suit its own purposes. The hinder feet are made in a manner similar to those of the bear, and, like that animal, the Tasmanian Devil is able to sit erect upon its hinder quarters, and to convey food to its mouth by means of its fore-paws, which it uses in a very adroit manner.

This animal is also known under the names of Ursine Dasyure and Ursine Opossum.

OF the animals which have been congregated into the genus Dasyures, four or five species are now admitted to be clearly separated from each other. In colour the Dasyures are extremely variable, so much so indeed, that it is hardly possible to find two individuals of the same species that are marked in precisely the same manner.

In the Common DASYURE the general colour of the fur is brown, of a very dark hue, sometimes deepening into positive black, diversified with many spots of white, scattered apparently at random over the whole of the body, and varying both in their position and dimensions in almost every individual. In some specimens the tail is washed with white spots similar to those of the body, but in many examples the tail is uniformly dark. In all the Dasyures this member is moderately long, but not prehensile, and is thickly covered with hair; a peculiarity which has caused zoologists to give the title of Dasyure, or hairy-tail, to these animals.

They are all inhabitants of Australasia, the common Dasyure being found numerously enough in New Holland, Van Diemen's Land, and some parts of Australia. The habits of all the Dasyures are so very similar that there is no need of describing them separately. They are all rather voracious animals, feeding upon the smaller quadrupeds, birds, insects, and other living beings which inhabit the same country. The Dasyure is said to follow the example of the two preceding animals, and to be fond of roaming along the sea-coasts by night in search of food.

The Dasyures are all nocturnal animals, and very seldom make voluntary excursions from their hiding-places so long as the sun is above the horizon. They do not, like the Tasmanian wolf and the

DASYURE.—*Dasyurus viverrinus.*

ursine Dasyure, lie hidden in burrows under the earth, or in the depths of rocky ground, but follow the example of the Petaurists, and make their habitations in the hollows of decayed trees.

The young of the Dasyures are, like those of all the animals of this order, extremely small. Their number is rather variable, but is usually from four to six. In this species the thumb of the hind-feet is entirely absent.

The PHASCOGALE, or TAPOA TAFA, as it is termed by the natives of the country which it inhabits, affords an excellent example of the little dependence that is to be placed on mere external appearance in judging the character of any living being.

In size, the Phascogale is small, hardly exceeding the house-rat of Europe in dimensions. The total length of this creature is about seventeen inches, the long, widely-formed tail occupying nine inches if measured to the point of the hairy tuft that decorates its extremity, and seven inches if denuded of its hairy covering.

The fur of this animal is long, soft, and woolly, and lies very loosely upon the skin, so that it is disturbed by every slight breath of air that may happen to pass over its surface. In colour it is a soft grey on the upper parts of the body, the head, and the outer faces of the limbs, the under portions of the body being white, and slightly washed with grey. A few black hairs are scattered sparingly over the body. In almost every specimen that has been captured, a dark line is seen to run from the nose towards the base of the skull.

The tail is clothed with fur of the same colour as that of the body for one-fifth of its length, but the remaining four-fifths are furnished with a bushy mass of long hair, each hair being about two inches in length. The colour of this graceful appendage is a jet black, which affords a very marked contrast to the light hues with which the body and limbs are tinged, and which gives to the animal a notably handsome aspect. The ears are rather large, and the head tapers rapidly towards the nose.

The general appearance of the Tapoa Tafa is that of a gentle, peaceable little animal, unlikely to do any harm, and well calculated to serve as a domesticated pet.

Never did animal or man hide under a specious mask of innocence a character more at variance with its mendacious exterior.

For the Tapoa Tafa is one of the pests of the colonists, a fierce, bloodthirsty, audacious creature, revelling in the warm flesh of newly-slaughtered prey, and penetrating, in search of food, into the very houses of civilized men. Its small size and sharply-pointed head enable it to insinuate itself through the crevices which are almost necessarily left open in fences and walls, and its insatiate appetite induces it to roam through the store-rooms in search of any animal substances that may have been laid up by the owners. Unless placed under lock and key, behind tightly-closed doors, provisions of various kinds are invaded by the Tapoa Tafa, for its powers of climbing are so great that it can ascend even a perpendicular wall, unless its surface be smooth and hard, so that its sharp curved claws can take no hold.

PHASCOGALE.—*Phascogale penicillata.*

Fortunately for the farmers, the Tapoa Tafa is not possessed of the chisel-shaped incisor teeth which enable the European rat to gnaw its way through opposing obstacles, so that a wooden door will afford a sufficient barrier against its depredations, providing it be closely fitting, and of solid material. It is said to be very destructive to poultry, and to penetrate by night into the fowl-houses, creeping towards its prey so silently that its presence is not detected, and slaying the inmates as they are slumbering quietly on their perches. Were its size equal to that of the Tasmanian wolf, the Phascogale would be an effectual bar to civilization in any district which it might frequent. In its wild state its food is of a mixed vegetable and animal nature, and in the stomach of one of these creatures was found a heterogeneous mass of insect remains, mixed with portions of certain fungi.

Not only is the Tapoa Tafa an object of destruction for the repeated acts of depredation which it commits in civilized dwellings, but it has also earned a renowned name among white and black men for the extraordinary energy with which it will defend itself when attacked. Small though it may be, and harmless though it may appear, it deals such fierce and rapid strokes with its sharp claws that it can inflict extraordinarily severe lacerations upon the person of its adversary. So celebrated is the animal for its powers of resistance, that not even the quick-eyed and agile-limbed native will venture to trust his hand within reach of the claws of an irritated Tapoa Tafa.

Night is the usual time for the Tapoa Tafa to leave its home and prowl about in search of food, but it is often seen by daylight, and appears to be equally vivacious at either time. It is always a most active animal, and chiefly arboreal in its habits, climbing trees and skipping among their branches with the agility of a squirrel. Its long tail may serve to act as a balance during these excursions, but as it is not in the least prehensile, it cannot afford assistance in the actual labour of passing from one branch to another.

Its home is generally made in the hollow trunks of the eucalypti, and in those dark recesses it produces and nourishes its young. It is very widely distributed over Australasia, being found in equal plenty upon plain or mountain, contrary to the usual habits of Australian animals, which are generally confined within certain local limits, according to the elevation of the ground or the character of the soil.

On account of the large tuft of black hair that decorates the tail, the Tapoa Tafa is in some works mentioned under the title of the "Brush-tailed Phascogale."

THE little animals which are grouped together under the title of Pouched Mice are tolerably numerous, the genus Antechinus comprising about twelve or thirteen species. They are spread rather widely over New South Wales and Southern Australia, and as they are prolific creatures, they are among the most common of the Australian quadrupeds. They are all of inconsiderable size, the greater number hardly exceeding the ordinary mouse in dimensions, though one or two species nearly equal a small rat in size.

Arboreal in their habits, they are among the most active of tree-loving quadrupeds, running up and down a perpendicular trunk with perfect ease, and leaping from

YELLOW-FOOTED POUCHED MOUSE.—*Antechinus flavipes.*

one branch to another with singular activity of limb and certainty of aim. They can even cling to the under side of a horizontal branch, and are constantly seen running round the branches and peering into any little crevice, precisely after the manner of the ordinary titmice among the birds. They can descend a branch with their heads downward, instead of lowering themselves tail foremost, as is generally the custom among tree-climbing quadrupeds, and traverse the branches with admirable rapidity and liveliness.

The YELLOW-FOOTED POUCHED MOUSE is a very pretty little creature, its fur being richly tinted with various pleasing hues.

The face, the upper part of the head, and the shoulders, are dark grey, diversified with yellow hairs, and the sides of the body are warmed with a wash of bright chestnut. The under parts of the body, the chin, and the throat are uniform white, and the tail is black. There is often a slight tufting of hair on the extremity of the tail. The total length of the animal is about eight inches, the head and body being rather more than four inches and a half in length, and the tail a little more than three inches.

The MYRMECOBIUS is remarkable for several parts of its structure, and more especially so for the extraordinary number of its teeth, and the manner in which they are placed in the jaw. Altogether, there are no less than fifty-two teeth in the jaws of an adult and perfect specimen of the Myrmecobius, outnumbering the teeth of every other animal, with the exception of one or two cetacea and the armadillo. There is no pouch in this animal, but the tender young are defended from danger by the long hairs which clothe the under portions of the body.

It is a beautiful little animal, the fur being of agreeable tints and diversified by several bold stripes across the back. The general colour of the fur is a bright fawn on the shoulders, which deepens into blackish-brown from the shoulders to the tail, the fur of the hinder portions being nearly black. Across the back are drawn six or seven white bands, broad on the back and tapering off towards their extremities. The under parts of the body are of a yellowish-white. The tail is thickly covered with long, bushy hair, and has a grizzled aspect, owing to the manner in which the black and white hairs of which it is composed are mingled together. Some hairs are annulated with white, red-rust, and black, so that the tints are rather variable, and never precisely the same in two individuals.

The length of the body is about ten inches, and the tail measures about seven inches, so that the dimensions of the animal are similar to those of the common water vole of Europe.

It is an active animal, and when running, its movements are very similar to those of the common squirrel. When hurried, it proceeds by a series of small jumps, the tail being elevated over its back after the usual custom of squirrels, and at short intervals it pauses, sits upright, and casts an anxious look in all directions before it again takes to flight. Although not a particularly swift animal, it is not an easy one to capture, as it immediately makes for some place of refuge, under a hollow tree or a cleft in rocky ground, and when it has fairly placed itself beyond the reach of its pursuers, it bids defiance to their efforts to drive it from its haven of safety. Not even smoke—the usual resort of a hunter when his prey has gone to "earth" and refuses to come out again—has the least effect on the Myrmecobius, which is either possessed of sufficient smoke-resisting powers to endure the stifling vapour with impunity, or of sufficient courage to yield its life in the recesses of its haven, rather than deliver itself into the hands of its enemies.

MYRMECOBIUS.—*Myrmecobius fasciatus.*

The food of the Myrmecobius is supposed chiefly to consist of ants and similar diet, as it is generally found inhabiting localities where ants most abound. For this kind of food it is well fitted by its long tongue, which is nearly as thick as a common black-lead pencil, and is capable of protrusion to some distance. In confinement, a specimen of the Myrmecobius was accustomed to feed on bran among other substances. It is known that in the wild state it will eat hay, as well as the "manna" that exudes from the branches of the eucalypti.

It is a very gentle animal in its disposition, as, when captured, it does not bite or scratch, but only vents its displeasure in a series of little grunts when it finds that it is unable to make its escape. The number of its young is rather various, but averages from five to eight. The usual habitation of the Myrmecobius is placed in the decayed trunk of a fallen tree, or, in default of such lodging, is made in a hollow in the ground. It is a native of the borders of the Swan River.

THERE are very few of the marsupiated animals which are more remarkable for their form, their habits, or their character, than the Opossums of America. They are nearly all admirable climbers, and are assisted in their scansorial efforts by their long, prehensile tails, which are covered with scales, through the interstices of which a few short black hairs protrude. The hinder feet are also well adapted for climbing, as the thumb is opposable to the other toes, so that the animal is able to grasp the branch of a tree with considerable force, and to suspend its whole body together with the additional weight of its prey or its young.

The VIRGINIAN, or COMMON OPOSSUM, is, as its name implies, a native of Virginia as well as of many other portions of the United States of America. In size it equals a tolerably large cat, being rather more than three feet in total length, the head and body measuring twenty-two inches, and the tail fifteen. The colour of this animal is a greyish-white, slightly tinged with yellow, and diversified by occasional long hairs that are white towards their base, but of a brownish hue towards their points. These brown-tipped hairs are extremely prevalent upon the limbs, which are almost wholly of the brown hue, which also surrounds the eye to some extent. The under fur is comparatively soft and woolly, but the general character of the fur is harsh and coarse. The scaly portion of the tail is white.

It is a voracious and destructive animal, prowling about during the hours of darkness, and prying into every nook and corner in hope of finding something that may satisfy the cravings of imperious hunger. Young birds, eggs, the smaller quadrupeds, such as young rabbits, which it eats by the brood at a time, cotton rats, and mice, reptiles of various kinds, and insects, fall victims to the appetite of the Virginian Opossum, which is often not content with the food which it finds in the open forests, but must needs insinuate itself into the poultry-yard and make a meal on the fowls and their eggs. When it has once determined on making such a raid, it can hardly be baffled in its endeavours by any defences except those which consist of stout walls and closely fitting doors; for it can climb over any ordinary wall, or thrust itself through any fence, so that there is but little chance of preventing it from making good its entrance into the precincts of the farmyard.

OPOSSUM.—*Didelphys Virginiana.*

Its proceedings are so admirably related by Audubon, that I can do no better than present the account in his own words, the words of one who has frequently been an eye-witness of the scene which he so graphically depicts :—

" Methinks I see one at this moment slowly and cautiously trudging over the melting snows by the side of an unfrequented pond, nosing as it goes for the fare its ravenous appetite prefers. Now it has come upon the fresh track of a grouse or hare, and it raises its snout and sniffs the pure air. At length it has decided on its course, and it speeds onwards at the rate of a man's ordinary walk. It stops and seems at a loss in what direction to go, for the object of its pursuit has either taken a considerable leap, or has cut backwards before the Opossum entered its track. It raises itself up, stands for a while on its hind-feet, looks around, snuffs the air again, and then proceeds; but now, at the foot of a noble tree, it comes to a full stand. It walks round the base of the large trunk, over the snow-covered roots, and among them finds an aperture, which it at once enters.

Several minutes elapse, when it reappears, dragging along a squirrel already deprived of life, with which in its mouth it begins to ascend the tree. Slowly it climbs. The first fork does not seem to suit it, for perhaps it thinks it might there be too openly exposed to the view of some wily foe, and so it proceeds, until it gains a cluster of branches, intertwined with grape-vines, and there composing itself, it twists its tail round one of the twigs, and with its sharp teeth demolishes the unlucky squirrel, which it holds all the while with its fore-paws.

The pleasant days of spring have arrived, and the trees vigorously shoot forth their leaves; but the Opossum is almost bare, and seems nearly exhausted by hunger. It visits the margin of creeks, and is pleased to see the young frogs, which afford it a tolerable repast. Gradually the poke-berry and the nettle shoot up, and on their tender and juicy stems it gladly feeds. The matin calls of the wild turkey-cock delight the ear of the cunning creature, for it well knows that it will soon hear the female, and trace her to her nest, when it will suck the eggs with delight.

Travelling through the woods, perhaps on the ground, perhaps aloft, from tree to tree, it hears a cock crow, and its heart swells as it remembers the savoury food on which it regaled itself last summer in the neighbouring farmyard. With great care, however, it advances, and at last conceals itself in the very henhouse.

Honest farmer! why did you kill so many crows last winter? aye, and ravens too? Well, you have had your own way of it; but now, hie to the village and procure a store of ammunition, clean your rusty gun, set your traps, and teach your lazy curs to watch the Opossum. There it comes! The sun is scarcely down, but the appetite of the prowler is here; hear the screams of one of your best chickens that has been seized by him! The cunning beast is off with it, and nothing now can be done, unless you stand there to watch the fox or the owl, now exulting on the thought that you have killed their enemy and your own friend, the poor crow. That precious hen under which you last week placed a dozen eggs or so, is now deprived of them. The Opossum, notwithstanding her angry outcries and ruffled feathers, has consumed them one by one; and now, look at the poor bird as she moves across your yard; if not mad, she is at least stupid, for she scratches here and there, calling to her chickens all the while.

All this comes from your shooting crows. Had you been more merciful or more prudent, the Opossum might have been kept within the woods, where it would have been satisfied with a squirrel, a young hare, the eggs of a turkey, or the grapes that so profusely adorn the boughs of our forest trees. But I talk to you in vain.

But suppose the farmer has surprised an Opossum in the act of killing one of his best fowls. His angry feelings urge him to kick the poor beast, which, conscious of its inability to resist, rolls off like a ball. The more the farmer rages, the more reluctant is the animal to manifest resentment; at last there it lies, not dead but exhausted, its jaws open, its tongue extended, its eyes dimmed; and there it would lie until the bottle-fly should come to deposit its eggs, did not its tormentor walk off. 'Surely,' says he to himself, 'the beast must be dead.' But no, reader, it is only ''possuming,' and no sooner has his enemy withdrawn than it gradually gets on its legs, and once more makes for the woods."

Besides the varied animal diet in which the Opossum indulges, it also eats vegetable substances, committing as much havoc among plantations and fruit-trees as among rabbits and poultry. It is very fond of maize, procuring the coveted food by climbing the tall stems, or by biting them across and breaking them down. It also eats acorns, beech-nuts, chestnuts, and wild berries, while its fondness for the fruit of the "persimmon" tree is almost proverbial. While feeding on those fruits it has been seen hanging by its tail, or its hinder paws, gathering the "persimmons" with its fore-paws, and eating them while thus suspended. It also feeds on various roots, which it digs out of the ground with ease.

Its gait is usually slow and awkward, but when pursued it runs with considerable speed, though in a sufficiently clumsy fashion, caused by its habit of using the limbs of the right and left side simultaneously in a kind of amble. As, moreover, the creature is plantigrade in its walk, it may be imagined to be anything but elegant in its mode of progress upon the ground. Although it is such an adept at " 'possuming," or feigning death, it does not put this ruse in practice until it has used every endeavour to elude its pursuers, and finds that it has no possibility of escape. It runs sulkily and sneakingly forward, looking on every side for some convenient shelter, and seizing the first opportunity of slipping under cover.

If chased by a dog, it takes at once to a tree, and unless the dog be accompanied by its master, only climbs to a convenient resting-place, above the limit of the dog's leaping powers, and there sits quietly, permitting the dog to bark itself hoarse, without troubling itself any further about so insignificant an enemy. If, however, as is generally the case, the dog be accompanied by human hunters, the unfortunate Opossum has but little chance of safety. For as soon as the creature is "treed," the quick, sharp bark of the dog conveys to its master the welcome tidings, and he immediately runs towards the point from whence proceeds the well-known voice of his dog.

Having reached the position of the enemy, he ascends the tree in chase of the Opossum, which begins to climb towards the highest branches, followed by its pursuing foe. At last it gains the very extremity of some branch, and holds on with tail and claws, while the man endeavours to dislodge it by shaking violently the bough to which it clings. For a time it retains its hold, but is soon wearied by the constant exertion, and falls heavily to the ground, where it is seized and despatched by the expectant dogs.

The negroes are especially fond of this sport, and look eagerly forward to the close of day when they have been promised a " 'possum-hunt," as a reward for good conduct. Not only do they very thoroughly enjoy the moonlight sport, with its exciting concomitants, but promise themselves a further gratification, after their return home, in eating the Opossums which have fallen victims to their skill. The flesh of the Opossum is white when cooked, and is considered to be remarkably good, especially when the animal is killed in autumn, for at that time of year it is extremely fat.

Although, from the great accession of fat in the autumn months, it might be thought a hybernating animal, it is found roaming the woods in search of food even in the coldest night of winter. Still the large amount of fat with which the body is loaded is calculated to give the animal greater powers of resisting hunger and the severity of the weather than would otherwise have been the case, and enables it to thrive upon the comparatively small amount of food which it can obtain during the season of intense cold.

It is not a gregarious animal, and even the members of the same family spread themselves widely apart when they are in the open air.

The Opossum, although so cunning in many respects, is singularly simple in others. There is hardly any animal which is so easily captured, for it will walk into the rudest of traps, and permit itself to be ensnared by a device at which an English rat would look with contempt. Strange mixture of craft and dulness; and yet one which is commonly found in all creatures, whether men or animals, that only possess cunning, and no observance at all. For there are none so prone to entangle themselves in difficulties as the over-artful. They must needs travel through crooked byeways, instead of following the open road, and so blunder themselves stupidly and sinuously into needless peril, from which their craftiness sometimes extricates them, it is true, but not without much anxiety and apprehension.

When captured it is easily tamed, and falls into the habit of domestication with great ease. It is, however, not very agreeable as a domestic companion, as it is gifted with a powerful and very unpleasant odour, which emanates from its person with great force, whenever the animal is irritated or excited.

The nest of the Opossum is always made in some protected situation, such as the hollow of a fallen or a standing tree, or under the shelter of some old projecting roots.

In forming an appropriate receptacle for her young, the Opossum is assisted by her fore-feet, which are well adapted for digging. The nest itself is composed of long moss and various dried leaves. Sometimes the creature has been known to usurp the domicile of some other animal, not without suspicion of having previously devoured the rightful owner. On one occasion a hunter sent a rifle-ball through a squirrel's nest, which was placed at some forty feet from the ground, and was surprised to see an Opossum fall dead on the ground. This creature has also been known to possess itself of the warm nest of the Florida rat.

When the young of the Opossum are born, they are transferred by the mother to her cradle-pouch, where they remain for some weeks. From repeated experiments that have been made on this animal, it is found that the transfer is made on the fifteenth day after the young have been called into existence, and that at that period they only weigh four grains, their total length being under an inch, the tail included. Their number is from thirteen to fifteen. After they are placed in the pouch, their growth is wonderfully rapid, for in seven days they have gained so much substance as to weigh thirty grains ; and even at this early period of their existence their tails exhibit the prehensile capacity, and are often found coiled round each other's bodies. In four weeks the little Opossums have gained sufficient strength to put their heads out of the pouch, and at the end of the fifth week they are able to leave it entirely for a short time.

Very great trouble was required in order to ascertain these particulars, as it was found that the Opossum was in the habit of hiding herself in her den until she had placed her young in the pouch, so that it was needful to search the cavity for these concealed females, and to watch their proceedings by night and day without intermission.

There are one or two circumstances in connexion with this subject that are well worthy of attention.

The young Opossums are not, as has been often asserted, mere helpless lumps of animated substances, without sense or power of determinate action, but are wonderfully active in proportion to their minute size and their undeveloped state. If placed upon a table, they can crawl about its surface, and are sufficiently hardy to retain life for several hours after their removal from the warm cradle in which their tender bodies were shielded from harm, and the maternal fount which poured a constant stream of nourishment into their tiny systems.

Another singular circumstance is, that when they are first placed in the pouch, they are blind and deaf, the eyes and ears being closed, and not opened until many days have elapsed. With partial blindness at the time of birth we are all familiar in the persons of kittens, puppies, and other little animals, but that the tender young of the Opossum should be deaf as well as blind, is truly singular. It appears that in the case of the kitten or puppy, the presence of light and the action of the atmosphere are needed in order to withdraw the obstacles that obstruct the sense of vision. In the young Opossum, how-ever, it seems that the action of the atmosphere is needed in order to render the ears sensitive to the sounds that are transmitted through its mediumship, but that in most cases the little creature requires the absence of light until the time comes for it to open its eyes as well as its ears.

What length of time elapses between the period of transmission into the pouch and the several opening of eyes and ears is not, I believe, as yet clearly ascertained, and would furnish an interesting subject for investigation. I would also suggest that the blood of the young animal be carefully examined in three of its stages, viz. just before it is born, immediately after being placed in the pouch, and after the period when the eyes and ears are opened, in order to ascertain whether any important change, chemical or otherwise, has been made in that liquid by the double action of air and light.

The CRAB-EATING OPOSSUM is not so large an animal as the Virginian Opossum, being only thi.   thirty-one inches in total length, the head and body measuring sixteen inches, a.   the tail fifteen. It can also be distinguished from the preceding animal by the darker hue of its fur, the attenuated head, and the uniformly coloured ears, which are generally black, but are sometimes of a yellowish tint.

CRAB-EATING OPOSSUM.—*Philander cancrivorus.*

The fur of the Crab-eating Opossum is long, and though rather woolly in texture, is harsh to the touch. From the peculiar colouring of the long hairs that protrude through the thick, close, woolly fur that lies next to the skin, the general tinting of its coat appears rather uncertain, and varies according to the portion which happens to be exposed to view at the time. These hairs are nearly white towards their base, but darken into sooty-black towards their extremities. The limbs and feet are black, and the head is a brownish-white. There is generally an indistinct dark line drawn over the forehead. The tail is covered with scales, interspersed with short hairs, and its basal half is black, the remainder being of a greyish-white. For the first three inches of its length it is densely clothed with sooty-black fur of the same tint as that upon the back, and the remainder of its length is covered with scales and short hair.

The Crab-eating Opossum is peculiarly fitted for a residence on trees, a s never seen to proper advantage except when traversing the boughs, or swinging ong the branches by means of its peculiarly prehensile tail. While it is engaged in arboreal wanderings, it always takes care to twine its tail firmly round the nearest object that is

1.   K K

capable of affording a firm hold, and thus secures itself against any unfortunate slip of its paws.

On the level ground its pace is slow, and its gait awkward. It is, however, seldom seen upon the ground, as it is unwilling to forego the advantages of its arboreal residence, except for the purpose of obtaining food. Like the Virginian Opossum, it feeds chiefly on animal food, such as the smaller mammalia, birds, reptiles, and insects, and is so fond of crustacea, that it has been called the Crab-eater from that predilection. As the crabs and other crustaceans on which it feeds are usually found upon low and marshy soils, the Crab-eating Opossum is in the habit of frequenting such localities, and may generally be found in their neighbourhood.

This animal is held in some estimation, as furnishing an agreeable meal to those who care for such diet, and its flesh is said by the initiated to resemble that of the hare. The young of the Crab-eating Opossum are, during their days of infancy, coloured very differently from the adult animal. When first they are born, they are entirely naked, but when they are large enough to leave the pouch, they are clothed with short silken hairs of a bright chestnut brown, which, after a while, fades into the dark brownish-black of the full-grown animal. In all cases the tinting of the fur is rather variable.

The Crab-eating Opossum is found very numerously in the Brazils, and is spread over the whole of tropical America.

MERIAN'S OPOSSUM.—*Philander Dorsigerus.*

THE beautiful little animal which is so well depicted in the engraving affords another instance of a marsupiated animal being devoid of a true pouch.

In MERIAN'S OPOSSUM there is no true pouch, and the place of that curious structure is only indicated by a fold of skin, so that during the infancy of its young, the mother is obliged to have recourse to that singular custom which has gained for it the title of "dorsigerus," or back-bearing. At a very early age, the young Opossums are shifted to the back of their mother, where they cling tightly to their mother's fur with their little hand-like feet, and further secure themselves by twining their own tails round that of the parent. The little group which is here given, was sketched from a stuffed specimen in the British Museum, where the peculiar attitude of mother and young is wonderfully preserved, when the very minute dimensions of the young Opossums are taken into consideration.

Many other species of Opossums are in the habit of carrying their young upon their backs, even though they may be furnished with a well developed pouch, but in the pouch-less Opossums the young are placed on the back at a very early age, and are retained there for a considerable period.

It is a very small animal, measuring when adult only six inches from the nose to the root of the tail, the tail itself being more than seven inches in length, thus exceeding the united measurement of the head and body. Its general appearance is much like that of a very large mouse, or a very small rat.

The fur of Merian's Opossum is very short, and lies closely upon the skin. On the upper portions of the body its colour is a pale greyish-brown, fading below into a yellowish-white. Round the eyes is a deep brown mark, which extends forwards in front of each eye, and forms a small dark patch. The forehead, the upper part of the head, the cheeks, together with the limbs and feet, are of a yellowish-white, tending to grey.

Towards the base, the tail is clothed with hair of the same texture and colour as that of the upper part of the body, but towards its extremity it becomes white. The habits of Merian's Opossum are similar to those of the Virginian and Crab-eating Opossums. Its native country is Surinam.

YAPOCK OPOSSUM.—*Cheironectes Yapock.*

LAST, and most singular of this group of animals, is the YAPOCK OPOSSUM, a creature which, abandoning the arboreal life in which its relations so much delight, shifts its residence to the river-banks, and passes an existence almost wholly aquatic.

It is a curious looking animal, and even by the bold markings with which its fur is diversified is easily distinguishable from any other Opossum. Upon the coat of this animal, the two contrasting hues of grey and sooty-black are so nearly balanced that it is almost impossible to choose either of them as the ground tint and the other as the accessory. We will, however, consider the lighter hue to form the ground tinting of the fur, and describe the animal accordingly.

The general hue of the body is a pale fawn-grey, with a very *watery* look about it, and set closely upon the skin. Four dark bands of sooty-black are drawn across the body in a peculiar, but extremely variable manner. The first band extends over the shoulders as far as the first joint of the fore limbs; the second passes saddlewise across the back, extending only half-way down the sides of the body; the third passes over the hinder quarters and traverses the greater portion of the thigh; while the fourth is reduced to a broad patch upon and above the insertion of the tail. Along the spine runs a broad black band, which consists of three dark patches, and spreads into a wide black patch upon the top of the head. The tail is dark for two-thirds of its length, and white for the remaining third.

In the young animal these peculiar markings are very strongly defined, for the pale greyish-fawn becomes almost white, and contrasts powerfully with the dead, sooty-black of the dark portions of the fur.

On a closer examination of the structure of the Yapock, we come upon certain peculiarities which distinguish the animal from any other of its relations, and give ample cause for placing it in a separate genus, if not in a separate family. Intended for an aquatic existence, and to gain its food in and about the waters, the Yapock is well fitted for its course of life by the structure of its feet. The hind-feet aref urnished with a membranous web that connects the toes together as far as the base of the claws, and serves as a paddle by which the creature may drive itself rapidly through the water, or as a rudder by which it may direct its course.

The fore-paws are endowed with great powers of grasping, and have a very hand-like aspect. They are webbed only as far as the first joint. One peculiarity in their form deserves notice. To a casual observer, the fore-feet of the Yapock appear to be furnished with six toes, the superabundant member being devoid of a nail. So close a resemblance does this structure bear to a real toe that it has been described as such by a very eminent naturalist. It is, however, nothing more than an unusual development of the pisiform bone, which supports a fold of the skin.

The under side of the feet is furnished with a large, rough, fleshy pad, and there are also large, rough pads upon the toes. The claws are small and weak, and the thumb-joint is not opposable to the others. The ears are moderate in size, sharp, and pointed, and the head tapers rapidly towards the nose. The entire aspect of the creature is aquatic, its elongated body and peculiarly shaped limbs being sufficient to proclaim it a skilled swimmer even if its webbed feet were not taken into consideration.

It is not a very large animal, its total length being only two feet, the head and body measuring rather more than ten inches. In some specimens the tail is more developed than in others, and measures as much as fifteen inches in length.

Another peculiarity of structure in the Yapock is the presence of large cheek-pouches, apparently similar in their use to those of certain monkeys. It is supposed that the object of these receptacles is to enable the animal to engage in a lengthened chase of the various aquatic animals on which it feeds, and to lay them up in store until it returns to shore laded with the produce of its watery toils. These cheek-pouches extend far backwards along the sides of the mouth, and seem to be capable of containing a large supply of food.

As may be supposed from its structure, the Yapock feeds principally on fish, aquatic insects, crustaceans, and other water-loving creatures. Powers of climbing would be useless for such an animal, and it is accordingly found that the Yapock is unable to ascend trees, but, as if in compensation for this deficiency, is a most admirable swimmer and diver. One of these animals was once taken in a fish "crawl," or conical basket, which affords ingress to the fish but does not permit them to get out again. The Yapock had evidently dived after a fish, followed it into the basket, and not being able to make good its retreat, had been ignominiously captured.

The residence of this animal is always near water, and is generally made in a hole that is tunneled close to the water side. It is a very rare animal, and comparatively few specimens have as yet been taken. On account of its aquatic propensities and the nature of its prey, it has been taken for an otter, and has been described under the title of the Demerara Otter. Buffon notices it under the name of *Petite Loutre de la Guyenne*. It is a native of Brazil, and is generally found by the banks of the smaller streams and rivers.

In VERY many respects, the marsupiated animals seem to take in Australia the places that are occupied in other countries by creatures of very different orders. For example, the Tasmanian wolf is clearly the representative of the true wolves that inhabit other parts of the earth, the kangaroos take the place of the jerboas, and so on. There is, however, one singular deficiency in the Australian fauna, which seems to be partially supplied by different members of this curious order.

Australia appears to be a country that would be peculiarly adapted for the monkey race, and we might reasonably expect to find many curious species of quadrumana inhabiting its deserts and vast forests. Yet, as far as is yet known, there is no member of the quadrumana to be found in that land. Many of the monkey's habits, however, are possessed by the indigenous marsupials, which seem to serve as the representatives of the quadrumana in Australia. Very many species among which we must look for these representative animals are arboreal in their habits, and, although they do not possess the true quadrumanous structure, are yet endowed with such hand-like extremities that they may fairly be considered as the analogues of these animals.

Perhaps the strangely brutalized form of humanity which exists in that wondrous land may have some connexion with the remarkable forms of animal life which are co-inhabitants of the same country ; for in truth, the aboriginal Australian is of so low a type that there seems to be little necessity for the existence of the quadrumana race to prevent too abrupt a transition from the bipedal to the quadrupedal form.

It is a strange race of humanity, and which, by its own showing, is on the point of vanishing from the face of the earth. The traditions of the Australian aborigines are most singular and prophetic. They say that their deity has been vanquished by that of the white man, that he is dethroned, and though still living is buried under the earth, existing only by permission. They even consider him to be inferior to the white man, and have consequently lost all feeling of reverence for so degraded a being. They have not even the paltry veneration that is inspired by fear, and seem to have reached the lowest depths of degradation to which religious feeling can fall. Some deference they still pay to the malevolent phantoms which are supposed to haunt the sepulchres of the dead, but when a race of mankind has fallen so low in spiritual knowledge as to think the white man superior to the black god, we can look for nothing less than total and speedy extinction.

Few religious creeds are there which have not so much of life in them that even when cast off by the more developed minds, they cannot be taken up by others and made vital and useful to them. But in the present instance it is impossible to conceive that any being possessed of human form and human intellect could avail himself of a creed so pitifully contemptible as to degrade its supreme deity below the level of created man.

GROUP OF SEALS.

PHÓCIDÆ, OR SEALS.

WE now arrive at a very wonderful series of animals, which, although they breathe atmospheric air like other mammalia, are yet almost entirely aquatic in their habits, and are never seen except in the water or its immediate vicinity. The first family of these aquatic mammalia is that which is formed of the animals which are popularly known by the name of SEALS.

The structure of their bodies shows that the Seals are intended to pass the greater portion of their existence in the water, for the body is elongated, and formed very much like that of a fish, while the limbs and feet are so unbiped that they greatly resemble fins, and are put to the same use.

In order to protect their bodies from the debilitating action of the element in which they live, they are thickly covered with a double fur, which, when immersed in water, is pressed tightly to the skin, and effectually throws off the moisture. In some Seals this fur is extremely valuable, and is largely employed as an article of commerce. The fur itself is kept constantly lubricated with a fatty matter secreted by the skin, and is thus rendered waterproof. The more effectually to defend the animal from the icy cold water in which it is often immersed, and from the ice-fields on which it loves to climb, a thick layer of fat is placed immediately below the skin, and being an excellent non-conductor of heat, serves to retain the internal heat through the severest cold. All the fat of the body seems to be pressed into this service, as there is comparatively little of the internal fat that is usually found plentifully in the mammalia.

Aided by the imperfectly developed limbs, the Seals are able to leave the water and to ascend the shore, where they are capable of proceeding with no small rapidity, though in a sufficiently awkward manner, their gait partaking equally of the character of a shuffle and a crawl. When moving in a direct line, without being hurried, they bend their spine in such a manner as to give them the appearance of huge caterpillars crawling leisurely along the ground; the spine is extremely flexible, so that the animal can urge itself through the water in a manner very similar to that which is employed by the fish.

Their clumsy, scuttling movements when on land form a curious contrast with the easy grace of their progress through the water. When the Seals swim, they drive themselves forward by means of their hinder feet, which are turned inward, and pressed against each other so as to form a powerful leverage against the water, as well as a rudder, by means of which they can direct their progress. They are also assisted in some measure by the fore-limbs, but these latter members are more employed upon land than in water, except perhaps for the purpose of grasping their young. On reference to the skeleton of the Seal on page 511, the peculiar formation of the Seal's limbs will be better understood than by the expenditure of a page of actual description alone.

When they desire to leave the water, they rush violently towards the shore, and by the force of their impulse shoot themselves out of the water, and scramble up the bank as fast as they can. On taking again to the water, they shuffle to the edge of the bank, and tumble themselves into the sea or river in a very unceremonious manner, gliding away as if rejoicing that they were once more in their proper element.

The food of the Seals consists chiefly of fish, but they also feed largely upon various crustacea, and upon molluscs. Their powers of swimming are so great that they are able to urge successful chase of the fish even in their native element, and it has several times happened that captive Seals have been trained to catch fish for the service of their owners.

The "whisker" hairs are extremely thick and long, and in many species are marked with a raised sinuous margin, which gives them the appearance of being covered with knobs. Their basal extremities are connected with a series of large nerves, similar to those of the lion's lip, which has already been figured on page 137, and it is very probable that this structure may aid them in the capture of their finny prey. The sense of smell is largely developed, and the tongue is rough, and slightly cleft at its extremity; the reason for this structure is not known.

The brain of the Seal is very large in proportion to the body, and, as might be expected from this circumstance, the creature is extremely intelligent, and is capable of becoming very docile when placed under the tuition of a careful instructor. The eyes are large, full, and intelligent, and the nostrils are so formed that they can be effectually closed while the creature is submerged beneath the surface of the water, and opened as soon as it rises for the purpose of respiration. At every breath the nostrils open widely, and seem to close again by means of the elasticity of the substance of which they are composed. The ears are also furnished with a peculiar structure for the purpose of resisting the entrance of water.

The true Seals are found only in the sea, and at the mouths of various large rivers, and are wonderfully abundant in the polar regions.   None of them are known to inhabit the tropical parts of the earth.   Several species have been known to occur upon our own shores, more especially on the more northern coasts, and the common Seal, *Phoca vitulina*, is found in great numbers around the northern British shores.

The teeth of the Seals are very remarkable, and admirably adapted for seizing and retaining the slippery prey.   The canine teeth are long, sharp and powerful, and the molar teeth are covered with long and sharp points of various sizes, so that when once caught in the gripe of these formidable weapons, there is but scant hope of escape for the fish.

The Seals are not very prolific animals; the number of their young family being seldom more than two, and often restricted to a single offspring.   As the young Seals would be unable, during the earlier portion of their existence, to battle with the sea waves, and to cling firmly to their parent while she afforded them their needful nourishment, the mother Seal retires to the shores when she is called upon to take upon herself the pleasing cares of maternity, and cherishes her young for a season on land, before she ventures to commit them to the waves.

Owing to the excessive shyness of disposition which characterize the Seals, and the wary caution with which they retire from the sight of mankind, their domestic habits are very little known.   Indeed, were it not that many specimens of the common Seal had been captured and tamed, we should have but little information on the manners or the habits of those curious animals.   There are many species of Seals, which have been separated into various genera by different authors upon different grounds.   Some, for example, found the generic distinction upon the absence or presence of external ears, others from the incisor teeth, and others from the molars and the general character of the skull.

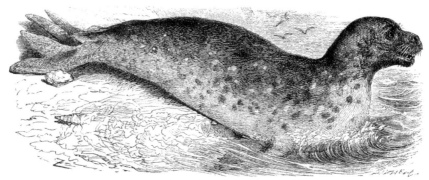

SEA LEOPARD.—*Léptonyx Weddéllii.*

The SEA LEOPARD, or LEOPARD SEAL, is distinguishable from the other Seals by means of its comparatively slender neck, and the wider gape of its mouth, which opens further backward than is generally the case among these animals.   The body is rather curiously formed, being slender at the neck and largest towards the middle, from whence it tapers rapidly to the short and inconspicuous tail.

The fore-paws are without any projecting membrane, and are largest at the thumb-joint, diminishing gradually to the last joint.   The claws are sharp and curved, and rather deeply grooved; their colour is black.   The hind-feet are devoid of claws and projecting membrane, and bear some resemblance to the tail-fin of a fish.   The colour of this Seal is generally a pale grey on the upper portions of the body, relieved with a number of pale greyish-white spots, which have earned for the animal the name of Leopard Seal. The external ears are wanting.

It may be observed in this place, that the situation of these organs is rather remarkable. The external orifice is not placed exactly over the passage that leads to the internal ear, but is situated below and a little behind the eyes, so that there is a tubular passage below the skin that seems to conduct the waves of sound towards the hidden organs of hearing. Partly on account of this structure, and partly because the Seals pass so much of their time below the surface of the water, it has been supposed that the sense of hearing will be little needed by them, and that it is not at all acute.

Yet, any one who has been accustomed to diving must have discovered that when the body is entirely submerged in the water, the auditory organs are very sensitive to sounds which are conveyed through the water, although not to those which are produced on land and are only transmitted through the upper atmosphere. For example, although when a man is entirely submerged he is unable to hear the loudest shouts that can be raised by persons on shore, his ears are almost painfully sensitive to any sound that is produced in the water and is transmitted through its mediumship. A stone thrown into the water, or a blow struck upon its surface, is heard with perfect distinctness, while the measured stroke of oars and their peculiar grinding roll in the rowlocks become perceptible to his ears long before the sound is audible to those who are on land.

We must be extremely cautious in offering any conjectures on the supposed efficiency or dulness of certain organs because we fancy that if we were placed under the same conditions our own organs would serve or fail us. In many cases these conjectural assertions, among which we may reckon many of Buffon's brilliant disquisitions, are found to be in direct contradiction to the real facts, and in all instances it is necessary to be exceedingly cautious lest we should overlook some circumstance which may entirely alter the whole aspect of affairs.

Very little is known of the habits of the Sea Leopard, which are probably much the same as those of the common Seal, as Captain Weddell, who first noticed this species, speaks of it casually as a well-known animal, merely mentioning that his men caught so many Leopard Seals, or that they secured so many Seal skins and so many Leopard Seal skins in the course of their hunt.

It does not appear to be a very large animal, as the average length of the largest specimens is scarcely ten feet. Around the largest part of the body, the circumference measures nearly six feet and a half, round the root of the tail about two feet three inches, and round the neck barely two feet. It was recorded by Captain Weddell to have been seen off the South Orkneys. Some specimens in the British Museum were taken off the eastern coast of Polynesia. As far as is yet known, these animals are only found in the Southern hemisphere.

The CRESTED SEAL is a very curious animal, being chiefly remarkable for the singular structure to which it is indebted for its title.

The head of the Crested Seal is broad, especially across the cranial region, and the muzzle is very short in comparison with that of the preceding animal. The teeth are also rather remarkable. The wonderful protuberance which decorates the head of this species with a projecting crest is confined to the adult males, and even in them is not always so conspicuously elevated as is represented in the figure. In the females and the young of both sexes it is hardly perceptible.

From the muzzle arises a cartilaginous crest, which rises abruptly over the head to the height of six or seven inches, and is keel-shaped in the middle. This crest seems to support the hood-like sac or cowl which covers the head, and is nothing but an extraordinary development of the septum of the nose, the true nostril opening at each side of it by oblong fissures. The sac is covered with short brown hair, and as it can be inflated or allowed to collapse at the pleasure of the owner, it presents a very grotesque sight.

The real object of this appendage is not known. Some writers lean to the opinion that it is intended to aid in some manner the sense of smell. This conjecture, however, seems to be worthless, as in that case the females and the young would equally need its assistance with the adult males.

Whatever may be the true purport of this crest, it is frequently of great service to the animal in moments of danger. It is well known that the Seals are peculiarly sensitive about the region of the nostrils, and that a comparatively slight blow upon the nose will suffice to stun a Seal that would be but little affected by the heaviest blows upon any other portion of its body. The Crested Seal, however, finds his air-filled helmet of truly invaluable service to him in deadening the force of any stroke that may be aimed at his nose ; for, as has already been mentioned, the nostrils are not placed at the extremity of the muzzle, but upon each side of it, and are consequently protected by the over-hanging head.

It has often happened that when the Seal-hunters have been engaged in the pursuit of their prey, they have laid several of these animals to all appearance senseless on the ground, awaiting the stroke of the knife that shall complete the victory. The animals, however, are but slightly stunned, and recovering from their temporary swoon, return to the conflict with such unexpected energy that their assailants are forced to have recourse to summary measures when engaged in the chase of these creatures.

CRESTED SEAL.—*Stemmatopus cristatus.*

The onset of an enraged Crested Seal is much to be dreaded, for the creature is marvellously fierce when its anger is roused, and its strength is very considerable. The teeth, too, are formidably powerful, and can inflict very dangerous wounds. In fighting, they can use their claws as well as their teeth. The males are always pugnacious animals, and during the season when they choose their mates are in the habit of fighting desperately among each other for the possession of some attractive female, and in these combats inflict severe lacerations. During these conflicts the two combatants express their mutual rage by emitting a torrent of loud, passionate, yelling screams, which are audible at a considerable distance.

It is a polygamous animal, one male ruling over a small herd of wives.

The fur of this animal is of some value, and great numbers of these skins are annually imported into Europe, where they are used for various purposes. To the Greenlander this Seal is of incalculable value, as he makes use of almost every portion of its body as well as of its skin. Of the fur he makes his thick, cold-resisting costume, and with the skin he covers those wonderful little boats in which he braves the fury of the ocean in search of his aquatic quarry. Of the stomach he makes air buoys, which he fastens to his lances, and which indicate the position of any Seal or other animal that he may strike, and also serve to tire the wounded prey, and enable the hunter to repeat his blow. Even the teeth are pressed into his service, and are used as convenient heads for his spears.

In the preparation of the Seal skin for civilized nations it is needful to remove the long coarse hairs, and to leave only the soft woolly fur adherent to the skin. The process is very simple, consisting in heating the skin, and then scraping it while hot with a wooden knife.

The colour of this creature's fur is, when adult, a dark blue-black upon the back, fading to a yellowish-white on the under portions of the body. A number of large grey patches are irregularly scattered over the body, and in the centre of each patch there is a dark spot. The head, the tail, and the feet, are black. In the young animal the colours are not of the same cast, being during the first year of a slate-grey upon the back and silvery-white below, darkening in the second year to a brownish-grey along the spine.

It is a moderately large animal, being from ten to twelve feet in length when adult, and stout in proportion.

The Crested Seal is found spread over the coasts of Southern Greenland, and is in the habit of reposing much upon ice islands, caring comparatively little for ordinary land. It also frequents the shores of Northern America. From September to March it is found in Davis's Straits, but leaves that locality for the purpose of producing and rearing its young, and returns again in June, together with its offspring, in a very bare and poor condition. About July it takes another excursion, and employs its time in recovering the health and strength which it had lost during the period of its former absence, so that in September it is very fat, and altogether in excellent condition.

By the native Greenlanders it is termed " Neitsersoak."

The natives of the localities which are inhabited by this animal are in the habit of employing two methods for their capture, the one being only a question of patience between the man and the Seal, and the other a fair measurement of human reason against sealish sagacity ; the former generally, but not always, proving the superior. The two modes are as follows.

The Seals are in the habit of making, or preserving in some way, certain round holes in the ice, which communicate with the water, and which serve them as doors through which they can enter or leave the water without being forced to crawl to the edge of the ice-field. It seems wonderful that the animals should be able to crawl up the steep and perpendicular sides of these holes, which are sometimes three or four feet in depth, but they manage to perform this feat with entire ease.

Taking advantage of these Seal-holes, the hunter shapes his course towards them, and according to the locality or the bent of his own genius, has recourse to one or other of the established methods by which Seals are killed. The easiest, but at the same time the tardiest and stupidest plan, is to build a kind of barricade of snow and ice at some distance from the Seal-holes, and to lie there concealed until the animal emerges from the sea, and makes its appearance upon the ice-field. As soon as it has travelled to some little distance from its spot of refuge, the hunter seizes the opportunity to inflict a fatal wound, and then uses his best endeavours to prevent his powerful prey from regaining its familiar element.

Should the Seal ever reach the ice-hole, the entire labour of the day is lost, for the unsuccessful hunter is not only disappointed by the escape of his intended victim, but has also the mortifiction of seeing every Seal upon the ice-field scouring towards the ice-holes, and disappearing therein, no more to venture upon open ice that day.

The second mode of Seal killing is much more sportsmanlike, and needs not the long and wearisome watch behind the icy barrier.

Leaving his sledge and dogs at a distance, but within convenient call, the cautious hunter takes his weapons, and proceeds silently and slowly towards the spot where he sees a Seal reposing itself upon the ice. As soon as he perceives the animal to betray signs of distrust, he drops prostrate upon the ice, and remains motionless until the Seal recovers from its alarm, and again composes itself to rest. From this moment, the man is obliged to cast away all human habits and movements, and while lying prostrate on the ice to imitate the actions of a Seal.

Taking care to remain motionless whenever he sees the Seal looking in his direction, he creeps gradually towards his intended prey, in hopes of getting between the ice-hole and the Seal, in which case the death of the latter is almost inevitable. If, however, the relative position of man, Seal, and ice-hole be such that this manœuvre becomes impossible, the hunter contrives to crawl up to the sleeping Seal, and with a single blow lays it lifeless upon the ice.

The COMMON SEAL is spread very widely over many portions of the globe, and is of very frequent occurrence upon our own coasts, where it is found in considerable numbers, much to the annoyance of the fishermen, who look upon it with intense hatred, on account of the havoc which it makes among the fish.

It is rather a handsome animal, with its beautifully mottled skin and large intelligent eyes, and although not so large as other species which are also found upon the British coasts, yields to none of them in point of beauty. The colour of its fur is generally of a greyish-yellow, sprinkled with spots of brown, or brownish-black, which are larger and more conspicuous along the back than upon the sides. The under portions of the body are of a much lighter hue. The feet are short, and the claws of the hinder feet are larger than those of the anterior limbs. The total length of the adult Seal is seldom more than five feet, the head being about eight or nine inches long.

This creature is wonderfully active both in water and on land, although its bodily powers are but awkwardly manifested when it is removed from the watery element in which it loves to roam. It is a persevering hunter of fish, chasing and securing them in a manner that greatly excites the wrath of the fishermen, who see their best captives taken away from them without the possibility of resistance. So cunning as well as active is the Common Seal, that one of these animals will coolly hang about the fishing grounds throughout the season, make itself familiar with all the turns and angles of the nets, and avail itself of their help in capturing the fish on which it is desirous to make a meal.

A crafty old Seal will sometimes continue this predatory mode of existence for a series of years, until his person becomes familiar to the fishermen, and will carry out his depredations with such consummate skill that the fishermen can find no opportunity for stopping his career with a rifle-bullet or a fish-spear. Seals have been known in this manner to haunt the salmon fisheries as long as the nets were down, and when the fishing season was over, and the nets had been removed, have been seen to ascend the rivers for some miles, in order to devour the spawning fish.

There is a curious tradition among the inhabitants of the Irish coast respecting the Seal, which constantly haunts the same spot through a series of many years.

They think that the animal is supernaturally protected from harm of any kind ;—that bullets will not strike him, however well the gun be aimed ; that steel will not enter his body, however keen the blade, or however strong the arm that urges it ; and that the long array of nets are powerless to retain so puissant a being in their manifold meshes. So after a while a Seal, if he be only bold and wary, may lead a luxurious life at the fishermen's cost, for no one will venture to attack an animal that bears a charmed life.

Fortunately for the Seals in general, they are not often visited by the wrath of those whom they rob, for there is a feeling prevalent among many fishermen that to kill a Seal is unlucky, and that such a deed would prevent the murderer from obtaining any more success at sea. This humane idea seems however to extend no further than the regular fishing grounds, for the chase of the Seal has long taken its place among the most valuable of commercial speculations, and is of extreme importance.

The general mode of securing these creatures is to land quietly, and to cut off the return of the terrified animals, which are quickly despatched by smart blows from a bludgeon across their nose. When driven to desperation, they fight savagely, and a single Seal is no mean antagonist for a man, provided that he is not a practical Seal-hunter. The creature has an awkward way of lying on its side, shuffling rapidly along, and scratching furiously with its fore-paws. And if its antagonist should endeavour to cut off its retreat, it will boldly fling itself upon him, and endeavour by the violence of its onset to bear him to the ground.

Should the beach be composed of pebbles or shingles, it is the hunter's best policy to face the animal, and trust to his cudgel for stunning the Seal as it comes along. For when the Seal is galloping along the beach after its own rapid but awkward fashion, it flings the stones behind it with such violence that the pursuer can hardly escape from receiving severe blows from these strangely-launched missiles. Some writers suppose that the Seal uses this mode of defence wittingly, but the general opinion on the subject is, that the peculiar gait of the animal is the cause of this Parthian assault, without any voluntary intention on the part of the Seal itself.

On the British coasts the chase of the Seal is but of local importance, but on the shore of Newfoundland it assumes a different aspect, and becomes an important branch of commercial enterprise, employing many vessels annually. In a successful season the number of Seals which are taken amount to many hundred thousand. A large quantity of oil is obtained from the bodies of the Seals, and is used for various purposes, while their skins are of considerable value either when tanned into leather or when prepared with the fur, and used for making various articles of dress and luxury.

SEAL.—*Phoca itulina*

On the British coasts Seal-shooting is much followed, and is thought to be a very exciting pursuit, requiring much steadiness of nerve as well as strength of body and quickness of eye.

So quick is the animal in its movements, that with the old flint-lock guns the sportsman could seldom succeed in killing a Seal; for at the flash of the powder in the pan the Seal would instantly dive below the surface, leaving the bullet to speed harmlessly over its head. With the present weapons the Seal may be readily killed, provided that the hunter be able to take a rapid and certain aim in spite of the rocking of his boat or the movements of the Seal. It is no easy matter to hit a Seal, even at the short range of a hundred yards, for the creature exposes but little of its person above the surface of the water, and if alarmed, does not remain exposed for more that a few seconds.

It is very tenacious of life, and requires a heavy missile to kill it upon the spot, the ordinary "pea rifles" being of little use for such a purpose. Should the Seal not be killed immediately, the sportsman will never obtain possession of his prey, and even when a well directed shot has instantaneously deprived the animal of life, it will often sink out of reach unless struck by a barbed fish-spear, which should always be kept in readiness for that purpose.

Much sea-craft is required for the chase of the Seal, as well as considerable knowledge of the animal and its habits. It is a remarkable fact that if the Seal be disturbed while the tide is ebbing, it will always make its way seawards; but that if it be alarmed while the tide is flowing, it will direct its course towards the land.

The Seal is also a good natural barometer, and by its movements indicates to a practised eye the forthcoming changes in the weather. Whenever an old Seal is seen rolling and tumbling along a bank, a storm of wind and rain is sure to ensue before many hours have passed.

This species of Seal, in common with several others, is mightily fond of musical sounds, and has often been known to follow a boat while some one was playing on a musical instrument. Some persons say that the Seal-shooters ought always to be accompanied by a piper in order to induce the Seals to approach, and assert that the animals prefer the sound of the bagpipes to that of any other instrument, an assertion which, if true, only shows that the Seal must be in the very earliest stage of musical feeling.

The Common Seal is very easily tamed, and speedily becomes one of the most docile of animals, attaching itself with strong affection to its human friends, and developing a beautifully gentle and loving nature, hardly to be expected in such an animal. Many of these creatures have been taken when young, and have been strongly domesticated with their captors, considering themselves to belong of right to the household, and taking their share of the fireside with the other members of the family. An interesting account of a tame Seal was lately sent to the *Field* newspaper, and runs as follows :—

" If taken young and treated kindly, the Seal will rival the dog in sagacity and affection for its master.

When a boy, I was presented by some fishermen with one apparently not more than a fortnight old, which in a few weeks became perfectly tame and domesticated, would follow me about, eat from my hand, and showed unmistakeable signs of recognition and attachment whenever I approached. It was fond of heat, and would lie for hours at the kitchen fire, raising its head to look at every new comer, but never attempting to bite, and would nestle close to the dogs, who soon became quite reconciled to their new friend.

Unfortunately the winter after I obtained it was unusually rough and stormy. Upon that wild coast boats could seldom put to sea, and the supply of fish became scanty and precarious. We were obliged to substitute milk in its place, of which the Seal consumed large quantities, and as the scarcity of other food still continued, it was determined, in a family council, that it should be consigned to its own element, to shift for itself.

Accompanied by a clergyman who took a great interest in my pet, I rowed out for a couple of miles to sea, and dropped it quietly overboard. Very much to our astonishment, however, we found that it was not so easy to shake it off. Fast as we pulled away it swam still faster after the boat, crying all the time so loudly that it might easily have been heard a mile away, and so pitifully that we were obliged to take it in again and bring it home, where, after this new proof of attachment, it lived in clover for several months, and I believe might still be in existence but for the untimely fate which most pets are doomed sooner or later to experience, and to which this one was no exception."

A somewhat similar story is told in Maxwell's "Wild Sports of the West," where may be found a very interesting and touching narrative of a tamed Seal, which lived for several years with a family, and which, although it was repeatedly taken out to sea in a boat and thrown overboard, always found its way back again to the house which it loved, even contriving to creep through an open window and to gain access to the warm fireside. The end of this poor creature was a sadly tragic one, and need not be narrated here.

In the same work is a very spirited account of another Seal adventure, in which the ludicrous element prevails, although it might have furnished material for tragedy.

A number of men had gone in a boat to the Sound of Achil, and having seen a Seal and her young one reposing on the sand, had borrowed an old musket, and set off to attack them. They succeeded in securing the cub before it could reach the sea, and tossed it into their boat. The mother Seal, however, inspired by maternal love, swam after the boat which contained her offspring, and could not be deterred from following the captors in the hope of rescuing her child. The men attempted several times to shoot the poor creature, but their ricketty weapon would not explode until it had been several times futilely snapped. At last, however, it performed its duty, and lodged the ball in the Seal's head.

The body was immediately lifted into the boat, when to the horror of the captors, the animal, which was only stunned, recovered its senses, and began a most furious attack upon its enemies, floundering about the boat with such energy that she nearly overset it, and snapping fiercely at the legs of her antagonists. The contest soon assumed a serious aspect, for the teeth of the angry animal were urged with such fury that they cut deeply into the oars with which her attacks were warded off, and if assistance had not speedily

SKELETON AND TEETH OF COMMON SEAL.

arrived, the result might have been of a very tragic nature. A gentleman, however, that happened to be sailing near the scene of combat, was attracted by the curious spectacle of a boat's crew engaged in such strange evolutions, and directing his course towards them, ended the combat by a rifle bullet.

Of late days, performing Seals have come into vogue under various titles, among which the "Talking Fish" is well known. These clever animals have been taught to perform sundry ingenious feats, requiring not only an intelligent mind to comprehend, but an activity of body to execute, apparently incompatible with the conformation of the animal.

They will bark at the word of command, their phocine tongue being very freely translated into the language of any country in which they may happen to be; they will rotate in their water tub with singular velocity; they will offer to kiss any one who is not afraid of their large wet hairy mouth, and in fine will go through many similar performances with great skill and seeming enjoyment of the attention paid to them by their visitors. For they are wonderfully fond of admiration, these Seals, and although very shy to those with whom they are unacquainted, and before they have become accustomed to the sight of strangers, are highly appreciative of the kind words and caresses which fall to their lot while they are going through their tricks.

Even the Seals at the Zoological Gardens, who are not put through any particular training, are not without their little coquetries with regard to the ever changing circle of visitors who stand around the railings which inclose their habitations, while their demonstrations of affection towards the keeper who attends to their wants are quite lively. They recognise him at a considerable distance, and shooting rapidly through the water, fling themselves upon the bank, scuttle to the iron fence and rear themselves against the rails in impatient greetings, long before he reaches the limits of their home. Perhaps these ebullitions of regard are not totally personal to the keeper, but are partly caused by certain pieces of fish which he carries with him. As soon as they receive the expected gift, they seize it between their teeth, and unless they happen to be hungry, become very playful, and execute all manner of aquatic gambols before they eat it, much after the manner of a cat with a mouse.

The specific name of Vitulina, or calf-like, has been given to the Common Seal, not because it presents a calf-like aspect, but because its voice is thought to bear some resemblance to the plaintive cry of a calf when separated from its mother.

The HARP SEAL derives its name from the very conspicuous manner in which its fur is coloured.

The general hue of the Harp Seal's coat is a whitish-grey, and upon that delicate tint are drawn two broad, semicircular bands of a deep black, their points nearly touching each other, and extending from the shoulders nearly to the root of the tail. These dark markings are thought to bear some resemblance to an ancient harp, and have given rise to the popular name by which this species is designated. The muzzle and fore part of the head is also black. This peculiar colouring, which seems to distinguish this animal from any other species of Seal, is however never seen except in the adult animal, and is not considered to be perfect until the creature has attained its fifth year. Very many animals are variously marked according to their ages, sex, and time of year, but there are few which undergo more decided changes than the Harp Seal ; changes so unexpected that they have caused the animal to be described under several different titles according to the particular coat which it happened at that time to wear.

In the first few months of its existence, the fur of the Harp Seal is white in colour and woolly in texture. At the expiration of a year the white changes to a greyish-cream. In the second year, the fur is entirely grey. In the third year, the grey is diversified with stripes of darker hues, and varying in number, dimensions, and position. The fourth year changes the stripes into spots, and in the fifth year, the semi-lunar black stripes make their appearance.

The Greenlanders designate the Harp Seal by different titles according to its years ; giving it the name of Atak, or Attarak, in its first year, Atteisiak in the second, Agletok in the third, Milektok, in the fourth, and Attarsoak, in the fifth.

The Harp Seal is found in great numbers upon the coasts of Greenland, Iceland, and other localities, and is rather peculiar in its mode of life. It does not much affect the shore after the fashion of many Seals, but prefers to take up its residence upon floating ice-islands, disregarding even the more solid fields of ice that are so much frequented by other animals belonging to the same family. It is found in considerable numbers, congregating together in herds under the leadership of a single chief, and constantly subject to his authority.

There is always a sentinel planted by every herd, whose duty lies in looking out for danger, and giving timely warning to the remainder of the herd. The Harp Seal seems withal to be rather a reckless being, in spite of this precaution, for it is easily approached by a cautious hunter, and can be destroyed with little difficulty. Sometimes it prefers to take its repose without trusting itself upon ice or land, and sleeps while floating on the surface of the sea. In this situation its slumbers are so profound that it has often been surprised and slain before it has awoke from its treacherous somnolence. The Harp Seal is moderate in size, being generally about seven or eight feet in length and proportionably stout. Some very large specimens, however, are said to reach the length of nine feet.

It is an extremely valuable animal both to Europe in general, and to the uncivilized inhabitants of Greenland. In this country we value the Harp Seal principally on account of the excellent oil which it furnishes, and which is allowed to be of a purer and better character than that which is obtained from other oil-producing animals. The creature is

HARP SEAL, OR ATAK.—*Phoca Grœnlándica.*

remarkably well furnished with the fatty substance from which the oil is obtained, and is therefore eagerly pursued, irrespective of the value which is set upon the fur.

The food of this animal is almost wholly of an animal nature, and consists chiefly of salmon and other fish, together with various molluscs and crustaceans. It however stands in great dread of other species of Seal, such as the Sea Lion and Sea Bear, and according to many accounts holds the spermaceti whale in awe, being chased by that formidable creature into the shallow waters of the shore. Twice in the year the Harp Seal indulges in a migration similar to that which has already been described when treating of the Sea Leopard. The young of this species are sometimes two in number, although the maternal Seal is often forced to content herself with a single child.

Like the Common Seal, the Atak is possessed of much intelligence, and is very capable of domestication.

Two of these animals which were placed in the zoological collection at the Jardin des Plantes, were at their first arrival extremely shy, and would avoid the person of man with every mark of terror. Yet in a very short time they became quite tame, and would voluntarily seek the caresses of those who had behaved kindly towards them. They also struck up a great friendship with two little dogs, and would permit their little playfellows to take all kinds of liberties with them, permitting the dogs to sit on their backs and bark, and not even resenting an occasional bite. They would even permit the dogs to take their food from their mouths, but if their relation attempted to act in like manner, a sharp combat immediately took place, the weaker being forced ultimately to succumb to superior might.

In cold weather, dogs and Seals were accustomed to huddle closely together for the sake of warmth, and when the dogs made their way out of the entrance, the Seals did their best to follow their little playfellows, caring nothing for the rough ground over which they were forced to pass.

This Seal has been several times seen upon the coasts of England, although generally in its immature state. Two young Harp Seals were taken in the mouth of the Severn, and others seem to have made their appearance off the Orkney Isles.

AMONG all the strange forms which are found among the members of the phocine family, there is none which presents a more terribly grotesque appearance than that of the WALRUS, MORSE, or SEA HORSE, as this extraordinary animal is indifferently termed.

The most conspicuous part of this animal is the head, with its protuberant muzzle bristling with long, wiry hairs, and the enormous canine teeth that project from the upper jaw.

1.   L L

WALRUS, OR MORSE.—*Trichecus Rósmarus.*

These huge teeth measure, in large specimens, from fourteen inches to two feet in length, the girth at the base being nearly seven inches, and their weight upwards of ten pounds each. In ordinary specimens, however, the length is about one foot. In some examples they approach each other towards their points, and in others they diverge considerably, forming in the opinion of some writers two distinct species. As, however, the relative position of these teeth varies slightly in every specimen that has yet been examined, the structure seems to be of hardly sufficient importance for the establishment of a separate species. The ivory which is furnished by these extraordinary weapons is of very fine quality, and commands a high price in the market.

By means of the great development of these teeth, which are necessarily buried deeply in the upper jaw, the muzzle is much enlarged in order to afford room for their sockets, and assumes the remarkable prominence that gives to the animal so ferocious an aspect. As there would be insufficient space for the nostrils if placed in their usual position, they are removed to a much higher locality than that which they occupy in other Seals, and open nearly vertically above the muzzle.

The other teeth of the Walrus are very small in comparison with the two canines of the upper jaw, and in number are exceedingly variable, according to the age of the animal. During the earlier years of its life, it is furnished with six incisors in each jaw, two canines, ten molars in the upper jaw and eight in the lower. But when the animal approaches maturity, the incisors all fall out, and even in a prepared skull, they come away together with the soft substances. The tooth which is sometimes considered to be the first molar of the lower jaw, is in reality the lower canine.

In order to accommodate itself to the position of the huge tusks, the lower jaw narrows rapidly towards its point, so as to pass easily between the canines. The food of the Walrus consists of small Seals, fish, shrimps, and various other animal substances, diversified with such vegetable diet as the sea can afford. It has been suggested that one object of the large tusks may be to drag the algæ from their hold upon the rocks.

A Walrus is a valuable animal, for even in this country its skin, teeth, and oil are in much request, while among the Esquimaux its body furnishes them with almost every article in common use. Among civilized men, the skin of the Walrus is employed for harness and other similar purposes where a thick and tough hide is required. The tooth furnishes very good ivory, of a beautiful texture, and possessing the advantage of retaining the white hue longer than ivory which is made from the elephant tusk. The oil is delicate, but there is very little to be obtained from each Walrus, the layer of fatty matter being scarcely more than a hand's-breadth in thickness. Among the Esquimaux the Walrus is put to a variety of uses. Fish-hooks are made from its tusks, its intestines are twisted into nets, its oil and flesh is eaten, and its bones and skin are also turned to account by these rude but ingenious workmen.

In former days, the chase of the Walrus was an easy matter, for the powerful brutes seemed to be so satisfied of their strength that they would permit their assailants to approach them closely, and to inflict fatal wounds without any opposition. Now, however, they have learned caution by many a bitter experience, and are extremely wary animals. They are tenacious of life, and dangerous antagonists, for although they seldom, if ever, commence an attack, they are most furious when opposed or wounded, and fight with marvellous energy. In the conflict the enormous tusks prove themselves to be truly formidable weapons, and have been known to pierce through the plankings of a boat. Even the polar bear stands in awe of these weapons, and has often been beaten off by an old Walrus on whom it had hoped to make a meal.

The Walrus is found in vast herds, which frequent the coasts of the arctic and antarctic regions, and which congregate in such numbers that their united roarings have often given timely warning to fog-bewildered sailors, and acquainted

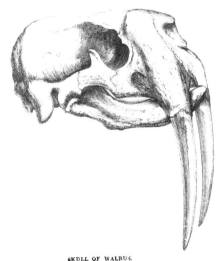

SKULL OF WALRUS.

them with the near proximity of shore. These herds present a curious sight, as the huge, clumsy animals are ever in movement, rolling and tumbling over each other in a strange fashion, and constantly uttering their hoarse bellowings.

As soon as a Walrus gets out of the water, it lies down on the shore and would not of its own free will stir from the spot on which it had first laid itself to repose. But another Walrus soon emerges from the sea, and as it cannot very well climb over its comrade, begins to butt him until he moves farther on, and makes room for the new comer. Others land in rapid succession, and the whole strand is soon full of life, for these unreflective creatures never think of taking a short walk inland, so as to secure a quiet berth at ease, but must needs lie down where they land, although they are sure to be disturbed by their comrades as they rise out of the sea. As many as seven thousand have been seen in a single herd, so that to attack one of these assemblies is no slight matter, for as soon as they take alarm, they all come scuttling towards the sea, tumbling over each other in their haste, and presenting a formidable front simply by the weight of their huge bodies.

In order to prevent the whole herd from making a simultaneous charge at their invaders, the hunters endeavour to disperse them by means of dogs trained to the business, and so to secure the animals as they fly affrighted in various directions. One such chase is technically termed a "cut," and if rightly conducted is so successful that at a single "cut" no less than fifteen hundred of these huge Seals have been taken. These chases take place at night.

The movements of the Walrus when on land are of a very clumsy character, as might be supposed from the huge, unwieldy body of the animal, and the evident insufficiency of the limbs to urge the weighty body forward with any speed. When this creature is hurried or alarmed, it contrives to get over the ground at a pace that, although not very rapid, is yet wonderfully so when the size of the animal is taken into account. The movement is a mixture of jerks and leaps, and the Walrus is further aided in its progress by the tusks. Should it be attacked, and its retreat cut off, the Walrus advances fiercely upon its enemy, striking from side to side with its long tusks, and endeavouring to force a passage into the sea. If it should be successful in its attempt, it hurries to the water's edge, lowers its head, and rolls unceremoniously into the sea, where it is in comparative safety.

The Walrus is possessed of the same docile and affectionate disposition as the other Seals, and has been more than once effectually tamed. One of these animals which was captured while young at Nova Zembla, and brought to England, was remarkably gentle in its demeanour, and learned many accomplishments from its owner. It had been so well instructed, that if taken in a boat, it would leap overboard at the word of command, chase and catch fish, and return to the boat bearing the fish in its mouth.

The number of young which the Walrus produces at a litter is seldom if ever more than one, and when newly born, the little animal is about the size of a yearling pig. Winter is the usual time of year for the appearance of the young, and the mother always repairs to the shore or to the ice-fields for the purpose of nourishing her family. The maternal Walrus is very attentive to her charge, and while in the water is very solicitous about its welfare, carrying it about under her fore-limbs, and defending it from any danger that may arise, regardless of her own safety in watching over that of her offspring. When a mother Walrus is surprised upon the shore, she places her young one upon her back, and hurries away to the sea, bearing her precious burden.

This animal attains to a very great size, so great, indeed, that its dimensions can hardly be appreciated except by ocular demonstration. A full-grown male Walrus is generally from twelve to fifteen feet in length, while there are many specimens that have been known to attain a still greater size. The skin is black and smooth, and is sparingly covered with brown hairs, which become more numerous on the feet. The eye is very small in proportion to the size of the animal, and after death sinks so completely into its socket that it cannot be seen except by an experienced observer. By pressure upon each side of the orbit, the eye suddenly starts forward, and becomes visible.

The Walrus has once or twice been seen off the British coasts, but is so very rare a visitant that any such occurrences can only be considered as exceptional to the general rule. The term Walrus literally signifies "whale horse," and the specific name, Rosmarus, is a Latinized form of the Norwegian word Rosmar, or "sea horse." The word Morse is slightly altered from the Russian Morss, or the Lapponic Morsk.

ANOTHER powerful and grotesque Seal now engages our attention. This is the ELEPHANT SEAL, or SEA ELEPHANT, so called not only on account of the strange prolongation of the nose, which bears some analogy to the proboscis of the elephant, but also on account of its elephantine size. Large specimens of this monstrous Seal measure as much as thirty feet in length, and fifteen or eighteen feet in circumference at the largest part of their bodies.

The colour of the Sea Elephant is rather variable, even in individuals of the same sex and age, but is generally as follows. The fur of the male is usually of a bluish-grey, which sometimes deepens into dark brown, while that of the female is darker, and variegated with sundry dapplings of a yellow hue. This animal is an inhabitant of the

SEA ELEPHANT.—*Morunga proboscidea.*

southern hemisphere, and is spread through a considerable range of country. It is extensively hunted for the sake of its skin and its oil, both of which are of very excellent quality, and, from the enormous size of the animal, can be procured in large quantities. It is not exclusively confined to the sea, but is also fond of haunting fresh-water lakes, or swampy ground, as is depicted in the engraving.

It is an emigrating animal, moving southwards as the summer comes on, and northwards when the cold weather of the winter months would make its more southern retreats unendurable. Their first emigration is generally made in the middle of June, when the females become mothers, and remain in charge of their nurseries for nearly two months. During this time the males are said to form a *cordon* between their mates and the sea, in order to prevent them from deserting their young charges. At the expiration of this time, the males relax their supervision, and the whole family luxuriates together in the sea, where the mothers soon regain their lost condition. They then seek the shore afresh, and occupy themselves in settling their matrimonial alliances, which are understood on the principle that the strongest shall make his choice among the opposite sex, and that the weakest may take those that are rejected by his conquerors, or none at all, as the case may be.

During the season of courtship the males fight desperately with each other, inflicting fearful wounds with their tusk-like teeth, while the females remain aloof, as quiet spectators of the combat. They are polygamous animals, each male being lord over a considerable number of females, whom he rules with despotic sway. When the victorious combatants have chosen their mates they are very careful about their safety, and refuse to quit them if they should be in any danger. Knowing this fact, the Seal-hunters always direct their attacks upon the females, being sure to capture the male afterwards. If they were to kill the male at first, his harem would immediately disperse and fly in terror, but as long as he lives they will continue to crowd round him.

Although these animals are of so great dimensions and bodily strength, and are furnished with a very formidable set of teeth, they are not nearly such dangerous antagonists as the walrus, and are most apathetic in their habits. When roused from

SEA LION.—*Otária jubáto.*

sleep they open their mouths in a threatening manner, but do not seem to think of using their teeth, and if they find that their disturbers do not run away, they take that office upon themselves, and move off deliberately for the water. As they proceed their huge bodies tremble like masses of jelly, in consequence of the fat with which they are so heavily laden. So plentiful is this fat, that a single adult male will furnish about seventy gallons of clear and scentless oil.

The extraordinary development of the nose, which gives so weird-like a character to the aspect of the Elephant Seal, is only found in the adult males, and even in them is not very perceptible unless the animal is alarmed or excited. While the creature is undisturbed, the nose only looks peculiarly large and heavy, as may be seen in the figures that occupy the background of the illustration on page 517. As soon, however, as the animal becomes excited, it protrudes this proboscis-like nose, blows through it with great violence, and assumes a very formidable appearance. The female is entirely destitute of this structure, and except for its enormous dimensions, might be mistaken for an ordinary Seal. In the male it does not make its appearance until the third year.

The Elephant Seal is easily tamed when taken young, and displays great affection towards a kind master. One of these animals was tamed by an English seaman, and would permit its master to mount upon its back, or to put his hand into its mouth without doing him any injury.

The teeth of this animal are very curious in their formation, especially the molar teeth, which are small, and pointed with a kind of mushroom-like apex. The canines are very large. The whisker hairs are very coarse and long, and are furnished with a raised margin, which gives them the appearance of being twisted like a screw. The food of the Elephant Seal is supposed to consist chiefly of cuttle-fish and sea-weed, as the remains of both these substances are generally found in the stomachs of those that are killed.

HARDLY less ferocious in aspect than the preceding animal is the SEA LION, of Kamtschatka and the Kurile islands.

It is of very large size, although not of such gigantic proportions as the sea elephant, measuring about fifteen feet in length, and weighing about sixteen hundred pounds. The colour of the male Sea Lion is a reddish-brown, which becomes paler after the animal is advanced in years. Upon the neck and shoulders is a heavy mass of stiff, curly, crisp hair, which bears some resemblance to the mane of the lion, and has earned for the creature the name of Sea Lion. The female is destitute of this mane, and her fur is sometimes chestnut and sometimes ochry-brown.

It is not exclusively confined to the localities above mentioned, but is sometimes seen off the coast of Northern America, in the month of July. During the autumn the Sea Lions are found in very great numbers upon the shores of Behring's Island, where they assemble for the purpose of rearing their young through the first few weeks of their life. They are polygamous, but restrict themselves to three or four females.

They are naturally quiet and peaceable animals, permitting the approach of mankind with great indifference, and suffering themselves to be roughly treated before they will condescend to move from the spot on which they may happen to be lying. The hunters make easy prey of these slothful animals, which are not so active as the elephant Seal, nor so fierce as the walrus. The females seem to be more apathetic respecting their cubs than is generally the case among Seals, and will frequently relinquish their offspring in their haste to escape from their human foes. The natives are in the habit of killing the Sea Lions by poisoned arrows, or by harpoons. As the wounded animal would be sufficiently strong to escape in spite of the harpoons, the native hunters attach the harpoon-line to a post firmly planted in the ground, and are thus enabled to delay the Sea Lion until they can inflict a fatal wound.

They are marvellously blatant animals, keeping up a continual chorus of vociferations as long as they are on land. The old males are the most noisy of the party, snorting discordantly, and roaring like magnified lions. The females answer by loud bleatings, and the young of both sexes add their voices in a less degree. The united cries of a large herd of Sea Lions are so deafening, that human senses are almost stunned by the clangorous uproar.

This species is said to feed upon fish and smaller Seals, being extremely dreaded by the latter animals, and ruling supreme in its own domains. The teeth of the Sea Lion are very singular in their shape and arrangement, the molars being furnished with sharp trenchant points, some of the incisors double-headed, and others long and pointed like canine teeth.

As the mane-clad neck and shoulders of the preceding animal have earned for it the title of Sea Lion, so the generally ursine aspect of the present species has gained the name of SEA BEAR.

It is not a very large animal, being hardly eight feet in length. As its limbs are larger and better developed than in the generality of the Seals, it can stand and walk in more active manner than any of the preceding members of the phocine family. The colour of its fur is very pleasing, the long hairs being of a greyish-brown, while the thick soft wool that lies next to the skin is reddish-brown. The fur is extremely soft and warm, and of high value as an article of commerce. When it is dressed by the furriers, the entire coating of long hairs is removed, the wool only being left adherent to the skin. Upon the neck and shoulders of the male animal there is a kind of mane, composed of rather stiff hairs about two inches in length, and of a grizzled aspect, the hairs themselves being jetty-black, and their tips white. The whole of the fur is thick and long, and does not lie closely to the body.

It is not so easily caught as the sea lion, for it is not only very active in the water, but can proceed upon land with such rapidity that a man who wishes to overtake an affrighted Sea Bear will be forced to exert his utmost speed before he can attain his object.

The Sea Bears are found in great numbers about Kamtschatka and the Kurile islands, and at the beginning of summer are so numerous as to blacken the banks on which they repose. Being polygamous, the males are extremely jealous, and will not suffer any strangers to approach the limits of his own family. The entire sea-beach is therefore mapped out, so to speak, in little domains, each belonging to a separate family, and guarded with the most jealous care. As the number of females over which a single male bears sway

SEA BEAR, OR URSINE SEAL.—*Arctocephalus ursinus.*

is about forty to fifty on an average, it will be seen that the family must be very extensive when the young are added to their number. From one hundred to a hundred and twenty is not at all an uncommon number for a single family of Sea Bears.

No family will allow the members of another household to crouch upon their territories, and it is very seldom that such an attempt is made. Sometimes, however, trespassers are detected, and then there is a general fight upon the beach, in which the animals of both sexes and all ages fight with great fury. They will not even permit a human being to encroach upon their territories, but advance upon him with such threatening cries and such menacing display of gleaming teeth that he is forced to make his escape as he best can. One traveller was so hard beset by these animals that he was fain to climb a rock which they could not surmount, and was watched by them for nearly six hours before he could make good his escape.

Sometimes an old Sea Bear is seen lying alone in solitary state, not permitting any living being to approach him, and continually uttering low, savage growls.

The males are very tyrannous in their behaviour to their wives, and treat the poor submissive creatures very cruelly. If a mother should happen to drop her cub as she is carrying it off, the male immediately turns upon her and bites her as a punishment for her offence. These animals seem to be very intelligent, and have a great variety of intonations, by which they can express their meaning so clearly that their language can even be understood by human ears. Their general voice is something like the lowing of a cow, but when they are wounded, they utter long plaintive cries like that of a suffering dog.

The food of this species consists of sea otters, small Seals, and other animals, which hold it in great terror. The Sea Bear, however, stands in considerable awe of the sea lion, and does not exercise the same indisputable sway as that animal. The name Arctocephalus is of Greek origin, and signifies "bear-headed."

GREENLAND WHALE.—*Balæna mysticetus.*

WHALES.

THE CETACEA, or WHALES, are more thoroughly aquatic than any other animals which have already been described, and are consequently framed in such a very fish-like manner that they have generally been considered as fishes by those who were but little acquainted with the animal kingdom. The entire livelihood of the Whale is obtained in the waters, and their entire structure is only fitted for traversing the waves, so that if they should happen to be cast upon the shore they have no means of regaining their native element, and are sure to perish miserably from hunger.

With the seals, the young are produced upon the land, and there nurtured until they have attained sufficient strength to enable them to cope successfully with the sea waves, and are, moreover, attended in their marine excursions by their mothers, who exercise a watchful guard over their offspring. But the young Whale knows no such terrestrial nurture, but is at once received into the bosom of the ocean, being capable from its very birth of accompanying its parent in her paths through the waves.

Although the Whales bear so close a resemblance to the fish, and are able to pass a considerable time below the water, they possess no gills through which they may respire and renew their blood through the agency of water, but breathe atmospheric air in the same manner as the other mammalia. If a Whale were to be detained below the surface of the water for too long a period it would be inevitably drowned, a fact which was once curiously exemplified by the death of a Whale which had entangled itself in a rope fastened to a dead and sunken Whale, and which was found drowned when the rope was drawn to the surface. No injury had been inflicted upon the animal, but it had not been able to disengage itself from the detaining cord in time to breathe, and was consequently suffocated.

When the Whales breathe, they are forced to rise to the surface of the sea, and there make a number of huge respirations, which are technically called "spoutings," because a column of mixed vapour and water is ejected from the nostrils, or "blow-holes," and spouts upwards to a great height, sometimes as much as twenty feet. In order to enable the animal to respire without exposing itself unnecessarily, the "blow-holes" are placed on the upper part of the head, so that when a Whale is reposing itself on the surface of the sea, there is very little of its huge carcase visible, except the upper portion of the head and a part of the back. The "spoutings" are made with exceeding violence, and can be heard to some distance.

The mode of respiration is, however, rather different from that of the generality of mammalia, being modified in order to meet the peculiar circumstances in which the animal is placed.

In nearly all the mammalia the movements of respiration take place in rather rapid succession, and are continuous in their action, and if they are checked for only a few minutes, the result is inevitably fatal. It is evident, however, that as the Whales are forced to seek their food in the depths of the ocean, and to remain for a considerable space of time below the surface, their respiration must be conducted on a different system. The mode which is adopted is truly one of the most marvellous contrivances that can be imagined, and which is so beautifully simple as well as profound that it raises our highest adoration of the unspeakable wisdom which planned it.

It is clear that the creature would not be able to take a supply of air into the depths of the ocean, and that another means must be found for oxygenizing the blood.

As, therefore, the animal is unable to breathe below the surface of the water, the difficulty is surmounted by furnishing it with a large reservoir of arterial blood, which is oxygenized during the short time that is occupied in the "spoutings," and which supplies the circulation until the Whale returns again to the upper regions for a fresh supply. The reserved blood is contained in a large mass of vessels which line the interior of the chest and the adjoining regions, and are capable of containing a sufficient amount of fresh blood to sustain life for a wonderfully long period.

As the Whales are in the habit of descending to very great depths—depths so profound, that if a piece of dry wood be equally deeply sunk it will be saturated with water, and will not float—their ears and nostrils must be guarded against the dangers that would arise from the penetration of the water into their cavities. There is consequently a beautifully simple and ingenious valvular structure, which perfectly answers this purpose, and firmly closes the external orifices in proportion to the depth to which the animal dives. The ear is remarkably small, and in some specimens is almost undiscernible. Some anatomists are of opinion that the Whales can hear by means of the communication of the ear with the mouth. As the spermaceti Whale is capable of communicating with its companions at a distance of several miles, it is evident that the sense of hearing must be better developed than would be the case if the creature were totally dependent for hearing on the external orifice; which must always be closed while under water, and which in many species is covered with the external integument.

The limbs of the Whales are so modified in their form that they can hardly be recognised by their external appearance alone as the limbs of a veritable mammal. In shape they closely resemble the fins of fish, and it is not until they are stripped of the thick skin which envelops them that the true limb is developed. The reader may see the bony structure of the Whale's fin by referring to the skeleton of the rorqual on page 529. The chief use of these organs seems to be that they assist the animal in preserving its position in the water, for the huge carcase rolls over on its back as soon as it is deprived of the balancing power of the fins. They are also employed for the purpose of grasping the young whenever the mother Whale is anxious for the safety of her offspring, but they are of little use in urging the animal through the water, that duty being almost entirely performed by the tail.

This member is very curious in its structure, for, as may be seen by reference to the rorqual skeleton, the Whales have no hinder limbs that may be modified into fins, as is the case with the seals, and are forced to depend solely on the soft structures for its powers of locomotion. The traces of hinder limbs are to be found in some little bones that lie loosely in the flesh, but they are of no real use, and are only representatives of the true limbs.

The tail of these animals is an enormously powerful organ, set transversely upon the body, and driving the creature forward by its powerful vertical sweeps. With such wonderful strength is the tail endowed, that the largest Whales, measuring some eighty feet in length, are able by its aid to leap clear out of the water, as if they were little fish leaping after flies. This movement is technically termed "breaching," and the sound which is produced by the huge carcase as it falls upon the water is so powerful as to be heard for a distance of several miles. The length of the tail is, in the larger Whales, about five or six feet, but it is often more than twenty feet in breadth. The substance of the tail is remarkably strong, being composed of three layers of tendinous fibres. When taken from the animal it is largely used in the manufacture of glue.

The skin of the Whales is devoid of hair, and is of a rather peculiar structure, as is needful to enable it to resist the enormous pressure to which it is constantly subjected at the vast depths to which the animal descends. The skin is threefold, consisting first of the scarf-skin, or epidermis; secondly, of the rete-mucosum, which gives colour to the animal; and thirdly, of the true skin, which is modified in order to meet the needs of the creature which it defends. The blubber, indeed, is nothing more than the true skin, which is composed of a number of interlacing fibres, capable of containing a very great amount of oily matter. This blubber is never less than several inches in thickness, and in many places is nearly two feet deep, and as elastic as caoutchouc, offering an admirable resistance to the force of the waves and the pressure of the water. In a large Whale the blubber will weigh thirty tons.

None of the Whales are able to turn their heads, for the vertebræ of the neck are fused together into one mass, and compressed into a very small space.

The GREENLAND WHALE, NORTHERN WHALE, or RIGHT WHALE, as it is indifferently termed, is an inhabitant of the Northern Seas, where it is still found in great abundance, although the constant persecutions to which it has been subjected have considerably thinned its numbers.

This animal is, when full-grown, about sixty or seventy feet in length, and its girth about thirty or forty feet. Its colour is velvety black upon the upper part of the body, the fins and the tail; grey upon the junction of the tail with the body and the base of the fins, and white upon the abdomen and the fore-part of the lower jaw. The velvety aspect of the body is caused by the oil which exudes from the epidermis, and aids in destroying the friction of the water. Its head is remarkably large, being about one-third of the length of the entire bulk. The jaw opens very far back, and in a large

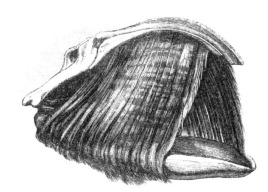

SKULL OF GREENLAND WHALE.
(To show the Whalebone.)

Whale is about sixteen feet in length, seven feet wide, and ten or twelve feet in height, affording space, as has quaintly been remarked, for a jolly-boat and her crew to float in.

The most curious part of the jaw and its structure is the remarkable substance which is popularly known by the name of Whalebone. This substance is represented in its natural position in the accompanying illustration, which is taken from a photographic portrait of the skeleton in the great Museum of Comparative Anatomy at the Jardin des Plantes.

The Whalebone, or baleen, is found in a series of plates, thick and solid at the insertion into the jaw, and splitting at the extremity into a multitude of hair-like fringes. On each side of the jaw there are more than three hundred of these plates, which in a fine specimen are about ten or twelve feet long, and eleven inches wide at their base. The weight of baleen which is furnished by a large Whale is about one ton. This substance does not take its origin directly from the gum, but from a peculiar vascular formation which rests upon it. These masses of baleen are placed along the sides of the mouth for the purpose of aiding the Whale in procuring its food and separating it from the water.

The mode of feeding which is adopted by the Whale is as follows. The animal frequents those parts of the ocean which are the best supplied with the various creatures on which it feeds, and which are all of very small size, as is needful from the size of its gullet, which is not quite two inches in diameter. Small shrimps, crabs and lobsters, together with various molluscs and medusæ, form the diet on which the vast bulk of the Greenland Whale is sustained. Driving with open mouth through the congregated shoals of these little creatures, the Whale engulphs them by millions in its enormous jaws, and continues its destructive course until it has sufficiently charged its mouth with prey. Closing its jaws and driving out through the interstices of the Whalebone the water which it has taken together with its prey, it retains the captured animals which are entangled in the Whalebone, and swallows them at its ease. The multitude of these little creatures that must hourly perish is so enormous, that the prolific powers of nature would seem inadequate to keep up a supply of food for the herds of Whales that inhabit the Northern Seas. Yet the supply is more than equal to the demand, for the sea is absolutely reddened for miles by the countless millions of living beings that swarm in its waters.

The Whale is an animal of very great value to civilized and to savage men. The oil which is procured in great quantities from its blubber and other portions of its structure is almost invaluable to us, while the bones and baleen find their use in every civilized land. To the natives of the polar regions, however, the Whale is of still greater value, as they procure many necessaries of life from various parts of its body, eat the flesh, and drink the oil. Repulsive as such a diet may appear to us who live in a comparatively warm region, it is an absolute necessity in these ice-bound lands, such oleaginous diet being needful in order to keep up the heat of the body by a bountiful supply of carbon.

Civilized beings, even though they may be living for the time in these northern regions, find themselves almost unable to join in the greasy banquet which so entirely delights the native palate. There are, however, some portions of the Whale which can be eaten without difficulty, and are rather palatable than otherwise. The skin of the Whale, when properly dressed, is of ebony blackness, and not at all attractive to the eye. But its flavour is quite agreeable, and is said to bear some resemblance to that of the cocoa-nut. When prepared for the table it is cut into little cubes like black dice. But the best part of the Whale is one that would hardly be expected to form an article of diet, namely the portion of the gums in which the roots of the baleen are still imbedded. The Tuskis call this substance their sugar, though its flavour is very like that of cream-cheese. One traveller who had been obliged, through motives of politeness, to take part in a native banquet, and who had been more than disgusted by the very remarkable dishes which were brought to table, became quite enthusiastic on the merits of Whale's skin and gum, acknowledging himself to be agreeably surprised by the former, and calling the latter article of diet, "perfectly delicious."

The chase of the Whale, its dangers and its excitement, are too well known to need description in these pages, and only as far as they form part of the animal's history will they be noticed.

In its character the Greenland Whale is inoffensive and timorous, and except when roused by the pain of a wound or by the sight of its offspring in danger, will always flee the presence of man. Sometimes, however, it turns fiercely upon the boat from which the fatal weapon has been launched, and with a single blow of its enormous tail—its only weapon—has been known to shatter a stout boat to fragments, driving men, ropes, and oars high into the air. It is a very affectionate animal, holding firmly to its mate, and protecting its young with a fearlessness that is quite touching to any one except a whaler, who takes advantage of the poor creature's natural affection to decoy the mother within reach of his harpoon.

As far as is yet known, the Greenland Whale produces only a single cub at a birth. When first born, the young Whale is without the baleen, depending upon its mother for its subsistence like any other young mammal. The maternal Whale keeps close to her offspring until the baleen is grown, and does not forsake it until it is capable of supporting itself. The young Whales, before the baleen has developed itself, are technically termed "suckers," and when the baleen is six feet in length, they are called by the name of "size."

The tongue of this Whale cannot be protruded from the mouth, as it is fixed throughout its entire length. It is very large, soft, and full of oil, so soft, indeed, that a man can make a depression deep enough to contain his closed fist by a tolerably strong pressure ; as I can testify by personal experience. The bones are porous and are very full of oil, the jawbones being so heavily charged with this valuable substance that they are removed from the animal, and so fastened in the rigging as to permit the oil to drain from them.

When the Greenland Whale is undisturbed, it generally remains at the surface of the water for ten minutes, and "spouts" eight or nine times. It then descends for a short time, from five to twenty minutes, and returns again to the surface for the purpose of respiration. But when harpooned, it dives to a very great depth, and does not return to the surface until half an hour has elapsed. By noticing the direction of the line which is attached to the harpoon, the whalers judge of the spot in which the creature will rise, and generally contrive to be so near their victim when it emerges that they can fix another harpoon, or strike it with a lance before it can again descend into the depths of the ocean.

Several species of the genus Balæna are found inhabiting the different oceans of our globe, such as the Western Australian Whale, the Cape Whale, the Japan Whale, the New Zealand Whale, the Scrag Whale, and others, of which the best known is the Cape Whale, or Southern Whale, as it is sometimes called.

This animal attains a considerable size, reaching the length of seventy feet when full grown, the length of its head being sixteen feet. It inhabits the Southern Ocean, and is often seen in the bays that adjoin the Cape of Good Hope in the months of June, July, and August, as the female is in the habit of frequenting these localities during the infancy of her young. The males are very seldom seen near their mates, so that out of sixty Cape Whales that were killed in False Bay only one was a male. The colour of this animal is a uniform black.

SEVERAL species of the HUMP-BACKED, or BUNCHED, WHALES are now known, although there is very great difficulty in deciding upon the distinctions that are needful for the founding of a species, in animals which are necessarily so far beyond our reach except on rare and limited occasions.

In all these animals the head is rather broad and flattened, and the throat and chest are marked with deep longitudinal folds or "reeves," as they are termed by Dudley in his account of the Bunch Whale. These folds are perceptible even on the sides, and extend as far as the fins. The hump or bunch is of no very great size, being only a foot or so in height, and hardly larger than a man's head. These animals may easily be distinguished from those of the succeeding genus by their shorter and more stout forms, the warty lip, and the large and rounded nose. The skull is about one-fourth of the entire length of the animal.

The species which is figured in the accompanying illustration is that of JOHNSTON'S HUMP-BACKED WHALE, a species which frequently attains very considerable dimensions, measuring from sixty to seventy feet in length. In spite, however, of its great size, it is not sought after by the whalers, and even if it should accidentally come across the course is seldom disturbed by them. Its oil, however, is said to be superior to that which is furnished by the Greenland Whale, and not much inferior to the oil of the Sperm Whale.

HUMP-BACKED WHALE.—*Megáptera longimana.*

It is an inhabitant of the Greenland seas, and is said to be found in greater profusion than any other species. It is furnished with baleen, but this substance is of no very great value, being short, and not splitting "kindly," like that of the Greenland Whale. When dry it takes a slight twist. When first born, the young of the Hump Whales are devoid of baleen, but a considerable number of rudimentary teeth are found in both jaws. The spout or blow-holes are situated on the top of the head, and not on the snout like those of the sperm Whale.

The name Megaptera signifies "great-finned," and is given to this genus on account of the large size to which the pectoral fin extends. This member sometimes measures as much as seventeen feet in length, being equal to the head, or about one-fifth the entire length of the body. When the integument is removed it is seen to be provided with only four fingers. The Latin specific name, longimana, signifies "long-handed." In colour it is white. The dorsal fins are placed rather low, and behind the middle of the body. This Whale is always infested with sundry parasitic animals belonging to the genera Diadema and Otion.

ALL the true carnivorous Whales are remarkable for the great proportionate size of the head. The PIKE WHALE, as may be seen from the illustration, belongs to this group of animals, and in some respects is not dissimilar to the Greenland Whale.

This animal is, however, not nearly so large as the preceding, being only about twenty-five feet in length when adult. It is furnished with baleen, but the plates are comparatively short, and of a slight pinkish hue. The volume of the mouth is made up by a development of the lower part of the mouth into a kind of huge pouch, which is capable of containing a very large volume of water and marine animals. The tongue is not tied down as in the Greenland Whale, but is free towards the apex, and almost as capable of movement as that of man.

It is a native of the seas that wash the coasts of Greenland, and is sometimes seen near Iceland and Norway, descending but rarely into warmer latitudes. The flesh of this animal is in some repute for its delicacy, and is therefore much coveted by the natives of these northern regions. They do not, however, attempt to harpoon the creature, on account of its great activity, but content themselves with inflicting severe wounds with their darts and spears, in the hopes that the wounded animal may die, and may in time be stranded on their coasts. The oil which it furnishes is said to be particularly delicate.

The Pike Whale feeds not only on the little creatures that form the food of the Greenland Whale, but chases and kills the active salmon and other fish. In the stomach of one of these animals have been found the remains of various fish, those of the dog-fish

PIKE WHALE.—*Balænóptera rostráta.*

being the most prevalent. The head of this species is elongated and rather flattened, and the throat and chest are furnished with very deep longitudinal folds, which are capable of dilatation to a great extent.

At the extremity of the snout there are eight distinct bristles, arranged in perpendicular rows on the top of each jaw. It has been called by a great number of names by different writers, and is mentioned by various authors under no less than seventeen distinct titles. The colour of this animal is black upon the upper parts of the body, and white on the abdomen, tinged with a reddish hue. The pectoral fin is almost entirely dark, but changes into white on its upper surface, near its base.

The name Balænoptera signifies "Finned-Whale," and is given to the animal on account of the size of the pectoral fins.

A GIANT among giant forms, the huge Rorqual roams the Arctic seas at will, seldom molested by the hunter, and scarcely ever captured.

The bulk of this animal is greater than that of any other Whale, as many specimens have been known to attain a length of more than one hundred feet, and one or two have reached the extraordinary length of one hundred and twenty feet. By inexperienced whalers it is sometimes mistaken for the Greenland Whale and harpooned, but is very seldom killed; for the creature is so remarkably active and fearless, that in many cases the aggressors have paid dearly for their error by a crushed boat and the loss of several lives. On one such occasion the Rorqual started off in a direct line, and at such a speed that the men lost their presence of mind and forgot to cut the rope that connected the

RORQUAL —*Physalus Boops.*

Whale with the boat. Making directly for a neighbouring ice-field, the Rorqual shot under it, and drew the boat with all its crew beneath the ice, where they disappeared for ever from the gaze of mankind.

Mr. Scoresby, desiring to secure one of these powerful animals, made preparations for the chase by employing very short lines, only two hundred fathoms in length, and attaching a buoy to their extremities in order to tire out the creature by the resistance which the buoy would offer to the water through which they would be dragged by the Whale.

Two Rorquals were struck, and in both cases the intended victims escaped. In the first instance, the Whale dived with such impetuous speed that the line snapped by the resistance of the buoy against the surface of the water, and in the second case the line only held together for a single minute, and was severed apparently by friction against the dorsal fin. A third Rorqual was afterwards harpooned through the error of the seamen, who mistook it for a Greenland Whale. As soon as it felt the sting of the harpoon, the animal dived with such rapidity that it carried nearly three thousand feet of line out of the boat in about a minute of time, and escaped by snapping the rope.

Not contenting itself with such mode of escape, the Rorqual will often turn fiercely upon the boats, and avenge itself by dashing them to pieces by repeated strokes of its fearful tail.

These belligerent qualities would make the whalers very cautious in dealing with such formidable foes, even if their capture were attended with profit equal to the bulk of their prey. But as it is found that the Rorqual is almost valueless when killed, the whalers permit it to pass unmolested, and turn their attention to more valuable quarry. The layer of blubber which encompasses the Rorqual is only about six or eight inches in thickness, and is very chary in yielding oil, a large Whale only furnishing at the best ten or fifteen tons, and sometimes scarcely a single ton of this valuable substance.

As the head of the Rorqual is not nearly so much arched as that of the Mysticetus, and the capacity of the mouth is more owing to the huge pouch of the lower jaws than to the form of the upper jaw, the baleen, or whalebone, is necessarily very short, scarcely reaching four feet in length. Even if its quality had been good, it would be of comparatively little value. Yet it is so coarse and "unkindly" that it is almost valueless for manufacturing purposes. Whalers would rejoice if this substance were of more value, as it is extremely plentiful in the Rorqual, the jaws being lined with five thousand distinct plates or "slabs" of baleen.

As the food of the Rorqual is not limited to the small animals which constitute the diet of the Greenland Whales, but consists also of various fish, it needs that the gullet should be larger than in that creature. In the stomach of a single Rorqual, six hundred large cod-fish have been found, together with a considerable number of pilchards. In order to procure a sufficiency of food for its vast bulk, the Rorqual often follows the shoals of migrating fish until it approaches the shores of Great Britain, where in many cases it prefers to take up its abode, hovering round the fishing-grounds, and swallowing whole boat-loads of herrings, pilchards, and other fish. One of these creatures haunted the Frith of Forth for a period of twenty years, and was popularly recognised under the title of the "hollie-pike," on account of a hole through its dorsal fin which had been perforated with a musket-ball.

Although the Rorqual may for a time support itself at the cost of our fishing-trade, it is nearly sure to fall a victim to its own temerity, and to be left by the returning tide, helplessly and ignominiously stranded on the shores. This is a season of great rejoicing among the fishermen, who flock to the fatal spot with their most deadly weapons, and avenge themselves of their losses by the slaughter of the giant robber. Even the "hollie-pike" himself fell a victim to his want of caution, and was at length stranded on the

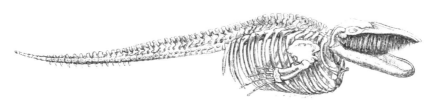

SKELETON OF RORQUAL.

shores of the very bay which he had haunted for so many consecutive years. The length of this animal was seventy-five feet.

Owing to the persevering manner in which the Rorqual follows its prey to our coasts, it is more frequently stranded upon the British shores than any other true Whale. One of these animals that was thus captured was ninety-five feet in length, and weighed two hundred and forty-nine tons. Its breadth was eighteen feet, the length of the head twenty-two feet. Each fin measured twelve feet six inches in length. The skeleton of this magnificent animal was preserved and mounted, and after the bones were dry, their united weight amounted to thirty-five tons. To procure the skeleton of so large an animal is no easy matter, for the preparation of a Rorqual that was only eighty-three feet in length occupied a space of three years.

The Laplanders, who find the bones and other portions of this animal to be of great service to them, unite in its chase, and employ a very simple mode of action. To harpoon such a being would be useless, so they content themselves with inflicting as many wounds as possible and leaving it to die. After the lapse of a few days the huge carcass is generally found dead upon the strand, and becomes the property of all those who have wounded it and can prove their claims by the weapons which are found in its body. The person who finds the stranded carcass is by law entitled to one-third of the value.

1.

The whalers appear to regard this animal with as much detestation as do the European fishermen, for the Greenland Whale has no love for the Rorqual, and seems to avoid the localities where this marine giant takes up its residence. It does not frequent the more icy seas, but prefers the clear waters. The spoutings of the Rorqual are very impetuous, as indeed are all its actions, and while engaged in respiration it shoots along the surface of the water at a velocity of four or five miles an hour instead of lying still during that process as is the custom with the Greenland Whale. The colour of this species is a dark-grey, tinged with blue.

The name Rorqual is derived from the Norwegian word, which signifies a "Whale with folds," in allusion to the deep longitudinal folds which lie along the under jaw and a considerable portion of the lower parts of the animal. The term Boöps is from the Greek, and signifies "ox-eyed," in allusion to the small rounded ox-like eyes of the Rorqual.

The Rorqual can be distinguished while in the water from the Mysticetus by the comparatively longer and more slender body and more cylindrical form, and by the fact of its possessing a dorsal fin. Its actions too are so peculiar as to mark it out to an experienced eye.

The animals which form the next little group of Cetacea are remarkable for their immensely large head with its abruptly terminated snout, and the position of the blow-hole, which is situated upon the fore part of the head, nearly at the tip of the snout.

They do not possess any baleen, but are armed with a most formidable set of teeth in the lower jaw, which fit into a series of conical depressions in the upper jaw. To a casual observer the upper jaw appears to be devoid of teeth, but on a closer examination it is found to possess a short row of them on each side, which are mostly placed nearer the interior of the jaw than the conical depressions already mentioned, but in some instances are found at the bottom of these cavities. The number of the teeth of the lower jaw is very variable, but the average in adult specimens is about fifty-two. The teeth are heavy, strong, and when the animal is young are rather sharply pointed, but become extremely blunt when worn by the attrition of a long course of service. In Europe the teeth of the CACHALOT, or SPERMACETI WHALE, are of no great value, being considered merely as marine curiosities, and often carved with rude engravings representing the chase of the animal from which they were taken, together with a very precise account of the latitude and longitude, and a tolerably accurate view of the vessel. In the South Sea Islands, however, these teeth are articles of the highest value, being thought worthy of dedication to the idol deities, or at least placed as rare ornaments in the king's house. So great is the conventional value of these teeth, that several wars have arisen from the possession of a Whale's tooth by an inferior and unfortunate chief who had discovered the rarity and meant to keep it.

The partly-hidden teeth of the upper jaw are about three inches in length, but they hardly project more than half an inch through the soft parts in which they are imbedded. In preparing the skull of the Spermaceti Whale these teeth are apt to fall out together with the softer parts, as their attachment to the jawbone is very slight. Eight of these teeth have been found on each side of the jaw,

The CACHALOT is one of the largest of the Whales, an adult male, or "old bull," as it is called by the whalers, measuring from seventy to eighty feet in length, and thirty feet in circumference. The head is enormously long, being almost equal to one-third of the total length. The term Macrocephalus is derived from two Greek words, signifying "long-headed," and has been given to the animal in reference to this peculiarity of structure. Upon the back there is a rather large hump, which rises abruptly in front and tapers gradually towards the tail. The colour of the Cachalot is a blackish-grey, somewhat tinged with green upon the upper portions of the body. Round the eyes and on the abdomen it is of a greyish-white.

This species is chiefly notable on account of the valuable substances which are obtained from its body, including oil and spermaceti. The oil is obtained from the blubber, which is not very thick in this animal, being only fourteen inches in depth on

SPERMACETI WHALE —*Catodon Macrocephalus.*

the breast and eleven inches on the other parts of the body, and is therefore not so abundant in proportion to the size of the animal as that which is extracted from the Greenland Whale. Its superior quality, however, compensates fully for its deficiency in quantity. The layer of blubber is by the whalers technically called the "blanket," probably in allusion to its office in preserving the animal heat.

The spermaceti is almost peculiar to a few species of the genus Catodon, and is obtained as follows.

The enormous and curiously formed head is the great receptacle of the spermaceti, which lies in a liquid, oily state, in two great cavities that exist in the huge mass of tendinous substance of which the head is chiefly composed. On reference to the skull of the Cachalot, the reader will observe that it dips suddenly over the eyes, and then is greatly prolonged. This portion of the skull is termed Neptune's chair by the sailors, and it is in Neptune's chair that the spermaceti is placed. When the Whale is killed and rowed to the ship's side, the head is cut off and affixed to tackles for the purpose of supporting it in a convenient position for the extraction of this valuable substance. A large hole is cut in the top of the head, and a number of sailors lower their buckets into the cavity and bale out the liquid matter.

When first exposed to the air it has a clear, oily appearance, but after it has been subjected to the action of the atmosphere for a few hours, the spermaceti begins to separate itself from the oil, and in a short time is sufficiently firm to be removed and put into a different vessel.

There is yet a considerable amount of oil mixed with the pure spermaceti, giving it a yellow, greasy aspect, which must be thoroughly removed before the spermaceti can assume its silky, crystalline appearance. The process of purifying it is rather a long and complicated one, consisting of various meltings and re-meltings, of squeezing through hair bags, and of treatment with a solution of potass. It is then sufficiently refined for commercial purposes, but if it should be required to be perfectly pure without any admixture of oil or extraneous substances, it is boiled in alcohol and is deposited in pearl-white laminated crystals, glistening with a silver sheen and separating easily into small scales.

The amount of spermaceti which is produced from the head of a single Whale is very large indeed. From a Cachalot that only measured sixty-four feet in length, and was therefore by no means a large one, twenty-four barrels of spermaceti and nearly one hundred barrels of oil were obtained.

Ambergris, that curious substance whose origin so long baffled the keenest inquirers, and which was formerly only found at rare intervals floating on the waves or cast upon the shore, is now often discovered within the intestines of the Cachalot, and is supposed to be a morbid secretion peculiar to the animal, and analogous to biliary calculi. Fifty pounds weight of this substance have been found in a single Whale, and on one occasion a single piece of ambergris of the same weight was discovered on the coast of the Bermudas by some sailors, who immediately deserted their ship and escaped to England with their valuable prize. The value of ambergris is rather variable, but it is always a costly article.

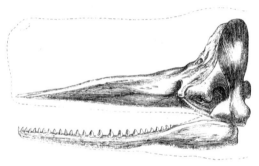

SKULL OF SPERMACETI WHALE.

It is seldom, if ever, found in young and healthy Cachalots, so that a ship may make a very successful whaling voyage, and yet return home without finding a single ounce of ambergris in all the Whales killed. Ambergris is generally employed as a perfume, and is prepared for the use of the purchaser by being dissolved in alcohol.

It sometimes happens that a stray Cachalot blunders into the shallow waters of the Bermudas, and being unable to discover the passage through which it passed, is caught like a mouse in a trap, and falls a ready victim to the intrepid and almost amphibious natives.

As soon as a Whale is discovered in this helpless situation, the populace is all astir and full of excitement at the welcome news. Boats are immediately launched, filled with men bearing guns, lances, and other destructive weapons, which would be of little use in the open sea, but are very effectual in the shoal waters of these strange islands. No sooner does the Whale feel the sharp lance in its body, than it dives with its ordinary velocity, forgetting that it is no longer in deep water, and strikes its head against the rocky bed of the sea with such unexpected force as to bring it to the surface half stunned. The hunters take advantage of its bewildered state to approach closely and to ply their deadly weapons with fatal effect. Some of these men are so cool and determined, that they will actually leap from their boats upon the Whale's back, and, setting their shoulders to the butt of the spear, urge the sharp blade by the weight of their bodies. The Whale soon yields up its life under such circumstances, and the huge carcass is brought to shore amid the shouts and congratulations of the spectators.

The fat and ivory of the slain animal are divided among the hunters who were actually engaged in the chase, but the flesh is distributed gratuitously to every one who chooses to apply for it. Every one who can own a barrow or basket, bears it to the scene of slaughter, and is at liberty to take as much Whale's flesh as he chooses. The connoisseurs in Whale's flesh assert that there are three qualities of meat in every Whale, the best resembling mutton, the second quality imitating pork, and the third resembling beef. Captain Scott, R.N., an eye-witness of this animated scene, and to whom I am indebted for the information, avers that there really is some semblance of these various meats, and that the "pork," when salted and barrelled, might be readily taken for the flesh of the veritable hog.

The same gentleman tells me that the leaping powers of the Cachalot are not in the least exaggerated, for that he has seen one of these animals spring to such a height out of the water, that the horizon could be seen under it, although the spectators were standing on the deck of a man-of-war. The Cachalot was about three miles from the ship at the time when it made its spring.

The Spermaceti Whale, when it is in the open seas, lives chiefly on the "squids," or cuttle-fish, which swarm in that ocean, and when it approaches land, feeds on various fish. It seems, however, to dislike the propinquity of the shore, and is very seldom taken in "soundings." It is a gregarious animal, being seldom seen alone, but in large herds, technically called "schools," and consisting of several hundred in number. The "schools" are generally divided into two bands, the one consisting of young males and the other of females. Each band of females is under the command of several large males, who exercise the strictest discipline over their harems, and will not permit any intruder to join their society. From their office, these leaders are called the schoolmasters.

At distant intervals a large overgrown Cachalot is seen roaming the ocean, apparently unconnected with any school. These solitary animals are, however, the "schoolmasters," which have laid down their authority for a short space of time, and are engaged in search of food. These animals generally fall easy victims to the harpooner, as they are very reckless in their conduct, and will permit a boat to approach them without much difficulty. Sometimes when struck they lie still and supine as if they did not feel the keen edge of the harpoon, and so afford time to the whalers to use their deadly lances at once instead of dragging them for miles over the waves.

Sometimes, however, a "large Whale" will become belligerent, and is then a most fearful antagonist, using its tail and its huge jaws with equal effect. One of these animals has been known to drive its lower jaw entirely through the plankings of a stout whaling boat, and another well-known individual destroyed nine boats in rapid succession. This formidable animal was at last killed, and in its carcass were found a whole armoury of harpoons and spears belonging to different ships. Not only boats, but even ships have been sunk by the attacks of an infuriated "old bull" Cachalot.

An American ship, the *Essex*, was thus destroyed by the vengeful fury of a Cachalot, which accidentally struck itself against the keel. The irritated animal, evidently thinking that the ship was a rival Whale, retired to a short distance and then charged full at the vessel, striking it on one side of the bows, and crushing beams and planks like straws. There were at the time only a few men on board, the remainder of the crew being in the boats engaged in chasing the Whales ; and when the poor men returned to their ship, they found her fast sinking, and only reached her in time to secure a scanty stock of provision and water. Husbanding these precious supplies to the utmost, they made for the coast of Peru, but all perished excepting three, who were almost miraculously rescued as they lay senseless in their neglected boat, which was drifting at large in the ocean, unguided by human hands.

Like the Greenland Whale, the Cachalot is an affectionate animal, and though constitutionally timid to a degree, is yet possessed of sufficient moral courage to come to the rescue of its wounded friends. If the harpooner strikes one female of the "school," her companions will not attempt to make their escape, but will swim anxiously round their suffering companion and remain in her vicinity until she is killed. Taking advantage of this trait of character, the whalers have often contrived, by sending a number of boats simultaneously, to secure almost every member of the "school." The young males however are far more selfish, and when one of them is struck, the others make off as fast as they can swim, leaving their wounded companion to shift for himself as he best can.

The natural timidity of the Cachalot is very remarkable, considering the gigantic size of the animal and the formidable array of teeth with which it is armed. Any strange object perceived by this creature throws it into a state of excited trepidation, during which time it performs several curious antics, and is said by the sailors to be "gallied." When uneasy, it has a strange habit of slowly sweeping its tail from side to side upon the surface of the water, as if feeling for the object that excited its terror.

When thoroughly frightened, and especially when roused to energetic action by the painful sting of the harpoon, the Cachalot darts along the surface of the water at an astonishing rate, its speed being often from ten to twelve miles per hour. As it proceeds in its rapid course, the alternate upward and downward strokes of the tail cause its head to sink and emerge alternately, producing that mode of swimming which is technically termed "head-out." As the lower part of the head is compressed into a kind of cut-water shape, there is less resistance offered to the water than if the creature swam entirely below the surface, as is its wont when undisturbed. It is conjectured that the enormous amount of oil and spermaceti which exists in the head of the Cachalot may be intended for the purpose of lightening the head, and enabling it to lie more easily upon the surface.

The "spoutings" of the Spermaceti Whale are very peculiar, and can be recognised at a distance of several miles. It generally lies still while spouting, but sometimes proceeds gently along the surface. Firstly the "hump" becomes perceptible as the animal rises, and at some forty or fifty feet distance the snout begins to emerge. From the extremity of the snout is ejected a continuous stream of water and vapour, which lasts for about three seconds, and is thrown forward at an angle of forty-five degrees.

The intervals of time between the "spoutings" are as regular as clockwork, and their number is always the same in the same individual. The snout sinks under water as each spouting is finished, and emerges for the next respiration. Sometimes a Cachalot is alarmed before "the spoutings are out," and dives below the surface. In such a case, the animal soon re-appears in another spot, and completes the number of the respirations. The interval of time between the spoutings is ten seconds in the "old bulls," and as the animal makes between sixty and seventy of these curious respirations, the time which is consumed in oxygenising the blood is ten or eleven minutes.

Having completed this business, the creature then lowers its head into the water, flings its tail in the air, and disappears into the far depths of the ocean, where it remains about an hour and ten minutes. The number and force of these "spoutings," together with the time which is consumed by respiration, and the period of the stay beneath the surface of the water, are extremely varied, according to the age, sex, and size of the individual.

The Spermaceti Whale does not seem to choose any particular portion of the year for the production of its young, but is found at all seasons in charge of its offspring. Moreover, young Whales, or "cubs," are found of all sizes and ages, simultaneously roaming the seas, either in company with their parents or turned loose upon the world to shift for themselves. There is but a single cub at a birth. The milk of the animal is exceedingly rich and thick, as indeed is the case with the milk of all Whales.

This animal is very widely spread over the world, as it is found in almost every portion of the aqueous portions of the globe with the exception of the Polar Seas. Several of these creatures have been discovered off our own coasts, and a few have been stranded on the beach. A Cachalot measuring fifty-four feet in length was driven ashore in the Frith of Forth in 1769, and its appearance off the Orkneys is said to be no very uncommon occurrence.

The BLACK-FISH has been separated from the genus Catodon, and placed in the genus Physeter, together with one or two other Whales, because, although they possess the huge truncated head and heavily armed lower jaw of the Cachalots, the spout-holes are removed from the extremity of the snout and placed upon the middle of the top of the head. These spout-holes are separate, but are covered with a common flap. The pectoral fin is moderate in size, being about four feet long, and rather triangular in its form, and the

BLACK-FISH.—*Physeter Tursio.*

dorsal fin is long, and slightly sickle-shaped. The head is remarkably large, and probably exceeds in length the fourth of the entire bulk.

This species is of considerable dimensions when adult, as it is known to measure from fifty to sixty feet in length. In the lower jaw is a bountiful supply of teeth, white, powerful, and conical. These teeth are very variable in number, in different species, varying altogether from twenty-two to forty-four. An equal number of cartilaginous sockets are placed in the upper jaw, into which the conical teeth are received when the mouth is closed. In the accompanying illustration of this species, the sockets of the upper jaw are engraved as if they were projecting teeth, instead of hollow cavities. The teeth that are placed in the middle of the jaw are larger and heavier than those of the front or base. Some of these teeth will exceed nine inches in length, and weigh more than eighteen ounces when perfectly dried.

The root of each tooth is hollow in the centre to the depth of several inches, and is so deeply buried in the jaw, that the projecting portion of the largest tooth rarely exceeds three inches. The teeth range from seven to nine inches in length. These teeth are very white and polished, are conical in their shape, tolerably sharp while the animal is young, but become blunt as the creature increases in years and dimensions.

The dimensions of one of these animals have been very accurately given by Sibbald.

In total length it measured between fifty-two and fifty-three feet, its girth at the largest part of the body was rather more than thirty-two feet, and as it lay on the ground the height of its back was twelve feet. The lower jaw was ten feet in length, and was furnished with forty-two teeth, twenty-one on each side. Each tooth was slightly sickle-shaped, and curved towards the throat. From the tip of the snout to the eyes was a distance of twelve feet, and the upper part of the snout projected nearly five feet beyond the tip of the lower jaw. The eyes were remarkably small, about the size of those of the common haddock. As may be supposed from the popular name of this animal, the colour of its skin is almost uniformly black. The throat is larger, in proportion, than that of other Whales. One of these animals was thrown ashore at Nice, in the month of November, 1736.

When the upper part of the head was opened, it was found to contain spermaceti, which lay in a mass of two feet in thickness in the usual locality. The blow-hole is graphically termed the "lum" or chimney.

In concluding this brief history of the Whales, it must be once more remarked, that, in spite of the earnest labours of many excellent observers, our knowledge of these wondrous creatures is as yet exceedingly rudimentary, and even the genera are not clearly ascertained. The native Greenlanders seem to possess a very large amount of information on this subject, and are extremely accurate in their knowledge of the various Whales and their habits. It has therefore been happily suggested, that succeeding voyagers should take advantage of this circumstance, and should use their best endeavours to extract from those illiterate, but very practical savages, the knowledge which they really possess.

ZIPHIUS.—*Ziphius Sowerbiensis.*

DOLPHINS.

THE MEMBERS of this family do not possess the enormous head which characterises the true whales, and have teeth in both jaws, although they are liable to fall out at an early age. The blow-holes are united together, so as to form a single lunate opening, which is set transversely on the crown of the head. When first born, the young Dolphins are remarkable for their very great proportional dimensions, being little less than one-fourth the size of the parent, and affording a wonderful contrast to the marsupials, whose young are of such minute proportions when first born.

From the circumstance that the lower jaws are only furnished with two teeth, the rare and curious animal which is represented in the accompanying illustration is sometimes scientifically termed the Diodon, or two-toothed animal. But as this generic title has already been appropriated to the urchin-fishes, the name has been more recently changed into Ziphius.

In the animals which belong to this genus, the spout-holes are placed upon the top of the head, the throat is furnished with two diverging furrows, and the teeth are only two in number, rather large in proportion to the skull, slightly curved and compressed, and are situated in the middle of the lower jaw. The pectoral fins are placed rather low, and their shape is oval, tapering towards their extremities. Sowerby's Ziphius is so called, because that well-known naturalist figured and described the animal in the British Miscellany. His description was founded upon a specimen that was cast ashore upon the estate of Mr. J. Brodie, in Elginshire. The skull of this individual was preserved by Mr. Sowerby in his museum, and after his death it was placed by Dr. Buckland in the Anatomical Museum at Oxford. As it is so valuable a specimen, it has been industriously multiplied by means of plaster casts, which have been distributed to various scientific institutions.

The length of the creature was sixteen feet, and its girth at the largest part of the body was eleven feet. The head is small, narrow, and pointed, and the lower jaw is longer, blunter, and wider than the upper jaw, so that when the mouth is closed, the lower jaw receives the upper. In the upper jaw there are two depressions corresponding with the teeth, and permitting the perfect closing of the mouth. The colour of the animal is black on the upper surface and grey below, and is remarkable for the pellucid and satin-like character of the skin, which reflects the rays of the sun to a considerable distance. On reference to the illustration, the reader will notice that the body is marked like watered silk. This effect is produced by a vast number of white streaks immediately below the skin, which are drawn irregularly over the whole body, and at a little distance appear as if they were made by means of some sharp instrument.

Nothing is known of the habits of this curious animal, which is unknown to science, except by means of the specimen above mentioned.

THE word NARWHAL is derived from the Gothic, signifying " Beaked-whale," and is a very appropriate term for the SEA UNICORN, as the animal is popularly entitled. The head of the Narwhal is round, and convex in front, the lower jaw being without teeth, and not so wide as the upper jaw. From the upper jaw of the Narwhal springs the curious weapon which has gained for the animal a world-wide reputation.

In the upper jaw of the young or the female Narwhal are found two small, hollow tusks, imbedded in the bone, which, in the female, are generally undeveloped throughout the whole of the animal's existence, but in the male Narwhal are strangely modified. The right tusk remains in its infantine state, excepting that the hollow becomes filled with bony substance ; but the left tusk rapidly increases in length, and is developed into a long, spiral, tapering rod of ivory, sometimes attaining to the length of eight or ten feet. The tusks are supposed to be formed by an excessive growth of the canine teeth, and not of the incisors, as might be supposed from the position which they occupy in the jaw.

The use of this singular tusk is very obscure, for if it were intended to serve some very important object, such as the procuring of food, it is evident that the females would need its aid as much as their companions of the opposite sex, for both sexes feed on the same food, and inhabit the same localities, at the same time. A very plausible conjecture has been offered, to the effect, that the " horn " is useful in the light of an auger, with which the animal is enabled to bore breathing-holes through the ice-fields, whenever it finds itself in want of air beneath those vast frozen plains. But this theory is equally liable to the objection, that the females want to breathe as much as the males, and would stand in equal need of so indispensable an apparatus.

That the " horn " is employed in some definite task, is evident from the fact, that its tip is always smooth and polished, however rough and encrusted the remainder of its length may be.

The male Narwhal may perhaps use the tusk as a weapon of war, wherewith to charge his adversaries, as a mediæval knight was wont to charge with shield on breast and lance in rest ; and if that be the case, the weapon is truly a terrible one. This conjecture derives some force from the fact, that a herd of these aquatic spearmen have been seen engaged in sportive pastime, crossing their ivory lances, and seeming to fence with them, as the white weapons clashed against each other. The play of animals, not to mention mankind, is almost invariably founded on the spirit of combativeness, and generally consists in a sham fight ; so that the Narwhal " horn " may probably be analogous to the tusks of boars and the horns of deer, and be given to the animal as an offensive weapon,

NARWHAL.—*Mónodon Monóceros.*

wherewith he may wage war with those of his own species and sex who arouse his feelings of jealousy, or would interfere with his supremacy.

The food of the Narwhal consists chiefly of marine molluscs and of occasional fish, but is found to be generally composed of the same kind of squid, or cuttle-fish, which supplies the gigantic spermaceti whale with subsistence. As the remains of several flat fish have been discovered in the stomach of the Narwhal, it was supposed by some authors that the animal made use of its tusk as a fish-spear, transfixing them as they lay "sluddering" on the mud or sand, after their usual fashion, thus preventing their escape from the toothless mouth into which the wounded fish are then received. However this may be, the force of the tusk is terrific when urged with the impetus of the creature driving through the water at full speed, for the whole combined power of the weight and velocity of the animal is directed along the line of the tusk. A Narwhal has been known to encounter a ship, and to drive its tusk through the sheathing, and deeply into the timbers. . The shock was probably fatal to the assailant, for the tooth was snapped by the sudden blow, remaining in the hole which it had made, and acting as a plug that effectually prevented the water from gaining admission into the vessel.

In some rare instances the right tusk has been developed instead of the left, and it is supposed that if the developed tooth should be broken, the right tusk becomes vivified, and supplies the place of the damaged weapon. One remarkable case is known where both tusks were almost equally developed, being rather more than ten inches in length; and another example is recorded of a Narwhal which possesses two long tusks, the one being seven feet five inches in length, and the other seven feet. These tusks diverge slightly from each other, as their tips are thirteen inches asunder, though there is only an interval of two inches between their bases. Both these specimens were females. Sometimes the female Narwhal possesses a spear like her mate, but this circumstance is probably the effect of age, which in so many creatures, such as the domestic fowl, gives to the aged female the characteristics and armature of the male.

As both these double-tusked Narwhals were females, it may be probable that they owed their unusual weapons to some peculiarity in their structure, which prevented them from becoming mothers, and forced the innate energies to expend themselves in the development of tusks instead of the formation of offspring. The tusks of male swine and other animals, the horns of male deer, the mane of male lions, and other similar structures, appear to be safety valves to the vital energies, which in the one sex are occupied in the continual formation of successive offspring, and in the other find an outlet in the development of tooth, horn, and hair, according to the character of the animal. In all probability, the health of the animal would greatly suffer if the calcareous and other particles which are deposited in the tusk were forced to remain in the system instead of being harmlessly removed from it and placed upon its exterior.

The ivory of the Narwhal's tusk is remarkably good in quality, being hard and solid, capable of receiving a high polish, and possessing the property of retaining its beautiful whiteness for a very long period, so that a large Narwhal horn is of no inconsiderable commercial value.

But in former days, an entire tusk of a Narwhal was considered to possess an inestimable value, for it was looked upon as the weapon of the veritable unicorn, reft from his forehead in despite of his supernatural strength and superhuman intellect. Setting aside the rarity of the thing, it derived a practical value from its presumed capability of disarming all poisons of their terrors, and of changing the deadliest draught into a wholesome beverage.

This antidotal potency was thought to be of vital service to the unicorn, whose residence was in the desert, among all kinds of loathsome beasts and poisonous reptiles, whose touch was death and whose look was contamination. The springs and pools at which such monsters quenched their thirst were saturated with poison by their contact, and would pour a fiery death through the veins of any animal that partook of the same water. But the unicorn, by dipping the tip of his horn into the pool, neutralized the venom, and rendered the deadly waters harmless. This admirable quality of the unicorn-horn was a great recommendation in days when the poisoned chalice crept too frequently upon the festive board ; and a king could receive no worthier present than a goblet formed from such valuable material.

Even a few shavings of unicorn-horn were purchased at high prices, and the ready sale for such antidotes led to considerable adulteration—a fact which is piteously recorded by an old writer, who tells us that "some wicked persons do make a mingle-mangle thereof, as I saw among the Venetians, being as I here say compounded with lime and sope, or peradventure with earth or some stone (which things are apt to make bubbles arise), and afterwards sell it for the unicorn's horn." The same writer, however, supplies an easy test, whereby the genuine substance may be distinguished from the imposition. "For experience of the unicorn's horn to know whether it be right or not ; put silk upon a burning coal, and upon the silk the aforesaid horn, and if so be that it be true, the silk will not be a whit consumed."

The native Greenlanders hold the Narwhal in high estimation ; for, independently of its value, it is welcomed on each succeeding year as the harbinger of the Greenland whale.

The Narwhal is, however, of the greatest service to the Greenlanders, for its long ivory tusk is admirably adapted for the manufacture of various household implements and of spear-heads, so that it is the sad fate of many a Narwhal to perish by means of the tooth that has been extracted from its near kinsman. It is easily killed, as it possesses no very great power of diving, and is soon tired out by means of the inflated buoys which are attached to the harpoon, and offer so great a resistance to the water. It seldom descends above two hundred fathoms below the surface, and when it again rises is so fatigued that it is readily killed by a sharp spear.

The oil which is extracted from the blubber is very delicate, but is not present in very great amount, as the coating of fatty substance is seldom more than three inches in depth. About half a ton of oil is obtained from a large specimen. The flesh is much prized by the natives, and is not only eaten in its fresh state, but is carefully dried and prepared over the fire.

The colour of this animal is almost entirely black upon the upper surface of the body, but is slightly varied by streaks and patches of a deeper tint. The sides fade into greyish-white, diversified with sundry grey marks, and the under portions of the body are white. The tints of the Narwhal are rather variable even in the same individual, which assumes different hues at different stages of its existence. There is no fin upon the back of the Narwhal, but its place is indicated by a fold or ridge of skin, which runs down the centre of the back, and in old specimens appears to have been subjected to hard usage. The pectoral fins are very small in proportion to the size of the animal, and appear to be of little service to the owner except for the purpose of preserving the balance of the body. In the upper jaw there are two other teeth beside the tusks, which are concealed in the gum and are supposed to be false molars.

The Narwhal is a gregarious animal, being seldom seen alone, and generally associating in little companies of fifteen or twenty in number. It seems to be gifted with a considerable amount of curiosity, as Sowerby mentions that several of these animals pursued the ship from some distance, diving below the strange monster and playing near the rudder. It is seldom found in southern latitudes, although it has two or three times been driven upon our coasts, but is seen in great numbers between the seventeenth and eighteenth degree of north latitude. The length of an adult Narwhal is about thirteen feet.

THE BELUGA, which is sometimes called the WHITE WHALE on account of the colour of its skin, is an inhabitant of the higher latitudes, being found in great numbers in Hudson's Bay and Davis' Straits, and is also known to frequent the northern coasts of Asia and America, being especially fond of the mouths of large rivers.

Although it has no love for the warmer seas, it has been found upon our own coasts, where it has fallen a victim to the wrath of the fishermen. A Beluga was seen almost daily in the Frith of Forth for nearly three months, taking advantage of the tide to pass up the Frith, and always securing its retreat before the water had sunk sufficiently to prevent it from escaping seawards. It was at last killed by means of spears and guns, and its body was very carefully examined by competent anatomists.

The head of the Beluga is short and rounded, the forehead being convex, and the lower jaw not so wide as the upper. Both jaws of this animal are well supplied with teeth, some of which have a tendency to fall out as the animal increases in years, and are generally wanting in the jaws of those specimens which are captured. The dorsal fin is absent, and the pectorals are tolerably large, thick, powerful, and rounded. The thick and powerful tail is bent under the body while the animal is swimming, and is used with such singular strength that the creature shoots forward with arrowy speed, whenever it is alarmed or excited.

The colour of this animal is generally a cream-white, but in some specimens the cream tint is dashed with red, and becomes a pale orange. When young, the Beluga is marked with brown spots, the general hue of the body being a slaty grey. The average length of an adult male is about eighteen or twenty feet. The eye of this animal is hardly larger than that of man, and the iris is blue. The food of the Beluga consists chiefly of marine fish, such as cod, haddocks, and flounders, which are easily caught by these active and voracious animals.

On account of their speed and agility, the whalers seldom attempt their capture, although their boldness is so great as to induce them to follow the boats in considerable numbers, and to play all kinds of antics within easy reach of a harpoon. As, however, they are so swift and agile as to elude the barbed steel, or to shake it from its hold if it should happen to strike them, the harpooner seldom runs the risk of losing time and patience in the chase of such a creature. Moreover, even were the animal fairly struck and secured, the blubber is not of sufficient value to repay the crew for their expenditure of time, labour, and personal risk.

The oil that is furnished by the Beluga is of very good quality, although small in quantity, and is sufficiently valuable to lead to the establishment of regular Beluga hunts in the great rivers of northern America, which they ascend for some distance in search of prey.

These hunts are often very successful, and furnish a large supply of oil and skin. As soon as a Beluga is seen in the river or inlet, its retreat is cut off by strong nets

BELUGA.—*Beluga Cátorlon.*

laid from bank to bank. A number of boats then start in chase of the animal, while others are stationed along the line of nets, and manned with well-armed crews. When alarmed by the boats, the Beluga makes for the sea, but is interrupted by the nets, which bar its farther progress seawards, and soon falls a victim to the bullets, spears, and other missiles which are rained upon it by its assailants. Sometimes the harpoon and rope are employed in this chase, and small specimens are occasionally taken by means of hooks baited with fish. Those Belugas which are taken in the St. Lawrence are seldom more than fifteen feet in length.

The skin of the Beluga is subjected to the process of tanning, and can be made into a peculiarly strong though soft leather, which is said to be able to resist an ordinary musket-ball. The flesh is held in some estimation, and is said to bear some resemblance to beef. Its oily flavour, however, which cannot easily be destroyed, would render it unpalatable to European palates. When prepared with vinegar and salt, it is thought to be equal to the best pork. The fins and tail are considered as the best portions.

It is a migrating animal, visiting the western coasts of Greenland at regular intervals, reaching that locality about the end of November. It swims in large herds, and is of exceeding value to the natives, who not only eat its flesh, and burn and drink its oil, but employ the sinews for thread, which may be made of any degree of fineness by splitting the tendons, and use the delicate internal membranes for windows to their huts. The coming of the Beluga is anxiously expected by the Greenlanders, as their provisions generally run short about the end of November, and are replenished by the flesh of their welcome visitor.

Most familiar of all the Dolphin fraternity is the well-known Porpoise, or Sea Hog, an animal which may be seen on any of our coasts, tumbling about on the waves, and executing various gambols in the exuberance of its sportive feelings.

Before steamboats came into general use, the Porpoises were constant attendants on the packet and passenger boats, sometimes pursuing the vessels from sheer curiosity, and at other times following in their wake in hopes of picking up the garbage that might be thrown overboard. In the present day, however, the Porpoises are so frightened at the paddles and screws, that they remain at a respectful distance from the boats, content themselves with keeping pace with the vessels for a course of some miles, and then drop astern to rejoin their companions.

The Porpoise is a very gregarious animal, herding together in large shoals, and sometimes swimming in "Indian file" as they shoot over the surface of the sea; just showing their black and glossy backs above the water, and keeping such excellent line that they seem to be animated by one spirit and one will.

As might be presumed from the formidable array of sharp teeth with which the jaws are studded, and which are so arranged that the upper and lower sets interlock when the animal closes its mouth, the food of the Porpoise consists entirely of animal substances, and almost wholly of fish, which it consumes in large quantities, much to the disgust of human fishermen. Herrings, pilchards, sprats, and other saleable fish, are in great favour with the Porpoise, which pursues its finny prey to the very shores, and, driving among the vast shoals in which these fish congregate, destroys enormous quantities of them. The fish are conscious of the presence of their destroyer, and flee before it in terror, often

PORPOISE, OR PORPESSE.—*Phocæna communis.*

flinging themselves into the certain death of nets or shallow water in their hope to escape from the devouring jaws of the Porpoise. Even salmon and such large fish fall frequent victims to their pursuer, which twists, turns, and leaps with such continuous agility that it is more than a match for its swift and nimble prey. Not even the marvellous leaping powers of the salmon are sufficient to save them from the voracious Porpoise, which is not to be baffled by any such impotent devices.

The Porpoise seems to keep closely to the coasts, and is seldom seen in mid-ocean. It appears to be a migratory animal, as the season of its disappearance from one locality generally coincides with its arrival on some other coast. It is very widely spread, appearing to inhabit with equal security the warm waters of the Mediterranean, the cool seas of our own coasts, or the icy regions of the high latitudes.

Many of these animals have been found in our rivers, which they have evidently ascended with the idea of varying their diet by a few meals on fresh-water fish, or in hope of chasing the salmon into their spawning grounds. The Thames was in former days noted for the number of Porpoises which have been seen in its waters, one animal having ascended the river until it passed London Bridge. It is now, however, a long time since a

Porpoise made its appearance in the "silver Thames;" for the combined influences of steamboats and sewerage afford a most effectual barrier to the presence in our beautiful river of any animal which could in any way remain in the sweet waters of the open sea. Other less maltreated rivers are often honoured with the visits of Porpoises; and it is seldom that a year passes away without several notices in local newspapers of Porpoises which have been seen or captured in fresh water.

The length of a full-grown Porpoise is extremely variable, the average being from six to eight feet. In this animal the very great size of the new-born young is very remarkable. A mother-Porpoise and her new-born offspring were taken in the Frith of Forth in 1838. The length of the mother was four feet eight inches, and her girth two feet ten inches; while the length of her offspring was two feet ten inches, and its girth sixteen inches. On the nose of the young Porpoise there are always two thick bristles, which fall out as the creature advances in age, and cause two depressions, which have erroneously been taken for nostrils. The colour of the Porpoise is a blue-black on the upper surface of the body, and a bright silvery white below; so that when the animal executes one of its favourite gyrations the contrasting tints produce a strange effect as they rapidly succeed each other. The iris of the eye is yellowish.

The word Porpoise is corrupted from the French term "Porc-poisson," i.e. Hog-fish, and bears the same signification as its German name, "Meerschwein."

When the skin of a Porpoise is removed from the body, a layer of white fat is seen lying upon the flesh, about an inch in depth. This fatty layer melts into oil when subjected to the action of heat, and is very fine and delicate in its quality. In common with the oil of other of the Porpoises, it contains a peculiar volatile acid, which can be separated from the fat by chemical agency, and is termed phocenine. The odour of this substance is very powerful, and its taste is acrid and aromatic. It does not congeal even when its temperature is lowered to fourteen degrees above zero, and its boiling point is higher than that of water.

The skin of the Porpoise is well suited for tanning, and can be manufactured into valuable leather. As it is naturally too thick and heavy for this process, it is planed down until it becomes partially transparent, and is then employed for covering carriages, as well as for some articles of apparel.

In former times the flesh of the Porpoise was valued very highly, and was looked upon as a regal dish, being cooked with bread-crumbs and vinegar. Before it is dressed it is very unpleasing to the eye, being dark-coloured, coarse-looking, and evidently too full of blood; and its flavour when cooked is said to be coarse and unpleasant. As the Porpoise was conventionally considered as a fish, in common with the otter, seal, and certain sea-fowl, by the ecclesiastical rulers of the land, its flesh was a great boon to those who cared not for a fish diet on the multiplied meagre days which studded the calendar, and at the same time were too reverential towards their ecclesiastical superiors to eat that which was openly considered as butchers' meat.

On examining the jaws of a Porpoise, we find them to be closely set with rather long, sharp, compressed, and formidable teeth, variable in number, but always multitudinous. There are seldom less than eighty teeth in a Porpoise's mouth, and in the jaws of the female specimen which has already been mentioned there were no less than one hundred and two teeth, twenty-six on each side of the upper jaw, and twenty-five on each side of the lower. In these animals there is no perceptible distinction between the teeth, incisors, canines, and molars being all alike.

THE head of the GRAMPUS is more rounded than that of the porpoise, and its forehead is more convex. There are several species of Delphinidæ which are called by the name of Grampus, the best known of which is the ordinary or common Grampus.

It is a decidedly large animal, an adult specimen measuring from twenty to thirty feet in length, and from ten to twelve in girth. The teeth are not nearly so numerous as in the porpoise, being only forty-four in total number, eleven at each side of each jaw. In shape they are somewhat conical, strongly made, and slightly curved. The colour of the Grampus is black on the upper part of the body, suddenly changing into white on the abdomen and part of the sides. There is generally a white patch of considerable size immediately above and rather behind the eyelid.

The name Grampus appears to be a corruption of the French word " Grand-poisson," just as porpoise is a transmuted form of " porc-poisson."

Although it sometimes wanders into more southern regions, its favoured home is in the northern seas that wash the coasts of Greenland and Spitzbergen, where it congregates in small herds. It is a very wolf in its constant hunger, and commits great havoc among the larger fish, such as the cod, the skate, and the halibut, caring little for the smaller fry. At times it is said to make systematic attacks on seals, by startling them from their slumber as they lie sunning themselves on the rocks or ice, and seizing them as the half-sleeping animals plunge instinctively into the sea. Even the smaller porpoises and dolphins fall victims to the insatiable appetite of the Grampus, as has been proved by the discovery of their remains in the dissected stomach of one of these animals.

It has been frequently seen on the British coasts, and on more than one occasion has been killed in the rivers which it had ascended in chase of its prey.

GRAMPUS.—*Delphinus Grampus.*

A Grampus was captured nearly opposite Greenwich Hospital in 1772, and was so swift and powerful, that after it had been struck with three harpoons, and covered with lance wounds, it twice dragged the boat from Blackwall to Greenwich, and once ran as far as Deptford, going at a rate of eight miles per hour against the tide. The struggles of the wounded animal were so formidable, that none of the boats could approach it. Several other specimens of this animal have been caught in the same river at different times, one being twenty-four feet in length, and another measuring more than thirty feet.

The Frith of Forth seems to be a favourite spot for these errant cetacea, which are evidently attracted by the salmon and other fresh-water fish which may be found in those waters.

It is said that the Grampuses are fond of amusing themselves by mobbing the Greenland whale, just as the little birds mob owls when they venture forth in the daytime, and that they persecute it by leaping out of the water and striking it sharply with their tails as they descend. The Americans, in consequence, have called it by the name of Thresher, or Killer. The sword-fish is reported to join the Thresher in this amusement, and to prevent the whale from diving by attacking it from below. Whatever credence may be given to the latter part of the story, the former is certainly true, and is corroborated by Captain Scott, who tells me that he has often seen the Thresher engaged in this strange amusement, springing high out of the water and delivering the most terrific blows with its tail on the object of its pursuit. For the co-operation of the sword-fish he does not vouch, but has remarked that the whale does not seek refuge in the ocean depths when thus persecuted, but makes short and hurried attempts to dive, seeming to be prevented from making its escape by some allied force beneath.

APART from the marvellous tales which were once rife concerning the beauty and accomplishments of the DOLPHIN, the animal is well worthy of notice, and needs no aid of fictitious narrative to enhance its value in the eyes of the naturalist or the observer.

The Dolphin is remarkable for the enormous number of teeth which stud its mouth, no less than forty-seven being found on each side of both jaws, the full complement being one hundred and ninety. In the head of one specimen were found fifty teeth on each side of each jaw, making a complement of two hundred in all. Between each tooth there is a space equal to the width of a single tooth, so that when the animal closes its mouth the teeth of both jaws interlock perfectly. All the teeth are sharply pointed and flattened, and slightly curved backward, so that the entire apparatus is wonderfully adapted for the retention of the slippery marine creatures on which the Dolphin feeds. Fish of various

DOLPHIN.—*Delphinus Delphis.*

kinds form the usual diet of the Dolphin, which especially delights in the flat fishes of our coasts, and often prowls about the shoals of herrings and pilchards that periodically reach our shores.

The Dolphin is not a very large animal, measuring, when full grown, from six to ten feet in length, seven feet being the usual average. Its colour is black upon the back, and silvery-white on the abdomen, while the flanks are greyish-white. There is a peculiar satin-like sheen upon the skin when the animal is submerged beneath the water or freshly removed from the sea, but which rapidly disappears as the skin becomes dry. The beautiful colours which have been said to play about the body of a dying Dolphin are not entirely mythical, but belong rightly to one of the fishes, the coryphene, or dorado, which is popularly called the Dolphin by sailors.

The eyes of the Dolphin are small, and are supplied with eyelids ; the pupil of the eye is heart-shaped. The ears have but a very minute external aperture, barely admitting an ordinary pin, so that its sense of hearing appears to be very limited.

In former days the flesh of the Dolphin was thought to be a very great luxury, so great, indeed, that a Dolphin was considered as a noble present to be made to the Duke of Norfolk by Dr. Caius, the well-known founder of the college bearing his name. As the

1.   N N

Dolphin, in common with the porpoise and all the cetaceans, was considered as belonging to the fishes, its flesh was a permitted diet upon maigre days, when all flesh meat was rigidly forbidden by ecclesiastic prohibition, and was served to table with a sauce composed of bread-crumbs, vinegar, and sugar. Now-a-days, however, the flesh of the Dolphin has fallen entirely into disrepute as an article of diet, and is not to be restored to its former station even by the force of prelatical discipline.

It is a lively and playful animal, and being remarkably active in its native element, is fond of gambolling among the waves, and engaging in various sports with its companions. Being of a very gregarious nature, it is seldom seen alone, but prefers to associate in little flocks or herds, and is in the habit of accompanying ships for considerable distances, hovering about the vessel and executing various strange manœuvres. Sometimes it falls a victim to its curiosity, and when paying too close a visit to the vessel is struck with the "grains," or barbed trident, which is kept on board in readiness for such an occasion, and is hauled struggling on deck, where it is soon deprived of life.

The formation of the Dolphin's brain is of such a nature that it indicates great intelligence on the part of its possessor, and goes far towards confirming some of the current reports on this subject. It is said that Dolphins have been tamed and taught to feed from the hand of their instructor, beside performing sundry feats at his bidding. That the seals are eminently capable of instruction is a well-known fact, and it is probable that the Dolphins may not be less endowed with intellectual powers.

From the peculiar shape of the snout and jaws, which are rather flattened and considerably elongated, the animal has derived its French titles of "Bec d'Oie" and "Oie de Mer," i.e. Goose-beak, or Sea-goose. The forehead is rather rounded, and descends suddenly towards the base of the "beak." The "beak" itself is about six inches in length in a moderately sized specimen, and is separated from the forehead by a small but distinct ridge. The Dolphin only produces a single young one at a time, and nurses her offspring with exceeding tenderness and assiduity.

The common Dolphin is found in the European seas, and in the Atlantic and the Mediterranean, and may possibly have a still wider range. There are Dolphins to be found near the coasts of Africa and America, but whether they belong to the same species as the common Dolphin is at present a mooted point.

BOTTLE-NOSED DOLPHIN.—*Delphinus Tursio.*

In the Bottle-nosed Dolphin there is not such an extraordinary array of teeth as in the preceding animal, their maximum number being one hundred, and their average about eighty-five. The average length of this animal is between seven and eight feet, although specimens have been taken which measured between ten and eleven feet in length.

The colour of the Bottle-nosed Dolphin is rather different from that of the common Dolphin. Its back is not of the same jetty hue, but is deeply tinged with purple, its flanks are dusky, and the under portions are greyish-white, and do not glisten with the pure silvery-white of the ordinary Dolphin of our coasts.

Although it is a rare animal, it has more than once been captured upon our coasts, one specimen having been taken in the river Dart in Devonshire, and another in the river at Portsea. Two more Bottle-nosed Dolphins, a mother and her young one, were caught upon the sea-coast near Berkely, where they had been seen for several days haunting the neighbourhood. The first of these specimens was captured when it had ascended the river about five miles, and was so powerful and active that it did not resign its life until it had fought for a space of four hours against eight men armed with spears and guns, and assisted by dogs. While struggling with its foes it bellowed loudly, making a sound like that of an enraged bull. This individual was more than eleven feet in length.

In many instances the teeth of the Bottle-nosed Dolphin are extremely blunt, a circumstance which was once thought to be peculiar to the species. Mr. Bell, however, proves to the contrary by the fact of possessing two skulls of Bottle-nosed Dolphins, in which the teeth are of the usual length, and as sharp as in the ordinary Dolphin. When the teeth are thus worn down, the creature is unable to interlock them rightly, as the narrow portion of the teeth has been ground down, and the interstices are too narrow to receive the wide stumps. The name of Blunt-toothed Dolphin has been given to this animal on account of the supposed normal shape of the teeth. The lower jaw of this species projects rather beyond the upper.

THERE is a curious animal belonging to this family, which inhabits the Ganges, and is known by the name of the Soosoo.

It is remarkable for the curious shape of its "beak," which is long, slender, compressed at the sides, and is larger at the extremity than in the middle. The number of its teeth is about one hundred and twenty. It is a swift and powerful, but at the same time a sluggish animal, appearing to partake largely of the curious mixture of sloth and energy which is found in the huge lizards that frequent the same river, and never caring to exert itself except in chase of its prey. Its colour is greyish-black upon the back, and white on the abdomen. The eye is wonderfully small, being only one-eighth of an inch in diameter in a Soosoo which measures four or five feet in length. There is no dorsal fin, its place being indicated by a small projection.

SÍRENIA.

THE small but singular group of animals that are classed together under the title of the SIRENIA, are so formed that anatomists have had much difficulty in deciding upon their proper position in the animal kingdom. Many parts of their structure exhibit so strong an affinity to the pachydermata, or thick-skinned mammalia, that they have been placed next to the elephants by some zoologists, while their fish-like form and aquatic habits have induced other writers to place them in the position which they now occupy in the British Museum. They feed chiefly on vegetable substances, and find the greater part of their subsistence in the thick herbage that edges the waters where they reside. Their nostrils are placed at the extremity of the muzzle, as is the case with most mammalia, and they are never employed as blow-holes, after the manner of the cetaceans.

THE MANATEE, or LAMANTINE, is a very strange-looking creature, appearing like a curious mixture of several dissimilar animals, the seal and the hippopotamus being predominant.

There are several species of Manatee, two of which are found in America and one in Africa, but always on those shores which are washed by the waters of the Atlantic Ocean.

The common Manatee is generally about nine or ten feet in length, and is remarkable for the thick fleshy disc which terminates the muzzle, and in which the nostrils are placed. It is found in some plenty at the mouths of sundry large rivers, such as the Orinoko or the Amazon, and feeds upon the algæ and other herbage which grows so plentifully in those regions. By some writers the animal is said to leave the water entirely, and to search for its food upon the land, but this assertion is now ascertained to be incorrect. It is, however, in the habit of crawling partly out of the water, and has a strange custom of elevating its head and shoulders above the surface in such a manner that it bears some resemblance to a human being.

The flesh of this animal is said to be well flavoured, and as the Manatee is ecclesiastically reckoned as a fish, together with the whales, seals, and other water-loving creatures, it is permitted as a lawful article of diet on fasting days. When properly salted and preserved by drying in the sun, the flesh of this animal will remain sweet for a whole year. The skin of the Manatee is in great request for the formation of sundry

MANATEE.—*Mánatus Austrális.*

leathern articles in which great strength is required, and the oil which is extracted from its fat is of excellent quality, and is free from the unpleasant rancid odour which characterises so many animal oils.

So valuable an animal is subject to great persecution on the part of the natives, who display great activity, skill, and courage in the pursuit of their amphibious quarry. The skin of the Manatee is so thick and strong that the wretched steel of which their weapons are composed,—the "machetes," or sword-knives, with which they are almost universally armed, being sold in England for three shillings and sixpence per dozen,—is quite unable to penetrate the tough hide. Nothing is so effectual a weapon for this service as a common English three-cornered file, which is fastened to a spear-shaft, and pierces through the tough hide with the greatest ease. The skin of the Manatee is so thick that it can be cut into strips like the too-celebrated "cow-hide" of America, which is manufactured from the skin of the hippopotamus. Before being dressed, the hide of the Manatee is thinly covered with rather stiff bristles.

The DUGONG may easily be distinguished from the manatee by the formation of the tail, which in the latter animal is rounded, but in the former is forked. These animals are found on the eastern coast of Africa and on the shores of the Indian Ocean.

In Ceylon the Dugong is exceedingly plentiful, and it also inhabits the northern coasts of Australia, where it is assiduously chased by the natives. The name of Sirenia, which is given to this group of animals, is chiefly owing to the peculiar form and habits of the Dugong, which has a curious custom of swimming with its head and neck above the surface of the water, so that it bears some grotesque resemblance to the human form, and might have given rise to the poetical tales of mermaids and sirens which have prevailed in the literature of all ages and countries. When the female Dugong is nursing her child, she carries it in one arm, and takes care to keep the head of her offspring, as well as her own, above the surface of the water, and thus presents a strangely human aspect. If alarmed, she immediately dives below the waves, and flinging her fish-like

DUGONG.—*Halicore Dugong.*

tail into the air, corresponds in no inadequate degree with the popular notions of mermaid form.

The usual haunts of the Dugong are at the mouth of rivers or similar spots, where the subaqueous algæ grow in greatest profusion, and it never seems to be found where water is more than three or four fathoms in depth. It is not so good a diver as the seals, not being furnished with the peculiar blood-reservoirs which enable those animals to survive beneath the water for so great a length of time ; and it is therefore unable to seek its food except in shallow waters. Whole herds of these animals may be seen sporting near the shores, diving at intervals to procure food, and rising again in order to breathe. They are most affectionate creatures, and if one of a pair be captured, the other falls an easy prey to the pursuers, as it refuses to leave the fatal spot, and will rather suffer itself to be killed than forsake even the dead body of its late partner.

There are several species of Dugong, which are all very similar in habits, although they vary in size. One species has been known to attain to the length of twenty-six feet.

The skull of these animals is very singularly formed, the upper jaw being bent downward over the lower jaw, and terminated by two large incisor teeth. It is supposed that the object of this structure is to assist the animal in gathering together and dragging up by the roots the algæ and other subaqueous vegetation on which it feeds.

The skin of the Dugong is capable of being manufactured into various useful articles, and the flesh is in some repute, being said to bear close resemblance to veal.

A THIRD genus of these herbivorous cetaceans is the RYTINA, which is supposed to be now extinct, the last known specimen having been killed in 1768, only twenty-seven years after the creatures were discovered.

The Rytina possessed no true teeth, and masticated its food by means of two bony plates, one of which was attached to the front of the palate, and the other to the lower jaw. It was a large animal, measuring about twenty-five feet in length, and nearly twenty feet in circumference. The Rytina was discovered in the year 1741 on an island in Behring's Straits ; and as the animals were large, heavy, and unarmed, they were most valuable in affording food to the unfortunate sailors who were shipwrecked upon that island, and were forced to abide there for the space of ten months. When the islands were visited by ships in search of sea-otters, which abounded in that locality, the crews found the Rytinas to be so valuable and so easy a prey that the entire race was extirpated in a few years.

The only account of the Rytina is that which was furnished by Steller, one of the shipwrecked party, who, undaunted by the terrible privations which he was forced to undergo, wrote an admirable description of the animal, which was afterwards published in St. Petersburg.

GROUP OF RODENT ANIMALS.

RODENTS.

THE RODENTS, or gnawing animals, derive their name from the peculiar structure of their teeth, which are specially fitted for gnawing their way through hard substances. The jaws of the Rodents are heavily made, and very large in proportion to the head, their size being not only needful for the support of the gnawing teeth, but for their continual development. There are no canines, but a wide gap exists between the incisors and the molars, which are nearly flat on their surfaces, and are well suited for grinding the soft substances on which these animals feed.

The structure of the chisel-edged incisor teeth is very wonderful, and may be easily understood by inspecting the teeth of a rat, mouse, hare, or rabbit.

As their teeth are continually worn out by the severe friction which they undergo continually, there must needs be some provision for renewing their substance, or the creature would soon die of starvation. In order to obviate this calamity, the base of the incisor teeth pass deeply into the jaw-bone, where they are continually nourished by a kind of pulpy substance from which the tooth is formed, and which adds fresh material in proportion to the daily waste. Sometimes it happens that one of the incisor teeth is broken or injured by some accident, so that it offers no resistance to its corresponding tooth in the opposite jaw. The result of such an accident is very sad to the sufferer, and is not unfrequently fatal in its termination. For the unopposed tooth, being continually increased by fresh substance from behind, is gradually pushed forward until it attains an enormous length, having sometimes been known to form a complete circle. Examples of these malformed teeth are of tolerably frequent occurrence, and specimens may be seen in almost every museum of comparative anatomy.

Something more is needed for the wellbeing of the animal than the mere growth of its teeth ; for unless their chisel-like edges were continually kept sharp, they would be of little use for cutting their way through the hard substances which the Rodents are in the habit of gnawing. This result is attained as follows :—

The enamel which covers the front face of the incisor teeth is much harder than that which is laid upon the remaining surfaces, while the dentine which makes up the solid mass of each tooth is also harder in front than behind. It is evident that when these teeth are employed in their usual task, the softer enamel and dentine are worn away very much more rapidly than the remainder of the teeth, so that the peculiar chisel-edge of the teeth is continually preserved. Following—perhaps unconsciously—the structure of these teeth, our cutlers have long been accustomed to make their axes on the same principle, a thin plate of steel being inclosed within two thick plates of iron, so that when the axe is used upon timber, the iron is continually worn away, leaving the plate of steel to project, and form a sharp cutting edge. These teeth are well represented in the engraving of the beaver's skull, to which the reader is referred.

In many species of Rodents, the front faces of the incisor teeth are tinted with a light orange red, or a reddish-brown, by means of a very thin layer of coloured enamel. In order to enable these teeth to perform their office rightly, the lower jaw is jointed so as to slide backwards and forwards.

The Rodent animals are widely spread over the entire globe, and are very numerous, comprising nearly one-third of the mammalia.

FEW animals are so well known or so thoroughly detested as the common BROWN RAT, or NORWAY RAT, as it is sometimes erroneously called.

It has spread itself over almost every portion of the globe, taking passages in almost every ship that traverses the ocean, and landing on almost every shore which the vessel may touch. Wherever they set their feet, the Brown Rats take up their abode ; and, being singularly prolific animals, soon establish themselves in perpetuity. They are marvellous exterminators of other "vermin," and permit none but themselves to be in possession of the domain which they have chosen. It is a well-known fact that they have driven away the black English Rat, and established themselves in its place with wonderful rapidity, having been accidentally brought to our coast by some vessel in which they had embarked, and found the English climate to suit them as a permanent residence. Some of these animals were purposely introduced into Jamaica, in order to extirpate the plantation Rats, which did such damage to the growing crops. They soon drove away the original "vermin;" but like the Saxons when invited to help the Britons, or like the man who was requested to aid the horse against the stag, were found to be moredangerous foes than the enemy whom they had overcome.

The Brown Rat is well fitted for its exterminating mission, as it is a fierce and dangerous animal, and can inflict very painful wounds with its long incisor teeth. An unarmed man would be quite impotent against the attacks of even a small party of old sewer Rats, while a large body of these animals would make but short work of any man, however well he might be armed. There is a wonderful power of combination in the Brown Rat, which enables it to act in concert with its companions, and renders it a truly formidable animal when it chooses to make a combined attack upon man or beast. A number of these animals have been known to attack a cat, and inflict such grievous injuries that the poor creature had to be killed as soon as its evil plight was discovered by its owner. Even a single Rat is no despicable antagonist, and, according to the observations of practical men, could beat off a ferret in fair fight, and would foil any but a properly trained dog.

It is an exceedingly voracious animal, eating all kinds of strange food, and not sparing its own species in times of scarcity. Like the wolves, the Rats will always fall upon and devour one of their companions if it should chance to be wounded, and excite their

BROWN RAT.—*Mus Decúmanus.*

carnivorous passions by the sight and smell of flowing blood. If a Rat should be caught by a foot or a leg in a steel trap, its former companions will often fly upon the poor captive and tear it to pieces, instead of endeavouring to effect its release. As in such instances the imprisoned limb is left in the teeth of the trap, it has been erroneously supposed that the Rat had severed its own limb in order to set itself free.

From some strange cause, the male Rats far outnumber the females, the proportion being about eight of the former to three or four of the latter. This disproportion of the sexes may possibly be caused by the cannibalistic habits of the Rat, the flesh of the female being more tender than that of the opposite sex. Whatever may be the cause, it is clear that the wider increase of these creatures is greatly checked by the comparative paucity of females.

There is scarcely a greater plague to the farmer, butcher, sailor, provision merchant, or poultry keeper than the Rat, whose mingled craft, daring, and perpetual hunger require the greatest watchfulness and the most elaborate precaution. The havoc which an army of Rats will make among the corn-ricks is almost incredible, while they carry on their depredation with so much secrecy that an unpractised eye would think the stacks to be sound and unharmed. Fortunately they can easily be dislodged from any rick by taking it down, and replacing it on proper "staddles," taking great care that no stray weeds or branches afford a foothold to these persevering marauders. While the rick is being rebuilt, no particular care need be taken to shake the Rats out of the sheaves, for, as they are thirsty animals, they will be forced to leap from the stack in search of water, and then will not be able to return.

Mice can subsist in a stack by means of the rain and dew which moisten the thatch, and may be often seen licking the straws in order to quench their thirst. But the Rats are less tolerant of thirst, and are forced to evacuate their premises. When mice and Rats are found inhabiting the same stack, the former animals reside in the upper parts, and the Rats in the lower.

Poultry of all kinds suffer sadly from these carnivorous creatures, which have a custom of invading the henroosts by night, and making prey of the fowls as they sit quietly sleeping on their perches. Birds are always indisposed to move during the hours of darkness, so that the cunning rodent finds no difficulty in carrying out its destructive intentions. Rabbit-fanciers have great cause to be indignant with the Rat, for when it once gains access to the hutches, the safety of the entire stock is in imminent danger. The only way to secure the survivors is to remove them at once to some spot which is made Rat-proof. Much of the damage which is done by Rats is laid upon innocent shoulders, the fox and the weasel being the ordinary scape-goats.

The audacity of these animals is really wonderful, especially when they have enjoyed an unmolested life. They have been known to enter a stable and nibble the horn away from the horses' hoofs, or to creep among dogs as they lay sleeping, and gnaw the callous soles of their feet. They have even been known to attack sleeping infants, and to inflict fearful damage before they were detected in their crime. The metropolitan butchers execrate the Rats very sincerely, as they are forced to remove every joint of meat as soon as their business is over for the day, and hang it up in some place which is so well protected that not even a Rat can gain access. Indeed, the black list of their misdemeanors is so extremely long that even a rapid enumeration of their crimes would more than occupy the entire space devoted to one animal.

Rats are not without their use, especially in large towns, which but for their never-failing appetites would often be in very sad case. Taking, for example, the metropolis itself, we find that the sewers which underlie its whole extent are inhabited by vast hordes of Rats, which perform the office of scavengers by devouring the mass of vegetable and animal offal which is daily cast into those subterranean passages, and which would speedily breed a pestilence were it not removed by the ready teeth of the Rats. So that, when kept within proper bounds, the Rat is a most useful animal, and will continue to be so until the drainage of towns is conducted in a different manner.

How to keep them to their own proper dominions is no easy task, as their sharp teeth can cut through almost any obstacle, and have been known even to grate away the corner of a particularly hard brick. It is found, however, that if these tunnels be stopped up with mortar or cement well studded with pieces of broken glass, they will not venture their teeth against such a barrier. Moreover, if a few table-spoonfuls of quick-lime be placed in the hole before it is stopped up, it will deter the Rats from coming in that direction, as the lime burns their feet.

Catching them in traps is by no means so easy a process as it appears to be, as the Rat is a very crafty animal, and is moreover gifted with so acute a nose that it can perceive the touch of a human hand upon a trap, and will keep aloof from so dangerous an article. In order to set a trap properly, it is needful to avoid touching it with the bare hand, and to wear thick gloves powerfully scented with aniseed, carraway, or other powerfully-smelling substance. Even in that case, the successful chase of the Rat requires such an accurate knowledge of the animal's habits, and needs so many precautions, that it is almost impossible for an amateur to be permanently successful in that line.

Although the Sewer and the Barn Rats belong to the same species, they are very different in aspect as well as in habits, the former being very much larger than the latter, and much fiercer in disposition. The Sewer Rats remain in their strange habitations during their whole lives, while the Barn Rats are in the habit of making annual migrations as soon as the spring season commences, some betaking themselves to the fields and hedge-rows, while others take up their abode on the river-banks, where they commit sad havoc among the fish.

During this temporary migration, the female Rats make their snug and comfortable nest in any sheltered spot; and before the autumnal season has fairly commenced add a considerable number of new members to the Rat family. It is a wonderfully prolific animal, beginning to breed at four months of age, and having three broods in the year, each brood being from eight to twelve or fourteen in number. When the autumn has set in, the emigrants return to their old quarters, marvellously increased in number.

The female Rat is a most affectionate mother, braving all dangers in defence of her young, and dashing boldly at any real or fancied foe who may happen to alarm her maternal sympathies. Unfortunately for her peace of mind, the paternal Rat is far from partaking of these tender affections, and if he condescends to pay a visit to his young family, only does so with the intention of eating them. Should the mother be at home, she shows such a defiant front, that he is fain to decamp from the cradle of his offspring, but if she should perchance happen to be absent from her charge, the result is tragical indeed.

Rats are very cleanly animals, always washing themselves after every meal, and displaying the greatest assiduity in making their toilet. They also exhibit considerable delicacy of palate wherever they find a sufficiency of provisions, although they are in no way nice in their diet when pressed by hunger. If, for example, a party of Rats discover an entrance into a butcher's store-house, they are sure to attack the best parts of the meat, utterly disdaining the neck, the shin, or other coarse pieces.

There is one peculiarity in the structure of the Rat which is worthy of notice. These animals are able not only to ascend a perpendicular tree or wall by the aid of their sharp, hooked claws, but also to descend head foremost with perfect ease. In order to enable them to perform this feat, their hind legs are so made that the feet can be turned outwards, and the claws hitched upon any convenient projections.

However unpromising a subject the Rat may appear, it has often been tamed, and is a very much more educatable animal than could be supposed. It will obey its master's commands with promptitude, and has been known to learn very curious tricks.

For further information on this subject the reader is referred to a work published by Messrs. Routledge and Co., entitled "The Rat," by James Rodwell, in which may be found an elaborate account of the animal and its habits, together with much curious and original information.

There is a well-known proverb that Rats always desert a falling house; in which aphorism there is really much truth. One curious example thereof I here offer to the reader.

On page 204 may be seen an account of a cat which had, by some mysterious intuition, migrated from a mill in which she had long lived, and to which she was greatly attached, and which was burned to the ground in a few hours after she had taken her departure. Pussy, it seems, was not the only animal which had been thus forewarned of impending danger, for the Rats also took alarm, and were actually seen upon their journey from their late habitation. They were about one hundred in number, and, starting from the mill some two hours before the fire broke out, proceeded in a compact body towards four stacks belonging to the landlord of the Commercial Inn, and there took up their abode.

A similar account of Rat prescience has been narrated to me by a spectator of the scene.

When H.M.S. *Leander* was brought into harbour after her voyage, in the year 1803, she was so infested with Rats that a wholesale destruction of these four-footed pests was rendered absolutely necessary, not only for the comfort of the crew, but for the very safety of the vessel. The entire contents of the ship were therefore landed on the wharf, a number of chafing-pans filled with lighted brimstone were placed between decks, and the hatches being battened down, the animals were soon stifled by the suffocating vapours. As soon as the preparation for this wholesale destruction commenced, the Rats took alarm, and endeavoured to make their way on shore by traversing the "warps," or ropes by which the vessel was made fast to the shore. Sentinels were accordingly placed by the warps, and furnished with sticks, so that as soon as a Rat came running along the ropes, it was speedily checked by a sharp blow, which struck it from its foot-hold, and knocked it dead or dying into the water, where it soon perished.

It is a curious fact that the Rats were all found lying dead in circles round the braziers, heaped thickly upon each other's bodies. They had instinctively run towards the spots which were comparatively free from vapour, as the heat of the burning coals forced the suffocating smoke to rise from the spot where it was generated.

The BLACK RAT derives it name from the colour of its fur, which is of a greyish-black, instead of the reddish-brown hue which tinges the coat of the brown Rat. The upper jaw projects considerably beyond the lower, and a number of long stiff hairs project through the ordinary fur. In size it is rather less than the above-mentioned animal, and the ears and tail are rather longer in proportion.

The Black Rat is found in all warm and temperate regions, and in England was in former days extremely numerous, although it has now been gradually driven away from its domains by the larger and more powerful intruder. It is not, however, so rare as is sometimes imagined, and may still be found by those who know where to look for it. According to Mr. Rodwell's theory, the manner in which the brown Rat has supplanted its black relation is not by war, but by love, the stronger males of the brown hue

BLACK RAT.—*Mus Rattus.*

carrying off the black females by force of superior strength, and thus by degrees merging the weaker black race into the powerful brown Rats. In France the two varieties—for the distinction of species really seems to be but doubtful—live together in perfect harmony, mixing freely with each other, and producing a curious kind of parti-coloured offspring.

The skins of these brown-black Rats are considered to be of some value, and they are accordingly pursued by the rat-catchers for the purpose of sale to the furriers. Even the brown Rat is not without its value in commerce, as the prepared skin is said to furnish the most delicate leather for the manufacture of the thumbs of the best kid gloves; and the fur is used as a substitute for beaver in the composition of hats.

Albino, or White Rats, are of no very uncommon occurrence; and when crossed with the black or brown species, their offspring is curiously pied with a darker or lighter hue, according to the colour of the parent.

"Yᴇ little vulgar MOUSE," as it is quaintly termed by old Topsel, is a truly pretty little creature, with its brown-grey back, grey throat and abdomen, soft, velvety fur, its little bright black bead-like eyes, and squirrel-like paws. A detailed description of so familiar an animal would be quite unnecessary, and we will therefore proceed to its habits and manners.

COMMON MOUSE.—*Mus Musculus.* (Brown, White, and Pied varieties.)

Like the rat, it frequents both town and country, doing an infinity of damage in the former, but comparatively little harm in the latter. In the country it attaches itself mostly to farmyards, where it gains access to the ricks, and when once firmly established, is not so easily dislodged as its larger relative the rat. However, if the rick be kept under cover, the Mice cannot make any lengthened stay, for the cover keeps off the rain, on which they chiefly depend for drink, and they are then obliged to leave the stack in search of water. If the rick be placed on staddles, it will be then safe from these little pests.

In the town they are not so objectionable as in the country, for they can only annoy the human inhabitants, and cannot inflict real damage upon them. They are bold little creatures in their way, although easily startled; and, if permitted to carry out their noisy sports undisturbed, run about an inhabited room with perfect nonchalance. The walls of many of the college rooms at Oxford are papered over canvas, and the Mice run scuffling and squeaking between the canvas and the plaster as if they were the legitimate owners of the place, and the tenants were only located there in order to cater for their benefit. Many a wall is riddled with holes that have been made by the irritated occupants making furious lunges with a toasting-fork—always unsuccessful, by the way—at the noisy little creatures as they scurry about behind the paper.

They are odd little animals, and full of the quaintest gamesomeness, as may be seen by any one who will only sit quite still and watch them as they run about a room which they specially affect. They are to the full as inquisitive as cats, and will examine any new piece of furniture with great curiosity.

Mice are very easily tamed, and, as far as my own experience goes, the common brown Mouse is more readily brought under subjection, and more docile, than the white or albino variety. I have kept many a set of Mice, brown, white, and mottled, and have always found them to be very susceptible of kindness. To tame a young brown Mouse is an easy task; but it must be remembered that, as all Mice are very cleanly animals, the strictest care is needful to rid their cage of all impurity. Their bedding should be constantly changed, and the false floor of their cage should be double, so that, while one is in use, the other is getting dry after being thoroughly washed.   Any soft substance, such as hay, cotton wool, or rags will suffice for their bedding ; but I have found that black cotton wool, or black " wadding," as it is sometimes termed, is fatal to Mice in the course of a single night.   Why it should be so I cannot venture to guess, but that such is the case I have had practical experience.

Mice are cunning creatures, and when they once have taken alarm at a trap, cannot be induced to put themselves within such peril, no matter how strong the inducement may be.   For a while it is possible to entrap them by changing the kind of bait as soon as they have begun to learn the result of eating that particular substance ; but in a few weeks the trap must be entirely removed until the animals have forgotten it.

It is a marvellously prolific animal, producing its young several times in the course of the year, and at a very early age.   The nests are made in any sheltered spot, and formed from any soft substance, such as rags, paper, or wool, that the mother can procure.   On taking up some boards in my own room, I once found a Mouse-nest nearly as large as a man's head, composed wholly of scraps of paper, and containing six or seven tiny red, semi-transparent mouselets, through whose little bodies one could almost see the substance of the nest on which they were lying.   Another Mouse-nest which I discovered, was made in an old disused harmonicon, which had been put away in a cupboard, and was filled by the Mice with empty nutshells, the refuse of a bag of hazel-nuts which had been placed in the same cupboard ; no very enviable bed, as one would fancy, and the reason for its construction not at all obvious.

Before closing this account of the common Mouse, a few words are due to the " Singing Mice," concerning whose musical accomplishments the scientific world is rather at issue, some persons thinking the song to be nothing more than a symptom of bronchitis, and others believing it to be voluntarily produced by the imitative powers of the performers.

In a letter to the *Field* newspaper, one of the correspondents gives a curious instance of " singing " which favours the former of these suppositions.   A Mouse had been caught in a trap with weak springs, and being half choked by the wire pressing on its neck, gave vent to a twittering or chirruping not unlike that of a small bird.   Other correspondents, however, who have met with examples of singing Mice, seem rather to incline to the opinion that the musical sound is produced by healthy animals, and is not owing to disease. A very interesting letter on this subject has been sent to me by the Rev. R. L. Bampfield, of Little Barfield, in Essex, and seems also to favour the latter supposition.   By the kind permission of the writer, I am enabled to present the account to the reader, and will leave him to come to his own conclusions on the subject.

" In a former residence of mine, some Mice took up their abode behind the wainscot in the kitchen.   From motives which few housekeepers would appreciate, we allowed them to remain undisturbed ; and most merry, cheerful little creatures they were.

It seemed to us that a young brood was being carefully educated ; but they did not learn all their accomplishments from their parents.   In the kitchen hung a good singing canary, and we observed that by degrees the chirp of the Mice changed into an exact imitation of the canary's song ; at least it was so with one, for though several attempted it, one considerably excelled the rest.   I am not sure that admiration of the music influenced them, for from the funny facetious way in which it was done, I should rather say it was out of mockery, or at least from a love of imitation.   Yet the result was very pleasing ; far inferior to the canary's note in volume, strength, and sweetness, it was, perhaps, superior to it in softness and delicacy.

Often have I listened to it with pleasure in the evening, when the canary was asleep with its head beneath its wing ; and more than once have I observed a kitchen-guest glance at the canary, then look round in some astonishment and say, ' Is that a bird, sir, singing ? '   One trustworthy person assured me that he too had had in his house a similar ' Singing Mouse.'   I have, therefore, little doubt that if a young family of Mice were brought up from the first close to a canary or some other songster, some of them would learn to sing."

I have also been favoured with an account of a young singing rat, which endeavoured to imitate the sounds produced by a piping-bullfinch and an ordinary goldfinch. In the first, the creature entirely failed, but was tolerably successful in its imitation of the mild notes of the goldfinch. The same animal would begin to sing if a melody were played in the minor key, but would give no response to the major. The fondness of Mice for music is already well known, and may afford some clue to their sensitiveness of ear. I believe, by the way, that the untaught cries of all the lower animals, whether they be quadrupeds or birds, are in the minor key.

SMALLEST, and perhaps the prettiest, of the British mammalia, the elegant little HARVEST MOUSE next claims our attention. The total length of this tiny creature is not quite five inches, its tail being nearly two inches and a half in length. The colour of its fur is a delicate reddish-brown, the base of each hair taking the darker tint, and the point warming into red, while the under parts of the abdomen are white. The line of demarcation between the brown and white is well defined.

The description which is given of the Harvest Mouse and its wonderful nest, by the Rev. Gilbert White, is so well known that it need only be casually mentioned. I have fortunately had opportunities of verifying his observations by means of a nest which was found in a field in Wiltshire by some mowers.

It was built upon a scaffolding of four of the rank grass-stems that are generally found on the sides of ditches, and was situated at some ten or eleven inches from the ground. In form it was globular, rather larger than a cricket-ball, and was quite empty, having probably been hardly completed when the remorseless scythe struck down the scaffolding and wasted all the elaborate labour of the poor little architect. The material of which it was composed was thin dry grass of nearly uniform substance, and its texture was remarkably loose, so that any object contained in it could be seen through the interstices as easily as if it had been placed in a lady's open-worked knitting basket. There was no vestige of aperture in any part of it, so that the method by which it was constructed seems quite enigmatical.

I am inclined to suppose that the little builder remained in its centre while engaged in its construction, and after weaving it around her, pushed her way out through the loosely woven wall, and re-arranged the gap from the outside. It may be that the nest is the joint work of both sexes, the one remaining inside and plaiting

HARVEST MOUSE.—*Micromys minútus.*

the grass, while her mate brings fresh material, and consolidates the work from the exterior.

Perhaps the young Mice, when snugly packed into their airy cradle, may be fed by the mother from the exterior, by making a temporary opening opposite each little one, and replacing the material when she proceeds to the next in succession. This is Mr. White's suggestion, and seems to be a very probable one. He also wonders how the little nest, which was entirely filled by the bodies of the eight young Harvest Mice that lay cradled in its embrace, could expand so as to accommodate itself to their increasing dimensions. This problem may be answered by the fact that the loose structure of the nest is precisely calculated for such extension, for the materials are so interwoven that the entire structure can be greatly expanded from the interior without losing its spherical shape. Such, at all events, was the case in my own specimen, and is probably so in all.

Very little is known of the habits of the Harvest Mouse in a wild state, except that it is destructive to corn whether stored in ricks or barns. It is also carnivorous, or rather insectivorous, to no small degree, [as was proved by Mr. Bingley, who kept one of these little creatures, and was accustomed to feed it with various insects. This propensity was discovered by mere accident, the Mouse springing with wonderful activity at a blue-bottle fly that happened to buzz against the wires of her cage. Taking the hint, Mr. Bingley caught the fly, and holding it against the wires, was pleased to see the little quadruped dart nimbly out of her hiding-place and take it from his fingers. She always preferred insects to vegetable food. The same observer noticed that the tail of this animal is prehensile.

Independently of its small size, the Harvest Mouse may be distinguished from a young ordinary Mouse by its short ears, narrow head, slender body, and less projecting eyes.

THE bold and elegant markings with which the fur of the BARBARY MOUSE is decorated, render it a very conspicuous animal, and when the creature is in captivity, always attract the attention of visitors who happen to pass before its cage.

BARBARY MOUSE.—*Golunda Bárbara.*

The dimensions of this animal are greater than those of the common Mouse, while they are smaller than those of the ordinary rat. Its colour is very pleasing, the ground tint of the fur being a rich brown, and the stripes of a whitish-yellow, verging by degrees into the white hue of the under portions of the body. These pretty creatures are tolerably hardy, and can endure an English climate as well as most animals which have been brought from a hot and arid to a cold and moist country. They run about their cage with considerable liveliness, sometimes diving among their bedding, and ever and anon poking their intelligent-looking little heads from among the hay, and tripping about as if pleased to exhibit their beautiful fur. As may be supposed from its title, the animal is a native of Barbary. It is not devoid of the carnivorous habits of its race, and even when bountifully supplied with food, has been known to develop its carnivorous nature into cannibalism, eating the body of one of its companions that had died while in the cage.

THE short, sturdy, stupid rodent which is so famous under the name of the HAMSTER is widely spread over many parts of Northern Europe, where it is an absolute pest to the agriculturists, who wage unceasing war against so destructive an animal. Before proceeding to the habits and character of the Hamster, a short description of its external appearance will be necessary.

The colour of its fur is a greyish-fawn on the back, deepening into black on the under portions of the body, and softening into a yellow hue upon the head and face. The otherwise uniform tinting of the fur is relieved by some patches of whitish-yellow upon the cheeks, shoulders, and sides. The creature is furnished with two large cheek-pouches, which are capable of containing a considerable amount of food, and which can be inflated with air at the pleasure of the animal. The length of the adult Hamster is about fifteen inches, the tail being only three inches long.

The Hamster is most destructive to the crops, whether of corn, peas, or beans, and when the autumn approaches, begins to plunder the fields in a most systematic manner, for the purpose of laying up a winter store of provisions. By dint of dexterous management, the animal fills its cheek-pouches with grain, pressing it firmly with its paws, so as to lose no space, and then carries off its plunder to its subterranean treasury, where it disgorges the contents of the pouches, and returns for another supply. The husbandmen are so well aware of this propensity that they search after the habitation of the Hamster after the harvest is over, and often recover considerable quantities of the stolen grain. The destructive capability of the animal may be gathered from the fact that a single Hamster has been known to hoard no less than sixty pounds of corn in its home, while a hundredweight of beans have been recovered from the storehouses of another specimen.

The skin of the Hamster is of some value in commerce, so that the hunters make a double use of a successful chase, for they not only recover the stolen property of the agriculturist, but gain some profit by selling the skins.

HAMSTER.—*Cricĕtus frumentarius.*

The burrow of the Hamster is a most complicated affair, and not very easy to describe. Each individual has a separate burrow, and not even in the breeding season do the male and female inhabit the same domicile. At some depth below the surface of the earth are several rather large chambers, communicating with each other by horizontal passages. In one of these chambers the creature lives, and in the others it places its store of provision. There are at least two entrances to each burrow, one being almost perpendicular, and the other sloping. Sometimes there are more than two entrances to the chambers, but there are never less than that number. The depth of the chambers is from three to five feet. Each burrow is only intended to serve for one season, and is abandoned at the end of winter.

As the Hamster is in the habit of throwing the excavated earth from the oblique burrow, technically called the "creeping-hole," its locality is discovered by means of the mound of loose earth which is heaped at its entrance. Eighty thousand of these animals have been killed in one year within a single district.

The Hamster is a very prolific animal, as appears from the fact that it still holds its own in spite of the constant persecution to which it is subjected by the agriculturists and the regular hunters. There are several broods in each year, the average number of each family being from seven to ten or twelve. As soon as the young Hamsters are able to

1. o o

shift for themselves, an event which occurs in a wonderfully short time, they leave the maternal home, and dig separate burrows.

The strangest part of the Hamster's character is its dull, unreasoning ferocity, which is utterly incapable of comprehending danger, and causes the animal to attack any kind of opponent, whether animate or not.   An irritated Hamster will fly upon a dog, a man, or a horse, without the least hesitation.   If a cart were to crush it, it would try to bite the wheel ; if a stone were to roll over it, it would turn upon the lifeless stone ; threaten it with a stick, and it fastens upon the senseless weapon with malign fury ; and when opposed by a bar of iron nearly red hot, it has been known to grasp the burning metal in its teeth, and to retain its hold in spite of the pain which it must have suffered.   This combative disposition leads it to fight desperately with its own species, caring nothing for sex or age ; and it has actually happened that when a pair of these animals have been placed together in a cage, the male has been killed and partly eaten by his disconsolate widow.

The food of this animal is chiefly vegetable, but is varied by animal diet, such as worms, insects, mice, small birds, lizards, frogs, and other such vermin.   It is a nocturnal animal, and achieves its robberies under cover of the darkness of night.   It can hardly be termed a true hibernating animal, as it is quite lively for a considerable portion of the winter, feeding on its ample stores for nearly two months, and becoming very fat by the combined influence of inactivity and good feeding.   Through a portion of the winter it becomes torpid, but awakes early in the spring, ready to renew its depredations in the fields.   During the spring and summer months its food consists chiefly of leaves and various herbage.

WATER RAT, OR WATER VOLE.—*Arvicola amphibius.*

THERE are many animals which have been saddled with a bad reputation merely on account of an unfortunate resemblance to another animal of really evil character. Among these misused innocents the WATER VOLE is very conspicuous, as the poor creature has been commonly supposed to be guilty of various poaching exploits which were really achieved by the ordinary brown Rat.

It is quite true that Rats are often seen on the river-banks in the act of eating captured fish, but these culprits are only the brown Rats which have migrated from the farmyards for the summer months, and intend to return as soon as autumn sets . in.   The food of the true Water Rat, or Water Vole, as it is more correctly named, is chiefly of a vegetable nature, and consists almost entirely of various aquatic plants and roots.   The common " mare's-tail," or equisetum, is a favourite article of diet with the Water Vole, and I have often seen it feeding on the bark of the common rush.   Many years ago

I shot a Water Vole as it was sitting upon a water-lily leaf and engaged in eating the green seeds ; and on noticing the kind of diet on which the animal was feeding, I determined to watch the little creatures with more care. My own testimony coincides precisely with those of other observers, for I never yet saw the true snub-nosed, short-eared, yellow-toothed Vole engaged in eating animal food, although the brown Rat may be often detected in such an act.

Many communications have been made to me on the subject, written for the most part by persons who have seen water-side Rats engaged in catching and eating fish, and have thought that the delinquents were the true Water Vole. Indeed, the Vole is allied very closely to the beaver, and partakes of the vegetarian character of that animal.

However guiltless the Water Vole may be of piscicapture, it is not altogether a harmless animal, for, independently of weakening the banks by its tunnels, it will sometimes leave the water-side and travel some little distance across the country in search of cultivated vegetables. One of these animals has been seen to cross a large field and enter a garden in which some French beans were growing. The Vole crept up the bean-stalks, and after cutting off several of the pods with its sharp and scissor-like teeth, picked them up and retraced its steps to its home.

The colour of the Water Vole is a chestnut-brown, dashed with grey on the upper parts and fading to grey below. The ears are so short that they are hardly perceptible above the fur. The incisor teeth are of a light yellow, and are very thick and strong. The tail is shorter than that of the common Rat, hardly exceeding half the length of the head and body. The average length of a full-grown Water Vole is thirteen inches, the tail being about four inches and three-quarters long. It is not so prolific an animal as the brown Rat, breeding only twice in the year, and producing from five to six young at a birth.

THE CAMPAGNOL, or SHORT-TAILED FIELD MOUSE, is even more destructive in the open meadows than the common grey mouse in the barns or ricks ; for not contenting itself with plundering the ripened crops of autumn, it burrows beneath the ground at sowing time, and devours the seed-wheat which has just been laid in the earth. Besides these open-air depredations, it make inroads into ricks and barns, and by dint of multitudinous numbers does very great harm. As its food is entirely of a vegetable nature, it does not enter human habitations, where it would find but a poor chance of a livelihood.

The colour of the Campagnol is ruddy brown on the upper surface of the body, and grey on the abdomen and chest. The ears are rounded and very small, closely resembling those of the water vole. The tail is only one-third the length of the body, and the total length of the animal is rather

CAMPAGNOL.—*Arvicola arvalis.*

more than five inches. As it belongs to the same genus as the water vole, and is very closely related to that animal, it sometimes goes by the name of Field Vole.

It is a very prolific animal, and its numbers are almost incredibly great in districts where no means have been taken for its destruction. Even in well-cultivated fields, whether of grass or corn, the Campagnol may be found in vast quantities by any one whose eyes are sufficiently accustomed to the task to distinguish the little creature from the earth on which it moves. At one time, when my eyes were in proper order, I have frequently gone into any grass-field at random, and amused myself by detecting the Field Mice as they crept through the grass blades, and endeavouring to watch them in their silent and almost imperceptible progress. They move so easily through the green herbage that they scarcely stir the blades; and are so similar in their colour to the earth as it shows between the leaves, that none but a practised eye can detect them. There is hardly any sign to tell of its presence, except an undefined sense of something red among the grass, which, unless it be immediately pounced upon, fades again into brown, and the thing is gone.

The Campagnol is a water-loving creature, and is oftener found in marshy ground than in meadows which are elevated above the level of the neighbouring lands and ditches. A dry summer is very trying for these animals, and a long-continued drought is fatal to hundreds of them.

The Field Vole carries its destructive powers even into woods and plantations, and is often the unknown cause by which some cherished young tree has drooped, withered, and died. These little animals are good burrowers, and are in the habit of digging into the ground, and nibbling the living roots of trees and shrubs. Sometimes the mice attack the bark, and, by completely stripping it from the circumference of the tree, destroy it as effectually as if it had been cut down with an axe.

THERE is another species of Field Mouse, in which the tail is much longer in proportion, and the dimensions are altogether smaller. This is the BANK VOLE, or BANK CAMPAGNOL, and must not be confounded with the Long-tailed Field Mouse, which is not a vole at all, but a veritable mouse.

AT uncertain and distant intervals of time, many of the northern parts of Europe, such as Lapland, Norway, and Sweden, are subjected to a strange invasion. Hundreds of

LEMMING.—*Myodes Lemmus.*

little, dark, mouse-like animals sweep over the land, like clouds of locusts suddenly changed into quadrupeds, coming from some unknown home, and going no one knows whither. These creatures are the LEMMINGS, and their sudden appearances are so entirely mysterious, that the Norwegians look upon them as having been rained from the clouds upon the earth.

Driven onwards by some over-powering instinct, these vast hordes travel in a straight line, permitting nothing but a smooth perpendicular wall or rock to turn them from their course. If they should happen to meet with any living being, they immediately attack, knowing no fear, but only urged by undiscriminating rage. Any river or lake they swim without hesitation, and rather seem to enjoy the water than to fear it. If a stack or a corn-rick should stand in their way, they settle the matter by eating their way through it, and will not be turned from their direct course even by fire. The country over which they pass is utterly devastated by them, and it is said that cattle will not touch the grass on which a Lemming has trodden.

These migrating hosts are accompanied by clouds of predaceous birds, and by many predaceous quadrupeds, who find a continual feast spread for them as long as the

Lemmings are on their pilgrimage. While they are crossing the rivers or lakes, the fish come in for their share of the banquet, and make great havoc among their columns. It is a very remarkable fact that the reindeer is often seen in chase of the Lemmings; and the Norwegians say that the deer is in the habit of eating them. This statement, however, seems to be rather of doubtful character. The termination of these extraordinary migrations is generally in the sea, where the survivors of the much-reduced ranks finally perish. Mr. Lloyd mentions that just before his visit to Wermeland, the Lemming had overrun the whole country. The primary cause of these strange migrations is generally thought to be hunger. It is fortunate for the country that these razzias only occur at rare intervals, a space of some ten or fifteen years generally elapsing between them, as if to fill up the places of those which were drowned or otherwise killed in the preceding migration.

The Lemming feeds upon various vegetable substances, such as grass, reeds, and lichens, being often forced to seek the last-named plant beneath the snow, and to make occasional air-shafts to the surface. Even when engaged in their ordinary pursuits, and not excited by the migratorial instinct, they are obstinately savage creatures. Mr. Metcalfe describes them as swarming in the forest, sitting two or three on every stump, and biting the dogs' noses as they came to investigate the character of the irritable little animals. If they happened to be in a pathway, they would not turn aside to permit a passenger to move by them, but boldly disputed the right of way, and uttered defiance in little sharp, squeaking barks.

The colour of the Lemming is dark brownish-black, mixed irregularly with a tawny hue upon the back, and fading into yellowish-white upon the abdomen. Its length is not quite six inches, the tail being only half an inch long.

THE common BEAVER has earned a world-wide reputation by the wonderful instinct which it displays, independently of its very great value in producing costly fur and perfume.

This animal is found in the northern parts of Europe and Asia, but is found in the greatest profusion in North America. In days long gone by, the Beaver was an inhabitant of our own island. It is generally supposed that there is but one species of true Beaver, and that the Beaver of Europe and Asia is specifically identical with that of America, even though certain small differences of fur and colouring may be discerned between them. The social Beaver of Northern America is a truly wonderful animal, displaying a singular mixture of reason and instinct, together with a curious absence of both on occasions. The best account of this animal is to be found in Audubon and Bachman's valuable work on the quadrupeds of North America, to which work I am indebted for the following particulars.

The Beaver lives in societies, varying considerably in number, and united together in the formation of works which may fairly be considered as belonging to the profession of the engineer. They prefer to make their habitations by small clear rivers and creeks, or close to large springs, although they sometimes take up their abode on the banks of lakes.

Lest they should not have a sufficient depth of water in all weathers and at all seasons, the Beavers are in the habit of building veritable dams, for the purpose of raising the water to the required level. These dams are composed of tree-branches, mud, and stones, and in order effectually to resist the action of the water, are about ten or twelve feet in thickness at the bottom, although they are only two feet or so wide at the summit. When the different parts of the stream run with varying velocity, the formation of the dam is really a triumph of engineering skill, for wherever the stream is gentle, the dam is built straight across it; but wherever the current runs smartly, the dam is curved so as to present a convex surface to its force. It often happens that when a dam has been made for some years, its dimensions become very large, in consequence of the trees and branches that are intercepted by it, and in process of time it sprouts thickly with vegetation, and even nurtures trees of some dimensions.

In forming the dam, the Beaver does not thrust the ends of the stakes into the bed of the river, as is often supposed, but lays them down horizontally, and keeps them in their place by heaping stones and mud upon them. The logs of which the dam is composed

BEAVER.—*Castor Fiber.*

are about three feet in length, and vary extremely in thickness. Generally, they are about six or seven inches in diameter, but they have been known to measure no less than eighteen inches in diameter. An almost incredible number of these logs are required for the completion of one dam, as may be supposed from the fact that a single dam will sometimes be three hundred yards in length, ten or twelve feet thick at the bottom, and of a height varying according to the depth of water.

Before employing the logs in this structure, the Beavers take care to separate the bark, which they carry away, and lay up for a winter store of food.

Near the dams are built the Beaver-houses, or "lodges," as they are termed; edifices as remarkable in their way as that which has just been mentioned. They are chiefly composed of branches, moss, and mud, and will accommodate five or six Beavers together. The form of an ordinarily sized Beaver's lodge is circular, and its cavity is about seven feet in diameter by three feet in height. The walls of this structure are extremely thick, so that the external measurement of the same lodges will be fifteen or twenty feet in diameter, and seven or eight feet in height. The roofs are all finished off with a thick layer of mud, laid on with marvellous smoothness, and carefully renewed every year. As this compost of mud, moss, and branches is congealed into a solid mass by the severe frosts of a North American winter, it forms a very sufficient defence against the attacks of the Beaver's great enemy the wolverene, and cannot readily be broken through, even with the help of iron tools. The precise manner in which the Beavers perform their various tasks is not easy to discern, as the animals work only in the dark.

Around the lodges the Beavers excavate a rather large ditch, too deep to be entirely frozen, and into this ditch the various lodges open, so that the inhabitants can pass in or out without hindrance. This precaution is the more necessary, as they are poor pedestrians, and never travel by land as long as they can swim by water. Each lodge is inhabited by a small number of Beavers, whose beds are arranged against the wall, each bed being separate, and the centre of the chamber being left unoccupied.

In order to secure a store of winter food, the Beavers take a vast number of small logs, and carefully fasten them under water in the close vicinity of their lodges. When a Beaver feels hungry, he dives to the store heap, drags out a suitable log, carries it to a sheltered and dry spot, nibbles the bark away, and then either permits the stripped log to float down the stream, or applies it to the dam.

Their teeth are wonderfully powerful and sharp, and their jaws are possessed of singular strength, as may be seen by the accompanying engraving of a Beaver's skull.

So sharp are their teeth, and with such address does the animal use them, that a tame Beaver has repeatedly been seen to take a potato or an apple in his fore-paws, to sit up on his hind feet, and by merely pressing the apple against his lower incisors, and manipulating it dexterously, to peel it as really as if the operation had been performed by human hands with the aid of a knife.

Not all the Beavers employ themselves in these united labours, for there are some which, like drones, refuse to take any part in the proceedings, and are technically called "Les paresseux," or the Idlers, by the Beaver-hunters. These animals make no dam and build no house, but content themselves with excavating long tunnels and taking up their abode therein. Several of these idlers inhabit the same burrow, and as they are always males, it is supposed that they must have

SKULL OF BEAVER.

been conquered in the contests which take place between most male animals while they are seeking their mates, and that they must have retired into comparative solitude until they have gained sufficient strength and courage to renew the fight. These idlers are gladly welcomed by the hunters, for they are easily caught, and a skilful trapper thinks himself ill-used if he does not capture every idler that he may meet.

We now must bestow a little time on the curious odoriferous substance which is called "castoreum" by the learned, and "bark-stone" by the trappers. This substance is secreted in two glandular sacs which are placed near the root of the tail, and gives out an extremely powerful odour.

To the castoreum the trapper is mostly indebted for his success, for the Beavers are strangely attracted by this substance, and if their nostrils perceive its distant scent, the animals will sit upright, sniff about in every direction, and absolutely squeal with excitement. Taking advantage of this curious propensity, the hunter always carries a supply of castoreum, in a closed vessel, and when he comes to a convenient spot for placing his trap, he sets the trap and then proceeds to manufacture his bait. This process is simple enough, consisting merely of taking a little twig of wood about nine inches long, chewing one end of it and dipping it in the castoreum. The trap is now laid so as to be covered by about six inches of water, and the stick arranged so that its perfumed tip projects from the water. Any Beaver which scents this bait will most certainly come to it, and will probably be captured in the trap.

Connected with this strange mode of baiting a trap, is a habit which has only recently been brought before the public by the researches of Messrs. Audubon and Bachman.

If two Beaver lodges are tolerably near each other, the inhabitants of the one lodge, which we will call lodge A, go to a little distance for the purpose of ridding themselves of the superabundant castoreum. The Beavers of lodge B, smelling the castoreum, go to the same spot, and cover the odoriferous substance with a thick layer of earth and leaves. They then place their own castoreum upon the heap, and return home. The inhabitants of lodge A then go through precisely the same process, until they have raised a mound some four or five feet in height.

To return to the baited trap spoken of in the last paragraph but one. If the Beaver which smells the bait is a young one, it will almost certainly be captured; but if it should chance to be an old and experienced animal, it will not only avoid capture, but render

COYPU RAT, OR RACOONDA.—*Myopótamus Coypus.*

the trap useless until it has been re-set. For instead of trying to get at the bait, it fetches quantities of mud and stones, heaps its load upon the trap until it has raised a small mound, and after placing its own superabundant castoreum upon the little hillock, goes away in safety.

In spite of their store of provisions, the Beavers become very thin during the winter months, so that they are in bad case when spring comes in. However, the succulent diet which they then find has a rapid and beneficial effect upon them, and by the beginning of autumn they are quite fat. By study of the Beaver's habits, the trappers are enabled to prognosticate the kind of weather which is likely to happen. For example, as it is well known that the Beavers always cut their winter's store of wood in good season, the fact of their early commencement of this labour shows that winter will be earlier than usual.

The colour of the long shining hairs which cover the back of the Beaver is a light chestnut, and the fine wool that lies next to the skin is a soft greyish-brown. The total length of the animal is about three feet and a half; the flat, paddle-shaped, scale-covered tail being about one foot in length. The flesh of the Beaver is eaten by the trappers, who compare it to flabby pork. The tail is something like beef marrow, when properly cooked, but it is too rich and oily to suit the taste of most persons. The female Beaver produces about three or four young at a litter, and the little creatures are born with open eyes.

THE COYPU RAT, or RACOONDA, as it is sometimes termed, is a native of Central America, where it is found in such great numbers that its beautiful fur is imported into Europe in very large quantities.

The colour of this animal is a light reddish-brown, the hairs being variegated with both tints, not unlike that of the beaver in character and general appearance. Indeed, the creature bears a great resemblance to a miniature beaver, with the exception of its tail, which is long and rounded, instead of being flattened like that of the true beaver. The incisor teeth are a light reddish-orange, and are very conspicuous even at some distance. The length of a full-grown Coypu is about two feet six inches, its tail being about fifteen inches long.

It is a quick and lively animal, and very amusing in its habits. It swims nearly, if not quite, as well as the beaver, using its webbed hind feet in much the same manner. It is wonderfully dexterous in the use of its fore-paws, which it uses as if they were hands, while it sits upright on its hinder paws and tail. I have often watched the funny antics of the Coypus in the Zoological Gardens, and have been much amused by the manner in

MUSQUASH, OR MUSK RAT, OR ONDATRA.—*Fiber Zibéthicus.*

which they traverse their domains, and examine everything that seems to be novel. If a tuft of grass is thrown to them, they pick it up in their fore-paws, shake it violently, in order to get rid of the earth that clings to the roots, and then, carrying it to the water-side, wash it with a rapid dexterity that might be envied by a professional laundress.

While swimming it looks very like a magnified water vole, and is remarkably quick and agile in its movements; but its gait on land is clumsy and awkward. It seems to be equally at home in salt and fresh water, inhabiting the banks of rivers or the shores of the sea creeks, according to the locality in which it is found, and living in burrows which it excavates along the banks. It is said to be a tolerably powerful animal, and to make no despicable resistance to the dogs which are employed in its chase. It is, however, naturally of a gentle disposition, and can be rendered very tame by those who bestow proper attention upon it.

The ONDATRA, MUSQUASH, or MUSK RAT, is a native of Northern America, where it is found in various places above the twentieth degree of north latitude.

The colour of this animal is a dark brown on the upper portions of its body, tinged with a reddish hue upon its neck, ribs, and legs, the abdomen being ashy grey; the tail is of the same dark hue as the body. In total length it rather exceeds two feet, of which measurement the tail occupies about ten inches. The incisor teeth are bright yellow, and the nails are white. The whole colouring of the animal is so wonderfully like the hue of the muddy banks on which it resides, that a practised naturalist has often mistaken the Ondatras for mere lumps of mud until they began to move, and so dispelled the illusion. The hinder feet of the Ondatra are well webbed, and their imprint on the soft mud is very like that of a common duck.

The food of the Ondatra in a wild state appears to be almost wholly of a vegetable nature; although, when confined in a cage, one of these animals has been seen to eat muscles and oysters, cutting open the softest shells, and extracting the inmates, and waiting for the hard-shelled specimens until they either opened of their own accord or died. Although the Ondatra is a clumsy walker, it will sometimes travel to some distance from the water-side, and has been noticed on a spot nearly three-quarters of a mile from any water. These animals have also been detected in ravaging a garden, which they had plundered of turnips, parsnips, carrots, maize, and other vegetables. The mischievous creatures had burrowed beneath them, bitten through their roots, and carried them away to their subterranean storehouses.   The maize they had procured by cutting the stalks near the level of the ground.

The Ondatra lives mostly in burrows, which it digs in the banks of the river in which it finds its food, but sometimes takes up its abode in a different kind of habitation, according to the locality and the soil.   In the stiff clay banks of rivers the Ondatra digs a rather complicated series of tunnels, some of them extending to a distance of fifteen or twenty yards, and sloping upwards.   There are generally three or four entrances, all of which open under water, and unite in a single chamber, where the Ondatra makes its bed. The couch of the luxurious animal is composed of sedges, water-lily leaves, and similar plants, and is so large as to fill a bushel basket.   On marshy ground, and especially if it be supplied by springs, the Ondatra builds little houses that rise about three or four feet above the water, and look something like small haycocks.

As the fur of the Ondatra is rather valuable, and the flesh is considered to be nearly as good as that of the wild duck, it is rather persecuted by the human inhabitants of the same land, as well as by the regular fur hunters.   If these creatures have taken up their abode in burrows, the hunters capture them by stopping up all the holes which they can reach, and intercepting the animals as they try to escape; but if the ground be marshy, and they live in houses or "lodges," a different plan is adopted.   Being armed with a four-pronged barbed spear, the hunter creeps quietly towards one of the houses, and with the full strength of his arm drives the barbed prongs completely through the frail walls, transfixing one or more of the inhabitants.   His companion, who is furnished with an axe, immediately hurls down the remainder of the wall, and secures the unfortunate victims who are held down by the merciless steel.

The habits of the Ondatra are very curious, and are admirably related by Messrs. Audubon and Bachman, in the work to which allusion has already been made :—

"Musk Rats are very lively, playful, animals when in their proper element, the water ; and many of them may be occasionally seen disporting themselves on a calm night in some mill-pond or deep sequestered pool, crossing and recrossing in every direction, leaving long ripples in the water behind them, while others stand for a few moments on little hurdles or tufts of grass, or on stones or logs, on which they can get a footing above the water, or on the banks of the pond, and then plunge one after the other into the water.   At times one is seen lying perfectly still on the surface of the pond or stream, with its body widely spread out, and as flat as can be.   Suddenly it gives the water a smart slap with its tail, somewhat in the manner of the beaver, and disappears beneath the surface instantaneously, going down head foremost, and reminding one of the quickness and ease with which some species of ducks and grebes dive when shot at.

At the distance of ten or twenty yards, the Musk Rat comes to the surface again, and perhaps joins its companions in their sports ; at the same time others are feeding on the grassy banks, dragging off the roots of various kinds of plants, or digging underneath the edge of the bank.   These animals seem to form a little community of social, playful creatures, who only require to be unmolested in order to be happy.

Should you fire off a fowling-piece while the Musk Rats are thus occupied, a terrible fright and dispersion ensues ; dozens dive at the flash of the gun, or disappear in their holes ; and although in the daytime, when they see imperfectly, one may be shot while swimming, it is exceedingly difficult to kill one at night.   In order to ensure success, the gunner must be concealed, so that the animal cannot see the flash, even when he fires with a percussion lock."

Traps are also largely employed for the destruction of this gentle but, unfortunately for itself, valuable animal.   The traps are so arranged, that when the creature is taken, and struggles to get free, it jerks the trap into the water, and is thus drowned.   If its companions discover it while still entrapped, they behave in the manner of the brown Rats, and tear their imprisoned companion to pieces.   If one of these animals is shot, and not immediately retrieved, the survivors surround the dead body of their companion, and carry it off to their homes from the reach of its murderer.   In character it is quiet and gentle, and although armed with such powerful teeth, makes no offensive use of them, even when handled by man for the first time.

BEAVER RAT, OR HYDROMYS.—*Hydromys chrysogaster.*

THE shy and retiring HYDROMYS, or BEAVER RAT, is not a very rare animal in its native country, but as, in addition to its natural timidity, it is nocturnal in its habits, it is but seldom seen by casual observers. It is a native of Van Diemen's Land, and is found inhabiting the banks of both salt and fresh water. It is an admirable swimmer and diver, reminding the spectator of the water vole of Europe. Like that animal, it has a habit of sitting upright, supported by its hind paws and tail, while it employs the fore-feet for the conveyance of food to its mouth.

The colour of the Beaver Rat's fur is as follows. The neck and upper parts of the body are of a dark rich brown, which is washed with a light golden hue along the sides of the face, shoulders, and the flanks, as far as the hind limbs. The under surface of the body is golden yellow, and has earned for the animal the name of "chrysogaster," which signifies "golden-bellied." The basal half of the tail is black, and the remaining moiety is white. In the engraving, the contrast of the colours is not sufficiently marked. The total length of the Beaver Rat is about two feet, the tail being the same length as the body. The hinder feet are webbed.

GROUND PIG.—*Aulacódus Swinderianus.*

The GROUND PIG is one of the links between the beavers and the porcupines, and has a considerable affinity with the latter animals.

It is found in many parts of Southern Africa, as well as on the coast of Guinea, where it is not at all uncommon. The hair of this animal is rather peculiar, and approximates closely to the quill-hairs of the true porcupines, being either flat and grooved above, or developed into flexile spines. The tail is but sparely covered with hair, and is rather short in proportion to the size of its owner. The hinder feet are only furnished with four toes, armed with large, rounded, and rather blunt claws. The ears are short and rounded.

The PORCUPINE has long been rendered famous among men by the extraordinary armoury of pointed spears which it bears upon its back, and which it was formerly fabled to launch at its foes with fatal precision.

This animal inhabits many parts of the world, being found in Africa, Southern Europe, and India. The spines, or quills, with which it is furnished, vary considerably in length, the longest quills being flexible, and not capable of doing much harm to an opponent. Beneath these is a plentiful supply of shorter spines, from five to ten inches in length, which are the really effective weapons of this imposing array. Their hold on the skin is very slight, so that when they have been struck into a foe, they remain fixed in the wound, and, unless immediately removed, work sad woe to the sufferer. For the quill is so constructed, that it gradually bores its way into the flesh, burrowing deeper at every movement, and sometimes even causing the death of the wounded creature. In Africa and India, leopards and tigers have frequently been killed in whose flesh were pieces of Porcupine quills that had penetrated deeply into the body, and had even caused suppuration to take place. In one instance, a tiger was found to have his paws, ears, and head filled with the spines of a Porcupine, which he had vainly been endeavouring to kill.

Conscious of its powers, the Porcupine is not at all an aggressive animal, and seldom, if ever, makes an unprovoked attack. But if irritated or wounded, it becomes at once a very unpleasant antagonist, as it spreads out its bristles widely, and rapidly backs upon its opponent. There are few horses which will face an irritated Porcupine ; and even the preliminary rustle of the quills with which a Porcupine generally prepares every attack, is sufficient to make an ordinary horse flee in terror. The rustling sound is produced by a number of hollow quills which grow upon the Porcupine's tail, and which, when that member is agitated, clash against each other with a sound very like the peculiar ruffling of a peacock's train.

The Porcupine is a nocturnal animal, seldom venturing out of its retreat as long as the sun is above the horizon, and is therefore not often seen even in the localities which it most prefers. It is said not to require the presence of water, but to quench its thirst by eating the succulent roots and plants which it digs out of the ground. Its food is entirely of a vegetable nature, and consists of various kinds of herbage, as well as of bark, fruit, and roots. This animal takes up its abode in deep burrows which it excavates, and in which it is supposed to undergo a partial hibernation.

As the spines of the Porcupines are of some commercial value, and are used for many purposes, the chase of the animal is rather popular in the countries which it inhabits, and derives a further interest from the fact that the Porcupine, although a timid creature, can make a very powerful resistance when it is driven to despair. In fighting, it depends wholly on its quills, and does not attempt to make the least use of its strong and sharp incisor teeth, which are able to cut their way through the hardest wood as if it were butter, and would inflict most dangerous wounds. So far, indeed, is it from making any use of these formidable weapons, that its first care is to protect its head, being probably led to that course of action by its fear for its nose, which is so sensitive that the animal is stunned by a comparatively slight blow on that organ.

It does not appear to be very susceptible of domestication, probably because it cannot find teachers who are sufficiently fearless of its quills to pay very close attention to it. With the exception of the hollow quills in the tail, the spines are encircled with alternate rings of black and white, producing a very rich contrast of colouring. The upper parts of the body are covered with hair instead of quills, and upon the head and neck there is a kind of crest, composed of very long stiff hairs, which can be erected or depressed at pleasure. Like the hedgehog, it can coil itself into a ball when it is surprised at a

PORCUPINE.—*Hystrix Cristáta.*

distance from its haven of refuge, and can present such an array of threatening spikes, that it is quite safe from any enemy excepting man. When, however, the animal is at peace it is capable of depressing the bristling spears, and can squeeze itself through an opening which would appear at first sight to be hardly large enough to permit the passage of an animal of only half its size.

The total length of the common Porcupine is about three feet six inches, the tail being about six inches long. Its gait is plantigrade, slow, and clumsy, and as it walks, its long quills shake and rattle in a very curious manner. Its muzzle is thick and heavy, and its eyes small and pig-like.

TUFTED-TAILED PORCUPINE.—*Atherúra Africána.*

The TUFTED-TAILED PORCUPINE is even a more singular animal than that which has just been described.

The quills which cover the body are very short in proportion to the size of the animal, and instead of preserving the rounded, bamboo-like aspect of the ordinary Porcupine-quills, are flattened like so many blades of grass. The tail is scaly throughout a considerable part of its length, but at the tip is garnished with a tuft of most extra-ordinary-looking objects, which can hardly be called hairs or quills, but, as Buffon remarks, look very like narrow, irregular strips of parchment. The colouring of the quills is rather various, but as a general rule, they are black towards the extremity and white towards the base. They are very sharply pointed, and are remarkable for a deep groove that runs along their entire length. Upon the head the quills are not more than one inch long, but on the middle of the body they reach four or even five inches. Among these quills there are a few long and very slender spines or bristles, which project beyond the others.

The Tufted-tailed Porcupine has been found at Fernando Po, and is an inhabitant of India and the Peninsula of Malacca.

The URSON, CAWQUAW, or CANADIAN PORCUPINE, is a native of North America, where it is most destructive to the trees among which it lives.

Its chief food consists of living bark, which it strips from the branches as cleanly as if it had been furnished with a sharp knife. When it begins to feed, it ascends the tree, commences at the highest branches, and eats its way regularly downward. Having finished one tree, it takes to another, and then to a third, always choosing those that run in the same line; so that its path through the woods may easily be traced by the line of barked and dying trees which it leaves in its track. A single Urson has been known to destroy a hundred trees in a single winter, and another is recorded as having killed some two or three acres of timber.

It is a tolerably quiet animal, and easily tamed; although subject to sudden fits of alarm at any strange object. One of these animals was so entirely domesticated, as to come voluntarily, and take vegetables or fruit from the hand of its master, and would rub itself against him after the manner of an affectionate cat. When irritated or alarmed, it has a curious habit of striking sharply with its tail, which is thickly set with short quills, and causing no small damage to the object of attack. In the work of Messrs. Audubon and Bachman is a very amusing little story of the manner in which the tame Urson above mentioned repelled an attack made upon it by a fierce dog.

"A large, ferocious, and exceedingly trouble-some mastiff, belonging to the neighbourhood,

CANADIAN PORCUPINE, OR URSON.—*Erethizon dorsatum.* had been in the habit of digging a hole under the fence, and entering our garden. Early one morning we saw him making a dash at some object in the corner of the fence, which proved to be our Porcupine, which had, during the night, made its escape from the cage.

The dog seemed regardless of all its threats, and probably supposing it to be an

COENDOO, OR BRAZILIAN PORCUPINE.—*Cercolabes prehensilis.*

animal not more formidable than a cat, sprang upon it with open mouth. The Porcupine seemed to swell up in an instant to nearly double its size, and as the dog pounced upon it, it dealt him such a sidewise blow with its tail, as to cause the mastiff to relinquish his hold instantly, and set up a loud howl in an agony of pain. His mouth, tongue, and nose were full of Porcupine quills. He could not close his jaws, but hurried, open-mouthed, off the premises. It proved to him a lesson for life, as nothing could ever afterwards induce him to revisit a place where he had met with such an unneighbourly reception. Although the servants immediately extracted the spines from the mouth of the dog, we observed that his head was terribly swelled for several weeks afterwards, and it was months before he finally recovered."

The victorious Urson did not long survive the affray, for as the summer weather approached, it betrayed unmistakeable signs of distress, and finally died of heat. A similar anecdote is recorded of an Urson, which took a sudden umbrage at the attentions of a person who was attempting to caress it, and unexpectedly dealt him such a blow with its tail that his offending right hand was instantly covered with wounds.

The Urson is not so fully defended with spines as the two preceding animals, but is covered with long, coarse, blackish-brown hair, among which the short pointed quills are so deeply set, that, except in the head, tail, and hinder quarters, they are scarcely perceptible. These spines are largely used by the American Indians in the decoration of their hunting-pouches, mocassins, and other articles, and after the quills are extracted, the remainder of the fur is sufficiently soft to be used for clothing. The flesh of the Urson is considered eatable, and is said to bear some resemblance to flabby pork.

The length of the Urson is not quite four feet, the head and body measuring rather more than three feet, and the tail about nine inches. The teeth are of a bright orange.

IN Southern America, the Porcupines find a representative in the COENDOO, an animal which is not only remarkable for its array of quills, but also for the prehensile power of its long tail.

As might be presumed, from the prehensile tail and the peculiarly armed claws, the Coendoo is of arboreal habits, finding its food among the lofty branches of trees. On the level ground it is slow and awkward, but among the more congenial boughs it climbs with great ease, drawing itself from branch to branch by means of its hooked claws; but seldom using its tail, except as an aid in descent. The food of this animal consists of leaves, flowers, fruit, bark, and the soft woody substance of young and tender branches, which it slices easily with its chisel-edged incisor teeth. During the summer months the Coendoo becomes extremely fat, and its flesh is then in great request, being both delicate in flavour and tender in character. The young of this animal are born in the month of September or October, and are very few in number.

The total length of the Coendoo is about three feet six inches, of which the tail occupies one foot six inches. Its nose is thick and blunt, like that of the common Porcupine, and the face is furnished with very long whisker-hairs of a deep black. The numerous spines which cover the body are parti-coloured, being black in the centre and white at each extremity. Their length is rather more than two inches on the back, an inch and a half on the fore-legs, and not quite an inch on the hinder limbs. A number of short quills are also set upon the basal half of the tail, the remainder of that organ being furnished with scales, and tapering to its extremity. The colour of the scales is black. The entire under surface of the tail is covered with similar scales, among which are interspersed a number of bright chestnut hairs. The abdomen, breast, and inner face of the limbs are clothed with dense, brown, coarse hairs.

It is a nocturnal animal; sleeping by day, and feeding by night.

THE two succeeding animals bear some resemblance to each other, but may be distinguished by the different shape of the head, and the structure of the feet and toes. These technical distinctions may be found in the list of generic differences which closes the first volume of this work. There are several species of Agoutis, the COMMON AGOUTI being considered as the type of the genus, and their habits being very similar.

The Agouti is a native of Brazil, Paraguay, Guiana, and other neighbouring countries, but its numbers have been considerably thinned in many spots where cultivation has been industriously carried on. In some of the Antilles, where it formerly swarmed, it is now nearly extirpated, and in St. Domingo is but rarely seen. It is a voracious animal, eating almost every kind of vegetable food, having, however, an unfortunate preference for those plants which have been reared under human superintendence. It is especially fond of roots, such as potatoes and yams, and is so destructive among sugar-canes that the planters are forced to wage a war of extermination against the Agouti before they can hope for a good crop. Very few of these animals are to be found in any spot where the sugar-cane has been cultivated to any extent. Besides plants and nuts, the Agouti eats various fruits, displaying a strong predilection for nuts. Like many of the rodent animals, it is capable of varying its diet with animal substances, and will seldom refuse a piece of meat if offered.

It is a tolerably swift animal, as might be supposed from the great comparative length of its hinder limbs, but does not appear to be capable of sustaining a long chase. Open country is on that account rather distasteful to the Agouti, who prefers wooded districts,

AGOUTI.—*Dasyprocta Agouti*

where it can find shelter without being forced to run for any considerable distance. When running, it bears some resemblance to the common hare, and, like that animal, is rather apt to overbalance itself when running down hill, and to roll for some yards before it can recover itself.

All its movements are sharp, quick, and active, and even while sitting upright and engaged in feeding itself by the assistance of its fore-paws, its head is continually being turned from side to side, and its bright eyes glance in every direction, in order to guard against a surprise. As it is a nocturnal animal, and spends the whole of the day in its dark hiding-place, its ravages take place under cover of night, and are the more difficult to be repelled. Its usual resting-place is in the cleft of a rock, or in the hollow of some decaying tree, where twenty or thirty of these animals may be found living amicably together.

In these dark recesses the young Agoutis are born, and are laid upon a soft bed of leaves, where they remain for a few weeks, and then sally out with their parents on their nocturnal expeditions. There are generally two broods in each year, and the number of young at a birth is from three to six.

The Agouti can be readily domesticated, but is in no great favour as a pet, because it is so fond of exercising its sharp teeth upon any article of furniture which may fall in its way, and will in a very few minutes cut its way through an ordinary wooden door. Moreover, it ill repays the trouble which has been taken in taming it, for it seems to lose all its amusing qualities when it is once placed in an inclosure and furnished with regular food. It appears hardly to be capable of distinguishing kindness from cruelty, and displays but little emotion at the presence of the person who brings its daily food. It is naturally a gentle creature, and when captured will not attempt to bite the hand that seizes it, but only gives vent to a piteous squeak as it feels itself made a prisoner. The flesh of the Agouti is white, and good-flavoured, and is thought to resemble a mixture of the hare and rabbit. In some countries which it inhabits it is commonly eaten, while in others a prejudice prevails against its use as an article of diet.

The name, Dasyprocta, which has been given to the genus, refers to the thick hair which falls over the hind quarters, and nearly conceals the little pointed stump of a tail.

1.   P P

The hair of this part of the body is a bright golden-brown, but on the back and sides the fur has a curious speckled aspect, on account of the black, brown, and yellow tints with which each hair is marked. On the greater part of the body the fur is only about one inch in length, but the golden-brown hair of the hinder parts is more than four inches long. In character it is coarse, though glossy.

Though all the species are furnished with powerful claws, the Agouti is incapable of climbing trees or digging burrows. It is said to have some idea of laying up provisions, and to hide any superabundance of food in some place of concealment. In size it exceeds the common rabbit, but does not equal the hare.

BETWEEN the agoutis and the pacas is placed the MARA, or Patagonian Cavy, as it is sometimes called, an animal which is remarkably swift for a short distance, but is so easily fatigued that it can be run down by a man on horseback. It is more tameable than the agouti, and is often kept in a state of domestication, being permitted to range the house and premises at will. It is generally found in couples, a male and his mate occupying the same "form." It does not seem to burrow, nor to keep so close to its retreat as the agouti, but is fond of crouching in a form like our common hare. It is about thirty inches in length, and about nineteen inches high at the crupper, which is the most elevated part of the animal. At the shoulder it hardly exceeds sixteen inches. The fur of this animal is soft and warm, and from the contrasting colours of black, white, and golden-brown, presents a very handsome appearance. Its scientific title is *Dolichôtis Patachônicus.*

SOOTY PACA.—*Cœlógenys Paca.*

The PACAS are remarkable for the extraordinary development of a portion of the skull, which gives to the entire head a very singular aspect. The cheekbone is enormously developed into a large, expanded mass of bone, concave and very rough on the exterior, and smooth and concave interiorly. This enlarged bone is so enormous that its lower edge descends below the lower jawbone, and hides a considerable portion of it. Closely connected with this curious structure is a cheek-pouch, for which no use has hitherto been discovered. There are also two large cheek-pouches which open into the mouth, and extend past the jaws into the neck. In consequence of this formation, the name of Cœlógenys, or "Hollow-cheek," has been given to this genus.

The Pacas inhabit Southern America, being mostly, if not entirely, restricted to the eastern portions of that country, and have also been found in some of the West Indian islands.

The DUSKY PACA is really a pretty animal, the rows of white spots which decorate its sides standing out in pleasing contrast to the rich black-brown hue with which the remainder of the fur is tinged. The throat and abdomen are white, and the lowermost of the four rows of white spots is often nearly merged into the white fur of the under portions of the body. The colouring is rather variable in different individuals. The paws are light flesh-colour, and the large full eyes are dark brown. The total length of this animal is about two feet.

In its native land it is quite as destructive as the agouti, and, like that animal, is a terrible foe to the sugar-canes, which are too frequently destroyed in great quantities by the nocturnal visits of the Paca. The aggrieved planters retaliate by making diurnal attacks on the Paca burrows when they know the animal will be at home ; and by stopping up two of the three entrances which lead to the secret chamber of the midnight robber, are enabled to dislodge the hidden animal from its retreat. When hard pressed, the Paca turns fiercely on its assailant, and fights desperately in defence of itself and its home.

It is an active animal, in spite of its clumsy looks, and not only runs with considerable speed, but is a good swimmer, and can jump well.

The favourite localities of the Pacas are in wooded districts, in marshy grounds, or near the banks of rivers. Their domiciles are excavated in the ground, but are at no great depth, and are remarkable for the admirable state of cleanliness in which they are preserved by the inhabitants. The burrows are often so shallow that their roofs cannot support any superincumbent weight, and will give way under the tread of man or horse.

When properly dressed by being scalded and roasted, the flesh of the Paca is much esteemed, although it is too rich and fat to please the palates of some persons. The fur is of little value, being short and harsh, so that the skin of the Paca is useless until it has been deprived of hair and tanned. The Paca is only moderately intelligent ; and when in captivity appears, like the agouti, to lose a great portion of the bright intelligence which characterises its wild nature. It appears to take great care of its fur, and is as fastidious in its toilet as the domestic cat, washing itself in the same manner, and combing itself carefully with the claws of its hind and fore feet.

FEW persons, on seeing a CAPYBARA for the first time, would be inclined to class it with the animals to which it is so nearly related. The great size, the harsh, coarse hair, more like the bristles of a hog than the soft, delicate fur, which clothes the generality of rodent animals, the hoof-like toes, and the heavy, clumsy bearing of the animal, are so swinish in appearance that any ordinary spectator might well imagine that he saw before him a very curious example of the wild-hogs. In allusion to the external resemblance which this animal bears to the swine, it has received the name of Hydrochœrus, or Water Hog. It is a native of Southern America, and has a rather wide range.

The Capybara is the largest of all the living rodent animals, rather exceeding three feet in total length, and being so bulkily made that when it walks its abdomen nearly touches the ground. The muzzle of this animal is heavy and blunt, the eyes are set high in the head, and are moderate in size, the tail is wanting, and the toes are partially connected together by a development of the skin. The colour of the Capybara is rather indeterminate, owing to the manner in which the hairs are marked with black and yellow, so that the general idea which its coat presents is a dingy, blackish-grey, with a tinge of yellow. The hairs are rather long, and fall heavily over the body. The incisor teeth are of enormous dimensions, and the molars are very curiously formed, presenting some analogy to those of the elephant.

It is a water-loving animal, using its webbed feet with great power, and fleeing instinctively to the stream when terrified by real or imaginary danger. It not only swims well, but is a good diver ; and when endeavouring to escape from a foe, always tries to evade its pursuer by diving as long as its breath will hold out, and only permitting the top of its head to appear above the surface when it rises for the purpose of respiration. As, however, it can remain under water for a space of eight or ten minutes, it finds no difficulty in escaping from any ordinary foe, if it can only gain the shelter of the welcome

CAPYBARA.—*Hydrochœrus Capybára.*

stream. The food of this animal is exclusively vegetable, and its curious teeth are needed in order to bruise the herbage on which it feeds into a mass sufficiently pulpy to enable it to pass through the very narrow throat.

The Capybara is a gregarious creature, being generally found in small herds upon the banks of the streams which they frequent. These animals are subject to considerable persecution at the hand of man and beast, as the flesh is remarkably good, and when properly treated can be preserved like ham or bacon. The jaguar preys largely on the Capybara, which is so large and fat that it affords a plentiful and succulent meal; and is so easily overcome that the jaguar finds no difficulty in supplying himself with a dinner. There is a kind of musky flavour about the flesh of the Capybara which is very attractive to some persons, but is equally repulsive to others. When startled, it utters a peculiar sound, something between a bark and grunt, in which an indefinite noise is produced, and a large amount of breath expended.

THE Cavies are well represented by the common GUINEA PIG.

Few animals have received less appropriate names than the Guinea Pig; for it is not a pig, but a rodent, and does not come from Guinea, but from Southern America. It is very easily tamed; for its disposition is so unimpressible and dull that it accommodates itself to change of locality without betraying any emotion, and seems hardly to be susceptible even of fear. Being a very pretty little creature, it is in some favour as a domestic pet; and as it is remarkably prolific, it very rapidly increases in numbers, if it is well defended from cold and preserved from damp, as without warmth and a dry habitation it soon dies. The food of the Guinea Pig is exclusively of a vegetable nature, and while feeding it generally sits on its hinder feet, and carries the food to its mouth with its fore-paws.

An idea of the extreme fecundity of this animal may be formed from the fact that it begins to breed at ten months of age, that each brood consists on an average of six or eight, and that in less than three weeks after the birth of the young family they are driven to shift for themselves, and the mother is then ready for another brood. The young Guinea Pigs are born with their eyes open, and covered with hair, and do not attain their full dimensions until they have reached the age of eight or nine months.

GUINEA PIG.—*Cávia Ajeráa*

The colour of the Guinea Pig is very variable; but is generally composed of white, red, and black, in patches of different size and shape in each individual. The bare portions of the skin are flesh-coloured, and the eye is brown. The animal is of little direct use to mankind, as its flesh is held in very low estimation, and its hair is so slightly attached to the skin that its coat is useless to the furrier. There was formerly a prevalent idea that rats had an especial antipathy to the Guinea Pig, and would not haunt any place where one of these animals was kept. Rabbit owners were therefore in the habit of placing a Guinea Pig in the same apartment with the hutches, in hopes of scaring away the rats, which are the chief enemies of tame rabbits. As, however, in several instances the Guinea Pigs were eaten by the rats instead of driving them from the premises, the custom has gradually fallen into deserved disrepute.

THE group of animals which is known by the name of Leporidæ, from the Latin word, *lepus*, a hare, is easily distinguishable from the other rodents by the peculiar dentition of the upper jaw. Usually there are only two incisor teeth in that jaw; but in the Leporidæ there are four incisors, a pair of smaller teeth being placed immediately behind the usual upper incisors.

The common HARE is known from the rabbit by the redder hue of its fur, the great proportionate length of its black-tipped ears, which are nearly an inch longer than the head; by its very long hind legs, and its large and prominent eyes. When full-grown it is of considerable size, weighing on the average about eight or nine pounds, and sometimes attaining the weight of twelve or even thirteen pounds. In total length it rather exceeds two feet, the tail being about three inches long. The colour of the common Hare is greyish-brown on the upper portions of the body, mixed with a dash of yellow; the abdomen is white, and the neck and breast are yellowish-white. The tail is black on the upper surface and white underneath, so that when the creature runs it exhibits the white tail at every leap. Sometimes the colour of the Hare deepens into black, and there are many examples of albino specimens of this animal.

It is popularly supposed to be a timid animal, and has therefore received the specific title of "timidus;" but it is really possessed of no small share of courage. According to a well-known English writer—not a sportsman—we malign the poor creature by stigmatizing it as cowardly or timid, because it runs away when it is hunted. Half a hundred horsemen, together with a pack of dogs, band together in pursuit of one defenceless Hare, which is likely to run away under such circumstances. There is hardly any animal, from an elephant or lion downwards, that would not run away in like manner ; and it is very unfair to brand the poor Hare with an offensive epithet because it does not attempt to fight a field of horsemen and a pack of hounds.

However disposed the Hare may be to flight, when matched against such overwhelming odds, she is really a courageous animal when more fairly dealt with.

A countryman had captured a young leveret in a furrow, and was proceeding to mark it by notching its ears, when he was interrupted in his work by the mother Hare, which flew at him with singular courage, and struck so fiercely with her feet that she tore his hands rather severely. Finding that she could not release her child, she stood within a few feet of the captor, and waited patiently until he liberated the little Hare, with which she went off. The Hare is a very pugnacious animal, and is in the habit of waging the most savage fights with those of its own species.

The very long and powerful hind legs of the Hare enable it to make prodigious bounds, and to cover a considerable space of ground at every leap. The hinder limbs are, indeed, of such great proportionate length that the animal does not walk, but proceeds by a series of hops or leaps. The Hare is so constituted that it never becomes fat, however rich and fertile may be the pasture in which it feeds, and is therefore enabled to run for a very great distance without being fatigued, as would be the case if its muscles were loaded with fat. It can also leap to a considerable height, and has been known to jump over a perpendicular wall of eight feet in height in order to escape from its pursuers.

It is a wonderfully cunning animal, and is said by many who have closely studied its habits to surpass the fox in ready ingenuity. Appearing to understand the method by which the hounds are enabled to track its footsteps, it employs the most crafty manœuvres for the purpose of throwing them off the scent. Sometimes it will run forwards for a considerable distance, and then, after returning for a few hundred yards on the same track, will make a great leap at right angles to its former course, and lie quietly hidden while the hounds run past its spot of concealment. It then jumps back again to its track, and steals quietly out of sight in one direction, while the hounds are going in the other.

The hare also displays great ingenuity in running over the kind of soil that will best suit the formation of her feet, and be most disadvantageous to her pursuers, and has been known, on more than one occasion, to break the line of scent most efficiently by leaping into some stream or lake, and swimming for a considerable distance before she takes to the land again. A Hare has been seen to brave the salt waters and tossing waves of the sea when closely pressed by the hounds, and to evade them by its bold ingenuity. Sometimes an old crafty Hare will baffle the hounds for a succession of seasons, until it is as familiar to the hunters as any of the dogs or horses, and makes the hounds so ashamed of their failures that they cannot be induced to chase it with any good will.

As may be supposed from the fact of its taking the water, the Hare is a good swimmer, and can sustain itself upon the surface for no inconsiderable time. One of these animals was seen to swim to an island which was at least a mile distant from the main land, and to perform its task right bravely. The clever animal actually waited upon the shore until slack water, when the tide is not running, and having ascertained this fact by frequently examining the rippling waves as they came curling over the beach, launched itself boldly upon the water, and swam rapidly to the nearest point of land.

Although possessed of a remarkably delicate sense of hearing, and furnished with very quick eyesight, the Hare seems to employ those senses upon objects which are behind her rather than on those in her front. On more than one occasion a Hare has been known to swerve in her course, and to run into the very midst of the hounds without having either seen or heard them.

The Hare does not live in burrows, like the rabbit, but only makes a slight depression in the ground, in which she lies so flatly pressed to the earth that she can hardly be distinguished from the soil and dried herbage among which she has taken up her

HARE.—*Lepus timidus.*

temporary abode. Although she has no definite home, the Hare is strongly attached to her "form," wherever it may be placed, and even if driven to a great distance by the hounds, contrives to regain her little domicile at the earliest opportunity. As the varying seasons of the year bring on their varied accompaniments of heat and cold, rain and drought, or clouds and sunshine, the Hare changes the locality of her "form," so as to be equally defended against the bitter frost and snows of winter, or the blazing rays of the noontide summer sun.

In countries where the snow lies deep in winter, the Hare lies very comfortably under the white mantle which envelops the earth, in a little cave of her own construction. She does not attempt to leave her form as the snow falls heavily around her, but only presses it backward and forward by the movement of her body, so as to leave a small space between herself and the snow. By degrees the feathery flakes are formed into a kind of domed chamber, which entirely incloses the inhabitant, with the exception of a little round hole which is preserved by her warm breath, and serves as a ventilating aperture. This air-hole is often the means of her destruction as well as of her safety, for the scent which issues from the aperture betrays her presence to the keen nostrils of the dogs which accompany the a solitary hare-hunter, and which are trained to search for these air-holes and stand sentinels over them until their master arrives and captures the hidden victim.

When "preserved" in great numbers, the Hare is a most troublesome neighbour to the farmer, as it does great damage to the crops of all kinds, eating the tender blades of wheat almost as soon as they peep through the earth, invading the garden, and even destroying great quantities of young trees, by nibbling the green bark from the whole of their circumference. These depredations can hardly be checked, as the animal lies quietly in its "form" during the daytime, and makes long nocturnal journeys in order to procure its food, so that the owner of the garden or field can have no clue to the home of the thief which has injured him.

It is a tolerably prolific animal, beginning to breed when only a year old, and producing four or five young at a litter. The young Hares, or "leverets," as they are technically termed, are born with their eyes open, and covered with hair. For the space of four or five weeks they remain under the care of their mother, but after that time they separate, and depend upon themselves for subsistence.

THE common Hare is not found in Ireland, but the Irish Hare, *Lepus Hibérnicus*, is extremely common in that country, and takes the place of the common *Lepus timidus*. It may be distinguished from its English relation by its shorter limbs, its round head, and short ears, which are not so long as the head. According to some writers, the Irish Hare is identical with the Alpine Hare, and ought to be ranked with that animal, under the title of *Lepus variábilis*, or Variable Hare, in reference to the annual blanching of its coat during the winter months.

RESEMBLING the hare in general appearance and in many of its habits, the RABBIT is readily distinguished from that animal by its smaller dimensions, its different colour, its shorter and uniformly brown ears, and its shorter limbs.

The Rabbit is one of the most familiar of British quadrupeds, having taken firm possession of the soil into which it has been imported, and multiplied to so great an extent that its numbers can hardly be kept within proper bounds without annual and wholesale massacres. As it is more tameable than the hare, it has long been ranked among the chief of domestic pets, and has been so modified by careful management that it has developed itself into many permanent varieties, which would be considered as different species by one who saw them for the first time. The little brown short-furred wild Rabbit of the warren bears hardly less resemblance to the long-haired, silken-furred Angola variety, than the Angola to the pure lop-eared variety with its enormously lengthened ears and its heavy dewlap.

In its wild state, the Rabbit is an intelligent and amusing creature, full of odd little tricks, and given to playing the most ludicrous antics as it gambols about the warren in all the unrestrained joyousness of habitual freedom. To see Rabbits at their best it is necessary to be closely concealed in their immediate vicinity, and to watch them in the early morning or at the fall of evening. No one can form any true conception of the Rabbit nature until he has observed the little creatures in their native home; and when he has once done so, he will seize the earliest opportunity of resuming his acquaintance with the droll little creatures.

To describe the manifold antics of a Rabbit warren would occupy the space which ought to be devoted to some twenty or thirty animals, and even then would be quite inadequate to the proposed task. They are such odd, quaint, ludicrous beings, and are full of such comical little coquetries and such absurd airs of assumed dignity, that they sorely try the gravity of the concealed observer, and sometimes cause him to burst into irrepressible laughter, to their profound dismay.

At one time they are gravely pattering about the doors of their subterranean homes, occasionally sitting upright and gazing in every direction, as if fearful of a surprise, and all behaving with the supremest gravity. Next moment, some one gets angry, and stamps his feet fiercely on the ground as a preliminary observation before engaging in a regular fight. Suddenly a whole party rush off at full speed, scampering over the ground as if they meant to run for a mile at least, but unexpectedly stop short at an inviting tuft of herbage, and nibble it composedly as if they had not run a yard. Then a sudden panic will flash through the whole party, and with a rush and a scurry every rabbit leaps into its burrow and vanishes from sight like magic. The spot that was so full of life but a moment since is now deserted and silent as if it had been uninhabited for ages; but in a few minutes one little nose is seen cautiously poked out of a burrow, the head and ears follow, and in a very few minutes the frightened Rabbits have come again into the light of day, and have recommenced their interrupted pastimes.

Few animals are so easily startled as the Rabbit, and with perfect good reason. For their enemies are found in so many directions and under such insidious guises, that they are well justified in taking every possible precaution for their safety. Sundry rapacious birds are very fond of young Rabbits, and swoop down unexpectedly from some unknown aërial region before the doomed creature can even comprehend its danger. Stoats and weasels make dreadful havoc in a warren, and even the domestic cat is sadly apt to turn poacher if a well-stocked warren should happen to be within easy distance of her home. Foxes are very crafty in the pursuit of young Rabbits, and dig them out of the ground in a very ingenious and expeditions manner ; while the common hedgehog is but too apt to indulge its carnivorous appetite with an occasional Rabbit.

The burrows in which the Rabbit lives are extremely irregular in their construction, and often communicate with each other to a remarkable extent.

From many of its foes, the Rabbit escapes by diving suddenly into its burrow ; but there are some animals, such as the stoat, weasel, and ferret, which follow it into its subterranean abode, and slay it within the precincts of its own home. Dogs, especially

RABBIT.—*Lepus cuniculus.*

those of the small terrier breeds, will often force their way into the Rabbit burrows, and have sometimes paid the penalty of their life for their boldness. The Rabbit has been seen to watch a terrier dog safely into one of the burrows, and then to fill up the entrance so effectually that the invader has not been able to retrace his steps, and has perished miserably beneath the surface of the ground.

When the female Rabbit is about to become a mother, she quits the ordinary burrows, and digs a special tunnel for the purpose of sheltering her young family during their first few weeks of life. At the extremity of the burrow she places a large quantity of dried herbage, intermixed with down which she plucks from her own body, so as to make a soft and warm bed for the expected occupants. The young Rabbits are about seven or eight in number, and are born without hair and with their eyes closed. Not until they have attained the age of ten or twelve days are they able to open their eyelids and to see the world into which they have been brought.

When domesticated, the female Rabbit is sometimes apt to eat her own young, a practice which has been considered as incurable. It seems, however, that the Rabbit acts in this apparently unnatural manner from very natural causes. It has long been the custom to deprive domestic Rabbits of water, on the plea that in a wild state they never drink, but obtain the needful moisture from the green herbage on which they feed. But in the open country, they always feed while the dew lies heavily upon every blade, which is never the case with the green food with which our domestic Rabbits are supplied. Moreover, we feed our Rabbits very largely on bran, pollard, oats, and other dry nourishment which they do not obtain in their normal state of freedom. The mother Rabbit instinctively licks her young when they are born, and is evidently liable to an exceeding desire for liquid nourishment which prompts her to eat anything that may assuage her burning thirst. A Rabbit, which had already killed and begun to eat one of her offspring, has been seen to leave the half-eaten body and to run eagerly to a pan of water which was placed in her hutch. It may easily be supposed that when an animal is obliged to afford a constant supply of liquid nourishment to her young, she is forced to imbibe a sufficiency of fluid to enable her to comply with the ever recurring demands of her offspring.

Rabbits are terribly destructive animals, as is too well known to all residents near a warren, and are sad depredators in field, garden, and plantation, destroying in very wantonness hundreds of plants which they do not care to eat. They do very great damage to young trees, delighting in stripping them of the tender bark as far as they can reach while standing on their hind feet. Sometimes they eat the bark, but in many cases they leave it in heaps upon the ground, having chiselled it from the tree on which it grew, and to which it afforded nourishment, merely for the sake of exercising their teeth and keeping them in proper order, just as a cat delights in clawing the legs of chairs and tables.

When the Rabbits have begun to devastate a plantation, they will continue their destructive amusement until they have killed every tree in the place, unless they are effectually checked. There are only two methods of saving the trees—one of killing all the Rabbits, and the other by making them disgusted with their employment. The latter plan is generally the most feasible, and can be attained by painting each tree with a strong infusion of tobacco, mixed with a sufficiency of clay and other substances to make it adhere to the bark. This mixture should be copiously applied to the first three feet of every tree, so that the Rabbit cannot find any portion of the bark that is not impregnated with the nauseous compound, and is an effectual preservative against their attacks.

In their normal state of freedom, Rabbits feed exclusively on vegetable food, but in domestication they will eat a very great variety of substances. Many of my own Rabbits were very fond of sweetmeats, and would nibble a piece of hardbake with great enjoyment, though they were always much discomposed by the adhesive nature of their strange diet, and used to shake their heads violently from side to side when they found themselves unable to disengage their teeth. They would also eat tallow candles, a fact which I discovered accidentally, by seeing them devour a candle-end that had fallen out of an old lantern. These curious predilections were the more unaccountable, because the animals were most liberally supplied with food, and were also permitted to run in the kitchen garden for a limited time daily, and to feed upon the growing lettuces, parsley, carrots, and other vegetables, as they pleased.

As a general fact, the Rabbit has a great antipathy to the hare, so that the two animals are seldom, if ever, seen in close proximity. The possibility of a hybrid progeny between the two species was, until late years, entirely denied. There are, however, several accidental instances of such a phenomenon, and in every case the father has been a Rabbit and the mother a hare. There are many examples of young Rabbits which possess much of the colouring and general aspect of the hare, but these are almost invariably the offspring of domesticated Rabbits which have been turned into a warren.

In its native state, the fur of the Rabbit is of nearly uniform brown, but when the animal is domesticated, its coat assumes a variety of hues, such as pure white, jetty black, pied, dun, slaty-grey, and many other tints.

The CHINCHILLA, so well known for its exquisitely soft and delicate fur, belongs to the group of animals which are known to zoologists under the title of Jerbóidæ, and which are remarkable for the great comparative length of their hinder limbs, and their long, hair-clothed tails.

The Chinchilla is an inhabitant of Southern America, living chiefly among the higher mountainous districts, where its thick silken fur is of infinite service in protecting it from the cold. It is a burrowing animal, digging its subterranean homes in the valleys which intersect the hilly country in which it lives, and banding together in great numbers in certain favoured localities. The food of the Chinchilla is exclusively of a vegetable nature, and consists chiefly of various bulbous roots, which it disinters by means of its powerful fossorial paws. While feeding, it sits upon its hinder feet, and conveys the food to its mouth with its fore-feet, which it uses with singular adroitness. It is a most exquisitely cleanly animal, as might be supposed from the beautiful delicacy of its fur, for we may always remark, that whenever an animal is remarkable for the colouring or the texture of its natural robes, it is always most assiduous in preserving them from any substance that might stain their purity or clog their fibres.

The fur of the Chinchilla is of a delicate clear grey upon the back, softening into a greyish-white on the under portions, and its texture is marvellously soft and fine. As the fur seems to be of two different qualities in animals that are brought from different parts of South America, it is supposed that there may be either two distinct species of this animal, or at least two permanent varieties, the hair of one being very much more delicate than that of the other. Besides being dressed and employed as a fur, the hair of the Chinchilla is so long and soft that it is well adapted for the loom, and has been manufactured into various fabrics where warmth and lightness are equally required.

CHINCHILLA.—*Chinchilla laniger.*

As the animal is very small, only measuring fourteen or fifteen inches in total length, the tail occupying nearly one-third of the measurement, many skins are employed in the manufacture of one article of ordinary dress, and the destruction of the Chinchilla is necessarily very considerable in order to supply the constant demand for this deservedly popular fur.

As far as is known, the Chinchilla is not a very intelligent animal, seeming to be hardly superior to the guinea pig in intellect, and appearing scarcely to recognise even the hand that supplies it with food.

The LAGÓTIS is distinguishable from the preceding animal by the structure of the fore-feet, which are only furnished with four toes, while those of the chinchilla possess five. The ears are very long in proportion to the head, and being somewhat similar to those of the hare, have gained for the animal the generic name of Lagotis, or Hare-eared.

The hinder limbs are long, and very much resemble those of the hare or rabbit; and the whole aspect of the creature partakes greatly of the leporine character. The coat is very like that of the hare in colour and texture, and is soft, long, and rather woolly, but as it is only slightly attached to the skin is valueless as a fur. The long ears are rounded at their extremities, and their margins are rolled inwards. The tail is so long that it forms a ready means of separating the Lagotis from the hares or rabbits, being quite as long as the body, and thickly covered with stiff hairs.

It is an inhabitant of Peru, and takes up its residence in the crevices of the rocky

LAGOTIS.—*Lagótis Cuviéri.*

localities among which it dwells. Although tolerably active, it appears to be possessed of little endurance, never attempting to escape by speed if it should chance to be alarmed, but diving at once into the welcome shelter of the nearest cranny. When wounded, they always seek the same retreat, so that unless they are killed by some instantly mortal injury, their bodies cannot be recovered by the hunter. The fur of this animal is so slightly attached to the skin that it comes away when handled. The flesh, however, is delicate and tender, and it is chiefly for the sake of its value as an article of food that the Lagotis is hunted.

The GERBOAS bear a curious resemblance to the kangaroos, not only in their general appearance, but in many of their habits. Like those animals, they leap over distances which are absolutely enormous when the size of their bodies is taken into consideration, they constantly sit upright in order to observe surrounding objects, their food is of the same nature, and they carry it to their mouths in a similar manner. Their fore-limbs are extremely short, while the hinder legs and feet are developed to a very great extent, and they are all furnished with a long, hair-clad tail, which serves to aid them in preserving their balance while shooting through the air.

One of the most familiar of these leaping rodents is the SPRING HAAS, or CAPE GERBOA, sometimes called, from its hare-like aspect, the CAPE LEAPING HARE.

It is a native of Southern Africa, and is found in considerable numbers upon the sides of mountains, where it inhabits certain burrows which it tunnels for itself in the ground. It prefers sandy ground for the locality of its habitation, and associates together in great profusion in favourable spots, so that the earth is completely honeycombed with its burrows. Being a nocturnal animal, it is rarely seen by daylight, seldom leaving its stronghold as long as the sun is above the horizon. The natives, who set some value on its flesh, take advantage of this habit, and being sure of finding the Spring Haas at home during the daytime, take their measures accordingly. Placing a sentinel at the mouth of the burrow, they force the inmate to evacuate the premises by pouring a deluge of water into the hole, and as it rushes into the open air, it is seized or struck down by the ready hand of the sentinel.

SPRING HAAS.—*Hélamys Capensis.*

Like the kangaroos, the Spring Haas prefers rough and rocky ground to a smooth soil, and displays such wonderful agility as it leaps from spot to spot, that it can baffle almost any foe by its mere power of jumping. At a single leap this creature will compass a space of twenty or thirty feet, and will continue these extraordinary bounds for a great distance. It is rather a mischievous animal, as, like the common hare, it is in the habit of making nocturnal raids upon the corn-fields and gardens, and escaping safely to its subterranean burrow before the sunrise.

With the exception of shorter ears, and the elongated hinder limbs, the Spring Haas is not unlike our common hare. The fur is of a dark fawn, or reddish-brown, perceptibly tinged with yellow on the upper parts, and fading into greyish-white beneath. In texture it is very similar to that of the hare. The tail is about as long as the body, and is heavily covered with rather stiff hairs, which at the extremity are of a deep black hue. Upon the fore-legs there are five toes, which are armed with powerful claws, by means of which the animal digs its burrows, while the hinder feet are only furnished with four toes, each of which is tipped with a long and rather sharply pointed claw.

THE Jerbóidæ find their best type in the common GERBOA of Northern Africa.

This beautiful and active little animal is hardly larger than an ordinary English rat, although its peculiar attitudes and its extremely long tail give it an appearance of greater dimensions than it really possesses. The general colour of its fur is a light dun, washed with yellow, the abdomen being nearly white. The tail is of very great proportionate length, is cylindrical in shape, and tufted at its extremity with stiff black hairs, the extreme tip being white. From various experiments that have been made upon this member and its use to the animal, it appears that the tail is of infinite service in preserving the proper balance of the body while the creature is flying through mid-air in its extraordinary leaps; for in proportion as the tail was shortened, the power of leaping diminished, and when it was entirely removed, the animal was afraid to leap at all. Such truncated specimens were almost deprived of all power of locomotion, for they could never preserve their balance as they rose upon their hinder feet, but rolled over on their backs. As the Gerboa rises from one of its huge bounds for the purpose of commencing

GERBOA.—*Dipus Ægyptius.*

a second leap, it curves its tail into the peculiar form which is represented in the engraving, but straightens it in its aërial course.

The Gerboa is a burrowing animal, and lives in society, so that it forms large natural "warrens" in those parts of the country where it takes up its residence. It is much hunted by the natives, who set some store by its rather unpalatable flesh, and is captured by stopping up as many burrows as can conveniently be reached, and killing the Gerboas as they rush affrighted from the open entrances. This is, indeed, almost the only successful mode of capturing these fleet and agile creatures; for if they can once leap away from the immediate vicinity of their pursuers, they scour over the ground with such wonderful speed that they can hardly be overtaken even by a trained greyhound.

Dry and sandy spots are in greatest favour with the Gerboa, which is better able to dig in such soils than in moist situations. Against the injurious effects of the hard and burning ground upon its feet it is guarded by a thick covering of stiff, bristly hairs, which defend the soles of the feet from injury, and, moreover, are useful in giving a firm hold upon the ground when the animal is in the act of making one of its extraordinary bounds. It is a lively and playful animal, delighting to bask itself in the sun near the entrance of its burrows, and to divert itself by occasional gambols with its companions. Although it makes these visits to the open air for the sake of enjoying the warm beams of the sun, the Gerboa is a nocturnal animal, and feeds only by night.

By the united powers of its teeth and claws it can drive its tunnel through impediments which would baffle any ordinary animal; for it can not only cut its way through the hardest sand, but is even able to gnaw a passage through the thin layer of stone which lies beneath the sand. The food of these animals consists chiefly of roots and similar substances, which it digs out of the earth, but it also feeds on various kinds of grain.

The generic term "Dipus," or two-footed, has been given to the true Gerboas because they press their fore-feet so closely to their breasts while they leap that they appear to be entirely destitute of those limbs, and only to possess the two long hind legs. All the animals that belong to this genus have five toes on their fore-feet and only three on the hinder feet. The hair of the tail is arranged in a double row, after the manner which is scientifically called "distichous."

THERE are many species of Jerboidæ inhabiting different countries, all of which are very similar in shape and habits. Among these may be noticed the ALACTÁGA, or Jumping Rabbit of Siberia, and the GERBILLES of Africa and India.

NEXT in order to the Gerboas is placed the small group of animals which are sufficiently familiar by the name of Dormice. This term signifies "Sleepy Mouse," and is most appropriate to the lethargic little creatures, which spend the greater part of their time in somnolency. One of these animals, the LOIRE, or FAT DORMOUSE, is celebrated in classical literature as being in great request among the luxurious Romans as an article of diet. For this purpose the Loire was carefully fattened, being placed in certain receptacles, which were called Gliraria, from the Latin word *glis*, which signifies a dormouse. The Loire is found in almost all the warmer portions of Europe, but is seldom seen at any great elevation above the level of the sea.

The LEROT, or GARDEN DORMOUSE, inhabits the same localities as the loire, but its flesh is not eatable like that of the Fat Dormouse, although it resembles it very nearly in every point but size.

The total length of this animal is rather more than eight inches, of which measurement the tail occupies three inches. The general colour of its fur is grey, deeply tinged with red upon the back, and becoming white upon the abdomen. Below the eye is a patch of black fur, which extends nearly to the ears. The tail of the Lerot is covered with short black hair, changing rather abruptly into white at its extremity. It has derived its title of Garden Dormouse from its annoying habit of entering gardens and making sad havoc of the choicest fruit; for it is an animal of great taste, and makes its selection among the ripest and best fruits with an accuracy of judgment that may be highly agreeable to itself, but is sincerely execrated by the owner of the garden. It is particularly fond of espalier-trained fruit-trees, and is much given to devouring the peaches when they are just in their bloom of rosy perfection.

The Lerot is not content with making these autumnal raids upon the gardens, but is sufficiently provident to lay up a store of food for the winter, and for that purpose to carry off corn, peas, and beans in no small quantity. Its winter nest is made in some convenient recess, where six or eight Lerots congregate, and pass the cold wintry months in a slumber

LEROT.—*Myoxus quercinus.*

which is almost unbroken, except by the needful occasional wakings for the purpose of taking food. In summer time it makes a temporary nest in hollow trees, holes in old walls, or in similar localities, and reposes during the daytime upon a bed of dried grass and leaves. Sometimes it is so bold that it will make its way into human habitations, and establish itself in the very home of the justly incensed owner of the garden.

The young Lerots enter upon their existence in the middle of summer, and grow with wonderful rapidity. They do not, however, become parents in their turn until the following year. The average number of young Lerots which are produced at a single birth is from four to six.

THE common DORMOUSE is abundantly found in many districts of England, as well as on the Continent, and is in great favour as a domestic pet.

The total length of this pretty little animal is rather more than five inches, the tail being two inches and a half long. The colour of its fur is a light reddish-brown upon the back, yellowish-white upon the abdomen, and white on the throat. These tints belong to the adult animal only, as in the juvenile Dormouse the fur is nearly of the same colour as that of the common mouse, the ruddy tinge only appearing on the head and sides. It is not until the little creatures have nearly completed a year of existence that they assume the beautiful hues of adult age. The tail is thickly covered with hair, which is arranged in a double row throughout its length, and forms a slight tuft at the extremity. The head is rather large in proportion to the body, the ears are large and broad, and the eye full, black, and slightly prominent.

DORMOUSE. –*Muscardinus avellanárius.*

The Dormouse is a nocturnal animal, passing the whole of the day in its warm and neatly constructed nest, which is generally built in the most retired spot of some thick bush or small tree. It is a very active little creature, leaping from branch to branch, and traversing the intricate mazes of the brushwood with such ready featness, that it can scarcely be taken by a human hand. Generally, when a Dormouse is captured, it is secured while sleeping in its nest, for during its slumbers it is so deeply buried in repose that it can be handled without offering resistance or attempting escape. The food of the Dormouse consists of various fruits and seeds, such as acorns, nuts, haws, and corn.

As the animal is one of the hibernaters, it is in the habit of gathering together a supply of dried food, to afford occasional nourishment during the long wintry months when it lies in its bed, imprisoned in the bands of irresistible sleep. Like many other hibernating animals, the Dormouse becomes exceedingly fat towards the end of autumn, and is therefore enabled to withstand the severity of the winter season better than if it retired into its home in only its ordinary condition. As soon as the weather becomes cold, the Dormouse retires into its nest, and there slumbers throughout the entire winter, waking up for a short period whenever a milder temperature breaks the severity of the frost, and after taking a little nourishment, sinking again into its former lethargy. Several interesting experiments have been made on this animal in connexion with the phenomenon which is termed hibernation, and with the same results as have already been mentioned when treating of the hedgehog and the bat.

This hoard of provisions is not gathered into the nest, which is solely employed for the purpose of warmth and concealment, but is hidden away in sundry convenient nooks and crannies, close to the spot where the nest is placed. Comparatively little of the store is eaten during the winter, unless, indeed, the weather should happen to be peculiarly mild, but it is of very great service in the earlier part of the spring, when the Dormouse is awake and lively, and there are as yet no fresh fruits on which it could feed.

The Dormouse is rather gregarious in its habits, so that whenever one nest is discovered, several others may generally be found at no very great distance. These nests are of considerable dimensions, being about six inches in diameter, and are composed of grass, leaves, and similar substances. The entrance to the nest is from above.

The young animals are generally three or four in number at a birth, and make their appearance about the end of spring, or the beginning of summer. It is probable that there may be a second brood towards the end of autumn, as Mr. Bell received from one locality in the month of September one half-grown Dormouse, which had evidently been born in the spring, and three very little specimens, which were apparently not more than a week or two old. They are born blind, but are able to see in a very few days, and in a remarkably short space of time become independent of their parents.

Like many other rodent animals, the Dormouse carries the food to its mouth with its fore-paws, while it sits upright on its hinder legs. It is also able to suspend itself by the hind-feet from any convenient branch, and may often be seen hanging in this manner, and eating as comfortably as if it were seated on firm ground. The Dormouse is not confined to England, but is spread over the whole of Southern Europe, and is common even in Sweden.

TAGUAN FLYING SQUIRREL.—*Pléromys Petaurista.*

THE beautiful and active group of animals of which our English Squirrel is so familiar an example, are found in almost every portion of the globe, and, with one or two exceptions, live almost exclusively among the branches of trees. In order to enable them to maintain a firm clasp upon the branches and bark, they are furnished with long, finger-like toes upon the fore-feet, which are armed with sharp curved claws.

In the Flying Squirrels, of which the TAGUAN is a good example, the skin of the flanks is modified in a method similar to that which has already been noticed in the Petaurists of Australia and the Colugo of Java. This skin is so largely developed, that when the animal is sitting at its ease, its paws but just appear from under the soft folds of the delicate and fur-clad membrane. When the creature intends to make one of its marvellous leaps, it stretches all its four limbs to their fullest extent, and is upborne

I. Q Q

through the air on the parachute-like expansion which extends along its sides. This animal is a native of India, where it is tolerably common.

It is rather a large species, as its total length is nearly three feet, the tail occupying about one foot eight inches, measured to the extremity of the long hairs with which it is so thickly clothed. The general colour of this animal is a clear chestnut, deepening into brown on the back, and becoming more ruddy on the sides. The little pointed ears are covered with short and soft fur of a delicate brown, and the tail is heavily clad with bushy hairs, greyish-black on the basal portions of that member, and sooty-black towards the extremity. The parachute membrane is delicately thin, scarcely thicker than ordinary writing-paper, when it is stretched to its utmost, and is covered with hair on both its surfaces, the fur of the upper side being chestnut, and that of the lower surface nearly white. A stripe of greyish-black hairs marks the edge of the membrane, and the entire abdomen of the animal, together with the throat and the breast, is covered with beautiful silvery greyish-white fur.

ASSAPAN.—*Sciuṙópterus Volucella.*

THERE are many other Flying Squirrels, belonging to different countries, but presenting very similar characteristics of form and character. They are all playful and lively animals, and engage in the most gamesome sports as they chase each other about the branches of the tree on which they have taken up their residence. Among these creatures we may record the names of the ASSAPAN, or Flying Squirrel of America, the POLATOUCHE of Siberia, and the RASOO of India.

THE true Squirrels possess no parachute flying membrane, as do the Flying Squirrels, nor are they furnished with cheek-pouches, as is the case with the Ground Squirrels of America.

One of the handsomest of the Squirrels is the JELERANG, or JAVAN SQUIRREL, a native of Java, part of India, and Cochin China. Its total length is about two feet, the tail and body being equal to each other in measurement. In colour it is one of the most variable of animals, so that it has been more than once described under different names. In the British Museum are several specimens of this animal, and all of them present many varieties in point of colour, while some are so very unlike each other that most persons would consider them to be separate species. Some specimens of this animal are pale yellow, while others are deep brown; in some the colour is tolerably uniform, while in others it is variously pied; but in all there seems to be a tolerably decided contrast between a darker and a lighter tint. From this circumstance it has sometimes been termed *Sciúrus bicolor*, or the two-coloured Squirrel.

In general, the darker hue prevails on the back and upper portions of the body, and the lighter tint is abruptly separated from it by a decided line of demarcation. The usual colour of the Jelerang is a dark brownish-black on the back, the top of the head yellowish, and the sides and abdomen golden yellow.

The Jelerang is rather common in the countries which it inhabits, and as it is very retiring in its habits, and dreads the proximity of mankind, it is not so mischievous a neighbour as is the case with the greater number of the Squirrels. It lives chiefly in the depths of the forests, and feeds upon the wild fruits that grow without any aid from the hand of mankind. It is easily tamed, and being an active, amusing animal, as well as possessed of a beautifully marked coat, is often domesticated among the inhabitants of the same country. The flesh of the Jelerang is thought to be very good, and is eaten by the natives.

The generic term *Sciúrus*, which is applied to all the animals that belong to this genus, is of Greek derivation, and signifies "shadow-tail," in allusion to the manner in which the Squirrels curl their bushy tails over their bodies, as if to shade them from the rays of the sun. The name *Ptéromys*, which is applied to the Taguan, is also taken from the Greek language, and signifies "winged-mouse."

EVERY one is familiar with the lively little English SQUIRREL, which makes the woods joyous with its active gambols, and is too often repaid for its gaiety by being captured and compelled to make sport for its owner within the narrow precincts of a wire cage.

This little animal is plentiful in many parts of England, and, indeed, is generally found wherever there is a tolerably large copse or a wood of moderate dimensions. In private grounds and parks it luxuriates, knowing instinctively that it may wander at its own will, unchecked and unharmed. Among the tree branches its powers of activity are absolutely surprising, for it will fling itself through such distances, and at such a height, that it seems likely to be dashed to pieces every instant. Yet it seldom or never makes a false step, and even if it should lose its foothold, it is not at all disconcerted, but spreads out its legs and bushy tail to their utmost expansion, so that it presents a large surface to the air, and comes quite lightly to the earth, even though it may have leaped from a considerable height.

On the ground it is not so much at its ease as when it is careering amid the branches of some large tree, and, as soon as it feels alarmed, always makes the best of its way towards the nearest tree trunk. Its gait is a kind of semi-gallop, and even when ascending a perpendicular tree stem, it maintains the same galloping movements, and ascends to a considerable height in a very small space of time.

JELERANG, OR JAVAN SQUIRREL.—*Sciúrus Javensis.*

To watch a little party of Squirrels in a tree is a most amusing occupation, but not very easily managed, as the little creatures are blessed with quick eyesight, and if they happen to spy any object which they fancy may be dangerous, they always keep themselves on the opposite side of the trunk or branches of the tree which they are traversing. So jealously do they guard themselves by the interposition of the branches, that it is most difficult to shoot one of these animals after it has once caught sight of the gunner. By dint of patience, however, it is possible to witness the whole proceedings of the merry little creatures, and to obtain a great fund of amusement by so doing.

Squirrel-hunting is always a great sport among boys, and is the more fascinating because the Squirrel is hardly ever captured in fair chase.

The only plan is to watch the animal until it has ascended an isolated tree, or by a well-directed shower of missiles, to drive it into such a place of refuge, and then to form a ring round the tree, so as to intercept the Squirrel if it should try to escape by leaping to the ground and running to another tree. The best climber is then sent in chase of the Squirrel, and endeavours, by violently shaking the branches, to force the little animal to loosen its hold and come to the earth. But it is by no means an easy matter to shake a Squirrel from a branch, especially as the little creature takes refuge on the topmost and most slender boughs which even bend under the weight of its own small body, and

SQUIRREL.—*Sciurus Europœus.*

can in no way be trusted with the weight of a human being. By dint, however, of perseverance, the Squirrel is at last dislodged, and comes to the ground as lightly as a snowflake. Hats, caps, sticks, and all available missiles are immediately flung at the luckless animal as soon as it touches the ground, and it is very probably struck and overwhelmed by a cap. The successful hurler flings himself upon the cap, and tries to seize the Squirrel as it lies under his property. All his companions gather round him, and great is the disappointment to find the cap empty, and to see the Squirrel triumphantly scampering up the trunk of some tree, where it would be useless to follow it.

During the hotter hours of the day the Squirrel is never seen, being quietly asleep in its lofty nest; but in the early morning, or in the cooler hours of the afternoon, it comes from its retreat, and may be seen leaping about the branches in search of the various fruits on which it feeds.

The nest of the Squirrel is an admirable specimen of natural architecture, and is almost invariably placed in the fork of some lofty branch, where it is concealed from the view of any one passing under the tree, and is out of the reach of any ordinary foe, even if its situation were discovered. Sometimes it is built in the hollow of a decayed bough, but is always admirably concealed from sight. In form it is nearly spherical, and is made of leaves, moss, grass, and other substances, woven together in so artistic a manner that it is impermeable to rain, and cannot be dislodged from its resting-place by the most violent wind. A single pair of Squirrels inhabit the same nest, and seem to consider some particular tree as their home, remaining in it year after year.

The female Squirrel produces about three or four young at a litter, the little ones being born in the middle of summer, and remaining under the care of their parents until the spring of the succeeding year, when they separate, and shift for themselves.

The food of the Squirrel is usually of a vegetable nature, and consists of nuts, acorns, wheat, and other fruits and seeds. Being a hibernating animal, the Squirrel is in the habit of laying up a winter store of provisions, and towards the end of autumn, while acorns and nuts are in their prime, becomes very busy in gathering certain little treasures, which it hides in all kinds of nooks, crevices, and holes, near the tree in which it lodges. The creature must be endowed with a very accurate memory, for it always remembers the spots where it has deposited its store of food, and even when the snow lies thickly upon the earth, and has covered the ground with a uniform white mantle, the Squirrel betrays no perplexity, but whenever it requires nourishment, goes straight to the hidden storehouse, scratches away the snow, and disinters its hidden treasures.

During the few last weeks of autumn, the Squirrel is quite in its element, paying daily visits to the nut-trees, and examining their fruit with a critical eye. Detecting intuitively every worm-eaten or defective nut, the Squirrel makes deliberate choice of the soundest fruit, and conveys it to the secret storehouse. Feeding abundantly on the rich products of a fruitful autumn, the Squirrel becomes very fat before the commencement of winter, and is then in its highest beauty, the new fur having settled upon the body, and the new hair having covered the tail with its plumy fringe.

The manner in which a Squirrel eats a nut is very curious. The little animal takes it daintily in his fore-paws, seats himself deliberately, and then carrying the nut to his mouth, cuts off the tip with his chisel-edged incisor teeth. He then rapidly breaks away the shell, and after carefully peeling the dry brown husk away from the kernel, eats it as if he had earned his little feast. Sometimes the food of the Squirrel is not limited to vegetable substances, as the animal possesses something of the carnivorous nature, and has been often found guilty of killing and eating sundry animated beings. Young birds, eggs, and various insects are eaten by the Squirrel, who has been detected in the very act of plundering a nest, and carrying off one of the young birds.

Although it is a most pretty and interesting animal, it is sometimes a very unpleasant neighbour, especially where there are plantations of young trees near the spot on which it has taken up its residence. It has a habit of nibbling the green and tender shoots as they sprout upon the topmost boughs, and often succeeds in stunting many a promising tree by its inveterate habit of exercising its teeth upon young wood.

The usual colour of the Squirrel's fur is a ruddy brown upon the back, and a greyish-white on the under portions of the body. It is, however, a most variable animal in point of colour, the tint of its fur changing according to the country which it inhabits. Even in England the ruddy fur is sometimes changed to grey during a severe winter, and in Siberia, it is generally of a bluish-grey. The feathery tufts of hair which fringe the ears are liable to great modifications, being very long and full in winter and in cold climates, and almost entirely lost during the hotter summer months of our own country.

It is easily tamed, and is in great request as a domestic pet. Let me here, however, warn the reader against purchasing the so-called tame Squirrels which are offered for sale in the streets. They appear at first sight to be very gentle, for they will permit themselves to be handled freely without displaying any signs of anger, and possess much of the quiet demeanour of a truly tame animal. But this quietude is almost invariably produced by a gentle dose of strychnine, which has the effect of reducing the poor creature to a state of non-resistance, and which, although it is always fatal in the end, is often sufficiently tardy in its operation to aid the vendor in completing his iniquitous sale. Those who desire to purchase a really tame Squirrel should also be careful to examine its mouth, for in some instances the incisor teeth are drawn, so that the poor animal is physically incapable of biting ; and in other cases, an old, yellow-toothed Squirrel is palmed off upon an incautious purchaser for a young animal.

THERE are so many species of the Squirrel tribe, that even a cursory notice of each animal would be wholly impracticable in a work of the present dimensions, and we must content ourselves with a brief description of those species which stand out more boldly from the rest, by reason of form, colour, or peculiar habits.

One of the most striking forms among the members of the genus Sciúrus is seen in the LONG-EARED SQUIRREL. This remarkable species is found in Borneo, and there is a tolerably good specimen in the collection of the British Museum. Although it is called the Long-eared Squirrel, its title is not due to the length of the ears, which are in reality hardly longer than those of an ordinary Squirrel, but to the very long hair-tufts with which those organs are decorated. The fringe of hair which adorns the ears is about two inches in length, of a glossy blackish-brown colour, and stiff in texture. The colour of the back and exterior of the limbs is a rich chestnut-brown, which fades into paler fawn along the flanks, and is marked by a single dark longitudinal stripe, extending from the fore to the hinder limbs. This dark band is narrow at each end, but of some

LONG-EARED SQUIRREL.—*Sciurus Macrótis.*

width in the centre. The inside of the limbs is a pale chestnut, and the paws are jetty black. The tail is remarkably bushy, reminding the spectator of a fox's "brush," and is generally of the same colour as the back, but grisled with yellowish-white hairs, which are thickly sown among those of the darker hue.

In length it is about two feet, of which the tail occupies one moiety. The word Macrótis is of Greek origin, and signifies Long-eared.

The BLACK SQUIRREL has most appropriately been named, for the whole of its fur, with very slight and variable exceptions, is dyed with the deepest jet.

Even the abdomen and under parts of the body, which in almost all quadrupeds are of a lighter hue than the back, are in the Black Squirrel of the same sable tinge, with the exception of a few small tufts of white hairs which are scattered at wide and irregular intervals. A few single white hairs are also sown sparingly upon the back, but are so few in number as to escape a mere casual glance. The tail is also slightly flecked with these white hairs. The total length of this animal is about two feet ten inches, the tail being about thirteen inches in length, measured to extremity of the fur. When the creature spreads its tail to its full width, it measures nearly five inches in diameter in the largest part.

The Black Squirrel is a native of many parts of Northern America, and is tolerably common in some localities, though very scarce in others. It is a curious fact, that it vanishes before the advent of the common northern Grey Squirrel, and in many instances has been driven from some of its private haunts and supplanted by the more powerful intruder. It seems to be rather a timid animal, as it has been observed to fly in terror when threatened with the anger of the Red Squirrel (*Sciúrus Hudsónius*). Despite its cowardice, it is rather a fierce creature when captured, biting savagely at its opponent, and is not very easily tamed. One of these animals which was partially domesticated, was always noted for its evil temper, and justified the opinion that had been formed of its disposition by biting a piece from a servant's hand as cleanly as if it had been cut with a chisel. The injury was of so severe a nature that the man was obliged to go into a hospital for some weeks.

When undisturbed in its native domains, it appears to be an active and lively animal, and is remarkable for a curious habit of suddenly ceasing its play and running to the water side to refresh itself before it recommences its sport. In drinking it does not lap after the manner of dogs and cats, but bends over the water, and thrusting its nose fairly beneath the surface, drinks a steady draught. After it has satisfied its thirst, it sits on its hind legs, and with its fore-feet carefully washes its face, occasionally dipping its paws into the water, as if to perform its ablutions in the most effectual manner.

BLACK SQUIRREL.—*Sciurus niger.*

The skin of the Black Squirrel is rather valuable, as it not only possesses the uniform jetty hue which is so universally admired in ornamental furs, but is also peculiarly smooth and glossy. As is generally the case with dark coated animals, the hairs are lighter towards the base, and partake of a slaty-blue tint.

ALL the preceding examples of the Squirrel tribe are remarkable for their extreme agility in climbing trees, traversing the branches, and making extraordinary leaps from one bough to another or from some elevated spot to the earth. The Ground Squirrels, however, are intended to abide on the earth, and are seldom known to ascend trees of any great height. As they possess cheek-pouches, they are placed in a separate genus, under the name of Támias, which is a Greek word, signifying a storekeeper, and are separate from the true Squirrels, which are not furnished with those appendages.

The HACKEE, or CHIPPING SQUIRREL, as it is sometimes termed, is one of the most familiar of North American quadrupeds, and is found in great numbers in almost every locality. It is a truly beautiful little creature, and deserving of notice both on account of the dainty elegance of its form, and the pleasing tints with which its coat is decked. The general colour of the Hackee is a brownish-grey on the back, warming into orange-brown on the forehead and the hinder quarters. Upon the back and sides are drawn five longitudinal black stripes and two streaks of yellowish-white, so that it is a most conspicuous little creature, and by these peculiar stripes may easily be distinguished from any other animal. The abdomen and throat are white. It is slightly variable in colour according to the locality in which it exists, and has been known to be so capricious of hue as to furnish specimens of pure white and jet black. As a fur it is extremely elegant, and if it were not quite so common would long since have taken nearly as high a rank as the sable or ermine.

The length of the Hackee is about eleven inches, the tail being about four inches and a half in length. It is, however, slightly variable in dimensions as well as in colour.

The Hackee is one of the liveliest and briskest of quadrupeds, and by reason of its quick and rapid movements, has not inaptly been compared to the wren. It is chiefly seen among brushwood and small timber; and as it whisks about the branches, or shoots through their interstices with its peculiar, quick, jerking movements, and its odd, quaint, little clucking cry, like the chip-chipping of newly-hatched chickens, the analogy between itself and the bird is very apparent. As it is found in such plenty, and is a bold little creature, it is much persecuted by small boys, who, although they are not big or wise enough to be entrusted with guns, wherewith to work the destruction of larger game, arm themselves with long sticks, and by dexterous management knock down many a Hackee as it tries to escape from its pursuers by running along the rail fences. Among boys the popular name of the Hackee is the "Chipmuck."

It is a burrowing animal, making its little tunnels in various retired spots, but generally preferring an old tree, or the earth which is sheltered by a wall, a fence, or a bank. The burrows are rather complicated, and as they run to some length, the task of digging the animal out of its retreat is no easy one. In the work of Messrs. Audubon and Bachman is given the following spirited narrative of an attack upon the home of some unfortunate Hackees. " This species is to a certain extent gregarious in its habits. We had in autumn marked one of its burrows which we conceived well adapted to our purpose, which was to dig it out. It was in the woods, in a sandy piece of ground, and the earth was strewed with leaves to the depth of eight inches, which we believed would prevent the frost from penetrating to any considerable depth. We had the place opened in January, when the ground was covered with snow about five inches deep. The entrance of the burrow had been closed from within. We followed the course of the small winding gallery with considerable difficulty. The hole descended at first almost perpendicularly for about three feet. It then continued, with one or two windings, rising a little nearer the surface until it had advanced about eight feet, when we came to a large nest, made of oak leaves and dried grasses. Here lay snugly covered three Chipping Squirrels.

" Another was subsequently dug from one of the small lateral galleries, to which it had evidently retreated to avoid us. They were not dormant, and seemed ready to bite when taken in the hand; but they were not very active, and appeared somewhat sluggish and benumbed, which we conjectured was owing to their being exposed to sudden cold from our having opened their burrow. There was about a gill of wheat and buckwheat in the nest; but in the galleries, which we afterwards dug out, we obtained about a quart of the beaked hazel nuts (*Córylus rostráta*), nearly a peck of acorns, some grains of Indian corn, about two quarts of buckwheat, and a very small quantity of grass seeds."

Whenever menaced by one of the numerous foes by which so defenceless and conspicuous an animal is sure to be surrounded, the Hackee makes at once for its burrow, and is there secured from the attacks of nearly every enemy. One foe, however, cares nothing for the burrow, but follows the poor Hackee through its windings, and never fails to attain its sanguinary object. This remorseless foe is the stoat, or ermine, one of which animals has been detected in entering a Hackee's burrow, where it remained for a few minutes, and then returned, licking its lips, and appearing highly satisfied with its proceedings. When the burrow was examined in order to ascertain the amount of slaughter which the stoat had performed, one female Hackee and five young were found lying dead in their home, the stoat having contented itself with sucking their blood, without deigning to eat their flesh.

From the principal burrow the Hackee drives several supplementary tunnels, in which it stows its stock of provisions. The general nature of this store, and the amount of treasure which is garnered within the burrows, may be gathered from the account which has just been quoted. When the Hackee carries off the beaked nuts into its cave, it goes through its work in a very business-like manner. Fearing lest the sharp "beak" of the nut may hurt its cheeks when it puts the fruit into its pouch, it bites off the sharp point, and then deliberately pushes it into one of the pouches with the assistance of its fore-paws. Another and another are similarly treated, and taking a fourth nut between its teeth, the Hackee dives into its burrow, packs away its treasures methodically, and then returns for another cargo. It is rather curious that it always carries four nuts at each journey. As the little creature goes along with its cheek-pouches distended to their utmost limits it has the most ludicrous aspect imaginable, its cheeks prodigiously swelled, and labouring most truly under an embarrassment of riches.

The Hackee moves into its winter quarters early in November, and, excepting occasional reappearances whenever the sun happens to shine with peculiar warmth, is not seen again until the beginning of spring. The young are produced in May, and there is generally a second brood in August. Their number is about four or five. The male Hackee is rather a pugnacious animal, and it is said that during their combats their tails are apt to snap asunder from the violence of their movements. It is undoubtedly true that those members

GROUND SQUIRREL, OR HACKEE.—*Tamias Lysteri.*

are wonderfully brittle, but whether they undergo such spontaneous amputation is not so certain.

Pretty as it is, and graceful as are its movements, it hardly repays the trouble of keeping it in a domesticated state; for its temper is very uncertain, and it is generally sullen towards its keeper. Although the food of the Hackee is mostly of a vegetable character, it is occasionally diversified with other substances; for the Chipping Squirrel, like his English relative, is occasionally carnivorous in his appetite. One of these animals was detected in the very act of robbing a bird's nest and devouring the callow young.

BETWEEN the squirrels and the marmots there are one or two intermediate links, one of which has already been noticed in Támias, and another is found in the genus Spermóphilus, to which the PRAIRIE DOG belongs.

The Prairie Dog, as it is popularly called, is found in very great plenty along the course of the Missouri and its tributaries, and also near the River Platte. It congregates together in vast numbers in certain spots where the soil is favourable to its subterranean habits of life and the vegetation is sufficiently luxuriant to afford it nourishment. The colour of this animal is a reddish-brown upon the back, mixed with grey and black in a rather vague manner. The abdomen and throat are greyish-white, and the short tail is clothed for the first half of its length with hair of the same tint as that of the body, and for the remaining half is covered with deep blackish-brown hair, forming a kind of brush. The

PRAIRIE DOG, OR WISH-TON-WISH.—*Spermóphilus Ludoviciánus.*

cheek-pouches are rather small, and the incisor teeth are large and protruding from the mouth. The length of the animal rather exceeds sixteen inches, the tail being a little more than three inches long. The cheek-pouches are about three-quarters of an inch in depth, and are half that measurement in diameter.

The Prairie Dog is a burrowing animal, and as it is very gregarious in its habits, the spot on which it congregates is literally honeycombed with its tunnels. There is, however, a kind of order observed in the "Dog-towns," as these warrens are popularly called, for the animals always leave certain roads or streets in which no burrow is made. The affairs of the community seem to be regulated by a single leader, called the Big Dog, who sits before the entrance of his burrow, and issues his orders from thence to the community. In front of every burrow a small heap of earth is raised, which is made from the excavated soil, and which is generally employed as a seat for the occupant of the burrow.

As long as no danger is apprehended, the little animals are all in lively motion, sitting upon their mounds, or hurrying from one tunnel to another as eagerly as if they were transacting the most important business. Suddenly a sharp yelp is heard, and the peaceful scene is in a moment transformed into a whirl of indistinguishable confusion. Quick barks resound on every side, the air is filled with a dust-cloud, in the midst of which is indistinctly seen an intermingled mass of flourishing legs and whisking tails, and in a moment the populous "town" is deserted. Not a "dog" is visible, and the whole spot is apparently untenanted. But in a few minutes a pair of dark eyes are seen gleaming at the entrance of some burrow, a set of glistening teeth next shine through the dusky recess, and in a few minutes first one and then another Prairie Dog issues from his retreat, until the whole community is again in lively action.

The title of Prairie Dog has been given to this animal on account of the sharp yelping sound which it is in the habit of uttering, and which has some resemblance to the barking of a very small and very peevish lapdog. Every time that it yelps it gives its tail a smart jerk. This peculiar sound is evidently employed as a cry of alarm; for as soon as it is uttered all the Prairie Dogs dive into their burrows, and do not emerge again until they hear the shrill whistle which tells them that the danger is past.

As it is so wary an animal it is with difficulty approached or shot, and even when severely wounded it is not readily secured, owing to its wonderful tenacity of life. A bullet that would instantly drop a deer has, comparatively, no immediate effect upon the Prairie Dog, which is capable of reaching its burrow, even though mortally wounded in such a manner as would cause the instantaneous death of many a larger animal. A tolerably large bullet through the brain seems to be the only certain method of preventing a Prairie Dog from regaining his strong-hold. The mode by which this animal enters the burrow is very comical. It does not creep or run into the entrance, but makes a jump in the air, turning a partial somersault, flourishing its hind legs and whisking its tail in the most ludicrous manner, and disappearing as if by magic. Scarcely has the spectator recovered from the ludicrous effect of the manœuvre when the animal begins to poke out his head again, and if not disturbed soon recommences his gambols.

The burrows of the Prairie Dog are generally made at an angle of forty degrees, and after being sunk for some little distance run horizontally, or even towards the surface of the earth. It is well known that these burrows are not only inhabited by the legitimate owners and excavators, but are shared by the burrowing owl and the rattlesnake. According to popular belief, the three creatures live very harmoniously together ; but careful observations have shown that the snake and the owl are interlopers, living in the burrows because the poor owners cannot turn them out, and finding an easy subsistence on the young Prairie Dogs. A rattlesnake has been killed near a burrow, and when the reptile was dissected, a Prairie Dog was found in its stomach.

Although it does not endure a domesticated life as well as many of the rodents, it is possessed of very great affection and courage, as is seen from the following anecdote. A hunter was engaged in shooting Prairie Dogs, and had succeeded in killing one animal, which was seated upon the little hillock in front of its burrow. A companion, which had not hitherto dared to expose itself to the hunter's fire, immediately issued from the same burrow, and seizing the body of its friend, dragged it into the hole. The hunter was so touched with this exhibition of true, loving feeling on the part of the little creature, that he never could be induced to shoot another Prairie Dog.

From the most recent accounts, it appears that the Prairie Dog does not hibernate, but that it is as fresh and lively during winter as in the heat of summer.

ANOTHER example of the genus Spermophilus may be found in the beautiful little creature which is scientifically known as Hood's Marmot, but more popularly as the Leopard Marmot.

This pretty little animal is about the same size as the hackee, and is remarkable for the brilliant and conspicuous manner in which its fur is diversified with contrasting hues. Along the back are drawn eight pale yellowish-brown bands, and nine dark brown bands of greater width. The five upper bands are marked with pale spots. The colouring is slightly variable, both in distribution and depth of tint, for in some specimens the dark bands are paler than in others, while in several specimens the pale spots have a tendency to merge altogether and form bands. The average length of this creature is nearly eleven inches, the tail slightly exceeding four inches in length. The cheek-pouches are moderate in dimensions. It is an inhabitant of Northern America.

This animal is said to be more lively and active than any of its relations, and to be remarkably fearless as it whisks about the neighbourhood of its home, uttering its sharp little cry of "Seek-seek-seek" continually. This cry is common to many of the Spermóphilus, and has given the name of Seek-seek to another species belonging to the same genus. It generally leaves its winter quarters at the beginning of spring, and roams about in search of a mate. At this time the males are very pugnacious, and engage in fierce contests for the possession of some favoured individual of the opposite sex. They are very heedless at this time of year, and can be easily caught in ordinary traps.

The burrow of the Leopard Marmot is generally driven perpendicularly into the ground, to the depth of four or nearly five feet ; but on the plains of the Upper Missouri, where the soil is sandy, and mixed with gravel, the burrow is almost horizontal, and lies

HOOD'S MARMOT.—*Spermóphilus Hoodii.*

barely one foot below the surface. The Leopard Marmot is rather a prolific animal, producing about eight or ten young at a litter. It is said to be destructive to gardens which may happen to be in the vicinity of its home, and is as capable of exercising its teeth upon an antagonist as on its food. Its bite is remarkably severe for so small an animal, and it is of such a sour disposition, that it is always ready to snap at those who attempt to capture or handle it.

BOBAC.—*A'rctomys Bobac.*

The BOBAC, or POLAND MARMOT, is one of the true Marmots, and is a native of parts of Northern Europe and Asia.

It is larger than the preceding animals, and appears to be of still greater dimensions owing to the full coat of thick hair with which it is profusely covered. The colour of this animal is a tolerably uniform grey-brown, slightly tinged with yellow, and having a "watered" appearance along the back. The length of the Bobac is rather more than twenty inches, the tail being about six inches long. The Bobac is a gregarious animal, living in small bands of thirty or forty in number, and being always found to prefer dry to moist soil. It does not seem to be fond of elevated situations, but generally takes up its residence on the sides of valleys, where the temperature is not so bleak as on the mountain-top.

Like many other burrowing animals, it lays up a store of provisions for the winter, and generally chooses well-dried hay for that purpose. So hard does the animal labour at amassing this treasure, that in a single burrow there is generally found as much hay as will suffice a horse for a night. It is slightly variable in colour, some specimens being more brown than others.

THE common MARMOT is about the size of an ordinary rabbit, and not very unlike that animal in colour. The general tint of the fur is greyish-yellow upon the back and flanks, deepening into black-grey on the top of the head, and into black on the extremity of the tail.

It is very common in all the mountainous districts of Northern Europe, where it associates in small societies. The Marmot is an expert excavator, and digs very large and rather complicated burrows, always appearing to reserve one chamber as a storehouse

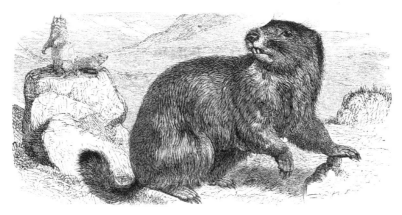

MARMOT.—*A'rctomys Marmotta.*

for the heap of dried grasses and other similar substances which it amasses for the purpose of sustaining life during the winter. The chamber in which the animal lives and sleeps is considerably larger than the storehouse, measuring, in some cases, as much as seven feet in diameter. The tunnel which leads to these chambers is only just large enough to admit the body of the animal, and is about six feet in length.

To these burrows the Marmot retires about the middle of September, and after closing the entrance with grass and earth, enters into the lethargic hibernating state, and does not emerge until the beginning of April. Like other hibernating animals, they are very fat just before they take up their winter-quarters, and as their fur is then in the best condition, they are eagerly sought after by the human inhabitants of the same country. The burrow of the Marmot is always dug in dry soil, and is seldom known to be at all above, or very much below, the line of perpetual snow. In these burrows the young Marmots are born, about three or four in average number. The burrow seems also a stronghold into which the Marmot can retire on the least alarm. It is so wary an animal that it always plants one of its number to act as a sentinel, and on the first symptom of danger, he gives the alarm cry, which is a signal for every Marmot to seek the recesses of its subterranean home.

The Marmot is a clumsy looking animal, and is not very active. Its movements are rather slow, and devoid of that brilliant activity which distinguishes the Leopard Marmot. Although it is easily tamed when taken young, it hardly repays the trouble of its owner, as it is a very unintellectual creature, and is ever too ready to use its powerful teeth upon the hand of any one who may attempt to handle or caress it. Naturally it is a timid animal, but when it finds itself unable to escape, it turns to bay and fights most desperately by means of the weapons with which its jaws are furnished.

At the end of the rodents are placed the singular animals which are grouped together under the title of Aspalácidæ, or Mole Rats, the word Aspalax, or Spalax, being the Greek term for a mole. The incisor teeth of these animals are extremely large, and project beyond the lips. The external ears are either wholly wanting or are of very small dimensions. The eyes are small, and in some species are concealed by the skin. The body is heavily and clumsily made, the tail is either very short or entirely absent, and the head is large and rounded.

SLEPEZ MOLE RAT.—*Spalax Typhlus.*

The common MOLE RAT, which is also known by its Russian name of SLEPEZ, is a native of Southern Russia, Asia Minor, Mesopotamia, and Syria. Like the ordinary mole, to which it bears no little external resemblance, it passes its existence in the subterranean tunnels which it excavates by means of its powerful claws. As it but seldom ventures into the light of day, it stands in no need of visual organs, but is compensated for their absence by the very large development of the organs of hearing. The place of the eyes is taken by two little round black specks, which lie under the fur-covered skin, so that even if they were sensitive to light, they would be unable to perceive the brightest rays of the noontide sun. The ears, however, are extremely large, and the hearing is exceedingly sensitive, so that the animal receives earlier information of danger through its sense of hearing than through that of sight, which latter faculty would indeed be useless in its dark abode. Sometimes the Slepez leaves the burrow and lies basking in the warm sunshine, but upon the least alarm, or unexpected sound, it plunges into its tunnel, and will not again make its appearance until it feels perfectly assured of safety.

Should it be unexpectedly attacked, it assumes an offensive attitude, and trusting to its delicate sense of hearing to inform it of the direction in which the foe is approaching, bites most savagely with its long chisel-like incisors.   While engaged in combat, or while threatening its adversary, it utters a sharp crying snort at short intervals.

The food of the Mole Rat is believed to be entirely of a vegetable nature, and it is in search of the various plants on which it feeds that it drives its long and complicated tunnels through the soil.   It is especially fond of roots, more particularly preferring those of a bulbous character, but will also feed on grain and different fruits, and is said to lay up a store of provisions in a subterranean chamber connected with its burrow.   The usual form of the Mole Rat's habitation and hunting-ground may be easily imagined.   A series of horizontal tunnels, or main roads, are driven through the ground at no great depth from the surface of the earth, and are connected with a number of chambers excavated at some depth, and with an endless variety of shallow passages which are made in the course of the animal's daily peregrinations in search of food.

The Russian peasants have an idea, that if any one will have the courage to seize a Slepez in his bare hands, permit the animal to bite him, and then squeeze it to death between his fingers, he will ever afterwards possess the power of curing goitre by the touch of his hands.   The general colour of the Slepez is a very light brown, slightly tinged with red in some parts, and fading into an ashen-grey in others.   Its total length is about ten or eleven inches, and the tail is wanting.   The head is broad, flat on the crown, and terminates abruptly at the muzzle.   The feet are short, and the claws small.

This animal is presumed to be the Blind Mole of the ancient Greek authors, and if so, affords another of the many instances where the so-called errors of the old writers on natural history have proved, on further acquaintance, to be perfectly correct.   The specific name Typhlus is a Greek word, signifying blind, and has been given to the Slepez on account of its absolute deprivation of eyes.

COAST RAT, OR SAND MOLE.—*Bathyergus Maritimus.*

THE incisor teeth of the COAST RAT, or SAND MOLE, are even larger in proportion than those of the preceding animal, and those of the upper jaw are marked by a groove running throughout their length.   The fore-feet are furnished with long and powerful claws, that of the second toe being the largest.   The eyes are exceedingly small, the external ears are wanting, and the tail is extremely short.

The Coast Rat is an inhabitant of the Cape of Good Hope and the coasts of Southern Africa, where it is found in tolerable profusion, and drives such multitudes of shallow tunnels that the ground which it frequents is rather dangerous for horsemen, and not at all pleasant even to a man on foot.   The burrows are made at so short a distance from the surface that the earth gives way under the tread of any moderately heavy animal. Mr. Burchell, the well-known African traveller, narrates that in traversing the great sand flats of Southern Africa he was often endangered by his feet sinking into the burrows of the Coast Rat, which had undermined the light soil in every direction. The animal is rather slow of foot upon the surface of the ground, but drives its subterranean tunnels with marvellous rapidity, throwing up little sandy hillocks at intervals, like those of the common mole. On account of this propensity it has received the name of *Zand Moll*, or Sand Mole, from the Dutch Boers who inhabit the Cape.

The colour of the Sand Mole is a uniformly light greyish-brown, rather variable in tinting. As it is very soft and full in texture, and can be obtained in great quantities, it might be profitably made a regular article of trade. The Sand Mole is as large as our ordinary wild rabbit, being about fifteen inches in total length, the tail measuring about three inches.

FUR COUNTRY POUCHED RAT.—*Saccophorus borealis.*

THERE has been much confusion between the two following animals, which have been by several authors considered to be identical with each other. They are, however, to be easily distinguished from each other by the deep longitudinal grooves which run along the upper incisors of the present animal, and the smooth or slightly grooved incisors of the succeeding species.

The FUR COUNTRY POUCHED RAT is a native of Canada, and is remarkable for the enormous size of the cheek-pouches. The colour of this animal's fur is generally of a pale grey washed with yellow, fading into a slaty-blue towards the base of each hair. The interior of the pouches, the abdomen, and the tail, are covered with white hair, that which lines the pouches being very short and fine. A dusky spot is observable behind each ear, the teeth are yellow and the claws white. The central claw of the fore-feet is almost deserving of the title of talon, as it is powerfully made and nearly half an inch in length. The total length of this animal is nearly ten inches, the tail measuring about two inches in length.

It is rather gregarious in its habits, associating together in moderately large bands, and undermining the ground in all directions. It is a vegetable feeder, preferring the bulbous roots of the quamash, or camas (*Scilla esculenta*), to any other diet, and is therefore called by some writers, the Camas Rat. This title is, however, given to several allied animals. It also feeds on nuts, roots, grain, and seeds of various kinds. The burrow of this animal is not very deep, but runs for a considerable distance in a horizontal direction, and along its course occasional hillocks are thrown up, by means of which it may be traced from the surface.

The CANADA POUCHED RAT is sometimes known by the name of "Mulo," and occasionally by that of "Gopher."

The incisor teeth of this animal are extremely long, and project beyond the lip, so as to be visible even at a profile view. The cheek-pouches are of great dimensions, measuring nearly three inches in depth, and reaching from the sides of the mouth to the insertion of the shoulder. They are lined with a soft covering of short fine hairs. The total length of the Canada Pouched Rat is about one foot, the tail being two inches long. The weight of an ordinary sized adult specimen is about fourteen ounces. In shape, it is heavily made and very clumsy, bearing no slight resemblance to the ordinary mole of our own country. Its fur is about half an inch in length upon the back, and much shorter upon the abdomen. Its colour is a reddish-brown on the upper parts of the body, fading into ashy-brown upon the abdomen, and the feet are white. The first third of the tail is clothed with short hair of the same colour as that of the back, but the remaining two-thirds are devoid of hairy covering.

This animal is a burrower, and is most destructive among plantations, as it is in the habit of eating the roots which happen to intercept the course of its tunnel, and has been known thus to destroy upwards of two hundred young trees in a few days and nights. Its ravages are not solely restricted to young plants, but are often extended to old and fullgrown fruit-trees. It continues its labour by day as well as by night, but is not readily discovered at its work, as it always ceases its labour at the least sound from above. The burrows of the Mulo are rather complicated, and are well described in the following extract from Audubon and Bachman.

"Having observed some freshly thrown up mounds in M. Chouteau's garden, several servants were called and set to work to dig out the animals if practicable alive; and we soon dug up several galleries worked by the Muloes, in different directions.

One of the main galleries was about a foot beneath the surface of the ground, except when it passed under the walks, in which places it was sunk rather lower. We turned up this entire gallery, which led across a large garden-bed and two walks into another bed, where we discovered that several fine plants had been killed by these animals eating off their roots just beneath the surface of the ground. The burrow ended near these plants under a large rose-bush. We then dug out another principal burrow, but its terminus was among the roots of a large peach-tree, some of the bark of which had been eaten off by these animals. We could not capture any of them at this time, owing to the ramification of their galleries having escaped our notice whilst following the main burrows. On carefully examining the ground, we discovered that several galleries existed that appeared to run entirely out of the garden into the open fields and woods beyond, so that we were obliged to give up the chase. This species throws up the earth in little mounds about twelve or fifteen inches in height, at irregular distances, sometimes near each other, and occasionally ten, twenty, or even thirty paces asunder, generally opening near a surface well covered with grass or vegetables of various kinds."

The burrow was probably sunk lower wherever it crossed a path, because the sense of hearing in this animal is so extremely acute, that it would be much annoyed by the continual sound of human footsteps immediately over its head.

Although it spends the greater part of its existence beneath the earth, it is frequently seen above the surface of the ground, as it resorts to the open air for the purpose of basking in the sun, or procuring leaves which have been brightened and vivified by the rays of the sun, as a change from the roots on which it chiefly depends for subsistence. When it revisits the regions of upper day, it emerges from the earth in some hitherto unbroken spot, pushing the soil upwards and causing a kind of miniature earthquake before it makes its appearance. Presently the head and shoulders of the animal emerge from the lump of earth, and shaking the loose mould from its fur, it draws itself entirely out of its burrow. It then runs forward for a yard or two, searching for food, nibbling off the green blades with its teeth, and stowing them into its cheek-pouches with the aid of its fore-paws. When it has filled the pouches, it runs back to the hole through which it had issued, and vanishes immediately from sight.

Should it be alarmed while out of its tunnel, it plunges precipitately into its strong-

1.   R R

hold, and drives an entirely new burrow in another direction, not venturing to entrust itself to that through which it had passed before it was alarmed.

The long and sharp incisor teeth are formidable weapons, and can be used with great effect upon an adversary. It is a sufficiently savage creature, and when captured or annoyed, bites fiercely in every direction, and squeals with rage. In captivity it is always employing these teeth upon every object that it can reach, and has even been detected in the act of endeavouring to cut its way through the wooden planks of the room in which it was placed. Two of these animals contrived to get into a pair of boots belonging to their owner, and not choosing to take the trouble of returning by the same aperture through which they had entered, cut a large hole in the toes, and so made their exit. They seemed to have a special liking for leather, as they afterwards gnawed to pieces the leathern straps which were dangling from a portmanteau that lay in the same room.

On the surface of the ground the Canada Pouched Rat is rather slow and clumsy in its movements, as its legs are short and ill fitted for such locomotion. So short indeed are its limbs, that if it be laid on its back, it has great difficulty in regaining its feet, but flounders about in almost total helplessness until it can seize a blade of grass, a twig, or similar object, by means of which it can draw itself into its normal attitude. In its tunnel, however, it proceeds with considerable activity, going faster than a man can walk, and being capable of running backwards or forwards with equal speed and ease.

The nest of the Mulo is not placed in one of the ordinary tunnels, but in a burrow dug specially for the purpose. It is about eight inches in diameter, globular in shape, and is made of dried herbage externally, and softly lined with hair plucked from the body of the female, and other appropriate substances. From the nest radiate a number of small galleries, which are again connected with smaller branch passages, and seem to conduct the animal to its feeding-grounds. It was formerly imagined that the Mulo was in the habit of filling its pouches with the excavated earth, and of emptying them at the mouth of burrows. This assertion is now disputed, for it is clearly ascertained that the creature only uses its cheek-pouches for the conveyance of its food. A little earth may perchance be imbedded together with the nuts and leaves, but the mistake has evidently arisen from the conduct of the natives, who, when they procure a skin of the Mulo, are accustomed to stuff the pouches with dry earth for the purpose of preserving them in their distended form.

The animal is found in many parts of Northern America, and has a very wide range.

The BAY BAMBOO RAT is one representative of the genus Rhizomys, of which there are several species.

This animal is a native of Nepal, Malacca, and China, and is very injurious to the bamboos, on the roots of which it feeds. In size it equals a rather small rabbit, and in colour it is of a uniform ruddy brown, slightly paler on the throat and abdomen. The long incisor teeth are faced with bright red enamel, which gives them a rather conspicuous appearance ; the tail is short and marked, and the claws are rather small. The head is of a peculiar form, which will be better understood from the engraving than by description alone.

THERE are several other genera belonging to this curious family, among which may be noticed the genus Ctenomys, containing the TUCUTUCO, a native of Magellan Straits. This is also a burrowing animal, and the peculiar name by which it is known has been given to it on account of the curious cry which it utters as it is engaged in its subterranean labours, and which is said to resemble the word " Tucutuco."

ON taking a retrospective view of the rodent animals, the reader will not fail to observe the frequency with which they reproduce some idea which is more fully mani- fested in other orders of the animal kingdom. The destructive idea is not more strongly developed in the lion than in the rat, which will attack and kill animals of much greater strength and bulk than itself. It is a truly bloodthirsty being, and will kill many a rabbit or fowl for the mere sake of sucking the hot blood as it pours from the fatal wound.

The tree-loving and agile squirrel plays the same part among the rodents as the monkey among the quadrumana; the flying squirrels have a close analogy to the colugo and the petaurists, and they again to the bats, which in their turn partake largely of the bird character and formation. The beaver and ondatra are evident reproductions of the aquatic idea, which is more thoroughly developed in the seals and whales, and is carried out to its greatest perfection in the fishes. The rodent capybara again, with its thick, coarse, bristly hair, heavy form, hoof-like claws, and water-loving propensities, is no indifferent representation of the pachydermatous water hog, which also may be looked upon as corresponding to the dugong and manatee. Lastly, the aspalacidæ, or rodent mole rats, are wonderfully similar to the true insectivorous moles, both in habit and formation of body.

In many instances this phenomenon is exhibited in the reverse order, the members of other groups exhibiting a tendency towards the rodent type. The aye-aye, for

BAY BAMBOO RAT.—*Rhizomys bâdius.*

example, a quadrumanous animal, displays so strong a resemblance to the squirrels, that it was long ranked together with those animals by systematic naturalists. The hyrax again, or klip-daas, a pachydermatous animal, and allied closely to the hippopotamus, is externally so rabbit-like in form, and even in the arrangement of its teeth, that it was as a matter of course placed among the rodents, until Cuvier's accurate eye discovered its true character. The insect-eating tupaias of Java, with their arboreal habits and long bushy tails, are so like the squirrels that the popular name of a squirrel and a tupaia is identical in the countries where they reside.

Thus, in this single order, we find external representatives of every idea which is embodied in the whole series of vertebrated animals, and cannot but notice the curious tendency which is found throughout the entire animal kingdom of each province to intersect several others, and to receive some of its privileges without detriment to its perfection. In no instance is the boundary of any single province defined with a clear line of demarcation, and in every case the outline is extremely irregular, sending out peninsulas into the neighbouring districts and receiving into its own territory some portion of another district. Sometimes these embodied ideas seem to bear some analogy to geological strata, and, after disappearing in one order of animals, to " crop out," so to speak, in another distant order, or even in another class or division.

All external objects are, in their truest sense, visible embodiments or incarnations of Divine ideas which are roughly sculptured in the hard granite that underlies the living and breathing surface of the world above ; pencilled in delicate tracery upon each bark flake that encompasses the tree trunk, each leaf that trembles in the breeze, each petal that fills the air with fragrant effluence ; assuming a living and breathing existence in the rhythmic throbbings of the heart-pulse that urges the life-stream through the body of every animated being ; and attaining their greatest perfection in Man, who is thereby bound, by the very fact of his existence, to outspeak and outact the Divine ideas, which are the true instincts of humanity, before they are crushed or paralysed by outward circumstances. Only thus can man be truly the image and likeness of God, only thus can the Divine ideas be truly manifested in him to the world. For just in proportion as he shrinks from speaking the truth that is in him, or from acting the good that is in him, so far he stifles the commencing outbirth of Divine power, and becomes less and less godlike.

Hence the necessity for the infinitely varied forms of animal life. Until man has learned to realize his own microcosmal being, and will himself develop and manifest the god-thoughts that are continually inbreathed into his very essential nature, it needs that the creative ideas should be incarnated and embodied in every possible form, so that they may retain a living existence upon earth.

This principle lies at the very root of all material formations. It is but obscurely shadowed in those portions of the creation which we term inanimate, but becomes more and more perceptible in every being in proportion as it assumes a more perfect form and a higher organization. In Man we see its very highest development, and recognise the absolute necessity of that great truth which has animated almost every form of theology upon the face of the earth, namely, the visible incarnation of Divinity in human form.

SCOTCH CATTLE.

OXEN.

In the large and important group of animals which now occupy our attention, the incisor teeth are entirely absent in the upper jaw, and are eight in number in the lower. There are six molars on each side of each jaw. The two middle toes of each foot are separate, and are furnished with hoofs instead of claws. From the frontal bones proceed two excrescences, which are generally armed with horns, particularly in the male animal. The structure of the stomach and gullet is very remarkable, and is employed in producing that peculiar action which is called "ruminating," or chewing the cud. Although the horns have in many varieties of domesticated Oxen been eradicated by a long course of careful management, they are always present in the wild species, and are permanently retained through life, instead of being annually shed like those of the deer. The peculiar

characteristics of the bovine skull are so well shown in the engraving, that further description is needless.

The Oxen, or Bóvidæ, as they are called, from the Latin word Bos, or Ox, are extremely difficult of systematic arrangement, as it is not easy to select any particular characteristic on which to base the distinctions of genus and species. Some writers have founded their arrangement upon the hoofs, others upon the muzzle, others upon the direction of the horns, and others upon the structure of their bony nucleus. Mr. Gray, in his elaborate elucidation of the Bóvidæ, considers that "the form of the horns affords the most natural character for subdividing them into groups," and employs other characteristics, such as the position of the knee, the beard of the male, and the formation of the muzzle, as means for further subdivision.

SKULL OF OX.

The DOMESTIC OX of Europe has been so modified in form, habits, and dimensions, by its long intercourse with mankind, that it has developed into as many permanent varieties as the dog, the pigeon, or the rabbit, and would in many cases be thought to belong to different species. Among the principal varieties of this animal may be noticed the Long Horned, the Short Horned, and the Polled or hornless breeds, and the Alderney cow, so celebrated for the quantity and quality of the milk which it daily furnishes. In almost every part of the world are found examples of the Ox, variously modified in order to suit the peculiar circumstances amid which they are placed, but in all instances they are susceptible of domestication, and are employed in the service of mankind.

There are few animals which are more thoroughly useful to man than the Ox, or whose loss we should feel more deeply in the privation of so many comforts. Putting aside the two obvious benefits of its flesh and its milk—both of which are so needful for our comfort that we almost forget to think about them at all—we derive very great benefit from its powers while living, and from many portions of its body when dead.

In many parts of England, Oxen are still employed in agricultural labour, drawing the plough or the wagon with a slow but steady plodding gait. The carpenter would find himself sadly at a loss were his supply of glue to be suddenly checked by the disappearance of the animal, from whose hoofs, ears, and hide-parings the greater part of that useful material is manufactured. The harness-maker, carriage-builder, and shoemaker would in that case be deprived of a most valuable article in their trade; the cutler and ivory turner would lose a considerable portion of the rough material upon which they work; the builder would find his best mortar sadly impaired without a proper admixture of cow's hair; and

LANCASHIRE BULL.

the practical chemist would be greatly at a loss for some of his most valuable productions if the entire Ox tribe were swept from the earth. Not even the very intestines are allowed to be wasted, but are employed for a variety of purposes, and in a variety of trades. Sometimes the bones are subjected to a process which extracts every nutritious particle out of them, and even in that case, the remaining innutritious portions of the bones are made useful by being calcined, and manufactured into the animal charcoal which has lately been so largely employed in many of the arts and sciences.

The best living example of the original British Ox is to be found in the celebrated white cattle of Chillingham.

The colour of these beautiful animals is a cream-white, with the exception of the ears and muzzle, the former of which are red, and the latter is black. Mr. Bell observes, that in every case of white cattle which have passed under his personal notice, the ears are marked with red or black, according to the breed. The white tint extends even to the horns, which are, however, tipped with black. They are rather slender in their make, and curve boldly upwards. As these Chillingham cattle are permitted to range at will through spacious parks in which they are kept, they retain many of the wild habits of their tribe, and are so impatient of observation that a stranger will generally find himself in a very unsafe position if he attempts to approach closely to the herd.

When they are alarmed or provoked at the intrusion of a strange human being within the limit of their territories, they toss their heads wildly in the air, paw the ground, and steadfastly regard the object of their dislike. If he should make a sudden movement, they scamper away precipitately, gallop round him in a circle, and come to another halt at a shorter distance. This process is continually repeated, the diameter of the circle

SHORT-HORN BULL.

being shortened at every fresh start, until the angry, yet half-frightened, animals, come so alarmingly close to the spectator, that he finds himself obliged to escape as he best can.

In performing these curious evolutions, they seem to be inspired by a mixture of curiosity, timidity, and irritation, which may be observed even in ordinary domestic cattle under similar circumstances.

On one occasion, when a herd of cattle were pressing upon me in a most uncomfortable manner, I owed my escape to early instruction in the art of the "acrobat." The herd, wholly composed of cows, was surrounding me with a very threatening aspect, and was advancing in such a manner that there was no mode of escape from their ranks. Seeing that a bold stratagem was the only resource, I ran sharply forward, and commenced rotating towards them in that peculiar method which is technically termed "turning a wheel," i.e. executing a series of somersaults on the hands and feet alternately. The cows were so terrified at the unknown foe who was attacking them in so extraordinary a manner, that they were panic-stricken, and galloped off at full speed, leaving me an easy escape before they had recovered from their surprise.

The domestic cow is too well known to need any detailed description of form and colour. Few persons, however, except those who have been personally conversant with these animals, have any idea of their intelligent and affectionate natures.

They are possessed of very susceptible feelings, and are remarkably sensitive to insulting or disrespectful conduct on the part of their inferiors. In a herd of cows, the senior animal is the leader in all things, and maintains a strict authority over her younger companions. Not a single member of the herd dares to leave or to enter the pasture until the leader has led the way, or even to take its food until she has decided whether she will

SUFFOLK BULL.

take possession of the banquet, or permit her inferiors to eat at peace. Should a younger animal commit a breach of etiquette by infringing any of the tacit rules which have been in force throughout Cowdom from time immemorial, the delinquent is butted at and punished until it returns to its allegiance.

To watch a calf through its various phases of existence is a most amusing employment. When the young animal is introduced for the first time into the farmyard, she is treated in the most supercilious manner by the previous occupants, who look with an air of supreme contempt upon the new comer. She is pushed aside by all her predecessors, and soon learns to follow humbly in the wake of her companions. She cannot even venture to take possession of a food-rack until all the others have begun their meal. So matters go on for a time, until she has attained a larger growth, and a younger calf is turned into the yard. She now in her turn plays the tyrant over the new comer, and receives no small accession of dignity from the fact of having a follower, instead of bringing up the rear in her own person. In process of time she makes her way to the head of the yard by virtue of seniority, and is then happy in the supreme rule which she enjoys.

Sometimes a three-parts grown heifer is introduced into a farmyard, and in that case, the new comer refuses to take her place below all the others, unless she is absolutely compelled to do so by main force. There is generally a considerable amount of fighting before such an animal finds her level, but when she has discovered her superiors and her subordinates, she quietly settles down in her place, and does not attempt to rise otherwise than by legitimate seniority.

As the Oxen, in common with the sheep, camels, giraffe, and deer, require a large amount of vegetable food, and are, while in their native regions, subject to innumerable disturbing causes that would effectually prevent them from satisfying their hunger in an ordinary manner, they are furnished with a peculiar arrangement of the stomach and digestive organs, by means of which they are enabled to gather hastily a large amount of food in any spot where the vegetation is luxuriant, and to postpone the business of mastication and digestion to a time when they may be less likely to be disturbed. The peculiarity of structure lies chiefly in the stomach and gullet, which are formed so as to act as an internal food-pouch, analogous in its use to the cheek-pouches of certain monkeys and rodents, together with an arrangement for regurgitating the food into the mouth at the will of the animal, previous to its mastication and digestion.

Owing to the absence of teeth in the upper jaw, the Ox is unable to cut or chew the grass as he feeds, and can only seize it between the lower incisor teeth and the upper jaw, so as to tear it by a movement of the head. The sound which is produced by this ripping or tearing process is familiar to all who have watched cows while grazing. As soon as the grass is taken into the mouth, it undergoes a slight rolling between the molar teeth, and is then swallowed, although it is not as yet in a fit state to be placed in the stomach, and there to be digested. The mode in which it undergoes that process is as follows.

The stomach and gullet are modified into four distinct compartments, one of which, called the paunch, is very much larger than the others, and is the receptacle into which the food is passed immediately after being swallowed. Here it remains comparatively unchanged until the animal is at rest, and ready to commence the process which is technically called "ruminating," and more popularly termed "chewing the cud." A small portion of the food then passes into the second compartment, which is lined with a series of hexagonal cells, not unlike the comb of the honey-bee, and is formed into little balls by being worked into the cells. From these cells the food is thrown into the mouth by a voluntary effort of the muscles, and is then subjected to a thorough mastication. Being again swallowed, it slips over the opening by which it had formerly passed into the paunch, and is received into the third compartment, technically called the "psalterium," or psalm-book, because it is lined with a number of thin longitudinal plates of membrane, which are thought to bear some resemblance to the leaves of a book. From thence it passes into the fourth compartment, which is the place where the business of digestion is carried on.

These different compartments of the stomach are familiar under the general name of tripe, and are popularly distinguished from each other as follows. The first compartment is called the paunch, and is lined with a vast number of little flattened projections of the membrane. In the paunch are found those curious concretions of hair and other substances which are known as hair-balls or bezoar stones.

The hair-balls are of various dimensions, a collection of them in my own possession varying from the size of a very large cricket-ball to a moderately sized marble. The hair is arranged most regularly in these balls, and all lies in the same direction, so that the axis on which the ball has revolved is plainly marked by the arrangement of the hair. In some of the balls the surface is covered with hair of different colours, some specimens being of a dark tint, while others are pure white. If the hair-ball be divided, its substance will be found to be of a spongy texture, affording considerable resistance to the knife, and requiring a strong and sharp blade to cut it neatly. Together with the hair is found a slight admixture of vegetable fibre. When first removed from the animal, these balls are wet and soft, receiving the impressions of the fingers unless handled with some care, but when they are quite dry, they are extremely light, hard, and strong, and tolerably elastic. Sometimes they are smooth on the exterior, which is then of a deep brown hue, and rather highly polished.

The second compartment is popularly called the "honeycomb," the "bag," or the "bonnet," and the third is termed the "monyplies," or "manyplus," on account of the membranous folds with which its interior is lined. The last stomach is generally termed the "red." In scientific language, the first compartment is called the "rumen," from which word is derived the term "ruminating;" the second is known under the name of "reticulum," or net; the third is called "omasus," or "psalterium," and the fourth is called the "abomasus," because it leads from the third compartment, or omasus.

Although the process of ruminating is mostly confined to the Ox and the other animals whose names have been already mentioned, it has, in more than one instance, been discovered in human beings.

In many parts of the world, such as the Pampas of America and the Australian colonies, vast herds of cattle roam the country as freely as if they were the original inhabitants. Although they are all sprung from domesticated cattle which have been permitted to run wild, or have escaped from their owners, they havere turned to the habits as well as the conditions of savage life, and can only be brought temporarily within the subjection of man by actual force. However free and uncurbed they may be, they are all private property, and except when of very tender age, are all branded with the name of their owner, burnt deeply into the skin. In detaching the unmarked cattle from the remainder of the herd, and bringing them safely to the enclosure where they are to receive the distinguishing brand of their proprietor, the cattle-drivers exhibit an extraordinary mixture of excellent horsemanship, great dexterity, cool patience, and fearless daring. Yet the man is sure to triumph over the beast at last, however cunning or powerful it may be, and before the poor animal has quite recovered from his surprise at finding himself mastered for the first time in his life, he has been captured, tied, branded, and set free again.

In Africa, the cattle are not only employed for the yoke, but are also educated for the saddle, and are taught to obey the bit as well as many horses. The bit is of very primitive form, being nothing more than a stick which is passed through the nostrils, and to which the reins are tied. One end of the stick is generally forked to prevent it from falling out of its place, and in guiding the animal, the rider is obliged to draw both reins to the right or left side, lest he should pull out the wooden bit. The saddle Oxen are not very swift steeds, their pace being about four or five miles an hour; and as their skin is so loosely placed on their bodies that the saddle sways at every step, their rider has no very agreeable seat. In training the Ox for the saddle, the teachers avail themselves of the aid of two trained Oxen, between whom the novice is tied, and who soon teach it the proper lesson of obedience.

The horns of this variety of the Ox are of marvellous length, having been known to exceed thirteen feet in total length, and nearly nine feet from tip to tip. The circumference of these enormous horns was more than eighteen inches, measured at their bases. One such horn is capable of containing upwards of twenty imperial pints. These weapons are not only long, but are sharply pointed, and are of so formidable a nature that a lion has been kept at bay during a whole night, not daring to leap upon an animal so well defended. As these horns might prove dangerous to the rider in case of the animal suddenly jerking its head, or flinging him forward by a stumble, the natives are in the habit of training them in various fashionable modes, by which the danger is avoided. Sometimes the horns are split into numerous ribbons, and curled fantastically in various directions; sometimes they are merely bent forwards and downwards; but the method most in vogue, is to cause them to swing loosely at each side of the head, their points towards the earth, and out of the way of the rider.

It is a remarkable fact that this Ox is in the habit of chewing dry bones whenever it finds them lying on the ground. The caribou, or American reindeer, is known to gnaw the fallen antlers of its companions, and probably with the same object.

Should the Ox turn out to be of a peculiarly savage disposition, he is soon conquered by having a heavy iron chain fastened round his neck. The continual weight which he is forced to carry whenever he moves, together with the jingling of the iron links, has such an effect upon his spirits, that he is forced to yield after a few days' trial. It is said that if the Ox is forced into the long grass which grows so luxuriantly in Southern Africa, it becomes alarmed, because it feels itself unable to see an approaching enemy, and is even terrified at the proximity of its own companions.

It is worthy of remark in the present place, that the skin of a white Ox is considered by the native tribes as an emblem of peace, and is analogous to the white bison hide which is displayed by the American Indians for similar purposes.

The Ox is also employed for draught in Southern Africa, and is used chiefly for

ZEBU.—*Bos I'ndicus.*

the purpose of drawing the wagons over the tracks which are by courtesy called roads. Although the wagons are remarkably light, and are built in such a manner as to take no harm by an occasional upset, the ground is so heavy, and the wheels sink so deeply, that a very large team of these cattle is required to draw the vehicle safely on its journey. Eight or ten yoke of oxen are frequently employed in drawing a single wagon. The conduct of the native drivers towards these poor beasts is cruel in the extreme, and deserving of the greatest reprobation. The "jambok," or whip, which these men employ is of very great length, and can be used with an effect that is perfectly terrible. Besides this more legitimate instrument, the Hottentot driver is in the habit of using various other methods of tormenting the poor beasts, and is absolutely ingenious in the refinements of his cruelty.

THE domestic cattle of India is commonly known by the name of Zebu, and is conspicuous for the curious fatty hump which projects from the withers. These animals are further remarkable for the heavy dewlap which falls in thick folds from the throat, and which gives to the fore part of the animal a very characteristic aspect. The limbs are slender, and the back, after rising towards the haunches, falls suddenly at the tail.

The Zebu is a quiet and intelligent animal, and is capable of being trained in various modes for the service of mankind. It is a good draught animal, and is harnessed either to carriages or ploughs, which it can draw with great steadiness, though with but little speed. Sometimes it is used for riding, and is possessed of considerable endurance, being capable of carrying a rider for fifteen hours in a day, at an average rate of five or six miles per hour. The Nagore breed is specially celebrated for its capabilities as a steed, and is remarkable for its peculiarly excellent action. These animals are very active, and have been known to leap over a fence which was higher than our five-barred gates, merely for the purpose of drinking at a certain well, and, having slaked their thirst, to leap back again into their own pasture. As a beast of burden, the Zebu is in great request, for it can carry a heavy load for a very great distance, though at no great speed.

BRAHMIN BULL.

The Zebu race has a very wide range of locality, being found in India, China, Madagascar, and the eastern coast of Africa. It is believed, however, that its native land is India, and that it must have been imported from thence into the other countries.

There are various breeds of Zebu, some being about the size of our ordinary cattle, and others varying in dimensions from a large Ox to a small Newfoundland dog. One of the most familiar of these varieties is the well-known Brahmin Bull, so called because it is considered to be sacred to Bramah.

The more religious among the Hindoos, scrupulously observant of the letter of a law which was intended to be universal in its application, but to which they give only a partial interpretation, indulge this animal in the most absurd manner. They place the sacred mark of Siva on its body, and permit it to wander about at its own sweet will, pampered by every luxury, and never opposed in any wish or caprice which it may form. A Brahmin Bull will walk along the street with a quaintly dignified air, inspect anything and anybody that may excite his curiosity, force every one to make way for himself, and if he should happen to take a fancy to the contents of a fruiterer's or greengrocer's shop, will deliberately make his choice, and satisfy his wishes, none daring to cross him. The indulgence which is extended to this animal is carried to so great a height, that if a Brahmin Bull chooses to lie down in a narrow lane, no one can pass until he gets up of his own accord.

Bishop Heber, in his well-known journal, mentions the Brahmin Bulls and the unceremonious manner in which they conduct themselves, and remarks that they are sometimes rather mischievous as well as annoying, being apt to use their horns if their caprices be not immediately gratified.

The BUFFALO is spread over a very wide range of country, being found in Southern Europe, North Africa, India, and a few other localities.

This animal is subject to considerable modifications in external aspect, according to the climate or the particular locality in which it resides, and has in consequence been mentioned under very different names. In all cases the wild animals are larger and more powerful than their domesticated relations, and in many instances the slightly different shape, and greater or lesser length of the horns, or the skin denuded of hairs, have been considered as sufficient evidences of separate species.

In India, the long, smooth-horned variety chiefly prevails, and is found in tolerable profusion. This animal frequents wet and marshy localities, being sometimes called the Water Buffalo on account of its aquatic predilections. It is a most fierce and dangerous animal, savage to a marvellous degree, and not hesitating to charge any animal that may arouse its ready ire. An angry Buffalo has been known to attack a tolerably-sized elephant, and by a vigorous charge in the ribs to prostrate its huge foe. Even the tiger is found to quail before the Buffalo, and displays the greatest uneasiness in its presence.

The Buffalo, indeed, seems to be animated by a rancorous hatred towards the tiger, and if it should come inadvertently on one of the brindled objects of its hate, will at once rush forward to the attack. Taking advantage of this peculiarity, the native princes are in the habit of amusing themselves with combats between tigers and trained Buffaloes. The arena is always prepared by the erection of a lofty and strongly-built palisade, composed of bamboos set perpendicularly, and bound together upon the outside. The object of this contrivance is, that the surface of the bamboo is so hard and slippery, that the tiger's claws can find no hold in case of an attempted escape.

The tiger is first turned into the arena, and generally slinks round its circumference, seeking for a mode of escape, and ever and anon looking up to the spectators, who are placed in galleries that overlook the scene of combat. When the tiger has crept to a safe distance from the door, the Buffalo is admitted, and on perceiving the scent of the tiger, it immediately becomes excited, its hairs bristle up, its eyes begin to flash, and it seeks on every side for the foe. As soon as it catches a glance of its enemy it lowers its head towards the ground, so that the tips of its horns are only a few inches above the earth, and its nose lies between its fore-legs, and plunges forward at the shrinking tiger. Were the latter animal to dare the brunt of the Buffalo's charge, the first attack would probably be the last; but as the tiger is continually shifting its position, the force of the onset is greatly diminished by the curve in the Buffalo's course.

As a general rule the Buffalo comes off the victor, for even when the tiger has gained an advantage, he does not follow it up with sufficient celerity, but permits his antagonist to regain his lost breath. The Buffalo, on the contrary, allows the tiger no breathing time, but continues his rapid charges without cessation, until he forces the tiger off his ground, and then with a rapid spring impales the foe on his horns. "Jungla," however, the celebrated fighting tiger, whose portrait may be seen on page 161, was invariably the conqueror in these combats, as he never tried to escape from the Buffalo, or to struggle with it, but quietly awaited its onset, and then, leaping nimbly aside from the deadly horns, dealt such a tremendous blow on the Buffalo's head with his herculean paw that he laid his antagonist dead on the ground.

It is generally supposed that the wild Buffaloes will destroy any tigers that may happen to approach their herds too closely. A wild adult male Buffalo, or Arnee as it is also called, is one of the largest of the Ox tribe, measuring no less than ten feet six inches from the tip of the nose to the root of the tail, and from six feet to six feet six inches in height at the shoulders. So confident are even the tiger-dreading herd-keepers of the prowess of their tamed animals, that they will ride them in search of pasture even when they know tigers to be in the near vicinity. One of these herds chanced to come across the spot where a tiger had been recently shot, and on perceiving the scent of the blood, they became powerfully excited, bellowed furiously, and at last charged in a body directly into a neighbouring covert, crushing everything that impeded their progress.

The Arnee lives in large herds, arranged after the manner of all bovine animals, the females and young being always placed in the safest spots, while the males post themselves

BUFFALO.—*Bubalus buffelus.*

in all positions of danger. These herds are never seen on elevated ground, preferring the low marshy districts where water and mud are abundant. In this mud they love to wallow, and when suddenly roused from their strange pastime, present a most terrible appearance, their eyes glaring fiercely from amid the mud-covered dripping masses of hair. Sometimes the Buffalo is said to fall a victim to its propensity for wallowing in the mud, and to be stuck so firmly in the oozy slime, as it dries under the scorching sunbeams of that burning climate, that it can be killed without danger. They generally chew the cud while they are lying immersed in mud or water.

Captain Williamson, in his work on "Oriental Field Sports," speaks thus of the Buffalo, and its mud-loving propensities :—

"This animal not only delights in the water, but will not thrive unless it have a swamp to wallow in. Then rolling themselves, they speedily work deep hollows, wherein they lie immersed. No place seems to delight the Buffalo more than the deep verdure on the confines of *jeels* and marshes, especially if surrounded by tall grass, so as to afford concealment and shade, while the body is covered by the water. In such situations they seem to enjoy a perfect ecstasy, having in general nothing above the surface but their eyes and nostrils, their horns being kept low down, and consequently hidden from view.

"Frequently nothing is perceptible but a few black lumps in the water, appearing like small clods, for the Buffaloes being often fast asleep, all is quiet; and a passenger would hardly expect to see, as often happens, twenty or thirty great beasts suddenly rise. I have a thousand times been unexpectedly surprised in this manner by tame Buffaloes, and once or twice by wild ones. The latter are very dangerous, and the former are by no means to be considered as innocent. The banks of the Ganges abound with Buffaloes in their wild state, as does all the country where long grass and capacious *jeels* are to be found. Buffaloes swim very well, or, I may say, float. It is very common to see droves crossing the Ganges and other great rivers at all seasons, but especially when the waters are low. At a distance one would take them to be large pieces of rock or dark-coloured wood, nothing appearing but their faces. It is no unusual thing for a boat to get into the thick of them, especially among reedy waters, or at the edges of jungles, before it is perceived. In this no danger exists; the Buffaloes are perfectly passive, and easily avoid being run down, so the vessel runs no danger."

The CAPE BUFFALO is quite as formidable an animal as its Indian relation, and much more terrible in outward aspect. The heavy bases of the horns, that nearly unite over the forehead, and under which the little fierce eyes twinkle with sullen rays, give to the creature's countenance an appearance of morose, lowering ill-temper, which is in perfect accordance with its real character.

Owing to the enormous heavy mass which is situated on the forehead, the Cape Buffalo does not see very well in a straight line, so that a man may sometimes cross the track of a Buffalo within a hundred yards, and not be seen by the animal, provided that he walks quietly, and does not attract attention by the sound of his footsteps. This animal is ever a dangerous neighbour, but when it leads a solitary life among the thickets and marshy places, it is a worse antagonist to a casual passenger than even the lion himself. In such a case, it has an unpleasant habit of remaining quietly in its lair until the unsuspecting traveller passes closely to its place of concealment, and then leaping suddenly upon him like some terrible monster of the waters, dripping with mud, and filled with rage. When it has succeeded in its attack, it first tosses the unhappy victim in the air, then kneels upon his body, in order to crush the life out of him, then butts at the dead corpse until it has given vent to its insane fury, and ends by licking the mangled limbs until it strips off the flesh with its rough tongue.

Many such tragical incidents have occurred, chiefly, it must be acknowledged, owing to the imprudence of the sufferer: and there are few coverts in Southern Africa which are not celebrated for some such terrible incident. Sometimes the animal is so recklessly furious in its unreasoning anger, that it absolutely blinds itself by its heedless rush through the formidable thorn-bushes which are so common in Southern Africa. Even when in company with others of their own species, they are liable to sudden bursts of emotion, and will rush blindly forward, heedless of everything but the impulse that drives them forward. In one instance, the leader of the herd, being wounded, dropped on his knees, and was instantly crushed by the trampling hoofs of his comrades, as they rushed over the prostrate body of their chief.

The flesh of the Cape Buffalo is not in great request even among the Kaffirs, who are in no wise particular in their diet. The hide, however, is exceedingly valuable, being used for the manufacture of sundry leathern implements where great strength is required without much flexibility. "Trek-tows," or the central leathern traces by means of which the draught oxen are harnessed to the wagons, are almost exclusively made of the Buffalo hide, as are also the numerous "rheims," or straps, which are in constant use about these curious wagons.

In South Africa, the Cape Buffalo, called by the Kaffirs "Inyati," or "Inthumba," plays much the same part as the arnee in India. Like that animal, it does much as it pleases, and fears no enemy but armed men. Even the lion dare not approach too closely to a herd of Cape Buffaloes, for with the cunning old bulls in front, and the cows and calves bringing up the rear, the lion would have but a poor chance against a general charge of such foes. Indeed, even in single combat, the lion would scarcely come off the conqueror.

The Cape Buffalo, although so terrible an animal, is not so large as the arnee, being little larger than an ordinary ox, but possessed of much greater strength. The strangely shaped horns are black in colour, and so large that the distance between their points is not unfrequently from four to five feet. On account of their great width at their bases,

CAPE BUFFALO.—*Bubalus Caffer.*

they form a kind of bony helmet, which is impenetrable to an ordinary musket-ball, and effectually defend their owner against the severe shocks which are frequently suffered by these testy animals.

I conclude this history of the Cape Buffalo with some personal reminiscences of the animal, which have been kindly placed at my disposal by Captain Drayson, R.A. :—

"The hide of this animal is a bluish-black in colour, and is so very tough that bullets will scarcely penetrate it if they are fired from a distance, or are not hardened by an addition of tin in the proportion of one to eight. It is of a fierce, vindictive disposition, and from its cunning habits is esteemed one of the most dangerous animals in Southern Africa. The Cape Buffalo is naturally a gregarious animal, but at certain seasons of the year the males fight for the mastery; a clique of young bulls frequently turn out an old gentleman, who then seeks the most gloomy and retired localities in which to brood over his disappointments.

These solitary skulkers are the most dangerous of their species; and although it is the nature of all animals to fly from man, unless they are badly wounded, or are intruded upon at unseasonable hours, these old hermits will scarcely wait for such excuses, but will willingly meet the hunter half-way and try conclusions with him.

Although frequently found in large herds on the plains, the Buffalo is principally a resident in the bush; here he follows the paths of the elephant or rhinoceros, or makes a road for himself. During the evening, night, and early morning, he roams about the open country and gorges, but when the sun has risen high, or if he has cause for alarm, the glens and coverts are sought; and amidst their shady branches he enjoys repose and

1. s s

obtains concealment. The 'spoor' of the Buffalo is like that of the common ox, the toes of the old bulls being very wide apart, whilst those of the young ones are close together ; the cow Buffalo's footprints are longer and thinner than the bull's, and smaller.

As these animals wander in the open ground during the night, and retreat to their glens during the day, their spoor may be taken up from the outside of the bush, and followed until the scent leads to the view. When the hunter comes near to his game, of which he should be able to judge by the freshness of the footprints, he should wait and listen for some noise by which to discover their position. Buffaloes frequently twist and turn about in the bush, and do so more especially just before they rest for the day.

I knew a Kaffir who carried about him the marks of a Buffalo's power and cunning. He was hunting Buffaloes one day in the bush, and came upon a solitary bull, which he wounded ; the bull bounded off, but the Kaffir, thinking him badly hurt, followed after at a run, without taking sufficient precautions in his advance. Now, dangerous as is a Buffalo when untouched, he is still more to be dreaded when hard hit, and should therefore be followed with the utmost caution.

The Kaffir had hurried on through the bush for a hundred yards or so, and was looking for the spoor, when he heard a crash close to him, and before he could move himself, he was sent flying in the air by the charge of the Buffalo. He fell into some branches and was thus safe, for the Buffalo was not satisfied with this performance, but wished to finish the work which he had so ably begun. After examining the safe position of his victim, he retreated.

The Kaffir, who had two or three ribs broken, reached his home with difficulty, and gave up Buffalo-shooting from that day.

It appeared that this cunning animal had retraced its steps after retreating, and had then backed into a bush, and waited for the Kaffir to pass.

A great sportsman at Natal, named Kirkman, told me that he was shooting Buffaloes when he was across the Sugela river on one occasion, and having wounded a bull, he was giving him his quietus, when the creature sent forth a sort of moan. Now the Buffalo always dies game, and rarely makes any other noise when hard hit. This moan was probably a signal ; and as such it was translated by the herd to which this animal belonged, as they suddenly stopped in their retreat, and came to the rescue. Kirkman dropped his gun and took to some trees, where he was in safety. Fortunate it was for him that timber happened to be near, as the savage herd really meant mischief, and came round his tree in numbers. When they found that he was safe from their rage, they retreated.

The vulnerable parts in a Buffalo are behind the shoulder, near the kidneys, or high up on the back. His head is so protected by his horny helmet, that a bullet does not easily find a vulnerable point in the forehead. I once met a Buffalo face to face in the bush, we were about three yards apart ; I fired at his forehead, aiming between the eyes. I know that my bullet struck true ; the Buffalo fell, but soon jumped up again, and scampered off. This was certainly a fair trial of lead versus horn, and horn had the best of it."

ANOTHER species of Buffalo is the ANOA, an inhabitant of the island of Celebes. This animal was formerly thought to belong rather more to the antelopes than to the oxen, but is now satisfactorily ascertained to be a member of the genus Bubalus. It is a small, but very fierce animal, and is with difficulty made prisoner. Some of these creatures, which were kept in confinement, killed in one night fourteen stags which were placed in the same inclosure. The horns of this animal are quite straight, and are set nearly in a line with the forehead. In length they equal the head, are boldly flattened in front, and are covered throughout their length with successive wrinklings. The Anoa is generally found among the more rocky localities of its native island.

The BANTENG, or JAVAN OX, possesses something of the homely aspect which belongs to the common domestic cattle. It is, however, a very strong, fleet, and active animal, inhabiting the wooded valleys of its native land, and living in small herds under the watchful guardianship of vigilant sentries.

BANTENG, OR JAVAN OX.—*Bibos Banting.*

This animal is rather variable in colour, according to its age and sex, the old bulls being of a blackish-brown colour, and the females a reddish-bay. Upon the hinder quarters is always a bold patch of white, the inside of the ears and the lips are of the same hue, and the lower half of the legs is white. It is a tolerably large animal, the height of an adult bull being about five feet six inches at the shoulder. In spite of its constitutional shyness and its dread of man, it is domesticated by the inhabitants of Borneo, and is employed for many useful purposes.

LARGEST of all the existing members of the Ox tribe, is the GAUR, or GOUR, an animal which may be easily recognised by the extraordinary elevation of the spinal ridge and the peculiarly white "stockings." The general colour of the Gaur is a deep brown, verging here and there upon black, the females being usually paler than their mates. The dimensions of the Gaur are very considerable, a full-grown bull having been known to measure six feet ten inches in height at the shoulders. The great height of the shoulder is partly owing to the structure of the vertebræ, some of which give out projections of sixteen inches in length.

The Gaur associates in little herds of ten, twenty, or thirty in number, each herd generally consisting of a few males and a great comparative number of the opposite sex. These herds frequent the deepest recesses of the forest, and in their own domains bear supreme rule, neither tiger, rhinoceros, or elephant daring to attack them. During the heat of noonday, the Gaurs are buried in the thickest coverts, but in the early morning, and after the setting of the sun, they issue from their place of concealment, and go forth to pasture on the little patches of open verdure that are generally found even in the

GAUR, OR GOUR.—*Bibos Gaurus.*

deepest forests. The watchfulness of this animal is extremely remarkable, as, independently of placing the usual sentries, the Gaurs are said to arrange themselves in a circle while at rest, their heads all diverging outwards, so as to preserve equal vigilance on every side. They may, however, be readily approached if the spectator be mounted on an elephant, as they seem to regard these huge animals without any suspicion or fear. In all probability, the imperturbable indifference with which they look upon the elephant is caused by the fact that the elephant is never used in Gaur-hunting, and, unless accompanied by human beings, never attempts to attack these animals.

The temper of the Gaur is naturally mild and equable, and, as a general fact, the herds are quite harmless. Solitary hermit Gaurs, however, are occasionally found, and are extremely irascible and vicious, concealing themselves in the deepest thicket, and suddenly springing at any unfortunate traveller. One of these animals has been known to drive his intended victim up a tree and watch at its foot for a space of twenty-four hours, only vacating its post when killed by the companions of the imprisoned hunter.

The voice of the Gaur is rather peculiar, being totally different from the bellow of an ordinary bull or the lowing of a cow, and partaking greatly of the nature of a grunt, or hoarse cough. The breath of the Gaur is even sweeter than that of the domestic cow, and is plainly perceptible at a distance of several yards. The skin of this animal is extremely thick, especially on the shoulders and hinder quarters. The hide of these parts of the animal will sometimes measure nearly two inches in thickness when it has been removed from the Gaur and permitted to contract during the process of drying. On account of its great strength, this portion of the skin is much esteemed for the purpose of being manufactured into shields. The flesh of the Gaur is said to be remarkably tender, and of good flavour.

AUROCHS.—*Bison Bonassus.*

ALTHOUGH some of the preceding animals have been popularly called by the name of Bison, they have no more right to that title than have the Bisons to the name of buffalo, which is so frequently bestowed upon them. All the true Bisons may be known by the short, crisp, woolly hair with which the body is covered, and which hangs in heavy masses over the head and shoulders of the male animal.

To the Bisons belongs the AUROCHS, or BONASSUS, the former name being a corruption of the word Auer-Ochs. It is also commonly known by the name of *Zubr.* This animal is now almost, if not exclusively, confined to the forest of Bialowikza, in Lithuania, a locality which is peculiarly suitable to its habits on account of the large marshy districts with which it abounds. In order that this magnificent animal may be preserved in perfection, it is protected by the most stringent forest laws. The Aurochs gives forth a powerful and curious odour, which is far from unpleasant, and partakes equally of the characters of musk and violet. This perfume is found to penetrate the whole of the body to a certain extent, but is exhaled most powerfully from the skin and hair which covers the upper part of the forehead. It is found in both sexes, but is much weaker in the cow than in her mate.

Although not so large as some of the preceding animals, standing only about six feet in height at the summit of the elevated shoulder, it is strongly and muscularly built, and is a terrible foe to any antagonist that may happen to arouse its ire. Like the buffalo, it has no fear of predaceous animals, a single Aurochs being supposed to be an overmatch for several hungry wolves. In general, it is shy, and fearful of the presence of man, slipping quietly away as soon as its acute senses perceive the symptoms of human neighbourhood; but if wounded or irritated, it fights most desperately, using its short, sharply-pointed horns, with terrible effect. In disposition it is said to be rather morose and untameable, never having been really domesticated and brought under the dominion of man, even when taken at a very early age.

It is a good swimmer, and is fond of dabbling in water, as well as of rolling itself in certain favoured mud-holes which it frequents. For this reason, it is generally found in thickets which border upon marshy land, ponds, or streams, and in consequence causes all experienced travellers to be very cautious how they approach such localities. Its food is various vegetable substances, and it is especially fond of lichens. In spite of its heavy and awkward look, it is sufficiently active and swift, running with considerable speed for a short time, but being unable to prolong the course for many miles. While running, it carries its head very low, placing the nose almost between the fore-feet.

THE American BISON looks at first sight like an exaggeration of the aurochs, the hair of the body being thicker, more woolly, and more closely curled; the mane, which hangs over the head and shoulders, actually reaching the ground, and the entire aspect of the animal more lowering.

This creature is only found in Northern America, never appearing north of lat. **33°**. It gathers together in enormous herds, consisting of many thousand in number, and in spite of the continual persecution to which it is subjected by man and beast, its multitudes are even now hardly diminished. The Bison is one of the most valuable of animals to the white hunter as well as to the aboriginal Red Indian, as its body supplies him with almost every necessary of life.

The flesh of the fat cow Bison is in great repute, being juicy, tender, and well-savoured, and possesses the invaluable quality of not cloying the appetite, even though it be eaten with the fierce hunger that is generated by a day's hunting. The fat is peculiarly excellent, and is said to bear some resemblance to the celebrated green fat of the turtle. The most delicate portion of the Bison is the flesh that composes the "hump," which gives to the animal's back so strange an aspect; and the hunters are so fond of this delicacy that they will often slay a magnificent Bison merely for the sake of the hump, the tongue, and the marrow-bones, leaving the remainder of the body to the wolves and birds. The pieces of hump-flesh that are stripped from the shoulders are technically called "fleeces," and sometimes weigh as much as a hundred pounds. The flesh of the Bison is also made of exceeding value to voyagers and travellers by being converted into "pemmican," a curious kind of preparation, which to the eye closely resembles tarred oakum, but which is composed of the dried fibres of Bison beef. "Jerked" beef is also made from this animal, the meat being cut into long thin strips, and hung in the sunshine until black, dry, and almost as hard as leather.

The hide is greatly valued by Indians and civilized men, for the many purposes which it fulfils. From this hide the Indian makes his tents, many parts of his dress, his bed, and his shield. For nearly the whole of these uses the skin is deprived of hair, and is so dressed as to be impervious to water, and yet soft and pliable. The shield is very ingeniously made by pegging out the hide upon the ground with a multitude of little wooden skewers round its edge, imbuing it with a kind of glue, and gradually removing the pegs in proportion to the consequent shrinking and thickening of the skin. One of these shields, although still pliable, is sufficiently strong to resist an arrow, and will often turn a bullet that does not strike it fairly.

Sometimes the Bison is the means of saving the hunter from the terrible death of thirst, for it oftentimes happens that the prairie-men find themselves parched with thirst in the midst of vast plains, without a drop of water in their vessels, and no stream within a long day's journey. Under these circumstances, they would inevitably die, were it not that they know how to have recourse to certain natural fountains which are never entirely empty. The Bison has the power of taking a large amount of water into its body, and depositing it in the "reticulum," or cells of the honey-comb department of the stomach, until it shall be needed for use. The hunters, therefore, are not long at a loss for materials wherewith to quench their thirst as long as a Bison is in sight, but slay the animal at once for the sake of the water which they know will be found in the usual situation.

BISON —*Bison Americanus.*

Vast quantities of Bisons are killed annually, whole herds being sometimes destroyed by the cunning of their human foes. The hunters, having discovered a herd of Bisons at no very great distance from one of the precipices which abound in the prairie-lands, quietly surround the doomed animals, and drive them ever nearer and nearer to the precipice. When they have come within half a mile or so of the edge, they suddenly dash towards the Bisons, shouting, firing, waving hats in the air, and using every means to terrify the intended victims. The Bisons are timid creatures, and easily take alarm, so that on being startled by the unexpected sights and sounds, they dash off, panic-struck, in the only direction left open to them, and which leads directly to the precipice. When the leaders arrive at the edge, they attempt to recoil, but they are so closely pressed upon by those behind them that they are carried forward and forced into the gulf below. Many hundred of Bisons are thus destroyed in the space of a few minutes.

A much fairer and more sportsmanlike method of hunting these animals is practised by red and white men, and consists in chasing the herds of Bisons and shooting them while at full speed. This sport requires good horsemanship, a trained steed, and a knowledge of the habits of the Bisons, as well as a true eye and steady hand. The hunter marks a single individual in the herd, and by skilful riding contrives to separate it from its companions. He then rides boldly alongside the flying animal, and shoots it from the saddle. In this method of shooting, the hunter requires no ramrod, as he contents himself with pouring some loose powder into the barrel, dropping a bullet from his mouth upon the powder, and firing across the saddle without even lifting the weapon to his shoulder. The Indians are very expert in this sport, and, armed with their little bows, will often give a better account of their day's sport than many a white hunter armed with the best firelock.

At certain seasons of the year, the whole Bison population becomes greatly excited about settling their matrimonial matters for the next twelvemonth, and dire are the quarrels among the bulls for favour in the eyes of their intended mates. Whole herds of these animals will run in a straight line for many miles, urged forward by some strange impulse, and being easily tracked, not only by the marks of their feet, which tear up the ground as if it had been ploughed, but by a succession of bull Bisons engaged in single combat, they having fallen out on the journey and halted to fight out their quarrel. A cloud of wolves always hangs about the skirts of these herds, as the cunning animals are well aware of the dangers that beset the infuriated wars of Bisons, and accompany them in hopes of pouncing upon some feeble or wounded straggler.

The Bison is remarkably fond of wallowing in the mud, and when he cannot find a mud-hole ready excavated, sets busily to work to make one for himself. Choosing some wet and marshy spot, he flings himself down on his side, and whirls round and round until he wears away the soil, and forms a circular and rather shallow pit, into which the water rapidly drains from the surrounding earth. He now redoubles his efforts, and in a very short time succeeds in covering himself with a thick coating of mud, which is probably of very great service in defending him from the stings of the gnats and other noxious insects which swarm in such localities.

In the summer, the Bison fares luxuriously, living on the sweet green herbage that always springs up after the prairie has been swept by the fires that are continually blazing in one part or another. In winter, however, it is often pinched with hunger as well as with cold, and would fare very badly, did it not instinctively employ its broad nose in shovelling away the snow and laying bare the grass that lies unhurt beneath the white covering. The nose of the Bison is admirably adapted for this purpose, being broad, strong, and tough, so that it can execute a work with ease which is most painful to ordinary cattle, and causes their noses to bleed sadly, from the unaccustomed friction. So severe, however, is the labour, that even old Bisons are often seen with their noses excoriated and bleeding from the effects of their toil. Mr. Palliser mentions that the common domesticated calves have been observed to hang about a Bison bull when he was engaged in shovelling away the snow, and to eat the herbage which he disclosed, without showing the least fear.

The Bison is a marvellously active animal, and displays powers of running and activity which would hardly be anticipated by one who had merely seen a stuffed specimen. The body is so loaded with hair that it appears to be of greater dimensions than is really the case, and seems out of all proportion to the slender legs that appear from under it and seem to bend beneath its weight. Yet the Bison is an enduring as well as a swift animal, and is also remarkably sure of foot, going at full speed over localities where a horse would be soon brought to a halt.

The YAK, or GRUNTING OX, derives its name from its very peculiar voice, which sounds much like the grunt of a pig. It is a native of the mountains of Thibet, and according to Hodson, it inhabits all the loftiest plateaus of High Asia, between the Altai and the Himalayas.

It is capable of domestication, and is liable to extensive permanent varieties, which have probably been occasioned by the climate in which it lives and the work to which it has been put. The Noble Yak, for example, is a large, handsome animal, holding its head proudly erect, having a large hump, extremely long hair, and a very bushy tail. It is a shy and withal capricious animal, too much disposed to kick with the hind feet and to make threatening demonstrations with the horns, as if it intended to impale the rider. The heavy fringes of hair that decorate the sides of the Yak do not make their appearance until the animal has attained three months of age, the calves being covered with rough curling hair, not unlike that of a black Newfoundland dog. The beautiful white bushy tail of the Yak is in great request for various ornamental purposes, and forms quite an important article of commerce. Dyed red, it is formed into those curious

YAK.—*Poéphagus grúnniens.*

tufts that decorate the caps of the Chinese, and when properly mounted in a silver handle, it is used as a fly-flapper in India under the name of a chowrie. These tails are carried before certain officers of state, their number indicating his rank.

The Plough Yak is altogether a more plebeian-looking animal, humble of deportment, carrying its head low, and almost devoid of the magnificent tufts of long silken hairs that fringe the sides of its more aristocratic relation. Their legs are very short in proportion to their bodies, and they are generally tailless, that member having been cut off and sold by their avaricious owner. There is also another variety which is termed the Ghainorik. The colour of this animal is black, the back and tail being often white. The natives of the country where the Yak lives are in the habit of crossing it with the common domestic cattle and obtaining a mixed breed. When overloaded, the Yak is accustomed to vent its displeasure by its loud, monotonic, melancholy grunting, which has been known to affect the nerves of unpractised riders to such an extent that they dismounted, after suffering half an hour's infliction of this most lugubrious chant, and performed the remainder of their journey on foot.

THE curiously shaped horns of the MUSK OX, its long woolly hair falling nearly to the ground in every direction, so as nearly to conceal its legs, together with the peculiar form of the head and snout, are unfailing characteristics whereby it can be discriminated from any other animal. The horns of the Musk Ox are extremely large at their base, and form a kind of helmet upon the summit of the forehead. They then sweep boldly downwards, and are again hooked upwards toward the tips. This curious form of the horns is only noticed in the male, as the horns of the female are set very widely apart on

MUSK OX —*Oribos moschatus.*

the sides of the forehead, and are simply curved. The muzzle is covered with hair, with the exception of a very slight line round the nostrils.

This animal is an inhabitant of the extreme north of America, being seldom seen south of the sixty-first degree of latitude, and ascending as high as the seventy-fifth. It lives, in fact, in the same country which is inhabited by the Esquimaux, and is known to them under the name of Oomingnoak. It is a fleet and active animal, and traverses with such ease the rocky and precipitous ground on which it loves to dwell, that it cannot be overtaken by any pursuer less swift than an arrow or a bullet. It is rather an irritable animal, and becomes a dangerous foe to the hunters, by its habit of charging upon them while they are perplexed amid the cliffs and crevices of its rocky home, thus often escaping unharmed by the aid of its quick eye and agile limbs. The hunters say that it is rather a stupid animal in some matters, and that it will not run away at the report of a gun, provided that it does not see the man who fired it, or perceive the smell of the powder. They believe that the Musk Ox takes the flash and the report to be only a species of thunder and lightning, and therefore does not think itself obliged to escape. The flesh of this animal is very strongly perfumed with a musky odour, very variable in its amount and strength. Excepting, however, a few weeks in the year, it is perfectly fit for food, and is fat and well flavoured.

The Musk Ox is a little animal, but owing to the huge mass of woolly hair with which it is thickly covered, appears to be of considerable dimensions. The colour of this animal is a yellowish-brown, deepening upon the sides.

GROUP OF SOUTH AFRICAN ANTELOPES.

ANTELOPES.

THE ANTELOPES form a large and important group of animals, finding representatives in many portions of the globe. Resembling the deer in many respects, they are easily to be distinguished from those animals by the character of the horns, which are hollow at the base, set upon a solid core like those of the oxen, and are permanently retained

GAZELLE.—*Gazella Dorcas.*

throughout the life of the animal. Indeed, the Antelopes are allied very closely to the sheep and goats, and, in some instances, are very goat-like in external form. In all cases the Antelopes are light and elegant of body, their limbs are gracefully slender, and are furnished with small cloven hoofs. The tail is never of any great length, and in many species is very short. The horns, set above the eyebrows, are either simply conical or are bent so as to resemble the two horns of the ancient lyre, and are therefore termed "lyrate" in technical language.

THE well-known GAZELLE is found in great numbers in Northern Africa, where it lives in herds of considerable size, and is largely hunted by man and beast.

Trusting to its swift limbs for its safety, the Gazelle will seldom, if ever, attempt to resist a foe, unless it be actually driven to bay in some spot from whence it cannot escape ; but prefers to flee across the sandy plains, in which it loves to dwell, with the marvellous speed for which it has long been proverbial. The herd seems to be actuated by a strong spirit of mutual attachment, which preserves its members from being isolated from their companions, and which, in many instances, is their only safeguard against the attacks of the smaller predaceous animals. The lion and leopard can always find a meal whenever they can steal upon a band of Gazelles without being discovered by the sentries which watch the neighbourhood with jealous precaution, for the Gazelles are too weak to withstand the attack of such terrible assailants, and do not even attempt resistance.

If, however, the insidious foe is detected by the eye or scent of the sentinel, his chance of a dinner is hopeless for a while. The alarm is instantly given by the animal whose acute senses have discovered the near presence of the dreaded enemy, and the whole herd immediately take to flight, skimming over the ground with such wondrous rapidity that neither lion nor leopard would be able to overtake their flying steps.

When opposed by less formidable enemies, the Gazelles can bid defiance to their assailants by gathering themselves into a compact circular mass, the females and the young being placed in the centre, and the outer circle being composed of the males, all presenting their horns towards the intruder. They then form a dense phalanx of sharply pointed weapons, arranged on strictly military principles, and being the prototype of the spear-phalanx of ancient warfare, and the "square" of more modern tactics.

ARIEL GAZELLE, OR CORA.

In this attitude the Gazelles will maintain their ground with considerable spirit and pertinacity, seeming to be entirely aware of the advantages which they derive from acting in concert, and oftentimes assuming the offensive as well as the defensive mode of action.

The eye of the Gazelle is large, soft, and lustrous, and has been long celebrated by the poets of its own land as the most flattering simile of a woman's eye. The colour of this pretty little animal is a light fawn upon the back, deepening into dark brown in a wide band which edges the flanks, and forms a line of demarcation between the yellow-brown of the upper portions of the body and the pure white of the abdomen. The face is rather curiously marked with two stripes of contrasting colours, one a dark black-brown line that passes from the eye to the curves of the mouth, and the other a white streak that begins at the horns and extends as far as the muzzle. The hinder quarters, too, are marked with white, which is very perceptible when the animal is walking directly from the spectator.

THERE is considerable difficulty in assigning the Antelopes to their proper position in the animal kingdom, and in many instances zoologists are sadly bewildered in their endeavours to ascertain whether a certain animal is entitled to the rank of a separate species, or whether it can only be considered as a variety of some species already acknowledged. Such is the case with the ARIEL GAZELLE, an animal which is now determined to be merely a variety of the preceding animal, and not entitled to take rank as an independent species.

This beautiful little creature is very similar to the Dorcas Gazelle in general appearance, but is much darker in all its tintings, the back and upper portions of the body being a dark fawn, and the stripe along the flanks almost black.

The Ariel is found in Syria and Arabia, and as it is not only a most graceful and elegant animal in appearance, but is also docile and gentle in temper, it is held in great estimation as a domestic pet, and may be frequently seen running about the houses at its own will. So exquisitely graceful are the movements of the Ariel Gazelle, and with such light activity does it traverse the ground, that it seems almost to set at defiance the laws of gravitation, and, like the fabled Camilla, to be able to tread the grass without bending a single green blade. When it is alarmed, and runs with its fullest speed, it lays its head back so that the nose projects forward, while the horns lie almost as far back as the shoulders, and then skims over the ground with such marvellous celerity that it seems rather to fly than to run, and cannot be overtaken even by the powerful, long-legged, and long-bodied greyhounds which are employed in the chase by the native hunters.

When the Gazelle is hunted for the sake of the sport, and not merely for the object of securing as many skins as possible, the falcon is called to the aid of the greyhound, for without such assistance no one could catch an Ariel in fair chase. As soon as the falcon is loosed from its jesses, it marks out its intended prey, and overpassing even the swift limbs by its swifter wings, speedily overtakes it, and swoops upon its head. Rising from the attack, it soars into the air for another swoop, and by repeated assaults bewilders the poor animal so completely that it falls an easy prey to the greyhound, which is trained to wait upon the falcon, and watch its flight.

When, however, the Gazelle is hunted merely for the sake of its flesh and skin, a very different mode is pursued.

Like all wild animals, the Gazelle is in the habit of marking out some especial stream or fountain, whither it resorts daily for the purpose of quenching its thirst. Near one of these watering-spots the hunters build a very large inclosure, sometimes nearly a mile and a half square, the walls of which are made of loose stones, and are too high even for the active Gazelle to surmount by means of its wonderful leaping powers. In several parts of the edifice the wall is only a few feet in height, and each of these gaps opens upon a deep trench or pit. The manner in which this enormous trap is employed is sufficiently obvious. A herd of Gazelles is quietly driven towards the inclosure, one side of which is left open, and being hemmed in by the line of hunters, the animals are forced to enter its fatal precincts. As the pursuers continue to press forward with shouts and all kinds of alarming noises, the Gazelles endeavour to escape by leaping over the walls, but can only do so at the gaps, and fall in consequence into the trenches that yawn to receive them. One after another falls into the pit, and in this manner they perish by hundreds at a time.

A very similar kind of trap, called the Hopo, is employed in Southern Africa, the walls of the inclosure being formed of trees and branches, and terminating in the pit of death. At the widest part the walls are about a mile asunder, and their length is about one mile. The pit at the extremity is guarded at its edges with tree-trunks, so as to prevent the sides from being broken down by the struggling animals in their endeavours to escape, and the plan is so successful, that sixty or seventy herd of large game are often captured in a single week.

The flesh of the Ariel Gazelle is highly valued, and is made an article of commerce as well as of immediate consumption by the captors. The hide is manufactured into a variety of useful articles. The Ariel is a small animal, measuring only about twenty-one inches in height at the shoulder. The JAIROU, or common Gazelle of Asia, which is so celebrated by the Persian and other Oriental poets, is ascertained to be a different species from the Dorcas, and may be distinguished from that animal by the general dimness of the marking, and the dark brown streak on the haunches. It is also known by the name of AHU, and DSHEREN. Several other species are now known to belong to the genus Gazella, among which we may mention the MOHR of Western Africa, the ANDRA of Northern Africa, and the KORIN, or KEVEL, of Senegal. The latter animal possesses no tufts of hair upon the knees. There is one animal, the CHIKARA, or RAVINE DEER of India, which is worthy of a passing notice, because it is by some authors supposed to belong to the Gazelles, and by others to form a separate genus, as is the case with the arrangement of the British Museum. This animal is also known under the titles of CHOUKA, GOAT ANTELOPE, and KALSIEPIE, or BLACK-TAIL.

The SPRING-BOK derives its very appropriate title from the extraordinary leaps which it is in the constant habit of making whenever it is alarmed.

As soon as it is frightened at any real or fancied danger, or whenever it desires to accelerate its pace suddenly, it leaps high into the air with a curiously easy movement,

SPRING-BOK.—*Antidorcas Euchore.*

rising to a height of seven or eight feet without any difficulty, and being capable on occasions of reaching to a height of twelve or thirteen feet. When leaping, the back is greatly curved, and the creature presents a very curious aspect, owing to the sudden exhibition of the long white hairs that cover the croup, and are nearly hidden by a fold of skin when the creature is at rest, but which come boldly into view as soon as the protecting skin-fold is obliterated by the tension of the muscles that serve to propel the animal in its aërial course.

The Spring-bok is a marvellously timid animal, and will never cross a road if it can avoid the necessity. When it is forced to do so, it often compromises the difficulty by leaping over the spot which has been tainted by the foot of man. The colour of the Spring-bok is very pleasing, the ground tinting being a warm cinnamon-brown upon the upper surface of the body, and pure white upon the abdomen, the two colours being separated from each other by a broad band of reddish-brown. The flesh of the Spring-bok is held in some estimation, and the hide is in great request for many useful purposes.

Inhabiting the vast plains of Southern Africa, the Spring-bok is accustomed to make pilgrimages from one spot to another, vast herds being led by their chiefs, and ravaging the country over which they pass as if they were quadrupedal and mammalian locusts. Thousands upon thousands unite in these strange pilgrimages, or "trek-bokken," as they are called by the Boers, and some faint idea of the moving multitudes that traverse the country may be obtained from the following description, written by Captain Cumming immediately after witnessing one of these migrations.

"For about two hours before the day dawned, I had been lying awake in my wagon, listening to the grunting of the bucks within two hundred yards of me, imagining that some large herd of Spring-boks was feeding beside my camp. But on my rising when it was clear, and looking about me, I beheld the ground to the northward of my camp actually covered with a dense living mass of Spring-boks, marching slowly and steadily along, extending from an opening in a long range of hills on the west, through which they continued pouring like the flood of some great river, to a ridge about a mile to the east, over which they disappeared. The breadth of the ground they covered might have been somewhere about half a mile.

I stood upon the fore-chest of my wagon for nearly two hours, lost in wonder at the novel and beautiful scene which was passing before me, and had some difficulty in convincing myself that it was reality which I beheld, and not the wild and exaggerated picture of a hunter's dream. During this time, their vast legions continued streaming through the neck in the hills, in one unbroken compact phalanx."

The wonderful density of these moving herds may be imagined from the fact, that a flock of sheep have been inextricably entangled among a herd of migrating Spring-boks, and carried along with them without the possibility of resistance or even of escape. Even the lion himself has been thus taken prisoner in the midst of a mass of these animals, and has been forced to move in their midst as if he belonged to their own order. Want of water is said to be the principal cause of these migrations, for they have been always observed to depart as soon as the district in which they live has been deprived of water, and to return as soon as the genial rains have returned moisture to the earth, and caused the green herbage to make its appearance. Dr. Livingstone, however, doubts whether the Spring-bok is a sufficiently thirsty animal to be driven into these migrations only by want of water, and thinks that there must be other causes.

They are extremely fond of the short tender grass as it springs from the earth, and the Bakalahari Kaffirs, taking advantage of this predilection, are in the habit of burning large patches of dry stubbly herbage for the sake of attracting the Spring-boks, who are sure to find out the locality, and to come and feed upon the short sweet grass that always makes its appearance on the site of burnt vegetation. Spring-boks are very seldom seen in the deep, rank grass, that is so plentiful in their native country, for they would not be able to raise their head above the tall blades, and to perceive the lion, leopard, or other enemy that might be crawling towards them under its shelter.

While engaged in these pilgrimages, the Spring-bok suffers sadly from many foes, man included, who thin their numbers along the whole of their march. Various beasts of prey, such as lions, leopards, hyænas, and jackals, hang around the skirts of the herd, and are always ready either to dash boldly among the moving mass, and to drag out some unfortunate animal which may happen to take their fancy; or to prowl in a crafty manner about the rear of the troop, in hopes of snapping up the weakly or wounded animals as they fall out of the ranks. The black and white inhabitants of Southern Africa also take advantage of the pilgrimages, and with guns and spears, which may be used almost indiscriminately among such multitudes of animals, without any particular necessity for a careful aim, destroy myriads of the Spring-boks, and load themselves with an ample supply of hides and meat.

There is a curious provision of nature for preserving the herds in proper condition. It is evident that as the animals move in a compact mass, the leaders will eat all the pasture, and those in the rear will find nothing but the bare ground, cut to pieces by the hoofs of their predecessors. The rearward animals would therefore soon perish by starvation, did not matters arrange themselves in a rather remarkable manner. The leading Spring-boks, having the choice of the best pasture, soon become so satiated and overloaded with food, that they are unable to keep pace with their eager and hungrily active followers, and so are forced to drop into the rear. The hindermost animals in the meantime are anxiously pushing forward in search of food, so that there is a continual interchange going on as the herd moves onwards, those in front dropping back to the rear, while those in the rear are constantly pressing forwards to take their place in front.

In size the Spring-bok is rather superior to the Dorcas gazelle, but may be immediately distinguished from that animal by means of the curious white patch of long hairs on the croup, which has already been described. Although the animal is so marvellously agile, the body is rather clumsily formed, and seems to be disproportionately large when contrasted with the slight and delicate limbs on which it is supported. While standing at rest, the Spring-bok may be recognised by the peculiar line of the back, which is more elevated at the croup than at the shoulders. The horns of this animal are much larger in the adult male than in the young or the female, and when full-grown are marked with eighteen or twenty narrow complete rings. The lyrate form of the horns is not so perceptible in the young Spring-bok as in the older animal, for until the creature has

PALLAH.—*Æpyceros Melampus.*

attained its full growth, the tips of the horns point forward, and only begin to turn inward as the animal increases in age. When the animal leaps into the air it curves its back, and exhibits the characteristic white patch upon the croup, and has, from this habit, received the name of Pronk-bok, or Showy Buck, from the Dutch colonists. The native name of the Spring-bok is Tsebe, a word that bears a remarkable resemblance to the Hebrew word Tsebi, which is supposed to signify the Dorcas gazelle.

The PALLAH, or ROOYE-BOK, is also an inhabitant of Southern Africa, where it is seen in large herds, almost rivalling in numbers those of the spring-bok.

It is a remarkably fine animal, measuring three feet in height at the shoulder, and being gifted with elegantly shaped horns and a beautifully tinted coat. The general colour of the Pallah is bay, fading into white on the abdomen, the lower part of the tail, and the peculiar disc of lighter coloured hairs which surrounds the root of the tail in so many Antelopes. There is a black semilunar mark on the croup, which serves as an easy method of distinguishing the Pallah from the other Antelopes. Its specific name, Melampus, is of Greek origin, signifying black-footed, in allusion to the jetty hue of the back of its feet.

The horns of this animal are of considerable length, often attaining to twenty inches, and are rather irregular in their growth. They are very distinctly marked with rings, and are lyrate in form, though not so decidedly as is the case with many other Antelopes.

The food of the Pallah is very similar to that of the spring-bok, and consists chiefly of tender herbage and the young twigs of the underwood among which it generally takes up its abode. It is hardly so timid an animal as the spring-bok, and will often allow strange creatures to approach the herd without much difficulty. It has a curious habit of walking away when alarmed, in the quietest and most silent manner imaginable, lifting up its feet high from the ground, lest it should haply strike its foot against a dry twig and give an alarm to its hidden foe. Pallahs have also a custom of walking in single file, each following the steps of its leader with a blind confidence ; and when they have settled the direction in which they intend to march, they adhere to their plan, and will not be turned aside even by the presence of human beings. It is not so plain-loving an animal as the spring-bok, but is generally found in or near the district where low brushwood prevails.

THE wide and comprehensive group of animals which includes the Antelopes is so extremely large that it is impossible in a work of the present dimensions to give illustrations even of the more important species, and we must content ourselves with but brief notices in any case.

The SASIN, or INDIAN ANTELOPE (*Antilope bezoártica*), is generally found in herds of fifty or sixty together, each herd consisting of one buck and a large harem of does.

It is a wonderfully swift animal, and quite despises such impotent foes as dogs and men, fearing only the falcon, which is trained for the purpose of overtaking and attacking them, as has already been related of the gazelle. At each bound the Sasin will cover twenty-five or thirty feet of ground, and will rise even ten or eleven feet from the earth, so that it can well afford to despise the dogs. As its flesh is hard, dry, and tasteless, the animal is only hunted by the native chiefs for the sake of the sport, and is always chased with the assistance of the hawk or the chetah, the former of which creatures overtakes and delays it by continual attacks, and the other overcomes by stealthily creeping within a short distance, and knocking over his prey in a few rapid bounds. It is a most wary animal, not only setting sentinels to keep a vigilant watch, as is the case with so many animals, but actually detaching pickets in every direction to a distance of several hundred yards from the main body of the herd.

The young Sasins are very helpless at the time of their entrance into the world, and are not able to stand upon their feet for several days, during which time the mother remains in the covert where her little one was born. As soon as it has attained sufficient strength, she leads it to the herd, where it remains during its life, if it should happen to be a doe, but if it should belong to the male sex, it is driven away from its companions by the leading buck, whose jealousy will permit no rivals in his dominions. Forced thus to live by themselves, these exiles become vigilant and audacious, and endeavour to attract mates for themselves from the families of other bucks.

The horns of this animal are large in proportion to the size of their owner ; their form is spiral, and they diverge considerably at their tips. From the base to the last few inches of the points, the horns are covered with strongly marked rings. In colour, the Indian Antelope is greyish-brown or black on the upper parts of the body, and white on the abdomen, the lips, breast, and a circle round the eyes. The outer sides of the limbs, together with the front of the feet and the end of the tail, are nearly black. Some of the oldest and most powerful males are so deeply coloured that their coats are tinted with the two contrasting hues of black and white, the fawn tint being altogether wanting. The height of this animal is about two feet six inches at the shoulder.

A VERY curious species of Antelopes is that which is scientifically known by the name of *Tetrácerus quadricornis,* both words bearing the same signification, viz. " four-horned." These singular animals are natives of India, where they are known under the titles of CHOUSINGHA, or CHOUKA, the last word being derived from the native term *chouk,* a leap, which has been given to the animal in allusion to its habit of making lofty bounds.

The front pair of horns are very short, and are placed just above the eyes, the hinder pair being much longer, and occupying the usual position on the head. The females are hornless. The colour of the Chousingha is a bright bay above and grey-white below, a few sandy hairs being intermixed with the white. The length of the hinder pair of horns is rather more than three inches, while the front, or spurious horns as they are sometimes termed, are only three-quarters of an inch long. The height of the adult animal is about twenty inches.

The GRYS-BOK, two females of which animal are represented in the accompanying illustration, is a native of Southern Africa, and is about the same size as the preceding animal, its height at the shoulder being between nineteen and twenty inches.

It is not very often found on the plains, but prefers to inhabit the wooded portions of the mountainous districts, and is an especially wary and vigilant creature, and endowed with great powers of speed. The colour of the Grys-bok is ruddy chestnut, largely intermixed with white hairs, which give it a stippled appearance, and have caused the Dutch Boers to term it the Grys-bok, or Grey-buck. The under portions of the body are

GRYS-BOK.—*Calótragus melanótis.*

not white, as is so often the case among the Antelopes, but are of a reddish-fawn. The ears are more than four inches in length, and from their conspicuously black tips have earned for the Grys-bok the scientific title of Melanotis, or black-eared. The hoofs are peculiarly small, sharp, and black, and the tail is so short that it barely protrudes beyond the hair of the hinder quarters.

The OUREBI is another of the many Antelopes which inhabit Southern Africa. For the following graphic description of its appearance and habits I am indebted to the kindness of Captain Drayson.

"Whilst many animals of the Antelope kind fly from the presence of man, and do not approach within a distance of many hundred miles of his residence, there are some few which do not appear to have this great dread of him, but which adhere to particular localities as long as their position is tenable, or until they fall victims to their temerity. It also appears as if some spots were so inviting, that immediately they become vacant by the death of one occupant, another individual of the same species will come from some unknown locality, and re-occupy the ground. Thus it is with the Ourebi, which will stop in the immediate vicinity of villages, and on hills and in valleys, where it is daily making hair-breadth escapes from its persevering enemy—man.

When day after day a sportsman has scoured the country, and apparently slain every Ourebi within a radius of ten miles, he has but to wait for a few days, and upon again

OUREBI.—*Scopóphorus Ourebi.*

taking the field he will find fresh specimens of this graceful little Antelope bounding over the hills around him. It is generally found in pairs, inhabiting the plains, and when pursued, trusts to its speed, seeking no shelter either in the bush or the forest. Its general habitation is among the long grass which remains after a plain has been burned, or on the sheltered side of a hill, among rocks and stones.

Its mode of progression, when alarmed or disturbed, is very beautiful. It gallops away with great rapidity for a few yards, and then bounds several feet in the air, gallops on, and bounds again. These leaps are made for the purpose of examining the surrounding country, which it is enabled to do from its elevated position in the air. Sometimes, and especially when any suspicious object is only indistinctly observed in the first bound, the Ourebi will make several successive leaps, and it then looks almost like a creature possessed of wings, and having the power of sustaining itself in the air. If, for instance, a dog pursues one of these Antelopes, and follows it through long grass, the Ourebi will make repeated leaps, and by observing the direction in which its pursuer is advancing, will suddenly change its own course, and thus escape from view. In descending from these leaps the Ourebi comes to the ground on its hind feet.

When first started, the Ourebi pursues over the ground a course somewhat similar to that which a snipe follows in the air. It dodges from side to side, leaps and rushes through the grass or over the plain with a lightning-like speed, and almost before the sportsman can get his gun ready, the Ourebi is scudding away at a distance of a hundred yards or so. Some sportsmen shoot this animal with buck-shot, and by walking through the long grass, and coming suddenly upon the creature in its lair, they pepper it with shot before it has time to get out of range. I tried this system for several days, but at length found that better and neater sport might be had by using a bullet instead of shot. As, moreover, the grass was in many places five feet in height, it would have prevented me from seeing the animals as they rushed off, whereas, from the back of my horse, I could look down upon the Ourebis as they moved out of their lairs. These animals are found in some parts of the colony of the Cape, and are very numerous in the plains about Natal.

They produce one fawn at a time, which can be easily caught with a good dog, and is particularly recommended as a table delicacy, when cooked with a proper mixture of fat.

The Ourebi does not carry off so heavy a charge of shot as the duyker or the reit-bok, and if wounded by a bullet, the sportsman is certain to secure his prize, provided that he watches the animal with care. When badly hit, they will frequently retire into long grass, and crouching low, will hide themselves from the casual observer. They will then creep away for several yards, and lie down behind a stone, ant-hill, or some similar cover. When the hunter passes and overlooks them, they will jump up and retreat as soon as his back is turned. Taking advantage of this practice, I always avoid looking directly at an Ourebi if I see it lying on the plain, and after taking 'bearings' of its position, I ride round the prostrate animal in circles, gradually drawing nearer and nearer, until it can be easily shot."

The Ourebi stands about two feet in height at the shoulder, and is about four feet in length. The horns of the full-grown male are about five inches long, straight, and pointed, and covered with bold rings at the base. The colour of the Ourebi is pale tawny above, and white below. The female is hornless.

ONE of the prettiest and most graceful of the Antelopes is the KLIPPSPRINGER, or KAINSI (*Oreótragus saltátrix*).

This "darling little Antelope," as Gordon Cumming terms it, almost equals the chamois in its agile traversing of the precipitous localities in which it takes up its residence. It is peculiarly formed for rocky ground, its hoofs being small, hard, sharply-pointed, and so formed that when the animal stands, its weight rests only on the tips of the feet. It may often be seen perched on some narrow point of vantage, standing like the chamois, with all its feet drawn closely together, and calmly surveying the prospect from a height which would prove instantly fatal were one of its feet to miss its hold. When startled, it dashes at once at the most precipitous rocks that are within reach, and bounds up their apparently inaccessible faces as if it were an India-rubber ball endowed with sudden vitality. The least projection serves it for a foothold, and its movements are so rapid, that a very few seconds serve to place it in safety from any other foe than a rifle-ball.

The Bechuanas have a curious idea that the cry of the Klippspringer is a kind of invocation to the pluvial powers of air, and are therefore in the habit of catching a number of these poor little creatures whenever they suffer from drought, and of making them cry continually by blows and pinches until rain falls. They rightly boast that it is a most infallible method of making rain, which is truly the case, as they never cease until the desired moisture is seen. In a similar manner the American Indians vaunt the potency of their sacred bison-dance for attracting the "buffalo," for if the bison herds do not make their appearance at the proper time, they commence the efficacious saltation, and never leave off until their scouts bring news of the desired herds.

The colour of the Klippspringer is dark brown, sprinkled largely with yellow, which gives to the coat a grizzled aspect. Each hair is yellow at the extremity, brown towards the end, and grey for the remaining part of its length. The tint of this animal is rather variable, according to the season of year and the age and sex of the individual. It is a very little creature, being scarcely twenty-one inches in height when full-grown. In general form it is not unlike the ibex kid of six or seven months old. The female is hornless.

THE graceful and slender-limbed MADOQUA (*Neótragus Saltiána*) is one of the tiniest of Antelopes, being hardly fourteen inches in height at the shoulder, and of most delicate proportions.

The little creature is so slightly made that it appears to be too fragile to resist the slightest breeze, or to brave the inclemency of the open air. Its legs are very long in proportion to the dimensions of the body, and hardly exceed a lady's finger in thickness. The colour of the Madoqua is remarkably beautiful, being a silvery-grey on the upper

DUYKER-BOK.—*Cephalopus mergens.*

parts of the body and outside of the limbs, deepening into warm chestnut-brown along the back, and becoming pure white on the abdomen, chest, inner surface of the limbs, and around the root of the tail. It is a native of Abyssinia, where it was discovered by Bruce, and is said to inhabit mountainous districts, where it lives in pairs.

The DUYKER-BOK, or IMPOON, has derived its name of Duyker, or Diver, from its habit of diving suddenly, when alarmed, into the heavy brushwood among which it lives, and of disappearing from the sight of the hunter. For the following valuable account of the animal I am again indebted to Captain Drayson's MS. notes on the fauna of Southern Africa, which he has kindly placed at my disposal.

"On the borders of the bush, the Antelope which is most commonly met is the Duyker, a solitary and very cunning animal.

If the sportsman should happen to overtake this buck, it will lie still, watching him attentively, and will not move until it is aware that it is observed. It will then jump up and start off, making a series of sharp turns and dives, sometimes over bushes, and at others through them. When it conceives that it is observed, it will crouch in the long grass or behind a bush, as though it were going to lie down. This conduct is, however, nothing but a ruse for the purpose of concealing its retreat, as it will then crawl along under the foliage for several yards, and when it has gone to some distance in this sly manner, will again bound away. It is therefore very difficult to follow the course of a Duyker, as it makes so many sharp turns and leaps, that both 'spoorer' and dogs are frequently baffled.

If the course of the buck can be watched, and the place discovered where it lies down after its erratic manœuvrings, it can be easily stalked by approaching it from the leeward side. One must, however, be a good shot to secure a Duyker with certainty, for the little creature is so tenacious of life that it will carry off a large charge of buck-shot without any difficulty, and the irregular course which it then pursues requires great perfection and quickness in shooting with a single ball.

The Duyker is not a very swift animal, and almost any ordinary dog can pull one down. An old pointer, which served me as a dog of all work, frequently caught and held a Duyker until I came to the death.

The flesh of this buck is celebrated for making good soup, and the skin for the thongs of the long waggon whips. As a general fact, the venison of South Africa is very inferior, being dry and tasteless, but to the epicure sportsman I can recommend the liver of all the small Antelopes as a great delicacy. The Dutch have an ingenious plan of improving the flavour of the native venison, by scooping a number of little holes in the meat before it is cooked, and pushing into the cavities bits of fat taken from the eland or the hippopotamus. This process is, indeed, a simple kind of 'larding,' and is very effectual in rendering the meat less dry.

As a general rule, a buck, or any animal, should be watched for some time after it has been fired at. It may be badly wounded and yet go away very freely at first, but after proceeding for a hundred yards or so, it usually comes to a halt, and gives evident tokens of distress. Many bucks which I thought had escaped my bullets I afterwards found to have been mortally wounded, and amongst them the Duyker was one which would frequently go off as if unharmed, though it had received a deadly hurt. Whenever the sportsman passes through long grass, or near low stunted bushes, he should be on the look out for a Duyker."

The height of the Duyker-bok is about twenty-one inches at the shoulder, but the animal is somewhat higher at the croup, where it measures nearly twenty-three inches. It may be distinguished from the other species belonging to the large genus in which it is placed, by a ridge upon the front surface of the horns, which runs through the four or five central rings with which the horns are marked, but does not reach either to the tip or to the base. The general colour of this animal is brown-yellow, fading into white on the abdomen and all the under parts, including the tail. The upper part of the tail is black, and there is a black streak running up the legs, and another on the nose.

The RHOODE-BOK, or NATAL BUSH BUCK (*Cephálopus Natalensis*), is, according to Captain Drayson's MS., "very common in the Natal forests, and although the advance of civilization and the spread of fire-arms among the Kaffirs must greatly reduce the numbers of these animals, still, from their watchful habits and selection of the most retired parts of large dense forests as their residence, they will remain much longer in their old quarters than those animals which inhabit plains and are destitute of any secure retreat. This buck is solitary, and rarely leaves the dense forests except in the evening or during rainy weather, when it seems to prefer feeling the rain-drops *au naturel* to receiving them second-hand from the bushes.

It is very amusing to watch the habits of this wary buck when it scents danger in the bush. Its movements become most cautious; lifting its legs with high, but very slow action, it appears to be walking on tip-toe among the briers and underwood, its ears moving in all directions, and its nose pointing up wind or towards the suspected locality. If it hears a sudden snapping of a branch or any other suspicious sound, it stands still like a statue, the foot which is elevated remains so, and the animal scarce shows a sign of life for near a minute. It then moves slowly onwards with the same cautious step, hoping thus to escape detection. If, however, it obtains a sight of danger, or clearly scents some foe, it gives a sharp sneeze, and bounds away through the forest, alarming all other animals in its progress.

The Kaffirs lay snares for this animal by making a noose which is held to the ground by a small peg, while the other end of the cord is fastened to the bent-down limb of a living tree. As soon as the head of the buck passes into the noose the peg is released, and the victim is jerked into the air by the recoil of the liberated branch, and so strangled. The Red Buck is about two feet high, its horns are about three inches long, straight and pointed, and its ears are rather large. The colour is a deep reddish-brown, which is difficult to distinguish in a dull day, but can be more easily seen on a bright sunny morning, especially if a gleam of light shines through the thick branches and alights on the ruddy coat of the animal."

" The BLUE-BUCK (*Cephalopus pygmœa*) is scarcely more than a foot in height, and about two feet long; it possesses small straight horns, about two inches in length closely annulated, its colour a dark blue or mouse tint.

The most practised eyes are required to discover this buck in the bush, as its colour is so similar to the gloom of the underwood that if it did not shake the branches in its progress it would be scarcely possible to see it.

Long after the sportsman has become sufficiently acquainted with bush-craft to secure with certainty one or two red bucks during a day's stalking, he would still be unable to bag the little Blue Buck.

Several times when I was with a Kaffir, who possessed eyes like those of an eagle, he would point, and with great excitement say, 'There goes a Blue-buck! there he is! there, there!' but it was of no use to me, I would strain my eyes and look to the spots pointed out, but could see no buck; and it was a considerable time before my sight became sufficiently quick to enable me to drop this little Antelope with any certainty."—*From Captain Drayson's MS.*

This animal must not be mistaken for the blau-bok (*Ægocerus leucophœus*), which will be mentioned and figured on a succeeding page. Although the two Antelopes are entirely distinct, even in external aspect, they have often been confused together in consequence of the Dutch name, Blau-bokje, having been given to both of them on account of the colour of their coat.

RIET-BOK, OR INGHALLA.—*Eleótragus arundináceus.*

PASSING by several interesting animals, for whose biography there is no space, we arrive at another of the South African Antelopes, the Riet-bok, or Reed-buck.

"This fine and handsome Antelope," writes Captain Drayson, "is found, as his name implies, principally among reeds or long grass. Few animals give the sportsman such a chance as the Riet-bok, for he usually lies concealed in the reeds or long grass until he is nearly trodden on, and when he does break away, he moves at a steady gallop for a short distance, and then stops to turn and look at his pursuers. As though to compensate for this incautious proceeding, the Riet-bok is gifted with a marvellous tenacity of life, and will frequently gallop very freely after a bullet has passed through his body. In very many cases his escape is but temporary, as he seeks some retired kloof in which he dies by inches, or his career is terminated during the night by a pack of hungry hyænas, who have tracked him for miles over his blood-stained spoor. But still he does sometimes retreat and recover after receiving very severe hurts.

When the Riet-bok is disturbed, he gives as he gallops off a kind of whistling sneeze, which is usually intended as a call for the doe. This whistle sometimes leads to his destruction, as the hunter may pass close to a hidden buck and not have seen it. Then, however, he usually jumps up and gallops away, giving this whistle, which at once attracts attention. The Riet-bok is very fond of young corn, and therefore the Kaffirs are most anxious to drive away or to kill any of these Antelopes which had chosen a retreat near their cornfields. On several occasions I won the eternal friendship of a whole village by shooting some trespassing bucks which had annoyed them for several weeks.

The height of this animal is about three feet, and its length nearly five feet. The horns are a foot in length, and covered with bold rings at the base ; the ears are six inches long. Its colour is ashy-grey above, and white beneath ; the female is rather smaller than her mate, and is destitute of horns. In the Kaffir language its name is Umseke."

The ÆQUITOON, KOB, or SING-SING, is a native of Western Africa, and is frequently found on the banks of the Gambia.

It is a large animal, equalling the common stag in dimensions, and bearing some resemblance to that animal in general aspect. The horns of the adult male are lyre-shaped, and covered with rings. The general colour of this animal is a pale brown, the entire under surface and inner faces of the limbs being white. There is no mane, and the tail is rather long, and covered with hair.

The WATER BUCK, or PHOTOMOK, is one of the handsome examples of the South African Antelopes.

It is a peculiarly timid animal, and when alarmed rushes at once towards the nearest river, into which it plunges without hesitation, and which it will cross successfully even when the stream is deep, strong, and rapid. The animals are probably induced to take to the water by their instinctive dread of the lion and leopard, which will never voluntarily enter the water, except under peculiar circumstances. The Water Bucks are generally found in small herds, which never wander far from the banks of some large river. The horns of this species are remarkable for their formation, being somewhat lyrate, bent back, and thrown forwards at their extremities. The tail is rather long, and is covered with long hairs towards its termination. The flesh of this animal is very powerfully scented, and is of so bad a flavour that none but a hungry Kaffir will eat it, and even he will not do so until forced by dire hunger. This peculiar scent is probably variable in potency according to the season of the year, as is the case with all perfumed animals. Captain Harris says that those which he has killed have been totally uneatable, not even the native palate being proof against the rank flavour. The scent extends to the skin, which exhales so powerful an odour that when Captain Harris was engaged in cutting off the head of a Water Buck which he had killed, the scent was so strong as to drive him repeatedly from his task.

The colour is brown, with the exception of a greyish-white oval patch round the base of the tail. The specific term, *ellipsyprymnus*, is given to the animal in reference to this elliptical mark. The female is without horns. The height of the adult male is about four feet six inches, and his horns are rather more than thirty inches in length.

THE two species which are placed in the genus Ægocerus bear a considerable resemblance to the ibex in the formation of the horns, which are of very great dimensions, large at the base, strongly ringed, and curved backwards towards the shoulders. The name Ægocerus is of Greek origin, and signifies Goat-haired, in allusion to the peculiar structure of these appendages.

The BLAU-BOK, or Blue Buck, as it is called, on account of the slaty-blue colour of its coat, is now a rather rare animal, although it was formerly common in many parts of Southern Africa.

WATER BUCK.—*Kobus ellipsyprymnus.*

It is a gregarious animal, living in little herds not exceeding ten or twelve in number, and preferring hills and slopes to level ground. Like the preceding animal, it exhales a powerful odour, which penetrates throughout its entire body, and which renders its flesh so unpalatable that it is never eaten as long as other food can be obtained. It is a swift and active creature, being remarkable for its speed even among the swift-footed Antelopes. There is a variety of this animal, called the DOCOI, which is found by the Gambia, and which is not quite of the same colour. The natives assert that the female never produces more than a single young one during her lifetime, for that the mother's horns grow so rapidly after the birth of the offspring, that they penetrate into her back and kill her. The Blaubok is about four feet in height, and the horns are nearly thirty inches in length.

ALTHOUGH the blau-bok is a truly handsome example of the Antelopes, it is surpassed in beauty by its congener, the SABLE ANTELOPE.

This truly magnificent creature is found in Southern Africa, but is never seen near the colony, as it is a very shy and crafty animal, and being possessed of great speed, is sure

BLAU-BOK, OR ETAAC.—*Ægóccrus leucophæus.*

to keep itself far aloof from civilization. Gordon Cumming's description of this animal is as follows.

"Cantering along through the forest, I came suddenly in full view of one of the loveliest animals which graces this fair creation. This was an old buck of the Sable Antelope, the rarest and most beautiful animal in Africa. It is large and powerful, partaking considerably of the nature of the ibex. Its back and sides are of glossy black, beautifully contrasting with the belly, which is white as driven snow. The horns are upwards of three feet in length, and bend strongly back with a bold sweep, reaching nearly to the haunches."

It lives in herds of no very great size, consisting mostly of ten or twelve does led by a single buck. As a general fact, the buck takes matters very easily, and trusts to the does for keeping a good watch and warning him of the approach of an enemy. Owing to the jealous caution of these female sentinels, the hunter finds himself sadly embarrassed when he wishes to enrich his museum with the horns of their leader, and if any of them should happen to take alarm, the whole herd will bound over the roughest ground with such matchless speed that all pursuit is hopeless. In Captain Cumming's well-known work, there is a most animated description of the proceedings of a herd of Sable Antelopes, and of the hunter's manœuvres in order to gain his point.

In the native dialect, the Sable Antelope is known under the name of Potaquaine. It is very tenacious of life, and will often make good its escape even though pierced entirely through the body with several bullets. It therefore fully tests all the powers of

SABLE ANTELOPE.—*Ægócerus niger.*

the hunter, and he who secures a specimen of an old male Sable Antelope may congratulate himself on possessing one of the noblest trophies of which a sportsman can boast.

We now arrive at some remarkably handsome animals, which are notable, not only for their dimensions, but for the size and beauty of their horns. The GEMS-BOK, or KOOKAAM, is a large and powerful member of the Antelope tribe, equalling the domestic ass in size, and measuring about three feet ten inches at the shoulder. The peculiar manner in which the hide is decorated with boldly contrasted tints, gives it a very peculiar aspect. The general hue is grey, but along the back, upon the hinder quarters, and along the flanks, the colour is deep black. A black streak also crosses the face, and passing under the chin, gives it the appearance of wearing harness. It has a short, erect mane, and long, sweeping, black tail, and its heavy horns are nearly straight from base to tip.

The long and sharply-pointed horns with which its head is armed, are terrible weapons of offence, and can be wielded with marvellous skill. Striking right and left with these

GEMS-BOK.—*Oryx Gazella.*

natural bayonets, the adult Gems-bok is a match for most of the smaller carnivora, and has even been known to. wage a successful duel with the lordly lion, and fairly to beat off its antagonist. Even when the lion has overcome the Gems-bok, the battle may sometimes be equally claimed by both sides, for in one instance, the dead bodies of a lion and a Gems-bok were found lying on the plain, the horns of the Antelope being driven so firmly into the lion's body, that they could not be extracted by the efforts of a single man. The lion had evidently sprung upon the Gems-bok, which had received its foe upon the points of its horns, and had sacrificed its own life in destroying that of its adversary.

In Captain Cumming's deservedly popular work on Southern Africa may be found the following notes concerning this animal.

"The Gems-bok was intended by nature to adorn the parched karroos and arid deserts of South Africa, for which description of country it is admirably adapted. It thrives and attains high condition in barren regions where it might be imagined that a locust could not find subsistence ; and burning as is the climate, it is perfectly independent of water, which, from my own observation and the repeated reports both of Boers and aborigines, I am convinced it never by any chance tastes. Its flesh is deservedly esteemed, and ranks next to that of the eland. At certain seasons of the year they carry a great quantity of fat, at which time they can more easily be ridden into.

Owing to the even nature of the ground which the Gems-bok frequents, its shy and suspicious disposition, and the extreme distances from water to which it must be followed, it is never stalked or driven to an ambush like the Antelopes, but is hunted on horseback, and ridden down by a long, severe, tail-on-end chase. Of several animals in South Africa which are hunted in this manner, the Gems-bok is by far the swiftest and the most enduring."

Although the Gems-bok is nearly independent of water, it stands as much in need of moisture as any other animal, and would speedily perish in the arid deserts were it not directed by its instincts towards certain succulent plants which are placed in those regions, and which possess the useful power of attracting and retaining every particle of moisture which may happen to settle in their vicinity. The most common and most valuable of these plants is a bulbous root, belonging to the Liliacea, called, from its peculiar property

ORYX.—*Oryx Leucoryx.*

of retaining the moisture, the Water-Root. Only a very small portion of the valuable plant appears above the ground, and the water-bearing bulb is so encrusted with hardened soil that it must be dug out with a knife. Several other succulent plants also possess similar qualities, among which may be noticed a kind of little melon which is spread over the whole of the great Kalahari desert.

RESEMBLING the gems-bok in many particulars, the Oryx can be easily distinguished from its predecessor by the shape of the horns, which, instead of being nearly straight, are considerably bent, and sweep towards the back in a noble curve.

It uses these horns with as much address as its near relative the Gems-bok, and if it should be lying wounded on the ground, the hunter must beware of approaching the seemingly quiescent animal, lest it should suddenly strike at him with its long and keenly-pointed horns, while its body lies prostrate on the earth. Should it be standing at bay, it is a very dangerous opponent, having a habit of suddenly lowering its head and charging forward with a quick, lightning-like speed, from which its antagonist cannot escape without difficulty.

The colour of this animal is greyish-white upon the greater part of its person, and is diversified by sundry bold markings of black and ruddy brown, which are spread over the head and body in a manner that can be readily comprehended from the illustration. The height of the Oryx is rather more than three feet six inches, and the long curved horns are upwards of three feet in length. These horns are set closely together upon the head, from whence they diverge gradually to their extremities. These weapons are covered with rings at their bases, but at their tips they are smooth and exceedingly sharp. Their colour is black.

ADDAX.—*Addax nasomaculatus.*

The Oryx is a native of Northern Africa, where it is known under several names; such as Abou-Harb, El-Walrugh, El-Bukras, Ghau-Bahrein, or Jachmur. It lives in herds of considerable size, and feeds mostly on the branches and leaves of a species of acacia.

The ADDAX is found in many parts of Northern Africa, and is formed by nature for a residence among the vast plains of arid sand which are spread over that portion of the globe.

These animals are not found living together in herds, but in pairs, and their range of locality seems to be rather wide. As they are intended for traversing large sandy regions, the feet are furnished with broad, spreading hoofs, which enable them to obtain a firm foothold upon the dry and yielding sand. The horns of this animal are long, and twisted after a manner that reminds the spectator of the Koodoo, an Antelope which will shortly be described and figured. Measured from the tip to the head in a straight line, the horns are about two feet three inches in length; but if the measurement is made to follow the line of the spiral, the length is obviously much greater. The distance between the tips is about the same as that from the tip to the base. From their roots to within a few inches of their extremities, the horns are covered with strong rings, arranged in an oblique manner, and some of them partially double. The spiral of the horns is as nearly as possible two turns and a half.

Upon the forehead there is a bunch or tuft of long hair, and the throat is also covered with a rather heavy mane of long hair, but there is no mane on the back of the neck. The muzzle and nose are rather peculiar, and bear some resemblance to the same parts of a sheep or goat. The general colour of the Addax is a milk-white, with the exception of the black patch of hair on the forehead, the brown-black mane, and a wash of reddish-brown, tempered with grey, which begins upon the head and envelops the shoulders and part of the back. The Addax is rather higher at the croup than at the shoulder, being about three feet seven inches high at the shoulder, and three feet eight inches at the croup. The horns are equally large and prominent in either sex, and at a little distance it is no easy matter to discriminate between the male and his mate.

GOAT-LIKE in aspect, and very hircine in many of its habits, the CHAMOIS is often supposed to belong to the goats rather than to the Antelopes.

It is, however, a true Antelope, and may be readily distinguished from any of its relations by the peculiar form of the horns, which rise straight from the top of the head for some inches, and then suddenly curve backwards, so as to form a pair of sharp hooks. Formerly, this animal was reported to employ these ornaments in aiding itself to ascend or descend the frightful precipices on which it dwells. This opinion is, however, entirely erroneous, the horns being intended for the same mysterious purpose which they serve when placed upon the head of the duyker, koodoo, or any other Antelope. In descending a precipitous rock, the Chamois is greatly aided by the false hoofs of the hinder feet, which it hitches upon every little irregularity in the stony surface, and which seem to retard its progress as it slides downwards, guided by the sharp hoofs of the fore-feet, which are placed closely together, and pushed well in advance of the body. Thus flattened against the rock, the Chamois slides downwards until it comes to a ledge broad enough to permit it to repose for a while before descending farther. In this manner the active creature will not hesitate to descend some twenty or thirty yards along the face of an almost perpendicular cliff, being sure to make good its footing on the first broad ledge that may present itself.

Although it is a very swift animal when upon level ground, and is unsurpassed in traversing the precipitous Alpine passes of its native home, it makes but a poor progress upon smooth ice, and in spite of its sharply pointed hoofs, slips and slides about upon the glassy surface as awkwardly as any ordinary animal.

The Chamois is one of the most wary of Antelopes, and possesses the power of scenting mankind at an almost incredible distance. Even the old and half-obliterated footmarks which a man has made in the snow are sufficient to startle the sensitive senses of this animal, which has been observed to stop in mid career down a mountain side, and to bound away at right angles to its former course, merely because it had come across the track which had been left by the steps of some mountain traveller. Like all animals which live in herds, however small, they always depute one of their number to act as sentinel. They are not, however, entirely dependent on the vigilance of their picket, but are always on the alert to take alarm at the least suspicious scent, sight, or sound, and to communicate their fears to their comrades by a peculiar warning whistle. As soon as this sound is heard, the entire herd take to flight. It is worthy of notice, that the sentinel must possess the power, not only of announcing danger to its friends, but also of indicating the direction from which it comes. Facts of this nature, of which there are abundance on record, prove that although the sounds of animal voices appear to us to be without definite signification, they yet possess the capability of communicating ideas to others of the same species.

When their attention is aroused by anything suspicious, they have a habit of gazing fixedly in the direction of the object which has excited their alarm, and will remain still, as if carved out of the very rock on which they stand, halting in one fixed attitude for an almost incredible length of time.

Their ears are as acute as their nostrils, so that there are few animals which are more difficult of approach than the Chamois. Only those who have been trained to climb the giddy heights of the Alpine mountains, to traverse the most fearful precipices with a quiet pulse and steady head, to exist for days amid the terrible solitudes of ice, rock, and snow, and to sustain almost every imaginable hardship in the pursuit of their game,— only these, or in very rare instances those who have a natural aptitude for the sport, and are, in consequence, soon initiated into its requisite accomplishments, can hope even to come within long rifle range of a Chamois when the animal is at large upon its native

CHAMOIS.—*Rupicápra Tragus.*

cliffs. There are many familiar tales of the Alpine hunts, and of the terrible privations and hair-breadth escapes of the huuters, but as these histories relate rather to the man than to the beast, we can only give them a passing reference.

The Chamois is, when captured young, capable of domestication, and is gifted with very amusing habits, and possessed of infinite curiosity, as is generally the case with all animals whose nervous system is peculiarly sensitive. The following account is taken from the pages of the " Annals of Sporting," and alludes to four of these Antelopes, a buck, a doe, and two kids, which were imported into England.

" Originally, they were the property of Mr. Lowther, of Wolvesey, who, during his *séjour* among the Alps, was determined to try the experiment of domesticating some of these creatures, hitherto considered, by the natives, the most difficult to tame of all animals in that mountainous region. He may be said to have succeeded ; for they were gradually familiarized to his chateau of Blonay, and to his domestics and people ; to whom, from their novel nature and peculiarities, they afforded much interest and amusement.

A *femme de chambre*, belonging to the establishment, they were most particularly attached to, and she to them ; on a certain day, one of them strayed away for some time, and much regret arose in consequence, but the maid was indefatigable in searching for her truant favourite, and was, at length, fortunate in reclaiming the wanderer ; who, on descrying her, trotted after her footsteps, obeying the enticing cry of ' *Lalotte, Lalotte,*' the name which had been given him.

They are stated, by Mr. Lowther, to have been particularly inquisitive and curious in their habits, prying into everything that was brought into the chateau ; the cook's basket, the tradesmen's wares, and the charcoal-man's bags. Of this last personage they were always indignant, and would pretend to make fight against him ; but on his approach, would leap, with surprising agility and to a great height, upon any wall, ledge, or projectment, that offered itself, and would sustain them, returning invariably to the charge when the man of fuel turned his back, or retrograded : indeed, they would never suffer themselves to be touched ; a finger not having yet reached them. They would admit of the hand being softly brought near their persons, but, immediately as it arrived within an inch of their head or body, they would vault, suddenly and lightly, from the proffered contamination. To the gardener and coachman they were amazingly gracious, and would, apparently, take great delight in seeing the one sow his seeds and delve his soil, and the other clean his carriages and groom his horses : in fact, they became quite pets with all parties, and seemed to forget their wild haunts and mountain dwellings, in the warmer and less terrific asylums of civilized man."

As the hind legs exceed the fore limbs in length, the Chamois is better fitted for the ascent of steep ground than for descending, and never exhibits its wonderful powers with such success as when it is leaping lightly and rapidly up the face of an apparently inaccessible rock ; taking advantage of every little projection to add impetus to its progress. Even when standing still, it is able to mount to a higher spot without leaping. It stands erect on its hind legs, places its fore-feet on some narrow shelf of rock, and by a sudden exertion, draws its whole body upon the ledge, where it stands secure.

The food of the Chamois consists of the various herbs which grow upon the mountains, and in the winter season it finds its nourishment on the buds of sundry trees, mostly of an aromatic nature, such as the fir, pine, and juniper. In consequence of this diet, the flesh assumes a rather powerful odour, which is decidedly repulsive to the palates of some persons, while others seem to appreciate the peculiar flavour, and to value it as highly as the modern gourmand appreciates the "gamey" flavour of long kept venison. The skin is largely employed in the manufacture of a certain leather, which is widely famous for its soft though tough character. The colour of the Chamois is yellowish-brown upon the greater portion of the body, the spinal line being marked with a black streak. In the winter months, the fur darkens and becomes blackish-brown. The face, cheeks, and throat are of a yellowish-white hue, diversified by a dark brownish-black band which passes from the corner of the mouth to the eyes, when it suddenly dilates and forms a nearly perfect ring round the eyes. The horns are jetty black and highly polished, especially towards the tips, which are extremely sharp. There are several obscure rings on the basal portions, and their entire surface is marked with longitudinal lines.

Several varieties of the Chamois are recorded, but the distinctions between them lie only in the comparative length of the horns and the hue of the coat. The full-grown Chamois is rather more than two feet in height, and the horns are from six to eight inches long.

The PRONG-HORNED ANTELOPE, or CABRIT, bears some resemblance to the Chamois, from which animal it may be known by a peculiar structure of the horns, which throw out a projecting point, or prong, just at the spot where the horns begin their backward curve. It is a native of North America, where it is sometimes called the Spring-Buck, to the great confusion of zoological neophytes. It is an active and vigorous animal, and cannot be easily overtaken by a horse unless its footsteps are hindered by a fall of snow. It is a gregarious animal, migrating at different times of the year. It inhabits the vast prairies of the Far West, and is there, under the popular name of the Antelope, an object of pursuit by bipedal and quadrupedal hunters. Its scientific title is *Antilocapra Americána.*

Of all the Antelopes, the GNOO presents the most extraordinary conformation. At the first sight of this curious animal, the spectator seems to doubt whether it is a horse, a bull, or an Antelope, as it appears to partake nearly equally of the nature of these three animals.

The Gnoos, of which there are several species, may be easily recognised by their fierce looking head, their peculiarly shaped horns, which are bent downwards and then upwards again with a sharp curve, by their broad nose, and long hair-clad tail. They live together in considerable herds, often mixing with zebras, ostriches, and giraffes, in one huge army of living beings. In their habits they are not unlike the wild cattle which have already been described. Suspicious, timid, curious of disposition, and irritable of temper, they display these mingled qualities in a very ludicrous manner whenever they are alarmed by a strange object.

"They commence whisking their long white tails," says Cumming, "in a most eccentric manner ; then, springing suddenly into the air, they begin pawing and capering, and pursue each other in circles at their utmost speed. Suddenly they all pull up together to overhaul the intruder, when some of the bulls will often commence fighting in

BRINDLED GNOO.—*Connóchetes Gorgon.*

the most violent manner, dropping on their knees at every shock ; then, quickly wheeling about, they kick up their heels, whirl their tails with a fantastic flourish, and scour across the plain, enveloped in a cloud of dust." On account of these extraordinary manœuvres, the Gnoo is called Wildebeest by the Dutch settlers.

The faculty of curiosity is largely developed in the Gnoo, which can never resist the temptation of inspecting any strange object, although at the risk of its life. When a Gnoo first catches sight of any unknown being, he sets off at full speed, as if desirous of getting to the farthest possible distance from the terrifying object. Soon, however, the feeling of curiosity vanquishes the passion of fear, and the animal halts to reconnoitre. He then gallops in a circle round the cause of his dread, halting occasionally, and ever drawing nearer. By taking advantage of this disposition, a hunter has been enabled to attract towards himself a herd of Gnoos which were feeding out of gunshot, merely by tying a red handkerchief to the muzzle of his gun. The inquisitive animals were so fascinated with the fluttering lure, that they actually approached so near as to charge at the handkerchief, and forced the hunter to consult his own safety by lowering his flag. The same ruse is frequently employed on the prairies of America, when the hunters desire to get a shot at a herd of prong-buck Antelopes.

Several experiments have been made in order to ascertain whether the Gnoo is capable of domestication. As far as the practicability of such a scheme was concerned, the experiments were perfectly successful, but there is a great drawback in the shape of a dangerous and infectious disease to which the Gnoo is very liable, and which would render it a very undesirable member of the cattle-yard. The animal is frequently infected with one of the Œstridæ, or Bot-flies, and suffers from them to such an extent that it ejects them from its nose whenever it snorts, an act which it is very fond of performing. Ordinary cattle have no love for the Gnoo, and on one occasion, when a young Gnoo of only four months old was placed in the yard, the cattle surrounded it and nearly killed it with their horns and hoofs.

U U 2

The colour of the ordinary Gnoo (*Connochetes Gnu*) is brownish-black, sometimes with a blue-grey wash. The mane is black, with the exception of the lower part, which is often greyish-white, as is the lower part of the tail. The nose is covered with a tuft of reversed hair, and there is a mane upon the chest. The BRINDLED GNOO may be distinguished from the common Gnoo, or Kokoon, by its convex and smooth face, the hair lying towards the nose, instead of being reversed. There is no mane upon the chest, and the brown hide is varied and striped with grey. It is higher at the withers than the Kokoon, and its action is rather clumsy. It is very local in its distribution, being found northwards of the Black River, and never being known to cross that simple boundary. It lives in large herds, and when observed, the whole herd forms in single file, and so flies from the object of its terror.

One of these animals, called in the interior the Blue Wildebeest, was captured by Cumming in a very curious manner. The animal had contrived to hitch one of his forelegs over his horns, and being thus incapacitated from running, was easily intercepted and killed. It had probably got into this unpleasant position while fighting. The Gnoo is about three feet nine inches high at the shoulders, and measures about six feet six inches from the nose to the root of the tail.

HARTEBEEST.—*Alcephalus Caäma.*

OF the genus Alcephalus, or Elk-headed, the HARTEBEEST, or LECAMA, is a good example.

This handsome animal may be easily known by the peculiar shape of the horns, which are lyrate at their commencement, thick and heavily knotted at the base, and then curve off suddenly nearly at a right angle. Its general colour is a greyish-brown, diversified by a large nearly triangular white spot on the haunches, a black streak on the face, another along the back, and a black-brown patch on the outer side of the limbs. It is a large animal, being about five feet high at the shoulder. Being of gregarious habits, it is found in little herds of ten or twelve in number, each herd being headed by an old male who has expelled all adult members of his own sex.

Not being very swift or agile, its movements are more clumsy than is generally the case with Antelopes. It is, however, very capable of running for considerable distances, and if brought to bay, becomes a very redoubtable foe, dropping on its knees, and charging forward with lightning rapidity. The Hartebeest is spread over a very large range of country, being found in the whole of the flat and wooded district between the Cape and the tropic of Capricorn.

The BUBALE, or BEKKER-EL-WASH, of Northern Africa (*Alcephalus Bubalis*), belongs to the same genus as the Hartebeest. It may be mentioned here, that the word "beest" is employed by the colonists in the same sense that it is used by British drovers; so that Wildebeest signifies "wild-ox," and Hartebeest "hart-ox."

SASSABY.—*Dámalis lunátus.*

The SASSABY, or BASTARD HARTEBEEST, as it is sometimes called, is by no means an uncommon animal, although some few years ago it was only known through the means of a mutilated skin.

The general colour of this animal is reddish-brown, the outer sides of the limbs being dark, and a blackish-brown stripe passing down the middle of the face. Sometimes the

BONTE-BOK.—*Dámalis pygarga.*

body is washed with a bluish-grey. It lives in small herds of six or ten, in the flat districts near the tropic of Capricorn, and is a most welcome sight to the wearied hunter when perishing with thirst. There are many Antelopes which are almost independent of water, and can quench their thirst by means of the moist roots and bulbs on which they feed. But the Sassaby is a thirsty animal, and needs to drink daily, so that whenever the hunter sees one of these animals he knows that water is at no great distance. It is rather persecuted by the hunters, as its flesh is in great esteem ; but as it soon becomes shy and wary, is not easily to be killed.

Concerning one of these animals, Cumming gives the following curious anecdote. " Having shot a Sassaby as I watched the water, he immediately commenced choking from the blood, and his body became swelled in a most extraordinary manner : it continued swelling with the animal still alive, until it literally resembled a fisherman's float, when the animal died of suffocation. It was not only his body that swelled in that extraordinary manner, but even his head, and legs down to his knees." The poor animal must have been shot through the lungs in such a manner that the air was forced by its efforts at respiration between the skin and flesh, until it assumed that puffy aspect.

THE regularly lyrate horns of the BONTE-BOK, or NUNNI, seem to distinguish it from its congener the sassaby.

The colour of the Bonte-bok is a purplish-red, the outside of the limbs deepening into a rich blackish-brown, and contrasting strongly with the white hair which appears upon the face, the haunches, and front of the legs. From the vividly contrasting tints of the coat,

KOODOO.—*Strepsiceros Kudu.*

it has derived the name of Pied Antelope, or White-faced Antelope. The female is not so highly coloured as the male, and the throat and under parts of the body are white. This animal is found in the district that borders the colony at the Cape of Good Hope, and lives in little herds of six or eight in number. Herds of much larger dimensions are said to be found in the more northern district. The height of the Bonte-bok is nearly four feet at the shoulder, and its length is about six feet, being thus inferior to the common stag in size. The horns are black in colour, and are furnished with a series of ten or twelve half-rings in their frontal surfaces. Their length is about fourteen or fifteen inches.

The BLESS-BOK (*Damalis albifrons*) has sometimes been confounded with the bonte-bok; there is, however, a marked distinction in the colour of the coat. The name, Bless-bok, or Blaze-buck, is given to this animal on account of the "blaze" of white upon the face, and is equally applicable to the bonte-bok.

By far the most striking and imposing of all South African Antelopes, the KOODOO, now claims our attention.

This truly magnificent creature is about four feet in height at the shoulder, and its body is rather heavily made, so that it is really a large animal. The curiously twisted horns are nearly three feet in length, and are furnished with a strong ridge or keel, which extends throughout their entire length. It is not so swift or enduring as the bless-bok, and can be run down without difficulty, provided that the hunter be mounted on a good horse, and the ground be tolerably fair and open. Its leaping powers are very great, for one of these animals has been known to leap to a height of nearly ten feet without the advantage of a run.

The Bushmen have a curious way of hunting the Koodoo, which is generally successful in the end, although the chase of a single animal will sometimes occupy an entire day. A large number of men start on the "spoor," or track, one taking the lead and the others following leisurely. As the leading man becomes fatigued he drops into the rear, yielding his place to another, who takes up the running until he too is tired. A number of women bearing ostrich egg-shells filled with water accompany the hunters, so that they are not forced to give up the chase through thirst. As the chase continues, the Koodoo begins to be worn out with continual running, and lies down to rest, thereby affording a great advantage to its pursuers, who soon come within sight, and force it to rise and continue the hopeless race. At last it sinks wearied to the earth, and falls an unresisting prey to its foes.

The flesh of the Koodoo is remarkably good, and the marrow of the principal bones is thought to be one of Africa's best luxuries. So fond are the natives of this dainty, that they will break the bones and suck out the marrow without even cooking it in any way whatever. The skin of this animal is extremely valuable, and for some purposes is almost priceless. There is no skin that will make nearly so good a "fore-slock," or whip-lash, as that of the Koodoo; for its thin, tough substance is absolutely required for such a purpose. Shoes, thongs, certain parts of harness, and other similar objects are manufactured from the Koodoo's skin, which, when properly prepared, is worth a sovereign or thirty shillings even in its own land.

The Koodoo is very retiring in disposition, and is seldom seen except by those who come to look for it. It lives in little herds or families of five or six in number, but it is not uncommon to find a solitary hermit here and there, probably an animal which has been expelled from some family, and is awaiting the time for setting up a family of his own. As it is in the habit of frequenting brushwood, the heavy spiral horns would appear to be great hindrances to their owner's progress; such is not, however, the case, for when the Koodoo runs, it lays its horns upon its back, and is thus enabled to thread the tangled bush without difficulty. Some writers say that the old males will sometimes establish a bachelor's club, and live harmoniously together, without admitting any of the opposite sex into their society.

It is a most wary animal, and is greatly indebted to its sensitive ears for giving it notice of the approach of a foe. The large, mobile ears are continually in movement, and serve as admirable conductors and condensers of sound. From the conduct of a young Koodoo that was captured by Mr. Anderson, and reared by him, the disposition of the animal appears to be gentle, playful, and affectionate. The little quadruped, which was taken at so tender an age that it was fed with milk from a bottle, became strongly attached to its owner, and was a most active and amusing little creature. Domestication to any extent, is, however, not very practicable, as the animal is, in common with the gnoo and the zebra, liable to the terrible horse sickness, which destroys so many of those useful animals.

The colour of the Koodoo is a reddish-grey, marked with several white streaks running boldly over the back and down the sides. The females are destitute of horns.

The ELAND, IMPOOFO, or CANNA, is the largest of the South African Antelopes, being equal in dimensions to a very large ox.

A fine specimen of an adult bull Eland will measure nearly six feet in height at the shoulders, and is more than proportionately ponderous in his build, being heavily burdened with fat as well as with flesh. Owing to this great weight of body, the Eland is not so enduring as the generality of the Antelopes, and can usually be ridden down without much trouble. Indeed, the chase of this animal is so simple a matter, that the hunters generally contrive to drive it towards their encampment, and will not kill it until it has approached the wagon so closely that the hunters will have but little trouble in conveying its flesh and hide to their wheeled treasure-house.

ELAND.—*Oreas Canna.*

The flesh of the Eland is peculiarly excellent ; and as it possesses the valuable quality of being tender immediately after the animal is killed, it is highly appreciated in the interior of South Africa, where usually all the food is as tough as shoe-leather, and nearly as dry. In some strange manner, the Eland contrives to live for months together without drinking, and even when the herbage is so dry that it crumbles into powder in the hand, the Eland preserves its good condition, and is, moreover, found to contain water in its stomach if opened. For its abstinence in liquids, the Eland compensates by its ravenous appetite for solid food, and is so large a feeder that the expense of keeping the animal would be almost too great for any one who endeavoured to domesticate the animal in England with any hope of profit.

The colour of the Eland is a pale greyish-brown, and the horns are nearly straight, spirally twisted, and of considerable size.

A variety of this animal, termed the Striped Eland, is sometimes, but rarely, seen. Some few years ago, when Colonel Faddy, R.A. was in Southern Africa, he shot several of these animals, and brought their skins home as trophies of success. Neither at the Cape nor in England was he believed when he described the animals which he had been fortunate enough to kill, and it was not until he produced the skins that his account was credited. The skins were presented to the institution attached to the Royal Artillery barracks at Woolwich, and may be seen in the museum.

For the following account of the Bosch-bok, I am again indebted to Captain Drayson's MS.

"The Black Bosch-bok is upwards of three feet in height, and five feet in length, very elegant, and stouter than the generality of Antelopes. The horns are a foot in length,

BOSCH-BOK.—*Tragélaphus sylvatica.*

nearly straight, and wrinkled near the base. The general colour is dark chestnut, black above, and marked with a streak of white along the spine, together with some white spots about the body. The ears are large and round. The female is without horns, smaller and lighter coloured. The animal is extremely watchful, and requires the perfection of bush-craft to be surprised.

These beasts are generally found in couples, male and female, although sometimes an old ram leads a hermit life. The Kaffirs frequently cautioned me about these solitary animals, but I never actually saw any signs of a ferocious disposition except when brought to bay, and under such circumstances even a rat will fight. I have heard that the tiger-bosch-katte (the serval) has been found dead in the bush, pierced by the horns of the Bosch-bok.

The wooded districts from the colony even to Delagoa Bay and some distance inland are the resorts of this Antelope. Although frequently passing from three to four days per week in the bush, I never saw more than a dozen black Bosch-boks, even though their spoor was imprinted on the ground in all directions, thus proving that they were numerous. Frequently I have heard the sharp crack of some twig as it snapped, in the distance, and upon approaching the spot have found that a Bosch-bok had retreated. Seldom by fair stalking can this crafty and wary Antelope be slain. The Kaffirs frequently form large hunting parties, and by 'spooring' their tracks and surrounding the bush in which they are concealed, drive them out and despatch them with assagais. This is, however, but a butcherly proceeding, and one which no true sportsman would follow. The Bosch-bok is so wary, so rare, and so beautiful an Antelope, that any one may feel delighted if he can fairly procure one or two specimens during his sporting career."

PASSING from Africa to Asia, we find a curious and handsome Antelope, partaking of many of the characteristics which are found in the Koodoo and the bosch-bok. This is the NYLGHAU, an inhabitant of the thickly wooded districts of India.

This magnificent Antelope is rather more than four feet high at the shoulders, and its general colour is a slate-blue. The face is marked with brown or sepia ; the long neck is

NYLGHAU.—*Portax tragocamēlus.*

furnished with a bold dark mane, and a long tuft of coarse hair hangs from the throat. The female is smaller than her mate, and hornless. Her coat is generally a reddish-grey, instead of partaking of the slate-blue tint which colours the form of the male. The hind legs of this animal are rather shorter than the fore-legs. Its name, Nylghau, is of Persian origin, and signifies " Blue Ox."

It does not seem to be of a social disposition, and is generally found in pairs inhabiting the borders of the jungle. There are, however, many examples of solitary males. It is a shy and wary animal, and the hunter who desires to shoot one of these Antelopes is obliged to exert his bush-craft to the utmost in order to attain his purpose. To secure a Nylghau requires a good marksman as well as a good stalker, for the animal is very tenacious of life, and if not struck in the proper spot will carry off a heavy bullet without seeming to be much the worse at the time. The native chiefs are fond of hunting the Nylghau, and employ in the chase a whole army of beaters and trackers, so that the poor animal has no chance of fair play. These hunts are not without their excitement, for the Nylghau's temper is of the shortest, and when it feels itself aggrieved, it suddenly turns upon its opponent, drops on its knees, and leaps forward with such astounding rapidity that the attack can hardly be avoided, even when the intended victim is aware of the animal's intentions.

Even in domesticated life the Nylghau retains its hasty and capricious temper, and though there may have been several successive generations born into captivity, the young Nylghaus display the same irritable temper as their parents. Its disposition is very uncertain and not to be depended upon. One of these animals which had been reared from a fawn by an officer, and was thought to be quite tame, turned suddenly upon its owner and attacked him with such ferocity that it left him dead. The males are much given to fighting, and in their duels they hurl themselves forward with such furious velocity that the shock of their contending heads seems to be sufficient to crush the skulls of both combatants. No one knows when a Nylghau will be offended, for it takes offence at the veriest trifles, and instantly attacks the object of its dislike. A captive Nylghau that once chose to feel insulted because a labourer happened to be passing near its domicile, dashed at the man with such violence that it completely shattered the wooden paling within which it was confined.

The Nylghau is not of very great value either to individual hunters or for commercial purposes. The hide is employed in the manufacture of shields, but the flesh is coarse and without flavour. There are, however, exceptions to be found in the "hump" of the male, the tongue, and the marrow bones; which are thought to be rather delicate articles of diet. Its gait is rather clumsy, but very rapid, and generally consists of a peculiar long swinging canter, which is not easily overtaken.

GOATS AND SHEEP.

CLOSELY allied to each other, the GOATS and the SHEEP can be easily separated by a short examination. In the Goats, which will first come under consideration, the horns are erect, decidedly compressed, curved backwards and outwards, and are supplied with a ridge or heel of horny substance in front. The males generally possess a thickly bearded chin, and are all notable for a powerful and very rank odour which is not present in the male sheep.

The JHARAL or JEMLAH GOAT (*Hemitragus Jemlaicus*) is a remarkably handsome animal, inhabiting the loftiest mountains of India, and traversing with ease the precipitous crags which are inaccessible to almost any wingless beings except themselves. Their strongholds, where they pass the night, and to which they fly when alarmed, are situated above the line of vegetation, and border upon the limits of perpetual snow. By day they descend to feed in little flocks of twenty or thirty in number, each flock being under the guidance of an old male, whose mandates they implicitly obey. They are shy and cautious animals, and the slightest unaccustomed sound is sufficient to send them towards their rocky fastnesses, ever and anon halting and looking back to examine the cause of their terror.

The hair of this animal is extremely long and coarse, hanging mane-like on each side of its head and neck. The general colour of the Jharal is a very pale greyish-fawn, diversified with a dark streak along the back, and a brown mark on the forehead and front of the legs. The horns are very curiously formed. They are very much depressed, and are very wide at the base, from whence they spread outwards, and then suddenly narrow into a point, which is curled so strongly inwards that the two points nearly meet above the neck. Upon their frontal edge are seven small distinct protuberances, becoming gradually obliterated as they are set higher upon the horns, and each creating a wrinkle which passes nearly round the entire horn. Their colour is greyish-buff.

OF the genus Capra, which includes several species, the IBEX or STEINBOCK is a familiar and excellent example.

This animal, an inhabitant of the Alps, is remarkable for the exceeding development of the horns, which are sometimes more than three feet in length, and of such extraordinary dimensions that they appear to a casual observer to be peculiarly unsuitable for an animal which traverses the craggy regions of Alpine precipices. Some writers say that these enormous horns are employed by their owner as "buffers," by which the force of a fall may be broken, and that the animal, when leaping from a great height, will alight on its horns, and by their elastic strength be guarded from the severity of a shock that

IBEX.—*Capra Ibex.*

would instantly kill any animal not so defended. This statement is, however, but little credited.

To hunt the Ibex successfully is as hard a matter as hunting the chamois, for the Ibex is to the full as wary and active an animal, and is sometimes apt to turn the tables on its pursuer, and assume an offensive deportment. Should the hunter approach too near the Ibex, the animal will, as if suddenly urged by the reckless courage of despair, dash boldly forward at its foe, and strike him from the precipitous rock over which he is forced to pass. The difficulty of the chase is further increased by the fact, that the Ibex is a remarkably endurant animal, and is capable of abstaining from food or water for a considerable time.

It lives in little bands of five or ten in number, each troop being under the command of an old male, and preserving admirable order among themselves. Their sentinel is ever on the watch, and at the slightest suspicious sound, scent, or object, the warning whistle is blown, and the whole troop make instantly for the highest attainable point. Their instinct always leads them upwards, an inborn "excelsior" being woven into their very natures, and as soon as they perceive danger, they invariably begin to mount towards the line of perpetual snow. The young of this animal are produced in April, and in a few hours after their birth they are strong enough to follow their parent.

The colour of the Ibex is a reddish-brown in summer, and grey-brown in winter; a dark stripe passes along the spine and over the face, and the abdomen and interior faces of the limbs are washed with whitish grey. The horns are covered from base to point with strongly marked transverse ridges, the number of which is variable, and is thought by some persons to denote the age of the animal. In the female the horns are not nearly so large nor so heavily ridged as in the male. The Ibex is also known under the name of BOUQUETIN.

GOAT.—*Hircus Ægágrus.*

The members of the genus Hircus may be distinguished from the ibex and the sheep by the peculiar formation of their horns, which are compressed, are rounded behind, and furnished with a well-developed keel in front. In some instances the keel is ragged, or appears like a series of knobs, but in all cases it is prominently conspicuous.

There are an enormous number of varieties of the common domestic GOAT, many of them being so unlike the original stock from which they sprang as to appear like a different species. For the present, we will turn to the common Goat of Europe, with which we are all so familiar. This animal is often seen domesticated, especially in and about stables, as there is a prevalent idea that the rank smell of the Goat is beneficial to horses. Be this as it may, the animal seems quite at home in a stable, and a very firm friendship often arises between the Goat and one of the horses. Sometimes it gets so petted by the frequenters of the stables, that it becomes presumptuous, and assaults any one whom it may not happen to recognise as a friend. Happily, a Goat, however belligerent he may be, is easily conquered if his beard can only be grasped, and when he is thus captured, he yields at once to his conqueror, assumes a downcast air, and bleats in a very pitiful tone, as if asking for mercy.

At the Cape of Good Hope, large flocks of these animals are kept, and are extremely sagacious, needing no goat-herd to watch them, and are altogether more wise than sheep. In the morning they sally out upon their foraging expeditions, and in the evening they voluntarily return. It is said that Goats are the only animals that will boldly face fire, and that their chief use in a stable is to lead the horses from the stalls in case of the stables being burned. Horses are such nervous, excitable animals, that when their dwelling has taken fire they cannot be induced to face the dreaded element, and must see some other animal lead the way before they will dare to stir. It is also said, and apparently with reason, that in case of fire, a horse may be easily removed from the scene of danger by harnessing him as usual, instead of trying to lead him out at once. The animal has learned to connect obedience and trustfulness with the harness, and while he bears the bit in his mouth, and the saddle or traces on his back, he will go wherever he may be led. Blindfolding the horse is another good method of inducing the animal to follow its guide without hesitation.

The Goat is, like several other domesticated animals, able to foretell stormy weather, and always contrives to place itself under shelter before the advent of a storm. The flesh of the Goat is not held in great estimation, and even that of the kid, which is comparatively tender and well-flavoured, has fallen into disrepute. The milk is, however, in some demand, being of a rather peculiar flavour, which is grateful to certain palates.

In its wild state, the Goat is a fleet and agile animal, delighting in rocks and precipitous localities, and treading their giddy heights with a foot as sure and an eye as steady as that of the chamois or ibex. Even in domesticated life, this love of clambering is never eradicated, and wherever may be an accessible roof, or rock, or even a hill, there the Goat may be generally found.

CASHMIR GOAT.

THE varieties of the Goat are almost numberless, and it will be impossible to engrave, or even to notice, more than one or two of the most prominent examples. One of the most valuable of these varieties is the celebrated Cashmir Goat, whose soft silky hair furnishes material for the soft and costly fabrics which are so highly valued in all civilized lands.

This animal is a native of Thibet and the neighbouring locality, but the Cashmir shawls are not manufactured in the same land which supplies the material. The fur of the Cashmir Goat is of two sorts; a soft, woolly under coat of greyish hair, and a covering of long silken hairs that seem to defend the interior coat from the effects of winter. The woolly under coat is the substance from which the Cashmir shawls are woven, and in order to make a single shawl, a yard-and-a-half square, at least ten Goats are robbed of their natural covering. Beautiful as are these fabrics, they would be sold at a very much lower price but for the heavy and numerous taxes which are laid upon the material in all the stages of its manufacture, and after its completion upon the finished article. Indeed, the English buyer of a Cashmir shawl is forced to pay at least a thousand per cent. on his purchase.

Attempts have been made to domesticate this valuable animal in Europe, but without real success. It will unite with the Angora Goat and produce a mixed breed, from which may be procured very soft and fine wool, that is even longer and more plentiful than that of the pure Cashmir Goat. As a commercial speculation, however, the plan does not seem to have met with much success.

There are at least forty acknowledged varieties of the Goat, among which may be mentioned the BERBURA, or RAM SAGUI, of India, a Goat which is remarkable for being destitute of beard and for the large dewlap which decorates the throat of the male. Its ears are very short, and its smooth fur is white, mingled with reddish-brown. The SYRIAN GOAT is notable for the extreme length of its ears, which hang downwards, and when the animal raises its head nearly touch the shoulders. The SPANISH GOAT is destitute of horns, and the MARKHUR, or SNAKE-EATING GOAT, of India and Thibet is celebrated for its large and exquisitely twisted horns, which are not dissimilar to those of the koodoo, only twisted in the opposite direction.

FROM time immemorial, the SHEEP has been subjected to the ways of mankind, and has provided him with meat and clothing, as well as with many articles of domestic use. The whole carcass of the Sheep is as useful as that of the ox, and there is not a single portion of its body that is not converted to some beneficial purpose. The animal as we now possess it, and which has diverged into such innumerable varieties, is never found in a state of absolute wildness, and has evidently derived its origin from some hitherto undomesticated species. In the opinion of many naturalists, the mouflon may lay claim to the parentage of our domestic Sheep, but other writers have separated the mouflons from the Sheep, and placed them in a different genus.

In many of its habits, especially in its rock-climbing propensities, it bears a strong resemblance to the Goats, to which animals it is closely allied. Whenever the flock can have access to elevated spots, they may always be seen perched upon the highest and most precipitous spots, and seem to take a curious pleasure in exposing themselves to the risk of being dashed to pieces. Mr. Bell mentions that he has seen a Sheep and her lamb perched nearly half-way down one of the lofty rocks that border the south-western side of the Isle of Wight. He was at first alarmed by the apparent danger in which the frail little creature was placed, but was re-assured by the boatman, who looked on the circumstance as nothing uncommon. Some of these Sheep will boldly descend the cliff in search of herbage until they nearly reach the sea level, and are in no way dismayed at the prospect of re-ascending the terrible cliffs down which they have come.

Although the Sheep is generally considered to be a timid animal, and is really so when forced into adverse circumstances and deprived of its wonted liberty, it is truly as bold an animal as can well be seen, and even in this country gives many proofs of its courage. If, for example, a traveller comes unexpectedly upon a flock of the little Sheep that range the Welsh mountains, they will not flee from his presence, but draw together into a compact body, and watch him with stern and unyielding gaze. Should he attempt to advance, he would be instantly assailed by the rams, which form the first line in such cases, and would fare but badly in the encounter. A dog, if it should happen to accompany the intruder, would probably be at once charged and driven from the spot.

Even a single ram is no mean antagonist when he is thoroughly irritated, and his charge is really formidable. Sheep differ from Goats in their manner of fighting; the latter animals rear themselves on their hind legs, and then plunge sideways upon their adversary, while the former animals hurl themselves forward, and strike their opponent with the whole weight as well as impetus of the body. So terrible is the shock of a ram's charge, that it has been known to prostrate a bull at the first blow. Nor is the Sheep only combative when irritated by opposition, or when danger threatens itself. A Sheep that had been led into a slaughter-house, has been known to turn fiercely upon the butcher as he was about to kill one of its companions, and to butt him severely in order to make him relinquish his grasp of its friend.

The Sheep does not seem to be so intelligent as the Goat, and has a curious habit of always following the individual who happens to be the leader, even though he should rush

SHEEP.—*Ovis Aries.* (Southdown.)

into danger. A herd of Sheep has jumped successively over the top of a precipice, merely because the leader happened to do the same thing. In the East, where the shepherds lead, not drive, the Sheep, they take advantage of this propensity, in managing the vast flocks over which they are set in charge. They have a few pet Sheep which will follow at their heels, and come to the musical call of the shepherd's flute. These privileged animals act as the leaders of the flock, and wherever they go, the rest will follow.

In the British Isles the Sheep breeds freely, producing generally one or two lambs every year, and sometimes presenting its owner with three lambs at a birth. One instance is on record of a wonderfully prolific ewe. She had hardly passed her second year when she produced four lambs. The next year she had five ; the year after that she bore twins ; and the next year five again. On two successive years she bore twins. Two out of the four and three out of the five were necessarily fed by hand.

We will now advert shortly to some of the principal breeds or varieties of the Sheep.

The SOUTHDOWN, which is figured at the commencement of this article, affords a good example of the short-wooled breed of domestic Sheep, and is valuable not only for the wool, but for the delicacy of the flesh. This breed derives its name from the extensive Southern Downs ; a range of grass-clad chalk hills which pass through Sussex, Surrey, and Kent. These downs are covered with a short sweet herbage, which is of great service in giving to the flesh of the animal its peculiarly delicate flavour. Multitudes of tiny snails are found upon almost every foot of the down-turf, and are thought by many agriculturists to be very efficacious in fattening and nourishing the animal. By careful crossing and good management, the horns of the Southdown Sheep have been abolished, and the vital energies which would have been expended in developing these appendages, are directed to the nourishment of the body and wool.

This valuable breed of Sheep is not confined to the southern downs of England, but has penetrated to every part of our island where the soil and grass are suitable for its welfare. The Wiltshire downs swarm with these Sheep, which have covered their entire extent with an elaborate interlacing system of Sheep-paths, understood by themselves, but very obscure to human senses. Hampshire, and other parts of England, are also in

1.   X X

LEICESTER SHEEP.

possession of the Southdown Sheep, which is often crossed successfully with some local breed. Indeed, this polled or hornless variety has superseded every horn-bearing breed throughout the kingdom, wherever it can find a habitable locality. In Scotland and elsewhere, the Southdown would not be able to live, as it is of too delicate a nature to withstand the severity of the terrible highland winter; so that the original horned breed still holds its place.

The Wiltshire Sheep have lost their horny armatures by continual crossing with the Southdown, and the result is that a remarkably fine variety has been produced, possessing greater dimensions, a lighter colour, and a finer fleece.

OWING to the very great number of the domestic varieties of the Sheep, amounting to nearly thirty distinct breeds, it will be impossible to give more than a mere outline of the most important among them. An example of the long-wooled variety is found in the LEICESTER SHEEP, under which general title are grouped six or eight sub-varieties of the same breed. This animal favours the low-lying level pasturages of the midland counties, and is not so fond of elevated spots as the Welsh and Southdown.

The most celebrated breed of Leicester Sheep is that which is known as the Dishley breed, and which was developed by the persevering energies of a single individual against every possible discouragement. Mr. Bakewell, seeing that the whole practice of Sheep-breeding was based on erroneous principles, struck out an entirely new plan, and followed it with admirable perseverance. The usual plan in breeding the old Leicester Sheep was to obtain a large body and a heavy fleece. Mr. Bakewell, however, thought that these overgrown animals could not be nearly so profitable to the farmer as a smaller and better proportioned breed; for the amount of wool and flesh which was gained by the larger animals would not compensate for the greater amount of food required to fatten them, and the additional year or eighteen months during which they had to be maintained.

His idea was, that three extra pounds of wool are not so valuable as ten or twelve pounds of meat, and that when the expense of keeping and feeding a Sheep for eighteen months is taken into consideration, the balance is certainly on the wrong side. He therefore set himself to improve the flesh, letting the wool take care of itself at first, and

MERINO, OR SPANISH SHEEP.

succeeded so admirably, that on the ribs of one three-year-old Leicester wether were found seven inches and one-eighth of solid fat, cut without any slope. Attention was then turned to the fleece, and by judicious selection and arrangement the two excellences of flesh and wool were combined in the same animal. It was found by experience, that Sheep which have an inordinately heavy fleece were slower in fattening than those whose coat was moderately thick, and that in consequence of the cost in keeping them for a longer period they do not pay the farmer so well as those which are heavy in body and moderately thick in fleece.

Of all the domestic varieties of this useful animal, the SPANISH, or MERINO SHEEP, has attracted the greatest attention.

Originally, this animal is a native of Spain, a country which has been for many centuries celebrated for the quantity and quality of its wool. The Merino Sheep, from whom the long and fine Spanish wool was obtained, were greatly improved by an admixture with the Cotswold Sheep of England, some of which were sent to Spain in 1464, and the fleece was so improved by the crossing, that the famous English wool was surpassed by that which was supplied by Spain.

The Merino Sheep is but of little use except for its wool, as, although its mutton is sufficiently good when fattened, it consumes so much food, and occupies so much time in the process of ripening, that it is by no means a profitable animal. The Merino is larger in the limbs than the ordinary English Sheep, and the male is furnished with large spiral horns. The female is generally hornless, but sometimes possesses these appendages on a very small scale. It is liable to bear a black fleece, the sable hue continually making its appearance, even after long and careful crossing. By good management the black tint has been confined to the face and legs, but is ever liable to come out in spots or dashes in the wool. There is always a peculiar hue about the face of a Merino Sheep, not easy to describe, but readily to be recognised whenever seen.

In Spain, the Merinos are kept in vast flocks, and divided into two general heads, the Stationary and the Migratory. The former animals remain in the same locality during the whole of their lives, but the latter are accustomed to undertake regular annual migrations. The summer months they spend in the cool mountainous districts, but as soon as the weather begins to grow cold, the flocks pass into the warmer regions of Andalusia, where they remain until April. The flocks are sometimes ten thousand in number, and the organization by which they are managed is very complex and perfect. Over each great flock is set one experienced shepherd, who is called the "mayoral," and who exercises despotic sway over his subordinates. Fifty shepherds are placed under his orders, and are supplied with boys and intelligent dogs.

Under the guardianship of their shepherds, the Merino Sheep, which have spent the summer in the mountains, begin their downward journey about the month of September; and after a long and leisurely march, they arrive at the pasture-grounds, which are recognised instinctively by the Sheep. In these pasturages the winter folds are prepared, and here are born the young Merinos, which generally enter the world in March, or the beginning of April. Towards the end of that month the Sheep begin to be restless, and unless they are at once removed, will often decamp of their own accord. Sometimes a whole flock will thus escape, and, guided by some marvellous instinct, will make their way to their old quarters unharmed, except perchance by some prowling wolf, who takes advantage of the shepherd's absence.

The very young lambs are not without their value, although they furnish no wool, for their skins are prepared, and sent to France and England, where they are manufactured into gloves, and called by the name of "kid."

Many attempts have been made to naturalize this most important animal, but with little success. At one time the Merino Sheep was in the highest repute, but as it does not combine the mutton-making with the wool-producing power, it has long ago been left unnoticed. On the Continent, however, the Merino Sheep has been most valuable, and by judicious crossings with the already existing varieties, has produced a number of very useful breeds. It is found that if a Merino be left untouched by the shears for two seasons, the wool will double its length, and be equally fine in texture. In one case, a half-bred Merino was clipped after having been put aside for a whole year, and it was found that her fleece weighed twenty-one pounds, the length of the "pile" being eight inches. The health of the Sheep appears to be uninjured by permitting the animal to retain its coat for two years.

The Australian Sheep, which roams the plains in such vast multitudes, and which furnishes so large a supply of wool to the world's commerce, owes much of its value to a cross with the Merino, several of which animals were imported into Australia by some far-seeing man of business.

A few words may fitly be spoken in this place upon the peculiar hair which decorates the Sheep, and which is called by the name of wool.

Wool is a very curious kind of hair, and may be recognised at once by any one who possesses a tolerable microscope. If a single hair of the Sheep's wool be subjected to a powerful lens, a vast number of serrations are seen, which, when carefully examined, resolve themselves into a series of notched ridges, which surround the hair closely. To use a familiar illustration, the hair bears a strong resemblance to a number of thimbles thrust into each other, and with their edges notched like so many saws. It is to this notched or jagged surface of the hair that the peculiar value of Sheep's wool is owing, for it is by means of these serrations that the hairs interlock with each other in that mode which is popularly termed "felting." If a handful of loose wool be taken and well kneaded the fibres become inextricably matted together, and form the substance which we term "felt." In a similar manner, when woollen thread is made into cloth, and subjected to the hard usage of its manufacture, the fibres of the different threads become so firmly adherent to each other that they never become unravelled when the cloth is cut or torn. The "felting" property is greatly increased by the propensity of woollen fibre to contract when touched by water. It is in consequence of this peculiarity that woollen fabrics will always shrink when they are wetted for the first time after their manufacture. The reader may naturally wonder why the wool does not become thus matted together when it is upon the Sheep's back, and subject to the influence of nightly dew and daily rain. The answer is, that the fleece is imbued with a peculiar secretion from the skin, which is technically called the "yolk," and which repels the action of water. Upon the quantity of this "yolk," the quality of the wool greatly depends.

The custom of annually depriving the Sheep of its wool by means of shears is of very ancient origin, and still holds its ground. But within a comparatively recent period, the poor creatures were even in this country barbarously stripped of their warm coats by main force, the operators grasping large handfuls of the wool and dragging it from the body. This operation was called "rowing," and those who are learned in old English ballad lore will remember many passages where reference to this cruel custom may be found. The Latin word for wool, "vellus," is derived from the verb "vellere," to pluck out, and evidently refers to the same custom. By that cruel mode of action, the Sheep owner was generally a bad economist, for the injury to the more delicate animals was so severe that their sensitive skins were unable to resist the effects of the weather, and the death of the poor creature was often the result.

The milk of the Sheep is not held in very much estimation, and is in these days almost invariably yielded to the lambs. It is, however, of very good flavour, but singularly rich, having, indeed, more of the consistency of true cream than the generality of the white liquid which passes under that name in the metropolis.

THE hardy, active, and endurant variety of the domestic Sheep which inhabit the Highlands of Scotland, partake in a great degree of the characters of the wild animal, and demand a specially trained shepherd to watch over them.

Pasturing together in enormous herds, and living upon vast ranges of bleak, hilly country, the light and active HIGHLAND SHEEP is a very intelligent and independent creature, quite distinct in character from the large, woolly, unintellectual animal that lives only in the fold, and is regularly supplied with its food by the careful hand of its guardian. It is very sensitive to atmospheric influences, and is so ready in obeying the directions of its own instinct, that a good shepherd when he first rises in the morning can generally tell where to find his Sheep, merely by noticing the temperature, the direction of the wind, and the amount of moisture in the air and on the ground. As the Highland Sheep is able to wander to considerable distances from its proper home, the shepherd is aided in his laborious task by several of those wonderful dogs whose virtues and powers have already been recorded in the course of this work.

Much of the disposition of the flock depends upon the temperament of the shepherd. An irritable or impatient man will speedily render his flock almost as unmanageable as his own temper, while he who is gentle and patient, though resolute and firm, will have his charge so thoroughly under control, that his very presence will, in many instances, cause them to do his bidding, even without calling in the aid of his dogs. One prolific source of trouble to the Scotch shepherd is a locomotive propensity which is inherent in Sheep, and which prompts them to quit their own ample boundaries and trespass upon those of their neighbours. Towards evening, when the flocks are inspected, this propensity becomes very annoying, and cannot be restrained by bad managers. Others, who understand the Sheep nature, and shape their conduct accordingly, will quietly move towards the boundary without being followed by the dogs, and by a series of gentle manœuvres entice the Sheep in the proper direction. In a very short time the shepherd establishes the custom, and whenever he moves towards the boundary, the Sheep instinctively recede.

The life of a Highland shepherd is necessarily one of great hardship, and is generally borne with admirable fortitude. In order that the man may feel a personal interest in the flock which is placed under his care, he is permitted to hold a property in a certain number of Sheep, which he may feed on his master's ground without payment. Sometimes he has the little flock of others to watch over as well as those of his own especial charge, so that the neighbours, far and few between as they are, can all have a fellow interest in the welfare of the Sheep. Each shepherd is generally in possession of a little flock consisting of ten to fifty or sixty Sheep, together with pasturage for a few cows.

HIGHLAND SHEEP.

When its intellectual faculties are developed by external circumstances, the Sheep is found to be a decidedly clever animal.

A lamb that belonged to one of my friends, was one of the oddest creatures that could be imagined, full of quaint and even grotesque humour, and cunning to a degree that was almost reprehensible. Excepting a monkey, the lamb was the greatest mimic which the house possessed, and would imitate everything and everybody in the most ludicrous manner. The great deficiency in its character was its utter want of self-reliance—a trait which may afford a clue to the extraordinary manner in which these animals will follow their leader. The creature seemed so dependent on the approbation of its human playfellows, and its disposition was so sensitive to praise or blame, that its mistress could hardly venture upon either course of conduct for fear of over-exciting the impetuous feeling of the animal. If blamed or scolded, it would shrink away into a corner, push its head out of sight, and appear quite overwhelmed with sorrow. But if, on the contrary, it were praised or patted, it became almost mad with excitement, rolling over and over like a ball, and even standing upon its head, an odd trick which it had contrived to acquire.

For music it possessed a discriminating ear, being delighted at brisk and lively airs, such as are set for polkas, quadrilles, and other dance-tunes; but abhorring all slow and solemn compositions. It had the deepest detestation for the National Anthem, and would set up such a continuous baa-baa as soon as its ears were struck with the unwelcome sounds, that the musician was fain to close the performance, being silenced by mirth if not by pity. Many of its pranks are fresh in the remembrance of its late owner, but I can only find space for a single anecdote :—

It was particularly fond of parsley, and ravaged the beds to such an extent that the gardener was forced to protect some of the coveted herb under a glass-shade. The creature soon discovered the treasure; and, nothing daunted by the supposed protection, broke the glass and ate the parsley, without damaging itself by the sharp fragments of the glass. "Bull's-eye" glass was then employed in the frames, and for a time seemed to protect the parsley; but after a while even the new frames were found broken, and the parsley gone. No one could conceive how the "innocent" lamb could have achieved such a feat, and a watch was consequently set upon it. Another frame was ·procured

BRETON SHEEP.

and set in the accustomed spot, so that it could be kept under surveillance. The lamb was soon seen to approach it, and after a careful inspection, walked away, and soon returned, bearing a tolerably large stone in its mouth. Rising on its hind legs, it brought the stone upon the glass with such force that the thick pane was shivered to fragments. It then laid down the stone, put its head through the opening, and quietly began to browse on the green herb which it so much coveted.

THE very small dimensions of the Welsh Sheep are sufficiently familiar to every frequenter of the metropolitan markets, on account of the small size of the delicately flavoured joints which are taken from the Welsh Sheep. There is, however, one variety of domesticated Sheep which is of such pigmy stature that even the Welsh animal rises into importance when compared with one of these curious little creatures. The variety in question is that which is known by the name of the Breton Sheep, and is of such wonderfully minute proportions that it irresistibly reminds the observer of the dwarfed oak-trees which are so prevalent among the Chinese.

A considerable number of these little animals have been lately imported into England, not for the sake of improving the British herds of Sheep, but merely as curious examples of the singular diversity of size and shape which can be assumed by a single species. If a Breton Sheep be placed by the side of a fair example of the Leicester breed, the difference in size would be much greater than that which is exhibited by the huge Flanders dray-horse and the diminutive Shetland pony.

IN several foreign breeds of the domestic Sheep there is a curious tendency to the deposition of fat upon the hinder quarters. This propensity is not valued in our own country, where the Sheep are almost invariably deprived of the greater portion of their tails by the hand of the shepherd, and in consequence is never developed. In some varieties, however, such as the steatopygous Sheep of Tartary, the fat accumulates upon the hinder quarters in such enormous masses that the shape of the animal is completely altered. The fat of this portion of the body will sometimes weigh between

AFFGHAN FAT-TAILED SHEEP.

thirty and forty pounds, and when melted down, will yield from twenty to thirty pounds of pure tallow. So, inordinate is the growth of the fat that the tail becomes almost obliterated, and is only perceptible externally as a little round fleshy button.

Some varieties present a different mode of producing fat, and deposit a large amount of fatty matter in the tail. Fat-tailed Sheep are found in every part of the world, and are much valued on account of the peculiarity from which they derive their name. The Syrian variety is remarkable for the enormous dimensions of the tail, which in highly fattened and carefully tended specimens will weigh from seventy to eighty pounds. So large, indeed, are the tails, and so weighty are they, that the shepherds are forced to protect them from the ground by tying flat pieces of board to their under surface. Sometimes they add a pair of little wheels to the end which drags on the ground, in order to save the animal the trouble of drawing the bare board over the rough earth. The fat which is procured from the tail is highly valued, and is used in lieu of butter, as well as to " lard " meat that would otherwise be unpleasantly dry and tasteless. It is also melted down and poured into jars of preserved meat, for the purpose of excluding the air. These Sheep are most carefully watched, and are generally fed by hand.

At the Cape of Good Hope a fat-tailed race of Sheep has long been prevalent, and is reared in flocks of considerable size, tended by Hottentot herdsmen. This is a very valuable animal, for it not only furnishes good mutton, together with great quantities of fat, but, when young, supplies its owner with beautifully soft and warm garments. The skins of the Cape Sheep are prepared for use by being cleaned, dressed, and sewn together ; and are of such excellent quality that they form a warmer coverlet than could be obtained from any other material. Their outward show is not at all inferior to their quality ; for they are so smooth and soft that few persons would guess that they had once formed part of the natural covering of a Sheep. The fat of these animals is mostly collected in the tail and hinder quarters, and is peculiarly soft when removed from the animal, being in an almost semi-fluid state. It is thought a great delicacy, and is also in great request for the manufacture of soap.

WALLACHIAN, OR CRETAN SHEEP.

The AFFGHAN FAT-TAILED SHEEP is remarkable not only for the extremely large and fatty tail, but for the delicate and silken texture of its wool. The coat of this animal is largely used in local manufactures, and a very considerable amount is also exported into neighbouring countries. Pelisses, caps, and carpets of various kinds are the chief articles into which this soft and valuable wool is manufactured. There are also several herds of Fat-tailed Sheep in different parts of India.

ONE of the most important of the ovine group, is the CRETAN, or WALLACHIAN SHEEP, remarkable for the enormous development and magnificent formation of its horns.

This splendid animal is a native of Western Asia and the adjacent portions of Europe, and is very common in Crete, Wallachia, and Hungary. The horns of the Wallachian Sheep are strikingly like those of the Koodoo, or the Addax, their dimensions being proportionately large, and their form very similar. The first spiral turn is always the largest, and the horns are not precisely the same in every specimen. As a general rule, they rise boldly upwards from the skull, being almost perpendicularly set upon the head; but in others, there is considerable variety in the formation of the spirals and the direction of the tips. In one specimen which was preserved in the gardens of the Zoological Society, the first spiral of the horns was curved downwards, and their tips were directed towards the ground.

The fleece of this animal is composed of a soft woolly undercoat, covered with and protected by long drooping hairs. The wool is extremely fine in quality, and is employed in the manufacture of warm cloaks, which are largely used by the peasantry, and which are so thick and warm that they defend the wearer against the bitterest cold. Even in

ARGALI.—*Caprovis Argali.*

the depth of winter the shepherd can safely lie on the ground wrapped in his sheep-skin mantle. For this purpose, the skin is dressed without removing the wool.

In a state of nature, all Sheep are furnished with a pair of horns, but in the cultivated races these ornaments generally become obliterated. A curious exception to this principle occurs in the many-horned varieties which are found in several parts of Asia, and which sometimes possess as many as three distinct pairs of horns. The additional or accessory appendages are slighter in their make than the true horns, and are generally placed on the upper parts of the head. Their tips almost invariably take an upward direction, while the true horns generally curl downward, and retain a portion of the tendency to a spiral form.

GIANTS among the ovine race, the MOUFLONS tower far above every other variety of the Sheep. These animals may be found in several portions of the world, several species being inhabitants of Asia, one of Sardinia and Corsica, and one of Northern America. Of these gigantic Sheep, the ARGALI of Siberia is the most conspicuous, as well for general dimensions, as for the enormous size of the horns.

The Argali is nearly as large as a moderately sized ox, being four feet high at the shoulders and proportionately stout in its build. The horns of a full-grown male Argali are very nearly four feet in length if measured along the curve, and at their base are about nineteen inches in circumference. They spring from the forehead, and after rising perpendicularly for a short distance, curve boldly downwards until they reach below the chin, when they recurve upwards and come to a point. The surface of the horns is covered with a series of deep grooves set closely together, and extending almost to the very extremities. Firmly as these weapons are fixed upon the animal's forehead, they are sometimes fairly broken off in the fierce conflicts which these creatures wage with each other when they fight for the possession of some desirable female. These broken horns

AOUDAD.—*Ammótragus Tragélaphus.*

are not suffered to lie unobserved on the ground, but are soon utilized by the foxes and other small mammalia which inhabit the same country, and converted at once into dwelling-houses, where they lie as comfortably as the hermit-crab in a whelk-shell. Man also makes use of these horns, by converting them into various articles of domestic economy.

It is a mountain-loving animal, being found on the highest grounds of Southern Siberia and the mountains of Central Asia, and not fond of descending to the level ground.

Its power of limb and sureness of foot are truly marvellous when the great size of the animal is taken into consideration. If disturbed while feeding in the valley, it makes at once for the rocks, and flies up their craggy surfaces with wonderful ease and rapidity. Living in such localities, they are liable to suffer great changes of temperature, and are sometimes wholly enveloped in the deep snow-drifts that are so common upon mountainous regions. In such cases they lie quietly under the snow in a manner similar to that which has already been related of the hare under the same circumstances, and are able to continue respiration by means of a small breathing-hole through the snow. For these imprisoned Argalis the hunters eagerly search, as the animal is deprived of its fleet and powerful limbs, and is forced ignominiously to succumb to the foe, who impales him by driving his spear through the snow into the creature's body. Like others of the same group, it is gregarious, and lives in small flocks.

ANOTHER example of the Mouflons may be found in the BIG-HORN, or ROCKY MOUNTAIN SHEEP, of California.

This animal is not at all uncommon in its native land, where it may be found in little troops of twenty or thirty in number, inhabiting the craggiest and most inaccessible rocks. From these posts of vantage they never wander, but are content to find their food upon the little knolls of green herbage that are found sprinkled among the precipices, without being tempted by the verdant expanse of the plains below. Before they became acquainted with the destructive powers of mankind, they were very fearless, and would curiously survey those who approached their lofty abodes. Now, however, they are peculiarly shy and suspicious, and at the sight of a man they blow their warning whistle, and immediately take refuge in the recesses of the rocks. When wounded, unless the injury is one that carries immediate death with it, the animal makes the best of its way into one of its retreats, and dying there, is useless to its slayer.

The flesh of the Big-horn is remarkably excellent, and is said to be superior to that of the native deer. When full-grown, a Big-horn measures about three feet six inches in height at the shoulders, and the horns are about the same length, thus preserving the same proportions of stature and length of horn as has already been noticed in the Argali of Siberia. The colour of the animal is extremely variable, changing according to the season of the year.

CLOSELY allied to the two preceding animals, the AOUDAD, or BEARDED ARGALI, may be easily distinguished from them by the heavy mane which commences at the throat and falls as far as the knees.

The Aoudad is a native of Northern Africa, and is a mountain-dweller, inhabiting only the loftiest and most inaccessible precipices. It is commonly found in the lofty woods of the Atlas mountains, where it disports itself with as much ease and absence of fear as if it were quietly standing on level ground. Like the argali and big-horn, it is remarkably active, as needs for an animal whose life is cast among the terrible precipices of the loftiest mountain ranges. The height of the Aoudad is rather more than three feet at the shoulder, so that it is a really large animal, although not of such gigantic proportions as the argali. The horns are about two feet in length. Round the fore-legs a quantity of long hair is placed, like ruffles, just above the knee, a peculiarity which has earned for the creature the French name of *Mouflon à manchettes*. It seems to be a lively but rather petulant animal, full of curiosity, and gentle in its disposition.

GIRAFFES.

TALLEST of all earthly dwellers, the GIRAFFE erects its stately head far above any animal that walks the face of the globe. It is an inhabitant of various parts of Africa, and is evidently a unique being, comprising in itself an entire tribe. The colour of the coat is slightly different in the specimens which inhabit the northern and the southern portions of Africa, the southern animal being rather darker than its northern relative.

The height of a full-grown male Giraffe is from eighteen to twenty feet, the female being somewhat less in her dimensions. The greater part of this enormous stature is obtained by the extraordinarily long neck, which is nevertheless possessed of only seven vertebræ, as in ordinary animals. Those bones are, however, extremely elongated, and their articulation is admirably adapted to the purpose which they are called upon to fulfil. The back of the Giraffe slopes considerably from the shoulders to the tail, and at first sight the fore-legs of the animal appear to be longer than the hinder limbs. The legs themselves are, however, of equal length, and the elevation of the shoulder is due to the very great elongation of the shoulder-blades. Upon the head are two excrescences which resemble horns, and are popularly called by that name. They are merely growths or developments of certain bones of the skull, somewhat similar to the bony cores on which the hollow horns of the oxen and antelopes are set. These quasi horns are covered with skin, and have on their summits a tuft of dark hair. On the forehead, and nearly between the eyes, a third bony projection is seen, occupying the same position that was traditionally accredited to the horn of the unicorn.

GIRAFFE.—*Giraffa Camelopárdalis.*

The singular height of this animal is entirely in accordance with its habits and its mode of acquiring food. As the creature is accustomed to feed upon the leaves of trees, it must necessarily be of very considerable stature to be able to reach the leaves on which it browses, and must also be possessed of organs by means of which it can select and gather such portions of the foliage as may suit its palate. The former object is gained by the great length of the neck and legs, and the latter by the wonderful development of the tongue, which is so marvellously formed that it is capable of a considerable amount of prehensile power, and can be elongated or contracted in a very wonderful manner.

Large as is the animal, it can contract the tip of its tongue into so small a compass that it can pass into the pipe of an ordinary pocket-key, while its prehensile powers enable its owner to pluck any selected leaf with perfect ease. In captivity the Giraffe is rather apt to make too free a use of its tongue, such as twitching the artificial flowers and foliage from ladies' bonnets, or any similar freak.

For grazing upon level ground the Giraffe is peculiarly unfitted, and never attempts that feat excepting when urged by hunger or some very pressing cause. It is, however, perfectly capable of bringing its mouth to the ground, although with considerable effort and much straddling of the fore-legs. By placing a lump of sugar on the ground, the Giraffe may be induced to lower its head to the earth, and to exhibit some of that curious mixture of grace and awkwardness which characterises this singular animal.

In its native country its usual food consists of the leaves of a kind of acacia, named the Kameel-dorn, or Camel-thorn (*Acacia giraffæ*). The animal is exceedingly fastidious in its appetite, and carefully rejects every thorn, scrupulously plucking only the freshest and greenest leaves. When supplied with cut grass, the Giraffe takes each blade daintily between its lips, and nibbles gradually from the top to the stem, after the manner in which we eat asparagus. As soon as it has eaten the tender and green portion of the grass, it rejects the remainder as unfit for camelopardine consumption. Hay, carrots, onions, and different vegetables form its principal diet while it is kept in a state of captivity.

The Giraffe is a gentle and playful animal, readily attaching itself to its companions or its keepers, and trying to attract attention by sundry little coquetries. It is full of curiosity, and seems to be greatly gratified by the advent of many visitors, whose costume and general appearance it investigates with an air of great interest. There is something peculiarly mild and pleasant in the full, round, dark eye of the Giraffe, whose gaze is really fascinating to those who feel attracted by a mild and gentle expression of soul. Even the ruthless hunter has felt himself overcome by the glances of the Giraffe's dark expressive eye, as the poor animal lay unresistingly and silently on the ground, watching its destroyer with reproachful but not vengeful gaze.

As far as is at present known, the Giraffe is a silent animal, like the eland and the kangaroo, and has never been heard to utter a sound, even when struggling in the agonies of death. When in its native land it is so strongly perfumed with the foliage on which it chiefly feeds, that it exhales a powerful odour, which is compared by Captain Cumming to the scent of a hive of heather honey.

Although an inoffensive and most gentle creature, it is not destitute of aggressive capabilities, and can defend itself against ordinary foes, such as the predaceous carnivora which inhabit the same land. In defending itself it does not bring its head within reach of its enemy, but delivers a shower of kicks with such lightness and celerity, that it has been known even to daunt the lion from the attack. When, however, the lion can steal unobserved upon the Giraffe, and especially when it unites with others of its own race in the pursuit of the huge prey, it brings down the Giraffe by dint of sheer bodily strength and sharpness of tooth and claw.

To man it falls an easy prey, especially if it can be kept upon level ground, where a horse can run without danger. On rough soil, however, the Giraffe has by far the advantage, as it leaps easily over the various obstacles that lie in its way, and gets over the ground in a curiously agile manner. It is not a very swift animal, as it can easily be overtaken by a horse of ordinary speed, and is frequently run down by native hunters on foot. When running, it progresses in a very awkward and almost ludicrous manner, by a series of frog-like leaps, its tail switching and twisting about at regular intervals, and its long neck rocking stiffly up and down in a manner that irresistibly reminds the observer of those toy birds whose heads and tails perform alternate obeisances by the swinging of a weight below. As the tail is switched sharply hither and thither, the tuft of bristly hairs at the extremity makes a hissing sound as it passes through the air.

The Giraffe is easily traced by its "spoor," or footmarks, which are eleven inches in length, pointed at the toe and rounded at the heel. The pace at which the animal has gone is ascertained by the depth of the impression, and by the scattering of disturbed soil along the path.

Besides the usual mode of hunting and stalking, the natives employ the pitfall for the purpose of destroying this large and valuable animal. For this purpose a very curiously constructed pit is dug, being about ten feet in depth, proportionably wide, and having a wall or bank of earth extending from one side to the other, and about six or seven feet in height. When the Giraffe is caught in one of these pits, its fore-limbs fall on one side of the wall, and its hind-legs on the other, the edge of the wall passing under its abdomen. The poor creature is thus balanced, as it were, upon its belly across the wall, and in spite of all its plunging, is unable to obtain a foothold sufficiently firm to enable it to leap out of the treacherous cavity into which it has fallen. The pitfalls which are intended for the capture of the hippopotamus and the rhinoceros are furnished with a sharp stake at the bottom, which impales the luckless animal as it falls; but it is found by experience that, in the capture of the Giraffe, the transverse wall is even more deadly than the sharpened pike.

In spite of the great size of the Giraffe, and its very peculiar formation, it is not nearly so conspicuous an animal as might be imagined. The long neck and dark skin of the creature are so formed that they bear a close resemblance to the dried and blasted stems of the forest trees. So close is the resemblance, that even the keen-eyed natives have been known to mistake trees for Giraffes, and *vice versâ*.

The Giraffe is generally found in little herds, sometimes only five or six in number, and sometimes containing thirty or forty members, the average being about sixteen. These animals are found of all sizes and both sexes, each herd being under the guidance of one old experienced male, whose dark chestnut hide and lofty head render him conspicuous above his fellows. These herds are always found either in or very close to forests, where they can obtain their daily food, and where they can be concealed from their enemies among the tree-trunks, to which they bear so close a resemblance.

As the hide of the Giraffe is enormously thick, the animal is not easily to be killed by the imperfect weapons with which the native tribes are armed, and does not readily yield its life even to the bullets of the white man. It is but seldom that a single shot has laid low one of these animals, and in these rare cases the balls were of heavy calibre and made of hardened metal. The flesh of the Giraffe is considered to be good, when rightly prepared, and its marrow is thought to be so great a delicacy that the natives eagerly suck it from the bones as they are taken from the animal. When cooked, it is worthy of a place on a royal table. The flesh is well fitted for being made into jerked meat. The thick, strong hide, is employed in the manufacture of shoe-soles, shields, and similar articles.

DEER.

THE characteristics by which the different groups of DEER are distinguished, as well as those which mark out the genus and species, are not at all self-evident, but are variously given by various zoologists. Most writers base their classification solely upon the horns, but as these ornaments are not to be found in every specimen, nor at every season, such a classification would evidently be impracticable in many cases. Moreover, the same species, or even the same individual, bears horns of quite a different aspect at different times of its life, while several species which are clearly distinct are furnished with closely similar horns. Bearing these difficulties in mind, Mr. Gray has judiciously employed several characteristics in his systematic arrangement of the Deer, and for that purpose has made use of the form and extent of the muzzle, the position and presence of glands on the hind legs, the general form of the horns, and the kind of hair which forms the fur.

From the antelopes the Deer are readily distinguished by the character of the horns, which only belong to the male animals, are composed of solid bony substances, and are shed and renewed annually during the life of the animal. The process by which the horns are developed, die, and are shed, is a very curious one, and deserves a short notice before we proceed to consider the various species of Deer which will be noticed in the present work. For a familiar instance, we will take the Common Stag, or Red Deer of Europe.

In the beginning of the month of March he is lurking in the sequestered spots of his forest home, harmless as his mate and as timorous. Soon a pair of prominences make their appearance on his forehead, covered with a velvety skin. In a few days these little prominences have attained some length, and give the first indication of their true form. Grasp one of these in the hand and it will be found burning hot to the touch, for the blood runs fiercely through the velvety skin, depositing at every touch a minute portion of bony matter. More and more rapidly grow the horns, the carotid arteries enlarging in order to supply a sufficiency of nourishment, and in the short period of ten weeks the enormous mass of bony matter has been completed. Such a process is almost, if not entirely, without parallel in the history of the animal kingdom.

When the horns have reached their due development, the bony rings at their bases, through which the arteries pass, begin to thicken, and by gradually filling up the holes, compress the blood-vessels, and ultimately obliterate them. The velvet now having no more nourishment, loses its vitality, and is soon rubbed off in shreds against tree-trunks, branches, or any inanimate object. The horns fall off in February, and in a very short time begin to be renewed. These ornaments are very variable at the different periods of the animal's life, the age of the Stag being well indicated by the number of "tines" upon his horns.

THE first group of Deer is that which includes the Deer of the snowy regions, and comprehends two genera, the Elk and the Reindeer.

The MOOSE or ELK is the largest of all the deer tribe, attaining the extraordinary height of seven feet at the shoulders, thus equalling many an ordinary elephant in dimensions. The horns of this animal are very large, and widely palmated at their extremities, their united weight being so great as to excite a feeling of wonder at the ability of the animal to carry so heavy a burden. It does not reach its full development until its fourteenth year. The muzzle is very large and is much lengthened in front, so as to impart a most unique expression to the Elk's countenance. The colour of the animal is a dark brown, the legs being washed with a yellow hue. It is a native of Northern Europe and America, the Moose of the latter continent and the Elk of the former being one and the same species.

As the flesh of the Elk is palatable, and the skin and the horns extremely useful, the animal is much persecuted by hunters. It is a swift and enduring animal, although its gait is clumsy and awkward in the extreme. The only pace of the Elk is a long, swinging trot; but its legs are so long and its paces so considerable, that its speed is much greater than it appears to be. Obstacles that are almost impassable to a horse, are passed over easily by the Elk, which has been known to trot uninterruptedly over a number of fallen tree-trunks, some of them five feet in thickness. When the ground is hard and will bear the weight of so large an animal, the hunters are led a very long and severe chase before they come up with their prey; but when the snow lies soft and thick on the ground, the creature soon succumbs to its lighter antagonists, who invest themselves in snow-shoes and scud over the soft snow with a speed that speedily overcomes that of the poor Elk, which sinks floundering into the deep snow-drifts at every step, and is soon worn out by its useless efforts.

It is as wary as any of the Deer tribe, being alarmed by the slightest sound or the faintest scent that gives warning of an enemy. As the Elk trots along, its course is marked by a succession of sharp sounds, which are produced by the snapping of the cloven hoofs, which separate at every step, and fall together as the animal raises its foot from the ground.

Generally, the Elk avoids the presence of man, but in some seasons of the year he becomes seized with a violent excitement, that finds vent in fighting with every living creature that may cross his path. His weapons are his horns and fore-feet, the latter being used with such terrible effect that a single blow is sufficient to slay a wolf on the

MOOSE, OR ELK.—*Alces Malchis.*

spot. The enormous horns form no barrier to his progress through the woods, for when the Elk runs, he always throws his horns well back upon his shoulders, so that they rather assist than impede him in traversing the forest glades. The Elk is a capital swimmer, proceeding with great rapidity, and often taking to the water for its own amusement in Africa. During the summer months of the year it spends a considerable portion of its time under water, its nose and horns being the only parts of its form which appear above the surface. Even the very young Moose is a strong and fearless swimmer.

The skin of the Elk is extremely thick, and has been manufactured into clothing that would resist a sword blow and repel an ordinary pistol ball. The flesh is sometimes dressed fresh, but is generally smoked like hams, and is much esteemed. The large muzzle or upper lip is, however, the principal object of admiration to the lovers of Elk flesh, and is said to be rich and gelatinous when boiled, resembling the celebrated green fat of the turtle.

When captured young, the Elk is very susceptible of domestication, and in a few hours will learn to distinguish its keeper, and to follow him about with playful confidence. If, however, the animal has attained to a moderate growth, it becomes fierce, surly, and dangerous. " In the middle of the night," says Audubon, speaking of a young captive Elk, " we were awakened by a great noise in the hovel, and found that as it had in some measure recovered from its terror and state of exhaustion, it began to think of getting home, and was much enraged at finding itself so securely imprisoned. We were unable to do anything with it, for if we merely approached our hands to the opening of the hut, it would spring at us with the greatest fury, roaring and erecting its mane in a manner that convinced

REINDEER.—*Tarandus Rangifer.*

us of the futility of all attempts to save it alive. We threw to it the skin of a deer, which it tore to pieces in a moment. This individual was a yearling, and about six feet high."

By careful attention, however, and good training, the Elk can be used as a beast of carriage or burden, and from its great size and power is extremely valuable in that capacity.

Two varieties of the REINDEER inhabit the earth ; the one, called the Reindeer, being placed upon the northern portions of Europe and Asia, and the other, termed the Caribou, being restricted to North America. We will first describe the European variety.

This animal is very variable in dimensions, specimens of very different height being in the British Museum. The colour is also variable, according to the season of year. In winter the fur is long, and of a greyish-brown tint, with the exception of the neck, hinder quarters, abdomen, and end of nose, which are white. In the summer, the grey-brown hair darkens into a sooty brown, and the white portions become grey.

In its wild state the Reindeer is a migratory animal, making annual journeys from the woods to the hills, and back again, according to the season. Their chief object in leaving the forests in the summer months appears to be their hope of escaping the continual attacks of mosquitoes and other insect pests that are found in such profusion about forest land. The principal plague of the Reindeer is one of the gad-flies, peculiar to the species, which deposits its eggs in the animal's hide, and subjects it to great pain and continual harassment. Even in the domesticated state the Reindeer is obliged to continue its migrations, so that the owners of the tame herds are perforce obliged to become partakers in the annual pilgrimages, and to accompany their charge to the appropriate localities.

The nature of the persecutions to which the Reindeer is continually subjected is well told by a correspondent to the *Field* newspaper :—" The herd looked very miserable, as I thought ; there is nothing of the antlered monarch about the Reindeer, but a careworn, nervous expression, which I do not wonder at, considering how they are bullied. There are creatures which sting them all over, and creatures which lay their eggs in their ears and nostrils, and make themselves comfortable under their skin ; and wolves, and gluttons, and dogs, and Laps—in short, I know of no animal so persecuted (barring a rat, and he has his revenge, and lives on the fat of the land), and nothing in return except snow, and moss which tastes like dry sponge."

The Laplanders place their chief happiness in the possession of many Reindeer, which are to them the only representatives of wealth. Those who possess a herd of a thousand or more are reckoned among the wealthy of their country ; those who only own a few hundreds are considered as persons of respectability ; while those who only possess forty or fifty are content to act as servants to their richer countrymen, and to merge their little herd in that of their employers. In the waste, dry parts of Lapland, grows a kind of white lichen, which forms the principal food of the Reindeer during winter, and is therefore highly prized by the natives. Although this lichen may be deeply covered with snow, the Reindeer is taught by instinct to scrape away the superincumbent snow with its head, hoofs, and snout, and to lay bare the welcome food that lies beneath. Sometimes the surface of the snow is frozen so firmly that the animal can make no impression ; and under these circumstances it is in very poor case, many of the unfortunate creatures dying of starvation, and the others being much reduced in condition.

The Reindeer is extensively employed as a beast of draught and carriage, being taught to draw sledges and to carry men or packages upon its back. Each Reindeer can draw a weight of two hundred and fifty or even three hundred pounds, its pace being between nine and ten miles per hour. There is, however, a humane law which prohibits a weight of more than one hundred and ninety pounds upon a sledge, or one hundred and thirty upon the back. It is a very enduring animal, as it is able to keep up this rate of progress for twelve or more hours together.

The eyes of the Reindeer are very quick, and his hearing also acute ; but his sense of smell is more wonderfully developed than either of the other senses.

The CARIBOU, or American variety of the Reindeer, is a large animal, measuring three feet six inches in height at the shoulder when adult. Although it is specifically identical with the European Reindeer, it has never yet been brought under the sway of man, and trained to carry his goods or draw his sledges. Should it be employed for these purposes, it would be a most valuable servant, for it is a very strong as well as an enduring animal, leading its pursuers a chase of four or five days, and often eventually making good its escape. A small herd of these animals was chased continually for a week ; and after tiring out their original hunters, lost two of their number by the bullets of some fresh hunters who took up the chase. Whenever practicable, the Caribou makes for the frozen surface of the lakes, and is then sure to escape, although the manner of doing so is ludicrously clumsy. Rushing recklessly forward, the Caribou will be suddenly startled by some object in its front ; and on attempting to check its onward career, falls on the ice in a sitting posture, and in that attitude slides for a considerable distance before it can stop itself. Recovering its feet, it then makes off in another direction, and gets over the ground with such celerity that the hunters always yield the chase whenever the animal gets upon the ice.

During the greater part of the year, the flesh of the Caribou is dry and tasteless, and when eaten seems to have no effect in satiating hunger. There is, however, a layer of fat, sometimes two or three inches in thickness, that lies under the skin of the back and croup in the male, and is technically termed the *depouillé*. This fatty deposit is so highly esteemed that it outweighs in value the remainder of the carcase, including skin and horns. The marrow is also remarkably excellent, and is generally eaten raw. When pounded together with the *depouillé* and the dried flesh it makes the best pemmican, a substance which is invaluable to the hunter. Even the horns are eaten raw while they

CARIBOU.—*Tarandus Rangifer.*

are young, soft, and in the velvet. The skin is very valuable, especially when taken from the young animal; and when properly dressed is an admirable defence against cold and moisture. With the addition of a blanket, a mantle of Caribou skins is an ample protection for any one who is forced to bivouac in the snow.

The Caribou lives in herds, which vary from ten to three hundred in number. As it is so valuable an animal, it is subject to great persecution at the hands of white and red hunters, who have very ingenious modes of trapping or stalking this wary and swift Deer. The most ingenious plan is that which is employed by the Esquimaux, who dig a large hole in the ground, about five feet in depth, and capable of holding several Deer. They then cover the aperture with a slab of ice or frozen snow, which is balanced on two pivots in such a way that when a Deer treads upon the treacherous floor it suddenly gives way, tilts him into the pit, and resumes its position in readiness for another victim. Another plan is to make a large inclosure, at least a mile in circumference, and to drive the Deer into its fatal precincts. The space within the inclosure is formed into numerous alleys, in each of which are long nooses, so that the Deer are caught and strangled as they move to and fro within the pound.

WE now come to the Deer which inhabit the warm or temperate regions of the world, and which include the greater portion of the family. The first on the list is the WAPITI, or CAROLINA STAG.

This magnificent animal is one of the largest of the Deer tribe, the adult male measuring nearly five feet in height at the shoulders, and about seven feet nine inches from the nose to the root of the tail. It is a native of North America, where it is popularly known under the name of the Elk.

The Wapiti lives in herds of variable numbers, some herds containing only ten or twenty members, while others are found numbering three or four hundred. These herds are always under the command of one old and experienced buck, who exercises the strictest discipline over his subjects, and exacts implicit and instantaneous obedience. When he halts, the whole herd suddenly stop, and when he moves on, the herd follow his example. There must be some method by which he communicates his orders to his followers, as the entire herd will wheel right or left, advance or retreat, with an almost military precision.

This position of dignity is not easily assumed, and is always won by dint of sheer strength and courage, the post being held against all competitors at the point of the horn. The combats that take place between the males are of a singularly fierce character, and often end in the death of the weaker competitor. An instance is known where a pair of these animals have perished in a manner similar to that which will be related of the carjacou, their horns having been inextricably locked together, causing the poor creatures to die a sad death of hunger and thirst. When attacked by the hunter, and wounded with a hurt that is not immediately mortal, the Wapiti will turn fiercely on his opponent, and fight with the reckless courage of despair.

Although the bucks display such courage in fighting for their spouses, they treat them very harshly when they have secured them, and always keep the poor creatures in constant fear. It is not until they have lost their horns that the does seem to lose the feelings of terror with which they regard their hard-hearted mates.

Even in captivity the male Wapiti retains its combative nature, as may be seen from the following anecdote, which is related in the work of Messrs. Audubon and Bachman.

" A gentleman in the interior of Pennsylvania, who kept a pair of Elks (Wapitis) in a large woodland pasture, was in the habit of taking pieces of bread or a few handfuls of corn with him when he walked in the inclosure, to feed these animals, calling them up for the amusement of his friends. Having occasion to pass through his park one day, and not having furnished himself with bread or corn for his pets, he was followed by the buck, who expected his usual gratification. The gentleman, irritated by the pertinacity with which he was accompanied, turned round, and picking up a small stick, hit the animal a smart blow ; upon which, to his astonishment and alarm, the buck, lowering his head, rushed at him, and made a furious pass with his horns.

Luckily, he stumbled as he attempted to fly, and fell over the prostrate trunk of a tree near which lay another log, and being able to throw his body between the two trunks, the Elk was unable to injure him, although it butted at him repeatedly, and kept him prisoner for more than an hour. Not relishing this proceeding, the gentleman, as soon as he escaped, gave orders to have the unruly animal destroyed."

The Wapiti is a good swimmer, and even when very young, will fearlessly breast the current of a wide and rapid river. Like many of the larger animals, it is fond of submerging itself under water in the warm weather, for the sake of cooling its heated

WAPITI.—*Cervus Canadensis.*

body, and of keeping off the troublesome insects. It is also a good runner, and although burdened with its large and widely branched horns, can charge through the forest haunts with perfect ease. In performing this feat, it throws its head well back, so that the horns rest on the shoulders, and shoots through the tangled boughs like magic. Sometimes a Wapiti will make a slight miscalculation in its leap, for Mr. Palliser saw one strike a small tree with its forehead so fiercely, that the recoil of the elastic trunk threw the Wapiti fairly on its back upon the ice of a frozen stream which it had just crossed.

The food of the Wapiti consists of grass, wild pea-vine, various branches, and lichens. In winter it scrapes among the snow with its fore-feet, so as to lay bare the scanty vegetation below. When alarmed or excited, it gives vent to its feelings in a peculiar loud whistling sound, which on a clear quiet day may be heard at the distance of a mile. While uttering this sound, the animal raises its head in a very peculiar manner, and seems to eject the cry by a kind of spasmodic jerk. The flesh of the Wapiti is in great favour among hunters, while the marrow-bones are prized as great dainties. The skin is also valuable, being employed in the manufacture of mocassins, belts, thongs, and other articles where strength and flexibility are required. The teeth are employed by the Indians in decorating their dresses; and a robe thus adorned, which is in the possession of Mr. Audubon, was valued by its manufacturers as equivalent to thirty horses. The horns are also employed for various useful purposes. It is a remarkable fact, that in no two individuals are the horns precisely alike.

The STAG, or RED DEER, is spread over many parts of Europe and Asia, and is indigenous to the British Islands, where it still lingers, though in vastly reduced numbers.

In the olden days of chivalry and Robin Hood, the Red Deer were plentiful in every forest; and especially in that sylvan chase which was made by the exercise of royal tyranny at the expense of such sorrow and suffering. Even in the New Forest itself the Red Deer is seldom seen, and those few survivors that still serve as relics of a bygone age, are scarcely to be reckoned as living in a wild state, and approach nearly to the semi-domesticated condition of the Fallow Deer. Many of these splendid animals are preserved in parks or paddocks, but they no more roam the wide forests in unquestioned freedom. In Scotland, however, the Red Deer are still to be found, as can be testified by many a keen hunter of the present day, who has had his strength, craft, and coolness thoroughly tested before he could lay low in the dust the magnificent animal, whose head with its forest of horns now graces his residence.

Formerly, the Stag was placed under the protection of the severest penalties, its slaughter being visited with capital punishment on the offender if he could be known and arrested. Indeed, a man who murdered his fellow might hope to escape retribution except by the avenging hand of some relation of the slain man, but if he were unfortunate or daring enough to dip his hands in the blood of a Stag, he could hope for no mercy if he were detected in the offence.

All the ancient works on hunting are filled with the praises of the Stag, which is belauded with a fluency of language and a fertility of expression that throw the modern sporting terminology completely into the shade. Every minute particular concerning the Stag itself, or the details of hunting, killing, cooking, and serving the animal is graced with its appropriate phrase, and if a gentleman should have perchance misplaced or omitted one of these ceremonious appellations, he would have been held in very low esteem by his compeers.

Although the Stag has been several times partially domesticated and trained to run in harness, it is a very capricious animal, and not a very safe servant. About the month of August the Stag always becomes very much excited, as that is the time when he seeks his mate, and during a space of three or four weeks the animal is testy and irritable in temper, and prone to attack with a kind of blind rage every other animal except a female of his own species. Comparatively tame Stags become dangerous at such a season, and have frequently assaulted those human beings to whom they were formerly attached. The sad death of a lady by an infuriated Stag is of recent occurrence, and may serve as a warning to persons who are ignorant of the strange fury that makes annual seizure of the animal's nature.

In the attack the Stag uses his fore-feet with as much force as the horns, and often with terrible effect, inasmuch as his opponent is seldom prepared for such a mode of action. The hard, pointed, sharp-edged hoofs of the creature become most formidable weapons in this mode of fighting, and are urged with such force and velocity that the coming blow can hardly be avoided. I once narrowly escaped an unexpected blow from a Stag's hoof. I had been feeding the animal with tufts of grass, and was stroking his neck and shoulders, when he suddenly reared up, and struck two blows with his fore-feet with such rapidity that although I was aware of his intention, and sprang backwards, the second stroke just reached one finger, and disabled it for some days.

STAG, OR RED DEER.—*Cervus Elaphus.*

The great speed of the Stag is proverbial, and needs no mention. It is an admirable swimmer, having been known to swim for a distance of six or seven miles, and in one instance a Stag landed in the night upon a beach which he could not have reached without having swum for a distance of ten miles. The gallant beast was discovered by some dogs as he landed, and being chased by them immediately after his fatiguing aquatic exploit, was overcome by exhaustion, and found dead on the following morning.

FALLOW DEER.—*Dama vulgári*.

The colour of the Stag varies slightly according to the time of year. In the summer the coat is a warm, reddish-brown, but in winter the ruddy hue becomes grey. The hind quarters are paler than the rest of the fur. The young Red Deer are born about April, and are remarkable for the variegated appearance of their fur, which is mottled with white upon the back and sides. As the little creatures increase in dimensions, the white marking gradually fades, and the fur assumes the uniform reddish-brown of the adult animal. For a short time after its birth the young Deer is helpless, and unable to escape even from a human pursuer, but it seems, nevertheless, to be possessed of much curious instinct, and to obey the mandates of its mother with instantaneous readiness. Mr. St. John mentions that he once saw a very young Red Deer, not more than an hour of age, standing by its mother and receiving her caresses. As soon as the watchful parent caught sight of the stranger, she raised her fore-foot and administered a gentle tap to her offspring, which immediately laid itself flat upon the ground, and crouched closely to the earth, as if endeavouring to delude the supposed enemy into an idea that it was nothing more than a block of stone.

The FALLOW DEER may readily be distinguished from the stag, by the spotted coat, the smaller size, and the spreading, palmated horns.

Whether it is indigenous to this country is an open point, but it is generally believed to be an importation from Southern Europe or Western Asia. It is never found in a truly wild state like the stag, but is largely kept in parks, and adds much to the beauty of the scene. There is hardly a more interesting sight than a herd of these graceful and active creatures, either lying calmly under the shadow of a broad clump of trees, or tripping along the sward under the guidance of their leaders, the old and sober proceeding at their peculiarly elastic trot, and the young fawn exerting all kinds of fantastic gambols by way of expressing the exuberance of youthful spirits. There is always one " master" Deer among them, who often couches alone in solitary state, apart from the rest of the herd, and only accompanied by a few chosen does whom he honours with his lordly preference.

In his absence, the herd is commanded and guided by the younger and less formidable bucks, but whenever he chooses to make his appearance among his subjects, his advent is always heralded by a general movement among the herd, the young bucks moving silently aside and making room for their monarch. Sometimes a more determined male will protest against such inglorious conduct, and will retain his post at the head of the herd. A threatening movement of the head is, however, generally sufficient to make him move slowly away from the place of honour, and in extreme cases, the offender against royal dignity is disdainfully swept aside by a blow from the horns of the master Deer. Not until he begins to fail in strength will the subordinate males venture to cross horns with one who has fought his way to the post which he holds, and whose prowess is too practically known to be questioned.

The colour of the Fallow Deer is generally of a reddish-brown, spotted with white, and with two or three white lines upon the body. There is, however, another variety which scarcely exhibits any of the white spots, and is of a deep blackish-brown.

The food of the Fallow Deer consists chiefly of grass, but it is very fond of bread, and will sometimes display a very curious appreciation of unexpected dainties. I have often seen them eat ham-sandwiches in spite of the mustard, and enjoy them so thoroughly that they pushed and scrambled with each other for the fragments as they fell on the ground. At Magdalen College, Oxford, where many Deer are kept, it used to be a common amusement to tie a crust to a piece of string, and let it down to the Deer out of a window. The animals would nibble the bread, and as it was gradually drawn aloft by the string, would raise themselves on their hind legs in order to reach it. But when the master Deer loomed in the distance, all retired, leaving him to eat the bread in solitary state. It was curious to see how a single Deer would contrive to take into her mouth the entire side of a " half-quartern" loaf, and though it projected on each side of her jaws, would manage, by dint of patient nibbling, to swallow the whole crust without ever letting it drop out of her mouth.

It is from the Fallow Deer that the best venison is procured, that of the stag being comparatively hard and dry. The skin is well known as furnishing a valuable leather, and the horns are manufactured into knife-handles and other articles of common use. The shavings of the horns are employed for the purpose of making ammonia, which has there-fore been long popularly known under the name of hartshorn. The height of the adult Fallow Deer is about three feet at the shoulders. It is a docile animal, and can be readily tamed. Indeed, it often needs no taming, but becomes quite familiar with strangers in a very short time, especially if they should happen to have any fruit, bread, or biscuit, and be willing to impart some of their provisions to their dappled friends.

The SAMBUR, or SAMBOO (*Rusa Aristótelis*), is an example of the Rusine Deer of Asia. It is a large and powerful animal, exceeding the red Deer in dimensions, and equalling that animal in activity and energy. The horns of the Samboo are set on a rather long footstalk, a snag projecting forwards just above the crown, and the tip simply forked. Its colour is a sooty-brown, with a patch of tan over the eyes, the feet, and by the root of the tail. The male possesses a rather full and dark mane. It is generally a savage and morose creature, being especially vicious when it is decorated with its powerful horns. In its native land it is a water-loving animal, and is generally found in low-lying forest land.

ANOTHER member of the Rusine Deer is the well-known AXIS, CHITTRA, or SPOTTED HOG DEER, of India and Ceylon.

The horns are not at all unlike those of the samboo, being placed on long footstalks, and simply forked at their tips. The colour of this pretty animal is rather various, but is generally a rich golden-brown, with a dark brown stripe along the back, accompanied by two series of white spots. The sides are covered with white spots, which at first sight appear to be scattered irregularly, but are seen on a careful inspection to be arranged in

AXIS DEER.—*Axis maculáta.*

oblique curved lines. There is also a white streak across the haunches. There are, however, many varieties of the Axis Deer, which differ in size as well as in colour. The height of the adult Axis is almost equal to that of the fallow Deer.

It does not appear to possess so much restless activity as is seen in many other Deer, and owing to its nocturnal habits, is but seldom seen by day. It frequents the thick grass jungles, preferring the low-lying lands, where a stream is within easy reach, and passing the greater part of the day asleep, in the deep shade of the heavy foliage. If disturbed, it flies off with great speed for a short distance, but does not appear to be capable of maintaining a long chase.

Of the Capreoline Deer, the common ROEBUCK is a familiar example.

This animal is smaller than the fallow Deer, being only two feet and three or four inches in height at the shoulder, but although so small, can be really a formidable animal, on account of its rapid movements and great comparative strength. Speaking of this animal, Mr. St. John makes the following remarks. After stating that when captured young it can readily be tamed, he proceeds to say :—

"A tame buck becomes a dangerous pet, for after attaining to his full strength, he is very apt to make use of it in attacking people whose appearance he does not like. They

ROEBUCK.—*Capreolus Capræa.*

particularly single out women and children as their victims, and inflict severe and dangerous wounds with their sharp-pointed horns. One day, at a kind of public garden near Brighton, I saw a beautiful but small Roebuck in an inclosure, fastened with a chain, which seemed strong enough and heavy enough to hold down an elephant. Pitying the poor animal, an exile from his native land, I asked what reason they could have for ill-using him, by putting such a weight of iron about his neck. The keeper of the place, however, informed me, that small as the Roebuck was, the chain was quite necessary, as he had attacked and killed a boy of twelve years old a few days before, stabbing the poor fellow in fifty places with his sharp-pointed horns. Of course I had no more to urge in his behalf."

Yet, according to some practical writers on the subject, the Roebuck will not turn upon its pursuer, even when wounded and brought to bay. It is not found in large herds like the fallow Deer, but is strictly monogamous, the single pair living together, contented with each other's society. The horns of this animal have no basal snag, and rise straight from the forehead, throwing out one antler in front, and one or two behind, according to the age of the individual. From the base of the horn to the first antler the horn is thickly covered with wrinkles. It is a most active little Deer, always preferring the highest grounds, thence forming a contrast to the fallow Deer, which loves the plains. It is seldom seen in England in a wild state, but may still be met in many parts of Scotland.

The colour of the Roebuck is very variable, but is generally as follows. The body is always of a brown tint as a ground hue, worked with either red or grey, or remaining simply brown. Round the root of the tail is a patch of pure white hair, and the abdomen and inside of the limbs are greyish white. The chin is also white, and there is a white spot on each side of the lips.

VIRGINIAN DEER, OR CARJACOU.—*Cariacus Virginianus.*

THE elegant and graceful CARJACOU, or VIRGINIAN DEER, is found in great numbers in North America, and is not only interesting to the naturalist on account of the beauty of its form, and the peculiarity of its habits, but is most valuable to the white and red hunters, as affording them an unfailing supply of food and clothing.

The Carjacou may be known by the peculiar shape of its horns, which, in the adult male, are of moderate size, bent boldly backwards, and then suddenly hooked forwards, the tips being nearly above the nose. There is a basal snag on the internal side, pointing backward, and several other snags on the posterior edge. The colour of this animal is extremely variable, being of a light reddish-brown in spring, slaty-blue in autumn, and dull brown in winter. The abdomen, throat, chin, and inner faces of the limbs are white. The fawn is a remarkably pretty little creature, the ruddy-brown fur being profusely decked with white spots, arranged in irregular lines, and sometimes merging into continuous stripes. The height of the adult animal is five feet four inches, measured from nose to root of tail.

It is a timid animal, and so easily scared that the sight of a child fills it with alarm, and urges it to seek refuge by flight. Yet, with a singular inconsistency, it hangs about the skirts of civilization, and refuses to be driven from its favourite spots by the presence of man, or even by the sound of fire-arms. Like the ourebi, it has a strong attachment to certain localities, and if driven from its resting-place on one day, it will surely be found on the next day within a few yards of the same spot. Sometimes it chooses its lair in close proximity to some plantation, and, after feasting on the inclosed vegetables, leaps over the fence as soon as its hunger is satiated, and returns to the spot which it had previously occupied. The animal, however, does not often lie in precisely the same bed on successive nights, but always couches within the compass of a few yards.

That the Carjacou is a good leaper has been already seen, and the experience of many eye-witnesses shows that it displays equal prowess in the water. It is a good swimmer, and is in the habit of venturing to the water-side in the warm weather, and immersing itself in the stream, in order to rid itself of the persecuting ticks and mosquitoes. In the work of Messrs. Audubon and Bachman is a rather amusing anecdote.

" We recollect an occasion, when on sitting down to rest on the margin of the Santel river, we observed a pair of antlers on the surface of the water, near an old tree, not ten steps from us. The half-closed eye of the buck was upon us ; we were without a gun, and he was therefore safe from any injury we could inflict upon him. Anxious to observe the cunning he would display, we turned our eyes another way and commenced a careless whistle, as if for our own amusement, walking gradually towards him in a circuitous route, until we arrived within a few feet of him. He had now sunk so deep in the water that an inch only of his nose and slight portions of his prongs were seen above the surface. At length we suddenly directed our eyes towards him and raised our hands, when he rushed to the shore, and dashed through the rattling cane-brake in rapid style."

The same author remarks, that the speed of the Carjacou, when swimming, is very considerable, the animal cleaving the water so rapidly that it can hardly be overtaken by a boat. As it swims, its whole body is submerged, the head only appearing above the surface. It is not only a swift but a very enduring swimmer, having been often seen crossing broad rivers, and swimming a distance of two miles. When hunted by hounds, the Virginian Deer has been known to baffle its pursuers by making for the sea-shore, taking boldly to the water, and swimming out to sea for a mile or more.

The male is a most pugnacious animal, and engages in deadly contests with those of his own sex, the prize being generally a herd of does. In these conflicts one of the combatants is not unfrequently killed on the spot, and there are many instances of the death of both parties in consequence of the horns interlocking within each other, and so binding the two opponents into a common fate. To find these locked horns is not a very uncommon occurrence, and in one instance three pair of horns were found thus entangled together, the skulls and skeletons lying as proofs of the deadly nature of the strife.

In those parts of the country where it is unable to visit the plantations, the Carjacou feeds on the young grasses of the plains, being fastidiously select in choosing the tenderest herbage. In winter it finds sustenance on various buds and berries, and in autumn it finds abundant banquets under the oaks, chestnuts, and beeches, revelling upon the fallen fruit in amicable fraternity with other quadrupeds and various birds. This variety of food does not render the animal fat at all times of the year, for excepting in the months of August, September, and October, the Carjacou is in very poor condition. It is then, however, very fat, and the venison is of remarkably fine quality. It is in October and November that the buck becomes so combative, and in a very few weeks he has lost all his sleek condition, shed his horns, and retired to the welcome shelter of the forest.

The sight of the Carjacou does not seem to be very keen, but its senses of scent and hearing are wonderfully acute. The slightest sound, even the snapping of a dry twig, will startle this wary animal, and the sense of smell is so acute that it is able to track its companions solely by means of the scent. It is a thirsty animal, requiring water daily, and generally visiting some stream or spring at nightfall. It is remarkably fond of salt, and resorts in great numbers to the saline springs, or " salt-licks," as they are popularly termed. The Deer do not drink the briny water, but prefer licking the stones at the edge where the salt has crystallized from the evaporation of the water.

When observed, the Carjacou leaps into the air like the bush-buck under similar circumstances, turning its head in every direction in order to detect the cause of its alarm, and then rushing away at full speed. Before it is accustomed to molestation, it starts from its lair long before the hunter can approach, but when it has frequently been harassed, it lies down, crouching to the ground, and endeavouring to escape the sight of its foe. Whenever it behaves in this manner it is easily outwitted, by riding or walking round the prostrate animal, and gradually lessening the circle, until it is within easy range.

When captured while young, the Carjacou is easily domesticated, and becomes even troublesome in its confident tameness. A pair of these animals that were kept by Mr.

Audubon were most mischievous creatures. They would jump into his study window, and when the sashes were shut would leap through glass and woodwork like harlequin in a pantomime. They ate the covers of his books, nibbled his papers, and scattered them in sad confusion, gnawed the carriage-harness, cropped all the choice garden plants, and finally took to biting off the heads and feet of the ducklings and chickens.

The skin of the Carjacou is peculiarly valuable to the hunter, for when properly dressed and smoked, it becomes as pliable as a kid glove, and does not shrivel or harden when subjected to the action of water. Of this material are formed the greater part of the native Indian's apparel, and it is also employed for various articles of civilized raiment.

As the Carjacou feeds, it always shakes its tail before it lowers or raises its head. So by watching the movement of the tail, the hunter knows when he may move towards his intended prey, and when he must lie perfectly quiet. So truly indicative of the animal is this habit, that when an Indian wishes to signal to another that he sees a Carjacou, he moves his fore-finger up and down. This sign is invariably understood by all the tribes of North American Indians.

THE Moschine Deer are readily known by the absence of horns in both sexes, the extremely long canine teeth of the upper jaw in the males, and the powerfully odorous secretion in one of the species, from which they derive their popular as well as their scientific title. There are at least eight or nine species of these curious animals.

The most celebrated of these little Deer, is the common MUSK DEER, which is a native of the northern parts of India, and is found spread throughout a very large range of country, always preferring the cold and elevated mountainous regions. The height of the adult Musk Deer is about two feet three inches at the shoulders; the colour is light brown, marked with a shade of greyish-yellow. Inhabiting the rocky and mountainous locations of its native home, it is remarkably active and surefooted, rivalling even the chamois or the goat in the agility with which it can ascend or descend the most fearful precipices. The great length of the false hoofs adds much to the security of the Musk Deer's footing upon the crags.

It is only in the male that the long tusks are seen, and that the perfume called musk is secreted. The tusks are sometimes as much as three inches in length, and therefore

KANCHIL, OR PIGMY MUSK.—*Trágulus pygmæus.*

project considerably beyond the jaw. In shape they are compressed, pointed, and rather sharp-edged. The natives say that their principal use is in digging up the kastooree plant, a kind of subterranean bulb on which the Musk Deer feeds, and which imparts the peculiar perfume to the odorous secretion. The musk is produced in a glandular pouch placed in the abdomen, and when the animal is killed for the sake of this treasure, the musk-bag is carefully removed, so as to defend its precious contents from exposure to the air. When securely taken from the animal, the musk is of so powerful an odour as to cause headache to those who inhale its overpowering fragrance. The affluence of perfume that resides in the musk is almost incredible, for a small piece of this wonderful secretion may remain in a room for many years, and at the end of that time will give forth an odour which is apparently not the least diminished by time.

On account of the value of the musk, the animal which furnishes the precious substance is subjected to great persecution on the part of the hunters, who annually destroy great numbers of these active little animals. The native hunters await the season of migration, while the Deer are forced to pass into more clement latitudes in search of subsistence, and beset their path with various traps, besides seizing every opportunity of destroying them by missiles. Although so good a leaper, and so well adapted for traversing the rocky crags of its native hills, the Musk Deer is not a very good climber, and descends slopes with great difficulty.

ANOTHER member of the Moschine group is the KANCHIL, or PIGMY MUSK (*Trágulus pygmæus*), a Deer which is found in the Asiatic islands, and which is as celebrated for its cunning as is the fox among ourselves.

This animal is not nearly so large as the musk Deer, and although somewhat similar in colour, may be distinguished by a broad black stripe which runs along the back of the neck, and forms a wide band across the chest. Instead of living in the cold and lofty mountain ranges which are inhabited by the musk Deer, the Kanchil prefers the thickly wooded districts of the Javanese forests. Like many other animals, the Kanchil is given to "possuming," or feigning death when it is taken in a noose or trap, and as soon as the successful hunter releases the clever actor from the retaining cord, it leaps upon its feet and darts away before he has recovered from his surprise.

CAMEL.—*Camélus Arábicus.*

The NAPU, or JAVA MUSK, inhabits Java and Sumatra, and without possessing the intellect of the Kanchil, is a very pleasing animal to the sight, and as it is readily domesticated, is well adapted to European menageries.

FROM the earliest times that are recorded in history, the CAMEL is mentioned as one of the animals which are totally subject to the sway of man, and which in eastern countries contribute so much to the wealth and influence of their owners.

There are two species of Camel acknowledged by zoologists, namely, the common Camel of Arabia, which has but one hump, and the Mecheri, or Bactrian Camel, which possesses two of these curious appendages. Of these two animals, the former is by far the more valuable, as it is superior to its two-humped relative in almost every respect. Admirably fitted, as are all animals, for the task which they are intended to perform, the Camel presents such wonderful adaptations of form to duty, that the most superficial observer cannot but be struck with the exquisite manner in which the creature has been endowed with the various qualities of mind and body which are needful under the peculiar circumstances amid which it dwells.

As the animal is intended to traverse the parched sand plains, and to pass several consecutive days without the possibility of obtaining liquid nourishment, there is an internal structure which permits the animal to store up a considerable amount of water for future use. For this purpose, the honeycomb cells of the "reticulum" are largely developed, and are enabled to receive and to retain the water which is received into the stomach after the natural thirst of the animal has been supplied. After a Camel has been accustomed to journeying across the hot and arid sand wastes, it learns wisdom by experience, and contrives to lay by a much greater supply of water than would be accumulated by a young and untried animal. It is supposed that the Camel is, in some way, able to dilate the honeycomb cells, and to force them to receive a large quantity of the priceless liquid.

A large and experienced Camel will receive five or six quarts of water into its stomach, and is enabled to exist for as many days without needing to drink. Aided by this internal supply of water, the Camel can satiate its hunger by browsing on the hard and withered thorns that are found scattered thinly through the deserts, and suffers no injury to its palate from their iron-like spears, that would direfully wound the mouth of any less sensitive creature. The Camel has even been known to eat pieces of dry wood, and to derive apparent satisfaction from its strange meal.

The feet of the Camel are well adapted for walking upon the loose, dry sand, than which substance is no more uncertain footing. The toes are very broad, and are furnished with soft, wide cushions, that present a considerable surface to the loose soil, and enable the animal to maintain a firm hold upon the shifting sands. As the Camel is constantly forced to kneel in order to be loaded or relieved of its burden, it is furnished upon the knees and breast with thick callous pads, which support its weight without injuring the skin. Thus fitted by nature for its strange life, the Camel faces the desert sands with boldness, and traverses the arid regions with an ease and quiet celerity that has gained for the creature the title of Ship of the Desert.

The Camel is invariably employed as an animal of carriage, when in its native land, and is able to support a load of five or six hundred pounds' weight without being over-loaded. The Arab will not willingly injure his Camel by placing too heavy a burden upon its back, but in India, and some other countries where the Camel has been naturalized and domesticated, its treatment is barbarous in the extreme. Hundreds of valuable animals are annually sacrificed on account of the covetousness of their owners, who know that they will receive payment for every Camel that falls upon the journey, and are consequently indifferent to the suffering and condition of those animals which they have nominally taken under their care.

The pace of the Camel is not nearly so rapid as is generally supposed, and even the speed of the Heirie, or swift Camel, has been greatly exaggerated. " In crossing the Nubian desert," says Captain Peel, " I paid constant attention to the march of the Camels, hoping it might be of some service hereafter in determining our position. The number of strides in a minute with the same foot varied very little, only from thirty-seven to thirty-nine, and thirty-eight was the average ; but the length of the stride was more uncertain, varying from six feet six inches to seven feet six inches. As we were always urging the Camels, who seemed, like ourselves, to know the necessity of pushing on across that fearful tract, I took seven feet as the average. These figures give a speed of 2·62 geographical miles per hour, or exactly three English miles, which may be considered as the highest speed that Camels, lightly loaded, can keep up on a journey. In general, it will not be more than two and a half English miles. My dromedary was one of the tallest, and the seat of the saddle was six feet six inches above the ground."

The speed of the Heirie is seldom more than eight or ten miles per hour, but the endurance of the animal is so wonderful, that it is able to keep up this pace for twenty hours without stopping. To back a Heirie at full speed is a terrible task, as the peculiar jolting trot at which the animal proceeds is so rough and irregular that it seems to dislocate every bone, and to shake the digestive organs almost out of their places. It is needful for any one who wishes to make a long journey on one of these animals to swathe himself tightly in bandages, in order to save himself from the ill effects of long continued jolting.

The gentle disposition and sweet temper of the Camel is quite as imaginary as its speed, for the creature is truly an ill-conditioned and morose beast, ever apt to bite, and so combative as to engage in terrible conflicts with its own species as soon as it is relieved of its load. Taking advantage of this disposition, the native chiefs will often amuse themselves by combats between fighting Camels, which are trained for the purpose, like the fighting tigers and buffaloes of India.

The true disposition of the Camel is told in a very spirited manner by the author of " Life among the Pandies."

"Invaluable he is, I admit; likewise hardy, capable of carrying enormous loads for great distances under a frightful sun, and generally admirably suited for the purpose to which he is put, namely, that of a baggage animal. But to say that a Camel is patient, to affirm that this great, grumbling, groaning, brown brute is either docile, meek, or sweet-tempered, is stating what is simply not the case; and I have no hesitation in saying, that never do I remember to have seen a Camel in a good humour, or otherwise than in open or moody hostility with the world at large; at least, if outward appearances are to be credited.

Watch him when he is being loaded; see his keeper struggling frantically with him, only succeeding in making him kneel down for the purpose by sheer force, and when down, only keeping him there by tying neck and fore-legs together tightly with a piece of string; hear him grumbling in deep, bubbling tones, with mouth savagely opened, and I think that then at least you will admit he is by no means in as amiable a frame of mind as one could wish. Observe him now that the process of loading is completed, and the string which held him in subjection loosened; up he rises, a great brown mountain, still groaning, still bubbling, and away he goes, madly dashing to and fro, and shaking off tables, portmanteaus, beds, furniture, and baggage in a scattered shower around him; and I think that even his stanchest admirers will allow, that neither at this moment is he in what one would call a pleasant humour.

Mr. Camel having, after some battling, been overcome and compelled to carry the load to which he so objected, but not until he has damaged it considerably, arrives when the march is over at the camping ground. It is then necessary to make him kneel down to have that load removed, grumbling as much as ever, in opposition as usual, beaten physically, but with soul unsubdued, and internally in a state of rebellion and mutiny, a sort of volcano ready at any moment to burst forth."

The "hump" of the Camel is a very curious part of its structure, and is of great importance in the eyes of the Arabs, who judge of the condition of their beasts by the size, shape, and firmness of the hump. They say, and truly, that the Camel feeds upon his hump, for in proportion as the animal traverses the sandy wastes of its desert lands, and suffers from privation and fatigue, the hump diminishes. At the end of a long and painful journey, the hump will often nearly vanish, and it cannot be restored to its pristine form until the animal has undergone a long course of good feeding. When an Arab is about to set forth on a desert journey, he pays great attention to the humps of his Camels, and watches them with jealous care.

Independently of its value as a beast of burden, the Camel is most precious to its owners, as it supplies them with food and clothing. The milk mixed with meal is a favourite dish among the children of the desert, and is sometimes purposely kept until it is sour, in which state it is very grateful to the Arab palate, but especially nauseous to that of a European. The Arabs think that any man is sadly devoid of taste who prefers the sweet new milk to that which has been mellowed by time. A kind of very rancid butter is churned from the cream by a remarkably simple process, consisting of pouring the cream into a goat-skin sack, and shaking it constantly until the butter is formed. The flesh of the Camel is seldom eaten, probably because the animal is too valuable to be killed merely for the sake of being eaten. Sometimes, however, in a season of great festivity, a rich Arab will slay one of his Camels, and calling all his friends and relations to the banquet, they hold high festival upon the unaccustomed dainty. The long hair of the Camel is spun into a coarse thread, and is employed in the manufacture of broad-cloths and similar articles. At certain times of the year, the Camel sheds its hair, in order to replace its old coat by a new one, and the Arabs avail themselves of the looseness with which the hair is at these times adherent to the skin, to pluck it away without injuring the animal.

In extreme cases, when the water has failed for many days, and the desert fountains are dried up, the Camel dies for the purpose of prolonging the life of its master, and yields up the store of water which is laid up in the cells of the stomach. The water thus obtained is of a light green colour, and very unpleasant to the palate; but when a man is dying of thirst he is not very particular as to the quality of the liquid which may save his life. Unpleasant though it be, this water is hardly more unpalatable than that

BACTRIAN CAMEL.—*Camélus Bactriánus.*

which is carried in leathern bags on the Camel's back, and which is not only heated by the rays of the fierce sun, but is strongly impregnated with a leathery flavour, and smells as if it were taken out of a tan-pit. The water which is taken from the Camel's stomach is even cooler than that which has been carried on its back, as the natural heat of the animal is not comparable to that which is produced by the continual rays of the burning desert sun.

The height of an ordinary Camel at the shoulder is about six or seven feet, and its colour is a light brown, of various depths in different individuals, some specimens being nearly black, and others almost white. The dromedary is the lighter breed of Camel and is chiefly used for riding, while the ordinary Camel is employed as a beast of burden. Between the two animals there is about the same difference as between a dray-horse and a hunter, the Heirie being analogous to the race horse.

The BACTRIAN CAMEL is readily to be distinguished from the ordinary Camel by the double hump which it bears on its back, and which is precisely analogous in its structure and office to that of the Arabian Camel.

The general formation of this animal; its lofty neck, raising its head high above the solar radiations from the heated ground; its valve-like nostrils, that close involuntarily if a grain of drifting sand should invade their precincts; its wide cushion-like feet, and its powers of abstinence, prove that, like its Arabian relative, it is intended for the purpose of traversing vast deserts without needing refreshment on the way. This species is spread through central Asia, Thibet, and China, and is domesticated through a large portion of

ALPACA LLAMA.—*Llama Pacos.*

the world. It is not so enduring an animal as the Arabian species, requiring a fresh supply of liquid every three days ; while the Arabian Camel can exist without water for five or even six days. It is employed by the Persians in a rather curious military capacity ; its saddle being furnished with one or two swivel guns, which are managed by the rider. The corps is called the Camel Artillery, and is of considerable value in the peculiar mode of fighting which is prevalent in the East.

The height of the Bactrian Camel is rather more than that of the Arabian species, and its colour is generally brown, which sometimes deepens into sooty black, and sometimes fades into a dirty white.

THE true camels are exclusively confined to the Old World, but find representatives in the New World in four acknowledged species of the genus Llama.

These animals are comparatively small in their dimensions, and possess no hump, so that they may easily be distinguished from the camels. Their hair is very woolly, and their countenance has a very sheep-like expression, so that a full-haired Llama instantly reminds the spectator of a long-legged, long-necked sheep. The feet of the Llamas are very different from those of the camels, as their haunts are always found to be upon rocky ground, and their feet must of necessity be accommodated to the ground on which they are accustomed to tread. The toes of the Llama are completely divided, and are each furnished with a rough cushion beneath, and a strong, claw-like hoof above, so that the member may take a firm hold of rocky and uneven ground.

Four species of Llamas are now acknowledged ; namely, the Vicugna, the Guanaco, the Yamma, and the Alpaca, each of which will be briefly described.

The VICUGNA is found in the most elevated localities of Batavia and Northern Chili, and is a very wild and untamable animal, having resisted all the attempts of the patient natives to reduce it to a state of domestication. It is extremely active and sure-footed in its mountain home, and being equally timid and wary, is seldom captured in a living state. It lives in herds near the region of perpetual snow, and in its habits bears some resemblance to the chamois. The short, soft, silken fur of this animal is very valuable, and causes the death of thousands of Vicugnas, which are slain by various methods merely for the sake of their coats. The colour of the Vicugna is a nearly uniform brown, tinged with yellow on the back, and fading into grey on the abdomen. Its height at the shoulder is about two feet six inches.

The GUANACO is spread over a very wide range of country, ranging over the whole of the temperate regions of Patagonia. The colour of this species is a reddish-brown, the ears and hind legs grey. The neck is long in comparison to the size of the body, and the height at the shoulder is about three feet six inches. .

The Guanaco lives in herds varying in number from ten to thirty or forty, but is sometimes seen in flocks of much greater numbers, resembling sheep, not only in their gregarious habits, but in the implicit obedience with which they rely upon their leader. Should they be deprived of his guardianship they become so bewildered that they run aimlessly from spot to spot, and can be easily destroyed by experienced hunters. It is a very wary and timid animal; but like many creatures of similar disposition, is possessed with so strong a feeling of curiosity that it can be attracted towards the hunter if he lies down on the ground and kicks his feet in the air. Even the reports of his rifle do not frighten the animals, who, says Darwin, consider them as part of the performance. Still, it is a quick-sighted and wary animal, and if it perceives a human being approaching its domicile, it sets up a shrill neighing scream, which is often the first intimation of its presence. The whole herd then set off into a rapid canter along the hill-side, and gain some elevated spot where they can feel themselves safe.

The Guanaco, in common with the other species, is rather short-tempered, and has a very unpleasant habit of displaying its anger by discharging a shower of half-digested food and saliva over the offender. Formerly, this salival discharge was thought to be acrid, and capable of raising blisters upon the human skin. This, however, is fortunately not the case, although the assault is eminently disagreeable, on account of the ill scent of the ejected liquid. In its wild state the Guanaco seems to have little or no idea of resistance, being easily held by a single dog until the hunter can come up and make sure of his prize. But in domesticated life, it seems to imbibe a spirit of combativeness, for it will kick with both hind legs, and deliver severe blows with the knees of those limbs. Among themselves, however, the males fight desperately, the cause of combat being generally some favoured and coveted female.

The Guanaco is wonderfully sure-footed upon rocky ground, and is also a good swimmer, taking voluntarily to the water, and swimming from one island to another. When near the sea, it will drink the salt water, and has often been observed in the act of drinking the briny waters of certain salt springs.

The YAMMA, or LLAMA, is of a brown, or variegated colour, and its legs are long and slender. In former days, this animal was the only beast of burden which was possessed by the natives, and it was largely used by the Spaniards (who described it as a sheep) for the same purpose. It is able to carry a weight of one hundred pounds, and to traverse about fourteen or fifteen miles per diem. As a beast of burden, it is now being rapidly supplanted by the ass, while the European sheep is gradually taking its place as a wool-bearer. The flesh of the Llama is dark and coarse, and is accordingly held in bad repute.

The ALPACA, or PACO, is, together with the last animal, supposed by several zoologists to be only a domesticated variety of the Guanaco. Its colour is generally black, but is often variegated with brown and white. The wool of this species is long, soft, silky, and extremely valuable in the commercial world. A herd of Llamas has been imported into Australia, and seems to have succeeded remarkably well, the yield of wool having been quite as rich as was hoped by the enterprising importer. It is a handsome and a gentle animal, and is only found in a domesticated state.

TARPAN, OR WILD HORSE.—*Equus Cabollus.*

HORSES.

THE HORSE has, from time immemorial, been made the companion and servant of man, and its original progenitors are unknown. It is supposed, however, that the Horse must have derived its origin from central Asia, and from thence have spread to almost every portion of the globe.

There are several countries, such as Tartary and Northern America, where the Horse runs wild, and has almost entirely reverted to its primeval state, thus affording an idea of the manners and customs of the Horse before it was subjected to the dominion of man. In Tartary, the Wild Horses are found in herds, consisting of many thousands in number, and are actuated by a wonderful spirit of discipline, each herd acting under the commands of a single leader, and executing his orders with military precision. The Tartars recruit their studs from these herds, capturing the best and strongest animals with the aid of a falcon, which is trained to settle on the Horse's head, and flutter its wings about his face so as to blind and detain him until the hunter comes up to secure his prize. The horses thus taken are coupled with the tame animals, and in a very short time learn to perform their share of the work, and to obey the orders of their master as implicitly as they once obeyed those of their quadrupedal leader.

Each herd is headed by an old experienced Horse, who holds his position by right of conquest, and loses his chieftainship if vanquished by any opponent. The young males are always excluded from these herds, and are forced to live solitary lives until they can attract some of the opposite sex, and set up an establishment on their own account. The colour of the Wild Horse of Tartary is red, with a black stripe along the back.

Not only do the Tartars ride their Horses, but they drink the milk and eat the flesh, so that a Horse-hunt is often conducted merely as a food-procuring expedition. From the milk the Tartars manufacture a peculiar sub-acid liquid, which they term "koumiss," and is made by permitting it to become sour, and then stirring the curd and milk violently with a large stick until it is forced into a homogeneous mass. From the same substance the Tartars make a fermented liquid. These Horses are very strong and hardy, and the breed is preserved in good condition by the custom which prevails among the Tartars of killing and eating the defective or weak foals, and preserving the strong and healthy for use. Being brought up with the family, the Tartar Horse is very gentle and familiar with its owners. When they are only a few months of age they are ridden by the children, but never backed by a man until they are five or six years old. They are then, however, severely treated, being forced to travel for several consecutive days, and to endure great privations of hunger and thirst.

ANOTHER well-known example of the Wild Horse is the MUSTANG of the American prairies.

This animal is congregated into vast herds, which are always under the guardianship of a single leader, who is able, in some wonderful manner, to convey his orders to all his subjects simultaneously. Although surrounded by various enemies, such as the puma, the wolf, and the jaguar, they care little for these ravenous and powerful carnivora, trusting in their united strength to save them from harm. There is no animal that will dare to face a troop of Wild Horses, which often entice the domesticated animals into their ranks, and carry them exultingly into the free plains.

The Mustang is always a strong and a useful animal, and is much sought after as a saddle-horse. To capture these wild creatures is a very difficult matter, and is generally managed by the help of the lasso, although the rifle is sometimes called into requisition in difficult cases. This latter plan, technically called "creasing," is never employed but by very accurate marksmen, as the difference of half an inch in the line of fire is sufficient either to miss the animal or to kill it on the spot. In "creasing" a Horse, the hunter aims so as to graze the skull just behind the ear, the sudden blow stunning the Horse for a few seconds, during which time the hunter pounces on the bewildered animal, and secures it before it has fairly recovered its senses.

The lasso is, however, generally employed for this purpose, and as it can be thrown with precision to a distance of thirty feet, is a terrible weapon in practised hands. This formidable instrument is very simple in construction, being a carefully plaited rope of green hide, one end being furnished with an iron ring, and the other extremity fastened to the saddle. When not in use, it is hung in coils upon a projection of the saddle, but when the hunter has his game in view, he throws the coils over his left arm, makes a slip-noose by means of the iron ring, and then grasping the ring and cord firmly in his left hand, so as to prevent the noose from slipping, he grasps the centre of the noose and the main cord in his right hand, and is then ready for action. Swinging the large noose, four or five feet in diameter, around his head, the weight of the iron ring giving a powerful impetus, the hunter is able to hurl the leathern cord to its full length, and with deadly aim. As the noose flies circling through the air it gradually contracts in diameter, so that the hunter is forced to accommodate the size of the loop to the distance of the object aimed at.

When fully caught, the Mustang is savage and furious at his discomfiture, and would speedily escape from his bondage but for the clever and simple method of subjection which is employed. The lasso being flung round its neck, the Horse nearly strangles itself by its plungings and struggles, and is soon reduced to stand still and gasp for breath. The hunter now dismounts from his Horse, and keeping his hands on the lasso, advances cautiously towards the captured animal, hauling the rope tight whenever it tries to escape. In a short time he works his way towards the creature's head, and seizing its muzzle in his hand, blows strongly into its nostrils. Overcome by some strange influence, the Horse immediately becomes quiet, and in a few hours can scarcely be distinguished from a regularly trained animal.

MUSTANG.

This mode of reducing the Horse to subjection is employed by the Comanche and neighbouring Indians, but the Gauchos, or inhabitants of the Pampas, manage in a different and far more cruel manner, the idea of humanity never entering the head of either Indian or white man. As soon as a Wild Horse is captured, its legs are suddenly pulled aside, and the poor animal falls prostrate on the ground. A Gaucho then seats himself on his head, while others gird a saddle tightly on his back, and force a bit into his mouth. The rider next stands astride the prostrate quadruped, which is then released from the weight upon its head. Up leaps the Horse, striving in vain to escape, for the Gaucho seats himself in the saddle as the animal rises, and is never to be shaken off as long as the Horse disobeys his will. However restive the poor creature may be, it soon exhausts itself by unavailing efforts, and becomes passively submissive. Sometimes a stubborn and determined animal refuses to move, and stands rooted to the spot on which it had fallen. The cruel spurs of the Gaucho, however, soon set it going, and in a very short time it is thoroughly subdued.

THE elegant, swift, and withal powerful Horses of which England is so proud, and which are employed in the chase or the course, owe their best qualities to the judicious admixture of the Arabian blood. The ARAB HORSE has long been celebrated for its swift limbs, exquisite form, and affectionate disposition ; the latter quality resulting, however, chiefly from the manner in which it is tamed.

There are several breeds of Arab Horses, only one of which is of very great value. This variety, termed the Kochlani, is so highly prized, that a mare of the pure breed can hardly be procured at any cost, and even the male animal is not easy of attainment. The

ARAB HORSE.

pedigree of these Horses is carefully preserved, and written in most florid terms upon parchment. In some cases, the genealogy is said to extend for nearly two thousand years. The body of the Arab Horse is very light, its neck long and arched, its eye full and soft, and its limbs delicate and slender. The temper of the animal is remarkably sweet, for as it has been born and bred among the family of its owner, it avoids injuring even the little children that roll about among its legs, as carefully as if they were its own offspring. So attached to its owner is this beautiful Horse, that if he should be thrown from its back, the animal will stand quietly by its prostrate master, and wait until he gains strength to remount.

The training of the Kochlani is not so severe as is generally imagined, for the presence of water and abundant pasturage is absolutely necessary, in order to rear the animal in a proper manner. Not until the strength and muscles of the animal are developed, is a trial permitted, and then it is truly a terrible one. When the mare—for the male animal is never ridden by the Arabs—has attained her full development, she is mounted for the first time, and ridden at full speed for fifty or sixty miles without respite. Hot and fainting, she is then forced into deep water, which compels her to swim, and if she does not feed freely immediately after this terrific trial, she is rejected as unworthy of being reckoned among the true Kochlani.

For the animals which will stand this terrible test the Arab has almost an idolatrous regard, and will oftentimes spare an enemy merely on account of his steed.

The RACE HORSE of England is, perhaps, with the exception of the foxhound, the most admirable example of the perfection to which a domesticated animal can be brought by careful breeding and training.

RACE HORSE.

Whatever may have been its original source, the Racer has been greatly improved by the mixture of Arab blood, through the means of the Godolphin and Derby Arabians. The celebrated Horse Eclipse was a descendant, on the mother's side, of the Godolphin Arabian, that wonderful animal which was rescued from drawing a cart in Paris, and which was afterwards destined to play so important a part in regenerating the breed of English racers. He was also descended, on his father's side, from the Darley Arabian. It is a remarkable fact, that both parents of this extraordinary animal were unappreciated by their owners, Marsk, his father, having been purchased for a mere trifle, and then permitted to run nearly wild in the New Forest. Spiletta, his mother, only ran one race, in which she was beaten, and Squirt, the father of Marsk, was actually saved by the intercession of a groom as he was being led to the slaughter-house.

Eclipse was never beaten, and his racing career extended only through seventeen months, and in that short period of time he won more than twenty-five thousand pounds. At his last race he was obliged to walk over the course, as no one dared enter a Horse against him. Ten years after that event, his owner, Mr. O'Kelly, was requested to sell him, and demanded the sum of twenty-five thousand pounds, an annuity of five hundred pounds a year, together with six of his offspring yearly. When he died, in 1789, he.was twenty-five years old, and had realized for his owner a princely fortune. His skeleton is now in the museum at Oxford. His shape was very remarkable, the hinder quarters being considerably higher than the shoulders, and his breathing was so thick that it could be heard at a considerable distance. He was originally purchased for seventy-five guineas, at the death of the Duke of Cumberland, by whom he was bred.

MANY thorough-bred Horses which are not suitable for the purposes of the turf are admirably adapted for the chase, and are trained for that purpose. The body of the Hunter should not be so long as that of the racer, and requires greater compactness, in order that he may not fatigue himself by taking too long a stride over ploughed land. A comparatively large foot is required, in order to save it from being destroyed by the rapid alternation of soft and hard ground which the animal is obliged to traverse, and which would batter a small contracted foot to such an extent as to render the Horse useless. The low shoulders of Eclipse would be very injurious in a Hunter, on account of the numerous and trying leaps which it is often called upon to perform.

The best bred Horses are generally the most affectionate and docile, although their spirit is very high, and their temper hot and quick. There are few animals which are more affectionate than a Horse, which seems to feel a necessity for attachment, and if his sympathies be not roused by human means, he will make friends with the nearest living being. Cats are great favourites with Horses, and even the famous Chillaby, called, from his ferocity, the Mad Arabian, had his little friend in the shape of a lamb, which would take any liberties with him, and was accustomed to butt at the flies as they came too near his strange ally. The Godolphin Arabian was also strongly attached to a cat, which usually sat on his back, or nestled in the manger. When he died, the cat pined away and soon followed her loved friend.

These examples are sufficient to show that the ferocity of these animals was caused by the neglect or ignorance of their human associates, who either did not know how to arouse the affectionate feelings of the animal, or brutally despised and crushed them. The Horse is a much more intellectual animal than is generally supposed, as will be acknowledged by any one who has possessed a favourite Horse, and treated it with uniform kindness.

There is no need for whip or spur when the rider and steed understand each other, and the bridle is reduced almost to a mere form, as the touch of a finger, or the tone of a voice, are sufficient to direct the animal. We are all familiar with the elephantine dray-horses that march so majestically along with their load of casks, and which instantaneously obey the singular sounds which continually issue from the throats of their conductors, and back, stop, advance, or turn to the right or left, without requiring the touch of a rein or the blow of a whip. The infliction of pain is a clumsy and a barbarous manner of guiding a Horse, and we shall never reap the full value of the animal until we have learned to respect its feelings, and to shun the infliction of torture as a brutal, a cowardly, and an unnecessary act. To maltreat a child is always held to be a cowardly and unmanly act, and it is equally cowardly and unworthy of the human character to maltreat a poor animal which has no possibility of revenge, no hope of redress, and no words to make its wrongs known. Pain is pain, whether inflicted on man or beast, and we are equally responsible in either case.

As an unprejudiced observer, with no purpose to serve, and without bias in either direction, I cannot here refrain from observing, that Mr. Rarey's method of bringing the Horse under subjection is a considerable step in the right direction, and a very great improvement on the cruel and savage method which is so often employed by coarse and ignorant men, and truly called "breaking." Having repeatedly witnessed the successful operations of that gentleman, in subduing Horses that had previously defied all efforts, I cannot be persuaded that it is a cruel process. The method by which it is achieved is now sufficiently familiar, and I will only observe, that the idea is a true and philosophical one. The Horse is mostly fierce because it is nervous, and bites and kicks, not because it is enraged, but because it is alarmed. Restore confidence, and the creature becomes quiet, without any desire to use its hoofs and teeth in an aggressive manner. It is clearly impossible to do so as long as the animal is at liberty to annihilate its teacher, and the strap is only used until the Horse is convinced that the presence of a human form, or the touch of a human hand, has nothing of the terrible in it. Confidence soon takes the place of fear, and the animal seems to receive its teacher at once into its good graces, following him like a dog, and rubbing its nose against his shoulder.

The ingenuity of the Horse is very considerable, and the creature will voluntarily perform acts that display a considerable amount of intellect. From a number of

HUNTER.

anecdotes relating to the intellectual powers of the Horse, I select the following, some of them entirely original, and others very little known.

An orchard had been repeatedly stripped of its best and ripest fruit, and the marauders had laid their plans so cunningly that the strictest vigilance could not detect them. At last the depredators were discovered to be a mare and her colt which were turned out to graze among the trees. The mare was seen to go up to one of the apple-trees and to throw herself against the trunk so violently that a shower of ripe apples came tumbling down. She and her offspring then ate the fallen apples, and the same process was repeated at another tree. Another mare had discovered the secret of the water-butt, and whenever she was thirsty, was accustomed to go to the butt, turn the tap with her teeth, drink until her thirst was satisfied, and then to close the tap again. I have heard of two animals which performed this feat, but one of them was not clever enough to turn the tap back again, and used to let all the water run to waste.

A careless groom was ordered to prepare a mash for one of the Horses placed under his care, and after making a thin, unsatisfactory mixture, he hastily threw a quantity of chaff on the surface and gave it to the Horse. The animal tried to push away the chaff and get his nose into the mash, but was unable to do so, and when he tried to draw the liquid into his mouth, the chaff flew into his throat and nearly choked him. Being baffled, he paused awhile, and then pulled a lock of hay from the rack. Pushing the hay through the chaff, he contrived to suck the liquid mash through the interstices until the hay was saturated with moisture. He then ate the piece of hay, pulled another lock from the rack, and repeated the process until he had finished his mash.

HACKNEY, OR ROAD HORSE.

LIKE the race Horse and the hunter, the HACKNEY or ROAD HORSE is obtained by judicious breeding, and is said by Mr. Youatt to be "more difficult to find than even the hunter or the courser. There are several faults that may be overlooked in the hunter, but which the Road Horse must not have. The former may start, may be awkward in his walk or even his trot, he may have thrushes or corns; but if he can go a good slapping pace, and has wind and bottom, we can put up with him and prize him. But the Hackney, if he is worth having, must have good fore legs and good hinder ones too; he must be sound on his feet, even-tempered, no starter, quiet in whatever situation he may be placed, not heavy in hand, and never disposed to fall on his knees. A Hackney is far more valuable for the pleasantness of his paces and his safety, good temper and endurance, than for his speed. We rarely want to go more than eight or ten miles an hour, and on a journey not more than six or seven. The fast Horses, and especially the fast trotters, are not even in their paces, and although they may perform very extraordinary feats, are disabled and worthless when the slower Horse is in his prime."

The same author, to whose valuable work on the Horse the reader is referred as a treasury of valuable information, proceeds to observe that pure blood is disadvantageous to a Hackney, as it gives small hoofs, slender legs, and a long stride, each of which qualities would be hurtful on the hard stony road. There should, however, be a spice of high breeding in the animal, the amount to be regulated by the country in which it lives and the work which it has to perform.

When properly managed and kindly treated, the Hackney is a most intelligent animal, displaying a singularly excellent memory. This extraordinary memory of the Horse has often proved serviceable to its owner, and in many instances has been made

AMERICAN TROTTER.

the means of saving his life. An ordinary Hackney had been ridden to a spot far from home, very difficult to find, and into which neither he nor his rider had previously been. Two years afterwards, the same journey was repeated, but at a distance of three or four miles from his destination the night closed in and the rain poured in torrents. Having entirely lost his way, the rider in despair flung the reins on his Horse's neck, and left him to his own desires. The intelligent animal proved himself equal to the trust which was reposed in him, and in half-an-hour drew up at the house which his master was visiting.

The power of the well-bred Hackney may be imagined from the following feat, recorded in the above mentioned work,—

"An English bred mare was matched to trot one hundred miles in ten hours ·and a half. She was one of those rare animals that could do almost anything as a hack, a hunter, or in harness. On one occasion, after having, in following the hounds and travelling to and from course, gone through at least sixty miles of country, she fairly ran away with her rider over several ploughed fields. She accomplished the match in ten hours and fourteen minutes, or deducting thirteen minutes for stoppages, in ten hours and a minute's actual work, and thus gained the victory. She was a little tired, and being turned into a horse-box, lost no time in taking her rest. On the following day she was as full of life and spirit as ever. The owner had given positive orders to the driver to stop at once on her showing decided symptoms of distress, as he valued her more than anything he could gain by her enduring actual suffering."

OUR Transatlantic brethren have long been celebrated for the excellence of their trotting Horses, and have succeeded in obtaining a breed of Horses that are intended

CLEVELAND, OR CARRIAGE HORSE.

exclusively for that pace. In America the trot is the only pace that is valued, and the energies of the animal are all directed to that single point. A good trotter is possessed of endurance as well as speed, for one of these animals trotted one hundred miles in ten hours and seven minutes, inclusive of thirty-seven minutes which were occupied in refreshment and stoppages, so that the actual time occupied was only nine hours and a half.

In the present times. when railways have taken the place of the old mail coaches, the regular Coach Horse is little needed, and has been metamorphosed into the handsome but less-enduring Carriage Horse.

A valuable Carriage Horse has a large admixture of good blood in him, and as he is required more for the sake of appearance than for steady, hard work, he is required to possess a high, strong action and proud bearing, well arched neck, and a light springy step. His speed is very considerable, and he can do a great amount of work, but he is not fitted for dragging heavy loads like his predecessors, nor can he endure a continuance of work, for several days in succession. The splendid action of the Carriage Horse, although it is very showy, and adds much to the magnificence of his appearance, is injurious to the welfare of his feet and legs, which are sadly damaged by being battered against the hard stones of the street pavements.

The name of Cleveland Horse is given to this animal because it derives its origin from the Cleveland Bay, a variety of the Horse that is largely bred at Cleveland, in the North Riding of Yorkshire, and which, when crossed with more or less thorough-bred animals, produces the best Carriage Horses in the world. Very great care is bestowed on this

SUFFOLK PUNCH.

important subject, and in the finest animals there is so much of the pure blood that, in the words of Mr. Youatt, "the Coach Horse is nothing more than a tall, strong, over-sized hunter." According to the same experienced author, the principal points in the Carriage Horse are substance well placed, a deep and well-proportioned body, bone under the knee, and sound, open, tough feet.

THE true, pure-blooded SUFFOLK PUNCH is now nearly extinct, having been so frequently crossed with other breeds that its individuality has been almost entirely lost.

The old Suffolk Punch, so called from its round, punchy form, is a wonderful animal for pulling, being built as if expressly for the purpose of dragging great weights with unflinching perseverance. A team of these Horses needs no incitement by the whip, but as soon as they hear the command of their driver, they fling their whole weight into the collar, and almost throw themselves on their knees in their anxiety to fulfil their task. They seem to be perfectly aware of their powers, and to be jealously tenacious of their supremacy, for even if they find after one or two efforts that the load resists their best endeavours, they do not refuse to exert themselves any further, as is often the case with draught Horses, but will persevere in pulling until they drop with fatigue. The low, heavy shoulder, and strong quarters of the Suffolk Punch are of infinite service in drawing the plough or the cart, and its hardy frame and determined disposition enable it to support a hard day's labour without being overcome.

These valuable characteristics have been employed in improving the breed of carriage Horses, for it is a wonderful fact, and one which cannot be too carefully considered, that mental traits are more enduring than bodily form, and that a crossed breed derives its true

1.   3 A

FLEMISH HORSE.

value, not so much from the outward form which is obtained by the cross, but from the mental characteristics that are transmitted through a series of generations. The reader may remember that in the case of the greyhound, a bull-dog cross was introduced in order to impart courage and determination to a breed that had sacrificed everything to speed, and that although the bull-dog form was totally eradicated in a few generations, the bull-dog spirit remained.

Thus with the Suffolk Punch. Some of the best carriage Horses have been obtained by crossing the Suffolk Punch with a thorough-bred hunter, so as to unite the excellences of the two animals, giving speed and rapid force to the draught Horse, and the power of pulling to the hunter.

AN elephant among Horses, the mixed Flemish and Black Draught Horse is familiar to all Londoners as drawing the heavy drays on which beer is conveyed from the breweries to the purchaser.

This enormous animal is really needed for his peculiar work, although a natural emulation that exists between the different firms leads them to rival each other in the size and magnificence of their dray Horses, as well as in the excellence of their beer. It is a general idea that the dray Horses derive their huge bulk from being fed on grains and permitted to drink beer, and that the draymen owe their large proportions and rubicund aspect to similar privileges. Such is, however, not the case, as the Horses are bred especially for the purpose, and the men are chosen with an eye to their jovial aspect. It would never answer for a brewer to keep a poor, wizened, starveling drayman, for the public would immediately lay the fault on the beer, and transfer their custom elsewhere.

CLYDESDALE CART HORSE.

The dray Horse is a very slow animal, and cannot be permanently quickened in his pace, even if the load be comparatively light. Its breast is very broad, and its shoulders thick and upright, the body large and round, the legs short, and the feet extremely large. The ordinary pace of the heavy Draught Horse is under three miles per hour, but by a judicious admixture of the Flemish breed, the pace is nearly doubled, the endurance increased, and the dimensions very slightly diminished. The great size of the dray **Horse** is required, not for the absolute amount of pulling which it performs, but for the need of a large and heavy animal in the shafts to withstand the extreme jolting and battering that takes place as the springless drays are dragged over the rough stones of the metropolis. And as a team of two or three small leaders and one huge wheeler would look absurd, it is needful to have all the Horses of uniform dimensions and appearance.

The genuine dray Horse is a noble beast, and it is very pleasant to see the kindly feelings which exist between them and their drivers. The long whip is carried upon the drayman's shoulders more as a badge of office than as an instrument of torture, and if used at all, it is gently laid upon the Horse's back, accompanied with some endearing language, which is very intelligible to the Horse, but not to be comprehended by ordinary human intellects.

ONE of the best Horses for ordinary heavy work is the CLYDESDALE CART HORSE, an animal which has derived its name from the locality where it was first bred. It is larger than the Suffolk Punch, and owes its origin to the Lanark Horse, crossed with the large Flemish breed. In temper it is docile, and it is possessed of enormous strength and great

SHETLAND PONY.

endurance. The pure breed is large and heavy, and is notable for a very long stride. When judiciously crossed with other breeds it produces offspring which are extensively employed in the carriage and for the saddle. The figure of the Clydesdale Cart Horse which accompanies this brief notice is a portrait of a remarkably fine animal named Prince Albert.

SEVERAL breeds of partially wild Horses are still found in the British islands, the best known of which is the SHETLAND PONY.

This odd, quaint, spirited little animal is an inhabitant of the islands at the northern extremity of Scotland, where it runs wild, and may be owned by any one who can catch and hold it. Considering its diminutive proportions, which only average seven or eight hands in height, the Sheltie is wonderfully strong, and can trot away quite easily with a tolerably heavy man on its back. One of these little creatures carried a man of twelve stone weight for a distance of forty miles in a single day. The head of this little animal is small, the neck short and well arched, and covered with an abundance of heavy mane, that falls over the face and irresistibly reminds the spectator of a Skye-terrier. It is an admirable draught Horse when harnessed to a carriage of proportionate size; and a pair of these spirited little creatures, when attached to a low lady's carriage, have a remarkably piquant and pretty appearance.

MAN has so long held the DOMESTIC ASS under his control, that its original progenitors have entirely disappeared from the face of the earth.

There are, as it is well known, abundant examples of wild Asses found in various lands, but it seems that these animals are either the descendants of domesticated Asses which have escaped from captivity, or mules between the wild and domestic animals. In size and general appearance the Ass varies greatly, according to the country which it inhabits, and the treatment to which it is subjected. The Spanish kind, for example, is double the size of the ordinary English Ass, and even the latter animal is extremely variable in stature and general dimensions. As a rule, the Ass is large and sleek-haired countries, and small and woolly-haired in the colder parts of the globe.

ASS.—*A'sinus vulgaris.*

Strong, surefooted, hardy, and easily maintained, the Ass is of infinite use to the poorer classes of the community, who need the services of a beast of burden, and cannot afford to purchase or keep so expensive an animal as a horse. In the hands of unthinking and uneducated people, the poor creature generally leads a very hard life, and is subjected to much and undeserved ill-treatment; not so much from deliberate cruelty as from want of thought. We often see the poor animal laden with a burden that is evidently beyond its powers, and continually urged forward by blows. Not long ago, I saw a poor donkey harnessed to a low cart in which were seated three full-grown women, one of whom was continually belabouring the animal with a thick stick. Presently they stopped, took up a fourth passenger, and again moved on in spite of all remonstrances on behalf of the unfortunate creature that was forced to drag so heavy a weight.

This cruel treatment is as impolitic as it is inhuman; for there are few animals which will better repay kindness than the Ass, or will develop better qualities.

Some years ago a very excellent movement was started by Captain Scott, R.N., for the purpose of ameliorating the condition of certain unfortunate donkeys which were employed in the conveyance of coal, and were in a most pitiable condition.

Several persons had attempted to remonstrate with the owners of the poor animals, and had only been insulted, without achieving any successful result. Captain S. however, struck out another line of conduct, and instead of abusing or persecuting those who treated their animals badly, he offered prizes to those who could produce the best and healthiest donkey. Several persons joined him in this most laudable undertaking, and they held quarterly meetings, at which the prizes were bestowed. A medal was also given to each successful competitor, and the association pledged themselves to employ no donkey-driver who could not produce a medal. The natural consequences followed. The public soon took up the idea, the medal-holders carried off all the trade, and the cruel and neglectful drivers were either forced to conform to the regulations of the society, or to betake themselves and their beasts elsewhere.

It is a very great mistake to employ the name of Ass or donkey as a metaphor for stupidity, for the Ass is truly one of the cleverest of our domesticated animals, and will lose no opportunity of displaying his capability whenever his intelligence is allowed to expand by being freed from the crushing toil and constant pain that are too often the concomitants of a donkey's life. Every one who has petted a favourite donkey will remember many traits of its mental capacities; for as in the case of the domestic fool of the olden days, there is far more knavery than folly about the creature.

One of these animals was lately detected in a most ingenious theft. A number of rabbits were kept in a little outhouse, and inhabited a set of hutches fastened to the wall. One day it was found that nearly all the store of oats had suddenly vanished from the outhouse without any visible reason. Next morning, however, the donkey who lived in an adjoining meadow was seen to open the gate which led into his field, and cautiously shut it after him. This conduct afforded a clue to the disappearance of the oats, and upon a careful search being made, his footmarks were traced along the path to the rabbit-house, and even on the ground among the hutches. It was very clear that the ingenious animal must have unlatched his own gate, unfastened the loop of the rabbit-house, finished all the oats, and have returned as he went, re-fastening all the doors behind him. In leaving the rabbit-house he must have backed out, as the place was not wide enough to permit him to turn.

He was very familiar with the children, and would permit three of them to ride on his back together. After a while the boys went to school, and some ponies were procured for the other members of the family, so that Sancho had a long holiday. When the boys returned from school, they mounted Sancho as usual for the purpose of having their ride. The cunning animal allowed them to seat themselves, and then coolly shook them off again. This process he repeated until they gave up the hopeless attempt, and Sancho gained his purpose.

That a donkey has more than once succeeded in beating off the attacks of a leopard by vigorous and rapid kicks of his hind-feet is well known, and an incident occurred some years ago which shows that the animal is as valiant in opposing dogs as in fighting leopards. A surly, ill-intentioned man, who possessed an equally surly bull-dog, set his animal at an unoffending donkey. The bull-dog, nothing loth, made at his intended victim and sprang at him. The Ass, however, cleverly avoided the dog's onset, seized him in its teeth, carried him to the river Derwent, near which the scene occurred, plunged him under water, and there lying down upon him, prevented him from regaining the surface, and fairly drowned his opponent.

Another Ass displayed a singular discrimination of palate, being celebrated for his love of good ale. At one road-side inn the landlady had been very kind in supplying the donkey with a glass of his loved beverage, and the natural consequence was, that the animal could never be induced to pass within a moderate distance of the spot without going for his beer. Neither entreaties nor force sufficed to turn his head in another direction, and his master was in such cases obliged to make the best of the matter, and permit the animal to partake of his desired refreshment. He had a curious knack of taking a tumbler of beer between his lips, and drinking the contents without spilling a drop of the liquid or breaking the glass. So curious a sight as a donkey drinking beer was certain to attract many observers, who testified their admiration by treating the animal to more beer. His head, however, was fortunately a strong one, for only once in his life was he ever seen intoxicated, and on that solitary occasion his demeanour was wonderfully decorous.

A petted donkey belonging to one of my friends was permitted to walk at large in the garden, on condition that he restrained himself from leaving the regular paths. Once or twice he had been seduced by the charms of some plant to walk upon the flower-beds, and had been accordingly drubbed by the gardener, who detected the robber by the marks of his footsteps, which were deeply imprinted in the soft mould. After a while the animal seemed to have reflected upon the circumstance which led to the discovery of his offence, and the next time that he walked upon the flower-beds, he scraped the earth over his foot-marks, and endeavoured to obliterate the traces of his disobedience. As, however, his hoofs were not very delicate tools, and his method of levelling anything but gentle, the marks were more conspicuous than before.

In the East, the Ass is used even more extensively than in Europe, and is generally employed for carrying burdens or for the saddle, the horse being used more for ostentation or for warfare than for the mere conveyance of human beings from one spot to another. The following account of donkey-riding in Cairo, by Bayard Taylor, gives a most vivid and animated description of the manner in which the Ass is employed in the East.

" To see Cairo thoroughly, one must first accustom himself to the ways of those long-eared cabs, without the use of which I would advise no one to trust himself in the bazaars. Donkey-riding is universal, and no one thinks of going beyond the Frank quarters on foot. If he does, he must submit to be followed by not less than six donkeys, with their drivers. A friend of mine who was attended by such a cavalcade for two hours, was obliged to yield at last, and made no second attempt. When we first appeared in the gateway of an hotel, equipped for an excursion, the rush of men and animals was so great, that we were forced to retreat until our servant and the porter whipped us a path through the yelling and braying mob. After one or two trials, I found an intelligent Arab boy named Kish, who for five piastres a day furnished strong and ambitious Donkeys, which he kept ready at the door from morning till night. The other drivers respected Kish's privilege, and thenceforth I had no trouble.

The donkeys are so small that my feet nearly touched the ground, but there is no end to their strength and endurance. Their gait, whether in pace or in gallop, is so easy and light that fatigue is impossible. The drivers take great pride in having high-cushioned, red saddles, and in hanging bits of jingling brass to the bridles. They keep their donkeys close shorn, and frequently beautify them by painting them various colours. The first animal I rode had legs barred like a zebra's, and my friend's rejoiced in purple flanks and a yellow belly. The drivers run behind them with a short stick, punching them from time to time, or giving them a sharp pinch on the rump. Very few of them own their donkeys, and I understood their pertinacity when I learned that they frequently received a beating on returning home empty-handed.

The passage of the bazaars seems at first quite as hazardous on donkey-back as on foot ; but it is the difference between knocking somebody down and being knocked down yourself, and one certainly prefers the former alternative. There is no use in attempting to guide the donkey, for he won't be guided. The driver shouts behind, and you are dashed at full speed into a confusion of other donkeys, camels, carts, water-carriers and footmen. In vain you cry out 'Bess' (enough), Piacco, and other desperate adjurations: the driver's only reply is, ' Let the bridle hang loose !' You dodge your head under a camel load of planks ; your leg brushes the wheel of a dust-cart ; you strike a fat Turk plump in the back; you miraculously escape upsetting a fruit stand; you scatter a company of spectral, white-masked women, and at last reach some more quiet street, with the sensations of a man who has stormed a battery.

At first this sort of riding made me very nervous, but presently I let the donkey go his own way, and took a curious interest in seeing how near a chance I ran of striking or being struck. Sometimes there seemed no hope of avoiding a violent collision, but by a series of the most remarkable dodges, he generally carried you through in safety. The cries of the driver running behind, gave me no little amusement. ' The howadji comes ! Take care on the right hand ! Take care on the left hand ! O man, take care ! O maiden, take care ! O boy, get out of the way ! The howadji comes !' Kish had strong lungs, and his donkey would let nothing pass him, and so wherever we went we contributed our full share to the universal noise and confusion."

The colour of the Ass is a uniform grey, a dark streak passing along the spine, and another stripe being drawn transversely across the shoulders. In the quagga and zebra these stripes are much more extended.

DZIGGETAI, OR KOULAN.—*A'sinus O'nager.*

The cross-breed between the horse and the ass, which is commonly known by the name of the MULE, is a very valuable animal for certain purposes, possessing the strength and power of the horse, with the hardiness and sure foot of the ass. The largest and most useful Mules are those which are produced by a male ass and a mare, the large Spanish Ass being the best for this purpose. In Spain and in many eastern countries the Mule is an animal of some importance, the parents being selected as carefully as those of the horse itself. The chief drawback in the rearing of this animal is that it is unproductive, and is incapable of continuing its species, so that there can be no definite breed of Mules, as of horses and asses.

THE Wild Asses are all celebrated for their extreme fleetness and sureness of foot, and among them the DZIGGETAI, KHUR, or KOULAN deserves especial mention.

This animal is so wonderfully swift that it cannot be overtaken even by a fleet Arabian horse, and if it can get upon hilly or rocky ground, it bids defiance to all wingless enemies. Not even the greyhound can follow it with any hope of success when it once leaves level ground. This great speed renders it a favourite object of chase with the natives of the countries which it inhabits ; and whether in Persia or India, it is held to be the noblest of game. Sometimes the falcon is trained to aid in the chase of the Wild Ass, but the usual method of securing this animal is to drive it towards rocky ground, and to kill it with a rifle bullet as it stands in fancied security upon some lofty crag.

It lives in troops, descending to the plains during the winter months, and returning to the cooler hills as soon as the summer begins to be unpleasantly warm. It is very common in Mesopotamia, and is always a most shy and wary, as well as swift animal.

QUAGGA.—*A'sinus Quagga.*

Each troop is under the command of a leader, who sways his subjects with unlimited authority, and takes upon himself to make all needful arrangements for their welfare.

The honour of success is not the only motive which urges the hunters to pursue the Dziggetai, for its flesh is remarkably excellent, and is universally thought to be one of the greatest dainties. The localities inhabited by this animal are Mesopotamia, Persia, the shores of the Indus, and the Punjâb. The colour of this animal is pale reddish-brown in the summer, fading into a grey-brown in the winter, and marked with a black stripe along the spine, becoming wider upon the middle of the back.

ANOTHER species of Wild Ass is the KIANG, or Wild Ass of Thibet, sometimes, but erroneously, called the Wild Horse of Thibet, because its noise resembles the neighing of that animal rather than the braying of the ass.

The Kiang inhabits the high table-lands of its native country, and is wonderfully fleet and active in traversing level or uneven ground. It is a rather large animal; a full-sized adult from Chinese Tartary measuring fourteen hands in height at the shoulder. It lives in little troops of eight or ten in number, and is found in districts where the cold is most intense, the thermometer falling below zero in the localities which are most frequented by them. As they pass their lives in such a climate, they are necessarily furnished with warm, woolly coats, which are of different colour and thickness according to the time of year. In the summer the fur is short, smooth, and of a light reddish-brown, but in winter the hair becomes long and rather woolly, and fades into a light grey brown. The legs too change the tinting, being straw-coloured in summer and whitish in winter. A broad black line is drawn along the back, but there is no transverse band across the shoulders, nor are their young marked with zebra-like stripes, as is the case with the young Dziggetai.

It is a swift and wary animal, fleeing in terror before the hunter, and yet stopping at intervals to gaze on the object of its alarm. Unless the hunter is very sure of his aim, he will not risk a shot, for the animals are so terrified by the report and the flash that they forget their curiosity in their fear, and gallop away at the best of their speed, which soon carries them out of danger. It is capable of domestication, and can be put in training like a horse or a domestic ass.

AFRICA produces some most beautiful examples of the Wild Asses, equalling the Asiatic species in speed and beauty of form, and far surpassing them in richness of colour and boldness of marking.

The QUAGGA looks at first sight like a cross between the common wild ass and the zebra, as it only partially possesses the characteristic zebra-stripes, and is decorated merely upon the hind and fore-parts of the body. The streaks are not so deep as they are in the zebra, and the remainder of the body is brown, with the exception of the abdomen, legs, and part of the tail, which are whitish-grey. The Quagga lives in large herds, and is much persecuted by the natives of Southern Africa, who pursue it for the sake of its skin and its flesh, both of which are in high estimation.

A NEARER approach to the true zebra is seen in the animal which is indifferently termed the DAUW, the PEECHI, or BURCHELL'S ZEBRA. This species is an inhabitant of Southern Africa, where it is found in large herds south of the Orange River. Unlike the wild ass of Asia, the Dauw keeps aloof from the rocky and hilly districts, and is only found on the plains, where it wanders in company with ostriches, various antelopes, and other strange comrades. The general appearance of this species bears a considerable resemblance to that of the zebra, from which animal it may be immediately distinguished by the colour, number, and extent of the dark stripes and bands. In the Dauw, the stripes are not so black as in the zebra, and instead of covering the entire body and limbs, they only extend over the head, neck, body, and the upper portions of the legs. The general colour of the fur is a pale-brown, becoming greyish-white upon the abdomen and inner faces of the limbs.

Like many other gregarious animals of Southern Africa, the Dauw is found to make periodical migrations, for the purpose of supporting itself with the food that has failed in its original district. In times of scarcity the Dauw, together with several species of antelope, visits the cultivated lands, and makes sad havoc among the growing crops. When rain has fallen, and the forsaken districts have regained their fertility, the Dauw leaves the scene of its plunder, and returns to its ancient pasturage.

The Dauw is capable of a partial domestication, and can be tamed to a considerable extent. It is, however, considered as possessing a tetchy and uncertain temper, and is of too obstinate a disposition to be of much use to man. By the Matabili and Bechuana Kaffirs it is called Peet-sey, and the Dutch colonists have given it the name of Bonte-quagga.

AMONG all the species of the Ass tribe, the ZEBRA is by far the most conspicuous and the most beautiful.

The general colour of the Zebra is a creamy white, marked regularly with velvety black stripes that cover the entire head, neck, body, and limbs, and extend down to the very feet. It is worthy of note, that the stripes are drawn nearly at right angles to the part of the body on which they occur, so that the stripes of the legs are horizontal, while those of the body are vertical. The abdomen and inside faces of the thighs are cream-white, and the end of the tail is nearly black. This arrangement of colouring is strangely similar to that of the tiger, and has earned for the animal the name of "Hippotigris," or Horse-tiger, among some zoologists, ancient and modern. The skin of the neck is developed into a kind of dewlap, and the tail is sparingly covered with coarse black hair. By the Cape colonists it is called "Wilde Paard," or Wild Horse.

At the best of times the flesh of the Zebra is not very inviting, being rather tough,

ZEBRA.—*A'sinus Zebra.*

coarse, and of a very peculiar flavour. The Boers, who call themselves by the title of "baptized men," think they would be derogating from their dignity to partake of the flesh of the Zebra, and generously leave the animal to be consumed by their Hottentot servants. When wounded, the Zebra gives a kind of groan, which is said to resemble that of a dying man.

In disposition the Zebra is fierce, obstinate, and nearly untameable. The efforts used by Mr. Rarey in reducing to obedience the Zebra of the Zoological Gardens are now matter of history. The little brindled animal gave him more trouble than the huge savages on whom he had so successfully operated, and it overset some of his calculations by the fact that it was able to kick as fiercely from three legs as a horse from four.

In its habits the Zebra resembles the dziggetai more than the dauw, as it is always found in hilly districts, and inhabits the high craggy mountain ranges in preference to the plains. It is a mild and very timid animal, fleeing instinctively to its mountain home as soon as it is alarmed by the sight of a strange object.

BETWEEN the zebras and the domestic ass several curious Mules have been produced, and may be seen in the collection of the British Museum. It is worthy of notice, that wherever a cross breed has taken place, the influence of the male parent seems to be permanently impressed on the mother, who in her subsequent offspring imprints upon them some characteristic of the interloper.

ELEPHANT.—*Elephas Indicus.*

PACHYDÉRMATA;

OR, THICK-SKINNED ANIMALS.

THE important family of the Elephantidæ includes, according to the catalogue of the British Museum, the Elephants, Tapirs, Swine, Hyrax, Rhinoceros, and Hippopotamus. All these animals, however different their aspect, are nearly related to each other by means of certain members of the family, which, although now extinct, have been recovered through the assistance of geological researches.

Of Elephants, two distinct species are found in different continents, the one inhabiting Asia, and the other taking up its residence in Africa. According to some zoologists, these animals belong to different genera, but the distinctions between the two creatures are not sufficiently determined to warrant such a suggestion. Although the Asiatic and African Elephants are very similar in external form, they may at once be distinguished from each other by the dimensions of the head and the size of the ear. In the Asiatic animal, the head is elongated, the forehead concave, and the ears of ordinary size, while in the

African Elephant the head is much shorter, the forehead convex, and the ears of enormous magnitude, nearly meeting on the back of the head, and hanging with their tips below the neck.

The molar teeth also afford excellent indications of the country to which their owner has belonged, for the enamel upon the surface of the teeth of the Asiatic Elephant is moulded into a number of narrow bands like folded ribands, while that of the African species is formed into five or six diamond or lozenge shaped folds. Indeed, each molar tooth seems to be composed of a number of flat, broad teeth, which are fastened closely together, so as to form a single large mass. Only a portion of each tooth is externally visible, the remainder being hidden in the jaw, and moving forward as the exposed portion is worn away. When the whole tooth is thus worn out, it falls from the jaw, and its place is taken by another which has been forming behind it. In this manner the Elephant sheds its molar teeth six or seven times in the course of its life. The tusks, however, are permanent, and are retained during the whole of the animal's existence. There are a pair of small "milk-tusks" when the Elephant is in its childhood, but these are soon shed and replaced by the true tusks. In the Indian Elephant only the males are furnished with tusks, and not every individual of that sex, whereas in the African species both sexes are supplied with these valuable appendages, those of the male being much larger and heavier than those of his mate.

The Elephant, whether Asiatic or African, always lives in herds, varying greatly in numbers, and being always found in the deepest forests, or in their near vicinity. Both species are fond of water, and are never found at any great distance from some stream or fountain, although they can and do make tolerably long journeys for the purpose of obtaining the needful supply of liquid. They have a curious capability of laying up a store of water in their interior, somewhat after the fashion of the camel, but possess the strange accomplishment of drawing the liquid supply from their stomachs by means of their trunks, and scattering it in a shower over their backs in order to cool their heated bodies. When drinking, the Elephant inserts the tip of his trunk into the stream, fills its cavities with water, and then, turning his trunk so as to get the extremity well into his throat, he discharges its contents fairly into his stomach, where it may be heard to splash by any one who is in near proximity to the animal.

The strangest portion of the Elephant's form is the trunk, or proboscis. This wonderful appendage is in fact a development of the upper lips and the nose, and is perforated through its entire length by the nostrils, and is furnished at its extremity with a kind of finger-like appendage, which enables the animal to pluck a single blade of grass, or to pick a minute object from the ground. The value of the proboscis to the Elephant is incredible; without its aid the creature would soon starve. The short, thick neck would prevent it from stooping to graze, while the projecting tusks would effectually hinder it from reaching any vegetables which might grow at the level of its mouth. And as it would be unable to draw water into its mouth without the use of the trunk, thirst would in a very short time end its existence.

As the trunk is required for so many purposes, it must needs be capable of extension, contraction, and of flexibility in every direction, as well as possessed of enormous strength. In order to effect these conditions, the trunk is composed of no less than fifty thousand distinct muscles, some of which run longitudinally along the axis of the proboscis, and others radiate from the centre to the circumference. When the trunk is at rest, its surface is covered with a series of thick, transverse wrinkles or corrugations, which become less distinct as this appendage is gradually stretched, and vanish entirely when it is extended to its full length. The little finger-like appendage at its tip is slightly different in shape in the two sexes.

In order to support the enormous weight of the teeth, tusks, and proboscis, the head is required to be of very large dimensions, so as to afford support for the powerful muscles and tendons which are requisite for such a task. It is also needful that lightness should be combined with magnitude, and this double condition is very beautifully fulfilled. The skull of the Elephant, instead of being a mere bony shell round the brain, is enormously enlarged by the separation of its bony plates, the intervening space being filled with a vast number of honeycomb-like bony cells, their walls being hardly thicker than strong paper, and their hollows filled during the life of the animal with a kind of semi-liquid fat or oil. The brain lies in a comparatively small cavity within this cellular structure, and is therefore defended from the severe concussions which it would otherwise experience from the frequency with which the animal employs its head as a battering-ram. It is easy to understand the difficulty of killing an Elephant by aiming at the head, for unless the shot be directed towards one of the apertures which lead to the brain, such as the eye, the ear, or the nostril, the bullet only enters the mass of bony cells, and does comparatively little damage. It is worthy of observation, that as the skull of the Asiatic and African Elephant is different in shape, a bullet which will destroy one animal might have little effect on the other.

In order to support the enormous weight which rests upon them, the legs are very stout, and are set perpendicularly, without that bend in the hinder leg which is found in most animals. There is no elongated cannon bone in the Elephant, so that the hind legs are without the so-called knee-joint. This structure, however, is of infinite use to the animal when it climbs or descends steep acclivities, a feat which it can perform with marvellous ease. It may seem strange, but it is nevertheless true, that localities which would be totally inaccessible to a horse are traversed by the Elephant with perfect ease.

In descending from a height, the animal performs a very curious series of manœuvres. Kneeling down, with its fore-feet stretched out in front, and its hinder legs bent backward, as is their wont, the Elephant hitches one of its fore-feet upon some projection or in some crevice, and bearing firmly upon this support, lowers itself for a short distance. It then advances the other foot, secures it in like manner, and slides still farther, never losing its hold of one place of vantage until another is gained. Should no suitable projection be found, the Elephant scrapes a hole in the ground with its advanced foot, and makes use of this artificial depression in its descent. If the declivity be very steep, the animal will not descend in a direct line, but makes an oblique track along the face of the hill. Although the description of this curious process occupies some time, the actual feat is performed with extreme rapidity.

Though the foot of an Elephant is extremely large, it is most admirably formed for the purpose which it is destined to fulfil, and does not, as might be supposed, fall heavily upon the ground. The hoof that incloses the foot is composed of a vast number of horny plates that are arranged on the principle of the common carriage-spring, and seem to guard the animal from the jarring shock of the heavy limb upon the soil. Those who for the first time witness the walk or the run of the Elephant, are always surprised at the silent ease of the creature's free, sweeping step. As there is no short ligament in the head of the thigh-bone, the hind foot is swung forward at each step, clearing the ground easily, but being scarcely raised above the surface of the earth.

Having thus given a short sketch of the characteristics which are common to both species of Elephants, I will proceed to a short account of the Asiatic animal.

The ASIATIC ELEPHANT bears a world-wide fame for its capabilities as a servant and companion of man, and for the extraordinary development of its intellectual faculties. Hundreds of these animals are annually captured, and in a very short period of time become wholly subjected to their owners, and learn to obey their commands with implicit submission. Indeed, the power of the human intellect is never so conspicuous as in the supremacy which man maintains over so gigantic and clever an animal as the Elephant. In all work which requires the application of great strength, combined with singular judgment, the Elephant is supreme ; but as a mere puller and hauler it is of no very great value. In piling logs, for example, the Elephant soon learns the proper mode of arrangement, and will place them upon each other with a regularity that would not be surpassed by human workmen. Sir Emerson Tennent mentions a pair of Elephants that were accustomed to labour conjointly, and which had been taught to raise their wood piles to a considerable height by constructing an inclined plane of sloping beams, and rolling the logs up the beams. The same writer, in his most valuable work on Ceylon, gives the following curious instance of intelligence in an Elephant :—

" One evening, while riding in the vicinity of Kandy, towards the scene of the massacre of Major Davie's party in 1803, my horse evinced some excitement at a noise which approached us in the thick jungle, and which consisted of a repetition of the ejaculation, *Urmph! urmph!* in a hoarse and dissatisfied tone. A turn in the forest explained the mystery, by bringing me face to face with a tame Elephant, unaccompanied by any attendant. He was labouring painfully to carry a heavy beam of timber, which he balanced across his tusks, but the pathway being narrow, he was forced to bend his head to one side to permit it to pass endways ; and the exertion and inconvenience combined, led him to utter the dissatisfied sounds which disturbed the composure of my horse.

On seeing us halt, the Elephant raised his head, reconnoitred us for a moment, then flung down the timber, and forced himself backwards among the brushwood, so as to leave a passage, of which he expected us to avail ourselves. My horse still hesitated : the Elephant observed, and impatiently thrust himself still deeper into the jungle, repeating his cry of *urmph!* but in a voice evidently meant to encourage us to come on. Still the horse trembled ; and, anxious to observe the instinct of the two sagacious creatures, I forbore any interference : again the Elephant wedged himself farther in amongst the trees, and waited impatiently for us to pass him, and after the horse had done so, tremblingly and timidly, I saw the wise creature stoop and take up his heavy burthen, turn and balance it on his tusks, and resume his route, hoarsely snorting, as before, his discontented remonstrance."

Another Elephant of Ceylon performed a feat of equal sagacity.

By profession he was a builder, and was employed in laying stones under the supervision of an overseer. Whenever he completed one course, he signalled to the overseer, who came and inspected his work, and after ascertaining that the task was properly performed, gave the signal to lay another course. On one occasion, the Elephant placed himself against a portion of the wall, and refused to move from the spot, when the overseer came to the part of the wall which his body concealed. The overseer, however, insisted on the animal's moving aside, and the Elephant, seeing that his ruse had failed, immediately set hard to work at pulling down the wall which he had just built, and which was defective in the spot which he had been attempting to conceal from the inspector's eye.

Although so valuable an animal for certain kinds of work, the Elephant is hardly so effective an assistant as is generally supposed. " The working Elephant," says Sir E. Tennent, " is always a delicate animal, and requires watchfulness and care ; as a beast of burden he is unsatisfactory ; for although in point of mere strength there is hardly any weight which could be conveniently placed on him that he could not carry, it is difficult to pack it without causing abrasions that afterwards ulcerate. His skin is easily chafed by harness, especially in wet weather. Either during long droughts, or too much moisture, his feet are liable to sores, which render him non-effective for months. Many attempts have been made to provide him with some protection for the sole of the foot, but from his extreme weight and mode of planting the foot, they have all been unsuccessful. His eyes are also liable to frequent inflammation. In Ceylon, the murrain among cattle is of frequent occurrence, and carries off great numbers of animals, wild as well as tame. In such visitations the Elephants suffer severely, not only those at liberty in the forest, but those which are carefully tended in the Government stables.

On being first subjected to work, the Elephant is liable to severe and often fatal swellings of the jaws and abdomen. On the whole, there may be a question as to the prudence or economy of maintaining a stud of Elephants for the purposes to which they are assigned in Ceylon. In the rude and unopened parts of the country—where rivers are to be forded, and forests are only traversed by jungle paths—their labour is of value in certain contingencies, in the carrying of stores and in the earlier operations for the construction of fords and bridges of timber. But in more highly civilized districts, and wherever macadamized roads admit of the employment of horses and oxen for draught, I apprehend that the services of Elephants might, with advantage, be probably reduced, if not altogether dispensed with." The able writer then proceeds to observe that if the peculiar constitution, irritability, and expensive maintenance of the Elephant be taken into consideration, the value of its labour will be found to be less than that of a good draught horse. The keep of an Elephant in Ceylon costs between six and seven shillings per diem, and the animal can only work, on an average, four days in each week ; while the keep of a powerful dray horse, which works five days in the week, is only half-a-crown per diem.

The general disposition of the Asiatic Elephant is gentle, but there are always some stray individuals that are not admitted into any herd, but live in solitary moodiness, and are termed "rogues," from their irritable temper. So gentle, indeed, is their nature that even when most irritated by wounds, they literally do not know how to kill their foe, even if he is lying at their mercy ; and there are many instances where hunters who have been chased and struck down by these animals have escaped without suffering any serious damage. The tusks are seldom employed as offensive weapons, and the Elephant has but little idea of directing them towards an adversary. A momentary pressure of the foot, or a blow with the tusk, would in any case be sufficient to cause death, but the animal seems to be scarcely aware of its own power, and often contents itself with kicking its prostrate foe from foot to foot, hustling him between the fore and hinder limbs in a very unpleasant manner. A little Indian Elephant, that had been much worried by wild boars, was accustomed to defeat them by receiving their charge, and then knocking them about from foot to foot until they were effectually disabled.

There are two modes of capturing the Asiatic Elephant, the one by pursuing solitary individuals and binding them with ropes as they wander at will through the forests, and the other by driving a herd of Elephants into a previously prepared pound, and securing the entrance so as to prevent their escape.

In the former method, the hunters are aided by certain trained females, termed "koomkies," which enter into the spirit of the chase with wonderful animation, and help their riders in every possible manner. When the koomkies see a fine male Elephant, they advance carelessly towards him, plucking leaves and grass, as if they were perfectly indifferent to his presence. He soon becomes attracted to them, when they overwhelm him with endearing feminine blandishments, and occupy his attention so fully that he does not observe the proceedings of the "mahouts," or riders. These men, seeing the Elephant engaged with the "koomkies," slip quietly to the ground, and attach their rope nooses to his legs, fastening the ends of the cords to some neighbouring tree. Should no suitable tree be at hand, the koomkies are sagacious enough to comprehend the dilemma, and to urge their victim towards some large tree which is sufficiently strong to withstand his struggles. As soon as the preparations are complete, the mahouts give the word of command to the koomkies, who move away, leaving the captive Elephant to his fate.

Finding himself deserted and bound, he becomes mad with rage, and struggles with all his force to get free. In these furious efforts, the Elephant displays a flexibility and activity of body that are quite surprising, and are by no means in accordance with the clumsy, stiff aspect of its body and limbs. It rolls on the ground in despair, it rends the air with furious cries of rage, it butts at the fatal tree with all its force, in hope of bringing it to the ground, and has been known to stand with its hind legs fairly off the ground, in its furious endeavours to break the rope. After a while, however, it finds its exertions to be totally useless, and yields to its conquerors. Formerly it was allowed to remain in its captivity until reduced by hunger, but as the ropes are apt to cut severely into the ankle, and to cause painful and dangerous wounds, the time of bondage is now shortened as much as possible, and the animal removed to another spot where ropes are needless. The koomkies afford invaluable assistance both in tying the animal and in leading him away from the tree to which he had been bound. One of these animals is reported to have gone on a solitary hunting expedition on her own account, and to have captured a fine male Elephant, which she tied to a tree with some iron chains.

The second mode of capturing Elephants is more complicated, and secures a greater number of beasts at a time, but as it necessarily includes the young, the old, and the vigorous of both sexes in the general seizure, its results are not so admirable as might be anticipated.

The inclosure into which the Elephants are driven is termed a "keddah," and is ingeniously constructed of stout logs and posts, which are supported by strong buttresses, and are so arranged that a man can pass through the interstices between the logs. When the keddah is set in good order, a vast number of hunters form themselves into a huge circle, inclosing one or more herds of Elephants, and moving gradually towards the inclosure of the keddah, and arranging themselves in such a manner as to leave the entrance towards the keddah always open. When they have thus brought the herd to the proper spot, a business which will often consume several weeks, the Elephants are excited by shouts, the waving of hands and spears, &c., to move towards the inclosure, which is cunningly concealed by the trees among which it is built. If the operation should take place at night, the surrounding hunters are supplied with burning torches, while the keddah is carefully kept in darkness. Being alarmed by the noise and the flames, the Elephants rush instinctively to the only open space, and are thus fairly brought within the precincts of the keddah, from which they never emerge again save as captives.

The terrified animals run round and round the inclosure, and often attempt a desperate charge, but are always driven back by the torch-bearers, who wave their flaming weapons, and discourage the captured animals from their meditated assault. At last the poor creatures are so bewildered and fatigued, that they gather together in the centre of the keddah, and are then considered to be ready for the professional Elephant-hunters. These courageous men enter the keddah either on foot or upon the backs of their koomkies, and contrive to tie every one of the captives to some spot from whence it cannot move. Most ingenious stratagems are employed by the hunters in this perilous task, the details of which may be found in many works on the subject.

When the natives hunt the Elephant merely for the sake of his ivory or his flesh, and do not care to take him alive, they achieve their object by stealing cautiously upon him as he dozes, and by gently tickling one of his hind-feet with a slight twig they induce him to lift the foot from the ground. As soon as he does so, the hunters, who are furnished with a mallet and a sharp wooden spike about eight inches in length, drive the spike into his foot, and effectually lame him with a single blow. He is then quite at their disposal, and is easily despatched. The flesh of the Elephant is thought to be very poor indeed ; but the heart, the tongue, the trunk, and the foot, are considered to be good eating if properly dressed.

The "points" of a good Elephant are as important in India and Ceylon as those of a horse in Europe. In a native work upon the Elephant, quoted by Sir E. Tennent, the points are given as follows :—"The softness of the skin, the red colour of the mouth and tongue, the forehead expanded and full, the ears large and rectangular, the trunk broad at the root, and blotched with pink in front, the eyes light and kindly, the cheeks large, the neck full, the back level, the chest square, the fore-legs short and convex in front, the hind quarters plump, five nails in each foot, all smooth, elastic, and round. An Elephant with all these perfections will impart glory and magnificence to the king."

The herds in which these animals congregate are not of very great size, containing only from ten to twenty or thirty individuals, and consisting, as is generally thought by men of practical experience, of members of the same family. This opinion is strengthened by the fact that certain physical peculiarities, such as the shape of the trunk or the head, have been found in every member of the same herd. Sometimes these herds will associate with each other for a time, but at the smallest alarm each little flock assembles together independently of the others. It is rather remarkable that a whole herd has never been known to charge a foe simultaneously. The leader generally faces the enemy, while the remainder of the herd manœuvre in his rear ; but that the entire herd should unite in a charge, is a circumstance never yet known to occur. The Asiatic Elephant will permit the temporary society of other animals, and may be seen at a fountain or feeding on an open space in close proximity to deer and wild buffaloes, neither animal displaying any aversion to or fear of the other.

In its general habits the Elephant is restless and irritable, or rather "fidgety," never remaining quite still, but always in motion in some way or other. At one time it will

1. 3 B

sway backwards and forwards, at another it will stoop and rise continually, or it will be getting sand or water and sprinkling it over its body, or it will pluck a leafy branch and wave it slowly and gracefully over its back. It is very fond of bathing, and has a curious predilection for drawing a mixture of mud and water into its trunk, and discharging it over its body. It is an admirable swimmer, and will cross large rivers with perfect ease. Sometimes it prefers walking on the bed of the river, merely protruding the tip of its proboscis above the surface for the purpose of breathing.

The Indian Elephant is employed more for purposes of state or for sport than for hard labour, and is especially trained for tiger-hunting. As there is a natural dread of the tiger deeply implanted in the Elephant's being, it is no easy matter to teach the animal to approach its brindled foe. A stuffed tiger-skin is employed for this purpose, and is continually presented to the Elephant until he learns to lose all distrust of the inanimate object, and to strike it, to crush it with his feet, or to pierce it with his tusks. After a while, a boy is put inside the tiger-skin, in order to accustom the Elephant to the sight of the tiger in motion. The last stage in the proceedings is to procure a dead tiger, and to substitute it for the stuffed representative. Even with all this training, it most frequently happens, that when the Elephant is brought to face a veritable living tiger, the fierce bounds, savage yells, and furious eyes of the beast are so discouraging, that he turns tail, and makes the best of his way from the spot. Hardly one Elephant out of ten will face an angry tiger.

The Elephant is always guided by a mahout, who sits astride upon its neck and directs the movements of the animal by means of his voice, aided by a kind of spiked hook, called the haunkus, which is applied to the animal's head in such a manner as to convey the driver's wishes to the Elephant. The persons who ride upon the Elephant are either placed in the howdah, a kind of wheelless carriage strapped on the animal's back, or sit upon a large pad, which is furnished with cross ropes in order to give a firm hold. The latter plan is generally preferred, as the rider is able to change his position at will, and even to recline upon the Elephant's back if he should be fatigued by the heavy rolling gait of the animal. The Elephant generally kneels in order to permit the riders to mount, and then rises from the ground with a peculiar swinging motion that is quite indescribable, and is most discomposing to novices in the art. Very small Elephants are furnished with a saddle like that which is used upon horses, and is fitted with stirrups. The saddle, however, cannot be conveniently used on animals that are more than six feet in height.

The size of Elephants has been greatly exaggerated, as sundry writers have given fourteen or sixteen feet as an ordinary height, and have even mentioned instances where Elephants have attained to the height of twenty feet. It is true that the enormous bulk of the animal makes its height appear much greater than is really the case. Eight feet is about the average height of a large Elephant, and nine or ten feet is the utmost maximum to which the creature ever attains.

It is rather remarkable that the Elephants should be so fond of intoxicating liquids as to be induced by the promise of porter, beer, wine, or spirits, to perform tricks which it would otherwise refuse to attempt. The natural food of the Elephant consists of grass and various leaves, which it plucks daintily with the tip of its trunk, and always beats against its fore-legs, in order to shake off the dust. While feeding, the Elephant never seems to be in a hurry, but eats deliberately, and often pauses in its meal, as if engaged in contemplation. In this country, the average daily food of an adult Elephant is one truss of hay, one truss of straw, a bushel of barley-meal and bran made into a mash, thirty pounds of potatoes, and six pints of water. In Ceylon, each Elephant employs two men in cutting leaves for its sustenance, and a very large animal would probably require the services of three leaf-cutters.

The general colour of the Elephant is brown, of a lighter tint when the animal is at liberty, and considerably deeper when its hide is subjected to rubbing with a cocoa-nut brush, and plenty of oil. Sometimes an albino or white Elephant is seen in the forests, the colour of the animal being a pinky-white, and aptly compared to the nose of a white horse. The King of Ava, one of whose titles is "Lord of the White Elephants," generally contrives to monopolize every White Elephant, and employs them for purposes of state, decorating them with strings of priceless gems, pearls, and gold coins, and lodging them in the most magnificent of houses, where their very eating-troughs are of silver.

Although the tame Elephant is usually gentle in his disposition, there are certain times in the year when he becomes greatly excited, and is sometimes so powerfully agitated, that he will attack anything that comes in his way, and has often been known even to assault his own keeper. Elephants in this condition are technically called "must" Elephants, and are carefully guarded as long as the paroxysm lasts. On one occasion, a mahout was forced to sit upon the animal's back for several days continuously, not daring to alight lest the infuriated animal should destroy him. As he sat upon the creature's back, it constantly endeavoured to pull him from his seat, but was held at bay by the sharp point of the "haunkus," which wounded his trunk whenever it threatened the mahout, and caused such pain that the animal was fain to desist from its deadly efforts. While in this state of excitement, the Elephant is largely employed as a combatant, being set to fight another "must" animal for the gratification of its owner. Very heavy wagers were often laid upon these combatants by their Eastern owners, and the fight was of a most terrific character. Each Elephant was mounted by his own mahout, who was furnished with a rope netting, to which he clung as the animals met in the deadly shock, in order to prevent himself from being flung off the creature's back. It is a remarkable fact, that the animal never interferes with a human being provided he is mounted upon an Elephant's back, and even the wild "rogue" Elephants do no harm to the men who come to ensnare them.

There are many breeds, or "castes," of the Asiatic Elephant, which are distinguished by certain technical terms.

The AFRICAN ELEPHANT is spread over a very wide range of country, extending from Senegal and Abyssinia to the borders of the Cape Colony. Several conditions are required for its existence, such as water, dense forests, and the absence of human habitations.

Although it is very abundant in the locality which it inhabits, it is not often seen by casual travellers, owing to its great vigilance, and its wonderful power of moving through the tangled forests without noise, and without causing any perceptible agitation of the foliage. In spite of its enormous dimensions, it is one of the most invisible of forest creatures, and a herd of Elephants, of eight or nine feet in height, may stand within a few yards of a hunter without being detected by him, even though he is aware of their presence. The only sure method of ascertaining the presence of Elephants is by listening for one sound which they are continually giving forth, and which they are unable to control. This peculiar noise resembles the bubbling of wine when poured from a bottle, and is caused by the large amount of water which is stored in their interior. This curious sound is emitted at regular intervals, and forms a sure criterion whereby to judge of the direction in which the creatures may be standing.

At the present day the African Elephant is never captured and domesticated, although there seems to be but little reason for such an omission. In the ancient times, this species was trained for the arts of war and peace as regularly as the Asiatic Elephant, and its present immunity from a life of captivity seems to be the result of the fears or laziness of the natives. The only object in possessing the African Elephant is to procure its valuable tusks and teeth, and to afford nourishment to the native tribes. Before the introduction of fire-arms among the Kaffir tribes, the Elephant was hunted by men armed with assagais, or spears, and after being unrelentingly pursued for several successive days, was at last forced to succumb under the multitudes of missiles which penetrated its body. Now, however, the musket-ball, however rude may be the weapon, does great service to the black hunter, and the Elephant is slain in far less time and in greater numbers than under the old system.

When wounded, the African Elephant is a most formidable animal, charging impetuously in the direction of the foe, and crashing through the heavy forest as if the trees were but stubble. In such a case, the best resource of the hunter is in his dogs, which bay round the infuriated animal, and soon distract his attention. The bewilderment

AFRICAN ELEPHANT.—*Loxodonta Africána.*

which the Elephant feels at the attacks of so small an animal as a dog is quite extraordinary. He does not seem to know what he is doing, and at one time will try to kneel on his irritating foes, or will even push down a tree in hopes of crushing them under its branches. This species is not so readily killed by a single ball as is its Asiatic relative, but instances are not wanting where an African Elephant has been slain by a single ball, which entered by the nostrils, and penetrated to the brain. In chasing this animal, the white hunter always prefers a gun with a very wide bore, as the execution which is done depends more upon the weight of the missile than on the accuracy with which it is sent. One great value of the heavy ball, of two or even three ounces in weight, is, that it will break the leg of the animal, and so render him at once helpless. The bone does not always give way at once when struck by so heavy a ball, but is sure to snap after the animal has made a few paces.

The most deadly gun for Elephant shooting seems to be a breech-loader, either double or single, and carrying a ball weighing not less than two ounces. All the apparatus of ramrod and powder-flask is thus rendered needless, and each charge being separately made up into a cartridge, is inserted into the breech, and is ready for use without the least delay. Very accurate shooting is of no great consequence in the pursuit of this giant game, as the hunter can always approach within a few yards of the animal, and deliver his fire from his horse's back, sheering off if the creature endeavours to charge. The bullets employed in Elephant shooting are always hardened with a mixture of one-eighth of tin or solder, and a steel-pointed bullet would probably be the most deadly missile that ever was employed for the purpose. The shell-bullets might also be used with terrible effect.

The Kaffirs are persevering Elephant-hunters, and are wonderfully expert in tracking any individual by the "spoor," or track, which is made by his footsteps. The foot of a male is easily to be distinguished by the roundness of its form, while that of the female is more oval, and the height of the animal is also ascertained by measurement of the foot-marks, twice the circumference of the foot being equal to the height at the shoulder. The mode by which the natives follow a single Elephant through all the multiplied tracks of his companions is very curious. The sole of each Elephant's foot is marked with certain wrinkles, which are never precisely alike in any two individuals, and may be compared to the minute depressions which are found on the human thumb, and which in more primitive times were employed as an expeditious mode of affixing a sign-manual, by being rubbed with ink and impressed upon the document. The black hunter, therefore, taking a piece of soft clay or earth, works it between his hands into a firm and smooth mass, resembling the footmark in shape, and with the point of a thorn traces upon it a chart of the lines which are found on the Elephant's foot. If he should become bewildered amid the multiplicity of footmarks, he has only to refer to his clay chart, and is guided against the possibility of mistaking one individual for another.

The death of a large Elephant is great matter of congratulation among the natives, who rejoice at the abundant supply of food which will fall to their share. Almost every portion of the animal is used by the Kaffirs, whose strong jaws are not to be daunted by the toughest meat, and whose accommodating palates are satisfied with various portions which would be rejected by any civilized being. Indeed, it seems to be a general rule among savages, that every part of an animal which is most repulsive to civilized tastes, is considered by the savage as a luxury, and in many cases thought too good to be spoiled by cooking. The flesh of the Elephant is dried in order to be formed into "biltongue," or jerked meat, and the fat is jealously preserved, being used in the decoration of the person and rubbed copiously over the head and body. Even the skin is of service to the natives, for beneath the hard, leather-like hide, there lies a tough inner skin, which is carefully removed in large sheets, and is made into vessels for the conveyance of water.

Some portions of the Elephant are, however, grateful even to European palates, and the foot, when baked, is really delicious. This part of the animal is cooked by being laid in a hole in the earth, over which a large fire has been suffered to burn itself out, and then covered over with the hot earth. Another fire is then built on the spot, and permitted to burn itself out as before, and when the place is thoroughly cool, the foot is properly cooked. The flesh of the boiled foot is quite soft and gelatinous, something resembling calf's head, and is so tender that it can be scooped away with a spoon. The trunk and the skin around the eye are also enumerated as delicacies, but have been compared by one who has had practical experience, as bearing a close resemblance to shoe-leather both in toughness and evil flavour.

The African Elephant is a most suspicious and wary animal, being very keen of scent and acute of hearing. So sensitive are the animal's olfactory faculties, that it can track a native by the scent of his footsteps, although perhaps it might find a difficulty in following the spoor of a shod and cleanly European. However close the Elephant may be, the pursued hunter is always safe if he can only climb a tree, for the animal never thinks of looking elsewhere than on the ground for its foe, and neither by scent nor vision directs its attention to the trees. While employed in thus trailing their enemies, it writhes the trunk into the most singular contortions, fully justifying the epithet of *anguimanus,* or snake-hand, which has so aptly been applied to that member.

The natives employ many methods of capturing Elephants, the pitfall being the most deadly. Even this insidious snare is often rendered useless by the sagacity of the crafty old leaders of the herds, who precede their little troops to the water, as they advance by night to drink, and carefully beating the ground with their trunks as they proceed, unmask the pitfalls that have been dug in their course. They then tear away the coverings of the pits, and render them harmless. These pitfalls are terrible affairs when an animal gets into them, for a sharp stake is set perpendicularly at the bottom, so that the poor Elephant is transfixed by its own weight, and dies miserably. Each pit is about eight feet long by four in width.

Whenever the Elephants approach the water at night, their advent may be at once known by the commotion that arises among the various animals which have also congregated around the pool for the purpose of slaking their thirst. " If the spring or pool," says Mr. Anderson, in his valuable work, "Lake Ngami," "be of small extent, all the animals present will immediately retire from the water as soon as they are aware of the presence of the Elephants, of whom they appear to have an instinctive dread, and will remain at a respectful distance until the giants have quenched their thirst. Thus, long before I have seen or even heard the Elephants, I have been warned of their approach by the symptoms of uneasiness displayed by such animals as happened to be drinking at the time. The giraffe, for instance, begins to sway his long neck to and fro ; the zebra utters sudden and plaintive cries ; the gnoo glides away with a noiseless step ; and even the ponderous and quarrelsome black rhinoceros, when he has time for reflection, will pull up short in his walk to listen : then turning round, he listens again, and if he feels satisfied that his suspicions are correct, he invariably makes off, giving vent to his fear or ire by one of his vicious and peculiar snorts. Once, it is true, I saw a rhinoceros drinking together with a herd of seven male Elephants ; but then he was of the white species, and, besides, I do not believe that either party knew of each other's proximity."

The ivory of the African Elephant is extremely valuable, and vast quantities are imported annually into this country. The slaughter of an Elephant is therefore a matter of congratulation to the white hunter, who knows that he can obtain a good price for the tusks and teeth of the animal which he has slain. A pair of tusks weighing about a hundred and fifty pounds will fetch nearly forty pounds when sold, so that the produce of a successful chase is extremely valuable. One officer contrived to purchase every step in the army by the sale of the ivory which he had thus obtained. On an average, each pair of tusks, taking the small with the great, will weigh about one hundred and twenty pounds.

There is an ingenious but a very cruel method of procuring ivory, which is employed by the Somali. The hunter contrives to crawl towards the Elephant as it is reposing, and with a single stroke of a very sharp sword nearly severs the principal tendon of the hind leg. At the time, the animal thinks little of the wound, evidently supposing it to be caused by the prick of a thorn. In order to rid himself of the supposed thorn, he stamps violently on the ground, and flings out the wounded limb, until the damaged sinew parts, and the Elephant is rendered incapable of locomotion. The hunters do not trouble themselves about the poor beast, knowing that he must soon die of hunger and thirst, as he cannot stir from the spot on which he was wounded. After a sufficient time has elapsed for putrefaction to have done its work, the hunters return to the spot, and easily draw the tusks from the skull. The tail is cut off, and evermore exhibited as a trophy of victory.

ONE of the links which unite the elephants to the swine and rhinoceros is to be found in the genus Tapírus. The animals which belong to this genus are remarkable for the prolonged upper lip, which is formed into a kind of small proboscis, not unlike that of the elephant, but upon a smaller scale, and devoid of the finger-like appendage at the extremity. Only two species are at present existing, but the fossil remains of many other species have been discovered, which, by the peculiar length of proboscis and general formation, seem to render the transition from the elephant to the swine less abrupt. The body is heavy and powerful, the skin thick and almost devoid of hair, and the tail is almost wanting.

TAPIR.—*Tapirus terrestris.*

The common or American TAPIR, sometimes called the Mbórebi, is a native of tropical America, where it is found in great numbers, inhabiting the densely wooded regions that fringe the banks of rivers. It is a great water-lover, and can swim or dive with perfect ease. Although a large animal, being nearly four feet in height, and very strongly made, it falls a victim to many destroyers, the jaguar being the most terrible of its enemies. It is said that when the jaguar leaps upon the Tapir's back, the affrighted animal rushes through the brushwood in hopes of sweeping away its deadly foe, and if it be fortunate enough to gain the river's bank, will plunge into the water, and force the jaguar, who is no diver, to relinquish his hold. The tough, thick hide, with which the Tapir is covered is of great service in enabling the animal to pursue its headlong course through the forest without suffering injury from the branches. When it runs, it carries its head very low, as does the wild boar under similar circumstances.

In disposition the Tapir is very gentle, and does not attack human beings except when wounded and driven to bay. It then becomes a fierce and determined opponent, and is capable of inflicting severe wounds with its powerful teeth. The hunter's dogs are often dangerously wounded by the teeth of the despairing Tapir. The voice of the Tapir is a curious shrill kind of whistling sound, which is but seldom uttered. The senses of the animal are very acute, and its sight, hearing, and scent appear to be equally sensitive. During the daytime it is seldom seen, preferring to lie quietly hidden in the deep underwood during the hotter hours of the day, and to emerge at night in order to obtain food and meet its companions. The nocturnal journeys which the Tapir will make are of considerable extent, and the animal proceeds straight onwards, heedless of bank or river, surmounting the one and swimming the other with equal ease. The food of the Tapir is generally of a vegetable nature, and consists of young branches and various wild fruits, such as gourds and melons.

KUDA-AYER, OR MALAYAN TAPIR.—*Tapirus Malayánus.*

The colour of the adult Tapir is a uniform brown, but the young is beautifully variegated with yellowish-fawn spots and stripes upon a rich brown-black ground, reminding the observer of the peculiar tinting of the Hood's marmot. The neck is adorned with a short and erect black mane. The Tapir can easily be brought under the subjection of man, and is readily tamed, becoming unpleasantly familiar with those persons whom it knows, and taking all kinds of liberties with them, which would be well enough in a little dog or a kitten, but are quite out of place with an animal as large as a donkey.

THE second species of Tapir is found in Malacca and Sumatra, and is a most conspicuous animal, in consequence of the broad band of white that encircles its body, and which at a little distance gives it the aspect of being muffled up in a white sheet.

The ground colour of the adult Malayan Tapir is a deep sooty-black, contrasting most strongly with the greyish-white of the back and flanks. The young animal is as beautifully variegated as that of the preceding species, being striped and spotted with yellow fawn upon the upper parts of the body, and with white below. There is no mane upon the neck of the Malayan Tapir, and the proboscis is even longer in proportion. In size it rather exceeds the preceding animal. In many of its habits the Malayan animal is exactly similar to the species which inhabits America, but it is said that although the Kuda-Ayer is very fond of the water, it does not attempt to swim, but contents itself with walking on the bed of the stream. Although a sufficiently common animal in its native country, it is but seldom seen, owing to its extremely shy habits, and its custom of concealing itself in the thickest underwood.

The hide of the Tapir is employed by the natives for several useful purposes, but the flesh is dry, tasteless, and not worth the trouble of cooking. The term Kuda-Ayer is a Malayan word, signifying "river-horse," and it is also known by the name of Tennu.

IN the SWINE, the snout is far less elephantine than in the preceding animals, and although capable of considerable mobility, cannot be curled round any object so as to raise it from the ground. Nor, indeed, is such a power needed, as the Swine employ the snout for the purpose of rooting in the earth, and of distinguishing, by its tactile powers, and the delicate sense of smell which is possessed by these animals, those substances which are suitable for its food.

In order to enable this instrument to perform its functions more effectually, it is furnished with a small bone, as is the case with the mole. Their form is heavy and massive, their neck and fore-quarters are very strong, and their heads are wedge-shaped, probably because in a wild state they inhabit dense bushes and thickets, and require this form of head and snout to enable them to pierce the tangled vegetation with ease. A wild boar will charge fearlessly at an apparently impenetrable thicket, and vanish into its interior as if by magic. The tusks, especially in the male, are largely developed, and are terrible weapons of offence, a boar being able to rip up a dog or a man's leg with a single blow of his tusks. When striking with these weapons, the boar does not seem to make any great exertion of strength, but gives a kind of wriggle with his snout as he passes his victim. In India, it is not uncommon for an infuriate wild boar to pursue some unfortunate native, to overtake him as he flies, and putting his snout between the poor man's legs, to cut right and left with an almost imperceptible effort, and to pass on his course, leaving the wounded man helpless on the ground.

There are many species as well as varieties of Swine, which are found in different parts of the earth, the first and most familiar of which is the DOMESTIC HOG of Europe.

This species is spread over the greater portion of the habitable globe, and was in former days common in a wild state even in England, from whence it has only been expelled within a comparatively late period. The chase of the wild boar was a favourite amusement of the upper classes, and the animal was one of those which were protected by the terribly severe forest laws which were then in vogue. The boar was usually slain with the spear, although the net or the arrow were sometimes employed in his destruction. In several continental countries the boar-hunt is still carried on, and by some more legitimate sportsmen is attacked solely with the spear. The chase is then a most exciting one, for the boar is a terrible antagonist, his charge is made with lightning swiftness, and together with his furious eyes and lips dripping with the foam, he is a sufficiently formidable foe to disconcert any one who is not possessed of good nerves and a steady hand. The animal has an awkward habit of swerving suddenly from his course, snapping at the spear-head and breaking it from the shaft. He also, when the hunter is on horseback, will charge at the horse instead of the rider, and rising on his hind legs, in order to give the blow greater force, will lay open the horse's flank and instantly disable it. There are, however, but few sportsmen of the present day who will restrict themselves to the use of the spear in boar-hunting, but employ the rifle in lieu of that weapon, so that the danger and excitement of the sport are almost entirely destroyed.

At the present time the wild Swine have ceased from out of England, in spite of several efforts that have been made to restore the breed by importing specimens from the Continent and turning them into the forests. There are, however, traces of the old wild boars still to be found in the forest pigs of Hampshire, with their high crests, broad shoulders, and thick, bristling manes. These animals are very active, and are much fiercer than the ordinary Swine.

Swine are very accommodating in their appetite, and will devour almost any vegetable or animal substance. Although more of a vegetable than an animal feeder, the Hog, whether wild or domesticated, will pick up any dead animal it may find, and will sometimes kill meat for itself. As a specimen of the carnivorous powers of the Swine, Buffon mentions that in the stomach of a wild boar opened by himself, he found part of the skin of a roebuck, and some feet of birds. Certain pig-keepers take a base advantage of the omnivorous qualities of the Hog, and instead of feeding their animals with such a vegetable diet as will produce a firm and sound flesh, maintain them on the worst kind of garbage, which they obtain at a cheap rate from slaughter-houses, and even force them

WILD BOAR.—*Sus scrofa.*

to eat the offal of their own species. The flesh of such ill-fed animals is always flabby and of ill-savour, and is also injurious to those by whom it is consumed.

In this country, the Hog is used not only for food, but for the sake of the hide, which, when prepared after a peculiar fashion, is found to make the best leather for saddles. The bristles which are so largely used in the manufacture of brushes are almost exclusively imported from the Continent.

Both to the Jews and the Mahometans the Hog is a forbidden article of diet, the latter prohibition being evidently in imitation of the former. In the Mosaical law the Hog is spoken of as an unclean animal that might not be eaten, although for what reason is not easy to ascertain, and the Rabbinical mandates which exercised such a potent sway over the people laid such a stress upon the interdict that they declared the animal itself to be a vile and foul beast, and pronounced a sentence of uncleanness against those who came in contact with a Hog or with anything which it had touched. It must be remarked, that the Egyptians, among whom the Hebrews had so long resided, held similar views of the Hog, and that might be in deference to their prejudices which they had contracted from their former masters. The Hebrews were taught in their law to hold the animal in the same light in which it had been regarded by those to whom they had been accustomed to look with reverence. By some persons it is thought that the flesh of the Hog is harmful to those who reside in hot countries; but even granting this to be the case—a matter which is by no means certain—it affords no clue to the cause why the Hog should have been held as a vile and unclean beast by the polished and learned Egyptians, who depicted so accurately the various animals found in their country, and employed them so largely in their symbolical literature.

In its wild and domesticated state, the Hog is a most prolific animal, producing from eight to twelve pigs twice in each year, when it is in full vigour and in good health. Gilbert White records a sow which, when she died, was the parent of no less than three hundred pigs.

We are rather apt to speak libellously of the Hog, and to ascribe to it qualities which are of our own creation. Although it is a large feeder, it really is not more gluttonous than the cow, the dog, or the sheep, for each of these animals will eat to repletion if furnished with a large amount of food, and will become inordinately fat in consequence of such high feeding.   In its wild state it is never found overloaded with fat, and, as has already been seen, is so active an animal that it can surpass a horse in speed, and is so little burdened with flesh that it can endure throughout a lengthened chase.   Neither is it naturally a dirty creature, for in its native woods it is as clean as any other wild animal.   But when it is confined in a narrow stye, without any possibility of leaving its curtailed premises, it has no choice, but is perforce obliged to live in a constant state of filth.

The Hog is also thought, and very wrongly, to be an especially stupid animal.   It appears stupid for the same reason that it appears to be gluttonous and dirty, merely because no attention has been paid towards developing its intellectual qualities, which have been left to exercise themselves in the narrow confines of the stye and on the daily supply of food.

When, however, its owner chooses to look upon the Hog as a living being, and not merely as a piece of animated pork or bacon, he finds that it is by no means the stupid animal that it has been supposed to be.   "Learned" pigs are familiar to us all, and though the animal does not display any very great amount of literature, it exhibits a capacity of observation and obedience which would hardly have been expected from so maligned an animal.

The senses of the Hog are wonderfully acute, and are capable of being turned to good purport.   So delicate is its sense of smell, that it has been trained to act as a pointer, and in this capacity acted its part so thoroughly, that it would often find birds which the dogs had missed.   "Slut," as this animal was called, was very fond of the sport, and would frequently walk a distance of seven miles in hopes of finding some one who was going out with a gun.   She would point at every kind of game with the curious exception of the hare, which she never seemed to notice.   Although she would willingly back the dogs, they were very jealous of her presence, and refused to do their duty when she happened to be the discoverer of any game, so that she was seldom taken out together with dogs, but was employed as a solitary pointer.   So sensitive was her nose, that she would frequently point a bird at a distance of forty yards, and if it rose and flew away, she would walk to the place from which it had taken wing, and put her nose on the very spot where it had been sitting.   If, however, the bird only ran on, she would slowly follow it up by the scent, and when it came to a stop, she would again halt and point towards it. She was employed in the capacity of pointer for several years, but was at last killed because she had become a dangerous neighbour to the sheep.

The Hog has also been trained to draw a carriage, a team of four Hogs having been driven by a farmer into the market-place of St. Alban's.   After driving once or twice round the market-place, he unharnessed his team, fed them, and in two hours put them again to his chaise, and drove them back to his house, a distance of two or three miles.   Absurd as the idea may seem, the Hog has been trained for the saddle as well as for harness. Another farmer, of Norfolk, laid a heavy wager that he would in one hour ride his boar pig from his own house to Wisbeach, a distance of four miles and a quarter.   He won his wager easily, accomplishing the distance in less than the given time.   The Hog seems to be a good leaper, for a livery-stable keeper, who petted a favourite pig, engaged that he could make his pig leap over a door four feet and a half in height.   In order to induce the animal to make the effort, he placed the door across the entrance to the stye, and laid a bounteous supply of favourite food within the inclosure.   A wild boar has been known to clear a paling nearly nine feet in height, and it is remarkably active in leaping across ravines.

There is a prevalent idea, that whenever the Hog takes to the water he cuts his own throat with the sharp hoofs of his fore-feet.   This, however, is by no means the case, for the animal is an admirable swimmer, and will often take to the water intuitively.   In one of the Moray Islands, three domestic pigs belonging to the same litter swam a distance of five miles ; and it is said that if they had belonged to a wild family, they would have swum to a much greater distance.

The flesh and fat of the Hog is especially valuable on account of its aptitude for taking salt without being rendered hard and indigestible by the process; and the various breeds of domesticated Swine are noted for their adaptation to form pork or bacon in the shortest time and of the best quality. A full account of the various English varieties, together with the mode of breeding them and developing their peculiar characteristics, may be found in many books which are devoted specially to the subject.

The WILD BOAR of India is reckoned by some naturalists to be a separate species, and deserves a few words on account of its superiority in size, strength, and swiftness, to the ordinary European Swine.

This animal is a sad plague to the agricultural population of India, as it makes terrible havoc among the crops, and is especially fond of frequenting the sugar-canes, eating them and chopping them into short lengths, which it forms into hut-like receptacles for its young. The Boar is a most fierce and savage animal, and if driven from the cane-brake, will rush at any man or animal that may be within his reach, and cut them terribly with his sharp tusks. Even the sow can do considerable damage with her teeth, but instead of ripping like her mate, she bites sharply and rapidly. When the animal is fairly roused, and takes to his heels, he puts the mettle of the swiftest and stanchest horse fairly to the test, and even on ground where the horse has all the advantage, he will frequently distance his pursuers, and regain his domicile in the cane-brake. Among the plantations are numbers of old disused wells, the sides of which have fallen in and were never properly filled up. In these wells the wild hog loves to lie, for the mouth of the well is so overgrown with thick verdure that the aperture is scarcely visible even to a person that stands on its brink, while from those who are not aware of its precise locality it is entirely hidden.

The spear is generally employed in Boar-hunting, or "pig-sticking," as the sport is familiarly termed, and is either thrown from the horse's back, or is held like a lance and directed so as to receive the animal's charge. When driven to bay, the Indian Boar is as savage an animal as can be imagined, as with flashing eyes and foaming mouth he dashes first at one and then another of the horsemen, sometimes fairly driving them from the spot, and remaining master of the field.

ONE of the most formidable looking of Swine is the BABYROUSSA of Malacca.

This strange creature is notable for the curious manner in which the tusks are arranged, four of these weapons being seen to project above the snout. The tusks of the lower jaw project upward on each side of the upper, as is the case with the ordinary boar of Europe, but those of the upper jaw are directed in a very strange manner. Their sockets, instead of pointing downwards, are curved upwards, so that the tooth, in filling the curvatures of the socket, passes through a hole in the upper lip, and curls boldly over the face. The curve, as well as the comparative size of these weapons, is extremely variable, and is seldom precisely the same in any two individuals. The upper tusks do not seem to be employed as offensive weapons; indeed, in many instances they would be quite useless for such a purpose, as they are so strongly curved that their points nearly reach the skin of the forehead. The female is devoid of these curious appendages.

From all accounts, the Babyroussa seems to be a very fierce and dangerous animal, being possessed of great strength, and able to inflict terrible wounds with the tusks of the lower jaw. A naval officer who had experienced several encounters with this creature, spoke of it with great respect, and seemed to hold its warlike abilities in some awe. The adult male Babyroussa is considerably larger than the boar of England, and the officer above mentioned told me that he had seen them as large as donkeys. It is a very good swimmer, and will take to the water for its own gratification, swimming considerable distances without any apparent effort.

The skin of the Babyroussa is rather smooth, being sparsely covered with short, bristly hairs. The object of the upper tusks is at present unknown, although certain old writers asserted that the animal was accustomed to suspend himself to branches by means of the appendage. The Babyroussa lives in herds of considerable size, and is found inhabiting the marshy parts of its native land.

BABYROUSSA.—*Babirussa Alfurus.*

The BOSCH VARK, or Bush Hog, of Southern Africa is a very formidable animal in aspect as well as in character, the heavy, lowering look, the projecting tusks, and the callous protuberance on the cheek, giving it a ferocious expression which is no way belied by the savage and sullen temper of the animal. The Bosch Vark inhabits the forests, and is generally found lying in excavations or hollows in the ground, from which it is apt to rush if suddenly disturbed, and to work dire vengeance upon its foe. In colour it is extremely variable, some species being of a uniform dark brown, others of a brown variegated with white, while others are tinged with bright chestnut. The young is richly mottled with yellow and brown. For the following account of the habits of the Bosch Vark I am indebted to Captain Drayson's MS.

"Where the locality is sufficiently retired and wooded to afford shelter to the bush bucks which I have mentioned, we may generally expect to find traces of the Bush Pig. His spoor is like the letter M without the horizontal marks, the extremities of the toes forming two separate points, which is not the case with the antelopes, at least very rarely so, the general impression of their feet being like the letter A with a division down the centre, thus ⋀.

The Bush Pig is about two feet six inches in height and five feet in length, his canine teeth are very large and strong, those in the upper jaw projecting horizontally; those in the lower upwards. He is covered with long bristles, and taking him all in all, he is about as formidable looking an animal, for his size, as can be seen.

The Bosch Varks traverse the forests in herds, and subsist on roots and young shrubs. A large hard-shelled sort of orange, with an interior filled with seeds, grows in great quantities on the flats near the Natal forests; this is a favourite fruit of the wild pigs, and they will come out of the bush of an evening and roam over the plains in search of windfalls from these fruit-trees.

The Kaffir tribes, although they refuse to eat the flesh of the domestic pig, will still feast without compunction on that of its bush brother.

In the bush I always found the Kaffirs disinclined to encounter a herd of these wild Swine, stating as their reason for doing so that the animals were very dangerous; they also said that the wounds given by the tusks of this wild pig would not readily heal.

The Berea bush of Natal was a favourite resort of these wild pigs, but although their spoor could be seen in all directions, the animals themselves were not so frequently encountered.

The Kaffirs are much annoyed by these wild pigs, which force a passage through the imperfectly made fences, and root up the seeds, or destroy the pumpkins in the various gardens. As a defence, the Kaffirs leave nice enticing little openings in different parts of their fences, and the pigs, taking advantage of these ready-made doorways, frequently walk through them, and are then engulfed in a deep pit in which is a pointed stake, and they

BOSCH VARK.—*Choiropótamus Africánus.*

are assagaied with great delight by the expecting Kaffirs, who are on the alert, and who hear the cries of distress from piggy himself.

The tusks are considered great ornaments, and are arranged on a piece of string and worn round the neck."

The VLACKE VARK, or EMGALLA, is even a more formidable animal in its aspect than the bosch vark. The general colour of the Vlacke Vark is a blackish hue upon the crown of the head, the neck, and upper part of the back, and dull brown upon the remainder of the body, except upon the abdomen, where it fades into a greyer hue. The tusks of an adult male are most terrible weapons, projecting eight or nine inches beyond the lips, and with them it has been known to cut a dog nearly in two with a single stroke, or to sever the fleshy parts of a man's thigh. It is a savage and determined opponent, and its charge is greatly to be dreaded. When chased, it presents a most absurd appearance, for it is naturally anxious to learn how much it has gained upon its pursuers, and is yet unable to look round, on account of its short neck and the large excrescence on each side of the face. The animal is therefore obliged to lift its snout perpendicularly in the air so as to look over its own shoulder; and as it always carries its tail stiff and upright when running, it has a most ludicrous aspect.

VLACKE VARK.—*Phacochœrus Æthiopicus.*

This animal is not devoid of sagacity, as was proved by Gordon Cumming: "I selected the old boar for my prey, and immediately separated him from his comrades. After ten miles of sharp galloping, we commenced ascending a considerable acclivity, where I managed to close with him, and succeeded in turning his head towards my camp. He now reduced his pace to a trot and regarded me with a most malicious eye, his mouth a mass of foam. He was entirely in my power, as I had only to spring from my horse and bowl him over. I felt certain of him, but resolved not to shoot as long as his course lay in the direction of my waggon. At length, surprised at the resolute manner in which he held for my camp, I headed him; when, to my astonishment, he did not in the slightest swerve from his course, but trotted along behind my horse like a dog following me. This at once aroused my suspicions, and I felt certain that the cunning old fellow was making for some retreat, so I resolved to dismount and finish him. Just, however, as I had come to this resolution, I suddenly found myself in a labyrinth of enormous holes, the haunt of the ant-bear. In front of one of them the wild boar pulled up, and charging stern foremost into it, disappeared from my disappointed eyes and I saw him no more. I rode home for my men; and returning, we collected grass and bushes, and tried to smoke him out, but without success."

The structure of the teeth in this animal is very curious, and will repay examination. Another species, the HALLUF or HAROJA (*Phacochœrus Æliani*), belongs to the same genus. This animal is sometimes known as the Æthiopian Wild Boar, or the Abyssinian Phacochœre.

AMERICA possesses a representative of the porcine group in the Peccaries, two species of which animals inhabit the Brazils.

The common PECCARY, or TAJAÇU, although it is of no very great dimensions, resembling a small pig in size, is yet as terrible an animal as the Wild Boar of India or the Phacochœre of Africa. Ever fierce and irritable of temper, the Peccary is as formidable an antagonist as can be seen in any land, for it knows no fear, and will attack any foe without hesitation. Fear is a feeling of which the Peccary is ignorant, probably because its intellect is not of a very high order, and it is unable to comprehend danger. Although the Peccary is a very harmless animal to outward view,

PECCARY.—*Dicotyles Tajaçu.*

being only three feet long and weighing fifty or sixty pounds, and its armature consists of some short tusks that are barely seen beyond the lips, yet these little tusks are as fearful weapons as the eight-inch teeth of the vlacke vark, for they are shaped like a lancet, being acutely pointed and double edged, so that they cut like knives and inflict very terrible wounds.

No animal seems to be capable of withstanding the united attacks of the Peccary, even the jaguar being forced to abandon the contest, and to shrink from encountering the circular mass of Peccaries as they stand with angry eyes and gnashing teeth ready to do their worst on the foe. In Webber's Romance of Natural History there is a very amusing account, too long to be quoted in this place, of the sudden consternation that was caused during a bear hunt by the charge of a herd of Peccaries, which came rushing over the very spot where the deadly struggle was being waged, scattering men, dogs, and bear in a common confusion. The singular courage of this animal seems, however, to be based in ignorance, for after a herd of Peccaries have been frequently assailed by the hunter, they appear to learn the power of their adversaries, and instead of charging at their opponents, make the best of their way to some place of concealment.

The usual resting-place of the Peccary is in the hollow of a fallen tree, or in some burrow that has been dug by an armadillo and forsaken by the original inhabitant. The hollow tree, however, is the favourite resort, and into one of these curious habitations a party of Peccaries will retreat, each backing into the aperture as far as he can penetrate the trunk, until the entire hollow is filled with the odd little creatures. The one who last enters becomes the sentinel, and keeps a sharp watch on the neighbourhood. The native hunters take advantage of this curious habit to immolate great numbers of these animals. There are two methods of Peccary killing, one by the gun and the other by the sword and pitchfork.

In the former method the hunter takes up his temporary abode in some concealed spot that commands the entrance of the tree or hole in which the Peccaries are known to sleep. As soon as the sentinel has assumed its post, the hunter takes a careful aim at the forehead, and kills it with a single ball. The wounded animal cautiously leaps from the cover, and its place is immediately taken by its successor. The hunter instantly reloads his rifle, and kills the second Peccary in like manner. In this way he will kill the entire family without giving the alarm. If the slain animal should not leap from the hollow, but fall dead at its post, the carcase is pushed out of the hole by the next in succession, who then assumes the part of sentinel without displaying any alarm. The other method requires the co-operation of two hunters, and is managed by one getting above the mouth of the hole and pinning the foremost Peccary to the ground with a pitchfork, while the other despatches it with a sword.

The food of the common Peccary is of a very varied character, and consists of fruits, seeds, grain, roots, reptiles, small birds and their eggs, and, indeed, of almost anything vegetable or animal which can be swallowed. The flesh of the Peccary is not of much value, as during many parts of the year it is wholly uneatable, on account of an odoriferous gland in the back, which taints the meat to such an extent that it cannot be eaten. The flesh of the male is at all times very unpleasant, but that of the female is in some months tolerably good, and has been compared to that of the hare. At the best, however, it is dry and insipid, as there is no fat or lard to be found in the Peccary. In all cases, the gland must be removed as soon as the animal is dead, for if it be permitted to remain but for a single hour, its effects will be perceptible throughout the entire body.

The common Peccary is not so harmful to the agriculturist as its large relation, and as it destroys such large numbers of reptiles, is probably rather beneficial than otherwise. The colour of the Peccary is a grizzled brown, with the exception of a white stripe that is drawn over the neck, and has earned for the animal the name of the Collared Peccary.

The TAGNICATE, or WHITE-LIPPED PECCARY, is larger than the preceding animal, assembles in larger herds, is fiercer in its disposition, and works more woe to the farmer.

The White-lipped Peccary derives its name from a band of white hairs that crosses the upper jaw, and covers nearly the whole of the lower. The colour of the adult animal is black-brown, flecked with a grey grizzle, but when young it is striped after the manner of the bosch-vark. A slight mane runs along its neck, and its ears are fringed with long and stiff hairs. It is a most mischievous animal, as it makes long marches over the country, ravaging the crops in its progress, and always choosing, with a perversely excellent taste, the best maize and grass. The cry of the Peccary is a sharp shrill grunt. When angry, the Peccary clashes its teeth smartly together, producing a sound which is recognisable at some distance, and is very useful to the hunters, as it serves to give timely notice of the animal's approach.

The generic name, Dicotyles, signifies " double-cupped," and is given to the animal on account of the peculiar open gland upon the back. This species is a good swimmer, and often crosses rivers of its own accord. As, however, it loses all its offensive powers while in the water, the Indians watch the opportunity, and by dashing among the floating animals, kill as many as they choose without any danger.

SEVERAL species of the RHINOCEROS are still inhabitants of the north, and several others have long been extinct, and can only be recognised by means of their fossilized remains. Of the existing species, two or three are found in various parts of Asia and its islands, and the remainder inhabit several portions of Africa. Before examining the separate species, we will glance at some of the characteristics which are common to all the members of this very conspicuous group.

The so-called horn which projects from the nose of the Rhinoceros is a very remarkable structure, and worthy of a brief notice. It is in no way connected with the skull, but is simply a growth from the skin, and may take rank with hairs, spines, or quills, being indeed formed after a similar manner. If a Rhinoceros horn be examined—the species of its owner is quite immaterial—it will be seen to be polished and smooth at the tip, but rough and split into numerous filaments at the base. These filaments, which have a very close resemblance to those which terminate the plates of whale-bone, can be stripped upwards for some length, and if the substance of the horn be cut across, it will be seen to be composed of a vast number of hairy filaments lying side by side, which, when submitted to the microscope, and illuminated by polarized light, glow with all the colours

1.   3 c

of the rainbow, and bear a strong resemblance to transverse sections of actual hair. At the birth of the young animal, the horn is hardly visible, and its full growth is the work of years.

As the horn is employed as a weapon of offence, and is subjected to violent concussions, it is set upon the head in such a manner as to save the brain from the injurious effects which might result from its use in attack or combat. In the first place, the horn has no direct connexion with the skull, as it is simply set upon the skin, and can be removed by passing a sharp knife round its base, and separating it from the hide on which it grows. In the second place, the bones of the face are curiously developed, so as to form an arch with one end free, the horn being placed upon the crown of the bony arch, so as to diminish the force of the concussion in the best imaginable manner. The substance of the horn is very dense, and even when it is quite dry, it possesses very great weight in proportion to its size. In former days, it was supposed to bear an antipathy to poison, and to cause effervescence whenever liquid poison was poured upon it. Goblets were therefore cut from this material, and when gorgeously mounted in gold and precious stones, were employed by Eastern monarchs as a ready means for detecting any attempt to administer a deadly drug.

The skin of the Rhinoceros is of very great thickness and strength, bidding defiance to ordinary bullets, and forcing the hunter to provide himself with balls which have been hardened with tin or solder. The extreme strength of the skin is well known both to the Asiatic and African natives, who manufacture it into shields and set a high value on these weapons of defence.

All the species of Rhinoceros are very tetchy in their temper, and liable to flash out into anger without any provocation whatever. During these fits of rage, they are dangerous neighbours, and are apt to attack any moving object that may be within their reach. In one well-known instance, where a Rhinoceros made a sudden dash upon a number of picketed horses, and killed many of them by the strokes of his horn, the animal had probably been irritated by some unknown cause, and wreaked his vengeance on the nearest victims. During the season of love, the male Rhinoceros is always vicious, and, like the elephant, the buffalo, and other animals in the like condition, will conceal himself in some thicket, and from thence dash out upon any moving object that may approach his retreat.

Sometimes the Rhinoceros will commence a series of most extraordinary antics, and seeming to have a spite towards some particular bush, will rip it with his horn, trample it with his feet, roaring and grunting all the while, and will never cease until he has cut it into shreds and levelled it with the ground. He will also push the point of his horn into the earth, and career along, ploughing up the ground as if a furrow had been cut by some agricultural implement. In such case it seems that the animal is not labouring under a fit of rage, as might be supposed, but is merely exulting in his strength, and giving vent to the exuberance of health by violent physical exertion.

The Rhinoceros is a good aquatic, and will voluntarily swim for considerable distances. It is very fond of haunting the river-banks and wallowing in the mud, so as to case itself with a thick coat of that substance, in order to shield itself from the mosquitoes and other mordant insects which cluster about the tender places, and drive the animal, thick-skinned though it may be, half mad with their constant and painful bites. In Sumatra, a curious result sometimes follows from this habit of mud wallowing, as may be seen from the following extract from the "Journal of the Indian Archipelago." "This animal, which is of solitary habits, is found frequently in marshy places with its whole body immersed in the mud, and part of the head only visible. The Malays call the animal 'Badak-Tapa,' or the recluse Rhinoceros. Towards the close of the rainy season they are said to bury themselves in this manner in different places; and upon the dry weather setting in, and from the powerful effects of a vertical sun, the mud becomes hard and crusted, and the Rhinoceros cannot effect its escape without considerable difficulty and exertion. The Semangs prepare themselves with large quantities of combustible materials with which they quietly approach the animal, who is aroused from his reverie by an immense fire over him, which, being kept well supplied by the Semangs with fresh fuel, soon completes his destruction, and renders him in a fit state to make a meal of."

INDIAN RHINOCEROS.—*Rhinoceros unicornis.*

In every species of Rhinoceros the sight appears to be rather imperfect, the animal being unable to see objects which are exactly in its front. The scent and hearing, however, are very acute, and seem to warn the animal of the approach of danger.

The Asiatic species of Rhinoceros are remarkable for the heavy folds into which the skin is gathered, and which hang massively over the shoulders, throat, flanks, and hind quarters. Upon the abdomen the skin is comparatively soft, and can be pierced by a spear which would be harmlessly repelled from the thick folds of hide upon the upper portions of the body. In the INDIAN RHINOCEROS this weight of hide is especially conspicuous, the skin forming great flaps that can be easily lifted up by the hand. In a tamed state the Rhinoceros is pleased to be caressed on the softer skin under the thick hide, and in the wild state it suffers sadly from the parasitic insects that creep beneath the flaps, and lead the poor animal a miserable life, until they are stifled in the muddy compost with which the Rhinoceros loves to envelop its body. The horn of the Indian species is large in width, but inconsiderable in height, being often scarcely higher than its diameter. Yet with this short, heavy weapon, the animal can do terrible execution, and is said, upon the authority of Captain Williamson, to repel the attack of an adult male Elephant.

The height of this animal when full-grown is rather more than five feet, but the average height seems scarcely to exceed four feet. In colour it is a deep brown-black, tinged with a purple hue, which is most perceptible when the animal has recently left its bath. The colour of the young animal is much paler than that of the mother, and partakes of a pinky hue.

The JAVANESE RHINOCEROS is not so large as its Indian relation, the skin-folds are much less conspicuous, and are arranged in a different manner. The hide, too, is covered with certain angular markings, interspersed with short hairs, and its limbs are proportionately longer and more slender. It is a nocturnal animal, seldom being seen by day, and issuing at night from its place of concealment for the purpose of feeding. Being a large and powerful beast, and happening to be very fond of several cultivated plants, such as the coffee and the pepper vine, it is apt to burst its way into the plantations, and to do considerable damage before it retires to its forest home. It seems to be more gentle and tractable than the common Indian Rhinoceros, and has been trained to wear a saddle, and to be guided by a rider.

The Sumatran species possesses two horns upon its nose, the first being tolerably long and sharp, and the second very thick, short, and pyramidal. The skin-folds are very slight in this animal; the hide is black in colour, rough in texture, and is covered with a thin crop of short bristly hairs. The neck is short and heavy, and the limbs are more clumsy than those of the Indian species. From all accounts it seems to be a very quiet creature, and to be held in no kind of dread, as an adult male has been seen to fly in terror before the attack of one of the native wild dogs. The head of this species is peculiarly long, a characteristic which is observable in the skull as well as in the living animal.

Of African Rhinoceroses four species are clearly ascertained, and it is very probable that others may yet be in existence. Two of the known species are black, and the other two white; the animals differing from each other not only in colour, but in form, dimensions, habits, and disposition. The commonest of the African species is the Borele, Rhinaster, or Little Black Rhinoceros, of Southern Africa; an animal which may be easily distinguished from its relations by the shape of the horns and the upper lip. In the Borele the foremost horn is of considerable length, and bent rather backward, while the second horn is short, conical, and much resembles the weapon of the Indian animal. The head is rather rounded, and the pointed upper lip overlaps the lower, and is capable of considerable extension.

The Borele is a very fierce and dangerous animal, and is more feared by the natives than even the lion. Although so clumsy in shape and aspect, it is really a quick and active creature, darting about with lightning speed, and testing the powers of a good horse to escape from its charge. Like many other wild animals, it becomes furiously savage when wounded, but it will sometimes attack a passenger without the least provocation. On one occasion an angry Rhinoceros came charging down upon a wagon, and struck his horn into the bottom plank with such force as to send the wagon forward for several paces, although it was sticking in deep sand. He then left the wagon, and directed his attack upon the fire, knocking the burning wood in every direction, and upsetting the pot which had been placed on the fire. He then continued his wild career in spite of the attempts of a native who flung his spear at him, but without the least effect, as the iron point bent against the strong hide.

The skin of this animal does not fall in heavy folds, like that of the Asiatic species, but is nevertheless extremely thick and hard, and will resist an ordinary leaden bullet, unless it be fired from a small distance. The skin is employed largely in the manufacture of whips, or jamboks, and is prepared in a rather curious manner. When the hide is removed from the animal it is cut into strips of suitable breadth and laid on the ground. These strips are then hammered for some time in order to condense the substance of the skin, and when they are dry are carefully rounded with a knife and polished with sandpaper. One of these whips will continue serviceable for several years. The horn of the Borele, from its comparatively small dimensions, is not so valuable as that of the other species, but is still employed in the manufacture of drinking-cups and sword-handles. Its value is about half that of ivory.

The food of the Black Rhinoceros, whether the Borele or the keitloa, is composed of roots, which the animal ploughs out of the ground with its horn, and of the young branches and shoots of the wait-a-bit thorn. It is rather remarkable that the black species is poisoned by one of the Euphorbiaceæ, which is eaten with impunity by the two white animals.

When wounded, the Black Rhinoceros is a truly fearful opponent, and it is generally

RHINASTER, OR BORELE.—*Rhinóceros bicornis.*

considered very unsafe to fire at the animal unless the hunter is mounted on a good horse, or provided with an accessible place of refuge. An old experienced hunter said that he would rather face fifty lions than one wounded Borele; but Mr. Oswell, the well-known African sportsman, always preferred to shoot the Rhinoceros on foot. The best place to aim is just behind the shoulder, as if the lungs are wounded the animal very soon dies. There is but little blood externally, as the thick loose skin covers the bullet-hole, and prevents any outward effusion. When mortally wounded the Rhinoceros generally drops on its knees.

It is at all times a rather savage beast, and is apt to quarrel with its own kind. Mr. Andersson mentions a curious battle of which he was an eye-witness, where four of these animals engaged furiously with each other. Two of them he contrived to shoot, and found that one was absolutely unfit for food, being covered with festering wounds which had been received in former encounters. The flesh of this animal is tolerably good, but that of the black species is rather tough, and possesses a bitter and unpleasant flavour, in consequence of the food on which the animal lives. The white species feeds almost exclusively on grass, and its flesh is remarkably good and tender. The Borele is a nocturnal animal, rousing himself from sleep at dark, and proceeding straightway to the nearest pool. Having refreshed himself, he takes long journeys in search of food, and returns to his temporary home soon after sunrise. When sleeping, he lies so still, that he may easily be mistaken for a fragment of dark rock.

As the eyes are set deeply in the head of the Rhinoceros, it is unable to see objects directly in its front if they are at any distance; its sight being hindered by the horns. But the hearing and scent of the creature are marvellously acute, and so wary is the animal, that even when feeding it will constantly halt, raise its ears, snuff the wind, and will not return to its occupation until its fears have been allayed.

KEITLOA, OR SLOAN'S RHINOCEROS.—*Rhinoceros Keitloa.*

The KEITLOA can readily be recognised by the horns, which are of considerable length, and nearly equal to each other in measurement. This is always a morose and ill-tempered animal, and is even more to be dreaded than the borele, on account of its greater size, strength, and length of horn. The upper lip of the Keitloa overlaps the lower even more than that of the borele; the neck is longer in proportion, and the head is not so thickly covered with wrinkles. At its birth the horns of this animal are only indicated by a prominence on the nose, and at the age of two years the horn is hardly more than an inch in length. At six years of age it is nine or ten inches long, and does not reach its full measurement until the lapse of considerable time.

The Keitloa is a terribly dangerous opponent, and its charge is so wonderfully swift, that it can hardly be avoided. One of these animals that had been wounded by Mr. Andersson, charged suddenly upon him, knocked him down, fortunately missing her stroke with her horns, and went fairly over him, leaving him to struggle out from between her hind legs. Scarcely had she passed than she turned, and made a second charge, cutting his leg from the knee to the hip with her horn, and knocking him over with a blow on the shoulder from her fore-feet. She might easily have completed her revenge by killing him on the spot, but she then left him, and plunging into a neighbouring thicket, began to plunge about and snort, permitting her victim to make his escape. In the course of the day the same beast attacked a half-caste boy who was in attendance on Mr. Andersson, and would probably have killed him had she not been intercepted by the hunter, who came to the rescue with his gun. After receiving several bullets, the Rhinoceros fell to the ground, and Mr. Andersson walked up to her, put the muzzle of the rifle to her ear, and was just about to pull the trigger, when she again leaped to her feet. He hastily fired and rushed away, pursued by the infuriated animal, which, however, fell dead just as he threw himself into a bush for safety. The race was such a close one, that as he lay in the bush he could touch the dead Rhinoceros with his rifle, so that another moment would probably have been fatal to him.

THE common WHITE RHINOCEROS (*Rhinoceros Simus*) is considerably larger than the two preceding animals, and together with the kobaoba, or long-horned white Rhinoceros, is remarkable for its square muzzle and elongated head. The foremost horn of this animal is of very considerable length, attaining a measurement of more than three feet when fully grown. The second horn is short and conical, like that of the borele. Fortunately for the human inhabitants of the regions where the White Rhinoceros dwells, its temper is remarkably quiet, and devoid of that restless irritability and sudden access of rage which is so distinguishing a quality of the two black species. Even when wounded it seldom turns upon its antagonist, but contents itself with endeavouring to make its escape. Sometimes, however, probably when it has its young to protect, it will assume the offensive, and is then even more to be dreaded than its black relatives. The following anecdote, which was related by Mr. Oswell, the hero of the tale, to Mr. Andersson, affords an instance of this rare display of combativeness:—

" Once as I was returning from an elephant chase, I observed a huge White Rhinoceros a short distance ahead. I was riding a most excellent hunter—the best and fleetest steed that I ever possessed during my shooting excursions in Africa—at the time ; but it was a rule with me never to pursue a Rhinoceros on horseback, and simply because this animal is so much more easily approached and killed on foot. On this occasion, however, it seemed as if fate had interfered.

Turning to my after-rider, I called out: 'By heaven ! that fellow has got a fine horn ! I will have a shot at him.' With that, I clapped spurs to my horse, who soon brought me alongside the huge beast, and the next instant I lodged a ball in his body, but, as it turned out, not with deadly effect. On receiving my shot, the Rhinoceros, to my great surprise, instead of seeking safety in flight, as is the habit of this generally inoffensive animal, suddenly stopped short, then turned sharply round, and, having eyed me most curiously for a second or two, walked slowly towards me. I never dreamt of danger. Nevertheless, I instinctively turned my horse's head away : but, strange to say, this creature, usually so docile and gentle—which the slightest touch of the reins would be sufficient to guide—now absolutely refused to give me his head. When at last he did so, it was too late ; for, notwithstanding the Rhinoceros had only been walking, the distance between us was so inconsiderable, that by this time I clearly saw contact was unavoidable. Indeed, in another moment I observed the brute bend low his head, and, with a thrust upwards, strike his horn into the ribs of the horse with such force as to penetrate to the very saddle on the opposite side, where I felt its sharp point against my leg.

The violence of the blow was so tremendous as to cause the horse to make a complete somersault in the air, coming heavily down on its back. With regard to myself, I was, as a matter of course, violently precipitated to the ground. Whilst thus prostrated, I actually saw the horn of the infuriated beast alongside of me ; but, seemingly satisfied with his revenge, without attempting to do farther mischief, he started off at a canter from the scene of action. My after-rider having by this time come up, I rushed upon him, and almost pulling him off his horse, leapt into the saddle ; and, without a hat, and my face streaming with blood, was quickly in pursuit of the retreating beast, which I soon had the satisfaction to see stretched lifeless at my feet."

THE flesh of the MUCHUCO, or MONOOHOO, as the White Rhinoceros is called by the natives, is apt to be rather tough, but is of good flavour. The best portions are those which are cut from the upper part of the shoulder and from the ribs, where the fat and the lean parts are regularly striped to the depth of two inches. If a large portion of the meat is to be cooked at one time, the flesh is generally baked in the cavity of a forsaken ant-hill, which is converted into an extempore oven for the occasion ; but if a single hunter should need only to assuage his own hunger, he cuts a series of slices from the ribs, and dresses them at his fire. The hide of the Monoohoo is enormously thick, and gives a novice no little trouble to get it from the body, as it is as hard as a board, and nearly as stiff. An adept, however, will skin the animal as quickly and easily as if it were a sheep.

The KOBAOBA, or Long-horned White Rhinoceros (*Rhinoceros Oswellii*) is much rarer than either of the preceding species, and is found far in the interior, mostly to the east of the Limpopo River. The peculiar manner in which this species carries its horns, makes it a very conspicuous animal. In all the other species, the horns are curved, and incline rather backward; but in the Kobaoba, the foremost horn is nearly straight, and projects forward, so that when the animal is running, the tip of the horn nearly touches the ground. Indeed, the extremity of an adult Kobaoba's horn is generally rubbed down on one side, owing to the frequency with which it has come in contact with the earth. The head of this and the preceding species is always carried very low, forming a singular contrast to the saucy and independent manner in which the borele carries his head.

The long horn of the Kobaoba sometimes exceeds four feet in length, and as it is almost straight, is most valuable for many purposes. The best, toughest, and straightest ramrods are manufactured from this horn, and I have seen one of these ramrods that was almost four feet long, even after being shaped and trimmed, so that the horn from which it was cut must have been still longer. The mother Kobaoba employs this horn for a very curious purpose, as was seen by Cumming. Whenever the mother and her young are abroad, the calf always takes the lead, and in this instance she guided her little one by pressing it against the calf's side. The horn is also used by the Kaffirs to make "knob-herries," or knob-headed sticks, which they can employ as clubs in hand-to-hand combat, or can throw with wonderful effect. A party of Kaffirs will often go out in chase of birds, armed with nothing but these knob-herries, which they will hurl with such force and precision that they generally return home loaded with game.

The four African species of Rhinoceros are not at all prolific animals, producing only one young one at a time, and, as far as is known, a considerable interval occurs between each birth. It is not a gregarious, neither does it appear to be a monogamous, animal. It seems, however, to find some gratification in the presence of others of its own species, and may be seen in little assemblies of eight or ten in number. These assemblies, however, cannot be termed flocks or herds, as their members are not under the command of a single leader, nor bound together by any common tie, and, when alarmed, each individual makes his escape as he best can. The skin is comparatively smooth, and devoid of hair, so that the animal bears some resemblance to an overgrown pig.

ONE of the most curious little animals in existence is the HYRAX, interesting not so much from its imposing external appearance, as for its importance in filling up a link in the chain of creation.

About as large as a tolerably sized rabbit, covered with thick, soft fur, inhabiting holes in the banks, possessing incisor-like teeth, and, in fine, being a very rabbit in habits, manners, and appearance, it was long classed among the rodents, and placed among the rabbits and hares. It has, however, been discovered in later years, that this little rabbit-like animal is no rodent at all, but is of one the pachydermata, and that it forms a natural transition from the rhinoceros to the hippopotamus. On a close examination of the teeth, they are seen to be wonderfully like those of the hippopotamus, their edges being bevelled off in a similar manner, and therefore bearing some resemblance to the chisel-edged incisors of the rodents. There are several species of Hyrax, one of which inhabits Northern Africa and Syria, while the other two are found in Abyssinia and South Africa.

The South African Hyrax is termed by the colonists KLIP DAS, or ROCK RABBIT, and is found in considerable plenty among the mountainous districts of its native land, being especially common on the sides of the Table mountain. It is largely eaten by the natives, who succeed in killing it in spite of its extreme wariness and activity. Among the crevices and fissures in the rock the Hyrax takes up its abode, and may often be seen sitting in the warm rays of the sun, or feeding with apparent carelessness on the aromatic herbage of the mountain side. It is, however, perfectly secure, in spite of its apparent negligence, for a sentinel is always on guard, ready to warn his companions by a peculiar shrill cry of the approach of danger. Sometimes the Hyrax is seen at a considerable height, but is often observed near the sea-shore, seated on rocks which are barely above high-water mark.

HYRAX, OR KLIP DAS.—*Hyrax Capensis.*

Besides mankind, the Hyrax has many foes, such as the birds of prey and carnivorous quadrupeds, and is destroyed in considerable numbers. The fore-feet of this animal are apparently furnished with claws like those of the rabbit, but on a closer inspection, the supposed claws are seen to be veritable hoofs, black in colour, and very similar to those of the rhinoceros in form. The Hyrax is an agile little creature, and can climb a rugged tree-trunk with great ease. It is rather hot in its temper, and if irritated, becomes highly excited, and moves its teeth and feet with remarkable activity and force.

The SYRIAN HYRAX is the animal which is mentioned under the name of "coney" in the Old Testament, and is found inhabiting the clefts and caverns of rocks. In its habits and general appearance it is very similar to the Cape Hyrax, and needs no farther description. Although it will bite fiercely when first captured, it is sufficiently docile in disposition, and soon learns to obey its keeper, towards whom it displays an affectionate disposition if it be rightly treated. The colour of both species is dark brown, but the Syrian animal can be distinguished from the Cape Hyrax by the presence of a great number of very long black hairs, which are thickly scattered over its body, and penetrate through the shorter fur. Its native name is Ashkoko.

THE last on the list of the pachydermatous animals is the well-known HIPPOPOTAMUS, or RIVER HORSE.

This enormous quadruped is a native of various parts of Africa, and is always found either in water or in its near vicinity. In absolute height it is not very remarkable, as its legs are extremely short, but the actual bulk of its body is very great indeed. The average height of a full-grown Hippopotamus is about five feet. Its naked skin is dark brown, curiously marked with innumerable lines like those on "crackle" china or old oil-paintings, and is also dappled with a number of sooty black spots, which cannot be seen except on a close inspection. A vast number of pores penetrate the skin, and exude a thick, oily liquid, which effectually seems to protect the animal from the injurious effects of the water in which it is so constantly immersed. I once spoiled a pair of gloves entirely by patting the male animal at present in the Zoological Gardens. The mouth is enormous, and its size is greatly increased by the odd manner in which the jaw is set in the head.

Within the mouth is an array of white, gleaming tusks, which have a terrific appearance, but are solely intended for cutting grass and other vegetable substances, and are seldom employed as weapons of offence, except when the animal is wounded or otherwise irritated. The incisor teeth of the lower jaw lie almost horizontally, with their points directed forwards, and are said to be employed as crow-bars in tearing up the various aquatic plants on which the animal feeds. The canines are very large and curved, and are worn obliquely, in a manner very similar to the rodent type of teeth. Their shape is a bold curve, forming nearly the half of a circle, and their surface is deeply channeled and ridged on the outer line of the curve, and smoother on the face. The entire tooth, when it has been removed from the animal and thoroughly dried, is covered with a series of fine, superficial cracks, which intersect each other diagonally with much regularity, being a veritable example of nature's "cross-hatching."

The tooth is very solid in its substance and close in its grain, and as it retains its colour under very trying circumstances, is admirably adapted for the manufacture of artificial teeth. Throughout the greater part of its length it is quite solid, but bears a conical hollow about three or four inches deep at the extremity which enters the socket. The extreme whiteness of the ivory obtained from the Hippopotamus' teeth renders it peculiarly valuable for the delicate scales of various philosophical instruments, and its natural curve adapts it admirably for the verniers of ship sextants. The weight of a large tooth is from five to eight pounds, and the value of the ivory is from twenty to twenty-five shillings per pound.

With these apparently combined teeth the Hippopotamus can cut the grass as neatly as if it were mown with a scythe, and is able to sever, as if with shears, a tolerably stout and thick stem.

Possessed of an enormous appetite, having a stomach that is capable of containing five or six bushels of nutriment, and furnished with such powerful instruments, the Hippopotamus is a terrible nuisance to the owners of cultivated lands that happen to be near the river in which the animal has taken up his abode. During the day it is comfortably asleep in its chosen hiding-place, but as soon as the shades of night deepen, the Hippopotamus issues from its den, and treading its way into the cultivated lands, makes sad devastation among the growing crops. Were the mischief to be confined to the amount which is eaten by the voracious brute, it would still be bad enough, but the worst of the matter is, that the Hippopotamus damages more than it eats by the clumsy manner of its progress. The body is so large and heavy, and the legs are so short, that the animal is forced to make a double track as he walks, and in the grass-grown plain can be readily traced by the peculiar character of the track. It may therefore be easily imagined that when a number of these hungry, awkward, waddling, splay-footed beasts come blundering among the standing crops, trampling and devouring indiscriminately, they will do no slight damage before they think fit to retire.

The aggrieved cultivators endeavour to protect their grounds and at the same time to make the depredators pay for the damage which they have done, by digging a number of pitfalls across the Hippopotamus paths, and furnishing each pit with a sharp stake in the centre.

When an animal falls into such a trap, the rejoicings are great, for not only is the ivory of great commercial value, but the flesh is very good eating, and the hide is useful for the manufacture of whips and other instruments. The fat of the Hippopotamus, called by the colonists "Zee-Koe speck," or Sea-cow bacon, is held in very high estimation, as is the tongue and the jelly which is extracted from the feet. The hide is so thick that it must be dragged from the creature's body in slips, like so many planks, and is an inch and a half in thickness on the back, and three quarters of an inch on the other portions of the body. Yet, in spite of its enormous thickness and its tough quality, it is quite pliable when seen on the living beast, and accommodates itself easily to all his movements.

The Hippopotamus is, as the import of its name, River Horse, implies, most aquatic in its habits. It generally prefers fresh water, but it is not at all averse to the sea, and will sometimes prefer salt water to fresh. It is an admirable swimmer and diver, and is able to remain below the surface for a very considerable length of time. In common with the elephant, it possesses the power of sinking at will, which is the more extraordinary when

HIPPOPOTAMI AT HOME.

the huge size of the animal is taken into consideration. Perhaps it may be enabled to contract itself by an exertion of the muscles whenever it desires to sink, and to return to its former dimensions when it wishes to return to the surface. It mostly affects the stillest reaches of the river, as it is less exposed to the current, and not so liable to be swept down the stream while asleep. The young Hippopotamus is not able to bear submersion so long as its parent, and is therefore carefully brought to the surface at short intervals for the purpose of breathing. During the first few months of the little animal's

HIPPOPOTAMUS, OR ZEEKOE.—*Hippopótamus amphibius.*

life, it takes its stand on its mother's neck, and is borne by her above or through the water as experience may dictate or necessity require.

There are various modes of hunting this mischievous but valuable animal, each of which is in vogue in its own particular region. The pitfalls above mentioned are universal throughout the whole Hippopotamus country, and lure many an animal to its destruction without needing any care or superintendence on the part of the men who set the snare. There is also the "down-fall," a trap which consists of a log of wood, weighted heavily at one end, to which extremity is loosely fixed a spear-head well treated with poison. This terrible log is suspended over some Hippopotamus path, and is kept in its place by a slight cord which crosses the path and is connected with a catch or trigger. As soon as the animal presses the cord, the catch is liberated, and down comes the armed log, striking the poisoned spear deep into the poor beast's back, and speedily killing it by the poison, if not from the immediate effects of the wound.

The white hunter of course employs his rifle and finds that the huge animal affords no easy mark, as unless it is hit in a mortal spot it dives below the surface and makes good its escape. Mortal spots, moreover, are not easy to find, or when found, to hit; for the animal soon gets cunning after it has been alarmed, and remains deeply immersed in the water as long as it is able, and when it at last comes to the surface to breathe, it only just pushes its nostrils above the surface, takes in the required amount of air, and sinks back again to the river bed. Moreover, it will often be so extremely wary, that it will not protrude even its mouth in the open water, and looks out for some reeds or floating substances which may cover its movements while breathing. As a general rule, it is found that the most deadly wound that can be given to a Hippopotamus is on the nose, for the animal is then unable to remain below the surface, and consequently presents an easy mark to the hunter. A heavy ball just below the shoulder always gives a mortal wound, and in default of such a mark being presented, the eye or the ear is a good place to aim at.

The most exciting manner of hunting the Hippopotamus is by fairly chasing and harpooning it, as if it were a whale or a walrus. This mode of sport is described very vividly by Mr. Andersson.

The harpoon is a very ingenious instrument, being composed of two portions, a shaft measuring three or four inches in thickness and ten or twelve feet in length, and a barbed iron point, which fits loosely into a socket in the head of the shaft, and is connected with it by means of a rope composed of a number of separate strands. This peculiar rope is employed to prevent the animal from severing it, which he would soon manage were it to be composed of a single strand. To the other end of the shaft a strong line is fastened, and to the other end of the line a float or buoy is attached. As this composite harpoon is very weighty it is not thrown at the animal, but is urged by the force of the harpooner's arm. The manner of employing it shall be told in Mr. Andersson's own words :—

" As soon as the position of the Hippopotami is ascertained, one or more of the most skilful and intrepid of the hunters stand prepared with the harpoons; whilst the rest make ready to launch the canoes, should the attack prove successful. The bustle and noise caused by these preparations gradually subside. Conversation is carried on in a whisper, and every one is on the *qui-vive*. The snorting and plunging become every moment more distinct; but a bend in the stream still hides the animals from view. The angle being passed, several dark objects are seen floating listlessly on the water, looking more like the crests of sunken rocks than living creatures. Ever and anon, one or other of the shapeless masses is submerged, but soon again makes its appearance on the surface. On, on, glides the raft with its sable crew, who are now worked up to the highest state of excitement. At last, the raft is in the midst of the herd, who appear quite unconscious of danger. Presently one of the animals is in immediate contact with the raft. Now is the critical moment. The foremost harpooner raises himself to his full height, to give the greater force to the blow, and the next instant the fatal iron descends with unerring accuracy in the body of the Hippopotamus.

The wounded animal plunges violently, and dives to the bottom; but all his efforts to escape are unavailing. The line or the shaft of the harpoon may break; but the cruel barb once imbedded in the flesh, the weapon (owing to the toughness and thickness of the beast's hide) cannot be withdrawn.

As soon as the Hippopotamus is struck, one or more of the men launch a canoe from off the raft, and hasten to the shore with the harpoon-line, and take a round turn with it about a tree, or bunch of reeds, so that the animal may either be ' brought up' at once, or, should there be too great a strain on the line, 'played' (to liken small things to great) in the same manner as the salmon by the fishermen. But if time should not admit of the line being passed round a tree, or the like, both line and ' buoy' are thrown into the water, and the animal goes wherever he chooses.

The rest of the canoes are now all launched from off the raft, and chase is given to the poor brute, who, so soon as he comes to the surface to breathe, is saluted with a shower of light javelins. Again he descends, his track deeply crimsoned with gore. Presently—and perhaps at some little distance—he once more appears on the surface, when, as before, missiles of all kinds are hurled at his devoted head.

When thus beset, the infuriated beast not unfrequently turns upon his assailants, and either with his formidable tusks, or with a blow from his enormous head, staves in or capsizes the canoes. At times, indeed, not satisfied with wreaking his vengeance on the craft, he will attack one or other of the crew, and with a single grasp of his horrid jaws either terribly mutilates the poor fellow, or, it may be, cuts his body fairly in two.

The chase often lasts a considerable time. So long as the line and the harpoon hold, the animal cannot escape, because the ' buoy' always marks his whereabout. At length, from loss of blood or exhaustion, Behemoth succumbs to his pursuers."

The Hippopotamus is a gregarious animal, collecting in herds of twenty or thirty in number, and making the air resound with their resonant snorts. The snort of this creature is a most extraordinary sound, and one that is well calculated to disturb the nerves of sensitive persons, especially if heard unexpectedly. The animals at the Zoological Gardens make the very roof ring with the strange unearthly sounds which they emit. In their native state it is very difficult to ascertain even approximately the number of a herd, as the animals are continually diving and rising, and never appear simultaneously above the surface of the water.

The creature is generally a harmless one, and need not be much dreaded. Sometimes, however, it becomes angry if molested in its watery home, and will then make a violent attack upon the object that has excited its anger. One of these animals, whose calf had been speared on the previous day, made at the boat in which Dr. Livingstone was sitting, and drove her head against it with such force that she lifted the forepart of the boat completely out of the water, capsized one of the black oarsmen fairly into the river, and forced the whole crew to jump ashore.

Although in its native river the female Hippopotamus is a most kind and affectionate mother, the tame animal does not display such excellent qualities. The female Hippopotamus in the Jardin des Plantes in Paris has twice been a mother, and twice has killed her offspring. On the last occasion she seemed to have been seized with a sudden fit of anger, for the marks of her teeth were only too plain on the poor little beast when its dead body was discovered, and her tusks had penetrated into its lungs. On the first occasion she killed it from sheer awkwardness ; and after carrying it about on her neck in the proper manner, she bruised it so severely in her clumsy efforts to teach her offspring the proper mode of getting out of the bath, that it never recovered from the hurts which it received.

The Hippopotamus has for years been extinct in Europe, but the fossil remains of the animal are found abundantly in the London clay, showing that in some remote age the Hippopotamus must have traversed the plains of England, and wallowed in its rivers. There is another species of Hippopotamus, which is smaller than that which has just been described, and is termed Hippopotamus Liberiensis. It is a native of Western Africa, and is remarkable for only having two incisors in the lower jaw.

DASÝPIDÆ.

THIS small but important family includes the Manis, the Armadillo, the Ant-eater, and the Platypus, or Duck-bill.

The Phatagin is one of the numerous species that compose the strange genus of Manis. All these animals are covered with a series of horny plates, sharp pointed and keen edged, that lie with their points directed towards the tail, and overlap each other like the tiles upon the roof of a house ; being the natural prototype of the metal scale-armour that was prevalent in the days of chivalry, and of the horn-scale bucklers that have been employed both in ancient and modern times. This defence of scales is not, however, entirely of a negative character, like the shell of the tortoise, but can be converted at will into a powerful weapon of offence towards all who come too hastily in contact with it. When the Manis is pursued, and is unable to escape, it rolls itself into a ball, after the manner of the hedgehog, so that the sharp-edged and acutely-pointed scales stand boldly outward, and can inflict very unpleasant wounds on the hand of man or the mouth of predaceous beast. The head is the most vulnerable part of the Manis, but as it always takes care to hide its head within the curve of the body, it has little fears on that score.

The fore-claws of the Phatagin are very large, and are employed for the purpose of tearing down the nests of the termite, or white-ant, as it is more popularly called, so as to enable it to feed upon the inmates, as they run about in confusion at the destruction of their premises. Ants, termites, and various insects are the favourite food of the Phatagin, which sweeps them up by means of its long and extensible tongue, caring nothing for their

PHATAGIN.—*Manis tetradactyla.*

formidable jaws, which are powerful enough to drive a human being almost distracted with pain. The claws are not only employed in destroying the nest of the termite, but in digging burrows for its own residence, a task for which they are well adapted by reason of their great size and strength, and the vigour of the limbs to which they are attached. As the limbs are short, and the claws very long, the pace of the Phatagin is very slow, and its tardiness is increased by the fact that the claws of the fore-feet are folded upon a thick, fleshy pad, and are therefore not at all adapted for locomotion.

The Phatagin is a native of Western Africa, and is of considerable dimensions, reaching five feet in average length, of which the tail occupies three feet. From the great length of the tail, it is sometimes called the LONG-TAILED MANIS.

The BAJJERKEIT, or SHORT-TAILED MANIS, is a native of various parts of India, and is also found in Ceylon. Of this species Sir Emerson Tennent gives the following short account. "Of the Edentates, the only example in Ceylon is the scaly ant-eater, called by the Singalese, Caballaya, but usually known by its Malay name of Pengolin, a word indicative of its faculty of 'rolling itself up' into a compact ball, by bending its head towards its stomach, arching its back into a circle, and securing all by a powerful hold of its mail-covered tail. When at liberty, they burrow in the dry ground to a depth of seven or eight feet, where they reside in pairs, and produce annually two or three young.

Of two specimens which I kept alive at different times, one from the vicinity of Kandy, about two feet in length, was a gentle and affectionate creature, which after wandering over the house in search of ants, would attract attention to its wants by climbing up my knee, laying hold of my leg by its prehensile tail. The other, more than double that length, was caught in the jungle near Chilaw, and brought to me in Colombo. I had always understood that the Pengolin was unable to climb trees, but the one last mentioned frequently ascended a tree in my garden in search of ants, and this it effected by means of its hooked feet, aided by an oblique grasp of the tail. The ants it seized by extending its round and glutinous tongue along their tracks. Generally speaking, they were quiet during the day, and grew restless as evening and night approached."

THE manis affords a curious example of scale-armour formed by nature, and a still more singular instance of natural plate-armour is found in the following little group of animals.

The ARMADILLOS are inhabitants of Central and Southern America, and are tolerably common throughout the whole of the land in which they live. The general structure of the armour is similar in all the species, and consists of three large plates of horny covering; one being placed on the head, another on the shoulders, and the third on the hind quarters. These plates are connected by a series of bony rings, variable in number, overlapping each other, and permitting the animal to move freely. Each plate and band is

BAJJERKEIT.—*Manis pentadáctyla.*

composed of a number of small plates, joined together, and forming patterns which differ in the various species. The whole of the animal, even to the long and tapering tail, is covered with these horny scales, with the exception of the upper part of the legs, which are concealed under the armour of the body, and need no other protection. At and soon after birth, the infant Armadillo is quite soft, like parchment, but the skin is marked in a similar manner to that of the adult animal, excepting that the hairs that protrude between the shelly plates are more numerous.

The common ARMADILLO, or POYOU, is about twenty inches in total length, the tail occupying some six or seven inches. It is very common in Paraguay, but is not easily captured, owing to its remarkable agility, perseverance, and wariness. Encumbered as it appears to be with its load of plate-armour, it runs with such speed that it can hardly be overtaken by a quick-footed man, and if it should contrive to reach its burrow, it can never be got out except by dint of hard work. Its hearing is exquisitely acute, and as during the daytime the creature never ventures very far from its home, it readily evades the attacks of every foe excepting man.

The natives, to whom time is of little value, employ a long but a sure process of obtaining the Armadillo after it has taken refuge in its home. In order to ascertain whether the animal is at home, they push a stick into the hole, and if a quantity of mosquitoes come buzzing out, it is a sure sign that the tenant is within. It seems very strange that the mosquitoes should attach themselves to an animal so well defended against their attacks, but such is nevertheless the case. Having ascertained the presence of the Armadillo, they push a stick into the hole, and sink a pit so as to catch the end of the stick. The stick is then pushed still farther, and another pit sunk, and so on, until the Armadillo is fairly captured.

The food of the Armadillo is nearly as varied as that of the swine, for there are few eatable substances, whether vegetable or animal, which the Armadillo will not devour, provided they are not too hard for its little teeth. Various roots, potatoes, and maize are among its articles of vegetable diet, and it also will eat eggs, worms, insects, and small reptiles of every description. Wherever wild cattle are slain, the Armadillo is sure to make its appearance in a short time, for the purpose of devouring the offal which the hunter leaves on the ground. It is not at all particular in taste, and devours the half-putrid remains with great eagerness, becoming quite fat upon the revolting diet.

ARMADILLO.—*Dásypus sexcinctus.*

As the Armadillo is a nocturnal animal its eyes are more fitted for the dark than for the bright glare of sunlight, which dazzles the creature, and sadly bewilders it. If it should be detected on the surface of the ground, and its retreat intercepted before it can regain its hole, the Armadillo rolls itself up as it best can, and tucking its head under the chest, draws in its legs and awaits the result. Even when taken in hand it is not without a last resource, for it kicks so violently with its powerful legs that it can inflict severe lacerations with the digging claws. The legs are wonderfully powerful in comparison with the dimensions of the animal. I have seen an Armadillo run about the ground with perfect ease, although it was carrying on its back three monkeys who had chosen to take their seats upon its mail-clad person. The Armadillo swims well, but does not enter the water from choice.

In spite of the unpleasant diet on which the animal feeds, its flesh is eaten by the natives, and is held by them in some estimation. It is, however, very rank and strong in flavour, and to European palates is rather disagreeable. The young of this animal are from six to eight in number.

ANOTHER curious species of Armadillo is the APARA, or MATACO (*Tatúsia tricincta*), which is often found on the Pampas. It is remarkable for the solid manner in which it is covered by its armour, there being only three bands in the centre of the body, the remainder of the creature being sheltered under the horny plates. When attacked, it can draw itself into a perfect ball, which is impervious to the teeth of predaceous animals, for it is too large to be taken into the mouth and cracked, and is so hard and smooth that the teeth glide harmlessly from its polished surface. The tail is very short, and, with the head, can be completely enveloped in the shell. On account of its shape when rolled together the Spaniards call it the "Bolita," or little ball.

The claws of this animal are feeble, and its legs weak, so that it is unable to burrow in the ground, and depends for defence totally on its coat of mail, which is the more required, as it is a diurnal animal.

The PEBA, or TATOUHOU (*Tatúsia septemcinctus*), is a native of Guinea, Brazil, and Paraguay, and is larger than either of the preceding species, being about thirty inches in total length, the slender and tapering tail being fourteen or fifteen inches long. Its colour is a very dark brown-black, from which circumstance it is sometimes called the Black

1.   3 D

Tatu. It is found in the open country, and is a good burrower. The natives seek it on account of the flesh, which is tender and well flavoured. In Messrs. Audubon and Buchanan's well-known work, is the following account of the Peba: "The Armadillo is not a fighting character, but, on the contrary, is more peaceable than even the opossum which will at times bite in a sly and treacherous manner quite severely.

A friend of ours, who formerly resided in South America, had a pet Armadillo in his bed-chamber, where it generally remained quiet during the day, but in the dark hours was active and playful. One night after he had gone to bed, the Armadillo began dragging about the chairs and some boxes that were placed round the room, and continued so busily engaged at this occupation that our friend could not sleep. He at length arose and struck a light, when, to his surprise, he found that boxes which he had supposed too heavy for such an animal to stir, had been moved and placed together, so as to form a sort of den or hiding-place in a corner, into which the animal retreated with great apparent satisfaction, and from whence it could only be drawn out after a hard struggle, and the receipt of some severe strokes from its claws.

But, in general, the Armadillo does not evince any disposition to resent an attack, and, in fact, one of them, when teased by a pet parrot, struck out with its claws only till pressed by the bird, when it drew in its head and feet, and, secure in its tough shell, yielded, without seeming to care much about it, to its noisy and mischievous tormentor, until the parrot left it to seek some less apathetic and more vulnerable object."

THE little PICHEY ARMADILLO (*Tatúsia minúta*) is only fourteen inches in length, the tail being four inches long. Like many of the African antelopes, it appears to be almost independent of water, and can live for months together without needing to drink. The food consists of various insects, small reptiles, and several kinds of roots, from the latter of which articles it hardly obtains the needful supply of moisture. It is a very active and rapid burrower, sinking below the ground with such celerity, that if a man on horse-back sees a Pichey scrambling over the ground, and wishes to secure it, he can hardly leap from his steed and stoop to take it up, before it has burrowed out of his reach. It also endeavours to escape observation by crouching closely to the ground, as if it were a stony pebble or lump of earth. Another example of the Armadillos is the TATOUAY (*Xenúrus unicinctus*). This animal is mostly remarkable for the undefended state of its tail, which is devoid of the bony rings that encircle the same member in the other Armadillos, and is only supplied with a coating of brown hair. For about three inches of the extremity the under side of the tail is not even furnished with hair, but is quite naked, with the exception of a few rounded scales.

THE last and largest of these animals is the TATOU, or GIANT ARMADILLO (*Priodonta gigas*).

This creature measures more than four feet six inches in length, the head and body being rather more than three feet long. It is as good a burrower as its relatives, and is so keen in its scent after the food which it loves, that the inhabitants of the same country are forced to line the graves of their departed friends with boards, in order to prevent the Tatou from exhuming and devouring them. The teeth are very remarkable, there being from sixteen to eighteen small molars on each side of the jaws. The tail is about seventeen inches long, and tapers gradually to a point from the base, at which spot it is nearly ten inches in circumference. This member is covered with regularly graduating horny rings, and when dried and hollowed, is used as a trumpet by the Botocudos. The Tatou is found in Brazil and Surinam.

NEARLY related to the armadillos is the remarkable little animal called the PICHICIAGO (*Chlamydóphorus truncatus*), a native of Chili, which looks like a mixture of the mole and the armadillo.

The top of the head, the back, and the hind quarters of the Pichiciago are covered with a shelly plate, which runs unbroken to the haunches, over which it dips suddenly, looking as if the creature had been chopped short by the blow of a hatchet, and a piece of shell

AARD VARK —*Orycléropus Capensis.*

stuck on the cut extremity. The remainder of the body is covered with long silken hair, very like that of the mole in its soft texture. It is a very little creature, scarcely surpassing the common English mole in dimensions, and living, like that animal, almost entirely below the surface of the earth. Its feet are formed for burrowing, and are most powerful instruments for that purpose, though they are not well fitted for rapid progress over the ground.

Its food consists, as far as is known, of worms, and other subterranean creatures, in addition to those which it may catch in its nocturnal expeditions into the open air. As is the case with the mole, and other subterranean animals, the eyes are of minute dimensions, and are hidden under the soft and profuse fur of the face.

The ANT-EATERS, as their name imports, feed very largely on ants, as well as on termites and various other insects, their long flexible tongue acting as a hand for the purpose of conveying food into the mouth. The tongue of the Ant-eater, when protruded to its fullest extent, bears some resemblance to a great red earth-worm, and as it is employed in its food-collecting task, it coils and twists about as if it possessed a separate vitality of its own.

The AARD VARK, or Earth-hog, is a native of Southern Africa, and is a very curious animal. The skin of the Aard Vark is not protected by scales or plates like those of the manis and the armadillo, but rather thinly covered with coarse bristly hair. Its length is about five feet, the tail being twenty inches long, and it is a very powerful creature, especially in the fore-limbs, which are adapted for digging, and are furnished with strong hoof-like claws at their extremities. These claws can be used with marvellous rapidity and force, and are employed for the purpose of destroying the dwellings of the ants on which the Aard Vark feeds, as well as for digging a burrow for its own habitation.

The burrows are not very deep, but are of tolerably large dimensions, and are often used, when deserted, as extempore tombs, to save the friends of the deceased from the trouble of digging a grave for their departed comrade. The creature makes its burrows with marvellous rapidity, and can generally dig faster with its claws than a man with a spade.

The Aard Vark is a nocturnal animal, and can very seldom be seen during the day-time. At night it issues from its burrow, and, making its way towards the ant-hills,

TAMANOIR, OR ANT-BEAR.—*Myrmecóphaga jubáta.*

begins its work of destruction. Laying its fore-feet upon the stone-like walls of these edifices, the Aard Vark speedily tears them down, and as the terrified insects run about in the bewilderment caused by the sudden destruction of their tenements, it sweeps them into its mouth with rapid movements of its long and extensile tongue. This member is covered with a tenacious glutinous secretion, to which the ants adhere, and which prevents them from making their escape during the short period of time that elapses between the moment when they are first touched and that in which they are drawn into the mouth.

THE remaining Ant-eaters possess no teeth whatever, and the aperture of their mouth is extremely small.

In its general habit and structure, the TAMANOIR, or GREAT ANT-EATER, or ANT-BEAR, is very similar to the preceding animal. It is, however, entirely destitute of teeth, possesses a wonderfully elongated and narrow head, and is thickly covered with long, coarse hair, which on the tail forms a heavy plume. The colour of this animal is brown, washed with grey on the head and face, and interspersed with pure white hairs on the head, body, and tail. The throat is black, and a long triangular black mark arises from the throat, and passes obliquely over the shoulders. There are four toes on the fore-feet, and five on the hinder. In total length it measures between six and seven feet, the tail being about two feet six inches long.

The claws of the fore-feet are extremely long and curved, and are totally unfitted for locomotion. When the animal is not employing these instruments in destroying, it folds the long claws upon a thick, rough pad which is placed in the palm, and seems to render the exertion of walking less difficult. As, however, the Ant-bear is forced to walk upon the outer edge of its fore-feet, its progress is a peculiarly awkward one, and cannot be kept up for any long time. Its mode of feeding is similar to that of the aard vark, which has

TAMANDUA.—*Tamándua tetradáctyla.*

just been detailed, and the creature seems to possess considerable grasping power in the toes of the fore-limbs, being able to pick up a small object in its paws. Though not a fighter, it can defend itself right well by means of these powerful instruments, and can not only strike with considerable violence, but when attacked by a dog or similar enemy, it clasps him in such a terrific gripe, that the half-suffocated animal is only too glad to be able to escape.

The Ant-bear is said to make no burrow, but to content itself with the shade of its own plumy tail whenever it retires to rest. While sleeping, the creature looks very like a rough bundle of hay, thrown loosely on the ground, for the hair of the mane and tail is so long and so harsh that it can hardly be recognised at the first glance for the veritable coat of a living animal. The eye of this creature has a peculiar and indescribably cunning expression. The Tamanoir is a native of Guinea, Brazil, and Paraguay.

The TAMANDUA possesses an elongated head, like that of the tamanoir, but the skull is not so extraordinarily long as in that animal, and the hair is short over the entire body. Indeed, the Tamandua looks like a small specimen of the tamanoir, which has been clipped from its neck to the tip of its tail. The colour of this species is much lighter than that of the tamanoir, and a black stripe passes over each shoulder. In size it is comparatively small, measuring, when full-grown, barely three feet and a half in total length.

It is a more active animal than the preceding species, and is a good climber of trees, which it ascends in search of the insects on which it feeds. The tail is long and tapering, and possesses something of the prehensile quality, though not so strongly as that of the little ant-eater, which will shortly be described. It is naked at the tip, but at the base is thickly covered with hair of the same short, coarse kind that is spread over the body. When young, its fur is a pale cinnamon.

The LITTLE ANT-EATER is a truly curious animal, possessing many of the habits of the two preceding animals, together with several customs of its own. The head of this creature is comparatively short; its body is covered with fine silken fur, and its entire length does not exceed twenty or twenty-one inches. The tail is well furred, excepting three inches of the under surface at the extremity, which is employed as the prehensile portion of that member, and is capable of sustaining the weight of the body as it swings from a branch. On looking at the skeleton, a most curious structure presents itself. On a side view, the cavity of the chest is completely hidden by the ribs, which are greatly flattened, and overlap each other so that on a hasty glance the ribs appear to be formed of one solid piece of bone. There are only two claws on the fore-feet and four on the hinder limbs.

The Little Ant-eater is a native of tropical America, and is always to be found on trees, where it generally takes up its residence, and where it finds its sustenance. It possesses many squirrel-like customs, using its fore-claws with great dexterity, and hooking the smaller insects out of the bark crevices in which they have taken unavailing refuge. While thus employed it sits upon its hind limbs, supporting itself with its prehensile tail. The claws are compressed, curved, and very sharp, and the little animal can use these instruments with some force as offensive weapons, and can strike smart blows with them. It is a bold little creature, attacking the nests of wasps, putting its little paw into the combs, and dragging the grubs from their cells.

Like its larger relations, it is nocturnal in its habits, and sleeps during the day with its tail safely twisted round the branch on which it sits. The generic name, Cyclothurus, signifies "twisted-tail," and is very appropriate to the animal.

LITTLE ANT-EATER.—*Cyclothurus didáctylus.*

THERE are few animals which have attracted such universal attention, both from scientific men and the reading world in general, as the MULLINGONG, DUCK-BILL, or PLATYPUS, of Australia. This little creature, the largest being but twenty-two inches in length, has excited more interest than animals of a thousand times its dimensions, on account of its extraordinary shape and singular habits. It is most appropriately called the Duck-bill, on account of the curious development of the intermaxillary bones, which are very much flattened and elongated, and their ends turned inwards in a kind of angular hook. The lower jaw is also lengthened and flattened, although not to such an extent as the upper, and the bones are covered with a naked skin.

In the stuffed and dried specimens the "beak" appears as if it were composed of the black leather taken from an old shoe, but in the living animal it presents a very different aspect, being soft, rounded, and of a pinky hue at its tip, mottled with a number of little spots. Dr. Bennett, to whom the zoological world is so much indebted for his researches into the habits of this curious animal, kindly showed me some excellent drawings, which gave a

DUCK-BILL, OR MULLINGONG.—*Plátypus Anatinus.*

very different idea of the animal from that which is obtained by the examination of stuffed skins. The beak is well supplied with nerves, and appears to be a sensitive organ of touch by means of which the animal is enabled to feel as well as to smell the insects and other creatures on which it feeds.

The Mullingong is an essentially aquatic and burrowing animal, and is formed expressly for its residence in the water, or under the earth. The fur is thick, soft, and is readily dried while the animal enjoys good. health, although it becomes wet and draggled when the creature is weakly. The opening of the ears is small and can be closed at will, and the feet are furnished with large and complete webs, extending beyond the claws in the fore limbs, and to their base in the hind legs. The fore-feet are employed for digging as well as for swimming, and are therefore armed with powerful claws rather more than half an inch in length, and rounded at their extremities. With such force can these natural tools be used, that the Duck-bill has been seen to make a burrow two feet in length through hard gravelly soil in a space of ten minutes. While digging, the animal employs its beak as well as its feet, and the webbed membrane contracts between the joints so as not to be seen. The hind-feet of the male are furnished with a spur, about an inch in length, curved, perforated, and connected with a gland situated near the ankle. It was once supposed that this spur conveyed a poisonous liquid into the wound which it made, but this opinion has been disproved by Dr. Bennett, who frequently permitted, and even forced the animal to wound him with its spurs, and experienced no ill consequences beyond the actual wound. The animal has the power of folding back the spur so as to conceal it entirely, and is then sometimes mistaken for a female.

The colour of the adult animal is a soft dark brown, interspersed with a number of glistening points which are produced by the long and shining hairs which protrude through the inner fur. Upon the abdomen the fur is a light fawn, and even softer than on the back. The under surface of the tail is devoid of hair—denuded, as some think, in forming its habitation—and the upper surface is covered with stiff, bristly hairs, brown towards the base and quite black at the extremity. The first coat of the young Duck-bill is always a bright, reddish-brown.

It can run on land and swim in water with equal ease, and is sufficiently active to be able to climb well. Some of the animals that were kept by Dr. Bennett were in the habit of ascending a perpendicular bookcase, performing this curious feat by placing their backs against the wall and the feet upon the shelves, and so pushing themselves upwards as a sweep ascends the chimney. Its pace is not very swift, but it gets over the ground with ease. The burrow in which the Mullingong lives is generally from twenty to forty feet in length, and always bends upwards, towards a sort of chamber in which the nest is made. This nest is of the rudest description, consisting of a bundle of dried weeds thrown carelessly together. The burrow has a very evil odour, which is unpleasantly adherent to the hand that has been placed within it.

Owing to the extremely loose skin of the Mullingong, it can push its way through a very small aperture, and is not easily retained in the grasp, wriggling without much difficulty from the gripe of the fingers. The loose skin and thick fur are also preventives against injury, as the discharge of a gun which would blow any other animal nearly to pieces, seems to take but little external effect upon the Duck-bill. The animal is, more-over, so tenacious of life, that one of these creatures which had received the two charges of a double-barreled gun, was able, after it had recovered from the shock, to run about for twenty minutes after it had been wounded.

The food of the Mullingong consists of worms, water insects, and little molluscs, which it gathers in its cheek-pouches as long as it is engaged in its search for food, and then eats quietly when it rests from its labours. The teeth, if teeth they may be called, of this animal are very peculiar, consisting of four horny, channeled plates, two in each jaw, which serve to crush the fragile shells and coverings of the animals on which it feeds. It seems seldom to feed during the day, or in the depth of night, preferring for that purpose the first dusk of evening or the dawn of morning. During the rest of the day it is generally asleep. While sleeping, it curls itself into a round ball, the tail shutting down over the head and serving to protect it.

The young Mullingongs are curious little creatures, with soft, short flexible beaks, naked skins, and almost unrecognisable as the children of their long-nosed parents. When they attain to the honour of their first coat, they are most playful little things, knocking each other about like kittens, and rolling on the ground in the exuberance of their mirth. Their little twinkling eyes are not well adapted for daylight, nor from their position can they see spots directly in their front, so that a pair of these little creatures that were kept by Dr. Bennett used to bump themselves against the chairs, tables, or any other object that might be in their way. They bear a farther similitude to the cat in their scrupulous cleanliness, and the continual washing and pecking of their fur.

At the present time—May, 1860—Dr. Bennett is endeavouring to accustom some Duck-bills to a life of confinement, with a view to their transportation to England. A very ingenious home has been constructed for them, precisely after the fashion of their own burrow. The chief difficulty lies in feeding them, for the Mullingong requires its food to be given at very frequent intervals, and soon perishes if not watched with the utmost care. The precise range of the animal is not satisfactorily ascertained, but it has never yet been seen in Southern Australia.

The ECHIDNA is found in several parts of Australia, where it is popularly called the hedgehog, on account of the hedgehog-like spines with which the body is so thickly covered, and its custom of rolling itself up when alarmed. A number of coarse hairs are intermingled with the spines, and the head is devoid of these weapons. The head is strangely lengthened, in a manner somewhat similar to that of the ant-eater, and there are no teeth of any kind in the jaws.

The food of the Echidna consists of ants and other insects, which it gathers into its mouth by means of the long extensile tongue. It is a burrowing animal, and is therefore furnished with limbs and claws of proportionate strength. Indeed, Lieutenant Breton, who kept one of these animals for some time, considers it as the strongest quadruped in existence in proportion to its size. On moderately soft ground it can hardly be captured, for it gathers all its legs under its body, and employs its digging claws with such extraordinary vigour that it sinks into the ground as if by magic. The hind-feet are employed by the animal for two purposes, i.e. locomotion and the offices of the toilet. There is a spur on the hind part of the male similar to that of the duck-bill. The flesh of the Echidna is very good, and is said to resemble that of the sucking-pig. There is

PORCUPINE ANT-EATER, OR ECHIDNA.—*Echidna Hystrix.*

another species of this curious animal, very similar in every respect except that of colour, which is of a darker brown, instead of the black and white which decorates the spines of the common Echidna. Its scientific title is *Echidna setosa.* The Echidna is tolerably widely spread over the sandy wastes of Australia, but has not been seen in the more northern portions of that country.

SLOTH.—*Cholœpus didáctylus.*

In the last group of the mammalia, we find a very remarkable structure, adapted to serve a particular end, and long misunderstood by zoologists. The common SLOTH, sometimes called the TWO-TOED SLOTH, is a native of the West Indies, where it is not very often seen, although it is not a very uncommon animal.

The peculiarity to be noticed in all the Sloths, of which there are several species, is, that they pass the whole of their lives suspended, with their backs downwards, from the branches of trees. The Sloth never gets upon a bough, but simply hooks his curved talons over it, and hangs in perfect security. In order to enable the animal to suspend itself without danger of falling, the limbs are enormously strong, the fore-legs are remarkable for their length, and the toes of all four feet are furnished with strong curved claws. Upon the ground the Sloth is entirely out of its element, as its limbs are wholly unadapted for supporting the weight of the body, and its long claws cannot be employed as adjuncts to the feet. The only manner in which a Sloth can advance, when he is unfortunately placed in such a position, is by hitching his claws into any depression that may afford him a hold, and so dragging himself slowly and painfully forward. On the trees, however, he is quite a different creature, full of life and animation, and traversing the branches at a speed which is anything but slothful. The Sloth travels best in windy weather, because the branches of trees are blown against each other, and permit the animal to pass from one tree to another without descending to the ground.

The food of the Sloth consists of leaves, buds, and young shoots. It appears to stand in no need of water, being satisfied with the moisture which clings to the herbage on which it feeds. In gathering the leaves and drawing the branches within reach, the Sloth makes great use of its fore-paws, which, however helpless upon the ground, can be managed with great dexterity. It is very tenacious of life, and is protected from any injury which it might receive from falls by the peculiar structure of its skull. In length it is about two feet.

The AI, or THREE-TOED SLOTH, is an inhabitant of South America, and is more common than the preceding animal, from which it can easily be distinguished by the third toe on its feet. The colour of this animal is rather variable, but is generally of a brownish-grey, slightly variegated by differently tinted hairs, and the head and face being darker than the body and limbs. The hair has a curious hay-like aspect, being coarse, flat, and harsh towards the extremity, although it is very fine towards the root. Owing to the colour and structure of the hair, the Ai can hardly be distinguished from the bough under which it hangs, and owes much of its safety to this happy resemblance; for its flesh is very good, and, in consequence, the poor creature is dreadfully persecuted by the natives, as well as by the white hunters. The cry of this creature is low and plaintive, and is thought to resemble the sound Ai. The head is short and round, the eyes deeply sunk in the head, and nose large and very moist.

AI OR THREE-TOED SLOTH.—*Bradypus tridactylus.*

The young of the Ai, as well as those of the other Sloths, cling to their mother as soon as they are born, and are carried about by her until they are able to transfer their weight from their parent to the branches. Several other species of Sloths are known to exist, but all are similar in appearance and habits.

COMPENDIUM OF GENERIC DISTINCTIONS.

EXPLANATION OF SIGNS AND TERMS.

TEETH.

INCISORS.—These are the teeth which are placed in the front of the jaw. They are inserted in the premaxillary bones, and in the corresponding portion of the lower jaw. They are termed "incisors," or cutting teeth, from the Latin word *incidere*, which signifies "to cut," even though their edges should not be formed for cutting.

CANINES.—These teeth are situated next to the incisors, and are inserted at or close to the suture of the premaxillary bones in the upper jaw. In the lower jaw, the canines are set opposite to those of the upper jaw, and when the mouth is closed pass in front of the crowns of the upper canines. They are called "canines," from the Latin word *canis*, a dog, because they are largely developed in the dogs.

PRÆMOLARS.—These teeth are situated behind the canines, and next to the true molars. The word " molar " is derived from the Latin *mola*, a mill, because these teeth serve to grind the food. Popularly they are called "grinders." In human subjects the præmolars are sometimes termed the "bicuspids," on account of the double cusp on their surfaces.

MOLARS.—These teeth are permanent, and are situated behind all the others. They are often not developed until comparatively late in life.

The DENTAL FORMULA is a concise mode of describing the number and positions of the various teeth, and is easily comprehended. The accompanying formula is that of Man :—

$$\text{I. } \frac{2-2}{2-2}, \text{ C. } \frac{1-1}{1-1}, \text{ P. } \frac{2-2}{2-2}, \text{ M. } \frac{3-3}{3-3} = 32.$$

In these formulas the upper figures refer to the teeth of the upper jaw, and the lower line to those of the lower jaw, while the short hyphen — serves to separate the right from the left side. In man, therefore, there are two incisor teeth on each side of the upper jaw, and the same in the under jaw ; one canine on each side of the upper jaw, and the same in the lower jaw ; two præmolars on each side of the upper jaw, and the same in the lower ; three molars on each side of the upper jaw, and the same in the lower : in all, thirty-two in number. The dentition is always presumed to be that of the adult animal.

CLASS I.—MAMMÁLIA.

ANIMALS possessed of vertebræ ; breathing atmospheric air by lungs ; heart with two auricles and two ventricles ; blood warm and red ; producing living young ; nurturing them by milk, which is secreted in the "mammary glands ;" skin covered with hair, spines or scales.

Order.—BÍMANA.

Hands and feet five-fingered, the nails all flat and broad. All the teeth even and close to each other, the molars equally enamelled. In this order there is but one species, namely Man,—Homo sapiens.

Teeth.—I. $\frac{2-2}{2-2}$, C. $\frac{1-1}{1-1}$, P. $\frac{2-2}{2-2}$, M. $\frac{3-3}{3-3}$ = 32.

Order.—QUADRÚMANA.

Hinder feet five-toed, the thumb opposable to the others ; fore-feet sometimes four-fingered, the thumb being absent. Molar-teeth equally enamelled ; with one exception, the Cheiromys, they possess incisor, canine, præmolar, and molar teeth. Skin covered with hair, with the exception of the palms of the hand, the face, and the callosities of the hinder quarters. Mammæ placed on the breast.

Family.—SIMIADÆ.

Teeth.—Molars, $\frac{5}{5}$, the false molars being tuberculate.

Nails rather flat or slightly rounded, and not pointed like claws. Fore-feet almost always five-toed. Thumb opposable. Tail never prehensile.

Genus.—Troglodýtes.

Teeth.—I. $\frac{2-2}{2-2}$, C. $\frac{1-1}{1-1}$, P. $\frac{2-2}{2-2}$, M. $\frac{3-3}{3-3}$ = 32.

Canines slightly elongated, and placed close to the incisors.

Head.—Muzzle rather short—Cheek-pouches none—Ears large and projecting.

Tail.—None.

Habitat.—Western Africa.

Genus.—Símia.

Teeth.—Canines much exceeding the others in length, and overlapping each other when the mouth is closed.—Two central incisors extremely broad.

Head.—Muzzle projecting very considerably—Ears small, and placed close to the head.—Cheek-pouches none.

Limbs.—Arms extremely long, the fingers resting on the ground when the animal stands erect.

Habitat.—Borneo and Sumatra.

Genus.—Siamanga.

Head small, and muzzle short—Cheek-pouches none. Throat furnished with a large air-pouch.

Limbs.—Arms extremely long—First and second fingers of the hands united as far as the middle of the second joint—Slight callosities on hinder quarters.

Habitat.—Sumatra.

Genus.—Hylóbates.

Head, throat, and limbs, resembling Siamanga, except that the fingers of the hand are all free. Many systematic naturalists consider the two genera to be really one, and that the Siamanga is only a species of Hylobates.

Habitat.—Malacca.

Genus.—Presbýtes.

Teeth.—Last molar of lower jaw with five tubercles.

Head.—Muzzle very slightly produced—rudiments of cheek-pouches.

Feet.—Elongated—Thumb of fore-feet very short.

Tail.—Extremely long, often surpassing the body.

Habitat.—India, China, &c. Only known in Asia.

Genus.—Cólobus.

Teeth ⎫
Head ⎬ As in Presbytes.
Tail ⎭

Feet.—Thumb of fore-feet altogether wanting, or only represented by a small tubercle.

Habitat.—Western Africa.

Genus.—Cercopithécus.

Teeth.—The last molar teeth of the lower jaw furnished with four tubercles.

Head.—Cheek-pouches large—Face rather long and rounded.

Tail.—Long, sometimes longer than the body.

Habitat.—Spread over the greater part of Africa.

Genus.—Cercocébus.

Teeth.—The last molar teeth of the lower jaw furnished with five tubercles, the others with four tubercles.

Head.—Muzzle more elongated than in Cercopithécus—Cheek-pouches large.

Tail.—Long, and not tufted.

Habitat.—Africa.

Genus.—Macácus.

Similar to Cercocebus, excepting that the tail is very varied in length, several species being almost destitute of the member, and others possessing it very slightly developed.

Habitat.—India, Sumatra, Japan, and the North of Africa.

Genus.—Silénus.

Similar to the genus Macacus, excepting the tail, which is furnished with a conspicuous tuft of hair.

Habitat.—India.

Genus.—Cynocéphalus.

Teeth.—Last molar of lower jaw furnished with one or two accessory tubercles, the others with four tubercles.

Head.—Face lengthened into a conspicuous snout, and abruptly terminated, the nostrils being placed at the extremity—Cheek-pouches large.

Tail.—Moderately long, and inserted high. In the Gelada it is furnished with a tuft, a peculiarity which has induced some writers to place the animal in a different genus.

Habitat.—Africa.

Genus.—Pápio.

Similar to Cynocephalus, excepting that the tail is extremely small, and set nearly perpendicularly to the line of the back.

Habitat.—Africa.

Family.—Cébidæ.

Nostrils very wide, separated by a broad septum, opening laterally. Tail long, and in most instances prehensile. Thumb of fore-hands totally distinct from the fingers. Cheek-pouches absent. Molar teeth comparatively small.

Genus.—Áteles.

Head.—Rounded and small.

Limbs.—Long and slender—Thumb of fore-hands wanting.

Tail.—Prehensile, naked below towards the tip.

Fur.—Long, stiff, and rather harsh.

Habitat.—Brazils.

Genus.—Brachýteles.

Head as in Ateles.

Limbs.—Thumb of fore-hands extremely small.

Tail.—Prehensile, and naked below towards the tip.

Fur.—Woolly.

Habitat.—Tropical America.

Genus.—Mycétes.

Head.—Rather pyramidal—A large beard on the cheeks and chin—Throat furnished with large, resonant pouch, formed by expansion of the hyoid bone.

Limbs.—Fore-feet five-fingered.

Tail.—Naked below towards the tip.

Habitat.—Tropical America.

Genus.—Cebus.

Head.—Rounded.

Tail.—Long, and entirely covered with hair.

Habitat.—Tropical America.

Genus.—Callithrix.

Teeth.—Incisors straight, the two middle being broad—Canines short, hardly exceeding incisors.

Tail.—Slender and rounded.

Habitat.—Brazils.

Genus.—Brachyúrus.

Teeth.—Incisors rather oblique, the lower being long—Canines large and stout—Molars small.
Tail.—Very hairy, shorter than body.
Habitat.—Guiana.

Genus.—Pithécia.

Teeth.—Like preceding genus.
Tail.—Equalling body in length.
Habitat.—Brazils.

Genus.—Nyctipithécus.

Teeth.—Lower incisors rather obliquely pointing forwards, two middle upper incisors broad—Canines moderate.
Head.—Ears small, and partially buried in hair—Eyes large, orbits very large.
Limbs.—Hind-feet longer than fore-feet.
Tail.—Longer than body.
Habitat.—Brazils.

Genus.—Jacchus.

Teeth.—Lower incisors long and rounded, rather convex externally—Præmolars with one tubercle in the outer margin—Molars with two tubercles.
Head.—Face short and blunt—nostrils wide.
Tail.—Long, and thickly furred.
Habitat.—Brazils.

Family.—LEMÚRIDÆ.

Teeth.—Upper incisors, 2—2, generally set in pairs, and separated from the canines by a small space ; lower incisors, either 2—2 or 1—1, often slightly projecting.
Limbs.—All the feet with five fingers, the fourth being the largest—Hind-feet larger than fore-feet—All the nails flat, excepting that of the second finger, which is narrow and curved.

Genus.—Lemur.

Teeth.—I. $\frac{2-2}{2-2}$, C. $\frac{1-1}{1-1}$, P. $\frac{3-3}{3-3}$, M. $\frac{3-3}{3-3}$ = 36.
Head.—Eyes large, and set closely together—Ears short and rounded.
Limbs.—First finger of fore-feet extremely short.
Tail.—Rather short.
Habitat.—Madagascar.

Genus.—Propithécus.

Teeth.—Upper incisors expanded towards the canines.
Habitat.—Madagascar.

Genus.—Loris.

Head.—Muzzle long and sharp, slightly directed upwards. Eyes extremely large.
Body and Limbs.—Slender and delicate.
Tail.—None.
Habitat.—Ceylon.

Genus.—Nycticébus.

Resembling Lemur, but having the tail extremely short.
Habitat.—Sumatra, Borneo, and Bengal.

Genus.—Gálago.

Teeth as Lemur.
Head.—Ears large and naked—Eyes large.

Limbs.—Tarsus elongated.
Tail.—Long, and thickly furred.
Habitat.—Madagascar, and various parts of Africa.

Genus.—Indris.

Teeth. —I. $\frac{2-2}{1-1}$, C. $\frac{1-1}{1-1}$, P. $\frac{2-2}{2-2}$, M. $\frac{3-3}{3-3}$ = 30.
Head.—Ears small and rounded.
Limbs.—Tarsus not elongated.
Habitat.—Madagascar.

Genus.—Társius.

Teeth.—I. $\frac{2-2}{2-2}$, C. $\frac{1-1}{1-1}$, P. $\frac{3-3}{3-3}$, M. $\frac{3-3}{3-3}$ = 36. Lower incisors oblique—False molars conic—Molars furnished with several sharp tubercles.
Head.—Eyes large—Ears rather large, very thinly supplied with hair.
Limbs—Hinder feet extremely long, with elongated tarsus.
Tail.—Very long, with tuft at the tip.
Habitat.—Borneo and Philippine Islands.

Family.—GALEOPITHÉCIDÆ.

There is only one genus in this family.

Genus.— Galeopithécus.

Teeth.—I. $\frac{2-2}{3-3}$, C. $\frac{1-1}{1-1}$, P. $\frac{2-2}{2-2}$, M. $\frac{3-3}{3-3}$ = 34. Some authors give the formula in a slightly different manner, as the teeth seem to be rather obscure :—I. $\frac{2-2}{2-2}$, C. $\frac{0-0}{1-1}$, P. $\frac{2-2}{2-2}$, M. $\frac{4-4}{4-4}$ = 34. The lower incisors are set pointing forwards, and are deeply notched on their crowns like the teeth of a comb.
Habitat.—Java, Borneo, Sumatra.

Family.—(?)

Genus.—Chéiromys.

Teeth.—I. $\frac{1-1}{1-1}$, C. $\frac{0-0}{0-0}$, M. $\frac{4-4}{3-3}$ = 18. Incisors pointed, compressed, and very sharp and powerful.
Head.—Rounded, and muzzle short and pointed.
Limbs.—Feet with five fingers—Fore-feet with toes long, the middle toe long and slender—Thumb of hind-feet with flat and broad nail.
Tail.—Long, and heavily furred.
The family in which this animal (the Aye Aye) ought to be placed is very doubtful, as is even the order to which it really belongs.

Family.—CHEIRÓPTERA.

The bones of the fore-limbs, and especially those of the fingers, much elongated, and sustaining a membrane of large dimensions, by means of which the animals fly in the air. The thumb-joint is not attached to the web, but is left free. It is furnished with a nail. The hinder-feet are small, and the toes furnished with sharp claws.

Genus.— Vampírus.

Teeth.—I. $\frac{2-2}{3-3}$, C. $\frac{1-1}{1-1}$, M. $\frac{6-6}{5-5}$ = 34. Incisors small, especially those of the lower jaw, and nearly contiguous at their bases—Canines large.

Head.—Nose with a double leaf-like membrane, one lying almost horizontally, and the other being erect—The ears are moderate, and the tragus is small and elongated.

Habitat.—South America.

Genus.—Rhinólophus.

Teeth.—I. $\frac{1-1}{2-2}$, C. $\frac{1-1}{1-1}$, P. $\frac{2-2}{3-3}$, M. $\frac{3-3}{3-3}$ = 32.

Or thus :—I. $\frac{0-0}{2-2}$, C. $\frac{1-1}{1-1}$, M. $\frac{4-4}{5-5}$, or $\frac{5-5}{5-5}$. Incisors small, and distinct from each other.

Head.—A complicated leaf-like membrane upon the nose, represented in the engraving on page 119—Ears large, without tragus.

Habitat.—Europe, Asia, Africa, and Australia.

Genus.—Barbastellus.

Teeth.—I. $\frac{2-2}{3-3}$, C. $\frac{1-1}{1-1}$, P. $\frac{2-2}{2-2}$, M. $\frac{3-3}{3-3}$ = 34.

Head.—Ears united at their bases, moderate in size—On the upper part of the muzzle is a depressed naked spot, in which the nostrils are set.

Habitat.—Europe.

Genus.—Plecótus.

Teeth.—I. $\frac{2-2}{3-3}$, C. $\frac{1-1}{1-1}$, P. $\frac{2-2}{3-3}$, M. $\frac{3-3}{3-3}$ = 36.

Head.—Ears very large, and united at their bases.

Habitat.—Europe.

Genus.—Noctilínia.

Teeth.—I. $\frac{2-2}{3-3}$, C. $\frac{1-1}{1-1}$, P. $\frac{2-2}{2-2}$, M. $\frac{3-3}{3-3}$ = 34.

Habitat.—Europe and Asia.

In Bell's *British Quadrupeds*, and in Van der Hoeven's *Handbook of Zoology*, Noctilinia is merged into the genus Vespertilio, together with Plecotus.

Genus.—Ptéropus.

Teeth.—I. $\frac{2-2}{2-2}$, C. $\frac{1-1}{1-1}$, P. $\frac{2-2}{3-3}$, M. $\frac{3-3}{3-3}$ = 34. Molars with flattened crowns, and a longitudinal groove.

Head.—Ears small, tragus none.

Limbs.—First finger of fore-paws with only three joints.

Tail.—None.

Habitat.—Indian Archipelago.

Order.—FERÆ.

Teeth.—Incisors always $\frac{3-3}{3-3}$—Canines large, strong, and pointed—Molars uniformly enamelled, with crowns more or less sharp, uneven, or tuberculated.

Family.—FÉLIDÆ.

Teeth.—I. $\frac{3-3}{3-3}$, C. $\frac{1-1}{1-1}$, P. $\frac{3-3}{3-2}$, M. $\frac{1-1}{1-1}$ = 30.

Limbs.—Feet digitigrade, soles of feet furnished with hairs.

Genus.—Felis.

Feet.—Fore-feet with five toes, hinder feet with four toes—Claws retractile.

Habitat.—Most parts of the world.

By some zoologists this genus is separated into four, namely :—Leo, Tigris, Leopardus, and Felis, but apparently on insufficient grounds.

Genus.—Lyncus.

Separated from Felis on account of the short tail, and pencils of hairs which tuft the ears.

Habitat.—Europe, Asia, and Africa.

In this genus are included Chaus, Caracal, and Lyncus.

Genus.—Gueparda.

Separated from Felis on account of the semi-retractile claws, larger limbs, and the short mane that runs along the neck and shoulders; and from Lyncus by the absence of the ear tufts and the long tail.

Habitat.—Asia and Africa.

Family.—VIVÉRRIDÆ.

Teeth.—Three præmolars on each side in the upper jaw, and either three or four in the lower.

Limbs.—Feet generally digitigrade—Claws often semi-retractile.

Glands.—Placed near junction of hinder limbs, secreting a substance of offensive odour.

In this family the Hyænas are placed by the best authorities.

Genus.—Hyæna.

Teeth.—I. $\frac{3-3}{3-3}$, C. $\frac{1-1}{1-1}$, P. $\frac{4-4}{3-3}$, M. $\frac{1-1}{1-1}$ = 34. On each side in the upper jaw is one tuberculate tooth.

Limbs.—Feet all with four toes.

Body.—Sloping from shoulder to tail.

Tail.—Short.

Habitat.—Asia and Africa.

In this genus Crocuta is included.

Genus.—Próteles.

Teeth.—Molars either $\frac{4-4}{4-4}$ or $\frac{5-5}{5-5}$, small and distant.

Limbs.—Fore-feet with five toes, the thumb being rather raised ; hind-feet with four toes.

Body.—Sloping like that of Hyæna.

Tail.—Rather short, and very bushy.

Habitat.—Southern Africa.

Genus.—Viverra.

Teeth.—I. $\frac{3-3}{3-3}$, C. $\frac{1-1}{1-1}$, P. $\frac{4-4}{4-4}$, M. $\frac{2-2}{2-2}$ = 40.

Limbs.—All the feet with five toes, the claws small and curved, the thumb-joint small and rather raised.

Habitat.—Africa, Asia, &c.

Genus.—Linsang.

Separated from preceding by its very slender and elongated body, its long legs, and very long whisker hairs.

Habitat.—Java and Nepâl.

Genus.—Genetta.

Separated from Viverra by its smaller size and longer tail.

Habitat.—Africa.

Genus.—Bássaris.

Separated from Viverra by its small, pointed head, long ears, and the different texture of its fur.

Habitat.—Mexico.

In the opinion of very many excellent zoologists this genus, together with its two predecessors, ought to be merged into the genus Viverra.

Genus.—Herpestes.

Teeth as in Viverra.
Head.—Ears small and rounded.
— —All with five toes, furnished with large, curved, ... sed claws.
— —Long and wiry, frequently annulated with different ...

... at.—Southern Europe, Asia, and Africa.
This genus includes Mungos and Urva.

Genus.—Cynictis.

Separated from the preceding genus because the hinder feet have only four toes. The limbs are rather longer, and the ears larger than in Herpestes.
Habitat.—Southern Africa.

Genus.—Crossarchus.

Teeth.—P. $\frac{2-2}{2-2}$, M. $\frac{3-3}{3-3}$.
Head.—Muzzle elongated, the nose resembling a proboscis—Ears small.
Limbs.—Feet with five toes, gait plantigrade.
Tail.—Rather shorter than body.
Habitat.—Western Africa.

—Suricáta.

P. $\frac{3-3}{3-3}$, M. $\frac{2-2}{2-2}$ = 36.

... s, furnished with long, curved,

... rather exceeding half the length of the body.
Habitat.—Southern Africa.

Genus.—Cynógale.

Teeth.—P. $\frac{4-4}{4-4}$, M. $\frac{2-2}{2-2}$, laniary teeth with tubercles.
Head.—Ears small—Muzzle elongated, blunt, and depressed—Whisker hairs remarkably long.
Limbs.—Feet with five short toes, gait plantigrade.
Tail.—Short.
Habitat.—Borneo.

Genus.—Paradoxúrus.

Teeth.—Molars as in Viverra—Laniary teeth thick and furnished with conical tubercles.
Limbs.—Feet with five toes conjoined by skin, the thumb not raised, gait plantigrade.
— —Long and cylindrical, mostly capable of being rolled, but not prehensile.
...at.—Africa and Asia.
...enus includes Nandínia and Páguma.

Genus.—Artictis.

Teeth.—I. $\frac{3-3}{3-3}$, C. $\frac{1-1}{1-1}$, P. $\frac{4-4}{4-4}$, M. $\frac{2-2}{2-2}$ = 40. Canines conical and compressed—Laniary teeth curiously tuberculated.
Head.—Ears furnished with pencil of long hairs; whisker hairs long.
Tail.—Nearly as long as body, prehensile, and heavily covered with hair at the base.
Habitat.—Sumatra, Borneo, and Java.
This genus is placed by V. der Hoeven among the Ursine animals.

Genus.—Cryptoprocta.

Limbs.—Feet with five toes and plantigrade, furnished with retractile claws.
Habitat.—Madagascar.
It is the opinion of many excellent zoologists that the genera of the Viverrine animals might be still further reduced.

Family.—CÁNIDÆ.

Teeth.—Molars either $\frac{6-6}{7-7}$, $\frac{7-7}{7-7}$, or $\frac{8-8}{8-8}$, but usually the former. Two or three on each side of both jaws tuberculated.
Limbs.—Fore-feet mostly with five toes, thumb raised; gait digitigrade, hinder feet with four toes.

Genus.—Canis.

Teeth.—I. $\frac{3-3}{3-3}$, C. $\frac{1-1}{1-1}$, P. $\frac{4-4}{4-4}$, M. $\frac{2-2}{2-2}$ = 42. Laniary tooth of upper jaw bi-lobed, with a small tubercle inside and rather forwards; the lower laniary divided into three portions.
Head.—Pupil of eye round.
Tail.—Moderate, covered with short hair.
Habitat.—All parts of the world.
Includes Cuon.

Genus.—Vulpes.

Separated from Canis by the oblong pupil of the eye, and the heavily brushed tail.
Habitat.—Most parts of the world.

Genus.—Otócyon.

Teeth.—Molars $\frac{8-8}{8-8}$. Laniary teeth less than the tuberculate; lower tuberculate with four sharp tubercles.
Head.—Ears very large, nearly as long as head, standing erect.
Tail.—Moderate, and covered with thick hair.
Habitat.—Southern Africa.

Genus.—Lycáon.

Separated from Canis because the fore-feet are furnished with only four toes.
Habitat.—Southern Africa.

Family.—MUSTÉLIDÆ.

Teeth.—Molars generally 4—4 or 5—5 in upper jaw, 5—5 or 6—6 in lower. On each side of both jaws there is a single tuberculate tooth.
Head.—Rather long, muzzle moderate and rounded, skull much elongated behind the eyes.
Limbs.—Feet with five toes.

Genus.—Putórius.

Teeth.—I. $\frac{3-3}{3-3}$, C. $\frac{1-1}{2-2}$, P. $\frac{3-3}{3-3}$, M. $\frac{1-1}{1-1}$ = 34.
Head.—Ears small and rounded.
Limbs.—Toes separated from each other.
Tail.—Moderate, of various lengths.
Habitat.—Europe and Asia.
Includes the Stoats, Weasels, and Polecats.

Genus.—Mustéla.

Teeth.—P. $\frac{4-4}{4-4}$.
Habitat.—Europe, Asia, and North America.
Includes Martes.

Genus.—Grisónia.

Teeth.—M. $\frac{4—4}{5—5}$,

Limbs.—Feet with soles naked, gait partly plantigrade.
Tail.—Hairy, partly "distichous."
Habitat.—Brazil and Guiana.
Includes Gálera.

Genus.—Mellívora.

Teeth.—I. $\frac{3—3}{3—3}$, C. $\frac{1—1}{1—1}$, P. $\frac{3—3}{3—3}$, M. $\frac{1—1}{1—1}$ = 32. Only
one tuberculate tooth on each side of upper jaw.
Limbs.—Fore-feet armed with large and powerful claws,
all the feet short, and the gait plantigrade.
Head.—Short.
Habitat.—Africa and Asia.

Genus.—Gulo.

Teeth.—I. $\frac{3—3}{3—3}$, C. $\frac{1—1}{1—1}$, P. $\frac{4—4}{4—4}$, M. $\frac{1—1}{2—2}$ = 38.
Head.—Ears short and rounded.
Limbs.—Gait nearly plantigrade.
Habitat.—Northern Europe, Asia, and America.

Genus.—Mephítis.

Teeth.—M. $\frac{4—4}{5—5}$. An accessory tubercle in the middle of
upper laniary tooth.
Tail.—Moderate, and thickly covered with long hair.
Habitat.—America.

Genus.—Mydaus.

Teeth.—I. $\frac{3—3}{3—3}$, C. $\frac{1—1}{1—1}$, P. $\frac{3—3}{3—3}$, M. $\frac{1—1}{2—2}$ = 32.
Head.—Muzzle much elongated, ears very small, and
buried in fur.
Limbs.—Fore-feet with large, compressed, and nearly
straight claws, gait nearly plantigrade.
Habitat.—Asia and Java.
Includes Arctonyx.

Genus.—Meles.

Teeth.—I. $\frac{3—3}{3—3}$, C. $\frac{1—1}{1—1}$, P. $\frac{4—4}{4—4}$, M. $\frac{1—1}{1—1}$ = 36. Lower
tuberculate teeth often missing.
Limbs.—Fore-feet with large digging claws, gait planti-
grade.
Tail.—Short.
Habitat.—Europe and Asia.

Genus.—Lutra.

Teeth.—I. $\frac{3—3}{3—3}$, C. $\frac{1—1}{1—1}$, P. $\frac{4—4}{3—3}$, M. $\frac{1—1}{2—2}$ = 36. Upper
laniary teeth very large.
Head.—Ears small, and set higher than the eyes.
Limbs.—Feet short and webbed, middle toe the largest.
Tail.—Moderate, rounded, but flattened beneath and
towards the tip.
Habitat.—Europe and Asia.

Genus.—Enhýdra.

Teeth.—I. $\frac{3—3}{2—2}$.

Head.—Ears set at the side of the head, and below the
eyes.

Limbs.—Feet webbed, hair covering fore-feet even to
claws, external toe of hind-feet the largest.
Tail.—Short.
Habitat.—Kamschatka and Northern America.

Family.—Úrsidæ.

Teeth.—Upper jaw with two tuberculate teeth on each
side, lower jaw with either one or two tuberculate teeth.
The laniary tooth resembling the tuberculates, the crown,
however, being flattened.
Limbs.—Feet all with five toes, gait plantigrade.

Genus.—Ursus.

Teeth.—I. $\frac{3—3}{3—3}$, C. $\frac{1—1}{1—1}$, P. $\frac{4—4}{4—4}$, M. $\frac{2—2}{3—3}$ = 42. Tuber-
culate teeth, $\frac{2—2}{2—2}$, the last in the upper jaw and last but
one in the lower being very large.
Head.—Ears small and erect; muzzle elongated, but blunt
at extremity, and very movable.
Tail.—Very short.
Habitat.—Europe, Asia, Africa, and America.
Includes all the Bears.

Genus.—Prócyon.

Teeth.—I. $\frac{3—3}{3—3}$, C. $\frac{1—1}{1—1}$, P. $\frac{4—4}{4—4}$, M. $\frac{2—2}{2—2}$ = 40. Tuber-
culate teeth, $\frac{2—2}{1—1}$. Laniary teeth of upper jaw with conic
tubercles.
Head.—Muzzle sharp.
Tail.—Moderate.
Habitat.—Northern America.

Genus.—Násua.

Teeth.—Molars smaller than those of Prócyon, but similar
in arrangement. Canines compressed, very sharp.
Muzzle.—Extremely elongated and movable.
Limbs.—Claws rather curved, long and compressed.
Tail.—Long.
Habitat.—Brazil and Surinam.

Genus.—Cercoleptes.

Teeth.—I. $\frac{3—3}{3—3}$, C. $\frac{1—1}{1—1}$, P. $\frac{3—3}{3—3}$, M. $\frac{2—2}{2—2}$ = 36.
Head.—Short, and face rounded; tongue long and very
flexible.
Tail.—Long and capable of being rolled round any object.
Habitat.—Guiana and Peru.

Genus.—Ailúrus.

Teeth.—Arrangement as in Cercoleptes. On each side of
upper jaw a false molar with one tubercle; two tuberculate
teeth on each side below.
Head.—Ears small and rounded.
Limbs.—Claws semi-retractile, curved, and compressed.
Tail.—Moderate, and very hairy.
Habitat.—Nepâl.

Family.—Tálpidæ.

Teeth.—Incisors variable in number; canines often wanting, their place being taken by false molars; molars with
sharp conical tubercles.
Feet.—Mostly with five toes, gait plantigrade.

Genus.—Talpa.

Teeth.—I. $\frac{3-3}{4-4}$, C. $\frac{1-1}{1-1}$, P. $\frac{4-4}{3-3}$, M. $\frac{3-3}{3-3}$ = 44. According to some, C. $\frac{0-0}{0-0}$, their place being occupied by the first molars.

Head.—Muzzle elongated, and blunt at extremity—Eyes hidden under fur, and very small.

Limbs.—Feet with five toes; fore-feet with sole turned backwards; claws very strong.

Tail.—Very short.

Habitat.—Europe.

Genus.—Scalops.

Teeth.—I. $\frac{3-3}{3-3}$, C. $\frac{1-1}{1-1}$, P. $\frac{4-4}{4-4}$, M. $\frac{3-3}{3-3}$ = 44. Two middle incisors of upper jaw large, and the others small.

Head.—Muzzle elongated, with a proboscis-like nose—Eyes minute.

Limbs.—Feet with five toes.

Tail.—Short, and thinly covered with hair.

Habitat.—Northern America.

Genus.—Chrysochlóris.

Teeth.—I. $\frac{3-3}{3-3}$ on either side; seven molars on each side in each jaw, having a space between them.

Head.—Muzzle elongated and naked—Eyes covered with skin.

Limbs.—Fore-feet with five toes, the fourth being small; the claw of the third toe powerful, curved, and broad; hinder feet with five toes.

Tail.—None.

Habitat.—Southern Africa.

Genus.—Astromyctes.

Teeth.—I. $\frac{3-3}{2-2}$, the lower projecting forward; seven molars on each side of upper jaw, and eight in the lower.

Head.—Muzzle elongated, with a curiously radiated extremity—Ears very small.

Limbs.—Feet with five toes.

Tail.—Moderate, sparsely covered with hair.

Habitat.—Northern America.

Genus.—Tupaia.

Teeth.—I. $\frac{2-2}{3-3}$, those of lower jaw projecting forwards, and the four central larger than the others; lower molars divided by a transverse groove; true molars, $\frac{4-4}{3-3}$.

Head.—Muzzle slender and elongated—Ears rather large.

Limbs.—Feet with five toes.

Tail.—Long, and thickly covered with hair, nearly "distichous."

Habitat.—India, Borneo, Sumatra.

Ptilocercus may be referred to this genus, from which it has been separated on account of the extraordinary tail.

Genus.—Macroscélides.

Teeth.—I. $\frac{3-3}{2-2}$, all small; seven molars on each side of upper jaw, and either eight or nine on each side of the lower. There are no true canines; true molars, $\frac{4-4}{3-3}$, or $\frac{4-4}{4-4}$.

Head.—Muzzle elongated into a slender proboscis, the nostrils being at the extremity—Eyes moderately large—Ears large, thickly covered with hair.

Limbs.—Fore-feet with five toes; hind-feet much larger than fore, and furnished with short, sharp, slight, and compressed claws.

Tail.—Long.

Habitat.—Africa.

Genus.—Sorex.

Teeth.—I. $\frac{3-3}{2-2}$, the upper being long, curved, and notched at their bases, the lower projecting almost horizontally. No true canines. Five small teeth in upper jaw between the incisors and true molars; the lower incisors serrate. True molars, $\frac{4-4}{3-3}$.

Head.—Muzzle lengthened and sharp—Eyes small, and ears broad.

Limbs.—Feet all with five toes.

Tail.—Moderate.

Habitat.—Europe.

Genus.—Cróssopus.

Teeth.—Only four small intermediate teeth in upper jaw, and the lower incisors not serrate.

Limbs.—Feet and toes edged with stiff hairs.

Habitat.—Europe.

Genus.—Solénodon.

Teeth.—I. $\frac{3-3}{3-3}$, M. $\frac{7-7}{7-7}$. No true canines; true molars, $\frac{4-4}{4-4}$. The middle incisors of upper jaw large and triangular, separated from the others by a narrow space. Two middle incisors of lower jaw small and narrow, next two long, conical, and grooved on the inside.

Head.—Upper jaw larger than the lower—Muzzle elongated, with a proboscis—Eyes very small—Ears round, and nearly naked.

Limbs.—Feet with five toes.

Tail.—Long and cylindrical, covered with scales for the greater part of its length.

Habitat.—St. Domingo.

Genus.—Gálemys.

Teeth.—I. $\frac{1-1}{2-2}$, those of upper jaw large, broad, and triangular; no true canines; true molars, $\frac{4-4}{3-3}$.

Head.—Muzzle elongated, with a slender, depressed proboscis.

Limbs.—Feet with five toes, and palmate.

Tail.—Long, compressed at tip, and scantily covered with hair.

Habitat.—South-eastern parts of Russia and the Pyrenees.

Genus.—Gymnúra.

Teeth.—I. $\frac{3-3}{3-3}$, C. $\frac{1-1}{1-1}$, P. $\frac{4-4}{4-4}$, M. $\frac{3-3}{3-3}$ = 44. Some consider the canines to be only false molars; two middle incisors of upper jaw large, two next small.

Head.—Muzzle elongated, and blunt at extremity—Ear round and naked.

Limbs.—Feet with five toes, three central toes largest.

Body.—Long bristles scattered among the fur.

Tail.—Rather long, scantily haired, and scaly

Habitat.—Malacca and Sumatra.

1.

Genus.—Erináceus.

Teeth.—I. $\frac{3-3}{3-3}$, P. $\frac{4-4}{2-2}$, M. $\frac{3-3}{3-3}$ = 36.

Head.—Muzzle rather elongated.

Limbs.—Feet with five toes.

Body.—Thickly covered with sharp quills or spines above, and with quills and hair below ; capable of contraction into a ball.

Tail.—Short.

Habitat.—Europe, Asia, and Africa.

Genus.—Centétes.

Teeth.—I. $\frac{2-2}{3-3}$, C. $\frac{1-1}{1-1}$, P. $\frac{3-3}{3-3}$, M. $\frac{3-3}{3-3}$ = 38. Sometimes I. $\frac{3-3}{3-3}$. Canines large, round, and conical, separated from other teeth by vacant space.

Head.—Muzzle elongated—Ears short and rounded.

Body.—Covered on upper surface with mixed spines and bristles.

Tail.—None.

Habitat.—Madagascar.

Family.—MACRÓPIDÆ.

Sub-family.—Phalangistina.

Skin of flanks developed into a parachute-like expansion, and affixed to the fore and hinder limbs. Hind feet with five toes, the thumb opposable to the others, and without a claw, the two next joined together as far as the claws.

Genus.—Acróbates.

Teeth.—I. $\frac{3-3}{1-1}$, C. $\frac{1-1}{1-1}$, P. $\frac{3-3}{3-3}$, M. $\frac{3-3}{3-3}$ = 36. The true molars furnished each with four acute cusps, premolars large and sharp pointed.

Hairs of tail stiff, and set in double row like the barbs of a feather.

Genus.—Petaurus.

Teeth.—I. $\frac{3-3}{1-1}$, C. $\frac{1-1}{1-1}$, P. $\frac{3-3}{3-3}$, M. $\frac{4-4}{4-4}$ = 40.

Tail hairy, but not prehensile, and extremely long.

Genus.—Petaurista.

Teeth.—Space between the molars and incisors occupied by two rudimentary minute teeth. M. $\frac{6-6}{5-5}$. The four last are true molars, and are furnished each with four pyramidical cusps, except the last tooth in the upper jaw, which only bears three cusps.

Genus.—Cuscus.

Teeth, as in Phalangista.

Tail.—Prehensile, destitute of hair except at the base, and covered with small tubercles.

Genus.—Phalangista.

Teeth.—Variable, incisors always $\frac{3-3}{1-1}$, and true molars always $\frac{4-4}{4-4}$. Molars, either $\frac{6-6}{6-6}$, $\frac{7-7}{7-7}$, or $\frac{7-7}{8-8}$. Inferior canines very small, and close to the incisors.

Tail.—Prehensile, and coloured with hair except at tip, and a naked stripe along the under side of the extremity.

Ears.—Elongated and triangular.

Genus.—Phascolarctos.

Teeth.—I. $\frac{3-3}{1-1}$, C. $\frac{1-1}{0-0}$, P. $\frac{1-1}{1-1}$, M. $\frac{4-4}{4-4}$ = 30. The crown of each true molar furnished with four angular pyramidical tubercles.

Toes of fore-feet in two sets, the one comprising the two inner, and the other the three outer toes.

Tail.—None.

Sub-family.—Macropína.

Hinder feet much longer than those of fore-limbs, furnished with four toes, the two inner toes being small, and connected together as far as the small claws.

Teeth.—Six incisors above, two below, lying nearly horizontally in the jaw, and projecting. Canines either wanting, or only in upper jaw, very close to the incisors. A considerable space between the canines and molars, which are $\frac{5-5}{5-5}$. The front molar has its crown narrow and compressed, but the others are furnished with two transverse tubercles.

Tail.—Long, covered with hair, but cannot be curled or twisted.

Genus.—Dendrólogus.

Teeth.—I. $\frac{3-3}{1-1}$, C. $\frac{0-0}{0-0}$, P. $\frac{1-1}{1-1}$, M. $\frac{4-4}{4-4}$ = 28. The two middle incisors of upper jaw hardly larger than the lateral.

Feet.—Hinder feet scarcely longer than fore-feet—Claws of fore-feet very strong, curved, and compressed—Fore-feet themselves larger than ordinary.

Tail.—Longer than body, powerful, and covered with hair.

Genus.—Mácropus.

Teeth.—Same as in preceding genus, but the two middle incisors of the upper jaw are equal in length to the other. The outermost on each side being broad.

Feet.—Hind-feet much longer than fore-feet—Claws, only of fore-feet, strong, curved, and compressed.

Tail.—Powerful, covered with hair, not so long as body.

Genus.—Halmatúrus.

Teeth.—Two middle incisors of upper jaw longer than the lateral.

Head.—Rather elongated. Muzzle, naked.

Feet.—Hinder far surpassing the fore-feet—Claws of fore-feet, flattish and strong.

Tail.—Shorter than the body, and covered with scales towards the tip.

Genus.—Petrógale.

Teeth.—Canines wanting ; upper incisors equal, but the front rather the longest, and slightly curved inwards ; hinder one hatchet-shaped, dilated towards the edge, and notched in the centre.

Head.—Muzzle bald.

Tail.—Cylindrical, furnished with a well-marked tuft at tip.

Genus.—Bettóngia.

Teeth.—Canines placed near the incisors, the space being about equal to one of the incisor teeth. Foremost compressed molar furnished with many vertical grooves ; true molars nearly square.

Head.—Short and broad.

Genus.—Hypsiprymnus.

Teeth.—Two middle incisors of upper jaw rather long, the two lateral incisors being small in proportion.

Feet.—Claws of fore-feet curved and compressed, the three middle claws very much longer than the two outer.

Tail.—Shorter than body, and slight.

Genus.—Lagorchestes.

Teeth.—Foremost upper incisor largest, and hinder the smallest. Behind the incisors, a very small canine. The last incisor has one vertical groove.

Genus.—Phascólomys.

Teeth.—I. $\frac{1-1}{1-1}$, C. $\frac{0-0}{0-0}$, P. $\frac{1-1}{1-1}$, M. $\frac{4-4}{4-4}$ = 24. Molars with flat crowns. A considerable interval between the incisors and the molars.

Feet.—Furnished with five toes, thumb-joint of hinder feet very short, and without a claw. The remaining claws powerful, and used for digging.

Tail.—Extremely short, only half an inch in length.

Sub-family.—Peramelina.

Very rat-like in general aspect.

Teeth.—Middle incisors of upper jaw not longer than the others.

Head.—Elongated, the snout being sharp, long, and pointed.

Feet.—Second and third toes of hinder feet joined as far as the claws—Thumb-joint of hinder feet very small.

Genus.—Perámeles.

Teeth.—I. $\frac{5-5}{3-3}$, C. $\frac{1-1}{1-1}$, P. $\frac{3-3}{3-3}$, M. $\frac{4-4}{4-4}$ = 48. Upper and outer incisors on each side separated from the others. Molars squared, with tubercles on the crown.

Feet.—Outer toe of fore-feet very short, and apart from the others—Thumb of hind-foot without a claw, sometimes entirely wanting.

Tail.—Rather short.

Genus.—Chœropus.

Teeth.—As in Perameles.

Feet.—Fore-feet with two toes, resembling those of swine; hind-feet without thumb-joint.

Tail.—Small and slight.

Sub-family.—Dasyurina.

Teeth.—I. $\frac{8}{6}$. Canines longer than incisors.

Feet.—Fore-feet with five toes; hinder feet either five or three-toed; thumb small, and without claw; second toe separated from third.

Tail.—Covered with hair, not prehensile.

Genus.—Parácyon (or Thylacínus).

Teeth.—I. $\frac{4-4}{3-3}$, C. $\frac{1-1}{1-1}$, P. $\frac{3-3}{3-3}$, M. $\frac{4-4}{4-4}$ = 46. External incisor on each side is the strongest—Canines very long, powerful, and sharply pointed—Last molars of upper jaw smaller than others; molars furnished with one large pointed cusp in centre, and two smaller lateral, one blunt cusp on inner side of crown.

Feet.—Fore-feet with five toes, the middle being slightly the longest—Hinder feet with four toes—Claws straight, strong, and blunt.

Tail.—Moderate, thick at root, covered with short hair.

Genus.—Diábolus.

Teeth.—I. $\frac{4-4}{3-3}$, C. $\frac{1-1}{1-1}$, P. $\frac{2-2}{2-2}$, M. $\frac{4-4}{4-4}$ = 42. Incisors arranged regularly without any interval, and of same length. Canines long and powerful. Grinding surface of upper molars triangular, the first having four sharp cusps, the second and third five, and the fourth three. All the molars of lower jaw covered with sharp cusps.

Head.—Short, and large in proportion.

Feet.—Thumb of hinder feet almost wanting.

Tail.—Short.

Genus.—Dasyúrus.

Teeth.—As in Diabolus, but not so strongly carnivorous.

Tail.—Long, and heavily covered with hair.

Genus.—Phascógale.

Teeth.—I. $\frac{4-4}{3-3}$, C. $\frac{1-1}{1-1}$, P. $\frac{3-3}{3-3}$, M. $\frac{4-4}{4-4}$ = 46. Two middle upper incisors longer than others, and separated from them by a narrow space; they are slightly curved, and projecting; the outermost incisors are the smallest. Canines not so large as in the preceding genera. The third præmolar of lower jaw is smaller than the others.

Tail.—Covered with short hair, and often tufted at extremity.

Genus.—Antechínus.

Teeth.—As in Phascogale, except that the two middle upper incisors are not larger than the others.

Tail.—Sparsely sprinkled with very short hairs, and very long.

Genus.—Myrmecóbius.

Teeth.—I. $\frac{4-4}{3-3}$, C. $\frac{1-1}{1-1}$, P. $\frac{3-3}{3-3}$, M. $\frac{6-6}{6-6}$ = 54. Incisors very small, pointed, and slightly compressed, separated from each other by considerable intervals—Canines hardly longer than præmolars—Molars very small, and separated from each other by a slight interval; covered with sharp, conical tubercles.

Head.—Pointed, skull very small.

Feet.—Hinder feet with four toes, the thumb being wanting, the claws curved, compressed, and sharp.

Tail.—Long, and very bushy.

Sub-family.—Didelphína.

Teeth.—Incisors always $\frac{5-5}{4-4}$.

Feet.—Furnished with five toes, the thumb of the hinder foot being broad, opposable to the other toes, and without a claw.

Genus.—Didelphys.

Teeth.—I. $\frac{5-5}{4-4}$, C. $\frac{1-1}{1-1}$, P. $\frac{3-3}{3-3}$, M. $\frac{4-4}{4-4}$ = 50. Two middle incisors of upper jaw rather longer than others, and separated from them by a slight interval—Upper canines stronger than the lower—Præmolars conical—Molars furnished with sharp cusps.

Head.—Long, gape of jaw very far back.

Pouch.—Tolerably developed.

Tail.—Long, covered with fur at the base, and with scales towards the extremity ; prehensile.

Genus.—Cheironectes.

Feet.—Toes of the hinder feet connected by a web. There is also a development of the pisiform bone, which supports a fold of the skin, and looks like a sixth toe.

Pouch.—Well developed.

Tail.—Longer than body.

Family.—PHÓCIDÆ.

Teeth.—Incisors variously deciduous ; molars with flattened crowns, or sometimes furnished with cusps.

Feet.—Furnished with five toes, short and palmate ; the hinder feet being turned backwards so as to approach each other.

Body.—Gradually tapering from the shoulders to the tail, which is very short and conical. Clothed with smooth hair, pressed firmly against the body.

Sub-family.—Phocina.

Teeth.—Incisors permanent ; molars $\frac{5-5}{5-5}$, or $\frac{6-6}{5-5}$.

Ears.—Very small, or wanting.

Genus.—Leptonyx.

Teeth.—I. $\frac{2-2}{2-2}$, C. $\frac{1-1}{1-1}$, P. $\frac{3-3}{3-3}$, M. $\frac{2-2}{2-2}$ = 32. Incisors much pointed ; molars divided into three long, conical, and slightly curved points.

Head.—Muzzle narrow and elongated.

Neck.—Long and tapering.

Limbs.—Nails very small, especially those of the hinder feet.

Genus.—Stemmátopus.

Teeth.—I. $\frac{2-2}{1-1}$, C. $\frac{1-1}{1-1}$, P. $\frac{3-3}{3-3}$, M. $\frac{2-2}{2-2}$ = 30. Incisors conical ; canines stout and large.

Head.—Adult male furnished with a large membranous and muscular sac, which is divided into two channels by a development of the septum of the nose.

Genus.—Phoca.

Teeth.—I. $\frac{3-3}{2-2}$, C. $\frac{1-1}{1-1}$, P. $\frac{3-3}{3-3}$, M. $\frac{2-2}{2-2}$ = 34. All the molars except the first furnished with double roots.

Genus.—Tríchecus.

Teeth.—Upper canines enormously developed, and without roots.

Head.—Muzzle tumid and protuberant, covered with thick bristles—External ears wanting.

Genus.—Morunga.

Teeth as in Stemmatopus.

Head.—Proboscis-like expansion of nose.

Genus.—Arctocéphalus.

Teeth.—I. $\frac{3-3}{2-2}$, C. $\frac{1-1}{1-1}$, P. $\frac{3-3}{3-3}$, M. $\frac{3-3}{2-2}$ (or rarely $\frac{2-2}{2-2}$) = 34. Four upper and middle incisors broad crowned, with groove, two others conical.

Limbs.—First toe of fore-feet longest, the middle hinder toes nearly equal ; membrane of hinder feet projecting like leathern straps beyond the toes.

Order.—CETE.

Teeth.—When present, conical and similar, the palate frequently furnished with baleen.

Body shaped like a fish.

Limbs.—Short and fin-shaped, the hinder pair forming a horizontal tail.

Skin.—Smooth and hairless, nostrils developed into blowing-tubes.

Family.—BALÉNIDÆ.

Nostrils two, palate with baleen, jaws without teeth.

Head.—Very large, equal to one-third the size of the body.

Genus.—Baléna.

No dorsal fin, abdomen smooth, baleen very long.

Teeth.—None in adult, only rudimentary in young.

Genus.—Megáptera.

Dorsal fin, abdomen furnished with longitudinal folds, pectoral fins long, equalling the head in length, baleen short, broad, and triangular.

Genus.—Balænóptera.

Dorsal fin sharp and falcate, abdomen and throat furnished with longitudinal folds, pectoral fins moderate, baleen short.

Genus.—Phýsalus.

Dorsal fins falcate, pectorals moderate, abdomen and throat with longitudinal folds. Blow-holes semi-lunar, divided from each other by a groove, and covered with a valve or flap. Baleen short.

Family.—CATODÓNTIDÆ.

Head very large, upper jaw apparently toothless, the lower jaw furnished with many conical teeth, which are received into cavities in the upper jaw. Blow-holes united, with a semi-lunar opening.

Genus.—Cátodon.

Dorsal hump rounded, blow-holes in front of the head.

Head.—Blunt at muzzle, skull much elongated.

Genus.—Physéter.

Upper jaw longest, blow-holes on the top of the head near the middle, separate, but covered with a common flap or valve. Dorsal fin high and falcate.

Family.—DELPHÍNIDÆ.

Head moderate—Both jaws furnished with teeth, which are frequently shed at an early age. Blow-holes united, forming a transverse semi-lunar opening on the top of the head.

Genus.—Zíphius.

Jaws tapering, upper jaw toothless, lower with two large, compressed teeth. Throat furnished with two diverging furrows. Lower jaw broad. Dorsal fin.

Genus.—Mónodon.

Few and early deciduous teeth in both jaws. Forehead convex. Upper jaw of male with one or two long, projecting teeth, spirally twisted. Dorsal fin.

Genus.—Belúga.

Both jaws with conical, deciduous teeth. Head rounded. No dorsal fin.

Genus.—Phocæna.

Both jaws with compressed and permanent teeth. Dorsal fin triangular, and placed in the middle of the back.

Genus.—Delphínus.

Head beaked, and rather convex in front. Dorsal fin fulcate in centre of back. Teeth many, small and conical.

Sub-order.—SIRENIA.

Body hairy. Muzzle with bristles. Two nostrils at extremity of snout. Fore-limbs like arms, hinder like fin or tail.

Genus.—Mánatus.

Teeth.—I. $\frac{1-1}{0-0}$, C. $\frac{0-0}{0-0}$, M. $\frac{8-8}{8-8} = 34$.

Tail or caudal fin rounded and oblong.
In adults the incisors are wanting, and in the young animal they are very small.

Genus.—Halícorë.

Teeth.—I. $\frac{1-1}{0-0}$, C. $\frac{0-0}{0-0}$, M. $\frac{5-5}{5-5}$.

Tail or caudal fin semi-lunar. Incisors are large in adults.

Genus.—Rhytína.

Teeth none, the jaws being furnished in their stead with a horny plate. Tail semi-lunar.

Order.—RODÉNTIA.

Teeth.—Two, long, curved, sharp-edged, rootless incisors in each jaw. Canines absent. Molars very few, and separated by a wide intervals from the incisors.
Feet furnished with claws.

Family.—MÚRIDÆ.

Teeth.—Lower incisors compressed and pointed. Molars generally six in each jaw.
Limbs.—Fore-feet with four toes, hind-feet with five.

Genus.—Mus.

Teeth.—I. $\frac{1-1}{1-1}$, P. $\frac{1-1}{1-1}$, M. $\frac{2-2}{2-2} = 16$. Incisors mostly smooth.
Limbs.—Fore-feet with four toes, and a nailed wart instead of a thumb. Hind-feet with five toes.
Habitat.—All lands.

Genus.—Cricétus.

Teeth.—Incisors smooth, lower compressed. Molars with tubercles.
Head.—Ears rounded. Cheek-pouches.
Limbs.—Feet as in Mus.
Tail.—Very short, and covered with hair.
Habitat.—Northern Europe.

Genus.—Arvícola.

Teeth.—I. $\frac{1-1}{1-1}$, P. $\frac{1-1}{1-1}$, M. $\frac{2-2}{2-2} = 16$. Molars curiously folded so as to form a double series of triangles on their crowns.
Head.—Ears rounded and very short.
Limbs.—Feet as in Mus, soles without hair.
Tail.—Rather short.
Habitat.—Europe.

Genus.—Myódes.

Teeth, as in Arvicola.
Head.—Ears not visible beyond the fur.
Limbs.—Feet with soles hairy, fore-feet with digging claws.
Tail.—Very short, covered with hair.
Habitat.—Norway and Sweden.

Sub-family.—Castorina.

Teeth.—Incisors covered with coloured enamel, and smooth in front. Molars with four folds of enamel.
Limbs.—Feet with five toes, hinder feet webbed.
Head.—Ears small and round.

Genus.—Castor.

Teeth.—I. $\frac{1-1}{1-1}$, P. $\frac{1-1}{1-1}$, M. $\frac{3-3}{3-3} = 20$.
Limbs.—Hind-feet entirely webbed.
Tail.—Wide, flat, and covered with scales.
Habitat.—North America and part of Europe.

Genus.—Myopótamus.

Teeth, as in Castor.
Limbs.—Only four toes of hind-feet webbed.
Tail.—Round and hairy.
Habitat.—Chili.

Genus.—Fiber.

Teeth.—I. $\frac{1-1}{1-1}$, P. $\frac{1-1}{1-1}$, M. $\frac{2-2}{2-2} = 16$.
Limbs.—Claws curved and flattened. Toes of hinder feet long, and edged with thick, stiff hairs.
Tail.—Moderate and compressed, a few short hairs appearing through the scales with which it is covered.
Habitat.—North America.

Genus.—Hýdromys.

Teeth.—I. $\frac{1-1}{1-1}$, P. $\frac{1-1}{1-1}$, M. $\frac{1-1}{1-1} = 12$.
Limbs.—Hinder feet partially webbed.
Tail.—Round, hairy, and large at the base.
Habitat.—New Holland and Van Diemen's Land.

Sub-family.—Echimjna.

Teeth.—Incisors straight, and abrupt at their extremities.

Genus.—Aulacódus.

Teeth.—Incisors broad and short, brown in front, three furrows in those of the upper jaw. Molars with four folds of enamel.

Limbs.—Fore-feet with thumb and outer toe short. Hind-feet with four toes.

Tail.—Rather short, slightly covered with hair.

Fur.—Composed of flattened and grooved hair, or grooved spines.

Habitat.—Southern Africa.

Family.—HYSTRÍCIDÆ.

Teeth.—Incisors smooth and large; molars with waving strips of enamel, always $\frac{4-4}{4-4}$.

Limbs.—Fore-feet with four toes, a wart instead of the thumb-joint.

Fur.—None, being replaced by strong and sharp spines.

Sub-family.—Hystricína.

Teeth.—Molars with undivided roots, set deeply in the bone; third molar placed under the anterior margin of the orbit.

Limbs.—Feet with soles smooth, but grooved.

Genus.—Hystrix.

Limbs.—Hind-feet with five toes.

Tail.—Not prehensile.

Habitat.—Parts of Europe, Africa, and Asia. Atherura is included in this genus.

Sub-family.—Cercolabina.

Teeth.—Molars with short, divided roots, set shallow in the bone. First molar placed under margin of orbit.

Limbs.—Feet with warty soles.

Genus.—Erethízon.

Tail.—Short and spined.

Limbs.—Hind-feet with five toes.

Fur.—Long hair, interspersed with short, sharp spines.

Habitat.—Canada.

Genus.—Cercólabes.

Tail.—Long and prehensile.

Limbs.—Hind-feet with four toes.

Habitat.—Brazils.

Sub-family.—Subungulâta.

Teeth.—Molars complex.

Limbs.—Fore-feet with four or five toes, hind-feet with three or four. Claws large and keeled above.

Genus.—Dasyprocta.

Head.—Lips cloven.

Limbs.—Fore-feet with four toes and a wart for the thumb; hind with three toes.

Tail.—Represented by a small, naked tubercle.

Fur.—Hair long on hind-quarters.

Habitat.—Brazils.

Genus.—Cœlógenys.

Head.—Lip cloven. Great development of zygoma, lined with a fold of skin. Cheek-pouches.

Limbs.—Fore-feet with four toes and nailed wart for thumb. Hind-feet with three toes.

Tail.—Very short.

Habitat.—Tropical America.

Genus.—Hydrochœrus.

Teeth.—Slight longitudinal groove on upper incisors. Molars without roots.

Head.—Thick, lips not cleft.

Limbs.—Feet partially webbed. Fore-feet with four toes, hinder with three.

Tail.—None.

Habitat.—Tropical America.

Genus.—Cavia.

Teeth.—Incisors smooth. Molars without roots, and curiously laminated.

Head.—Ears short and rounded.

Limbs.—Feet cloven; fore-feet with four toes, hinder with three; feet short.

Tail.—None.

Habitat.—Brazils.

Family.—LEPÓRIDÆ.

Teeth.—Four incisors in upper jaw, a pair of these teeth being placed behind the two usual incisors. Molars without roots, formed of two laminæ.

Limbs.—Fore-feet with five toes, hinder with four; soles hairy.

Tail.—Short, or absent.

Genus.—Lepus.

Teeth.—I. $\frac{2-2}{1-1}$, P. $\frac{2-2}{2-2}$, M. $\frac{3-3}{3-3} = 26$.

Head.—Ears long.

Limbs.—Hind-legs longer than fore-limbs.

Tail.—Short, and curved upward.

Habitat.—Europe, Asia, Africa, and America.

Family.—JERBOIDÆ.

Limbs.—Fore-feet short. Hind-feet long, and formed for leaping.

Tail.—Long, and thickly haired.

Sub-family.—Chinchillina.

Teeth.—Incisors smooth. Molars rootless, and composed of narrow laminæ.

Genus.—Chinchilla.

Teeth.—M. $\frac{4-4}{4-4}$, three laminæ in each.

Head.—Ears large, rounded, and scantily haired; whisker hairs very long.

Limbs.—Fore-feet with five toes, hind with four.

Habitat.—Peru.

Genus.—Lagótis.

Head.—Ears long, like those of the hare.

Limbs.—Fore-feet with four toes.

Habitat.—Peru.

Sub-family.—Dipína.

Teeth.—Molars with roots, and complex.

Genus.—Helamys.

Teeth.—I. $\frac{1-1}{1-1}$, P. $\frac{1-1}{1-1}$, M. $\frac{3-3}{3-3}$ = 20. Incisors smooth and broad. Molars with crown, divided into two portions by a fold of enamel.

Head.—Ears long.

Limbs.—Fore-feet with five toes, sharp, long claws; hinder with four toes, much elongated.

Habitat.—South Africa.

Genus.—Dipus.

Teeth.—I. $\frac{1-1}{1-1}$, P. $\frac{1-1}{1-1}$, M. $\frac{2-2}{2-2}$ = 16. Upper incisors grooved, all slender and sharp. Molars with roots.

Head.—Ears short.

Limbs.—Fore-feet with five toes, hind-feet with three.

Tail.—Long, covered with hair set in double row.

Habitat.—Part of Europe and Egypt.

Sub-family.—Myoxina.

Teeth.—Incisors smooth and compressed. Molars $\frac{4-4}{4-4}$, with roots, and with transverse bands on the crown.

Genus.—Myoxus.

Head.—Ears moderate, rounded, and covered with short, fine hair. Whisker hairs, long.

Limbs.—Fore-feet with four toes, and a wart for thumb, without claw. Hind-feet with five toes.

Tail.—Long, and thickly haired.

Habitat.—Europe.

Sub-family.—Sciurina.

Teeth.—Incisors smooth, brown or orange-coloured in front. Molars complex. M. $\frac{5-5}{4-4}$.

Limbs.—Fore-feet with four toes, and clawed wart for thumb.

Genus.—Sciuropterus.

Teeth.—I. $\frac{1-1}{1-1}$, P. $\frac{2-2}{1-1}$, M. $\frac{3-3}{3-3}$ = 22. Molars with tubercles.

Tail.—Rather short and flat.

Development of skin along sides so as to form a flying membrane.

Habitat.—India, North America, and Siberia.

Genus.—Sciurus.

Skin.—Not expanded along the sides.

Head.—Cheek-pouches none.

Habitat.—Europe, Asia, Africa, and America.

Genus.—Támias.

Head.—Cheek-pouches.

Feet.—Shorter than those of the true squirrels.

Tail.—Shorter than the body.

Habitat.—North America.

Genus.—Arctomys.

Teeth.—I. $\frac{1-1}{1-1}$, P. $\frac{1-1}{1-1}$, M. $\frac{4-4}{3-3}$ = 22. Incisors smooth and rounded. Molars with tubercles set transversely on the crown.

Head.—Cheek-pouches none.

Limbs.—Fore-feet with four toes, and nailed wart for thumb; hind-feet with five toes.

Tail.—Short, covered with long hair.

Habitat.—Northern Europe and America.

Genus.—Spermophilus.

Head.—Cheek-pouches.

Tail.—Moderate. General form more slender than Arctomys.

Habitat.—Northern Europe and America.

Family.—ASPALÁCIDÆ.

Teeth.—Incisors very long, and visible outside the mouth. Molars $\frac{4-4}{4-4}$ or $\frac{3-3}{3-3}$.

Head.—Ears none, or very small.

Limbs.—Front, five toes, cloven.

General form, thick, heavy, and clumsy.

Genus.—Spalax.

Teeth.—Molars $\frac{3-3}{3-3}$, complex, and small.

Head.—Very flat, and abrupt at muzzle. Eyes hidden under the skin. Ears, none externally.

Limbs.—Feet short, with small claws.

Tail.—None.

Habitat.—Europe.

Genus.—Bathyergus.

Teeth.—I. $\frac{1-1}{1-1}$, P. $\frac{1-1}{1-1}$, M. $\frac{3-3}{3-3}$ = 20. Upper incisors with deep groove, a stripe of enamel across the crown.

Head.—Eyes very small. Ears none externally.

Limbs.—Large digging claws on fore-feet, the claw of second toe the largest.

Habitat.—Cape of Good Hope.

Genus.—Saccóphorus.

Teeth.—Incisors with deep longitudinal groove—Molars $\frac{4-4}{4-4}$, rootless.

Head.—Ears very small, and rounded. Eyes very small. Very large cheek-pouches. nearly retractile when empty.

Limbs.—Three middle claws of fore-feet long, the third the longest.

Tail.—None.

Habitat.—Canada.

N.B. Diplostoma may be separated from Saccophorus by the smooth incisor teeth.

Genus.—Rhizomys.

Teeth.—Incisors broad, smooth, and red in front; molars $\frac{3-3}{3-3}$.

Head.—Broad, short, and abruptly terminated in front. Ears very small. Eyes also small.

Limbs.—Fore-feet four-toed, with clawed wart for thumb.

Habitat.—Malacca and China.

Genus.—Cténomys.

Teeth.—I. $\frac{1—1}{1—1}$, P. $\frac{1—1}{1—1}$, M. $\frac{3—3}{3—3}$ = 20. Molars decreasing in size from first to last; rootless and simple.

Order.—UNGULÁTA.

Teeth.—Incisors and canines often absent in one or both jaws. Molars all similar, when present.
Limbs.—Toes large, covered with hoofs.

Furcípeda.

Two middle toes large and equal.

Family.—BÓVIDÆ.

Teeth.—I. $\frac{0—0}{4—4}$ or $\frac{3—3}{4—4}$, M. $\frac{6}{6}$.
Head.—Mostly horns on frontal bones.
Limbs.—Two middle toes separate.
Gullet and stomach, complex.

Tribe I.—BOVINA.

Horns developed into permanent sheath, set upon the bony "core" or process from the frontal bone.

Sub-tribe.—Bóveæ.

Teeth.—Incisors nearly equal, and projecting slightly outwards.
Horns.—Smooth, bent outward, and curved upwards at tip.
Head.—Nose broad, nostrils at side.
Limbs.—Knee below the middle of fore-leg.

Horns not ridged or knobbed.

Genus.—Bos.

Horns.—Cylindrical and conical, curved upwards and outwards.
Head.—Frontal and facial portions of skull equal.
Dorsal ridge distinct.
Habitat.—Nearly all the world.

Genus.—Búbalus.

Horns.—Depressed or angular at base.
Head.—Forehead convex.
Habitat.—Africa and Asia.

Genus.—Bibos.

Horns depressed at base.
Shoulders very high, on account of the processes of the dorsal vertebræ.
Habitat.—Asia.

Genus.—Bison.

Horns round, and rather depressed at base; lateral, and curved upwards and outwards.
Head.—Muzzle short and rather wide.
Body covered with short crisp hair, longer on the head, neck, and shoulders. Dewlap none.
Habitat.—Europe and North America.

Genus.—Poéphagus.

Horns nearly cylindrical, curved outward.
Nose hairy; muzzle narrow and bald between nostrils.
Tail moderate, with thick, long hair.
Habitat.—Thibet.

Genus.—Óvibos.

Horns (of male) very broad at base, bent downwards over sides of face, and hooked upwards at tip. Those of the female smaller, and their bases farther apart.
Nose all hairy.
Tail short, and hidden by long hair of hind quarters.
Habitat.—North America.

Sub-tribe.—Antilopeæ.

I. Antelopes of the Field. Nostrils without hair inside.
Horns lyrate, sometimes conical; set over eyebrows.
Limbs slight, and hoofs small.
Tail short, with long hairs at base.

Genus.—Gazella.

Horns.—Moderate, lyrate; nose, tapering. Females hornless. Crumen (or tear-bag, situated below the eyes) distinct.
Habitat.—Part of Asia and Africa.

Genus.—Antidorcas.

Known by expansile white streak across back. Crumen small.
Habitat.—Southern Africa.

Genus.—Æpýceros.

Horns rather long, wide, and spreading. Tuft of black hair on posterior.
Habitat.—South Africa.

Genus.—Antílopë.

Horns erect, slightly spiral. Crumen large.
Habitat.—India.

Genus.—Tetrácerus.

Male with four horns, straight, and conical; female hornless. Muzzle large. Crumen longitudinal.
Habitat.—India.

Genus.—Calótragus.

Horns erect, slight, and tapering; female hornless. Crumen arched. Knees tuftless.
Habitat.—Southern Africa.

Genus.—Scopóphorus.

Horns slight and tapering; female hornless. Crumen transverse. Knees tufted.
Habitat.—Southern Africa.

Genus.—Oreótragus.

Horns slight and tapering; female hornless. Crumen transverse. Hoofs square, high, and contracted.
Habitat.—Southern Africa.

Genus.—Neótragus.

Horns short and conical; female hornless. Crumen large. Muffle none. Crown crested.
Habitat.—Abyssinia.

Genus.—Cephálopus.

Horns short and conical, set far back. Muffle large. Crumen represented by double series of pores. Crown crested.
Habitat.—Africa.

Genus.—Eleótragus.

Horns conical and diverging; bent forward at tips. Nose conical. Crumen none. Crown not crested.
Habitat.—Africa.

Genus.—Kobus.

Horns nearly lyrate, tips slightly recurved; female hornless. Mane on sides of neck. Crumen none.
Habitat.—South Africa.

Genus.—Ægócerus.

Horns recurved. Females horned. Compressed mane running down nape of neck; tuft of hair over crumen.
Habitat.—South Africa.

Genus.—Oryx.

Horns very long and slender, straight or slightly curved. Crumen none. Mane on nape of neck.
Habitat.—South Africa.

Genus.—Addax.

Horns long and spiral. Tuft of hair over crumen. Long hair on forehead and on throat. No mane on nape of neck.
Habitat.—North Africa.

Genus.—Rupicapra.

Horns erect, hooked abruptly backward at tips. Nose hairy.
Habitat.—Parts of Europe.

Genus.—Connóchetes.

Horns broad at base, bent downward and outwards on sides of head, then recurved at tip. Tail long and hairy from base.
Habitat.—South Africa.

Genus.—Alcéphalus.

Horns lyrate, thick at base, then suddenly bent backwards, nearly at right angles; set on upper edge of frontal bones. Tuft of hair on crumen. Muzzle broad. Muffle small and moist.
Habitat.—South Africa.

Genus.—Dámalis.

Horns lyrate and diverging. Muzzle rather broad. Muffle small and moist. Crumen without hair-tuft.
Habitat.—Africa.

Horns ridged.

Sub-tribe.—Strepsicéreœ.

Horns spiral, inclining backward. Crumen distinct, and nostrils near each other in front. No beard on chin of male.

Genus.—Strepsíceros.

Horns spiral, with bold keel or ridge. Short mane on neck.
Limbs equal.
Habitat.—South Africa.

Genus.—Óreas.

Horns spirally keeled, but nearly straight. Short mane on neck.
Limbs equal.
Habitat.—South Africa.

Genus.—Portax.

Horns short, almost triangular. Muffle large and moist. Hind-limbs shorter than fore-legs.
Habitat.—Asia.

Sub-tribe.—Capreœ.

Forehead convex, chin of males mostly with beard. Horns compressed, curved backward and outwards, with keel in front. Males with strong odour.

Genus.—Hemítragus.

Horns nearly triangular, compressed, heavily knobbed in front. Male without beard. Muffle naked.
Habitat.—Nepal.

Genus.—Capra.

Horns (of male) very large, heavily wrinkled and knobbed, and nearly square. Smaller in female. Muffle hairy.
Habitat.—Parts of Europe, Asia, and Africa.

Genus.—Hircus.

Horns triangular and compressed, slightly keeled and knobbed in front. Muffle hairy.
Habitat.—Europe, Asia, and Africa.

Sub-tribe.—Oveœ.

Forehead flat or concave. Horns spiral; females often hornless. Hoofs triangular, and shallow behind. Males not odorous.

Genus.—Ovis.

Crumen large; tail long. Skin with thick woolly coat, or flattened hair.
Habitat.—Europe, Asia, and Africa.

Genus.—Caprovis.

Crumen large. Tail very short. Wool hidden under thick hair.
Habitat.—Siberia, many parts of Asia and California.

Genus.—Ammótragus.

Crumen none. Tail long, and forehead concave. Neck heavily maned beneath. Chin not bearded.
Habitat.—North Africa.

Tribe.—GIRAFFÍNA.

Horns covered with hairy skin, tufted with hair at the tips.

Genus.—Giraffa.

Neck exceedingly elongated, back sloping. Lips not grooved, and totally hairy. Tongue very extensile. Tail long, with tuft of hair at extremity.
Habitat.—Africa.

Tribe.—CERVÍNA.

Horns, when present, shed and renewed annually.
Teeth.—Incisors wanting in upper jaw. False hoofs large.

Sub-tribe.—Alceœ.

Muzzle broad and hairy. Small bald muffle between nostrils. Horns large and palmed, without any basal snag near crown.

Genus.—Alces.

Neck short and thick; hair thick and brittle. Mane on throat. Hind-legs with tuft of hair above middle of metatarsus.
Habitat.—Northern Europe and America.

Sub-tribe.—Rangerine Deer.

Horns with large basal snag near crown. No naked muffle.

Genus.—Tarandus.

Muzzle hairy; crumen with pencil of hairs.
Habitat.—Northern Europe and America.

Sub-tribe.—Elaphine Deer.

Muzzle tapering, with bald, moist muffle, separated from muzzle by a hairy band. Horns with basal snag. Tuft of hair on hind leg, above middle of metatarsus.

Genus.—Cervus.

Horns round and erect, medial snag in front dividing into branches at tip (one or two branches on middle of front of beam). Crumen large. Hoofs narrow, triangular, and compressed.
Habitat.—Europe, Asia, America, and Africa.

Genus.—Dama.

Horns round below and expanded above; branched on hinder edge. Crumen large. Hoofs like Cervus. Fur spotted in summer.
Habitat.—Europe and Asia.

Sub-tribe.—Rusine Deer.

Horns with anterior basal snag. Muffle not separate from muzzle, and set high. Hair tuft in hind legs, as in Elaphines.

Genus.—Axis.

Horns set on rather long footstalks. Fur reddish, and spotted white at all seasons. Tail and ears rather long.
Habitat.—India.

Sub-tribe.—Capreoline Deer.

Horns without basal snag. Crumen small.

Genus.—Capréolus.

Horns small, erect, and round; slightly branched, with short footstalk. Tail none. Tuft on hind legs slightly above middle of metatarsus. Outer incisors of lower jaw very narrow, two central wide above.
Habitat.—Europe and Northern Asia.

Genus.—Caríacus.

Horns round and arched, central internal snag, tips bent forward, lower branches on hinder edge. Tail moderate, lower part dark, upper pale.
Habitat.—North America.

Tribe.—MOSCHÍNA.

Horns none. Upper incisors none. Hinder edge of metatarsus without hair. False hoofs large. Male with odoriferous gland.

Genus.—Moschus.

Muffle naked. Crumen none. Canine teeth of males extremely long.
Habitat.—Thibet and Nepal.

Genus.—Trágulus.

Throat and chin partially hairless. Hinder edge of metatarsus rather callous.
Habitat.—Parts of Asia.

Tribe.—CAMELÍNA.

Incisor teeth $\frac{1-1}{3-3}$. Upper lip hairy, but naked in front, and elongated. Canines in each jaw. Neck long. Legs long. Toes two, callous beneath, the hoofs only covering their upper surfaces.

Genus.—Camélus.

Back humped. Molar teeth $\frac{6-6}{6-6}$, the foremost being conical, like canines, and separated from the others. Toes broad, soles not divided.
Habitat.—Africa.

Genus.—Lama.

Back without hump—no conical molar teeth $\frac{5-5}{5-5}$. Toes long, soles separate.
Habitat.—South America.

Family.—ÉQUIDÆ.

Two middle toes united, and covered with a common hoof. No false hoofs. Incisor teeth $\frac{3-3}{3-3}$, C. $\frac{1-1}{1-1}$, M. $\frac{6-6}{6-6}$. Neck maned.

Genus.—Equus.

Tail entirely covered with long hair. All the legs with wart on inner side.
Habitat.—Europe, Asia, and Africa. Also naturalized on plains of America.

Genus.—Ásinus.

Tail with long hair only at extremity. Hind-leg without inner wart. Neck maned.
Habitat.—All the Old World.

Family.—ELEPHÁNTIDÆ.

Feet either ungulate, or furnished with flat and angular nails. Molars in both jaws, with very broad crowns ; incisors and canines sometimes absent. Skin mostly very thick, and generally with scanty hair.

Sub-family.—*Elephantina.*

Teeth.—I. $\frac{1-1}{0-0}$, very long and projecting—Canines none —Molars extremely long, with elongated crowns. Nose produced into proboscis. Feet with five toes.

Genus.—Elephas.

Teeth.—M. $\frac{2-2}{2-2}$, formed of a series of laminæ, succeeding each as they are worn.
Habitat.—Asia and Africa.

Sub-family.—*Tapirina.*

Incisor, canine, and molar teeth in both jaws. Three or four toes on fore-feet, three on hind-feet. Nose developed into a small proboscis.

Genus.—Tapírus.

Teeth.—I. $\frac{3-3}{3-3}$, C. $\frac{1-1}{1-1}$, P. $\frac{4-4}{3-3}$, M. $\frac{3-3}{3-3}$ = 42.
Fore-feet with four toes. Tail very short.
Habitat.—Asia and America.

Sub-family.—*Suína.*

Feet mostly with four toes, hinder feet sometimes with three toes. Nose abruptly truncated, not forming proboscis. Tail short, or almost absent.

Genus.—Sus.

Teeth.—I. $\frac{3-3}{3-3}$, or $\frac{2-2}{3-3}$, lower incisors directed forward. Canines of lower jaw directed upward. Molars with tubercles.
Feet with four toes. Tail short.
Habitat.—Nearly the whole world.

Genus.—Phacochœrus.

Teeth.—Molars with $\frac{3-3}{3-3}$, or $\frac{5-5}{5-5}$, according to age. Canines very large. A large wart under each eye.
Habitat.—Africa.

Genus.—Dicótyles.

Teeth.—I. $\frac{2-2}{3-3}$, M. $\frac{6-6}{6-6}$, with tubercles. Upper canines directed downward, and not projecting. Tubercle for tail. Hind-feet with three toes. Odoriferous gland in back.
Habitat.—Brazils.

Sub-family.—*Rhinocerina.*

Teeth.—Canines none, molars mostly $\frac{7-7}{7-7}$. One or more "horns" on nose and forehead. Feet with three toes. Skin very thick, and hanging in folds.

Genus.—Rhinóceros.

Upper lip rather extensile, and very mobile. "Horn" composed of aggregated longitudinal fibres.
Habitat.—Asia and Africa.

Genus.—Hyrax.

Teeth.—I. $\frac{1-1}{2-2}$, canines none, molars six or seven on each side of each jaw. Fore-feet with four toes, hind with three. Hoofs small and flat, somewhat resembling claws. Tubercle for tail.
Habitat.—Asia and Africa.

Sub-family.—*Hippopotamina.*

Teeth.—I. $\frac{2-2}{2-2}$, the two lower projecting forward. Canines large. Molars in adult $\frac{6-6}{6-6}$
Feet with four toes. Short hoofs. Tail short.

Genus.—Hippopotamus (*as the Sub-family*).

Habitat.—Southern Africa.

Order.—EDENTATA.

Teeth, none in forepart of jaws, sometimes wholly wanting. When present they are not enamelled, and are rootless. Feet furnished with strong curved claws.

Family.—DASÝPIDÆ.

Teeth, when present, small and similar. Head produced and snout long and narrow.

Sub-family.—*Manina.*

Teeth none. Body and tail covered with horny, sharp-edged scales, overlapping each other. Tongue round, and very long. Tail long.

Genus.—Manis (*as the Sub-family*).

Habitat.—Asia and Africa.

Sub-family.—*Dasypina.*

Teeth small and cylindrical, rootless in both jaws. Body covered with rows of scales, arranged in bands ; hair between the scales and bands.

Genus.—Dásypus.

Feet with five toes. Tail short. Body very convex.
Habitat.—South America.

Sub-family.—*Myrmecophagina.*

Teeth none. Body covered with thick, coarse hair. Tail long.

Genus.—Oryctéropus.

Teeth.—I. $\frac{0-0}{0-0}$, C. $\frac{0-0}{0-0}$, M. $\frac{7-7}{6-6}$, or $\frac{5-5}{5-5}$ when aged.
Their form is cylindrical and their crowns flat.
Limbs.—Fore-feet with four toes, hind with five. Claws very powerful adapted for digging.
Tail moderate and covered with hair.

Genus.—Myrmecóphaga.

Fore-feet with four toes, hind with five.
Habitat.—South America.

Genus.—Tamandua.

Distinguished from Myrmecóphaga by the shorter head and plumeless tail.

Genus.—Cyclothúrus.

Fore-feet with two toes, hind with four. Nose not so long as in preceding genus. Ribs very broad and flat, overlapping each other.
Habitat.—South America.

Sub-family.—*Ornithorhyncina.*

Teeth horny, or none. Hind-feet in males with hollow spur. Feet with five toes, and short. Snout long, and covered with naked skin.

Genus.—Plátypus.

Teeth $\frac{2-2}{2-2}$, flat and horny, without fangs. Snout flattened like duck's bill. Lower jaw shorter and narrower. Body covered with soft hair. Tail broad and flattened.
Habitat.—Australia.

Genus.—Echidna.

Teeth none. Snout long, slender, and rather pointed ; and very little mouth. Tongue long and extensile. Feet with large, curved claws. Tail short. Body covered with spines, mixed with hairs.
Habitat.—Australia.

Family.—BRADÝPIDÆ.

Head flat and short. Legs long, and furnished with large, curved, compressed claws.

Genus.—Cholœpus.

First molar tooth long, and like a canine. Fore-feet with two toes. Tail none.
Habitat.—West Indies.

Genus.—Brádypus.

Teeth of adult.—M. $\frac{5-5}{4-4}$, separate and cylindrical. Ears very short. Fore-feet with two or three toes, hind with three, joined as far as the claws. Tail none, or very short.
Habitat.—South America.

INDEX.

END OF MAMMALIA.

R. CLAY, PRINTER, BREAD STREET HILL.